The chemistry of
alkenes

Volume 2

THE CHEMISTRY OF FUNCTIONAL GROUPS

A series of advanced treatises under the general editorship of
Professor Saul Patai

The chemistry of alkenes (published 2 volumes)
The chemistry of the carbonyl group (published)
The chemistry of the ether linkage (published)
The chemistry of the amino group (published)
The chemistry of the nitro and nitroso groups (published in 2 parts)
The chemistry of carboxylic acids and esters (published)
The chemistry of the carbon-nitrogen double bond (published)

The chemistry of
alkenes

Volume 2

Edited by
JACOB ZABICKY

Institute for Fibres and Forest Products Research, Jerusalem
and
The Weizmann Institute of Science, Rehovoth, Israel

1970
INTERSCIENCE PUBLISHERS
a division of John Wiley & Sons
LONDON – NEW YORK – SYDNEY – TORONTO

First published 1970 John Wiley & Sons Ltd.
All Rights Reserved. No part of this publication
may be reproduced, stored in a retrieval system,
or transmitted, in any form or by any means
electronic, mechanical photocopying, recording
or otherwise, without the prior written permission of the Copyright owner.

Library of Congress Catalogue Card No. 64–25218

ISBN 0 471 98050 1

Made and printed in Great Britain by
William Clowes and Sons, Limited, London and Beccles

To my parents
far away and yet so close

Contributing authors

Jean-Francois Biellmann	Faculté des Sciences, Strasbourg, France.
Marvin Charton	Pratt Institute, Brooklyn, N.Y., U.S.A.
K. J. Crowley	Trinity College, Dublin, Ireland.
J. H. Goldstein	Emory University, Atlanta, Georgia, U.S.A.
Morton A. Golub	Ames Research Center, National Aeronautics & Space Administration, Moffett Field, California, U.S.A.
Henri Hemmer	Société Nationale des Pétroles d'Aquitane, Lacq, France.
Akira Kasahara	Yamagata University, Yamagata, Japan.
Jacques Levisalles	Faculté des Sciences, Nancy, France.
A. G. Loudon	University College, London, England.
Allan Maccoll	University College, London, England.
Kenneth Mackenzie	University of Bristol, Bristol, England.
P. H. Mazzocchi	University of Maryland, College Park, Maryland, U.S.A.
G. G. Meisels	Department of Chemistry, University of Houston, Texas, U.S.A.
Sekio Mitsui	Tohoku University, Sendai, Japan.
Herman G. Richey, Jr.	Pennsylvania State University, University Park, Pennsylvania, U.S.A.
V. S. Watts	Emory University, Atlanta, Georgia, U.S.A.

Foreword

The main purposes for a second volume dealing with the chemistry of the carbon-carbon double bond, were to cover the aspects that were left out in the previous volume because of authors having failed to deliver their contributions and to widen its coverage.

The topics of electrophilic additions and biological aspects of the double bond seem to have a fate of their own as, again, the commitment to write these chapters was not fulfilled by a second group of authors.

The material included in the present volume can be classified into three main groups:

General-theoretical: to this group belongs a chapter on n.m.r. spectroscopy.

Chemical behaviour: this is dealt with in chapters on rearrangements, hydrogenation, olefinic ions, complexes with transition metals, and the effects of various types of radiation.

Related compounds: these are treated in chapters on olefinic polymers and cyclopropanes.

I am indebted to Dr. M. Levin, director of the Institute of Fibres and Forest Products Research, for enabling me to carry out part of the present work during my stay in that Institute. Finally, I would like to thank Prof. Saul Patai for his help and guidance, and my wife for her constant encouragement and patience.

JACOB ZABICKY

Jerusalem and Rehovoth, 1968

The Chemistry of the Functional Groups
Preface to the series

The series 'The Chemistry of the Functional Groups' is planned to cover in each volume all aspects of the chemistry of one of the important functional groups in organic chemistry. The emphasis is laid on the functional group treated and on the effects which it exerts on the chemical and physical properties, primarily in the immediate vicinity of the group in question, and secondarily on the behaviour of the whole molecule. For instance, the volume *The Chemistry of the Ether Linkage* deals with reactions in which the C—O—C group is involved, as well as with the effects of the C—O—C group on the reactions of alkyl or aryl groups connected to the ether oxygen. It is the purpose of the volume to give a complete coverage of all properties and reactions of ethers in as far as these depend on the presence of the ether group, but the primary subject matter is not the whole molecule, but the C—O—C functional group.

A further restriction in the treatment of the various functional groups in these volumes is that material included in easily and generally available secondary or tertiary sources, such as Chemical Reviews, Quarterly Reviews, Organic Reactions, various 'Advances' and 'Progress' series as well as textbooks (i.e. in books which are usually found in the chemical libraries of universities and research institutes) should not, as a rule, be repeated in detail, unless it is necessary for the balanced treatment of the subject. Therefore each of the authors is asked *not* to give an encyclopaedic coverage of his subject, but to concentrate on the most important recent developments and mainly on material that has not been adequately covered by reviews or other secondary sources by the time of writing of the chapter, and to address himself to a reader who is assumed to be at a fairly advanced post-graduate level.

With these restrictions, it is realized that no plan can be devised for a volume that would give a *complete* coverage of the subject with *no* overlap between chapters, while at the same time preserving the readability of the text. The Editor set himself the goal of attaining *reasonable* coverage with *moderate* overlap, with a minimum of cross-references between the chapters of each volume. In this manner, sufficient freedom is given to each author to produce readable quasi-monographic chapters.

The general plan of each volume includes the following main sections:

(a) An introductory chapter dealing with the general and theoretical aspects of the group.

(b) One or more chapters dealing with the formation of the functional group in question, either from groups present in the molecule, or by introducing the new group directly or indirectly.

(c) Chapters describing the characterization and characteristics of the functional groups, i.e. a chapter dealing with qualitative and quantitative methods of determination including chemical and physical methods, ultraviolet, infrared, nuclear magnetic resonance, and mass spectra; a chapter dealing with activating and directive effects exerted by the group and/or a chapter on the basicity, acidity or complex-forming ability of the group (if applicable).

(d) Chapters on the reactions, transformations and rearrangements which the functional group can undergo, either alone or in conjunction with other reagents.

(e) Special topics which do not fit any of the above sections, such as photochemistry, radiation chemistry, biochemical formations and reactions. Depending on the nature of each functional group treated, these special topics may include short monographs on related functional groups on which no separate volume is planned (e.g. a chapter on 'Thioketones' is included in the volume *The Chemistry of the Carbonyl Group*, and a chapter on 'Ketenes' is included in the volume *The Chemistry of Alkenes*). In other cases, certain compounds, though containing only the functional group of the title, may have special features so as to be best treated in a separate chapter as e.g. 'Polyethers' in *The Chemistry of the Ether Linkage*, or 'Tetraaminoethylenes' in *The Chemistry of the Amino Group*.

This plan entails that the breadth, depth and thought-provoking nature of each chapter will differ with the views and inclinations of the author and the presentation will necessarily be somewhat uneven. Moreover, a serious problem is caused by authors who deliver their manuscript late or not at all. In order to overcome this problem at least to some extent, it was decided to publish certain volumes in several parts, without giving consideration to the originally planned logical order of the chapters. If after the appearance of the originally planned parts of a volume, it is found that either owing to non-delivery of chapters, or to new developments in the subject, sufficient material has accumulated for publication of an additional part, this will be done as soon as possible.

It is hoped that future volumes in the series 'The Chemistry of the Functional Groups' will include the topics listed below:

The Chemistry of the Alkenes (published in two volumes)
The Chemistry of the Carbonyl Group (Volume 1 published, Volume 2 in preparation)
The Chemistry of the Ether Linkage (published)
The Chemistry of the Amino Group (published)
The Chemistry of the Nitro and Nitroso Group (published in two parts)
The Chemistry of Carboxylic Acids and Esters (published)
The Chemistry of the Carbon–Nitrogen Double Bond (published)
The Chemistry of the Cyano Group (in press)
The Chemistry of the Amides (in press)
The Chemistry of the Carbon–Halogen Bond
The Chemistry of the Hydroxyl Group (in press)
The Chemistry of the Carbon–Carbon Triple Bond
The Chemistry of the Azido Group (in preparation)
The Chemistry of Imidoates and Amidines
The Chemistry of the Thiol Group
The Chemistry of the Hydrazo, Azo and Azoxy Groups
The Chemistry of Carbonyl Halides (in preparation)
The Chemistry of the SO, SO_2, $-SO_2H$ and $-SO_3H$ Groups
The Chemistry of the $-OCN$, $-NCO$ and $-SCN$ Groups
The Chemistry of the $-PO_3H_2$ and Related Groups

Advice or criticism regarding the plan and execution of this series will be welcomed by the Editor.

The publication of this series would never have started, let alone continued, without the support of many persons. First and foremost among these is Dr. Arnold Weissberger, whose reassurance and trust encouraged me to tackle this task, and who continues to help and advise me. The efficient and patient cooperation of several staff-members of the Publisher also rendered me invaluable aid (but unfortunately their code of ethics does not allow me to thank them by name). Many of my friends and colleagues in Jerusalem helped me in the solution of various major and minor matters and my thanks are due especially to Prof. Y. Liwschitz, Dr. Z. Rappoport and Dr. J. Zabicky. Carrying out such a long-range project would be quite impossible without the non-professional but none the less essential participation and partnership of my wife.

The Hebrew University, SAUL PATAI
Jerusalem, ISRAEL

Contents

1. Nuclear Magnetic Resonance Spectra of Alkenes — 1
   V. S. Watts and J. H. Goldstein
2. The Properties of Alkene Carbonium Ions and Carbanions — 39
   Herman G. Richey, Jr.
3. Alkene Rearrangements — 115
   Kenneth Mackenzie
4. Hydrogenation of Alkenes — 175
   Sekio Mitsui and Akira Kasahara
5. Alkene Complexes of Transition Metals as Reactive Intermediates — 215
   Jean-François Biellmann, Henri Hemmer and Jacques Levisalles
6. Photochemistry of Olefins — 267
   K. J. Crowley and P. H. Mazzocchi
7. The Mass Spectrometry of the Double Bond — 327
   A. G. Loudon and Allan Maccoll
8. The Radiolysis of Olefins — 359
   G. G. Meisels
9. Polymers containing C=C bonds — 411
   Morton A. Golub
10. Olefinic Properties of Cyclopropanes — 511
    Marvin Charton

Author Index — 611

Subject Index — 651

CHAPTER **1**

Nuclear Magnetic Resonance Spectra of Alkenes

V. S. WATTS and J. H. GOLDSTEIN

Emory University, Atlanta, Georgia, U.S.A.

I.	INTRODUCTION	2
	A. Chemical Shifts	2
	B. Coupling Constants	3
II.	SPECTRA OF ETHYLENE AND ITS SIMPLE DERIVATIVES	4
	A. Ethylene	4
	B. Vinyl Halides	5
	C. Vinyl Ethers and Sulphides	7
	D. Vinyl Formate, Acrolein, and Vinylacetylene	8
	E. Derivatives Containing Other Elements (Si, Hg, Al, Sn)	9
	F. Alkyl Substituted Ethylenes	12
III.	MULTIPLY SUBSTITUTED ETHYLENES	15
	A. Chemical Shifts	15
	B. Coupling Constants	18
IV.	COMPOUNDS WITH MORE THAN ONE DOUBLE BOND	18
	A. 1,3-Butadiene and Derivatives	19
	B. Cumulenes	21
V.	SPECIAL EFFECTS	22
	A. Medium Effects on n.m.r. Parameters	22
	B. Hindered Rotation About Single Bonds in Alkenes	26
VI.	APPLICATIONS TO COMPLEX SYSTEMS	31
	A. Cyclic Monoolefins	31
	B. Cyclic Compounds with more than One Double Bond	32
	C. Norbornenes and Norbornadienes	33
	D. Miscellaneous Applications	34
VII.	REFERENCES	35

I. INTRODUCTION

A. Chemical Shifts

The structure of high-resolution n.m.r. spectra can be accounted for by a phenomenological Hamiltonian of the form

$$H = \sum_i \nu_i I_z(i) + \sum_{ij} J_{ij} \vec{I}(i) \cdot \vec{I}(j)$$

where ν_i is the chemical shift of the i-th nucleus, J_{ij} the coupling constant between nuclei i and j, and $\vec{I}(i)$, $I_z(i)$ are the vector spin angular momentum operator and its z-component, respectively. The chemical shift may be expressed as

$$\nu_i = \frac{\gamma_i}{2\pi}(1 - \sigma_i)H_0$$

where γ_i is the magnetogyric ratio, σ_i is the dimensionless shielding (screening) constant for nucleus i, and H_0 is the applied external magnetic field.

For simple molecules shielding constants may be evaluated by the use of perturbation or variational calculations[1-4]. More complex molecules can be treated only in qualitative or semi-quantitative terms because of the lack of adequate wave functions and the partial cancellation of long-range diamagnetic and paramagnetic contributions[5].

For complex molecules the screening constant may be subdivided into four separate terms[6] representing contributions from: (*1*) diamagnetic intraatomic currents, (*2*) paramagnetic intraatomic currents, (*3*) other atoms, (*4*) interatomic currents. If local diamagnetic contributions to the screening of vinyl compounds (item *1*) are dominant, then substituent effects on the vinylic proton shifts should relate to the inductive ability of the substituent groups and to their ability to conjugate with the double bond in mesomeric structures. In some cases the proton shieldings are also strongly affected by magnetic fields caused by the anisotropic susceptibilities of the substituents (item *3*). Interatomic currents would be expected to affect the shieldings with a phenyl substituent and paramagnetic effects might be important for substituents like nitrogen.

In addition to the above effects, it has been suggested[7,8] that proton shieldings may be modified by a direct electrostatic field from a substituent.

B. Coupling Constants

The theory of nuclear spin–spin couplings is based upon the Hamiltonian developed by Ramsey, who subsequently showed with the aid of perturbation theory that couplings could be expressed as the sum of four contributions[9]. Of these, the Fermi contact term is believed to be dominant for proton–proton interactions and for other cases such as coupling between bonded ^{13}C and H. Most of the efforts at theoretical estimation of coupling constants have therefore been based upon this contribution. In these calculations both valence-bond and molecular-orbital approaches have been employed. Among the difficulties encountered are the lack of suitable precise wave functions and the necessity for employing approximations, such as the average energy approximation.

In the valence-bond method, as developed by Karplus and others[10,11], the calculated coupling constant reflects the extent to which deviations from perfect pairing occur. The coupling between non-bonded nuclei N and N' is directly related to the coefficients determining the weight of structures involving a long bond between these nuclei[12,13]. The molecular-orbital theory, as originally developed by McConnell, led to expressions for the coupling constant in terms of the bond order between the coupled atoms. The inability of the molecular-orbital theory to account for negative couplings has been attributed to the omission of configuration interaction in the calculations[14]. More recent developments of the molecular-orbital theory of couplings by Pople and Santry[15] make it possible to avoid the average energy approximation. While these and other recent developments in coupling theory are promising, it is still very difficult to account quantitatively for observed coupling values in diverse and complex molecules.

Coupling falls off rapidly with the number of intervening bonds in saturated molecules. However, in conjugated systems where the perfect-pairing approximation breaks down, longer range couplings occur. Thus, the study of proton spin-couplings permits a critical evaluation of electron delocalization in organic molecules[16].

The long-range couplings encountered in unsaturated molecules have been interpreted in terms of a π-electron contact mechanism, similar to that used to explain the hyperfine structure in the e.s.r. spectra of organic free radicals[17-20]. The contribution to the coupling as obtained from a second-order perturbation treatment is

$$J_{NN'}^{\pi} = \beta^2 Q^2 \eta_{NN'}^2 / h\Delta E$$

where β is the Bohr magneton, Q the hyperfine coupling constant, $\eta_{NN'}$ the π bond-order between carbon atoms N and N' and ΔE is the average excitation energy[17]. This coupling interaction between the protons and the π orbitals arises from σ-π exchange terms in the Hamiltonian. Since the hyperfine splitting constants appropriate for directly bonded protons (**1**) and for methyl group protons (**2**) are of similar magnitude but opposite signs[21-23],

$$\overset{\cdot}{C}-H \qquad \overset{\cdot}{C}-CH_3$$
$$(1) \qquad\qquad (2)$$

the π contribution to coupling should remain approximately constant in magnitude but change sign if a methyl group is substituted for a directly bound proton[16,24]. Such substitution thus allows estimation of the π-electron contributions to couplings between protons in olefinic systems.

In general the range of coupling constants for ethylenic compounds has been found to be approximately $J_{trans} = +12$ to $+18$ Hz, $J_{cis} = +4$ to $+12$ Hz, and $J_{gem} = -3.5$ to $+2.0$ Hz. The chemical shifts cover relatively a much larger range from almost 3·33 to 6·66 ppm with respect to TMS. It appears that with the exception of long-range coupling, the coupling constants in unsaturated molecules are primarily dependent upon electron withdrawal in the sigma framework. Chemical shifts, on the other hand, are more intimately related to the electron distribution of the π system.

II. SPECTRA OF ETHYLENE AND ITS SIMPLE DERIVATIVES

Spectra of important singly-substituted ethylenes have been considered in some detail, since they can provide a rational basis for applications to more complex molecules.

A. Ethylene

The p.m.r. spectrum of the prototype alkene, ethylene, consists of a single line due to the chemical equivalence of the four protons. The only information obtainable from the proton spectrum is, therefore, the chemical shift of the hydrogens. The various couplings have been obtained from spectra of the isotopically substituted compounds $H_2C\!=\!^{13}CH_2$ and $H_2C\!=\!CHD$[25-27]. For the first compound, the carbon-13 as well as the proton spectrum has been analysed. The couplings obtained are given in Table 1 together with those obtained for acetylene and ethane for comparison.

TABLE 1. Coupling constants for ethylene, acetylene and ethane[a]

Compound	J_{HH}	J_{CC}	J_{CH}	J_{CCH}
Ethane	8.0	34.6	125.0	−4.8
Ethylene	2.3 gem 11.5 cis 19.1 trans	67.2	156.2	−2.4
Acetylene	9.8	170.6	248.7	49.7

[a] All values in Hz; data taken from Ref. 27.

Consideration of the experimental results in terms of current theoretical treatments shows that the Fermi contact term[9] can be considered as the most important one in determining the magnitude of the coupling constants in ethylene. On this basis it has been inferred that ^{13}CH couplings depend linearly on the s-character of the hybrid carbon orbital, bonding to the hydrogen atom[28-30]. Assuming, further, that in acetylene this orbital is a pure sp hybrid, the value of J_{CH} in ethylene indicates an orbital with 31.5% s-character[26]. The chemical shift of the ethylene protons with respect to TMS as solvent (~30%) and internal reference, is 5.29 ppm[25].

B. Vinyl Halides

N.m.r. spectra of the vinyl halides including their ^{13}CH satellite spectra have been completely characterized and analysed[31]. The parameters for these compounds and for a number of other monosubstituted ethylenes are listed in Table 2. In addition to their utility for interpreting spectra of more complex derivatives, these data have been utilized in efforts to correlate n.m.r. parameters with other molecular or substituent properties.

When the chemical shifts for the chloro, bromo, and iodo compounds are corrected for the estimated diamagnetic anisotropy contribution of the CX bond, they correlate rather well with both substituent electronegativity and the quantity μ/R, where μ is the molecular dipole moment and R is the C—X bond length[31]. The quantity μ/R can be considered as a rough index of the charge transferred in the molecule. The fluoride shifts do not correlate with electronegativity in the above series, but this is not surprising as the fluoride shifts were not corrected for anisotropy.

In the series CH_2:CHX (X = F, Cl, Br, O, C) satisfactory straight lines are obtained for graphs of J_{cis} and J_{gem} vs. J_{trans} suggesting that

TABLE 2. N.m.r. Parameters for some monosubstituted ethylenes[a]

Substituent	Conc. (%) solvent	ν_1	ν_2	ν_3	J_{cis}	J_{trans}	J_{gem}	Ref.
F	11·30 TMS	388·72	258·71	281·65	4·72	12·67	−3·21	31
Cl	10·92 TMS	371·08	316·22	324·20	7·25	14·78	−1·32	31
Br	6·56 TMS	381·35	350·21	345·10	7·25	15·11	−1·62	31
I	8·13 TMS	389·09	387·39	371·36	7·91	15·83	−1·21	31
OCH_3	10(vol.) TMS	386·0	232·7	241·8	7·0	14·1	−2·0	32
SCH_3	10(vol.) TMS	380·7	304·5	290·4	10·3	16·4	−0·3	32
CN	dilute C_6H_{12}[b]	328·10	346·2	358·0	11·65	17·89	1·20	33
CHO	8·78 TMS	375·62	376·87	367·62	10·00	17·50	1·00	34
OCHO	—	—	—	—	6·3	13·8	−1·7	35
C≡CH	30(vol.) $CHCl_3$	—	—	—	11·5	17·3	2·05	36
$CH(CH_3)_2$	10(mole) CCl_4	—	—	—	10·4	17·3	1·6	37
H	70(vol.) TMS	318·5	318·5	318·5	11·4	18·8	2·0	25

[a] All values in Hz; shifts are downfield from and relative to, internal TMS at 60 MHz. Subscripts 1, 2, and 3 refer to positions α, β-trans and β-cis to the substituent, respectively.
[b] Cyclohexane.

the same substituent properties affect the various couplings in a similar manner[37]. All three couplings have been found[37,38] to decrease linearly with increasing electronegativity. Similarly, a good correlation exists between electronegativity and the sum of the ethylenic proton couplings. This has been demonstrated for over a hundred compounds of the type CH_2:CHX with a range of $J_{gem} + J_{cis} + J_{trans} = 14$ to 50 Hz[39]. Such correlations indicate that the greatest contribution to the coupling is probably transmitted through the sigma framework. The scatter in these plots is sufficient to indicate, however, that other factors may be important in determining the substituent effect on the couplings.

On the other hand, the differences between the α and β chemical shifts have been shown[37] to vary almost linearly with Taft's resonance parameters, σ_R. This suggests that changes in the π-electron distributions are influential in determining the differential chemical shifts $|\nu_\alpha - \nu_\beta|$. Also, the shifts of the α-protons in the vinyl halides do not parallel the substituent electronegativities[31] in contrast with the situation in methyl and ethyl halides[40,41]. These observations may be explained in terms of contributions from the customary resonance structures $H_2\bar{C}$—CH=$\overset{+}{X}$.

In the vinyl halides and most other monosubstituted ethylenes the α-protons resonate at lower fields than the β-protons. This has been explained as due to the short-range nature of inductive effects combined with resonance interaction between the double bond and lone-pair electrons on the substituent[37]. Both of these mechanisms result in lower field resonance of the α-compared with the β-protons.

In acrylonitrile the situation is reversed with the β-protons occurring at lower field. This observation can be adequately explained in terms of the additional shieldings due to diamagnetic anisotropy of the CN triple bond. These additional shieldings have been calculated, assuming the centre of anisotropy to be at the centre of the C≡N bond, as 1·13 ppm (α), 0·29 ppm (*cis*), and 0·38 ppm (*trans*) all to low field[42]. When corrections for this effect are applied to the chemical shifts of acrylonitrile, the α shift does occur at lower field than the β shifts.

C. Vinyl Ethers and Sulphides

Contributions from resonance structures are particularly important in the case of —OR and —SR substituents[32,37]. The chemical shifts for the β-protons in vinyl ethers lie approximately 1·9–2·1 ppm above ethylene[43]. Anisotropy effects on the β-proton shifts due to oxygen are expected to be smaller than those due to the cyano group. Since inductive effects would operate in the opposite direction, it seems likely that lone-pair conjugation of oxygen with the unsaturated system, and the resulting flow of charge into the β-position, is responsible for most of the observed upfield shift. Such conjugation appears to be considerably less important in sulphides than in ethers. For example, the β-protons of methyl vinyl sulphide are only 0·30–0·70 ppm above ethylene[32], in accord with n.m.r. data for vinyl halides, which indicate that the smaller the halogen the greater is the capability for lone-pair conjugation[44]. The α-proton shifts do follow the order of electronegativities and since inductive effects are expected to be short range, it would seem that this is the controlling factor here.

It is interesting to note that the β-vinyl protons in methyl vinyl sulphide couple with those of the methyl group, but there is no evidence of such coupling with the α-proton. These couplings are 0·4 Hz to the *cis* proton and 0·2 Hz to the *trans* proton. The sulphur atom can interact with unsaturated structures as if it were pseudo-ethylenic in character. The mechanism of such interaction possibly involves structures in which the methyl group donates charge to the unoccupied d orbitals of sulphur through hyperconjugation. Structures such as $CH_2{=}CH{-}\bar{S}{=}\overset{+}{C}H_3$ are energetically much more favourable with sulphur than with oxygen. Coupling between methyl and β-vinyl protons is not observed in methyl vinyl ether. It has been suggested [32,45] that the lack of coupling between the methyl group and α-proton in methyl vinyl sulphide might be of geometrical origin (angle dependence).

Ethyl vinyl ether and isobutyl vinyl ether have also been analysed [32]. Neither the chemical shifts nor the couplings in these systems vary significantly from those for methyl vinyl ether. It appears that these groups have little effect on the capacity of oxygen to conjugate with the vinyl system, and that the oxygen atom effectively buffers any inductive effect from the above groups.

D. Vinyl Formate, Acrolein, and Vinylacetylene

These three compounds are considered together, in spite of obvious structural differences, because they all exhibit long-range coupling, over four or five bonds. For the first two compounds this coupling is to the carbonyl proton and for vinylacetylene it is to the acetylenic proton. The reported values (in Hz) are, for couplings to the α, *trans*, and *cis* protons, respectively: vinyl formate, ±0·6, ±1·7 and ±0·8 [35]; acrolein, 7·40, −0·30, and 0·00 [34]; vinylacetylene, −2·1, −0·8, and 0·7 [36,46]. These long-range couplings are probably transmitted via the π-electron system.

The relative magnitudes of these long-range couplings in vinyl formate are interesting, both of the five-bond couplings being of greater magnitude than the four-bond coupling. This together with the relative magnitudes of the *cis* and *trans* long-range couplings has led to the suggestion that angular factors may be important in the π as well as σ contributions to spin–spin coupling [35].

Table 2 shows that the β chemical shifts in acrolein occur at rather low field strengths, relative to other monosubstituted ethylenes. This indicates the possible importance of contributions from resonance structures of the type $H_2C^+{-}CH{=}CH{-}O^-$ [34].

Data on the ^{13}C chemical shifts are also available for a number of monosubstituted ethylenes[47]. For seventeen compounds of the type $C_{(\beta)}H_2$—$C_{(\alpha)}HX$, the chemical shifts of $C_{(\alpha)}$ were found to cover a range of 67·5 ppm., while those of $C_{(\beta)}$ were found to cover a range of 57·5 ppm. No simple relations were found between the carbon shifts and the corresponding proton shifts. This is not unexpected in view of the different geometrical orientations with respect to the substituent. Such differences in orientation give rise to corresponding differences in effects due to 'neighbour anisotropy' or intramolecular dispersion forces.

However, a rough linear correlation exists between $^{13}C_{(\alpha)}$ shifts for vinyl compounds and the corresponding $^{13}C_{(1)}$ shifts of the similarly substituted carbon atom in phenyl compounds. A similar correlation holds for the $^{13}C_{(\beta)}$ shifts in vinyl compounds and the corresponding *ortho* shifts in phenyl compounds. Such correlations must be due to similar inductive, resonance, and neighbour effects in the two sets of compounds, resulting from similar electronic structures and molecular geometries. Each of the carbon atoms in both types of compounds utilizes three sp^2 σ hybrid orbitals for bonding, and contributes one p orbital to a π bond. Thus, the neighbour anisotropy and intramolecular dispersion effects should be comparable[47].

A plot of $\nu_{C(\beta)}$ (vinyl) vs. ν_{ortho} (phenyl) has a slope about twice that of the analogous plot for the $^{13}C_{(\alpha)}$ shifts. This is believed to arise from differences in the importance of resonance interactions in the two systems. In vinyl compounds the excess positive and negative charge from mesomeric interaction is confined to one carbon, while in phenyl compounds it is spread over two *ortho* and one *para* positions[47].

No correlation has been found between vinyl and phenyl $^{13}C_{(\alpha)}$ shifts and those in corresponding acetyl compounds. The lack of correlation seems to be related to differences in the inductive and resonance effects of the =O fragment as compared to the =CH$_2$ fragment, probably as a result of differences of polarity in these fragments[47].

E. Derivatives Containing Other Elements (Si, Hg, Al, Sn)

The vinylic n.m.r. parameters are rather markedly affected by bonding with elements such as Si, Hg, Al, Sn. Table 3 compares the shifts in compounds of this type with those of 1-hexene.

The proton chemical shift for ethylene in dilute TMS solution is −211 Hz at 40 MHz[48]. Thus, alkyl substitution as in 1-hexene

TABLE 3. Chemical shifts for vinyl groups bonded to Si, Hg, Al, and Sn[a]

	cis	trans	α	methyl	Ref.
1-Hexene	−197·6	−195·1	−228·7	−80·0[b]	48
ViMe$_3$Si	−225·2	−234·9	−244·4	−2·5	48
Vi$_2$Me$_2$Si	−226·6	−237·2	−244·2	−5·0	48
Vi$_4$Si	−230·8	−242·9	−244·5	—	49
Vi$_2$Hg	−181·0	−202·6	−234·7	—	50
Vi$_3$Al(etherate)	−167·1	−181·4	−191·5	—	50
Vi$_4$Sn	−165·9	−184·1	−191·6	—	50

[a] All values in Hz at 40 MHz, relative to internal TMS.
[b] Terminal methyl group in 1-butene.

displaces each β-proton upfield by 13–16 Hz. In view of the fact that carbon is appreciably more electronegative than silicon ($x_\text{C} = 2\cdot 5$ and $x_\text{Si} = 1\cdot 8$ on the Pauling scale), an even greater upfield displacement would be expected on the basis of naive inductive considerations alone[48,49]. The vinyl shifts of these silanes are in fact seen to be considerably downfield from those in 1-hexene.

The p.m.r. spectra of metal-organic compounds (Table 3) are often characterized by a reduction of the $|\nu_\alpha - \nu_\beta|$ values as compared to those found in ordinary organic compounds (Table 2). This is due to the tendency of metal atoms to increase the electronic shielding around adjacent groups. The large difference in shielding of the mercury and tin compounds in Table 3 cannot be explained in terms of inductive differences, since the Pauling electronegativity of both is 1·9. It has been suggested[50] that a possible explanation lies in the known variation of diamagnetic susceptibilities for the metal ions, being in the order $\text{Hg}^\text{II} > \text{Sn}^\text{IV} > \text{Al}^\text{III}$.

Also of interest is the fact that the *trans* proton is shifted considerably downfield compared to the *cis* proton (Table 3). The downfield shifts of the vinyl protons in metal substituted compounds and the larger effect at the *trans* compared with the *cis* position, can be explained in terms of d_π–p_π bonding between the metal atom and the vinyl group.

In the series $\text{R}_3\text{SiCH}=\text{CH}_2$ (where R = chloro, alkyl, or a substituted phenyl group)[51], an approximate correlation has been noted between the electronic nature of R and the *trans* proton chemical shift. This is consistent with a variable d_π–p_π resonance effect. Such resonance results when charge is donated from the p_π orbitals of the vinyl system into vacant d_π orbitals on the metal atom, as in **3 ↔ 4 ↔ 5**.

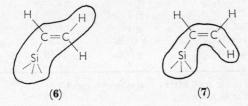

Preferential *trans* delocalization in such resonance structures is consistent with the hypothesis of the minimum bending of molecular orbitals. The smaller curvature of a molecular orbital **6** encompassing the silicon atom and the β-*trans* proton, could presumably make it more important in the ground state than **7**[51].

Table 4 shows some of the vinyl proton–proton couplings obtained in metal substituted ethylenes. These couplings are considerably larger than those encountered in the usual vinyl derivatives ($J_{cis} \sim 10$ Hz, $J_{trans} \sim 16$ Hz, Table 2). In the compounds shown in Table 4 the sum of the coupling constants is ~ 40 Hz. Commonly, this sum in vinyl derivatives is ~ 20 Hz for compounds with substituents capable

TABLE 4. Couplings in vinyl compounds bonded to Si, Hg, Al and Sn[a]

	J_{cis}	J_{trans}	J_{gem}	Ref.
ViMe$_3$Si	14·6	20·4	3·8	48
Vi$_2$Me$_2$Si	14·6	20·2	3·7	48
Vi$_4$Si	14·6	20·4	3·6	49
Vi$_2$Hg	13·1	21·0	3·5	50
Vi$_3$Al(etherate)	15·3	21·4	6·3	50
Vi$_4$Sn	13·2	20·3	3·7	50

[a] All values in Hz.

of donating lone-pair electrons and ~30 Hz for compounds with substituents that do not possess this capacity[48].

Table 5 lists the X—H couplings for a variety of vinyl compounds. Linear correlation is observed between the coupling constants and the

TABLE 5. X—H couplings in some vinyl compounds[a]

	$J_{X-H_{cis}}$	$J_{X-H_{trans}}$	$J_{X-H_{gem}}$
^{19}F(CH=CH$_2$)	±19·8	±52·7	±84·6
^{29}Si(CH=CH$_2$)$_4$	±8	±22[c]	
^{31}P(CH=CH$_2$)$_3$	±13·62	±30·21	±11·74
^{117}Sn(CH=CH$_2$)$_4$	±86·1	±174·1	±91·2
^{119}Sn(CH=CH$_2$)$_4$	±90·4	±183·0	±96·0
	±90·4[b]	±183·1	±98·3
^{199}Hg(CH=CH$_2$)$_2$	±159·5	±296·4	±128·3
	±159·6[b]	±295·5	±128·5
^{207}Pb(CH=CH$_2$)$_4$	±161·7	±330·1	±212·4
[^{205}Tl(CH=CH$_2$)]$^{2+}$	±1806	±3750	±2004
[^{205}Tl(CH=CH$_2$)$_2$]$^{+}$	±805	±1618	±842

[a] All values in Hz at 60 MHz unless otherwise stated.
[b] Measured at 40 MHz.
[c] Estimated from α-chlorotrivinylsilane.
[Reprinted from Reference 52, *J. Phys. Chem.*, **68**, 1240 (1964). By permission of the American Chemical Society.]

atomic number of the substituent atom within a given group in the periodic table. The increases in X—H coupling parallel the expected increases in electron densities of ns valence electrons surrounding the nucleus, $|\psi_{ns}(0)^2|$, indicating, at least qualitatively, that the contact contribution predominates in these couplings[52].

F. Alkyl Substituted Ethylenes

The n.m.r. spectra of alkyl substituted ethylenes are of interest for their implications with regard to the controversial topic of hyperconjugation. The chemical shift of ethylene at 40 MHz with TMS as solvent and internal reference is −211·5 Hz[53]. Under similar conditions, the proton α to the methyl group in propylene resonates 17·8 Hz to lower field than the ethylene protons, while the β-protons are both at higher field, the *trans* by 16·5 Hz and the *cis* by 13·0 Hz[53]. An analogous low-field shift for protons in the α-position occurs with methyl substitution in alkanes.

An inductive or hyperconjugative loss of charge to the remainder of the system by the methyl group, would result in a high-field rather than the observed low-field shift in such systems. It is possible that the low-field shift of a proton α to a methyl group, both in the alkanes and propylene, may be due to the difference in magnetic anisotropy of the C—C and C—H bonds[53-55].

It seems unlikely that anisotropic shielding by the methyl substituent would be as much as 15 Hz at the β-positions, especially in view of the fact that this is just the order of magnitude of the shift of the much nearer α-proton. An inductive loss of charge to the system by the methyl group would result in the observed direction of shift at the β-position, but inductive effects are presumably too short range to produce the magnitude of shift observed. It appears, therefore, that the high-field shift experienced by the β-protons as compared to ethylene, is most plausibly explained in terms of a hyperconjugative mechanism. This simple picture does not account for the difference in the two β shifts. Such a difference is predicted, however, if the charge displacement is parallel to the direction of the molecular dipole moment.[56]

It has been shown that for a number of methyl substituted ethylenic derivatives, the effect of methyl substitution on the chemical shifts of the ethylenic protons is approximately constant. Values for the methyl substituent effects are -17.8 Hz for α, $+16.5$ Hz for β (*trans* to CH_3) and $+13.0$ Hz for β (*cis* to CH_3). Table 6 shows how the shifts predicted for methylated ethylenes, obtained by applying these corrections to the parent compounds (the compound with a hydrogen atom in place of a methyl group), compare with actual observed shifts. Although the methyl substituent effect is seen to show some dependence upon the nature of the second substituent, it is constant enough to be a useful aid in making spectral assignments and choosing initial values of spectra parameters for the purpose of fitting n.m.r. spectra.

Theoretical analyses have been carried out for the spectra of propylene, 1-butene and 1-hexene. The vinylic chemical shifts are quite constant throughout the series: for 10 mole per cent solutions in CCl_4 the variations are only ± 1.70 Hz (α), ± 0.54 Hz (*trans*) and ± 0.72 Hz (*cis*). The couplings obtained are listed in Table 7.

The long-range couplings between the methyl and the vinyl protons in propene are -1.75 Hz (*cis*) and -1.33 Hz (*trans*)[56]. Both the magnitude and sign of these couplings are in accord with expectations based on a hyperconjugative π interaction mechanism. Since the predicted attenuation factor for π-electron coupling when a bonded

TABLE 6. Predicted and observed shifts in methyl substituted vinyl compounds[a]

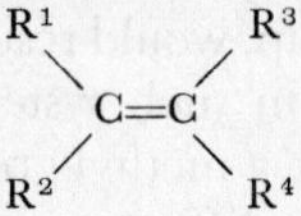

$$\begin{array}{c} R^1 \\ \diagdown \\ R^2 \end{array} C=C \begin{array}{c} R^3 \\ \diagup \\ R^4 \end{array}$$

R^1	R^2	R^3	R^4	Proton	Predicted	Observed
H	H	CN	CH_3	1	−222.2	−225.8
				2	−218.6	−221.6
CH_3	H	H	CN	3	−206.0	−208.7
				2	−256.5	−261.3
CH_3	H	CN	H	4	−202.5	−206.6
				2	−249.4	−251.0
H	H	Cl	CH_3	1	−200.0	−201.0
				2	−198.5	−199.8
Cl	H	H	CH_3	2	−234.8	−235.3
				3	−234.3	−232.8
H	H	Br	CH_3	2	−221.0	−218.3
				1	−213.7	−211.1
Br	H	H	CH_3	2	−241.5	−238.1
				3	−248.0	−243.3
H	H	CH_3	CH_3	1,2	−182.0	−183.5
CH_3	H	CH_3	H	2,4	−212.8	−213.4
CH_3	H	H	CH_3	2,3	−216.3	−210.0

[a] From Ref. 53. Values in Hz at 40 MHz, relative to internal TMS.

TABLE 7. Couplings for alkyl ethylenes $(H_2C_{(\beta)}=C_{(\alpha)}HCH_2R)$[a]

Coupling protons	J		
	R = H	CH_3	C_3H_7
cis $H_{(\alpha)}$—$H_{(\beta)}$	10.02	10.32	10.23
trans $H_{(\alpha)}$—$H_{(\beta)}$	16.81	17.23	17.03
gem $H_{(\beta)}$—$H_{(\beta)}$	2.08	1.96	2.23
Me-$H_{(\alpha)}$	6.40	6.22	6.55
Me-cis $H_{(\beta)}$	−1.75	−1.66	−1.51
Me-trans $H_{(\beta)}$	−1.33	−1.26	−1.18

[a] In Hz.
[Reprinted from Reference 56, *J. Am. Chem. Soc.*, **83**, 231 (1961). By permission of the American Chemical Society.]

proton is replaced by a methyl group, that is when a three-bonds coupling is substituted by a four-bonds coupling, has a value of -1, the above values indicate π contributions to the *cis* and *trans* couplings in ethylene of $+1\cdot7$ and $+1\cdot3$ Hz, respectively[57].

III. MULTIPLY SUBSTITUTED ETHYLENES

A. Chemical Shifts

A number of studies of geometrical isomerism about double bonds have been carried out using n.m.r. spectroscopy[58-63]. Table 8 lists

TABLE 8. Differential proton and ^{13}C shifts in *cis* and *trans* CHX=CHX[a,b]

X	H	^{13}C
Cl	$-0\cdot08$	$-1\cdot9$
Br	$-0\cdot38$	$-7\cdot0$
I	$-0\cdot76$	$-17\cdot1$
CO$_2$Me	$+0\cdot525$	—
CO$_2$Et	—	$+5\cdot0$

[a] Values from Ref. 59.
[b] $\delta cis - \delta trans$, in p.p.m.

the differential proton and ^{13}C shifts for a number of symmetrical disubstituted ethylenes. These data have been explained in terms of contributions from the anisotropic magnetic polarizability of the C—X bonds[64] according to the equation

$$\sigma_i = (1-3\cos^2\theta)(X_L - X_T)/3R^3$$

The C—X bonds are assumed to be axially symmetric with X_L and X_T, the longitudinal and transverse magnetic susceptibilities respectively; R is the length of the line from nucleus i to the centre of the anisotropic group; and θ is the angle between this line and the C—X axis. Structural factors (bond lengths, interbond angles) are assumed to be the same in the *cis* and *trans* structures.

Because of the R^{-3} factor, the anisotropy contribution to the differential shifts should be decidedly greater for the *trans* isomer. If the magnetic dipole is placed at the X nucleus, for halogen bonds this leads to small positive shifts for the *trans* structures. If the paramagnetic term is dominant the effect should parallel the polarizabilities of the X atoms, i.e., $\sigma_I > \sigma_{Br} > \sigma_{Cl}$, as is observed. The sign inversion for the esters studied was ascribed to the fact that for the C=O group $X_T > X_L$ (T is now perpendicular to the plane of the trigonal carbon orbitals).

However, the above explanation fails for the ^{13}C shifts since, for identical bond lengths and structural angles, the carbon nucleus is in an identical magnetic environment in both isomers. For this case it has been proposed that the differential shifts arise from the greater steric interference in the *cis* isomer[59]. This explanation is also applicable to the differential proton shifts. It is assumed that, for the molecules of Table 8, there are appreciable contributions to the ground-state structures from the mesomeric forms **8** and **9**, where X is

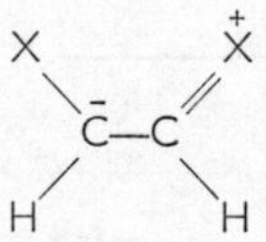

capable of mesomeric electron release. Effective participation of **8** or **9** requires planarity of these structures, which becomes difficult to attain in the *cis* isomers as the size of the substituent group rises. Since **8** would exhibit high field shifts, reflecting the greater shielding of C and H relative to the uncharged structures, and conversely for **9**, the observed differential shifts are accounted for. The order in which these appear for the halogens is explained in terms of differences in the magnetic anisotropy of the C—X and C=X$^+$ bonds.

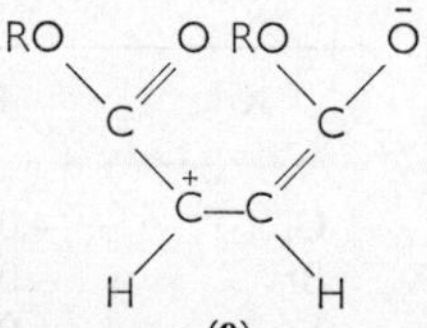

It is interesting that in the 2-substituted propenes, where steric interference of the type described above is absent, the differential shieldings are of opposite sign from those in the 1,2-disubstituted ethylenes. For these 2-substituted propenes the values of $\delta_{cis} - \delta_{trans}$ are 0·00 (Cl), 0·20 (Br), and −0·55 (CO_2Me), all in p.p.m., with *cis* and *trans* referring to the 2-substituent[58].

It has recently been suggested that proton shifts in olefinic systems may be estimated by the use of additive increments (Z_i) for substituents[65]:

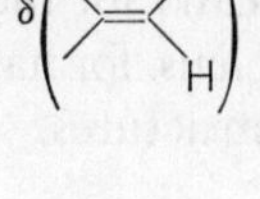

$$\delta\left(\!\!{>}\!\!=\!\!{<}_H\right) = \delta_{ethylene} + \sum Z_i$$

$$= 5\cdot 28 + \sum Z_i$$

1. Nuclear Magnetic Resonance Spectra of Alkenes

where δ is in ppm with respect to internal TMS. The incremental shifts for a large number of substituents are given in Table 9. Shifts calculated in this manner agree with the observed values with a standard deviation of 0·15 ppm.

Correlation charts for chemical shifts have proved useful in determining structures and configurations of alkenes. Dietrich and

TABLE 9. Values for incremental substituent shifts (Z_i)

Substituent[a] R	Z_i for R		
	gem	cis	trans
—H	0	0	0
—alkyl	0·44	−0·26	−0·29
—alkyl-ring	0·71	−0·33	−0·30
—CH$_2$O, —CH$_2$I	0·67	−0·02	−0·07
—CH$_2$S	0·53	−0·15	−0·15
—CH$_2$Cl, —CH$_2$Br	0·72	0·12	0·07
—CH$_2$N	0·66	−0·05	−0·23
—C≡C	0·50	0·35	0·10
—C≡N	0·23	0·78	0·58
—C=C alone	0·98	−0·04	−0·21
—C=C conj	1·26	0·08	−0·01
—C=O alone	1·10	1·13	0·81
—C=O conj	1·06	1·01	0·95
—COOH alone	1·00	1·35	0·74
—COOH conj	0·69	0·97	0·39
—COOR alone	0·84	1·15	0·56
—COOR conj	0·68	1·02	0·33
—CHO	1·03	0·97	1·21
—CNO	1·37	0·93	0·35
—COCl	1·10	1·41	0·99
—OR, R aliph	1·18	−1·06	−1·28
—OR, R conj	1·14	−0·65	−1·05
—OCOR	2·09	−0·40	−0·67
—aromat	1·35	0·37	−0·10
—Cl	1·00	0·19	0·03
—Br	1·04	0·40	0·55
—NR$_2$, R aliph	0·69	−1·19	−1·31
—NR$_2$, R conj	2·30	−0·73	−0·81
—SR	1·00	−0·24	−0·04
—SO$_2$	1·58	1·15	0·95

[a] The values given for 'R conj' apply when either R or the ethylene group is conjugated with another unsaturated group. The term 'alkyl-ring' indicates that substituent R forms a ring with the double bond.

[Reprinted from Reference 65, *Helv. Chim. Acta*, **49**, 164 (1966), by permission.]

Keller[66] have prepared several bar charts showing proton shift ranges for various molecular environments. Approximately 1500 proton shift values from over 700 compounds were used in preparing these charts. Stehling and Bartz[67] have also prepared correlation charts for both olefinic and non-olefinic protons in a number of alkenes. These charts were derived from the spectra of 60 alkenes. These authors describe several examples of structure determination using the correlation charts and also report coupling constant ranges for a series of alkenes.

B. Coupling Constants

In many instances coupling constants are probably more reliable than chemical shifts in determining geometrical configurations. As a general rule, for both mono- and disubstituted ethylenes $J_{trans} > J_{cis}$. A fair correlation has been shown to exist between both of these couplings and the sum of the electronegativities of the substituents[68]. The following additivity relations have been proposed for vicinal couplings in substituted ethylenes[68,69]:

$$H_2C{=}CHR \quad J_{cis} = 11\cdot 7\,(1-0\cdot 34\,\Delta E)$$
$$J_{trans} = 19\cdot 0\,(1-0\cdot 17\,\Delta E)$$
$$RCH{=}CHR' \quad J_{cis} = 11\cdot 7\,(1-0\cdot 36\,\Sigma\,\Delta E)$$
$$J_{trans} = 19\cdot 0\,(1-0\cdot 17\,\Sigma\,\Delta E)$$

where ΔE is the difference in electronegativity between R and hydrogen.

The variation in coupling constant with substituent electronegativity is about five times as great in the ethylenes as in the corresponding substituted ethanes. This sensitivity, together with the approximately additive substituent effect described above, accounts in large measure for the usefulness of coupling constants as a structural guide. Spin–spin couplings are also much less influenced, in general, by such complicating factors as magnetic anisotropy or solvents. (Solvent effects on couplings are not negligible in some cases, as will be discussed later, but these effects are of no consequence for *cis* and *trans* H-H couplings in the ethylenes.)

IV. COMPOUNDS WITH MORE THAN ONE DOUBLE BOND

Only the structurally simplest compounds will be considered here, specifically some butadienes and cumulenes. These are the compounds whose spectra have been studied in the greatest detail.

A. 1,3-Butadiene and Derivatives

By using deuterated analogues to obtain initial coupling values, the n.m.r. spectrum of butadiene has been completely analysed. The high resolution spectrum of this molecule consists of about 90 peaks in two different spectral regions. The lower field region corresponds to the two interior protons, and the higher field region to the four terminal protons. The n.m.r. parameters determined for butadiene and some of its derivatives are given in Table 10.

The chemical shift of the interior protons in butadiene occurs considerably downfield from ethylene (~ 346 Hz from cyclohexane at 60 MHz). A similar shift is observed for the $C_{(2)}$-proton in propylene, a possible explanation for which could be based on the difference in anisotropy of the C—H and C—C bonds. Application of a similar anisotropy correction to butadiene (60 Hz to high field) as determined by an empirical method[70], brings the chemical shift of the interior protons to very nearly the ethylene value. The uncertainty in the amount of double-bond character in the interior C—C bond of butadiene renders the quantitative value of this correction somewhat tenuous.

The chemical shifts in butadiene would be expected to reflect somewhat the π-electron distribution. Naive molecular-orbital calculations appear to support this expectation[71]. A good correlation has been found to exist between the logarithm of the rate constant for Diels–Alder addition to maleic anhydride of the dienes in Table 10 (as well as isoprene and chloroprene) and the chemical shift of the proton at the B position[71]. This proton is probably the least affected by the anisotropies of the substituents, and hence could be expected to reflect more adequately the relative electron densities at this position for the compounds studied. The values for 2-*t*-butyl-1,3-butadiene and 2-methoxy-1,3-butadiene do not correlate as well as the other dienes. This is not unexpected for the former compound since it occurs predominantly as a non-transoid structure at room temperature. From its chemical shift one would predict a smaller rate constant for this compound than is observed. This is understandable since the activation energy for the reaction of 2-*t*-butyl-1,3-butadiene would not contain a term for conversion from the *s-trans* to *s-cis* form. The non-conformity of 2-methoxy-1,3-butadiene is probably related to the strong conjugation between the vinyl system and the methoxy group, as evidenced by the extremely high-field values for the chemical shifts of the terminal protons β to the methoxy group.

TABLE 10. P.m.r. parameters for 1,3-butadiene and derivatives[a]

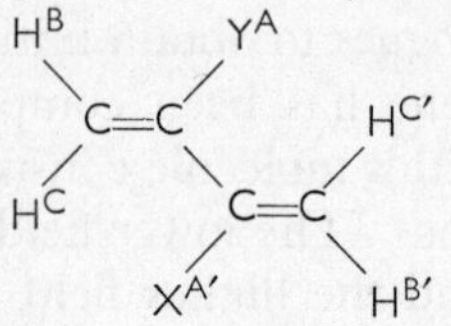

$$\begin{array}{c}H^BY^A\\ C=C\\ H^CH^{C'}\\ C=C\\ X^{A'}H^{B'}\end{array}$$

Parameter	X:H Y:H	CH$_3$O H	(CH$_3$)$_3$C H	Cl Cl	CH$_3$ CH$_3$
ν_A	−289.7	−276.5	−294.7	—	−26.4[b]
$\nu_{A'}$	−289.7	−129.5[b]	+21.4[b]	—	−26.4[b]
ν_B	−217.3	−214.9	−210.2	−249.2	−208.0[c]
$\nu_{B'}$	−217.3	−158.1[d]	−212.2	−249.2	−208.0[c]
ν_C	−223.5	−242.9	−232.5	−273.8	−213.1[c]
$\nu_{C'}$	−223.5	−156.5[d]	−199.0	−273.8	−213.1[c]
$J_{AA'}$	10.41	—	—	—	
J_{AB}	10.17	10.82	10.80	—	
$J_{AB'}$	−0.86	0.00	±1.00[e]	—	
J_{AC}	17.05	17.27	17.00	—	
$J_{AC'}$	−0.83	0.00	±0.40[e]	—	
$J_{A'B}$	−0.86	—	—	—	
$J_{A'B'}$	10.17	—	—	—	
$J_{A'C}$	−0.83	—	—	—	
$J_{A'C'}$	17.05	—	—	—	
$J_{BB'}$	1.30	1.50	0.00	1.90	
J_{BC}	1.74	1.87	2.30	−1.70	
$J_{BC'}$	0.60	0.57	0.00	0.55	
$J_{B'C}$	0.60	0.51	0.00	0.55	
$J_{B'C'}$	1.74	−1.90[f]	1.70	−1.70	
$J_{CC'}$	0.69	0.51	0.00	0.59	

[a] Table reprinted from Ref. 71. Spectral parameters expressed in Hz at 60 MHz with respect to internal cyclohexane.

[b] Chemical shift of the methyl group.

[c] No analysis was performed. The values given here refer to the centres of rather broad peaks (∼4 Hz).

[d] Accurate to within 0.4 Hz.

[e] The signs cannot be determined by the method employed here. However, they are probably negative.

[f] Assumed.

[Reprinted from Ref. 71, *J. Mol. Spectry.*, **12**, 76 (1964). By permission of Academic Press, Inc.]

B. Cumulenes

Since all of the protons in allene are equivalent, its p.m.r. spectrum consists of a single line from which only the chemical shift can be obtained. In the vapor phase allene exhibits a shift approximately 0·84 p.p.m. to higher field than the protons in ethylene. This could arise largely from the diamagnetic circulation in the cyclindrically symmetrical π-electron distribution around the digonally hybridized central carbon atom (see **10**). An equivalent magnetic shell expansion based on the semiclassical ring current model with two electrons localized in $2p$ orbitals whose effective radius is estimated from Slater functions, accounts for 2/3 of the observed difference[72]. The proton–proton coupling cannot be obtained from the proton spectrum but has

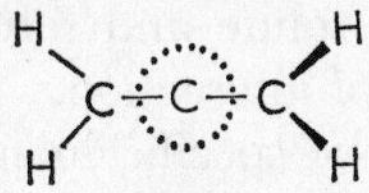

$$\underset{H}{\overset{H}{>}}C\!\!\div\!\!C\!\!\div\!\!C\underset{H}{\overset{H}{<}}$$

(**10**)

been determined from the ^{13}CH satellite pattern to be 7·0 ± 0·1 Hz[72]. Such a coupling is much larger than observed in similar single bonded systems, undoubtedly as a result of contributions from the π system. Such contributions could originate from structures of the type **11**.

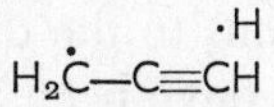

$$H_2\overset{\cdot}{C}\!\!-\!\!C\!\!\equiv\!\!C\overset{\cdot H}{H}$$

(**11**)

N.m.r. parameters have also been determined for a number of substituted allenes[73–80]. Values for the H-H coupling in several monosubstituted allenes are 6·3 ± 0·1 (Cl), 6·1 ± 0·1 (Br and I) and 6·67 ± 0·05 (CH$_3$). Utilizing such data, an additive substituent effect has been demonstrated for the four bond H-H coupling in CHR=C=CHR' and in the five bond CH$_3$-H coupling in CHR=C=C(CH$_3$)R' compounds[74].

N.m.r. studies of butatriene, CH$_2$=C=C=CH$_2$, have been hindered by its tendency to polymerize. However, by cooling the compound to −55°C, the rate of polymerization becomes sufficiently slow to provide satisfactory spectra. Both the long-range *cis* and *trans* H-H couplings were determined to be 8·95 Hz. This compares with Karplus' calculated value of 7·8 Hz for the π coupling in butatriene. The ^{13}CH couplings were found to be 170·9 Hz as compared with 168·2 Hz for allene and 156·4 Hz for ethylene. No simple relationship is evident among the chemical shifts of the three compounds[81].

V. SPECIAL EFFECTS

A. Medium Effects on N.m.r. Parameters

The sensitivity of chemical shifts to the liquid medium (solvent and concentration effects, exclusive of contributions due to bulk susceptibility of the solvent) is well known. Recent studies have shown that some coupling parameters in unsaturated systems are also influenced by the medium and should not, therefore, be described as 'constants'[82-86]. The resultant dependence of n.m.r. spectral patterns on solvent and concentration will naturally vary a good deal, reflecting the particular characteristics of various spin systems. Nevertheless, the topic is of considerable intrinsic interest with relation to the characteristic properties of the solute and solvent molecules under consideration and their mode of interaction.

These interactions may be specific in nature, involving preferred orientations and points of attraction between solvent and solute molecules. This introduces perturbations of electronic distributions and anisotropic magnetic influences which no longer average to zero. Both factors can affect chemical shifts and the former is capable of modifying coupling values.

In many recent studies of medium-dependent couplings it has been possible to relate the variations to the dielectric properties of the solvent[83]. These observations have been accounted for in terms of the 'reaction-field' theory[87] which may be explained in the following way: the electric dipole of a solute polarizes the surrounding medium (regarded as a continuum) and this polarization induces a secondary and reinforcing electric field, the reaction field, at the dipole. The extent to which this secondary field polarizes the solute dipole and alters the nuclear shieldings depends on the polarizability of the solute molecule. The functional relationship between the contribution to the shielding, σ, of a solute proton and the factors determining the reaction field, depends on the model chosen for the shape of solvent cavity containing the solute molecule. For a spherical cavity[87]

$$\sigma = -10^{-12} X \cdot \frac{2(\varepsilon - 1)(n^2 - 1)}{3(2\varepsilon + n^2)} \frac{\mu}{\alpha} \cos\theta$$

and for an ellipsoidal cavity[88]

$$\sigma = -10^{-12} X \cdot \frac{3\mu}{abc} \xi_a [1 + (n^2 - 1)\xi_a] \frac{\varepsilon - 1}{\varepsilon + \beta} \cos\theta$$

where ε is the dielectric constant of the solvent; n is the refractive index of the solute for the sodium D line; μ is the permanent dipole moment of the solute molecule and α is its polarizability; θ is the angle between the X—H axis and the direction of the reaction field; a, b, and c are the axes of the ellipsoid; ξ_a is a shape factor[89]; $\beta = n^2 \xi_a/(1 - \xi_a)$ and X is a constant. Values of X vary somewhat depending on the model but are generally around 3·0.

Concentration and solvent studies of unsaturated solutes have shown their n.m.r. parameters to be functions of both the above. The magnitude of observed effects depends on both the polarity of the solute and the solvent molecules. The greater the polarity of the solute, the more marked are the concentration and solvent effects. In general increasing the dielectric constant of the solvent results in chemical shifts moving toward lower-field strengths. The reaction field chemical shift should be linear in $(\varepsilon - 1)/(2\varepsilon + n^2)$ for a spherical cavity.

TABLE 11. Solvent effects on the n.m.r. parameters of α-chloroacrylonitrile[a,b]

Solvent	v_1^∞	v_2^∞	J^∞	ε^c
Dimethyl sulphoxide	−395	−406	−3·24	45
Nitromethane	−379	−385·5	−3·04	35·10
Acetonitrile	−377	−384	−3·00	35·10
Dimethylformamide	−396	−407	−3·19	35·05
Methyl alcohol	−380	−387	−2·93	30·61
Acetone	−386	−394·5	−3·07	19·75
4-Heptanone	−383	−392	−2·94	11·72
Bromoethane	−370	−376·5	−2·56	8·78
Iodoethane	−373·5	−380	−2·52	7·42
Chloroform	−368	−374	−2·42	4·59
Bromoform	−374	−381·5	−2·46	4·23
Cyclohexane	−352	−357	−1·96	1·97
Tetramethylsilane	−351·5	−357	−1·96	1·89

[a] N.m.r. parameters are in Hz at 60 MHz. Chemical shifts are relative to internal TMS. v_1 refers to proton *cis* to the chlorine atom.

[b] Parameters extrapolated to infinite dilution in indicated solvents.

[c] At 35°C. From the American Institute of Physics Handbook (McGraw-Hill Book Company, Inc., New York, 1957), 1st. ed., pp. 134–142; and *Natl. Bur. Std.* (*U.S.*) Circ. 589. Where given at another temperature they were corrected to 35°C.

[Reprinted from Ref. 83, *J. Chem. Phys.*, **42**, 228 (1965). By permission of American Institute of Physics.]

The effects of medium on the parameters of a highly polar compound, α-chloroacrylonitrile, are given in Table 11. Table 12 shows the effects observed in monosubstituted ethylenes.

TABLE 12. Solvent effects on the n.m.r. parameters for vinyl compounds[a]

Substituent	Solvent[b]	ν_α	ν_{trans}	ν_{cis}	J_{cis}	J_{trans}	J_{gem}
CN	neat	−343·79	−364·62	−372·19	11·75	17·92	0·91
	CH	−328·10	−346·20	−358·00	11·65	17·89	1·20
	DMF	−364·02	−375·75	−383·25	11·81	17·88	0·96
F	CH	−388·57	−258·80	−281·42	4·70	12·68	−3·06
	DMF	−413·40	−276·73	−291·71	4·63	12·47	−3·39
Cl	neat	−374·82	−321·07	−326·77	7·14	14·79	−1·48
	CH	−371·12	−316·08	−323·72	7·16	14·78	−1·28
	DMF	−395·02	−334·29	−334·47	6·95	14·64	−1·67
Br	neat	−385·88	−356·57	−348·23	7·15	15·01	−1·80
	CH	−391·90	−350·30	−344·69	7·27	15·10	−1·59
	DMF	−404·70	−367·24	−354·97	7·00	14·87	−2·05
I	neat	−391·72	−394·35	−373·84	7·81	15·89	−1·47
	CH	−389·40	−386·35	−370·65	7·83	15·82	−0·88
	DMF	−408·02	−401·42	−379·57	7·78	15·82	−1·52

[a] Parameters are in Hz at 60 MHz. Chemical shifts are referenced relative to internal TMS.
[b] The values of all parameters in cyclohexane (CH) and dimethylformamide (DMF) have been extrapolated to infinite dilution.
[Reprinted from Ref. 83, *J. Chem. Phys.*, **42**, 228 (1965). By permission of American Institute of Physics.]

The difficulty in estimating reaction-field contributions to n.m.r. parameters lies in the corrections necessary for other factors contributing to the magnitude of these parameters. One such attempt at correcting for the effect of other factors on the chemical shift involves the simultaneous study of *cis* and *trans* isomers. The dichloro and dibromoethylenes have been studied in this manner[90]. All four have proton spectra consisting of a single peak. The *cis* forms are dipolar and should experience a reaction field, whereas the *trans* forms are non-polar and should be unaffected by changes in the dielectric constant of the medium (to first order in the reaction field). Since the shapes of the corresponding *cis* and *trans* forms are rather similar, it may be anticipated that magnetic and dispersion interactions with any one medium are very nearly the same. Subtraction of the chemical shift of the *trans* from the *cis* form should isolate the contribution of the reaction field to the total proton shift of the *cis* form[90]. When the shifts so obtained in water–dioxane solutions are plotted against the dielectric function of such solutions, good correlation is obtained. The ^{13}CH couplings of the dichloro and dibromoethylenes are also found to be sensitive to concentration and solvent effects[85].

Medium effects on $J_{^{13}\text{CH}}$ and ν_H are most pronounced at positions α to the substituent, suggesting that a more specific and localized form of solvent–solute interaction, influencing the halogen atom, may be involved. Such interaction could be dipolar in nature and not entirely unrelated to the general dielectric properties of the solution. Variations in the C—Cl stretching frequencies of cis- and trans-dichloroethylene have been explained in terms of a combination of dipolar interaction and dielectric effects in polar solvents[91]. Table 13

TABLE 13. Comparison[a] of the solvent dependence of n.m.r.[b] and IR[c] parameters[d] in some halogenated ethylenes

Compound	$\Delta \nu_\text{H}$	ΔJ_CH	$\Delta \tilde{\nu}_\text{C–H}$	$\Delta \tilde{\nu}_\text{C–X}$
cis-$C_2H_2Cl_2$	20·47	2·97	9·6	9·7
trans-$C_2H_2Cl_2$	19·60	2·40	16·8	10·5
1,1-$C_2H_2Cl_2$	13·80	0·73	14·2	3·5
cis-$C_2H_2Br_2$	16·35	2·70	19·9	6·2
trans-$C_2H_2Br_2$	16·30	2·00	19·6	8·5

[a] From Ref. 85.
[b] Values in Hz; concentrations are 50 mole per cent.
[c] $\Delta\tilde{\nu}$ values in cm^{-1}; concentrations are 4–9 mole per cent.
[d] Δ(parameter) = |parameter (DMF)–parameter (CH)|.
[Reprinted from Ref. 85, *J. Mol. Spectry.*, **17**, 348 (1965). By permission of Academic Press, Inc.]

shows how changes in $J_{^{13}\text{CH}}$ and the chemical shift for the dichloro- and dibromoethylenes correlate with changes in the C—H and C—X stretches. With certain types of solvents, those containing particularly strong donor centres, specific interactions with the protons of unsaturated molecules become important[92]. If the interactions are strong enough to hold the solute and solvent molecules in a fixed geometrical arrangement for a period of time, long by comparison with that required for an n.m.r. transition, the anisotropy in the solvent molecule will affect the shifts of the solute protons. This was patent when the effects of nonaromatic with aromatic solvents were compared on the highly polar solute molecule α-chloroacrylonitrile[92]. The additional downfield shift experienced by the solute protons in aromatic solvents can be understood in terms of an equilibrium between complexes such as **12** and **13**.

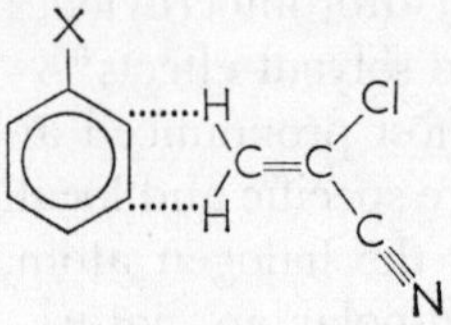

(12) 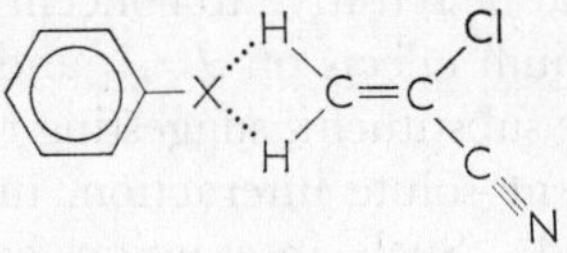   (13)

If the lifetimes of the above complexes are sufficiently long on the n.m.r. time scale, the effect of the anisotropy of the aromatic ring on the proton shifts can be calculated for configuration **12** by assuming the separation between the ring and protons equal to the sum of the Van der Waal's radii involved. The contribution of the aromatic anisotropy to the solute proton shifts is decreased as interactions of type **13** increase in relative importance. It is possible in some cases to correct for this anisotropy contribution with reasonable success[92].

Since the magnitude of solvent shifts varies with the geometrical position of a proton in a solute molecule, solvent effects can be used to increase the separation in chemical shift between protons, thereby simplifying spectral analysis[93].

B. Hindered Rotation About Single Bonds in Alkenes

The existence of distinguishable rotational conformers about a C—C bond adjacent to a double bond, has been demonstrated through substituent series, and solvent and temperature studies of the n.m.r. spectra of the compounds concerned[94–98]. The microwave spectrum of propylene indicates that its most stable conformation is that in which a methyl proton is eclipsed with the vinyl methylene group[99], but rapid interconversion averages the couplings to the methyl protons to a single value.

The vicinal proton–proton coupling constants for protons on the C_{sp^2} – C_{sp^3} bond have been determined to be the same within experimental error for propene, 1-butene and 1-hexene[100]. This observation is in agreement with 2 sets of postulates: (*1*) 1-butene and 1-hexene may exist predominantly as different conformers and the H—C_{sp^2} – C_{sp^3}—H coupling constant is independent of the H—C—C—H angle; or (*2*) the coupling constant depends on the H—C—C—H angle but the conformers of the two molecules are about equally populated[96]. The second alternative was confirmed by substituting the tetrahedral carbon with alkyl groups large enough to interfere sterically with the terminal vinylic methylene group. The attendant changes in the H—C_{sp^2}—C_{sp^3}—H coupling are most reasonably explained in terms of population changes of probable rotamers[96]. Values for the couplings of interest are given in Table 14.

TABLE 14. Coupling parameters in some hindered olefins (R^2R^3C=CH^1—$CHR^4R^{4'}$)[a]

R^2	R^3	R^4	$R^{4'}$	$J_{1,2}$	$J_{1,3}$	$J_{1,4}$	$J_{2,3}$	$J_{2,4}$	$J_{3,4}$
H	H	CH_3	CH_3	10·37	17·22	6·41	1·74	−1·17	−1·43
H	H	H	i-C_3H_7	10·13	17·02	7·00	2·05	−1·15	−1·43
H	H	H	t-C_4H_9	10·02	17·10	7·46	2·37	−0·94	−1·32
H	H	t-C_4H_9	t-C_4H_9	9·97	17·01	10·65	2·63	−0·10	−0·63
CH_3	CH_3	t-C_4H_9	t-C_4H_9	−1·25[b]	−1·25[b]	11·37	—	—	—

[a] From Ref. 96. Values in Hz.
[b] J_{H-CH_3}.

In the 3-monoalkylpropenes and 3,3-dialkylpropenes J for H—C_{sp^2}—C_{sp^3}—H, assumed to be a function of angle independent of substituents, may be expressed in terms of characteristic values for vicinal H-H coupling constants between protons in particular conformational arrangements and the relative populations of these conformations[96]. Thus, if the three probable stable rotamers 14–16 are present

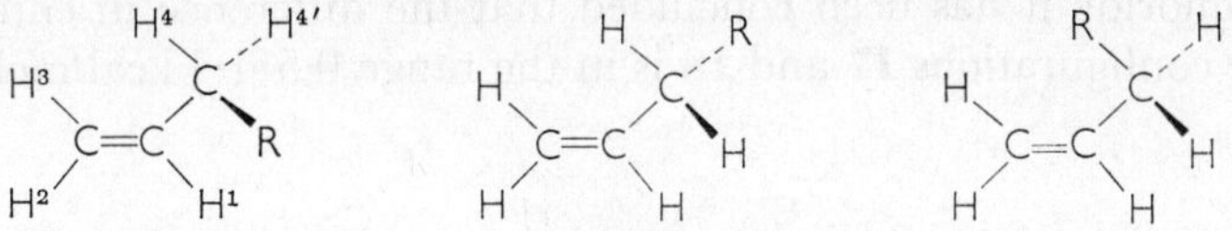

(14)   (15)   (16)

in the proportions $x:x:(1-2x)$, the observed coupling $J_{1,4}$ may be related to the characteristic values for coupling between *gauche* hydrogens, J_g, and *trans* hydrogens, J_t, by the expression

$$J_{1,4} = xJ_t + (1 - x)J_g$$

If R is some atom capable of coupling with H, then

$$J_{1,R} = 2xJ_{g,R} + (1 - 2x)J_{t,R}$$

An expression of the same form would describe the H—C_{sp^2}—C_{sp^3}—H coupling in the 3,3-substituted propenes. Such expressions have been used, together with the assumptions that propene and 1-butene exist essentially as 1:1:1 mixtures of rotamers and that 3-t-butylpropene exists essentially in two forms, to calculate an upper limit of 3·9 Hz for J_g and a lower limit of 11·1 Hz for J_t.

For a number of allyl compounds the coupling parameters have been determined[97,98]. By analogy with propene the important conformers are again considered to be **14–16**, and the observed coupling will be given by

$$J = \frac{P_a}{2} J_t + \left(P_c + \frac{P_a}{2}\right) J_g$$

where $P_a/2$, $P_b/2$ and P_c are the populations of rotamers **14, 15** and **16** respectively, with $P_a = P_b$. Using observed values of $J_{1,4}$ after correction for the effect of substituent electronegativity, E, according to the equation

$$J = 8·4 - 0·4\, E$$

and the values of $J_g = 3·7$ Hz and $J_t = 11·54$ Hz[96], the populations of the conformers in several allylic compounds have been calculated. They are $P_a = 0·78$ (allyl chloride), 0·95 (allyl bromide), 1·0 (allyl iodide), 0·74 (allylbenzene) and 0·44 (allyl cyanide). Equal distribution in the three forms would give $P_a = 0·66$. The above conclusions are consistent with the variation in the long-range coupling.

The fluoropropenes present the possibility of calculating populations and equilibrium constants from both the H-H and H-F couplings. From a temperature study of the H-H and H-F couplings in allylidene fluoride it has been concluded that the difference in enthalpy for the configurations **17** and **18** is in the range 0·5–1·4 kcal/mole.

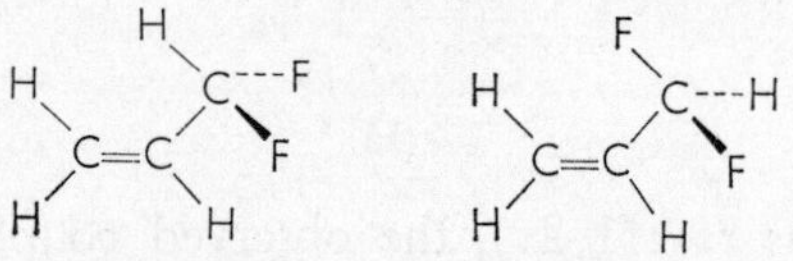

(17)   (18)

Another interesting temperature study of rotational isomerism has been carried out using compounds containing no hydrogen. These were compounds of the type CF_2=CFY in which Y was CF_3, CF_2Cl, CF_2Br, CF_2I and COF. The F-F couplings were studied and conclusions very similar to those obtained from studies of H-H couplings in systems with F replaced by H, were reached[101].

Values for the long-range H-H couplings in compounds of the type H_2C=CY—CH_2X have been found to depend markedly on solvent polarity[95]. The long-range coupling in a system of this type is expected to be determined principally by the π contribution discussed earlier

$$J^\pi = 2{\cdot}1 \times 10^{-15} \sum_T \frac{a_H a_{H'}}{\Delta E} \tag{1}$$

where the sum is taken over the triplet (T) states. The hyperfine splitting constant for the allylic proton, $a_{H'}$, depends on the square of the cosine of the dihedral angle, ϕ, between the π axis of the central carbon and the CCH plane. Thus, the observed *cis*, 1,3 coupling constant should show the approximate angular dependence $J = J_0 \cos^2 \phi$. If more than one methylene proton is involved the coupling must be averaged over both the bonds and the motions. The following classical expression has been used for $H_2C{=}CY{-}CH_2X$ systems

$$\bar{J}_2 = J_0 \frac{\int_0^{2\pi} \exp[-V(\phi)/kT](\tfrac{1}{2} + \sin^2 \phi) d\phi}{2\int_0^{2\pi} \exp[-V(\phi)/kT] d\phi}$$

by making two simplifying assumptions: (*1*) that J_0 as derived from equation (1) is unaffected by allylic substitution in propenes and (*2*) that the potential function $V(\phi)$ can be replaced without serious error by a delta-type function which is infinite everywhere except at the angles ϕ_N and ϕ_P, corresponding to the two potential minima indicated by the doubled C=C, C—C, stretching and =CH$_2$ bending modes in the i.r. spectra of the compounds studied.

Under the above assumptions J_P (i.e. J for the angle ϕ_P) is inferred to be $\approx 0{\cdot}7$ Hz and a corresponding maximum value of $2{\cdot}1$ Hz is inferred for J_N. The dependence of these values on the dihedral angles involved, lead to **19** and either **20** or **21** as the stable conformers[95].

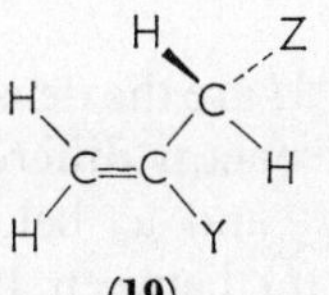

(**19**) 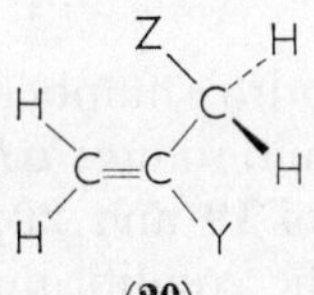   (**20**)

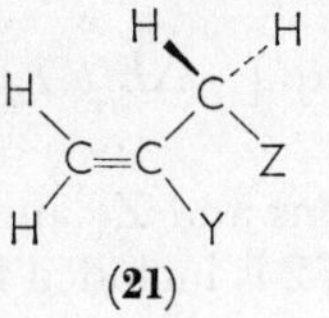

(**21**) 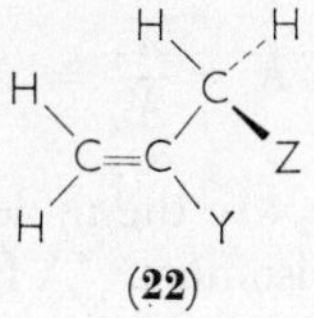   (**22**)

Conformer **21** may be eliminated on the basis of the direction of the solvent effect. Conformers **20** and **22** would be expected by analogy with propylene, however, electrostatic repulsion between the halogens might be expected to alter **22** to some extent in the direction of **19**.

The qualitative trends of substituent effects on $\bar{J}_2$ in the 2,3-disubstituted propenes may be rationalized in terms of structures **19** and **20** as follows: For negatively polarized 2-substituents, both increasing electronegativity and decreasing size across a series Z = I, Br, Cl in the 3-substituent should favour **20** over **19** accounting for the observed order $\bar{J}_2(Z = Cl) > \bar{J}_2(Z = Br) > \bar{J}_2(Z = I)$ for Y = Cl, Br. If the Z substituent is fixed and negatively polarized, then increasing electronegativity of the Y substituent should favour **20** while decreasing size might favour **19**. Since the polarizations, as indicated by bond dipoles, are nearly equal for Y = Cl and Y = Br, then the latter effect should predominate, leading again to the observed order $\bar{J}_2(Y = Br) > \bar{J}_2(Y = Cl)$. If the Y substituent is positively polarized, as in the case of methyl groups, both steric and electrostatic effects should favour **19**, and values of $\bar{J}_2$ close to the predicted minima have been observed in both polar and non-polar solvents[95].

The added potential energy of the molecular system due to a solvent reaction field in the Onsager treatment, may be given by

$$E_s = -[2(\varepsilon - 1)/(2\varepsilon + 1)](\mu^2/a^3)$$

where ε is the solvent dielectric constant, μ the solute dipole moment and a the radius of a spherical cavity surrounding the solute molecule. The energy difference between **19** and **20** in any solvent may be approximated as

$$\Delta E = \Delta E_0 - \left(\frac{8\pi Nd}{3M}\right)[(\varepsilon - 1)/(2\varepsilon + 1)](\mu_1^2 - \mu_2^2)$$

where N is Avogadro's number; d and M are the density and molecular weight of the liquid solute; ΔE_0 is the energy difference between isolated molecules of **19** and **20**; and μ_1 and μ_2 belong to **19** and **20** respectively. The equilibrium constant between **19** and **20** may be expressed as

$$K = \frac{X_1}{X_2} = 2\left(\frac{Z_1}{Z_2}\right) \exp\left[-\Delta E/RT\right]$$

where X_1 and X_2 are the mole fractions and Z_1 and Z_2 are partition functions for the isomers. A factor of 2 is included for **19** to allow for the equivalent conformation obtained on rotation through 180°. Such a calculation[95] yields an equilibrium constant of 1·905 for 2,3-dichloropropane in benzene and an energy difference between the two isomers of 914 cal/mole.

VI. APPLICATIONS TO COMPLEX SYSTEMS

Because of the now almost universal use of n.m.r. spectroscopy as a structural aid, any attempt at comprehensive coverage of applications to alkene systems within present space limitations would be impossible. The applications covered in this section represent a sampling from the relatively recent literature.

A. Cyclic Monoolefins

The variation of coupling parameters with ring size in cyclic olefins has been the subject of several studies[102-104]. Chapman has shown that the vinylic coupling in *cis* cyclic olefins varies appreciably with ring size but is not too dependent on substituents[104]. Ranges of values found for 5-, 6-, 7-, and 8-membered rings were: 5·4–7·0 Hz, 9·9–10·5 Hz, 9·7–12·5 Hz, and 11·8–12·8 Hz, respectively. Similar observations were also made by Laszlo and von R. Schleyer[103]. Smith and Kriloff[102] have reported coupling values for a number of unsubstituted cyclic monoolefins (**23**). These data reveal the variation of $J_{5,6}$ with ring size and suggest correlations between $J_{1,5}$ and $J_{1,6}$ with dihedral angle.

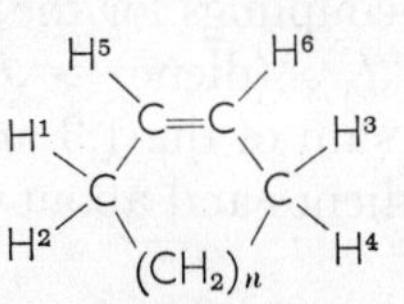

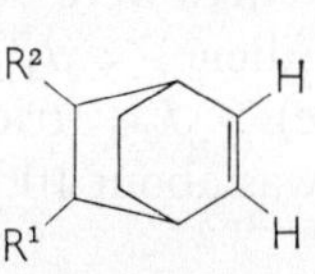

Roll and colleagues[105] have applied double-resonance (decoupling) techniques in their study of some 5,6-disubstituted bicyclo[2·2·2]oct-2-enes (**24**). The vinylic protons were assigned and an allylic coupling of 1·5 Hz was established. Decoupling methods were also used by Jefford and coworkers[106] in investigating the n.m.r. spectra of a series of substituted bicyclo[3·2·1]oct-2-enes. It was shown that although the skeleton permits the occurrence of three kinds of long-range couplings (W-plan, homoallylic, and allylic), only the first two are actually observed. Hanna and Harrington[107] have investigated the long-range couplings in the spectrum of 4-vinylidenecyclopentene in relation to the Karplus theory of π-electron contributions to spin couplings. A larger splitting is predicted here than for the case of methylallene. The spectra of 33 stereoisomeric alkyl cyclohexenes have been studied by Lippmaa and coworkers[108] who found the protons to be at lower field in the *cis* isomers as compared with the *trans* structures.

The conformational mobility of some 1,3-dithiepenes (**25**) and 1,3-dioxepenes (**26**)

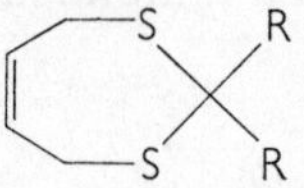

(**25**) 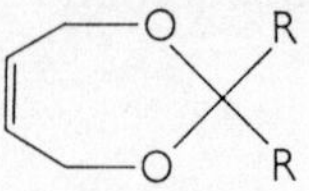   (**26**) 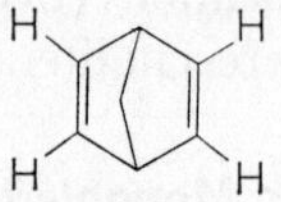   (**27**)

have been investigated by Freibolin and coworkers[109]. The chair and twist forms of these rings exhibit characteristic differences which permit the determination of their conformation by n.m.r. methods.

B. Cyclic Compounds with more than One Double Bond

An early investigation of bicycloheptadiene (**27**) was reported by Mortimer[113] working at 40 MHz. An unusually small coupling of 3·45 ± 0·1 Hz was observed for the two pairs of olefinic protons.

Guenther and Hinrichs[110] have reported parameters for a number of substituted cyclic 1,3-hexadienes and 1,3,5-heptatrienes. The following differences were noted in the couplings for the two types of systems: $J_{1,2}$ (diene) < $J_{1,2}$ (triene); $J_{1,3}$ (diene) > $J_{1,3}$ (triene); and $J_{3,4}$ (diene) > $J_{3,4}$ (triene). The sum of the 1,3 and 1,4 coupling constants was about 10 Hz for the dienes and about 6 Hz for the trienes.

Anet[111] has obtained n.m.r. spectral evidence for non-planarity of cycloheptatriene at $-150°$C. At this temperature two bands were observed in the methylene region. An activation energy of 6·3 kcal/mole was reported for the inversion process. Jensen and Smith[112] have also studied the low-temperature ($-170°$C) spectrum of cycloheptatriene and 7-deuterocycloheptatriene in CF_3Br and concluded that the former is a mixture of rapidly interconverting conformers. In the latter the conformer with ^2H *syn* to the ring is the more stable form. The results suggest that in cyclohexane the axial position would be preferred for deuterium. The structure and conformation of *cis,cis,cis*-1,4,7-cyclononatriene have been confirmed by Untch and Kurland[114], using proton–proton spin decoupling to be **27a**. The arguments leading to the acceptance of this conformation

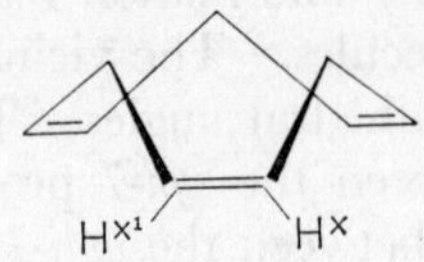

(27a)

are: (*a*) the AB quartet of the decoupled methylene protons in the low-temperature spectra verify that there is only one type of methylene hydrogen pair, thus eliminating the alternative saddle conformation; (*b*) the single sharp peaks in the high-temperature spectra prove the magnetic equivalence of the olefinic protons and that of the methylene protons; (*c*) the H^X and H^{X^1} patterns observed at low temperature when each type of methylene proton is irradiated in turn are decidedly different, thus substantiating the magnetic equivalence of the olefinic protons; (*d*) the irradiating frequency range necessary completely to decouple H^A and H^X, H^{X^1} is quite narrow. Chemical shift values for cyclopentadiene, 1-methylcyclopentadiene and cycloheptatriene have been reported by Strohmeier and coworkers[115].

Couplings of 7·4 to 11 Hz have been reported between the 1- and 4-protons of unsymmetrically substituted 1,4-dihydrobenzenes but the stereochemistry of these compounds is still in some doubt[116]. In the case of 1,4-dihydrobenzoic acid and 1,4-dihydrobiphenyl there is no chemical shift difference between the two CH_2 protons, which also couple equally to the methine proton. This may be associated with a boat–boat interconversion process between **28** and **29**.

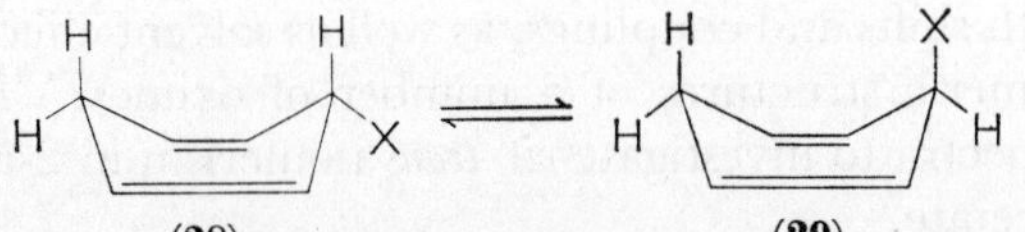

(28)   (29)

(X = Ph, COOH)

C. Norbornenes and Norbornadienes

Davis and Van Auken[117] have used decoupling methods to obtain complete analyses of the n.m.r. spectra of three *endo-exo* pairs of 2-substituted norbornenes. The *endo-endo* vicinal couplings are smaller than the *exo-exo* vicinal couplings. The bridge proton *syn* to the double bond is not always at higher field than the *anti* bridge proton.

Laszlo and von R. Schleyer[118] have used ^{13}C-H satellites and selective solvent effects to assign all protons and determine all the couplings for several norbornene derivatives. The existence of virtual coupling reduces the information readily available from the usual spectra of these molecules. The vicinal couplings were found not to depend solely on dihedral angles. There are indications of long-range couplings between the *syn*-7 proton and the *endo*-5 and *endo*-6 protons, as well as between the *anti*-7 proton and the olefinic protons (2 and 3). Allylic (1,3) and olefinic (1,2) couplings were found to have the same sign. The ^{13}C-H couplings varied regularly with the C—^{13}C—C angle.

Snyder and Franzus[119] have studied a number of 7-substituted norbornenes and norbornadienes and established criteria for configurational assignment of the 7-substituent with respect to the double bond. These criteria are: (*1*) bridge H couples to *anti* vinylic H by 0·8 Hz; (*2*) the sum of vicinal and allylic couplings between bridgehead H and vinylic protons is greater when *syn* to bridge H; and (*3*) bridge H is more shielded when *syn* to double bond. These long-range couplings and the delocalization of the bridge C orbitals are related to abnormally fast solvolytic reactions in these systems.

D. Miscellaneous Applications

Slomp and Wechter[120] have used n.m.r. spectra in characterizing the structures of oximes of α,β-unsaturated carbonyl compounds. The method is based on the interaction between the oxime OH with H atoms on adjacent carbon atoms. For the oximes of isophorone it was shown that OH is near the olefinic H in the *syn* form, and near the equatorial methylene H in the *anti* form. Karabatsos and coworkers[121] have used both shifts and couplings, as well as solvent effects, to determine the isomeric structures of a number of oximes. Tidd[122] has used n.m.r. spectra to investigate *cis–trans* isomerism in 2-methylpent-2-enyl thioacetate.

Stothers and coworkers[123] have recently analysed the vinylic p.m.r. spectra of 23 substituted styrenes. The series studied includes twelve *meta*- and *para*-substituted and ten *ortho*-substituted compounds. The effects of steric inhibition of conjugation on the vinylic proton parameters are considered in detail, and the vinylic proton shifts are compared with the corresponding vinylic ^{13}C shifts. Matsouka[124] has calculated the β-proton shieldings for styrene and for a number of other derivatives of ethylene. Reasonably good results were obtained for styrene in accord with the planar model for the molecule. Deviations in the case of bromoethylenes may be attributed to anisotropy effects caused by d orbital deformation.

N.m.r. spectra have been useful in investigating the structures of the adducts of addition reactions of unsaturated compounds. Bystrov and coworkers[125] have described such an application for the reaction between substituted cyclopentadienes with maleic anhydride. Weir and Hyne[126] have used n.m.r. evidence to support a proposed cyclic structure as the product of the base-catalysed dimerization of alkylidenemalononitriles.

Ouellette and coworkers[127] have carried out dilution studies to determine the chemical shift at infinite dilution of the hydroxyl proton in some 1-vinylcyclohexanols. The results were used to establish conformational preferences in these structures.

VIII. REFERENCES

1. H. L. Anderson, *Phys. Rev.*, **76,** 1460 (1949).
2. M. J. Stephen, *Proc. Roy. Soc. A (London)*, **243,** 264 (1958).
3. T. P. Das and R. Bersohn, *Phys. Rev.*, **115,** 897 (1959).
4. H. F. Hameka, *Z. Naturforsch.*, **14a,** 599 (1959).
5. N. F. Ramsey, *Phys. Rev.*, **86,** 243 (1952).
6. A. Saida and C. P. Slichter, *J. Chem. Phys.*, **22,** 26 (1954).
7. M. J. Stephen, *Mol. Phys.*, **1,** 223 (1958).
8. A. D. Buckingham, *Can. J. Chem.*, **38,** 300 (1960).
9. N. F. Ramsey, *Phys. Rev.*, **91,** 303 (1953).
10. M. Karplus and D. H. Anderson, *J. Chem. Phys.*, **30,** 6 (1959).
11. M. Karplus, *Rev. Mod. Phys.*, **32,** 455 (1960).
12. E. Hiroike, *Progr. Theoret. Phys. (Kyoto)*, **21,** 943 (1959).
13. E. Hiroike, *J. Phys. Soc. Japan*, **15,** 270 (1960).
14. H. M. McConnell, *J. Chem. Phys.*, **30,** 126 (1959).
15. J. A. Pople and D. P. Santry, *Mol. Phys.*, **8,** 1 (1964).
16. R. J. Hoffman, *Arkiv. Kemi*, **17,** 1 (1960).
17. H. M. McConnell, *J. Mol. Spectry.*, **1,** 11 (1957).
18. H. M. McConnell, *J. Chem. Phys.*, **30,** 126 (1959).
19. M. Karplus, *J. Am. Chem. Soc.*, **82,** 4431 (1960).
20. R. A. Hoffman and S. Gronowitz, *Arkiv. Kemi*, **16,** 563 (1960).
21. A. Forman, J. N. Murrell and L. E. Orgel, *J. Chem. Phys.*, **31,** 1129 (1959).
22. W. D. Phillips and R. E. Benson, *J. Chem. Phys.*, **33,** 607 (1960).
23. A. D. McLachlan, *Mol. Phys.*, **1,** 233 (1958).
24. R. A. Hoffman and S. Gronowitz, *Acta Chem. Scand.*, **13,** 1477 (1959).
25. G. S. Reddy and J. H. Goldstein, *J. Mol. Spectry.*, **8,** 475 (1962).
26. R. M. Lynden-Bell and N. Sheppard, *Proc. Roy. Soc. A (London)*, **269,** 385 (1962).
27. D. M. Graham and C. E. Holloway, *Can. J. Chem.*, **41,** 2114 (1963).

28. J. N. Shoolery, *J. Chem. Phys.*, **31,** 1427 (1959).
29. N. Muller and D. E. Pritchard, *J. Chem. Phys.*, **31,** 768 (1959).
30. N. Muller and D. E. Pritchard, *J. Chem. Phys.*, **31,** 1471 (1959).
31. R. E. Mayo and J. H. Goldstein, *J. Mol. Spectry.*, **14,** 173 (1964).
32. R. T. Hobgood, G. S. Reddy and J. H. Goldstein, *J. Phys. Chem.*, **67,** 110 (1963).
33. V. S. Watts and J. H. Goldstein, *J. Chem. Phys.*, **42,** 228 (1965).
34. A. W. Douglas and J. H. Goldstein, *J. Mol. Spectry.*, **16,** 1 (1965).
35. T. Schaefer, *J. Chem. Phys.*, **36,** 2235 (1962).
36. R. Hirst and D. M. Grant, *J. Am. Chem. Soc.*, **84,** 2009 (1962).
37. C. N. Banwell and N. Sheppard, *Mol. Phys.*, **3,** 351 (1960).
38. J. C. Muller, *Bull. Soc. Chim. France*, 1815 (1964).
39. T. Schaefer, *Can. J. Chem.*, **40,** 1 (1962).
40. A. A. Bothner-By and C. Naar-Colin, *J. Am. Chem. Soc.*, **80,** 1728 (1958).
41. J. N. Shoolery and B. P. Dailey, *J. Am. Chem. Soc.*, **77,** 3977 (1955).
42. G. S. Reddy, J. H. Goldstein and L. Mandell, *J. Am. Chem. Soc.*, **83,** 1300 (1961).
43. G. S. Reddy and J. H. Goldstein, *J. Am. Chem. Soc.*, **83,** 2045 (1961).
44. E. B. Whipple, W. E. Stewart, G. S. Reddy and J. H. Goldstein, *J. Chem. Phys.*, **34,** 2136 (1961).
45. H. S. Gutowsky, M. Karplus and D. M. Grant, *J. Chem. Phys.*, **31,** 1278 (1959).
46. E. I. Snyder, L. J. Altman and J. D. Roberts, *J. Am. Chem. Soc.*, **84,** 2004 (1962).
47. G. E. Maciel, *J. Phys. Chem.*, **69,** 1947 (1965).
48. R. T. Hobgood, J. H. Goldstein and G. S. Reddy, *J. Chem. Phys.*, **35,** 2038 (1961).
49. R. T. Hobgood and J. H. Goldstein, *Spectrochim. Acta*, **19,** 321 (1963).
50. D. W. Moore and J. A. Happe, *J. Phys. Chem.*, **65,** 224 (1961).
51. R. Summitt, J. J. Eisch, J. T. Trainor and M. T. Rogers, *J. Phys. Chem.*, **67,** 2362 (1963).
52. S. Cawley and S. S. Danyluk, *J. Phys. Chem.*, **68,** 1240 (1964).
53. G. S. Reddy and J. H. Goldstein, *J. Am. Chem. Soc.*, **83,** 2045 (1961).
54. J. Tillieu, *Ann. Phys.*, **2,** 471, 631 (1957).
55. P. T. Narasimhan and M. T. Rogers, *J. Chem. Phys.*, **31,** 1302 (1959).
56. A. A. Bothner-By and C. Naar-Colin, *J. Am. Chem. Soc.*, **83,** 231 (1961).
57. R. A. Hoffman and S. Gronowitz, *Arkiv Kemi*, **16,** 471 (1960).
58. L. M. Jackman and R. N. Wiley, *J. Chem. Soc.*, 2881 (1960).
59. G. B. Savitsky and K. Namikawa, *J. Phys. Chem.*, **67,** 2754 (1963).
60. C. A. Reilly, *J. Chem. Phys.*, **37,** 456 (1962).
61. M. H. Gianni, E. L. Stogryn and C. M. Orlando, Jr., *J. Phys. Chem.*, **67,** 1385 (1963).
62. H. Hageveen, G. Maccagnani and F. Taddei, *Rec. Trav. Chim.*, **83,** 937 (1964).
63. R. A. Beaudet and J. D. Baldeschwieler, *J. Mol. Spectry.*, **9,** 30 (1962).
64. H. M. McConnell, *J. Chem. Phys.*, **27,** 226 (1957).
65. C. Pascual, J. Meier and W. Simon, *Helv. Chim. Acta*, **49,** 164 (1966).
66. M. W. Dietrich and R. E. Keller, *Anal. Chem.*, **36,** 258 (1964).
67. F. C. Stehling and K. W. Bartz, *Anal. Chem.*, **38,** 1467 (1966).

68. P. Laszlo and P. v. R. Schleyer, *Bull. Soc. Chim. France*, 87 (1964).
69. C. N. Banwell and N. Sheppard, *Disc. Faraday. Soc.*, **34,** 115 (1962).
70. J. H. Goldstein and G. S. Reddy, *J. Chem. Phys.*, **38,** 2736 (1963).
71. R. T. Hobgood, Jr. and J. H. Goldstein, *J. Mol. Spectry.*, **12,** 76 (1964).
72. E. B. Whipple, J. H. Goldstein and W. E. Stewart, *J. Am. Chem. Soc.*, **81,** 4761 (1959).
73. E. B. Whipple, J. H. Goldstein and L. Mandell, *J. Chem. Phys.*, **30,** 1109 (1959).
74. D. F. Koster and A. Danti, *J. Phys. Chem.*, **69,** 486 (1965).
75. M. L. Martin, G. J. Martin and P. Caubere, *Bull. Soc. Chim. France*, 3066 (1964).
76. E. I. Snyder and J. D. Roberts, *J. Am. Chem. Soc.*, **84,** 1582 (1962).
77. S. Alexander, *J. Chem. Phys.*, **32,** 1700 (1960).
78. M. P. Simonnin and G. Pourcelot, *Compt. Rend.*, C **262,** 1279 (1966).
79. G. T. Jones, M. Randic and J. J. Turner, *Croat. Chem. Acta*, **36,** 111 (1964).
80. H. Walz and P. Kurtz, *Z. Naturforsch.*, **18b,** 334 (1963).
81. S. G. Frankiss and I. Matsubara, *J. Phys. Chem.*, **70,** 1543 (1966).
82. V. S. Watts, G. S. Reddy and J. H. Goldstein, *J. Mol. Spectry.*, **11,** 325 (1963).
83. V. S. Watts and J. H. Goldstein, *J. Chem. Phys.*, **42,** 228 (1965).
84. P. Bates, S. Cawley and S. S. Danyluk, *J. Chem. Phys.*, **40,** 2415 (1964).
85. V. S. Watts, J. Loemker, and J. H. Goldstein, *J. Mol. Spectry.*, **17,** 348 (1965).
86. P. Laszlo and H. J. T. Bos, *Tetrahedron Letters* 1325 (1965).
87. A. D. Buckingham, *Can. J. Chem.*, **38,** 300 (1960).
88. P. Diehl and R. Freeman, *Mol. Phys.*, **4,** 39 (1961).
89. I. G. Ross and R. A. Sack, *Proc. Phys. Soc. D (London)*, **63,** 893 (1950).
90. F. Hruska, E. Bock and T. Schaefer, *Can. J. Chem.*, **41,** 3034 (1963).
91. H. E. Hallam and T. C. Ray, *Trans. Faraday Soc.*, **58,** 1299 (1962).
92. V. S. Watts and J. H. Goldstein, *J. Mol. Spectry.*, **21,** 260 (1966).
93. T. Schaefer and W. G. Schneider, *Can. J. Chem.*, **38,** 2066 (1960).
94. E. B. Whipple, J. H. Goldstein and G. R. McClure, *J. Am. Chem. Soc.*, **82,** 3811 (1960).
95. E. B. Whipple, *J. Chem. Phys.*, **35,** 1039 (1961).
96. A. A. Bothner-By, C. Naar-Colin and H. Gunther, *J. Am. Chem. Soc.*, **84,** 2748 (1962).
97. A. A. Bothner-By and H. Gunther, *Disc. Faraday Soc.*, **34,** 127 (1962).
98. A. A. Bothner-By, S. Castellano and H. Gunther, *J. Am. Chem. Soc.*, **87,** 2439 (1965).
99. D. R. Herschbach and L. C. Krishner, *J. Chem. Phys.*, **28,** 728 (1958).
100. A. A. Bothner-By and C. Naar-Colin, *J. Am. Chem. Soc.*, **83,** 231 (1961).
101. K. C. Ramey and W. S. Brey, Jr., *J. Chem. Phys.*, **40,** 2349 (1964).
102. G. V. Smith and H. Kriloff, *J. Am. Chem. Soc.*, **85,** 2016 (1963).
103. P. Laszlo and P. v. R. Schleyer, *J. Am. Chem. Soc.*, **85,** 2017 (1963).
104. O. L. Chapman, *J. Am. Chem. Soc.*, **85,** 2014 (1963).
105. D. B. Roll, B. J. Nist and A. C. Huitric, *J. Pharm. Sci.*, **56,** 212 (1967).
106. C. W. Jefford, B. Waegell and K. Ramey, *J. Am. Chem. Soc.*, **87,** 2191 (1965).
107. M. W. Hanna and J. K. Harrington, *J. Phys. Chem.*, **67,** 940 (1963).

108. E. Lippmaa, S. Rang, O. Eisen and J. Puskar, *Eesti NSV Tead. Akad. Toim., Fuus. Mat. Tehnikatead. Seer.*, 615 (1966).
109. H. Friebolin, R. Mecke, S. Kabuss and A. Luettringhaus, *Tetrahedron Letters*, 1929 (1964).
110. H. Guenther and H. H. Hinrichs, *Tetrahedron Letters*, 787 (1966).
111. F. A. L. Anet, *J. Am. Chem. Soc.*, **86,** 458 (1964).
112. F. R. Jensen and L. A. Smith, *J. Am. Chem. Soc.*, **86,** 956 (1964).
113. F. S. Mortimer, *J. Mol. Spectry.*, **3,** 528 (1959).
114. K. G. Untch and R. J. Kurland, *J. Mol. Spectry.*, **14,** 156 (1964).
115. W. Strohmeier, E. Lombardi and R. M. Lemmon, *Z. Naturforsch.*, **14a,** 106 (1959).
116. L. J. Durham, J. Studebaker and M. J. Perkins, *Chem. Commun.*, 456 (1965).
117. J. C. Davis, Jr. and T. V. Van Auken, *J. Am. Chem. Soc.*, **87,** 3900 (1965).
118. P. Laszlo and P. v. R. Schleyer, *J. Am. Chem. Soc.*, **86,** 1171 (1964).
119. E. I. Snyder and B. Franzus, *J. Am. Chem. Soc.*, **86,** 1166 (1964).
120. G. Slomp and W. J. Wechter, *Chem. and Ind. (London)*, 41 (1962).
121. G. J. Karabatsos, R. A. Taller, and F. M. Vane, *J. Am. Chem. Soc.*, **85,** 2326, 2327 (1963).
122. B. K. Tidd, *J. Chem. Soc.*, 3909 (1963).
123. Gurudata, J. B. Stothers and J. D. Taiman, *Can. J. Chem.*, **45,** 731 (1967).
124. S. Matsuoka, Shinichi, *Sci. Repts. Kanazawa Univ.*, **10,** 1 (1965).
125. V. F. Bystrov, A. U. Stepanyants, V. A. Mironov, *Zh. Obshch. Khim.*, **34,** 4039 (1964).
126. M. R. S. Weir and J. B. Hyne, *Can. J. Chem.*, **42,** 1440 (1964).
127. R. J. Ouellette, K. Liptak and G. E. Booth, *J. Org. Chem.*, **31,** 546 (1966).

CHAPTER 2

The Properties of Alkene Carbonium Ions and Carbanions

HERMAN G. RICHEY, JR.

The Pennsylvania State University, University Park, Pennsylvania, U.S.A.

I. INTRODUCTION	40
II. VINYL IONS	42
A. Introduction	42
B. Cations	43
1. Stability	43
2. Geometry	48
3. Formation and reactions	48
C. Anions	49
1. Stability	49
2. Geometry	53
3. Formation and reactions	56
III. ALLYL IONS	56
A. Introduction	56
B. Cations	58
1. Stability	58
2. Geometry	61
3. Formation and reactions	63
C. Anions	67
1. Stability	67
2. Geometry	70
3. Formation and reactions	77
IV. NON-CONJUGATED IONS	77
A. Introduction	77
B. Cations	83
1. 5,6-Double bonds	83
2. 6,7- and 4,5-Double bonds	93
3. 3,4-Double bonds	95
C. Anions	101
V. ACKNOWLEDGEMENTS	106
VI. REFERENCES	107

I. INTRODUCTION

The properties of carbonium ions and carbanions are significantly influenced by proximate alkene functions. This chapter on ions containing alkene groups concentrates particularly on stability and on the closely interrelated subjects of geometry and bonding. When these aspects of the ions are understood, their chemical behaviour often is predictable from our experience with their saturated counterparts. The formation and chemical reactions of an ion will be discussed only to the extent that they present unusual features or have received little previous review.

The nature of the effect exerted by the alkene function is dictated by its distance from the formally charged carbon. In vinyl ions (**1**), the formal charge is placed directly on a carbon of the double bond. In

$$\underset{(1)}{\overset{\diagdown}{\underset{\diagup}{C}}=\overset{+(-)}{\underset{}{C}}-} \qquad \underset{(2)}{\overset{\diagdown}{\underset{\diagup}{C}}=\overset{|}{\underset{}{C}}-\overset{\diagup}{\underset{\diagdown}{C}}+(-)} \qquad \underset{(3)}{\overset{\diagdown}{\underset{\diagup}{C}}=\overset{|}{\underset{}{C}}-\overset{|}{\underset{|}{C}}-\overset{\diagup}{\underset{\diagdown}{C}}+(-)}$$

allyl ions (**2**), the charge is conjugated with a double bond. However, in the formally non-conjugated ions (**3**), any major interaction must depend on bonding across intervening saturated carbon atoms.

It seems more natural to consider together the cations and anions of these categories than to attempt to discuss together the different ions of like charge. The section on each category will begin with a brief discussion of the predictions about the properties of the cation and anion that can be made on the basis of current theory. Separate subsections will then summarize experimental observations relevant to the properties of the carbonium ions and the carbanions.

In this brief chapter it is impossible to refer to all pertinent studies, and consequently selection has been unavoidable. However, an effort is made to refer to other reviews that consider these ions. Other important topics relevant to the considerations of this chapter, but omitted for lack of space, include the effects exerted on carbonium ions and carbanions by aryl and alkyne functions, often similar to those exerted by alkene functions, and the behaviour of the related free radicals, often relevant to the prediction of properties for the carbonium ions and carbanions.

When possible, information obtained by direct studies of stable solutions of the ions is emphasized. However, where the ions have yet to be prepared as long-lived species, properties must be inferred from kinetic and product studies of chemical reactions. The usual

2. The Properties of Alkene Carbonium Ions and Carbanions

assumption will be made throughout this chapter that properties of transition states that lead to and from reactive intermediates bear considerable resemblance to the intermediates themselves[1].

The properties and reactions of carbonium ions[2,2a] and carbanions[3] can be influenced significantly by interactions with the counterions and with the solvent. This chapter will often skirt these complexities that, although important, certainly are not unique to the categories of ions considered here. However, some interactions are so significant in considering the data related to carbanions that they are discussed briefly below, rather than in subsequent sections.

It would be ideal to discuss carbanions having a minimum of specific bonding to a particular cation, or at least to have such bonding remain constant throughout the series of carbanions to be considered. However, this ideal is rarely approached and some studies are concerned with species that clearly are not carbanions at all.

Ions of sodium, potassium, and the heavier elements of group IA should generally have the weakest interactions with carbanions[4]. Some experiments to be discussed involve such organometallic compounds as transient intermediates, but the extent to which they ionize in solution is not well established. The insolubility of these compounds in hydrocarbons and their rapid attack on the more polar solvents in which they might have greater solubility are responsible for the paucity of information concerning their nature in solution[4,5,6]. Some compounds of these metals with resonance stabilization by groups such as aryl can be studied in solution. The properties of even these resonance stabilized carbanions can be significantly affected by ion-pairing with a cation[7]. Even less is known about the nature of organometallic compounds of calcium and the heavier elements of group IIA[4,8,9].

Organolithium and Grignard reagents are sometimes the most polar species for which certain information is available. Therefore, the nature of these reagents in solution and particularly the degree to which they approximate carbanions must be considered.

Alkyllithium reagents exist in hydrocarbon solutions as discrete tetramers or hexamers, the more highly branched alkyl groups apparently favouring tetramer formation. The electron-deficiency of the lithium is met in these structures by formation of 'electron-deficient' bonds to several carbons. In more polar solvents such as ethers and amines, interactions with the solvent are significant and the polymeric structures are degraded, though probably not significantly to monomers. Lithium reagents with significant resonance stabilization by groups such as aryl generally have the properties expected for ionization to carbanions, though ion-pairing and solvent interactions remain significant. The problem of the structures of alkyllithium compounds has recently been reviewed [10].

Though Grignard reagents possess polar C—Mg bonds, extensive ionization does not occur except to carbanions that have very large resonance stabilizations. The structures of Grignard reagents and their degree of association in solution have long been matters of controversy. However, it is now thought that Grignard reagents are monomeric in dilute solutions in tetrahydrofuran. The bromides and iodides are monomeric at low concentration, even in ether, though apparent molecular weight increases with increasing concentration and in some cases approaches that of a dimer. The chlorides are largely dimeric, even at lower concentrations. The dominant monomeric species probably is RMgX, though this species must be in equilibrium with R_2Mg.

$$2\,RMgX \rightleftharpoons R_2Mg + MgX_2$$

The ethereal solvents in which Grignard reagents ordinarily are studied become integral parts of the reagent by bonding strongly to the magnesium. The structural problem of Grignard reagents has been reviewed recently [11].

II. VINYL IONS

A. Introduction

The formally charged carbon in a vinyl cation or anion also is a carbon of the double bond. The nature of this carbon, disubstituted instead of trisubstituted as are carbons that bear or share the charge in the carbonium ions and carbanions more commonly considered, has a dominant influence on the properties of vinyl ions.

A linear geometry (4) in which the charged carbon has sp hybridization and a bent geometry (5) in which this carbon has sp^2 hybridization are extremes to be considered for these ions.

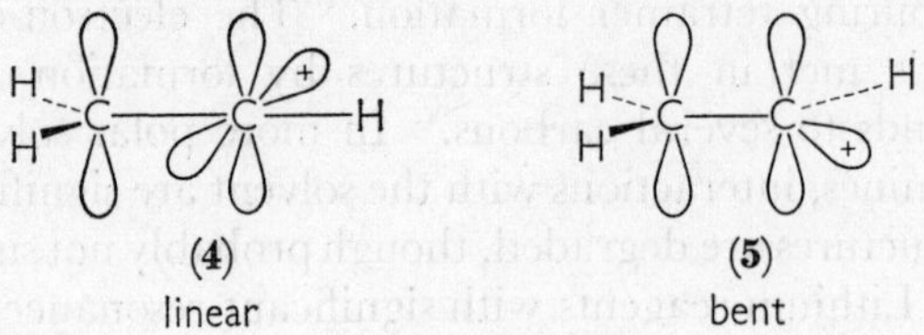

(4) linear

(5) bent

A linear geometry should be more favourable for the cation. In this geometry, the sp bonding orbitals utilize all of the s character and the bonding electron pairs are as close to the nucleus and yet as remote from each other as possible. The vacant orbital, p as in saturated cations, uses none of the s character. Alteration of a linear vinyl cation to a bent geometry requires promotion of one-third of an electron from an s to a p orbital. Charge density in the unsubstituted vinyl cation has been calculated using an extended Hückel method[12].

$$\begin{array}{cc} \overset{+0.218}{H}\diagdown & \overset{+0.529}{C}\text{---}H \\ H\diagup\underset{-0.116}{C}\!=\!\!=\!\!\underset{+0.150}{C} & \end{array} \qquad \begin{array}{cc} H\diagdown & \overset{+0.496}{H}\diagup \\ \underset{-0.231}{C}\text{---}C & \diagdown H \\ H\diagup & \end{array}$$

The disubstituted carbon of a linear vinyl cation is predicted to have a slightly greater electron deficiency than the trisubstituted carbon of the ethyl cation.

A bent geometry is predicted for the vinyl anion on the basis of similar reasoning. Such an anion has the nonbonded electron pair in an orbital (sp^2) of maximum s character. The isoelectronic species (6) containing carbon-nitrogen double bonds are well known to

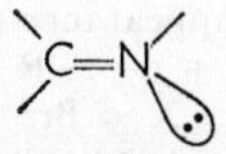

$$\diagdown\!\!\!\diagup C\!=\!N\diagdown\!\!\!\diagup\!\!:$$
(6)

assume this geometry. A vinyl anion is expected to be more stable (relative to neutral precursors) than are saturated carbanions; the unshared electron pair in a vinyl anion is favourably placed in an orbital having greater s character and therefore closer to the nucleus than the sp^3 orbital presumably occupied by the unshared electron pair of a saturated carbanion.

B. Cations

The properties of vinyl cations are discussed extensively in recent reviews[13,14].

I. Stability

Scant systematic consideration was devoted to vinyl cations as reaction intermediates until quite recently. A widely accepted impression that they were much less stable than saturated cations and not likely to be encountered as reaction intermediates may have been partly responsible. In fact, vinyl cations (**8**) probably are more stable

$$RCH_2^+ \quad < \quad RC^+{=}CH_2 \quad < \quad RCH^+CH_3$$
$$(7) \qquad\qquad (8) \qquad\qquad (9)$$
increasing stability →

than the corresponding primary cations (**7**), though somewhat less stable than the corresponding secondary cations (**9**). Reference is made here to thermodynamic stabilities of the ions relative to suitable precursors.

Mass spectral studies provide the most direct evidence concerning stability. Data for formation of the vinyl cation and other simple ions by impact of electrons on neutral precursors are summarized in Table 1. The energy needed to form the vinyl cation is intermediate between the energies needed to form the methyl and ethyl cations in the equivalent reaction process. It seems probable that the energies of the cations fall in the order of the energies needed to generate

TABLE 1. Appearance potentials of cations and ionization potentials of radicals

R^+	Appearance potential for $RX + e^- \rightarrow R^+ + 2e^- + X\cdot$				Ionization Potential of $R\cdot^d$ (e.v.)
	$X = Cl^a$ (e.v.)	$X = Br^a$ (e.v.)	$X = I^a$ (e.v.)	$X = H$ (e.v.)	
CH_3^+	13.5	13.2	12.36	14.39[b]	9.95
$CH_2{=}CH^+$	12.81			14.00[c]	9.45
$CH_3CH_2^+$	12.2	11.4	11.0	12.80[b]	8.78
$CH_3CH_2CH_2^+$					8.69
$CH_2{=}CHCH_2^+$				11.95[b]	8.16
$(CH_3)_2CH^+$	11.08	10.45	9.74	11.73[b]	7.90
$CH_2{=}\overset{\underset{\displaystyle CH_3}{\mid}}{C}CH_2^+$			9.40		8.03
$CH_3CH{=}CHCH_2^+$			9.15		7.71
$(CH_3)_3C^+$	10.27				7.42

[a] From the compilation by A. Maccoll in *The Transition State*, Special Publication No. 16, The Chemical Society, London, 1962, p. 159.

[b] From F. H. Field and J. L. Franklin, *Electron Impact Phenomena and the Properties of Gaseous Ions*, Academic Press, Inc., New York, 1957, Appendix, Part 1.

[c] A. G. Harrison and F. P. Lossing, *J. Am. Chem. Soc.*, **82**, 519 (1960).

[d] Selected from the tabulation of values in F. P. Lossing in *Mass Spectrometry* (Ed. C. A. McDowell), McGraw-Hill Book Co., Inc., New York, 1963, Chap. 11.

2. The Properties of Alkene Carbonium Ions and Carbanions

them, in spite of factors that may somewhat influence the relative values of the energies of the precursors[14].

Solutions of vinyl cations with lifetimes sufficient to permit their observation by spectral methods have not been prepared. As a consequence, estimates of the stabilities of vinyl cations by measurements of equilibria between the ions and neutral precursors are not available. However, the rates of those reactions in which formation of vinyl cations can be implicated as the rate-determining step can be used to assess their stabilities. The results of such studies described below are consistent with the suggestion that vinyl cations are somewhat less stable than equivalently substituted saturated cations.

Vinyl cations are certainly intermediates in some electrophilic additions to alkynes. Evidence for the intermediacy of vinyl cations is particularly compelling for reactions of alkynes with strong acids in polar solvents. Recent mechanistic investigations have included studies of hydrations of arylalkynes (**10**)[15,16], arylpropiolic acids (**11**)[13,16], and alkynyl ethers[17,18] and thioethers (**12**)[18,19], and of

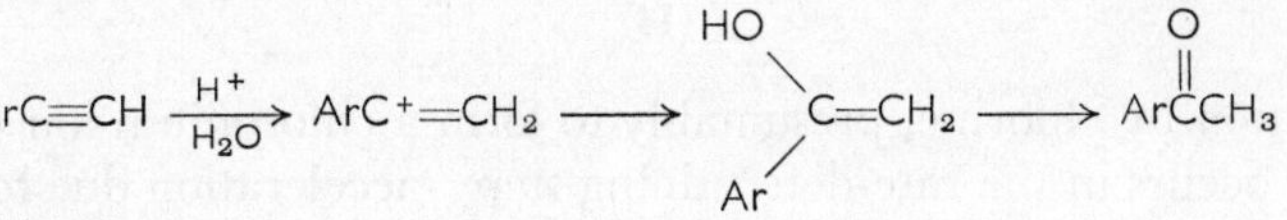

$$ArC\equiv CH \xrightarrow[H_2O]{H^+} ArC^+=CH_2 \longrightarrow \underset{Ar}{HO}C=CH_2 \longrightarrow ArCCH_3$$

(**10**)

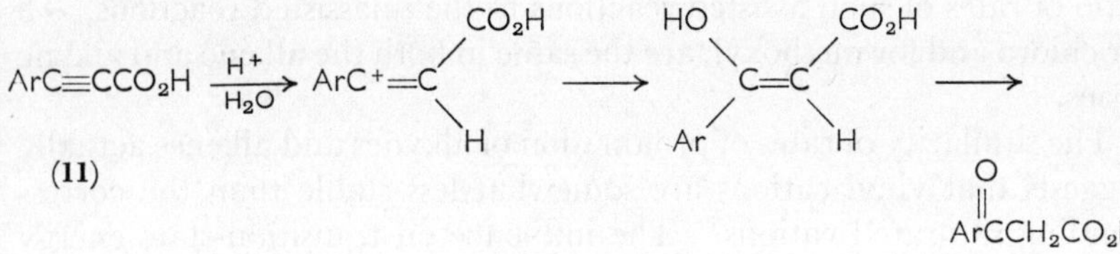

$$ArC\equiv CCO_2H \xrightarrow[H_2O]{H^+} ArC^+=C\underset{H}{\overset{CO_2H}{}} \longrightarrow \underset{Ar}{HO}C=C\underset{H}{\overset{CO_2H}{}} \longrightarrow$$

(**11**)

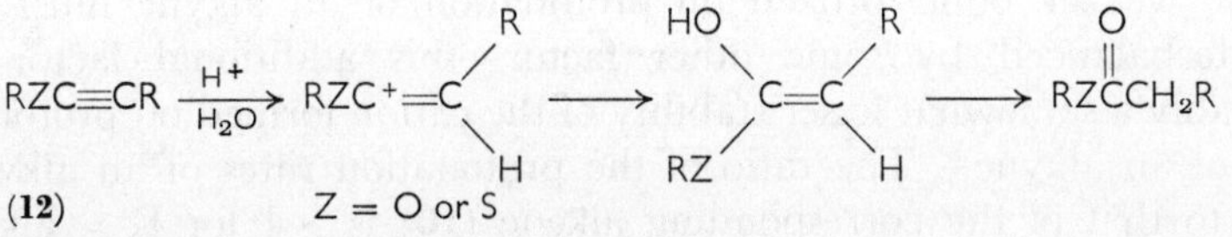

$$ArCCH_2CO_2H$$

$$RZC\equiv CR \xrightarrow[H_2O]{H^+} RZC^+=C\underset{H}{\overset{R}{}} \longrightarrow \underset{RZ}{HO}C=C\underset{H}{\overset{R}{}} \longrightarrow RZCCH_2R$$

(**12**) Z = O or S

$$EtC\equiv CEt \xrightarrow{CF_3CO_2H} EtC^+=C\underset{H}{\overset{Et}{}} \longrightarrow$$

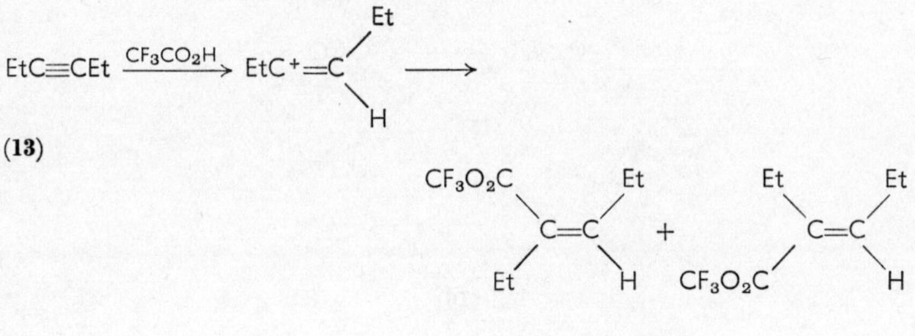

(**13**)

$$\underset{Et}{CF_3O_2C}C=C\underset{H}{\overset{Et}{}} + \underset{CF_3O_2C}{Et}C=C\underset{H}{\overset{Et}{}}$$

addition of trifluoroacetic acid to aliphatic alkynes (**13**)[20,21]. Protonation to form a vinyl cation is the probable rate-determining step for all of these reactions. In fact, the characteristics of these additions are remarkably similar to those of the similar additions to the corresponding alkenes, additions for which rate-determining proton addition is well established.

A striking example of the similarities between the additions to alkynes and to alkenes is provided by the nearly identical effects exerted on additions to 1-pentyne[20] and 1-pentene[22] by a diverse group of substituents at $C_{(5)}$. The similarity of substituent effects is particularly remarkable for those reactions that involve more than simple addition. Formation of the major product in the addition to 5-chloro-1-pentyne, for example, involves migration of the chlorine.

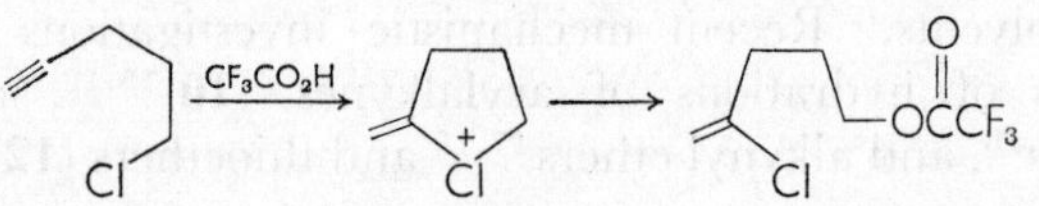

(**14**)

Participation by chlorine, presumably to form a chloronium ion (**14**), probably occurs in the rate-determining step—acceleration due to the chloro substituent is inferred from analysis of the kinetic data[20]. The ratio of rates of such assisted reactions to the unassisted reactions, ∼5 for chloro and for methoxyl, are the same in both the alkene and alkyne series.

The similarity of rates of protonation of alkynes and alkenes actually suggests that vinyl cations are somewhat less stable than the corresponding saturated cations. The influence on transition-state energy of the weaker bond broken on protonation of an alkyne must be counterbalanced by some other factor—this additional factor is probably a somewhat lesser stability of the cation formed on protonation of an alkyne. The ratio of the protonation rates of an alkyne (**15**) to that of the corresponding alkene (**16**) is ∼1 for R = alkyl,

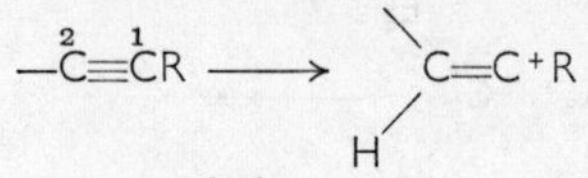

(**15**)

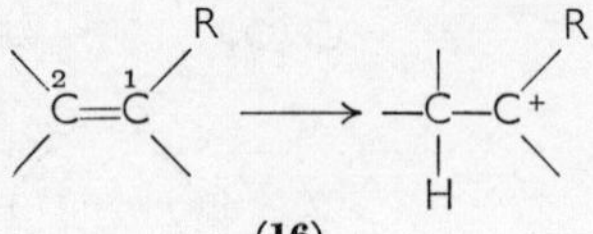

(**16**)

2. The Properties of Alkene Carbonium Ions and Carbanions

~10 for R = aryl, and ~10^2 for R = alkoxyl in the reactions with strong acids referred to above.

Carbonium ion stabilizing substituents at $C_{(1)}$ in **15** exert large accelerating effects on the rates of formation of vinyl cations by protonation of alkynes[14]. However, in accord with the expectation that relatively little positive charge is placed at $C_{(2)}$, methyl substituents at that position have only small effects on the rates. In fact, a methyl at $C_{(2)}$ in **10** or **12** (Z = O) actually is rate-retarding, an indication that its stabilization of the cation is less than of the parent alkyne.

Vinyl cations must also form as intermediates in solvolysis reactions of alkynes such as **17**[23] and **20**[24] that furnish cyclization products.

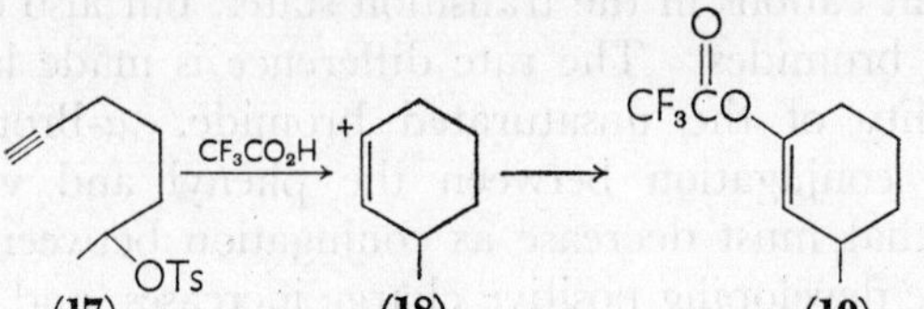

(17) → [CF$_3$CO$_2$H] → (18) → (19)

OTs = p-toluenesulphonate

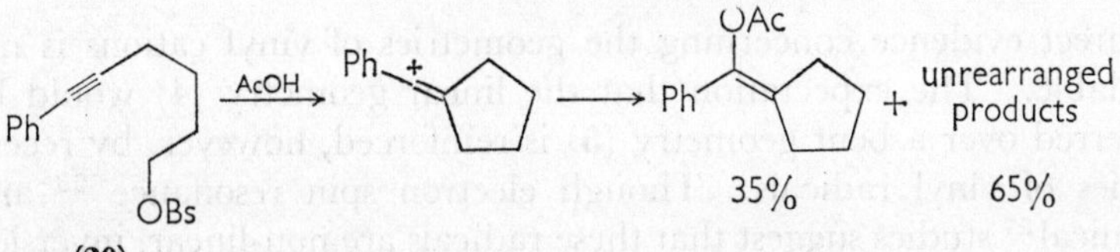

(20) → [AcOH] → 35% + 65% unrearranged products

OBs = p-bromobenzenesulphonate

These reactions are analogous to those of compounds with similarly placed double bonds that are discussed in Section IV.B. The ability of cyclization to compete successfully with other available reaction paths is another indication that the ease of formation of vinyl cations from electrophilic additions to alkynes is comparable to that of the formation of alkyl cations from alkenes.

Vinyl cations also have been formed in solution by loss of a nucleophilic group from appropriate vinyl compounds. For example,

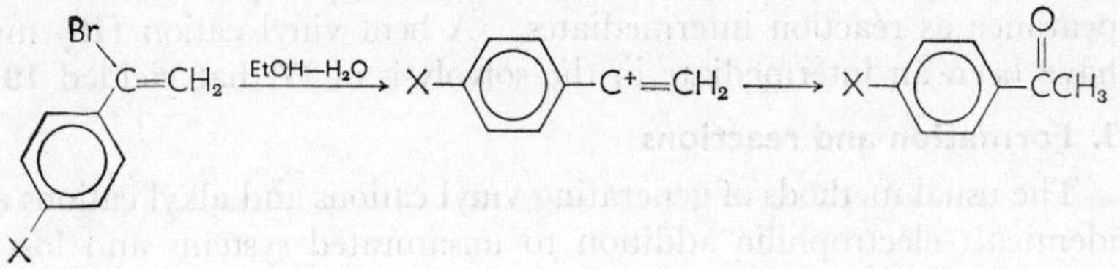

[Br-C(Ar)=CH$_2$] → [EtOH–H$_2$O] → X–C$_6$H$_4$–C$^+$=CH$_2$ → X–C$_6$H$_4$–COCH$_3$

hydrolyses of *para* substituted α-bromostyrenes have the characteristics of unimolecular nucleophilic displacements[25]. In fact, the reactions resemble those of the corresponding saturated α-arylethyl bromides. Therefore, formation of vinyl cations must be the rate-determining step.

The 1-phenylvinyl cation probably is less stable than the 1-phenylethyl cation, but by an amount smaller than the energy difference between the transition states leading to their formation from the corresponding bromides. The hydrolysis rate in 80% aqueous ethanol at 100° of α-phenylethyl bromide is $\sim 10^8$ that of α-bromostyrene[14]. The rate difference reflects not only different stabilities of the incipient cations in the transition states, but also different stabilities of the bromides. The rate difference is made larger by the greater stability of the unsaturated bromide. α-Bromostyrene is stabilized by conjugation between the phenyl and vinyl groups, stabilization that must decrease as conjugation between the phenyl group and the developing positive charge increases, and perhaps also by conjugation between the bromo and vinyl groups.

2. Geometry

Direct evidence concerning the geometries of vinyl cations is not available. The expectation that the linear geometry (**4**) would be preferred over a bent geometry (**5**) is reinforced, however, by recent studies of vinyl radicals. Though electron spin resonance [26] and chemical[27] studies suggest that these radicals are non-linear, inversion cannot have an energy barrier exceeding a few kcal/mole. Therefore,

$$\begin{array}{c}H\\ \diagdown \\ \end{array} \begin{array}{c}H\\ \diagup \\ \end{array} \qquad \qquad H$$

the linear geometry, through which a radical must pass during inversion, cannot be much less stable than the bent geometry. Factors favouring a linear structure should be more important in the cation than in the radical.

Bent vinyl cations, though thought to be less stable than linear vinyl cations, apparently are not of such high energy to preclude their appearance as reaction intermediates. A bent vinyl cation (**18**) must have been an intermediate in the solvolysis of **17** that yielded **19**[23].

3. Formation and reactions

The usual methods of generating vinyl cations and alkyl cations are identical: electrophilic addition to unsaturated systems and loss of

2. The Properties of Alkene Carbonium Ions and Carbanions

nucleophilic groups from suitable derivatives. Vinyl cations arise not only in electrophilic additions to alkynes, as seen in examples above, but also in similar reactions of some allenes [14,28].

The reactions of vinyl cations also are analogous to those of alkyl cations. The common reactions of both classes are the reverse of the reactions leading to their formation: addition of nucleophiles (including as nucleophiles carbon–carbon multiple bonds and neighbouring groups) and β-elimination of protons or other groups. Migration of a group to the formally positive carbon from an adjacent carbon, a reaction characteristic of alkyl cations, probably also occurs readily in vinyl cations. For example, the acid-catalysed decomposition of a

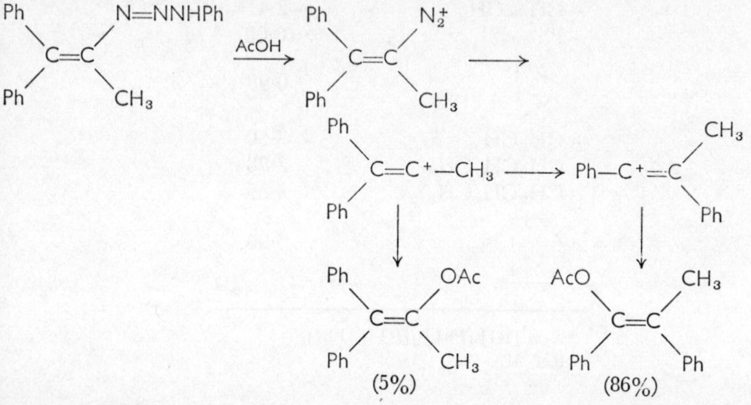

$$\underset{Ph}{\overset{Ph}{>}}C=C\underset{CH_3}{\overset{N=NNHPh}{<}} \xrightarrow{AcOH} \underset{Ph}{\overset{Ph}{>}}C=C\underset{CH_3}{\overset{N_2^+}{<}} \longrightarrow$$

$$\underset{Ph}{\overset{Ph}{>}}C=C^+-CH_3 \longrightarrow Ph-C^+=C\underset{Ph}{\overset{CH_3}{<}}$$

$$\underset{Ph}{\overset{Ph}{>}}C=C\underset{CH_3}{\overset{OAc}{<}} \qquad \underset{Ph}{\overset{AcO}{>}}C=C\underset{Ph}{\overset{CH_3}{<}}$$
$$(5\%) \qquad\qquad (86\%)$$

vinyltriazene that leads to a rearranged product must involve formation of a vinyldiazonium cation followed by its decomposition to a vinyl cation [29].

C. Anions

Vinyl organometallic compounds have been reviewed.[30]

I. Stability

That vinyl anions are considerably more stable than the corresponding saturated ions, in accord with expectation, is indicated by the available experimental evidence.

Equilibrium data relevant to stabilities of lithium and magnesium reagents are available from studies of exchange of organic groups between the organometallic compounds and less polar compounds. To interpret the results in terms of carbanion stabilities, the assumption is made that at equilibrium the most stable incipient carbanion will be associated with the most electropositive element.

Equilibrium data for the exchange of organic groups between lithium

$$\text{RLi} + \text{R'I} \rightleftharpoons \text{RI} + \text{R'Li}$$

reagents and iodides in ether or in ether-pentane solutions are recorded in Table 2[31]. The values of log K, which decrease as RLi is favoured

TABLE 2. Equilibrium constants[a] for RLi + PhI ⇌ RI + PhLi in ether or ether-pentane at $-70°$ [b]

R	log K
—CH=CH$_2$	-2.41
—Ph	(0.00)
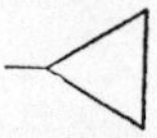 —◁	0.98
—CH$_2$CH$_3$	3.50
—CH$_2$CH$_2$CH$_3$	3.88
—CH$_2$CH(CH$_3$)$_2$	4.59
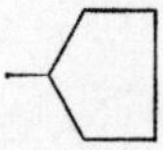 —⬠	6.90

[a] K = [RI] [PhLi]/[RLi] [PhI].
[b] Ref. 31.

at equilibrium, are referred to phenyl as a standard. Since carbon and iodine have similar electronegativities (both 2·5 on the Pauling scale)[32], the stabilities of the iodides are probably relatively insensitive to the nature of the organic group, and differences in log K are due to different stabilities of the lithium reagents. It is hoped that differences in degree of aggregation or interaction with solvent do not significantly affect the relative stabilities—in fact, the observed order is qualitatively that anticipated for the inherent abilities of the organic groups to bear negative charge.

The vinyl and phenyl reagents, in which lithium is associated with sp^2 hybridized carbons, are decidedly more stable than the alkyl reagents. That cyclopropyllithium is more stable than other alkyllithium reagents is expected since the external atomic orbitals of cyclopropyl groups, on the basis of calculations[33] as well as physical evidence[34], are thought to have hybridization intermediate between sp^2 and sp^3. Of those reagents which have lithium associated with carbons that are essentially sp^3, the secondary are less stable than the primary, in accord with the expected destabilization by the electron releasing inductive effect of alkyl groups.

Similar data for exchange of organic groups between magnesium

$$RMg- + R'Hg- \rightleftharpoons RHg- + R'Mg-$$

and mercury compounds in dimethoxyethane are recorded in Table

TABLE 3. Equilibrium constants[a] for $RMg- + PhHg- \rightleftharpoons RHg- + PhMg-$ in dimethoxyethane at $33°$[b]

R	log K
—C≡CPh	≪ −3·0
—CH$_2$Ph	−0·7
—CH$_2$CH=CH$_2$	−0·4
—Ph	(0)
—CH=CH$_2$	0·3
—◁	0·7
—CH$_3$	1·8
—CH$_2$CH$_3$	4·0
—CH$_2$CH(CH$_3$)$_2$	4·3
—CH(CH$_3$)$_2$	>6·0

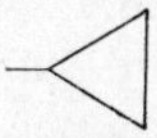

[a] $K = [RHg-][PhMg-]/[RMg-][PhHg-]$.
[b] Ref. 35.

3[35]. It is assumed that the effects of the organic groups are less on the relatively non-polar mercury compounds than on the polar magnesium compounds, and that differences in log K are therefore due principally to different stabilities of the magnesium reagents. Stability again decreases in the expected manner as s character decreases: $sp > sp^2 >$ cyclopropyl $> sp^3$. The allyl (see Section III.C) and benzyl groups, with their additional possibilities for resonance and inductive stabilization, form more stable magnesium compounds than do other alkyl groups. Stabilities of the alkylmagnesium compounds decrease as expected in the series methyl > primary > secondary.

The similarity between the series of values of log K obtained from the lithium–iodine and magnesium–mercury exchange studies reinforces the hope that these values are related to carbanion stabilities. Vinyl, though certainly more stabilizing than primary alkyl groups, unfortunately is the only group whose position seems significantly different in the two series of equilibria. Its placement in the magnesium series, as somewhat less stabilizing than phenyl, is more in accord with the results of kinetic acidity experiments that are described below. Its somewhat different placement in the lithium series may be due to the incursion of effects on relative stabilities of different degrees of aggregation of the organolithium compounds. It has been suggested that vinyllithium is predominantly associated even in tetrahydrofuran—in that solvent, vinyllithium addition to 1,1-diphenylethylene[36] and the metallation of triphenylmethane[37] both exhibit fractional kinetic orders in vinyllithium. Metallations of triphenylmethane by methyl- and n-butyllithium also exhibit fractional orders in lithium reagent, but metallations by allyl-, benzyl-, and phenyllithium are first order in lithium reagent[37]. Addition of lithium bromide greatly decreases the reactivity of vinyllithium as an initiator for the polymerization of styrene in tetrahydrofuran[38]—this is probably caused by ready complexation with lithium bromide and suggests that vinyllithium may also complex with itself.

Similar equilibrium studies of the more carbanion-like vinyl derivatives of even more polar metals are not available, but their stabilities relative to the corresponding hydrocarbons can be inferred from 'kinetic acidity' measurements[39,40]. The equilibrium acidity of a weak hydrocarbon acid (RH) cannot be measured easily, but the

$$\text{RH} + \text{B}^- \underset{k_2}{\overset{k_1}{\rightleftarrows}} \text{R}^- + \text{BH}$$

rate of proton removal (k_1) is more readily accessible by studying base-catalysed hydrogen isotope exchange between the hydrocarbon and the solvent. The ionization constants of acids (K_{RH}) sometimes are related to k_1 in the manner stated in the Brønsted catalysis law,

$$\log k_1 = \alpha \log K_{RH} + \text{constant},$$

α being a proportionality constant having a value between 0 and 1[39,40]. Adherence to this relation between $\log k_1$ and $\log K_{RH}$ depends on structural changes in the acid having proportional effects on the transition state for proton removal and on the carbanion. This condition is most likely to be met when RH is a weak acid compared to HB[39-41]. Then the transition state for proton removal most closely resembles the anion[1], and values of k_2 all approach a similar diffusion-controlled limit.

Relative rates of removal of tritium atoms from hydrocarbons using caesium cyclohexylamide in cyclohexylamine appear in Table 4[39,42,43].

2. The Properties of Alkene Carbonium Ions and Carbanions

TABLE 4. Relative rates of tritium-hydrogen exchange catalysed by caesium cyclohexylamide in cyclohexylamine at 25° [a]

Compound	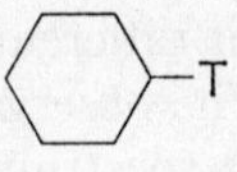 ⌬—T	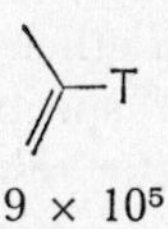 ⟩=⟨—T	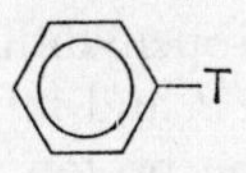 ⌬(aryl)—T
Relative Rate	1	9×10^5	9×10^7

[a] Ref. 39, 42 and 43.

These rates are thought to follow the Brønsted catalysis law with a value of α approaching 1. The equilibrium acidity of a secondary vinyl proton is therefore at least 10^6 greater than that of a secondary proton at a saturated carbon. The acidity of a phenyl proton is only about 10^2 that of a secondary vinyl proton and therefore probably about the same as that of a primary vinyl proton. Differences of similar magnitudes between rates of exchange of hydrogens on vinyl, aryl and saturated carbons are found in qualitative studies of exchange between hydrocarbons and ND_3 catalysed by KND_2[44].

2. Geometry

Vinyl organometallics are found to have non-linear geometries.

Studies of the stereochemistry of their reactions, as well as spectral studies, indicate that vinyllithium reagents exist in *cis* and *trans* forms that are stable for long periods of time[30,45]. The energy barrier for *cis–trans* isomerization, a process that would involve passing through a linear geometry, must be high. For example, the high overall retention of configuration observed in the products obtained from *cis*- and from *trans*-1-bromopropene indicates that both reactions, the

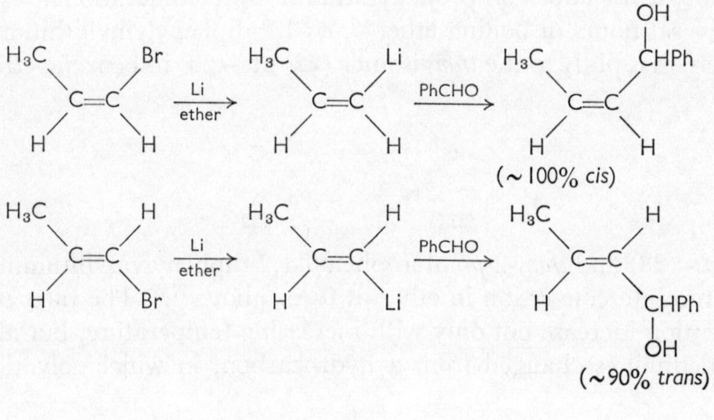

$$\underset{H}{\overset{H_3C}{>}}C=C\underset{H}{\overset{Br}{<}} \xrightarrow[\text{ether}]{Li} \underset{H}{\overset{H_3C}{>}}C=C\underset{H}{\overset{Li}{<}} \xrightarrow{PhCHO} \underset{H}{\overset{H_3C}{>}}C=C\underset{H}{\overset{CHPh-OH}{<}}$$
(~100% *cis*)

$$\underset{H}{\overset{H_3C}{>}}C=C\underset{Br}{\overset{H}{<}} \xrightarrow[\text{ether}]{Li} \underset{H}{\overset{H_3C}{>}}C=C\underset{Li}{\overset{H}{<}} \xrightarrow{PhCHO} \underset{H}{\overset{H_3C}{>}}C=C\underset{CHPh-OH}{\overset{H}{<}}$$
(~90% *trans*)

formation of a lithium reagent and its addition to benzaldehyde, occur with high stereospecificity[46]. It is now known that both reactions proceed with retention as indicated[47]. In accord with the configurational stability deduced from reactions, the *cis* and *trans* isomers of propenyllithium in ether exhibit different proton magnetic resonance[47] and infrared[48] spectra. The observation of an ABC pattern in the proton magnetic resonance spectrum of unsubstituted vinyllithium in ether[49] also indicates a non-linear structure, in which inversion is slow relative to the time scale (~ 0.01 sec) of the proton magnetic resonance experiment.

Stereospecificity observed in reactions such as those of the *cis-* and *trans*-2-bromo-2-butenes[50] is proof that Grignard reagents also have

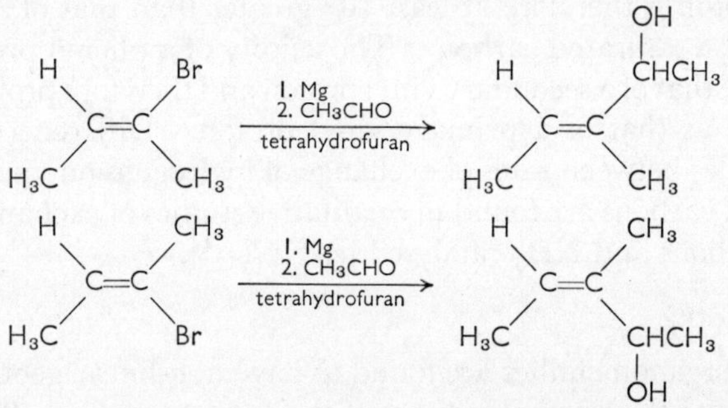

considerable configurational stability in solution[50,51,52]. *Cis* and *trans* isomers of Grignard reagents also exhibit discrete proton magnetic resonance spectra[51]. The ABC pattern exhibited by the unsubstituted vinyl Grignard reagent[53,54,55] indicates that inversion also does not occur in this reagent within the characteristic proton magnetic resonance time.

Though *cis-* and *trans*-propenyllithium are configurationally stable for several hours in boiling ether[46], *cis*-1,2-diphenylvinyllithium (**21**) isomerizes rapidly to the *trans* isomer (**22**) at $-20°$ in benzene-ether[56],

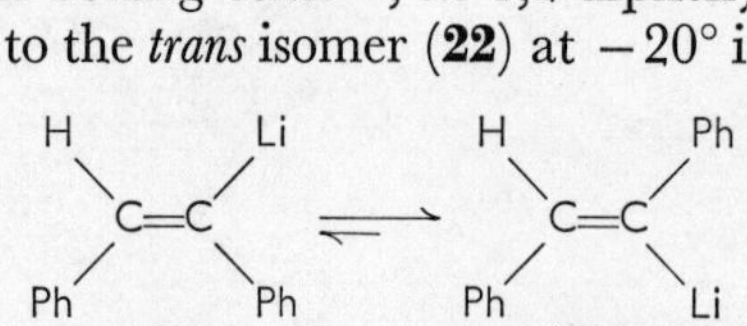

and *cis-* (**23**) and *trans*-2-*p*-chlorophenyl-1,2-diphenylvinyllithium (**24**) undergo interconversion in ether at 0° or above[57]. The rates of isomerization increase not only with increasing temperature, but also as the medium is changed from a hydrocarbon, in which solvation of

2. The Properties of Alkene Carbonium Ions and Carbanions

$$\underset{(23)}{\underset{p\text{-}ClH_4C_6}{Ph}\!\!>\!\!C\!\!=\!\!C\!\!<\!\!\underset{Li}{Ph}} \quad \rightleftharpoons \quad \underset{(24)}{\underset{p\text{-}ClH_4C_6}{Ph}\!\!>\!\!C\!\!=\!\!C\!\!<\!\!\underset{Ph}{Li}}$$

lithium ions would be small, to solvents in which significant solvation is possible[58]. Isomerization, slow in hydrocarbons, is more rapid in ether–hydrocarbon mixtures and even more rapid in tetrahydrofuran. It was suggested that isomerization involves ionization of the partially covalent C—Li bond followed by isomerization of the vinyl anion through a linear transition state or intermediate (25)[46]. The resonance contribution shown in 25b would lower the energy of the linear

$$\underset{(25a)}{>\!\!C\!\!=\!\!\bar{C}\!\!-\!\!\bigcirc} \quad \longleftrightarrow \quad \underset{(25b)}{>\!\!C\!\!=\!\!C\!\!=\!\!\bigcirc^{\!-}}$$

geometry for an anion when α-aryl groups are present. Relief of the steric strain associated with two *cis* aryl groups could also be a factor promoting the more rapid isomerization of these aryl derivatives.

Similar spectral observations and stereochemical studies of reactions of long-lived derivatives are not available for vinyl species associated with even more polar metals. Investigations of base-catalysed vinyl hydrogen isotope exchange and *cis–trans* isomerization of olefin isomers are relevant to the problem of geometry, but cannot be interpreted unambiguously.

For example, Hunter and Cram have found that hydrogen isotope exchange at the vinyl position of *cis*-stilbene catalysed by potassium *t*-butoxide in *t*-BuOD is $\sim 5 \times 10^3$ faster than the isomerization to *trans*-stilbene[59]. The high stereospecificity of exchange requires that different vinyl intermediates, not rapidly interconverted, must be formed from *cis*- and from *trans*-stilbene. It is possible that these intermediates are non-linear (26 and 27) and that capture of the

$$\underset{(26)}{\underset{Ph}{H}\!\!>\!\!C\!\!=\!\!C\!\!<\!\!\underset{Ph}{:}} \qquad \underset{(27)}{\underset{Ph}{H}\!\!>\!\!C\!\!=\!\!C\!\!<\!\!\underset{:}{Ph}}$$

hydrogen isotope from solvent is more rapid than inversion. Alternatively, it is possible that the intermediates are linear, an arrangement favoured here by conjugation with the phenyl groups, but differ in

the geometric relationship between the ions in intimate ion pairs
(**28** and **29**); the hydrogen isotope addition from one of the *t*-butyl

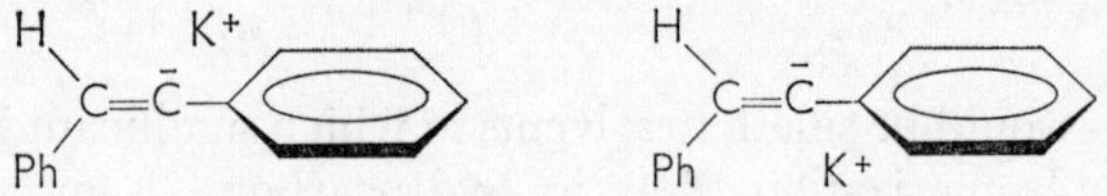

$$\underset{Ph}{\overset{H}{>}}C=\bar{C}-\underset{}{\overset{K^+}{\bigcirc}}\qquad \underset{Ph}{\overset{H}{>}}C=\bar{C}-\underset{K^+}{\overset{}{\bigcirc}}$$

(**28**) (**29**)

alcohol molecules surrounding the potassium ion might occur more
rapidly than processes that would interconvert the isomeric ion pairs.
It is known from Cram's studies that saturated carbanions conjugated
with aryl groups, though probably planar, can participate in reactions
that proceed with significant stereospecificity, presumably because of
similar formation of isomeric ion pairs[45]. For example, hydrogen
isotope exchange at the benzyl position of 2-phenylbutane catalysed
by potassium *t*-butoxide in *t*-butyl alcohol proceeds with considerable
retention of configuration[60].

Exchange without significant isomerization is also observed for *cis*-
and *trans*-1,2-dichloroethylene in D_2O to which sodium methoxide
has been added[61].

3. Formation and reactions

The preparations and reactions of polar vinyl organometallic compounds do not differ fundamentally from those of their saturated
counterparts. A general review discusses some of their reaction
chemistry[30], and a detailed review is concerned specifically with preparations and reactions of vinyl Grignard reagents[62]. The extensive
recent developments in the chemistry of α-halovinyl organometallics
have been reviewed[63].

III. ALLYL IONS

A. Introduction

To think of allyl cations and anions as simply vinyl substituted ions
(**30**) is misleading. An allyl ion is described by the resonance theory

$$\overset{}{\diagup\!\!\!\diagdown} + (-) \longleftrightarrow \overset{}{\diagdown\!\!\!\diagup} + (-)$$

(**30**) (**31**)

as a hybrid receiving equal contributions from the equivalent contributing structures **30** and **31**[64]. The simple molecular orbital theory describes an allyl ion as possessing a π-system constructed from three atomic p orbitals and containing two electrons in the cation and four electrons in the anion[65].

The ions are predicted on the basis of either description to have identical terminal carbons (in the absence of unsymmetrical substitution), carbon–carbon bonds intermediate in properties between those of carbon–carbon single and double bonds, and stabilities greater than would be expected for 'vinyl' substituted ions (**30**). The bonding stabilization should be maximal when the ions are planar, and a considerable barrier to rotation around the carbon–carbon bonds should exist, since rotation requires the destruction of one of the partial bonds. The resonance theory predicts equal division of the charge between the terminal carbons, and simple Hückel molecular orbital calculations[66] lead to the same conclusion. A delocalization energy of $0.828\ \beta$ is predicted for both the allyl cation and anion by simple Hückel calculations; though the appropriate value of β, the resonance integral, is uncertain[67], the delocalization energy should be 20 kcal/mole or greater.

More elaborate molecular orbital calculations[68] have taken account of electron-repulsion effects and other deficiencies of the simple Hückel calculations, and valence bond calculations[69] have included a structure that involves bonding between the terminal allylic carbons. In addition, allyl ions have been described by a 'non-pairing' treatment[70] and by an extended Hückel method[12,71]. These more elaborate approaches generally predict that some of the charge will be on the central carbon, though less than on the terminal carbons, but lead otherwise to predictions qualitatively similar to those of the simpler theories. It has been predicted on the basis of extended Hückel calculations that the barrier to rotation around the carbon–carbon bonds should be greater in the anion than in the cation[71].

The cyclopropenyl cation (**32**) appears to be an allyl cation, the ends of which have been brought into close proximity. However, this cation must be regarded as a true aromatic system rather than a simple allyl cation. This completely conjugated, monocyclic, and planar system containing two π-electrons fits Hückel's '$4n + 2$' rule[72].

(32)

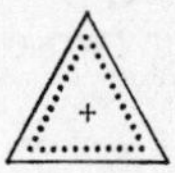

Experimental observations are in accord with a large delocalization energy, calculated to be 2β[72]. Even acyclic allyl cations could have some bonding between the terminal carbons, and it is predicted that cyclopropenyl character should increase as the C—C—C angle is decreased by geometric constraints[73,74]. Cyclopropenyl anions may have 2 of the 4 π-electrons in antibonding molecular orbitals[72] and are predicted[72] and found[74a] to have less delocalization energy than acyclic allyl anions.

B. Cations

A recent review has been concerned with the properties of allyl cations[75].

I. Stability

In accord with theoretical predictions, experiments indicate that allyl cations are considerably more stable than the corresponding saturated cations.

Measurements in solution of the protonation equilibria between allyl cations and the corresponding dienes (or diene mixtures) are recorded in Table 5. Similar data for equilibria between saturated

TABLE 5. Equilibria between allyl cations and dienes, and ultraviolet spectra of allyl cations[a]

	Cation	pK[b]	λ_{max}[c] (mμ)	log ϵ[c]	Ref.
33	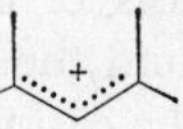	−5·9	305	4·03	76
34	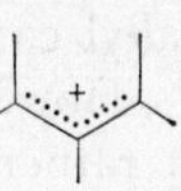	−7·1	307	3·98	76

(*Table continued*)

2. The Properties of Alkene Carbonium Ions and Carbanions

Table 5 contd.

	Cation	pK^b	λ_{max}^c (mμ)	log ϵ^c	Ref.
35	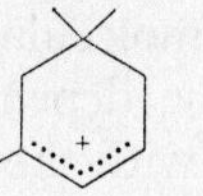	−6·8	319	>3·7	76
36	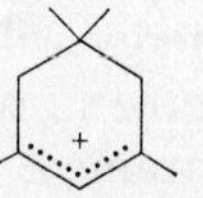	−3·2	314	3·96	76
37	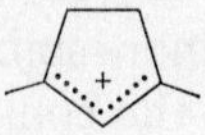	−1·9	275	4·04	76
38	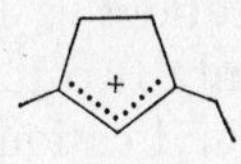	−2·4	278	4·07	76
39	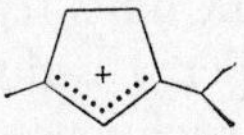	−2·4	280	4·03	76
40	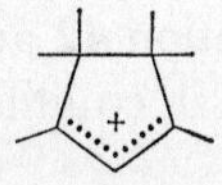	−1·8	291	4·13	76
41	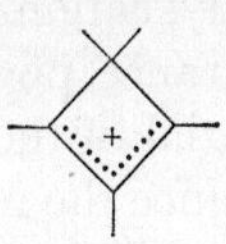		245	3·46	74
42	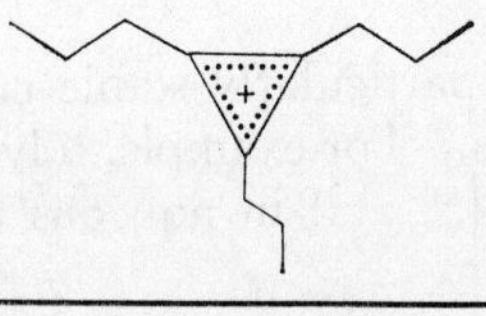	7·2	<185		77

[a] Spectra reported for the unsubstituted allyl cation and other simple alkyl-substituted allyl cations [J. Rosenbaum and M. C. R. Symons, *J. Chem. Soc.*, 1 (1961)] are now thought to be due to other species.

[b] Calculated from the equation pK = H_0 + log ([BH$^+$]/[B]). All determinations are in aqueous sulphuric acid solutions except that of **42** which is in 50% aqueous acetonitrile. For those cations in equilibrium with a mixture of isomeric dienes, these values refer to the equilibrium with the diene mixture, not with one of its components.

[c] In sulphuric acid solutions except the spectrum of **41** which is of a dichloromethane solution of an AlCl$_4^-$ salt.

cations and alkenes with which these values could be compared are not available. However, a pK of $-15\cdot5$ for the t-butyl cation-isobutene equilibrium has been estimated by extrapolation of comparisons of known equilibrium and solvolysis rate data[78]. It should be noted that efforts to correlate pK differences with molecular orbital predictions are hazardous. The pK values may be dependent on medium because of varying interactions of the ions with the solvent, and will be influenced by any difference in the relative energies of the precursors. In addition, the actual stabilization energy will be less than the calculated delocalization energy to the degree to which an ion is unable to achieve the favoured geometry or is strained. Nevertheless, it does seem apparent that allyl cations with four (**33, 34,** and **36–40**) and three (**35**) terminal alkyl substituents are more stable than the t-butyl cation, particularly when it is noted that the conjugative stabilization of the diene precursors of the alkyl cations is lost on protonation. The stabilities of allyl cations with only two terminal alkyl substituents should also exceed that of the t-butyl cation, since the effect of a terminal methyl group is about 3–4 pK units (compare **35** and **36**). The large pK observed for cyclopropenyl cation **42** is in accord with assigning a much greater stability to such ions. However, this pK may not be comparable to those of the other cations; a difference between the large strains present both in cation **42** and its precursor could be significant[77], and the cation may be in equilibrium with an alcohol rather than a diene[75].

A somewhat larger stabilization for the unsolvated allyl cation in the gas phase than in solution is indicated by the appearance potential data in Table 1. The value for the ionization potential of the allyl radical must not reflect the full stability of the cation, since the radical itself has a large resonance stabilization[79]. A large stabilizing effect of a terminal methyl group is again noted.

The conclusion that the allyl cation is particularly stable can be drawn also from rates of solvolysis reactions. For example, solvolytic reactivities of **43** in 50% aqueous ethanol[80] and in aqueous acetic

$$CH_2{=}CHCCH_3 \underset{Cl}{\overset{CH_3}{|}}$$

(**43**)

acid[81] at 25° are more than one hundred times those of t-butylchloride[82,83] in spite of the rate-retarding inductive effect of the alkenyl group on the transition states.

2. Geometry

The available evidence indicates that the preferred geometry of allyl cations is planar and that the energy barrier to rotation around the C⸺C bonds is sizable.

The effect of structure on pK provides strong evidence that planar structures lead to the greatest stabilization. The data in Table 5 indicate a striking increase in allyl cation stabilities in the order: acyclic < cyclohexenyl < cyclopentenyl. The alkylated linear cation (**33**) cannot attain planarity because of steric interference of

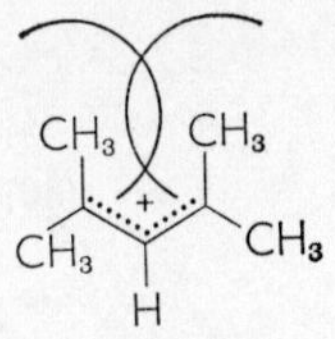

(**33**)

methyl substituents. The delocalization energy must therefore be reduced. Note that **34** is actually less stable than **33**. The additional methyl group should lead to even greater steric hindrance to the achievement of planarity. The cyclopentenyl cations must be planar, and the allyl system in cyclohexenyl cations must achieve planarity much more easily than in acyclic allyl cations.

Convincing evidence for high rotational barriers is provided by observation of discrete proton magnetic resonance absorptions for '*cis*' and '*trans*' groups. Spectra of some typical cations are summarized in Table 6. The discrete absorptions seen for *cis* and *trans* methyls in **33** and **45** at room temperature indicate that these groups do not become equivalent even at that temperature on the time scale (~ 0.01 sec) of the proton magnetic resonance observations.

Similar qualitative conclusions that *cis* and *trans* allyl cations do not interconvert readily have been drawn from solvolysis studies. For example, the crotyl alcohols formed initially from hydrolyses of

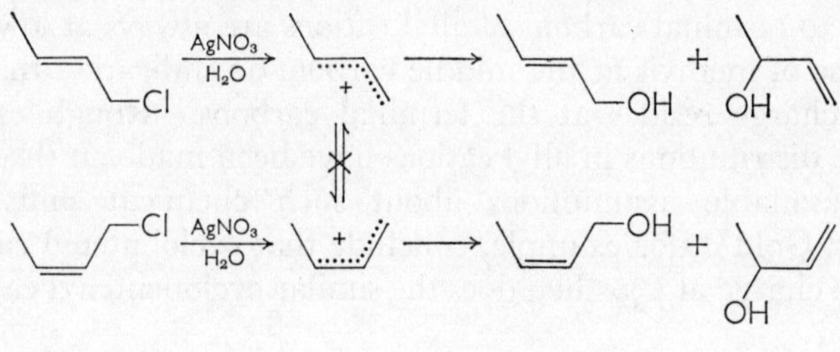

TABLE 6. Proton magnetic resonance spectra of allyl cations[a]

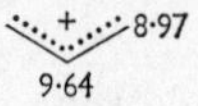

(structures of cations 44, 33, 34, 45, 36, 46, 41 with chemical shifts)

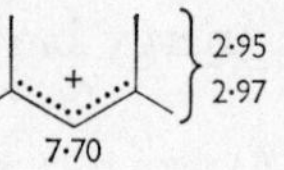

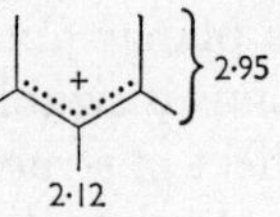

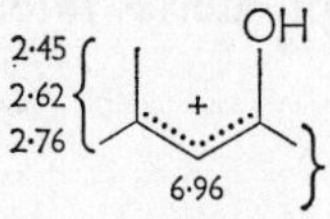

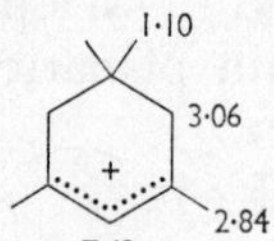

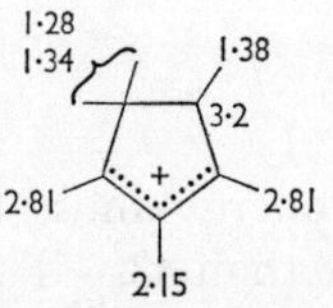

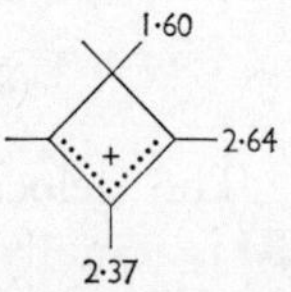

[a] Chemical shifts are expressed in ppm downfield from tetramethylsilane.

[b] Spectrum of an SbF_6^- salt in SO_2–SbF_5 at $-60°$; the reference is external tetramethylsilane [G. A. Olah and M. B. Comisarow, *J. Am. Chem. Soc.*, **86**, 5682 (1964)].

[c] Spectrum in sulphuric acid solution at room temperature; the reference is internal tetramethylammonium chloride assumed to absorb at 3·10[84].

[d] The stereochemistry of this ion is unknown.

[e] Spectrum of an $AlCl_4^-$ salt in CH_2Cl_2 at room temperature; the reference is internal tetramethylsilane[74].

cis- and trans-crotyl chlorides with aqueous silver nitrate are thought to preserve more than 99% of the geometric configurations of the starting materials[85].

The proton magnetic resonance absorptions of methyl groups attached to terminal carbons of allyl cations are always at lower field than those of methyls at the middle carbon, an indication that more positive charge resides at the terminal carbons. Rough estimates of charge distributions in allyl cations have been made on the basis of some reasonable assumptions about such chemical shifts[69,74,86]. Katz and Gold[74], for example, conclude that cyclobutenyl cation **41** has more charge at $C_{(2)}$ then does the similar cyclopentenyl cation **46**.

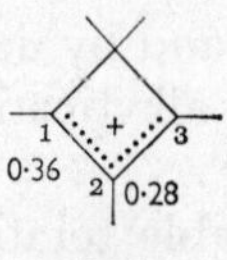

(41) 0.36, 0.28 (46) 0.40, 0.20

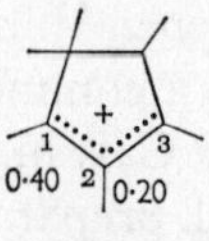

Ultraviolet spectra of allyl cations, some of which are listed in Table 5, are in accord with the theoretical predictions. The energy of $\sqrt{2}\,\beta$ predicted by simple Hückel molecular orbital theory for the $\pi \to \pi^*$ absorption corresponds to about 300 mμ using a reasonable spectroscopic value[87] of β. The simple Hückel molecular orbital theory leads to the prediction of a much greater energy (3β) for the absorption of the cyclopropenyl cation, again in accord with observation. The progressive decrease in wavelength in the series cyclohexenyl > cyclopentenyl > cyclobutenyl may be a reflexion of increasing cyclopropenyl character as the terminal carbons of the allylic system are brought closer together. In fact, comparison of the ultraviolet spectrum of **41** with those of other allyl cations led to the prediction that the 1,3 π-resonance integral ($\beta_{1,3}$) in the Hückel molecular orbital treatment of **41** must be $\sim 0.33\beta$[74]. A portion of the increased stability of cyclopentenyl cations over cyclohexenyl cations (Table 5) might also be due to such increased 1,3 bonding.

The infrared spectrum of **36**[88] exhibits its most intense absorption at 1533 cm^{-1}, and similar absorptions[78] are noted for other allyl cations. The position of these absorptions, between the regions typical for carbon–carbon single and double bond stretching absorptions, is in accord with their assignment to a stretching vibration of the allyl system. The considerable intensity is also reasonable since a vibration in an allyl cation could be associated with an unusually large change in dipole moment.

3. Formation and reactions

Allyl cations often are formed and react in ways common to their saturated counterparts. Formation of allyl cations as transient reaction intermediates by loss of nucleophilic groups, and reaction by recombination with the nucleophile, addition of another nucleophile, or loss of a proton to form a diene are most common. The intricacies of the mechanisms of such reactions, in which ion-pairing and solvation effects often are significant, have been reviewed by DeWolfe and Young in this series[89] as well as in an earlier review[90]. Allylic rearrangements that involve allyl cation intermediates have been reviewed by MacKenzie[91] in this series and by de la Mare[92]. Addition to dienes, also a common route to alkyl cation intermediates, is reviewed briefly by Cais in this series[93].

Recent successful preparations of stable solutions of allyl cations have used fundamentally the same methods, though of necessity always employing systems in which nucleophilic activity is low. The allyl cations studied in sulphuric acid solutions disappear by processes first-order in the cation and first-order in its diene precursor, presumably because of dimerization, perhaps followed by further reactions[76,84]. Therefore, an ion can be stable in concentrated acid, where diene concentration is low, but disappear rapidly at lower acidities, where both the ion and its precursor diene (or dienes) are present in significant concentrations. For example, it has been calculated that a 0·1 M solution of the equilibrium mixture of **37** and the corresponding diene mixture has a half-life of 2 days in 96% sulphuric acid or in aqueous solution at pH = 5, but a half-life of only 0·02 sec in 35% sulphuric acid, where the ion and dienes are present in equal concentrations[84]. Therefore, successful preparation of a solution of an allyl cation from a diene requires that protonation be rapid, so that at no time is a sizable concentration of precursor in contact with the ion.

Allyl cations have also proved to be products of a variety of complex rearrangement processes. The alkyl shifts, hydride transfers, alkylations, dealkylations, protonations, and deprotonations that must be involved in these rearrangements are not unique to the formation of allyl cations. Nevertheless, these rearrangements are reviewed below because they do emphasize the stability that is achieved in reaching an allyl cation. Some of these reactions have been observed in strong acid solutions and might not be noted under more ordinary reaction conditions where ion lifetimes are short.

Of the most general significance is the observation that a variety of olefins or saturated alcohols when dissolved in sulphuric acid form complex mixtures of cyclopentenyl cations and of alkanes[94]. In

1-butanol, 2-butanol,
t-butyl alcohol, isobutene, $\xrightarrow{96\% H_2SO_4}$
2,2,4-trimethylpentenes

$\begin{cases} 50\% \text{ cyclopentenyl} \\ \text{cations, mainly} \\ C_{10}-C_{18} \\ + \\ 50\% \text{ alkanes, mainly} \\ C_4-C_{18} \end{cases}$

same reactants $\xrightarrow{75\% H_2SO_4}$ dimers, trimers, etc. of C_4H_8

75% sulphuric acid, the well-known dimerization to produce iso-octenes is observed; the dimeric products are largely protected from further alkylation in that medium by separating as a second phase, though ultimately trimers and higher polymers are formed. In more concentrated acid, the residence time of the dimeric cations is sufficient to permit the sequences of steps that are needed to rationalize the formation of the cyclopentenyl cations. A hydride transfer must occur at some stage in these disproportionation reactions. A more direct formation of allyl cations by simple hydride transfer occurs when tertiary alcohols, such as **47,** possessing skeletons identical to

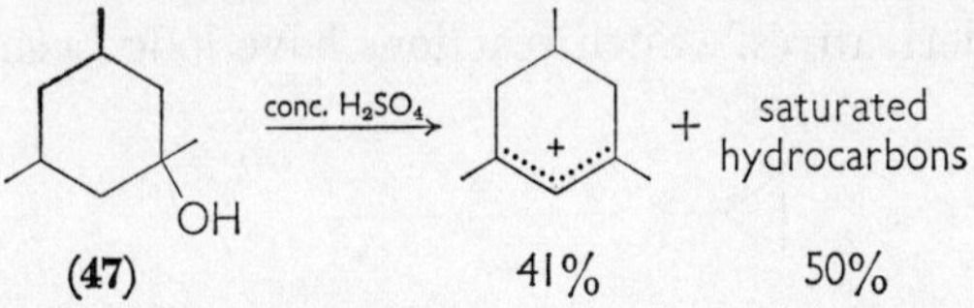

those of relatively stable allyl cations, are dissolved in acid[95].

Allyl cations are also formed by isomerizations of other cations. Several examples have been observed of isomerizations of pentadienyl cations, such as **48**[96], to cyclopentenyl cations[96-101]. These reactions

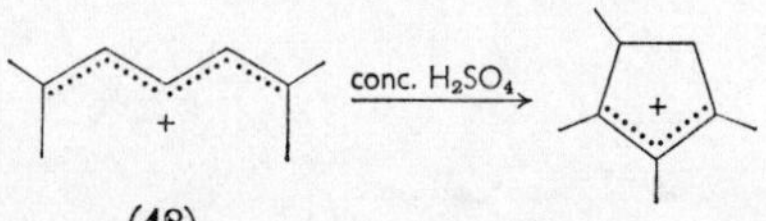

involve skeletal rearrangement as well as cyclization. Some attempts to prepare strong acid solutions of cyclopropylcarbonium ions, from **49**, for example[102], have led to allyl cations[102-104]. Presumably

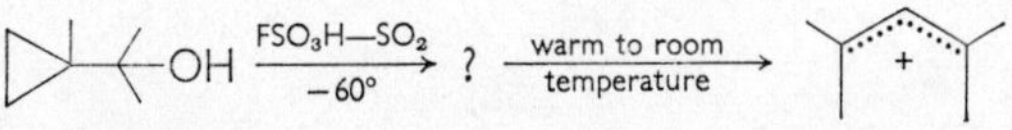

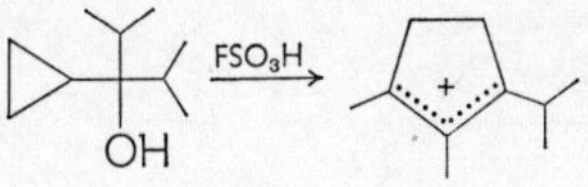

cyclopropylcarbonium ions are transient intermediates in these reactions, since they are formed from closely related precursors under identical conditions[102-104]. Formation in this manner of cyclopentenyl cations, such as **50**[103], involves oxidation as well as rearrangement[103,104]. Allyl cations are also formed by dissolving some bicyclic alcohols[105,105a], such as fenchol[105] (**51**), in sulphuric acid.

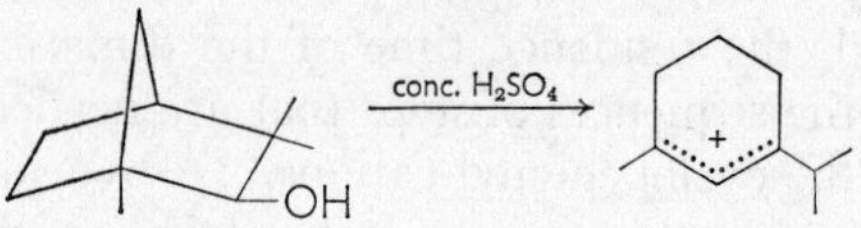

$$\text{(51)} \xrightarrow{\text{conc. } H_2SO_4} \text{80\%}$$

Allyl cations are formed as transient intermediates in reactions of cyclopropyl derivatives. Such reactions have long been known, but

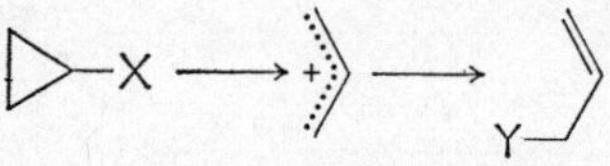

$$\triangleright\!-X \longrightarrow \overset{+}{\triangleright} \longrightarrow \searrow_{Y}$$

are of particular current interest because of the predictions made by Woodward and Hoffman about the stereochemical path of the 'electrocyclic transformation' that is involved[106]; they concluded that disrotatory (**52**) rather than conrotatory (**53**) motion should be

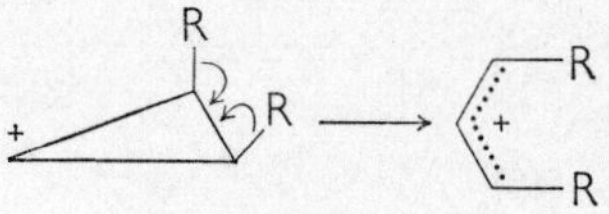

(**52**)

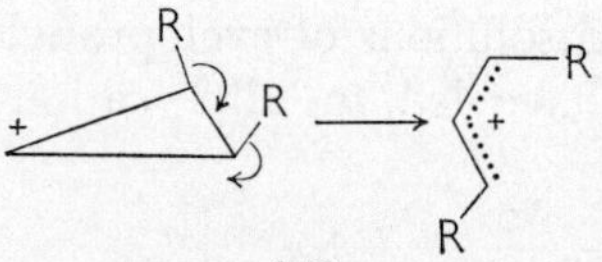

(**53**)

favoured, and, furthermore, that the disrotatory motion that involves inward (**54**) rather than outward (**55**) rotation of the groups *cis* to the leaving group is preferred when the cyclopropyl cation is formed

(**54**)   (**55**)

2. The Properties of Alkene Carbonium Ions and Carbanions

by ionization of X. Recent experiments are in accord with these predictions[107].

Isomerizations of one allyl cation to another provide further information about the relative stabilities of different allyl cations. These isomerizations involve not only shifts of substituent alkyl groups[96,98-101,108], as seen in **40 → 56**[108], but also obvious alterations

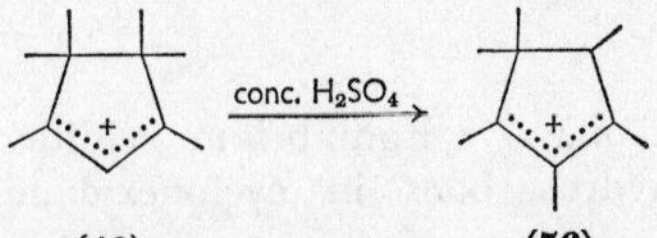

(40)   conc. H₂SO₄ →   (56)

of the substituent alkyl groups[100,101], as seen in **57 → 58 ⇌ 59**[100],

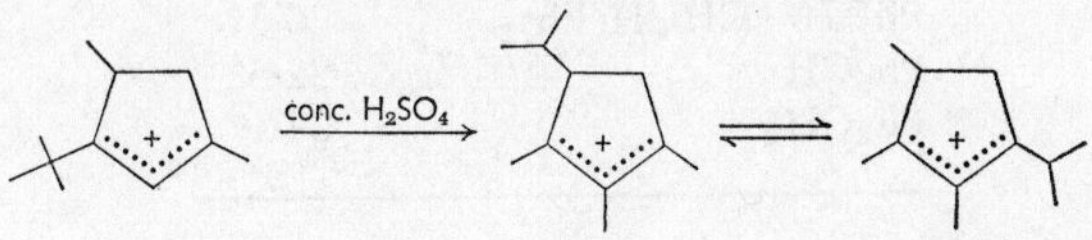

(57)   conc. H₂SO₄ →   (58)   ⇌   (59)

and rearrangements of cyclohexenyl to cyclopentenyl cations, such as **60 → 61**[105]. Mechanistic details of such isomerizations have been studied[100,101,108].

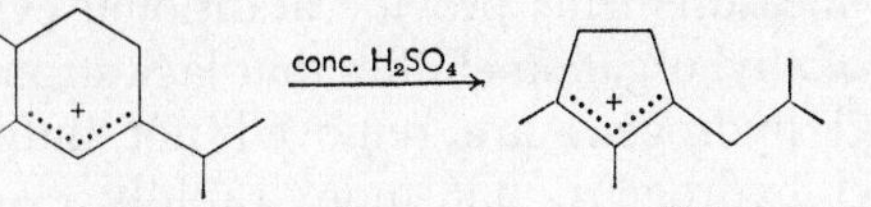

(60)   conc. H₂SO₄ →   (61)

C. Anions

I. Stability

Allyl anions are considerably more stable than their saturated counterparts. Equilibrium data in Table 3 (Section II.C.1) suggest that the allyl Grignard reagent has a stability comparable to benzyl and far greater than ethyl or even methyl. Isomerization of 1-sodio-[109] and 1-potassio-1-dodecene[110] to their allyl isomers demon-

$$CH_3(CH_2)_8CH_2CH=CHM \longrightarrow CH_3(CH_2)_8CH\text{-----}\bar{C}H\text{-----}CH_2 \; M^+$$

strates that the allyl organometallics are even more stable than their vinyl isomers. Other available equilibrium data are for allyl systems which are conjugated with particularly effective stabilizing groups and clearly ionized to carbanions[111,112]. The data of Table 7 indicate that 1,3,3-triphenylpropene is much more acidic than model compounds[111]. The pK_a values, obtained by measurements of equilibria between two hydrocarbons and the corresponding caesium

$$RH + R'^-M^+ \rightleftharpoons R^-M^+ + R'H$$

or lithium salts, are thought to be the correct values in aqueous solution.

TABLE 7. Equilibrium acidities of hydrocarbons in cyclohexylamine[a]

Hydrocarbon	pK_a
PhCH=CHC$\underline{H}$(Ph)$_2$	26·5[b]
Ph$_3$CH	32·5[c]
Ph$_2$CH$_2$	34·1[c]

[a] Ref. 111.
[b] Equilibrium experiments done with lithium salts.
[c] Equilibrium experiments done with caesium salts.

Kinetic acidity measurements provide additional evidence for the relative stabilities of allyl organometallics that lack strongly stabilizing substituents. Allyl hydrogens are removed not only much more rapidly than alkyl hydrogens, but more rapidly even than vinyl hydrogens. For example, the relative rates of isotopic exchange of allylic hydrogens catalysed by potassium amide in ammonia are known from qualitative studies to be much greater than those of vinyl hydrogens[44].

Reprotonation of the intermediate of such a reaction can lead, of course, to olefin isomerization. Such isomerization reactions are

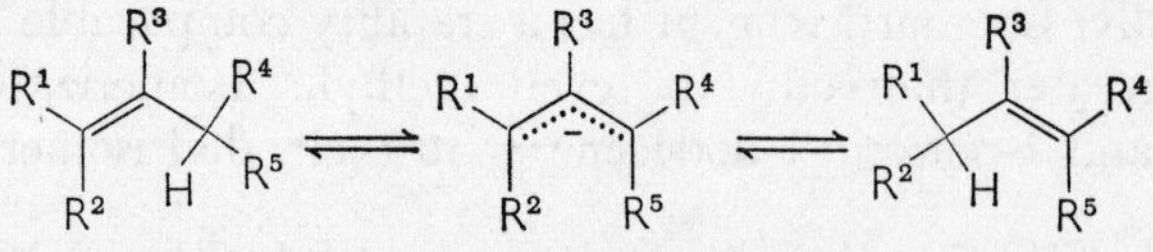

reviewed by MacKenzie in an earlier chapter of this book[91] and the details of their mechanisms have been reviewed[113].

Some of the rates observed in a thorough study[114–118] of olefin

2. The Properties of Alkene Carbonium Ions and Carbanions 69

isomerizations catalysed by potassium *t*-butoxide in dimethyl sulphoxide are listed in Table 8. It has been established that the rate-determining step of such an isomerization is proton removal[115]; therefore, the rates should reflect the carbanion stabilities. Comparison of the relative rates of **62, 64,** and **65** shows that the effect of a terminal methyl group on allyl anion formation is decelerating, as expected, though surprisingly small in comparison to the effect of a

TABLE 8. Relative rates of isomerization of terminal olefins in dimethyl sulphoxide catalysed by potassium *t*-butoxide at 55°[a]

	Olefin	Relative Rate[b]
62		(1·0)
63		1·4 × 10⁵[c]
64	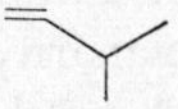	0·24
65	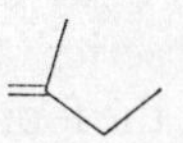	0·15
66		0·57
67	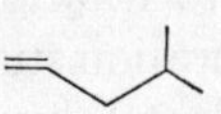	0·18
68		0·0072
69	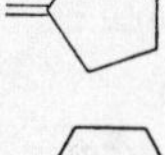	0·091
70	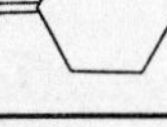	2·7
71	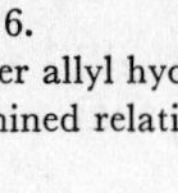	1·2
72		0·0026

[a] Ref. 116.
[b] Rate per allyl hydrogen.
[c] Determined relative to 1-butene at 30°.

non-terminal methyl. It is also noted in isomerizations involving more stable carbanions that effects of methyl groups, and of phenyl groups as well, are less than in carbonium ion chemistry[119]. Comparison of the relative rates of **63** and **62** indicates that the accelerating effect of vinyl is large. The additional stabilization provided by a vinyl group, even when attached to the already stable allyl system, obviously is still considerable.

2. Geometry

In accord with expectation, it appears that allyl anions are planar in the absence of steric restraints. The effect of structural variation on the rates of isomerization listed in Table 8 provides strong evidence for a preference for planarity. It is reasonable that the greatest stabilization of the transition state for proton removal will be found when its geometry can approach that preferred for an allyl anion. The decrease in rates observed in the series **62** > **66** > **67** > **68** can be mostly attributed to increasing steric interference by the alkyl groups, hindering the achievement of planarity by the anions and perhaps also inhibiting their solvation[116]. A large decrease is also noted in the equivalently substituted series **70** > **71** > **69** > **72**; achieving a planar geometry is easier with the four- and five-membered ring systems than with the six-membered ring or acyclic systems.

The existence of an energy barrier to rotation around the C⋯C bonds is demonstrated by several equilibration studies of olefins[120-122] An investigation of potassium *t*-butoxide catalysed isomerization and

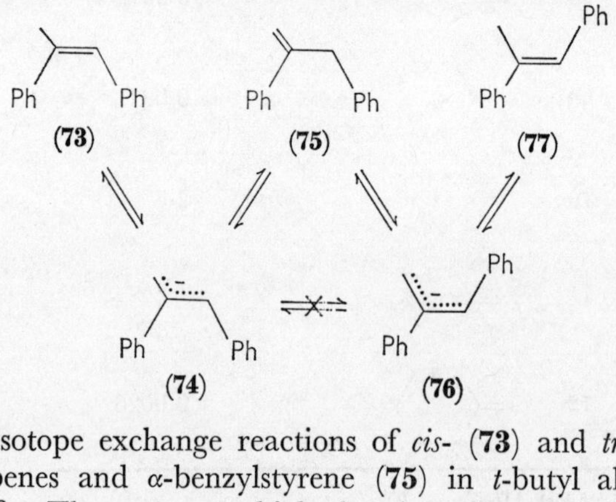

hydrogen-isotope exchange reactions of *cis*- (**73**) and *trans*- (**77**) α-methylstilbenes and α-benzylstyrene (**75**) in *t*-butyl alcohol is an example[122]. The rates at which the *cis* and *trans* olefins form the equilibrium mixture, the composition of that mixture, and the rates at which each olefin undergoes isotopic exchange have all been determined. Analysis of these data indicates that direct isomerization of the *cis* and *trans* olefins into one another is not detectable; rather the isomerization proceeds through α-benzylstyrene. Therefore, there must be two discrete intermediates (**74** and **76**) that are not interconverted.

The analysis of such data leads to estimates of the relative stabilities of isomeric carbanion intermediates. The decrease in stability in each

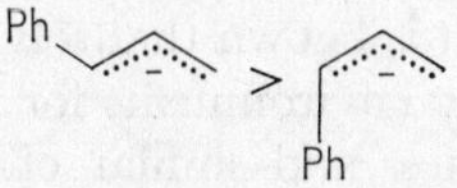

Ref. 120

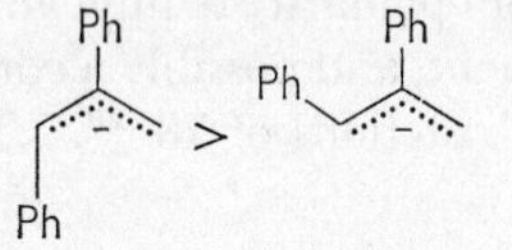

Ref. 122

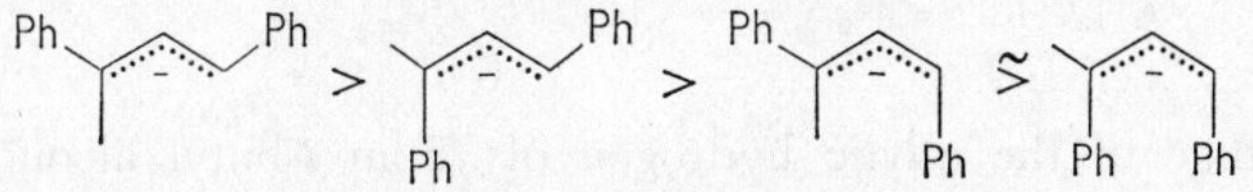

Ref. 121

series is attributed to increasing difficulty in achieving planarity because of steric interference by the substituents. It has been concluded that 1,3-interactions of substituents in these ions are generally more significant than 1,2-interactions.

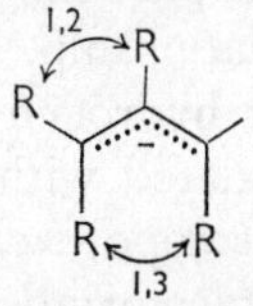

Isomerizations by a variety of base-solvent systems of 1-alkenes that lead to alkyl substituted allyl anions furnish *cis*-2-olefins much more rapidly than they do the more stable *trans*-2-olefins[117,118,123]. The *cis*–*trans* ratios initially observed exceed those at equilibrium by factors of $\sim 10^2$ [117,118]. Even reactions of the butenyl Grignard reagent with a variety of proton donors lead to more *cis*-2-butene than exists in the equilibrium mixture[124]. The more rapid formation of *cis* than of *trans* olefins has been attributed to greater stability of *cis* than of *trans* alkyl substituted anions[118]. A greater stability for the *cis* alkyl anions has been rationalized by consideration of dipolar interactions[125]

cis / trans (R positions shown)

and of other factors[113], and predicted for the butenyl anion by extended Hückel calculations that consider the bonding of the allyl π-system with orbitals of the methyl group[71].

Allyl anions when fully formed and free of inhibiting steric effects probably prefer a planar geometry so that maximum delocalization can be achieved. However, it is known that solvation or ion-pairing effects can provide asymmetric environments for anions generated as transient reaction intermediates, and similar effects might actually lead to some distortion from planarity of allyl anions [113]. Illustrations of asymmetry of environment, and possibly a consequent asymmetry of the ion, are provided by reactions of **78**[126]. The rates of hydrogen

$$\underset{(78)}{CH_3\overset{Ph}{\underset{|}{C}}HCH=CH_2} \rightleftharpoons \underset{(79)}{CH_3\overset{Ph}{\underset{|}{C}}=CHCH_3}$$

isotope exchange of the allylic hydrogen of **78** in *t*-butyl alcohol catalysed by potassium *t*-butoxide and in ethylene glycol catalysed by potassium ethylene glycoxide are considerably slower than isomerization of **78** to **79**. The greater rate of isomerization indicates that, at least in the limited lifetime of the allyl anion, a special relationship with the solvent molecule formed by hydrogen abstraction allows considerable chance for recapture of the particular hydrogen that was lost in the abstraction step. This relationship, which may be due to hydrogen bonding by this solvent hydrogen to the terminal carbons of the allyl anion, might be associated with a distortion of the anion. The rate of allylic hydrogen isotope exchange in *t*-butyl alcohol is more than ten times the rate of racemization, exchange occurring with predominant retention of configuration[126]. In contrast, hydrogen isotope exchange in ethylene glycol is only 0·7 times the rate of racemization, exchange occurring with an excess of inversion. These stereochemical results, though not unique to allyl systems[45], are another indication of the significance of unsymmetrical influences on reactions of anions.

The positions of the ultraviolet maxima, listed in Table 9, of allyllithium reagents **80–82**, are in accord with the energy ($\sqrt{2}\,\beta$) predicted by simple Hückel molecular orbital theory for the absorption of the allyl anion. The large extinction coefficients are consistent

TABLE 9. Ultraviolet spectra of allylic organometallic compounds

	Compound	λ_{max} (mμ)	log ε	Solvent[a]	Ref.
80	 ⌒⌒ Li+	315	3·66	THF	127
81[b]	 ⌒⌒ Li+	291	3·79	90% THF–10% ether	127
82	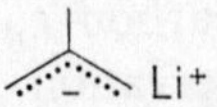 ⌒⌒ Li+	330		THF	128
83	 ⌒⌒ Na+	~315[c]		THF	129
84[b]	 Ph⌒⌒ Li+	395	4·38	THF	127
85[b]	 Ph⌒⌒ K+	420	4·34	ammonia	130
86	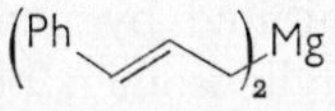 (Ph⌒⌒)$_2$Mg	252	4·30	ether	131

[a] THF is tetrahydrofuran and ether is ethyl ether.
[b] Stereochemistry not known with certainty.
[c] This value is estimated from the published spectrum. An additional but weaker absorption was observed at 375 mμ, but it has been suggested[127] that this absorption may have been due to an impurity.

with the assumption that these absorptions are due to π–π* transitions. Therefore, it seems reasonable to assume that allyl anions are present in these dilute solutions in tetrahydrofuran, though, of course, the anions may exist in ion pairs or other clusters. The near identity of the extinction coefficients of the absorptions of **80** and **81** is good evidence that the ionization of allyllithium (**80**) is essentially complete. If some significant portion of the allyl groups of **80** were present in non-ionic form, then addition of the electron-releasing methyl group should lead to decreases in the ion concentration and apparent extinction coefficient of **81**.

The spectra of the phenylallyl lithium and potassium compounds (**84** and **85**) can reasonably be attributed to the 1-phenylallyl anion. In contrast, however, the spectrum of dialkylmagnesium compound **86** and the similar spectrum of the corresponding Grignard reagent (prepared from the bromide) are essentially identical to those of *trans*-propenylbenzene and other model compounds containing a phenyl group conjugated with only an olefinic linkage[131]. The absence of absorption at longer wavelengths suggests that ions, even tightly held in ion pairs, are not present in significant concentrations in these solutions. The unsubstituted allyl Grignard reagent, lacking the stabilizing phenyl group, should be even less prone to ionize.

The reported infrared spectra seem to be generally consistent with the conclusions drawn from ultraviolet spectra. Nujol mulls of solid allyllithium, -sodium, and -potassium are reported[132] to show a common spectrum, the most striking feature of which is an intense band at ~ 1535 cm^{-1}. This band was assigned to the carbon–carbon unsymmetrical stretching of an allyl anion. The observation of similar spectra for allyl organometallic compounds of three metals and the position of the intense absorption, intermediate between ordinary C—C and C=C stretching bonds but similar in intensity and position to those exhibited by allyl cations (Section III.B.2), are in accord with this assignment. Mulls of isobutenylsodium (1520 cm^{-1}) and cyclohexenylsodium (1522 cm^{-1}) also show similar absorptions[132]. The two bands, at about 1525 and 1560 cm^{-1}, exhibited by pentenylsodium have been assigned to the *cis* and *trans* allylic isomers possible

for this anion[132]. A strong absorption at ~ 1540 cm^{-1} also observed for allyllithium in ether[133–135], tetrahydrofuran[134,135] and other solvents[134,135] suggests that the allyl anion is present in solution as well as in the solid. In contrast, solutions of allylmagnesium reagents show strong absorptions with frequencies much nearer to the normal C=C stretching region and sensitive to solvent and to the other group associated with magnesium[135,136]. For example, absorptions are exhibited by the Grignard reagent prepared from allyl bromide[135] at 1587 cm^{-1} in ether and 1570 cm^{-1} in tetrahydrofuran, by the Grignard reagent prepared from allyl chloride[136] at 1580 cm^{-1} in ether and 1565 cm^{-1} in tetrahydrofuran, and by diallylmagnesium[136] at 1577 cm^{-1} in ether. Therefore, it seems reasonable to assign structures that are more covalent to the magnesium reagents.

The proton magnetic resonance spectra of allyllithium[53,134,135,137] and of allyl Grignard reagents[53,134,138,139] are very similar. Both show an AX$_4$ pattern with $J = 12$ cps and nearly identical chemical shifts. In ethyl ether, for example, the quintet of relative area 1 is at 6·38 (ppm downfield from tetramethylsilane) for the Grignard[138,139] and 6·46 for the lithium[134] reagent, and the doublet of relative area 4 is at 2·50 for the Grignard and 2·40 for the lithium reagent. The absorption of the four methylene hydrogens, but not that of the lone central hydrogen, is far upfield from that ordinarily observed for hydrogens at a trigonal carbon, in accord with the presence of considerable negative charge at the terminal carbons of the allyl system.

The near identity of the proton magnetic resonance spectra of allyllithium and allyl Grignard reagents is surprising. Since the ultraviolet and infrared data suggest that their nature in solutions is different, however, the similarity in the spectra may be only coincidental. The proton magnetic resonance spectra of crotyllithium (81)[140] and the crotyl Grignard reagent[141] are also similar to one another, and the chemical shifts are those expected for concentration of considerably more negative charge at the primary than at the secondary allylic carbon.

The AX_4 patterns exhibited by the allyl reagents is inconsistent with the presence of only an ionic structure or of only one covalent structure. An A_2B_2X pattern would be expected for an ionic structure (87) since

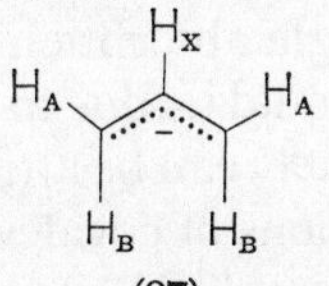

$$H_A \quad H_X \quad H_A$$
$$H_B \quad H_B$$
(87)

the terminal hydrogens are of two kinds (A and B). Rotation around the C⁝⁝⁝C bonds would have to be rapid (< 0.001 sec) for H_A and H_B to appear equivalent in the proton magnetic resonance spectra, but this is certainly in accord neither with the chemical evidence nor with the large barrier to rotation expected in an allyl anion. A covalent structure (88) should exhibit an even more complex $ABCX_2$ pattern.

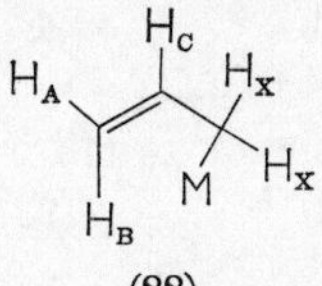

$$H_A \quad H_C \quad H_X$$
$$H_B \quad M \quad H_X$$
(88)

A predominance of allyl anion in solution is consistent with an AX_4 spectrum if the ion is in equilibrium ($89 \rightleftharpoons 90$) with low concentrations of covalent species in which, of course, H_A and H_B rapidly

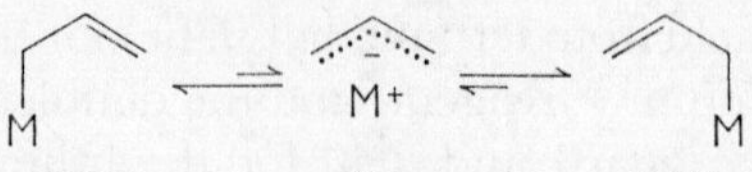

(90a) (89) (90b)

become equivalent. A concentration of covalent species **90** too low to be detectable by spectral means could account for the equivalence of H_A and H_B in the proton magnetic resonance spectrum, and probably is responsible for the AX_4 pattern exhibited by allyllithium. The formation of the pentadienyl anion (**91**) by ionization of pentadienyllithium makes more probable the assumption that allyllithium pre-

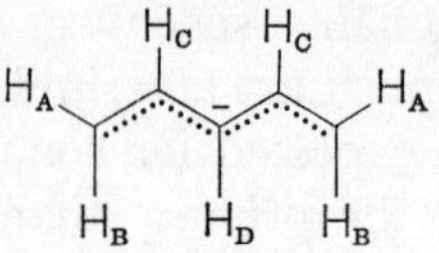

(91)

dominantly ionizes. The proton magnetic resonance spectrum of pentadienyllithium in tetrahydrofuran at 15° is that expected for an anion (**91**) and exhibits discrete absorptions for H_A and H_B[142]. These absorptions collapse to a single absorption at higher temperature; the equilibration of H_A and H_B could be due to more rapid rotation around the C═C bonds, but more likely can be attributed to more rapid equilibration with low concentrations of covalent species. The delocalization energy calculated by simple Hückel theory is less for the allyl anion than for the dienyl anion[66]. Therefore, it is reasonable that the concentration of covalent form in equilibrium with the allyl anion is somewhat higher, permitting more rapid equilibration of H_A and H_B in the allyl anion (**87**) than in the pentadienyl anion (**91**). A covalent structure is consistent also with an AX_4 pattern if it is assumed that the metal can exchange rapidly from one end of the system to the other (**92** ⇌ **93**), either by an inter- or intramolecular process. Proton

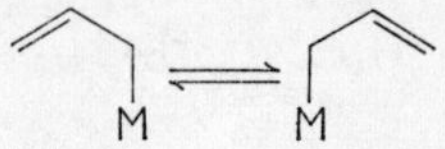

(92) (93)

magnetic resonance studies by Roberts and his coworkers[143] make this an attractive explanation for the spectrum of the allyl Grignard reagent. That the Grignard reagent prepared from **94** has chemical shifts anticipated for structure **95** rather than **96** is reasonable because

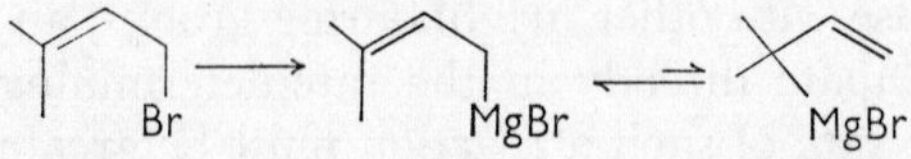

$$\text{(94) } \underset{\text{Br}}{\diagdown\!\!\diagup\!\!\diagdown} \longrightarrow \underset{\text{(95) MgBr}}{\diagdown\!\!\diagup\!\!\diagdown} \rightleftharpoons \underset{\text{(96) MgBr}}{\diagdown\!\!\diagup\!\!\diagdown}$$

the negative charge is concentrated on a primary rather than a tertiary carbon[143]. However, though only one methyl absorption is observed at room temperature, this singlet splits into two equal peaks at lower temperatures. Presumably at low temperature, discrete absorptions are seen for the methyls of covalent **95**; at higher temperatures, the rate of formation of **96** increases, though its concentration remains small, permitting the methyls to equilibrate rapidly enough to exhibit an average absorption.

3. Formation and reactions

The modes of formation and the reactions of allyl organometallics have been qualitatively little different from those of saturated organometallics. Allyl anions, like saturated anions, do not undergo the multitude of rearrangements involving 1,2-alkyl shifts at some stage which were noted with allyl cations. The counterpart in anion chemistry of the rearrangement of cyclopropyl to allyl cations has been observed only in systems that lead to particularly stable aromatic anions[143a]—even cyclopropylsodium is reported to be stable to ring opening[144]. An unsymmetrically substituted allyl anion reacts at different rates at the two ends—the factors influencing these rates have been reviewed[113] and also considered in recent publications[120,121].

IV. NON-CONJUGATED IONS

A. Introduction

Any major influence exerted on the rates or products of a reaction by a non-conjugated double bond must be a result of formation at some stage of the reaction of a bonding interaction involving the double bond. Purely inductive interactions, though certainly present, are small for double bonds insulated from a reaction centre by one or more carbon atoms.

A double bond that becomes bonded or partially bonded to a non-contiguous reaction centre is acting as a 'neighbouring group'[145,146]. A wealth of information is available about a variety of neighbouring groups in reactions involving cationic, but not anionic, intermediates.

As is often the case with other neighbouring groups, an olefinic function could participate directly in the rate-determining step for ion formation. The rate of such a reaction must be greater than in the absence of participation—the participation of the double bond exerts a 'driving force' upon the ion-formation step and the reaction is 'anchimerically'[147] assisted or accelerated. Observation of such rate acceleration is one criterion for participation. Alternatively, an olefinic function could participate in a reaction, but only following rate-determining formation of an ion. The rate of such a reaction would be influenced only by the small inductive effect of the double bond, though the products might reflect the participation. The rate of an available reaction pathway not involving double bond participation imposes a level which the rate of a participation pathway must approach or exceed if participation is to play a significant role in the reaction. The practical limit to the observation of the effects of participation that is imposed by the rates of other reaction processes is analogous to the limit to the observation of acidity imposed by the 'levelling effect' of a solvent. The importance of neighbouring group interactions generally is found to depend on the ring size of the transition state or intermediate that is formed in the order: $4 \ll 3, 4 \ll 5 \simeq 6$, and $7 < 6$. Consideration of a double bond as a neighbouring group is complicated, however, by the possibility of interaction at either end.

The possibilities for participation by a double bond are illustrated for a particular carbonium ion system that will be discussed later.

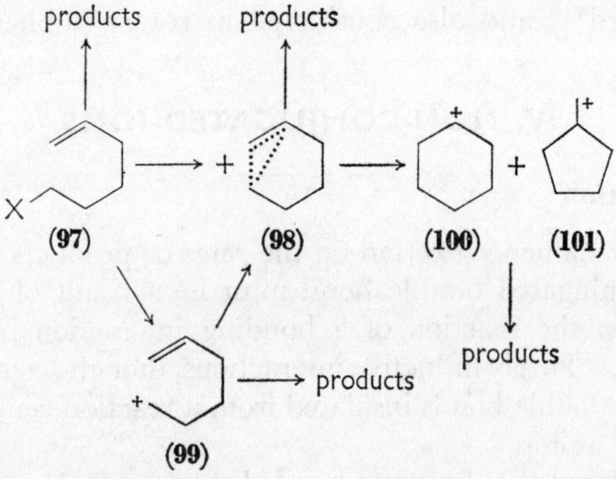

Participation by the double bond would lead to a transition state or intermediate that for the moment will be represented as **98**. This species could form either directly from **97** in the rate-determining step

2. The Properties of Alkene Carbonium Ions and Carbanions 79

or subsequent to the rate-determining formation of cation **99**. Alternatively, **97** or **99** could react to form products without direct involvement of the double bond. Some of the steps above might be reversible and several could occur simultaneously. Similar schemes for interaction of the double bond can be envisioned for reactions of carbonium ions generated in ways other than by loss of a nucleophile and for reactions of carbanions.

Participation in non-conjugated systems may lead through fleeting transition states and intermediates to cyclic ions, such as **100** and **101**, in which the double bond has disappeared. A cyclic ion such as **100** could also form from a saturated precursor (**102**). The composition

(97)   (100)   (102)

of the product mixture that results from a participation path will depend on the extent to which the ionic intermediates from which the products are formed have become identical to the ions formed from saturated cations. However, the significance of a path involving participation relative to paths in which the neighbouring group is not involved depends only on the energy of the transition state for participation relative to the energies of other transition states or intermediates. Therefore, the nature of the transition state for participation is significant in determining the fraction of the reaction that will proceed by a participation path, even though the products from that path may be the same that would have formed from a saturated precursor.

Assigning structures to the cations formed from saturated precursors sometimes raises problems that complicate the consideration of their relationship to the cationic intermediates formed by participation. Structure **100** is usually considered to be an adequate representation of the fully formed cyclohexyl cation, though in principle one could imagine this cation to be a hybrid (**98**) to which structures **99** and **101** might make minor contributions. However, such hybrid structures have been widely considered to be important for some of the cations that will be relevant to the subject of this section.

For example, the norbornyl cation is common to **103** and **104**. The experimental basis for proposals of delocalized structures for this ion is well known and extensively reviewed[148,149,150]. In the ions formed

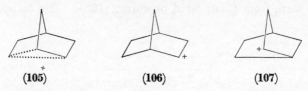

from norbornyl systems (**104**), $C_{(1)}$ and $C_{(2)}$ either are equivalent or are rapidly interchanged by a process that can have only a very low activation energy. Widely considered structural possibilities for the norbornyl cation are the symmetrical **105**, a resonance hybrid of **106**

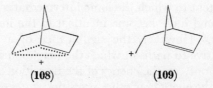

and **107**, and **108**, which includes an additional resonance contribution from **109**. A rapid equilibration between the asymmetric ions

that are represented as **106** and **107** has also been proposed. Symmetrical structure **105** (or **108**) is either more stable or only slightly less stable than **106** ⇌ **107**; equilibration of **106** and **107** would involve passing through **105** which therefore could not be of much higher energy.

The consideration of transition states or intermediates in carbonium ion reactions of systems containing double bonds therefore is entangled with the problem of "nonclassical"[151] carbonium ions. Structural problems, such as that mentioned briefly for the norbonyl system (**105**), can be raised for a number of systems that will be mentioned in this Section. However, this Section will usually consider only the reactions of olefinic starting materials. The reader is referred to other sources for experimental results, often voluminous, obtained from reactions of the related saturated starting materials.

There is a theoretical basis for believing that participation in cationic reactions may, at least initially, involve significant bonding to both carbons of the double bond, as indicated in structure **98**. Simple molecular orbital calculations suggest that bonding by a positive carbon to both carbons of a double bond leads to greater stabilization than does bonding to only one carbon. In contrast, bonding by a negative or radical carbon leads to greatest stabilization if it is only to one carbon of the double bond.

A simple model (**111**) for the 3-buten-1-yl system (**110**) was considered by Simonetta and Winstein in the first relevant calculation[152].

(**110**)   (**111**)

It was assumed that p orbitals at $C_{(1)}$, $C_{(3)}$, and $C_{(4)}$ are involved in pure π-bonding between $C_{(3)}$ and $C_{(4)}$ but part σ- and part π-bonding between $C_{(1)}$ and $C_{(3)}$. A maximum stabilization energy (delocalization energy minus strain energy) of about 6 kcal/mole was calculated by varying the angle $C_{(1)}$—$C_{(2)}$—$C_{(3)}$. Using a similar model, Howden and Roberts[153] calculated a stabilization energy of about 4 kcal/mole. However, inclusion of σ- and π-bonding between $C_{(1)}$ and $C_{(4)}$ led to a greater stabilization energy, ~ 11 kcal/mole, for the cation but reduced stabilization energies for the anion and radical. Similar calculations by Piccolini and Winstein[154] found greater delocalization energy for **112** by considering $C_{(2)}$–$C_{(5)}$ and $C_{(2)}$–$C_{(6)}$ rather than just $C_{(2)}$–$C_{(6)}$

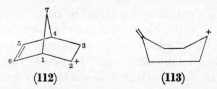

(**112**)   (**113**)

bonding. Similarly, the calculated stabilization energy for **113** was greatest for a geometry permitting considerable bonding to both carbons of the double bond[154]. Similar calculations for other ions have only considered the possibility of bonding to both carbons of the double bond[154,155]. It has been pointed out that the carbon bonded to the double bond carbons might be expected to have hybridization nearer sp^3 than the sp^2 used in these calculations[156]. These molecular orbital calculations resemble those for the cyclopropenyl system[72]. The assembly of atomic p orbitals in **111** would be similar

4+C.A. 2

to that in a cyclopropenyl cation (**32**) if $C_{(1)}$ could be located symmetrically with respect to $C_{(3)}$ and $C_{(4)}$. As mentioned previously (Section III.A), the cyclopropenyl cation has aromatic stability, but the corresponding anion does not.

The results of calculations based on a different model, considered in connection with 1,2-rearrangements, also suggest that bonding to both carbons of the double bond should be favourable in cations but unfavourable in anions and radicals. The model considers the relative energies of **114** and **115**, the latter representing a transition

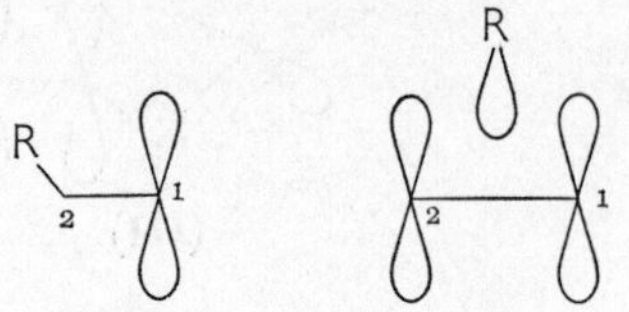

(**114**)   (**115**)

state or intermediate in a 1,2-rearrangement of R. It is assumed that the relevant orbitals at $C_{(1)}$ in **114** and at $C_{(1)}$ and $C_{(2)}$ in **115** are p, that R in **115** has sp^3 hybridization, and that bonding between $C_{(1)}$ and $C_{(2)}$ in **115** is π. Species **114** resembles the approach of R to only one carbon and **115** the approach of R to both carbons of a double bond, though in **114** and **115** such bond formation is considered to be fully developed. Zimmerman and Zweig found that **115** would be more stable than **114** for the cation but less stable for the radical and anion[157]. Streitwieser reached the same conclusion on the basis of a related calculation[158].

Stabilization of a cation is calculated to be greatest if the bonding to both carbons of the double bond is not only significant but equal in systems whose geometry and substitution make such symmetry possible. By extended Hückel molecular orbital calculations, Hoffmann has predicted that **116** is most stable if $C_{(7)}$ is located symmetrically with

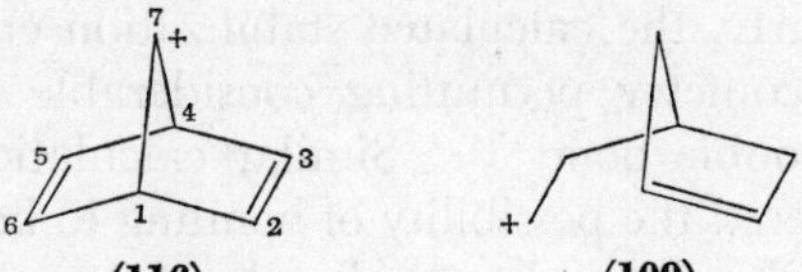

(**116**)   (**109**)

respect to $C_{(2)}$ and $C_{(3)}$[159]. It is also predicted that $C_{(7)}$ is nearer to one of the double bonds than to the other. A similar calculation suggests that the cation formed from approach of the positive carbon of **109** to the double bond is most stable if the approach is symmetrical with respect to the carbons of the double bond[160].

B. Cations

Examples of double bond participation in carbonium ion reactions are too numerous to be reviewed comprehensively in this brief section. A review discusses some of the variety of systems in which such participation is noted[161].

I. 5,6-Double bonds

The formation of 2-norbornyl derivatives in solvolyses of the arylsulphonates of structure **117** demonstrates that the double bond

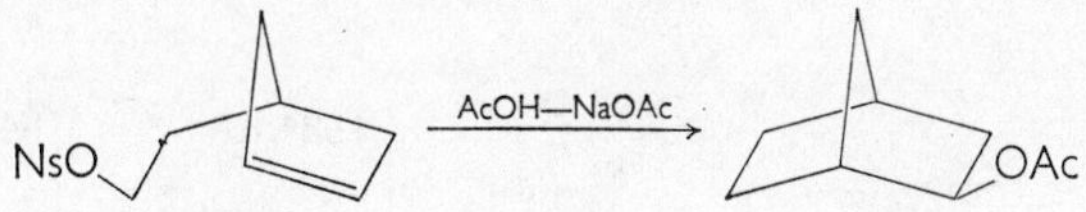

NsO— (**117**) —AcOH—NaOAc→ (**118**) —OAc

ONs = *p*-nitrobenzenesulphonate

participates at some stage in the reactions[162,163]. The particularly thorough studies of **117** and related systems provide a good introduction to the role played by non-conjugated double bonds in carbonium ion reactions.

Participation by the double bond in the rate-determining step is shown (Table 10) by the greater rate of acetolysis of **117** than of its saturated analogue (**119**). The participation by the double bond is actually somewhat more important than indicated by the rate ratio of 95—in the absence of participation, the inductive effect of the double bond should slightly decrease the solvolysis rate of **117**, to perhaps ∼ 0·7 times that of its saturated analogue[164].

Other mechanistic features of this reaction are in accord with direct participation by the double bond in the rate-determining step. The solvolysis rates of unsaturated arylsulphonates (**117**) exhibit a different dependence on solvent variation than do the rates of saturated arylsulphonates (**119**)[163,164]. The ratio $k_{\text{unsaturated}}/k_{\text{saturated}}$ varies from 5·8 for *p*-toluenesulphonates in 50% aqueous ethanol at 70° to 640 for *p*-nitrobenzenesulphonates in the much less nucleophilic formic acid at 25°[164]. Solvent nucleophilicity is less important for the unsaturated systems, as expected for reactions involving internal nucleophilic attack by the double bond. The rate (Table 10) of the

TABLE 10. Relative rates of acetolyses of *p*-nitrobenzenesulphonates at 60°

	Compound[a]	Relative rate	Ref.
117	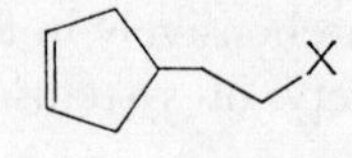	95	162
119	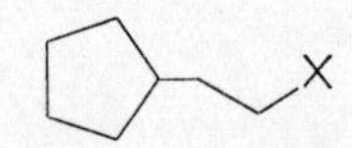	(1·0)	162
120	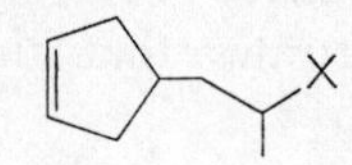	299[b]	164
121	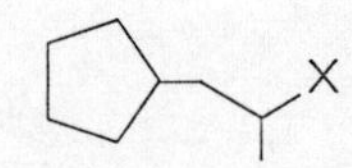	138[b]	164
122	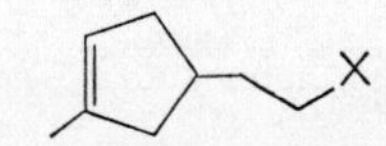	663[c]	165
123	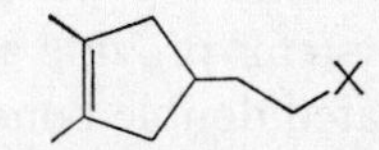	3650[c]	165
124	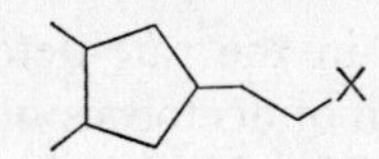	1·09[d]	165
125	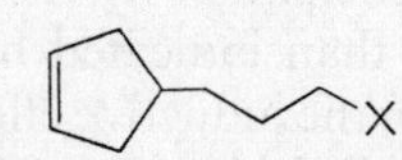	1·4[d]	166
126	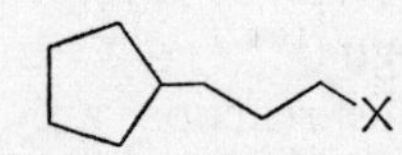	1·8[d]	166
127	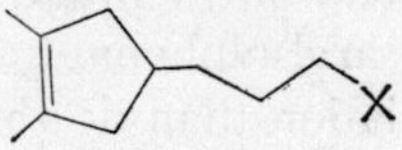	7·6[c]	166

[a] X = *p*-nitrobenzenesulphonate.
[b] Comparison made at 54·4°. Rate for comparison substance extrapolated from other temperatures.
[c] Interpolated from data at other temperatures.
[d] Extrapolated from data at other temperatures.

unsaturated secondary p-nitrobenzenesulphonate (**120**) is only 3·1 times that of the corresponding primary compound (**117**). This is in accord with the experience that anchimeric assistance by a neighbouring group is less for ionization of secondary than of primary compounds[145]. A much larger acceleration by the methyl would be expected for a solvolysis without participation, and a factor of 138 actually is observed for the similar rate comparison of the corresponding saturated compounds (**121** and **119**). The magnitude of the rate-decrease caused by substitution of α-deuteriums in solvolysis of **117** is not characteristic of solvolyses of primary arylsulphonates, but is reasonable if internal nucleophilic attack by the double bond is involved[167].

That the double bond carbons bear equal or nearly equal charge in the transition state, consistent with symmetrical approach as illustrated in **108**, is suggested by the cumulative effects of methyl substituents at the double bond. Insertion of one methyl group into **117** (to give **122**) leads to a rate enhancement by a factor of 7, and of another methyl (to give **123**) to an additional enhancement of 5·5. Bartlett and Sargent have noted that to the degree that the transition state for cyclization is symmetrical, it is reasonable to expect equivalent symmetry in the initial intermediate ion[165]. It is conceivable, though, that the initial intermediate ion in the participation reaction is not identical with the 'norbornyl' cation that forms directly from norbornyl starting materials. The rate-accelerating effect of a methyl substituent at the double bond (in **122**) is greater than that of an α-methyl substituent (in **120**). This difference suggests that in the transition state for ion formation, more charge may be at the carbon atoms of the original double bond than at the carbon from which the leaving group is departing[164].

The norbornyl products obtained from participation reactions[162,163] have the *exo* configuration, as noted in the formation of only **118** from **117**[162]. This configuration, particularly consistent with product formation from an intermediate such as **105**, is observed for products of similar reactions of norbornyl derivatives. A significant amount of unrearranged, as well as of cyclic product, is obtained from the modestly accelerated acetolysis of **120**[164]. It is probable that the unrearranged product arises entirely from a non-participation path. The product composition agrees with that predicted by assuming that the accelerated and non-accelerated portions of the reaction lead to the unrearranged and cyclic products, respectively.

Evidence for participation is noted in solvolyses of acyclic arylsulphonates with 5,6-double bonds. For example, acetolysis of

5-hexen-1-yl *p*-nitrobenzenesulphonate (**128**) is slightly accelerated[168]. The proportion of cyclic product observed in the presence of urea[169] or sodium acetate[168,169] is consistent with cyclization being associated

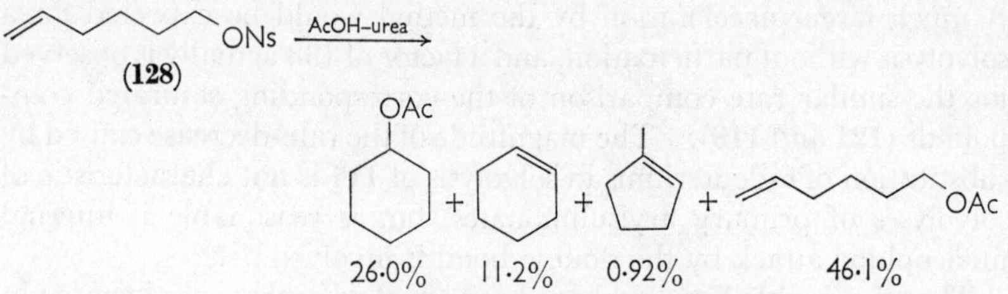

$$\text{(128)} \quad \diagdown\!\!\diagup\!\!\diagdown\!\!\diagup\!\!\diagdown\!\!\diagup\text{ONs} \xrightarrow{\text{AcOH-urea}}$$

cyclohexyl-OAc + cyclohexene + methylcyclopentene + hex-5-enyl-OAc

26·0%   11·2%   0·92%   46·1%

ONs = *p*-nitrobenzenesulphonate

solely with the accelerated portion of the reaction (if a suitable correction is made for the rate-decreasing effect of the double bond on the unassisted reaction). The ratio of cyclic to acyclic products is greater in formolysis[170] than in acetolysis of **128**. A higher proportion of participation in formic acid than in the more nucleophilic acetic acid was noted also for **117**. Formolysis of **129** at 75° is ~20 times

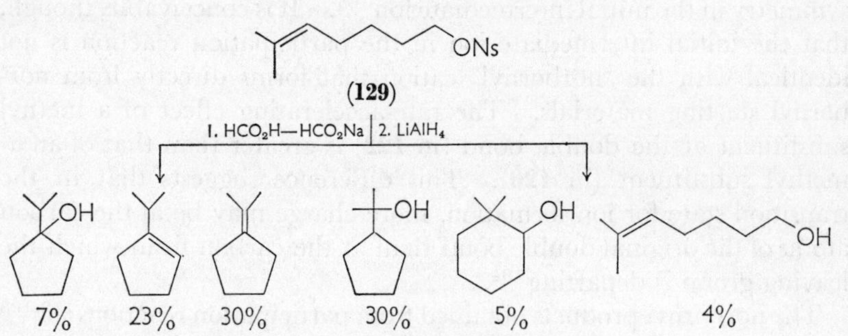

$$\text{(129)} \quad (CH_3)_2C=CH\text{-}CH_2\text{-}CH_2\text{-}CH_2\text{-ONs}$$

1. HCO$_2$H–HCO$_2$Na   2. LiAlH$_4$

7%   23%   30%   30%   5%   4%

ONs = *p*-nitrobenzenesulphonate

faster than that of the n-hexyl ester and the products are largely cyclic[171]. The amount of unrearranged acyclic product is that expected if it is formed by the small amount of unassisted reaction. The cyclic products, except for 2,2-dimethylcyclohexanol, are those of closure to five-membered rings. The cyclohexanol is actually thought to arise from isomerization of five-membered ring products, since it is found in increasing amounts at their expense as the reaction period is lengthened[171].

The intermediate cations from which products are formed cannot be identical in the acetolyses of 5-hexen-1-yl (**128**) and cyclohexyl *p*-

nitrobenzenesulphonates. The ratio of cyclohexene to cyclohexyl acetate is 0·43 in the acetolysis of **128** but 3·6 in an identical acetolysis of cyclohexyl *p*-nitrobenzenesulphonate[168]. Differences in ionic intermediates generated from related cyclic and acyclic compounds have also been noted in other systems related to **128**[172]. The intermediates could differ in the structures of the cations themselves, that from the acyclic starting material resembling **98** but that from the cyclohexyl material resembling **100**[168]. However, the intermediates will be formed with different positional relationships between the cation and the *p*-nitrobenzenesulphonate counterion[168]. Since product formation may occur before the positional relationships have time to become identical, this factor may also influence the product ratios and conceivably could be more important than differences in cation structure[173]. The dependence of the reactions of short-lived cationic intermediates on the history of their formation is well known in other situations[2,2a,146].

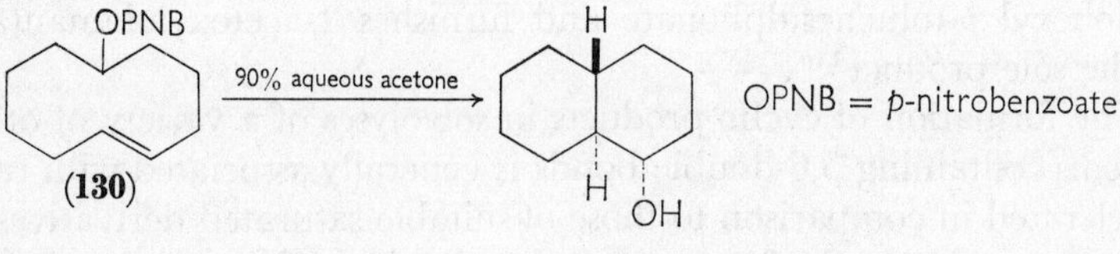

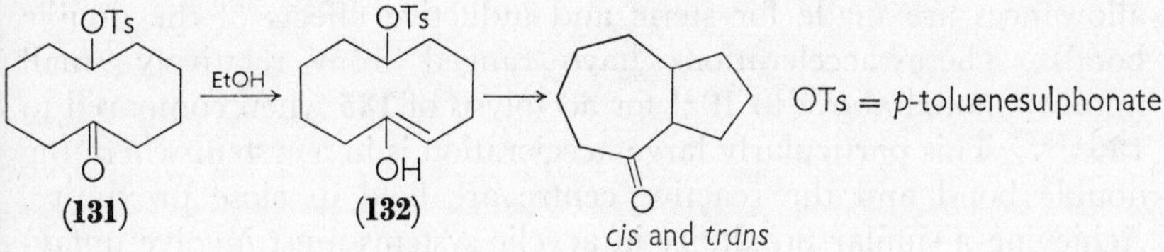

Cyclizations of **128** and **129**, unlike the cyclizations involving the symmetrical double bond of **117**, can lead to the formation of products with either five- or six-membered rings. The particular ring size noted can be rationalized as being due to the influences on the transition states for product formation of the stabilities of the possible extreme product-forming ions (**100** and **101** in the reaction of **128**). The predominant formation of six-membered rather than five-membered ring products from **128** can be attributed to the greater stability of a cyclohexyl (**100**) than a cyclopentymethyl cation (**101**); both the lesser strain of a cyclohexane than a cyclopentane ring and the greater stabilization of a secondary than a primary cation contribute to the difference in stabilities. The formation of five-membered ring products from **129** is ascribed to the greater stabilization of a tertiary cation overcoming the greater strain of a five-membered ring. The ability of substituents to alter the direction of cyclization is noted also by comparison of the solvolyses of **130**[174] and **131**[175] that were studied by Goering and his coworkers. The enol (**132**) was demonstrated to be the reactive species in the solvolysis of **131**, so the altered direction of addition can be attributed to the carbonium ion stabilizing ability of the hydroxyl substituent.

Evidence for participation of 5,6-double bonds has been noted in reactions of a variety of other cyclic and fused ring systems, **133** and **134**, for example. The rate of acetolysis of **133** is > 30 times greater

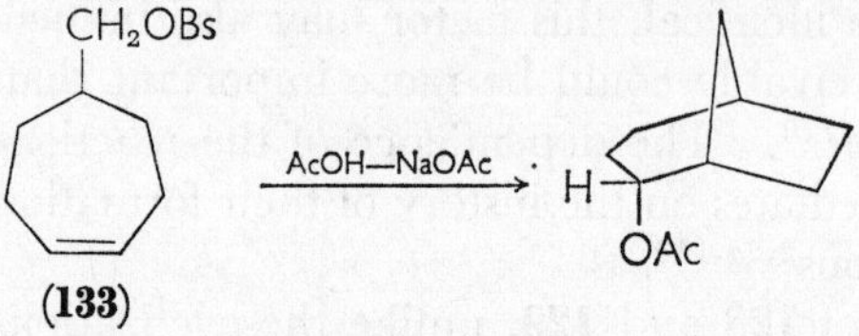

(**133**) → AcOH—NaOAc →

OBs = p-bromobenzenesulphonate

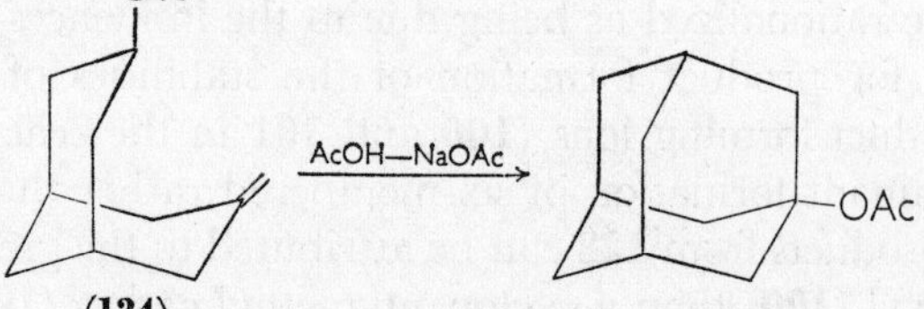

(**134**) → AcOH—NaOAc →

OTs = p-toluenesulphonate

than that of the corresponding saturated compound, and the stereochemistry of the product is that expected for its formation from a bridged ion[176]. Acetolysis of **134** is 13,500 times faster than that of cyclohexyl p-toluenesulphonate and furnishes 1-acetoxyadamantane as the sole product[177].

The formation of cyclic products in solvolyses of a variety of other systems containing 5,6-double bonds is generally associated with rates accelerated in comparison to those of suitable saturated derivatives, if allowances are made for steric and inductive effects of the double bonds. These accelerations have ranged from relatively small numbers noted above to 10^{11} for acetolysis of **135** when compared to **136**[178]. This particularly large acceleration is in a system where the double bond and the reactive centre are held in close proximity. Achieving a similar proximity in acyclic systems must involve unfav-

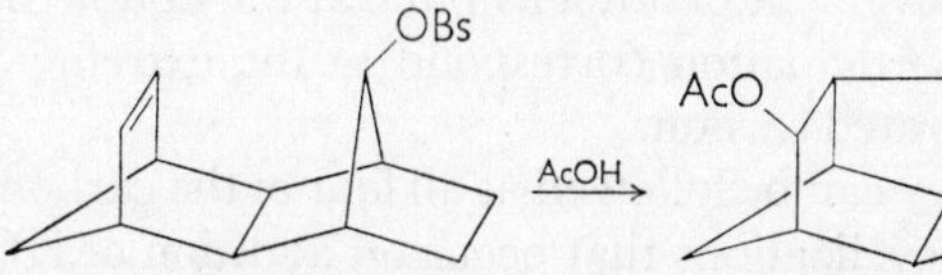

(135)

OBs = p-bromobenzenesulphonate

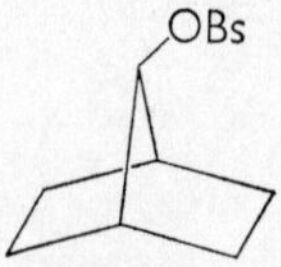

(136)

ourable strain and entropy factors which will decrease the overall stabilization of the intermediate. In addition, loss in the transition state of some of the unusually large strain of the double bond of **135** may also be a significant accelerating factor.

Sulphuric acid solutions of **137** with R = Ph and of the solid

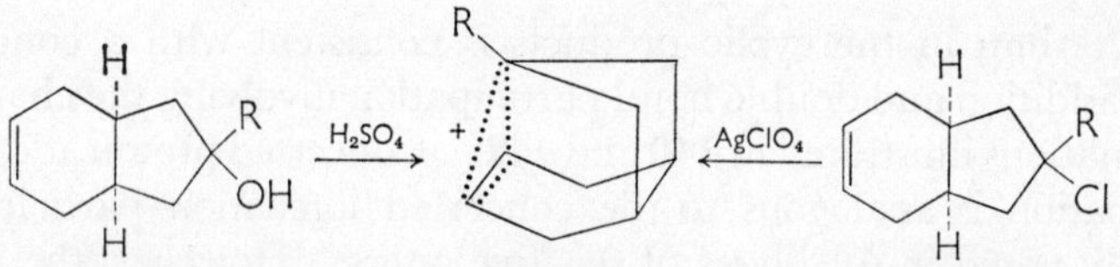

(137) (138) (139)

perchlorate prepared from **139** exhibit essentially identical long wavelength ultraviolet absorptions that were attributed to a long-lived cation to which structure **138** was assigned[179]. Similarly, **137** with R = CH_3 was reported to form the corresponding methyl substituted ion[179]. The long lifetimes of these ions in sulphuric acid are surprising in view of recent experience with a variety of carbonium ions[2,2a,78].

The extent to which participation of the double bond at a developing cationic centre is concerted with the reaction of a nucleophile at the double bond itself may be important in determining product stereochemistry. As one possible extreme, a cation might be fully formed without specific interaction with a nucleophile and only then undergo reaction with a nucleophile to form products. As the other extreme, reaction with a nucleophile could be so concerted with the participation by the double bond, that a cation might never be present. These extremes for the involvement of a nucleophile at the double bond in a reaction involving participation correspond to the extreme S_N1 and S_N2 processes at saturated carbon.

The stereochemistry can be followed at all four of the carbons undergoing change in the cyclizations that occur on addition of HCO_2D to **140** and to **142**[180]. The stereochemistry of the methyl groups and

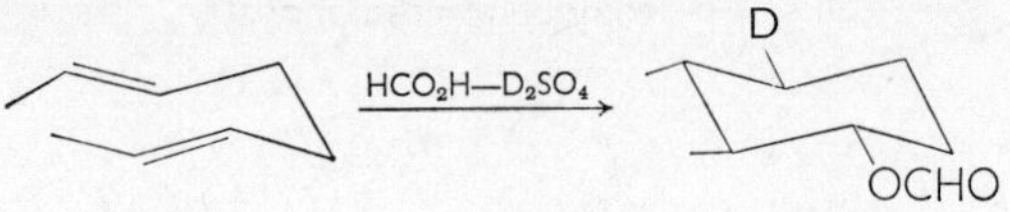

(140) → (141) 37–40%

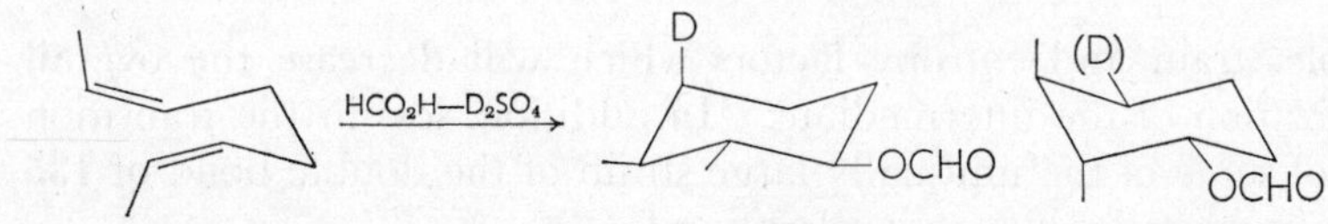

(142) → (143) 10% (144) 1%

the deuterium in the cyclic products is consistent with a concerted proton addition and double bond participation involving the chair-like conformations illustrated in **140** and **142**. Concerted proton addition–participation is analogous to the concerted ionization–participation generally noted in solvolyses of reactive esters. However, the major cyclization product (**143**) formed from **142** is not that expected for concerted addition of the nucleophile, formic acid. Its formation could be explained by conformational isomerization of an initially formed cyclic cation **145** to the more stable **146**, followed by the

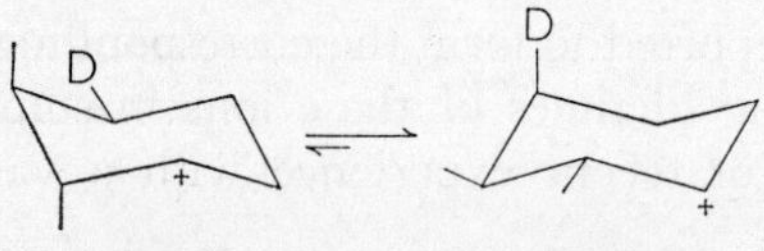

(145) ⇌ (146)

expected equatorial attack on **146**. The minor product (**144**) could form by a completely concerted reaction or from **145**. The product (**141**) obtained from **140** presumably is formed by the same mechanism, though it would be anticipated for either concerted addition of the nucleophile or addition of the nucleophile to the most stable cyclohexyl cation.

A variety of reactions have been studied in systems, such as **147** and

2. The Properties of Alkene Carbonium Ions and Carbanions

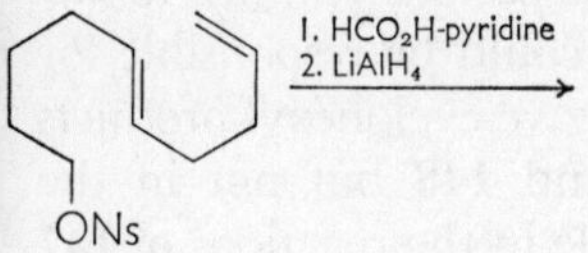

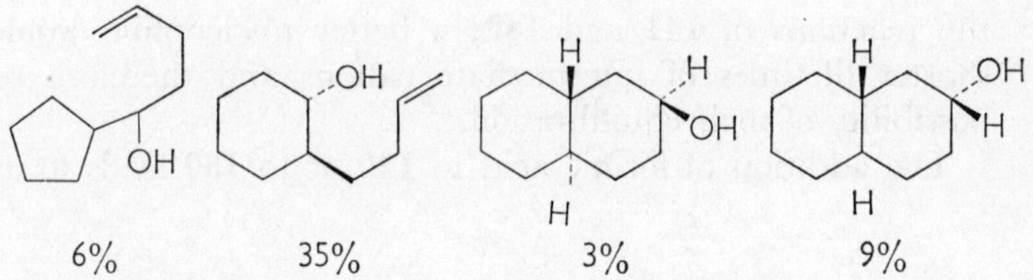

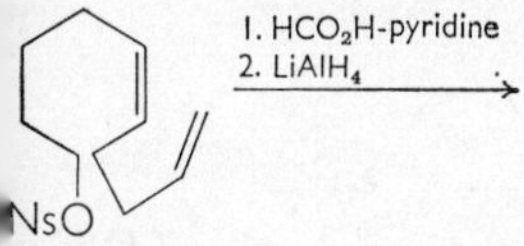

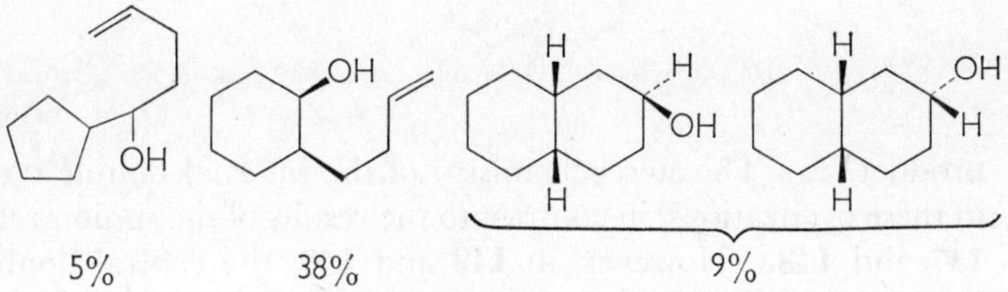

148, in which action of an additional double bond as a nucleophile leads to formation of a second ring. The solvolysis of **147** in 80% aqueous formic acid containing sodium formate is somewhat faster than that of 5-hexen-1-yl *p*-nitrobenzenesulphonate (**128**)[170]. In accord with this kinetic evidence for participation, the products obtained from **147** and from its *cis* isomer (**148**) in formic acid-pyridine are largely cyclic[181]. That both product mixtures contain a cyclopentyl product as well as cyclohexyl products is not surprising in view of the effects already noted on ring size due to alkyl substituents on 5,6-double bonds. Attack by the solvent leading to butenylcyclohexanols or by the terminal double bonds leading to decalols cannot proceed through a common intermediate in the reactions of **147** and **148** since these reactants furnish products of different stereochemistries. The results are compatible either with attack by the nucleophile (formate or terminal double bond) on cationic intermediates that still reflect the stereochemistries of the precursors, or with attack by the nucleophile being concerted with the participation by the internal double bond. Isomeric cationic intermediates could be cyclohexyl cations that differ in conformation or bridged cations (similar to **98**) that differ in configuration. Several factors could be responsible for the observations that reaction with solvent to give cyclohexyl products was stereospecific in the reactions of **147** and **148** but not in the reaction of **142**. For example, the nucleophile in the reactions of **147** and **148** may be formate instead of the less nucleophilic formic acid in the reactions of **141** and **142**; a better nucleophile would lead to shorter lifetimes of intermediate cations and therefore reduce the possibility of their equilibration.

The addition of formic acid to **149** or to **150** leads to a common product[182]. The stereochemistry of the internal double bond is lost in these cyclizations, in contrast to the results of the similar reactions of **147** and **148**. However, in **149** and **150**, the central double bonds have an additional alkyl substituent and the terminal double bonds are less nucleophilic due to the electron withdrawing carboxyl groups, factors that could prolong the lifetimes of cationic intermediates and permit their interconversion; alternatively, both trienes could initially form a common monocyclic diene whose subsequent cyclization would be responsible for the observed product.

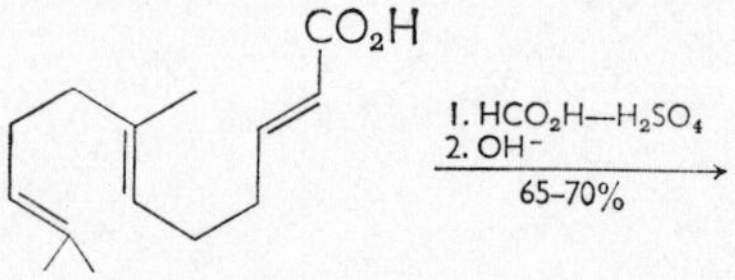

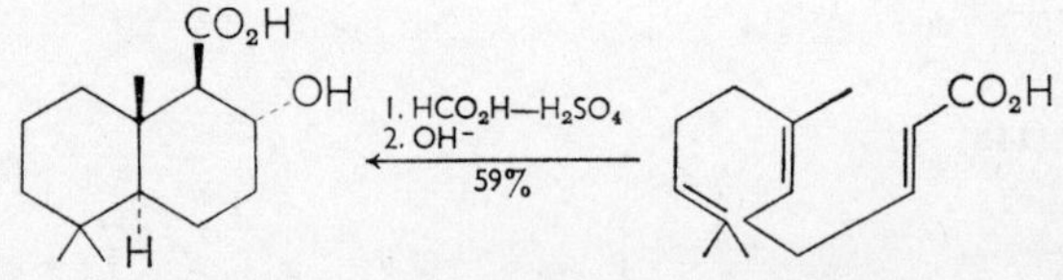

The factors determining the involvement of nucleophiles in participation reactions are not yet well formulated. However, in favourable cases, the stereoselectivity of product formation can be high, even in reactions involving more than two cyclizations. For example, the cyclization of **151** produces an excellent yield of isomers of structure **152**, possessing the *trans,anti,trans*-dodecahydrophenanthrene skeleton[183].

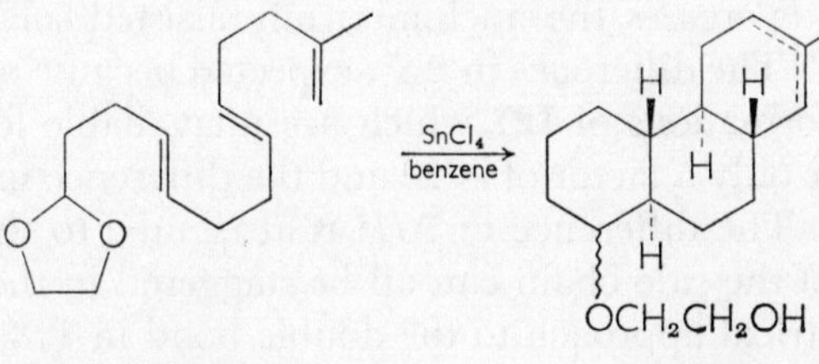

(151)   (152)

The synthetic value of cationic cyclizations is obvious. The reactions described in this section have often involved cation generation by solvolysis of a reactive ester, since this is convenient for kinetic studies. However, a variety of other ways of generating cations, two of which are seen in examples above, are useful for synthesis.

Cationic cyclizations involving 5,6-double bonds are also of great interest because of their relation to cyclizations of polyolefins that play an important role in the biogenesis of many terpenoid compounds[184]. Indeed, the impetus to much of the chemical work with these systems, particularly those in which multiple cyclizations are possible, has been to ascertain the degree to which the high stereoselectivity of biological cyclizations might be due to the intrinsic requirements of cationic cyclizations rather than to conformational control of the reactants by enzymes.

2. 6,7- and 4,5-Double bonds

Participation in acyclic systems by 6,7-double bonds is less important than by 5,6-double bonds. Formolysis of **153** is not noticeably accelerated and furnishes only ~1% of cyclic (cycloheptyl) product[170],

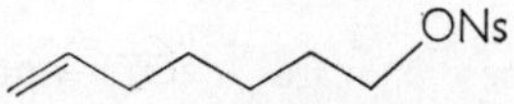

ONs = p-nitrobenzenesulphonate

(153)

in contrast to the evidence for significant participation in an identical reaction of **128**, the homologue with one carbon less. Similarly, **125** undergoes acetolysis without acceleration (Table 10) or formation of cyclic products, in marked contrast to the acetolysis of **117**[165]. The two added methyl groups in **127** do lead to sufficient participation by the double bond to be detected by a modest rate acceleration and by the formation of cyclized as well as of uncyclized products[166]. Comparison of **127** and **123** indicates that the elongation of the chain from ethyl to propyl decreases the anchimerically assisted solvolysis by a factor of ~600. The difference in $\Delta S^{\ddagger}$, expected because of the larger number of conformations of **127** which are unavailable for reaction, is responsible for only a factor of ~10 and the difference in $\Delta H^{\ddagger}$ for a factor of ~60. The difference in $\Delta H^{\ddagger}$ is attributed to the fact that the hydrogens of the side chain can all be staggered in the transition state for symmetrical approach to the double bond in **123**, but not in **127**[166]. However, participation by 6,7-double bonds is significant in reactions of **154**[185] and related systems[185,186] and of **155**[187].

(154)

(155)

OTs = *p*-toluenesulphonate

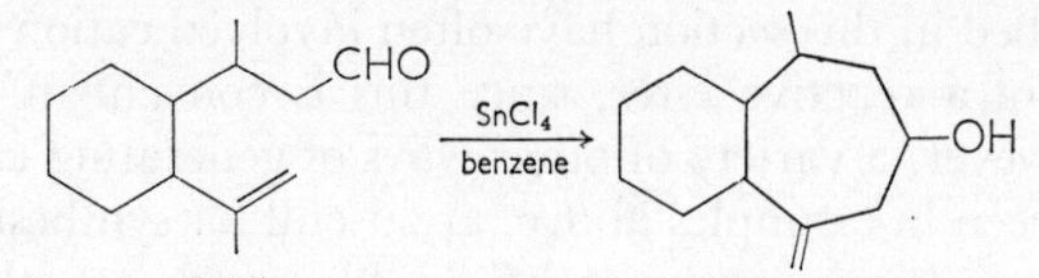

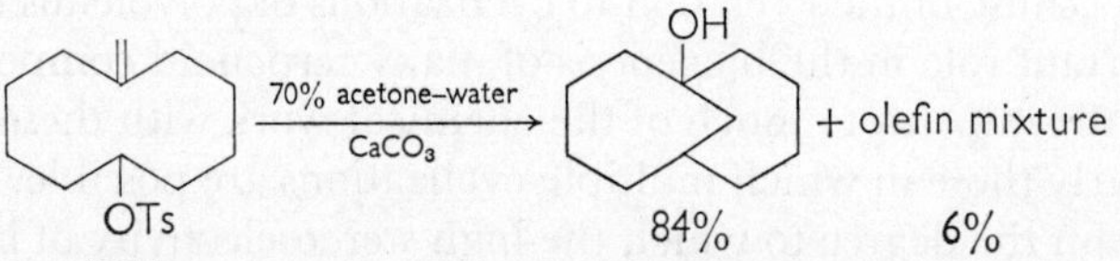

Participation in acyclic systems by 4,5-double bonds is also less important than by 5,6-double bonds. For example, participation is not noted in acetolyses[168] and formolyses[170] of **156** and **157**[188], in

(156) **(157)**

ONs = *p*-nitrobenzenesulphonate OTs = *p*-toluenesulphonate

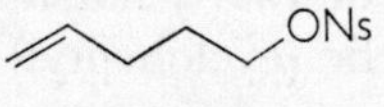

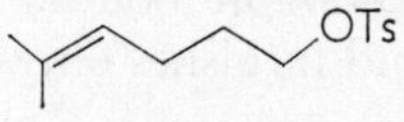

contrast to the observation of acceleration and cyclized products in similar reactions of **128** and **129**. However, the product mixture obtained from acetolysis of **158** indicates that participation is important in this reaction[189].

It is clear that participation in acyclic systems is more favourable by 5,6- than by 4,5- or 6,7-double bonds. The same delocalization energy is potentially available to the transition states of reactions of any of these systems. However, achieving a geometry that will permit obtaining a substantial portion of this delocalization energy will involve unfavourable strain and entropy factors. Attaining a favourable

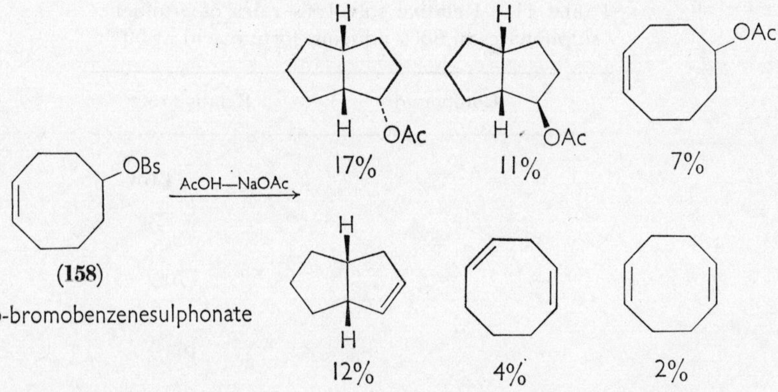

2. The Properties of Alkene Carbonium Ions and Carbanions

[Reaction scheme: compound (158) with OBs group, treated with AcOH—NaOAc, gives products: bicyclic with H/OAc cis 17%, bicyclic with H/OAc 11%, cyclooctenyl OAc 7%, bicyclic alkene 12%, cycloocta-1,3-diene 4%, cycloocta-1,5-diene 2%]

OBs = p-bromobenzenesulphonate

geometry, perhaps one that permits a relatively symmetrical distribution of charge between the carbons of the former double bond, apparently is least costly for systems with 5,6-double bonds.

3. 3,4-Double bonds

A variety of reactions that involve participation of 3,4-double bonds have been subjected to intensive study. Systems (**159**) with 3,4-double

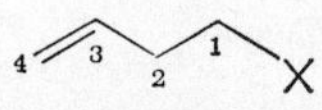

$$\overset{4}{\diagup}\overset{3}{\diagdown}\overset{2}{\diagup}\overset{1}{\diagdown}X$$

(**159**)

bonds were designated 'homoallylic' by Winstein, to emphasize the possibility of a bonding interaction between a reaction centre and a double bond separated by a methylene group[152]. This term is often now used more generally to refer to systems that contain even more remote double bonds.

Participation in acyclic systems by 3,4-double bonds is comparable to that by 5,6-double bonds. For example, solvolysis of 3-butenyl p-toluenesulphonate (**160**) in 98% aqueous formic acid leads largely to cyclic products and the rate (Table 11) is 3·7 times that of the corresponding saturated sulphonate[190]. Solvolysis of **161** at 100° in

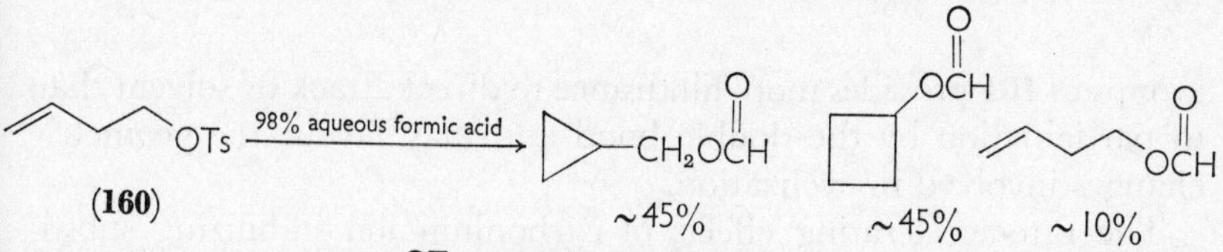

[Reaction: (160) CH₂=CHCH₂CH₂OTs + 98% aqueous formic acid → cyclopropylmethyl formate (~45%) + cyclobutyl formate (~45%) + 3-butenyl formate (~10%)]

OTs = p-toluenesulphonate

TABLE 11. Relative solvolysis rates of *p*-toluene-sulphonates in 98% aqueous formic acid at 50°[a]

Compound[b]	Relative rate
 CH₃CH₂CH₂CH₂—X	(1·0)[c]
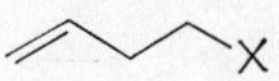 CH₂=CHCH₂CH₂—X	3·7[c]
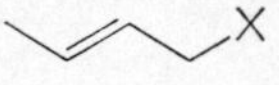 *trans*-CH₃CH=CHCH₂CH₂—X	770[d]
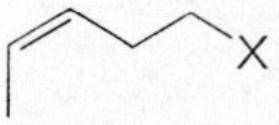 *cis*-CH₃CH=CHCH₂CH₂—X	165[d]
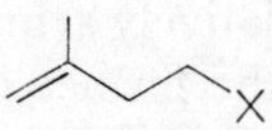 CH₂=C(CH₃)CH₂CH₂—X	12[d]
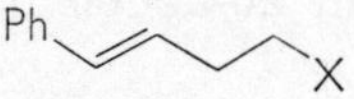 PhCH=CHCH₂CH₂—X	350[e]
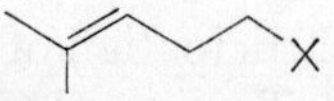 (CH₃)₂C=CHCH₂CH₂—X	16,500[e]

[a] Ref. 191.
[b] X = *p*-toluenesulphonate.
[c] Ref. 190.
[d] The rate was measured in a solvent mixture of 10% pyridine in 98% formic acid. The relative rate figure is based on the assignment of a relative rate of 3·7 to 3-butenyl *p*-toluenesulphonate in that medium.
[e] The relative rate figure is based on the assignment of a relative rate of 770 to *trans*-3-penten-1-yl *p*-toluenesulphonate in 98% formic acid.

acetic acid is ~60 times as fast as that of the corresponding saturated compound[192]. Even though participation generally is less significant in acetic acid than in formic acid, the steric effect of the methyl

CH₃CH=CHC(CH₃)₂OBs   OBs = *p*-bromobenzenesulphonate
(**161**)

groups in **161** provides more hindrance to direct attack by solvent than to participation by the double bond and may favour the geometric changes involved in cyclization.

The rate-accelerating effects of carbonium ion stabilizing substituents at the unsaturated carbons of **160** are shown by the relative formolysis rates in Table 11. The effects of methyl substituents suggest that though both carbons of the double bond become more positive in the transition state, much more charge accumulates at $C_{(4)}$ than at $C_{(3)}$. The smaller effect of phenyl than of methyl is contrary to the usually greater ability of a phenyl substituent to stabilize a positive charge. However, the smaller effect of phenyl, which has been noted in like situations[193], can be attributed in part to a greater stabilization of the starting alkene by phenyl than by methyl. Moreover, in situations where the bonding in a transition state is part σ and part π, the balance of the electron-withdrawing inductive and electron-releasing conjugative effects of phenyl may lead to less electron release than is usually found[191].

The transition state for the solvolysis of **160**, having positive charge both at $C_{(3)}$ and $C_{(4)}$, can be represented by **162**, a hybrid receiving

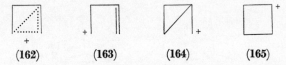

(162)   (163)   (164)   (165)

contributions, though certainly far from equal, from **163, 164**, and **165**. However, it is not clear if the products from reactions of systems with 3,4-double bonds form directly from ions best represented as **162**, or instead, following isomerization to other ions. The formolysis of **160** furnishes a product mixture similar to those obtained from reactions of unsubstituted cyclopropylmethyl and cyclobutyl derivatives[194,195], and may involve the same cationic intermediates, at least by the time that product is formed. Therefore, the consideration of the structures of cationic intermediates in reactions of systems with 3,4-double bonds is inextricably related to the unresolved problem of the proper assignment of structures to the cations formed from cyclopropylmethyl[194,195] and cyclobutyl[195] systems. Proposals for structures have included the suggestion that reactions of cyclopropylmethyl, cyclobutyl, and 3-buten-1-yl compounds lead directly to **162**, the bicyclobutonium ion, as a common intermediate[196]. However, more recent evidence suggests[195] that there may instead be discrete, though sometimes rapidly equilibrating, cyclopropylmethyl and cyclobutyl cations. These cations may be most stable when they have the geometries illustrated in **166**[194,195] and **167**[195], which if correct, may differ from the geometries of the transition states for participation in systems with 3,4-double bonds.

Evidence for participation has been encountered in studies of a

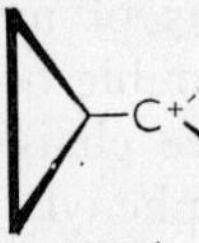

(166)  (167)

variety of cyclic and fused ring homoallylic systems. Many of these systems have geometries which favour bonding to $C_{(3)}$ more than to $C_{(4)}$. The extensively studied cholesteryl[197] (**168**) and 5-norbornenyl (**169**)[150] systems are examples in which the formation of cyclic products

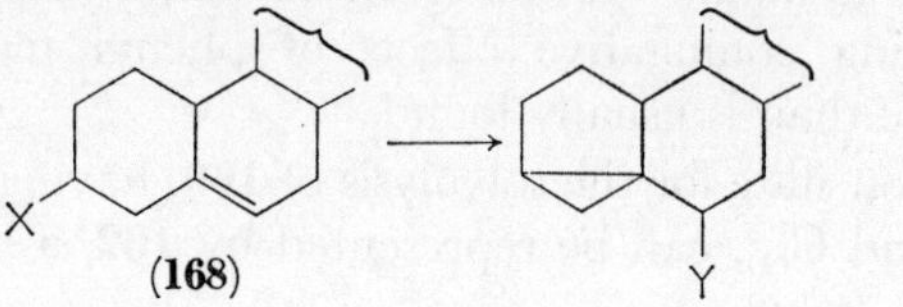

(168)

(169)

makes participation obvious. The acetolysis of **170**[198] is an example

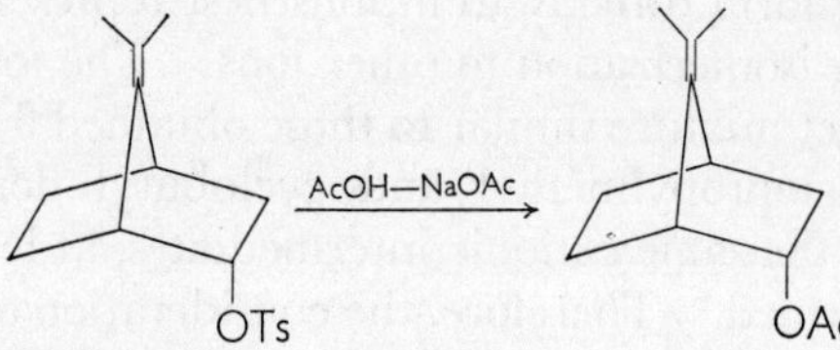

(170) → AcOH—NaOAc → (171)

OTs = p-toluenesulphonate

in which participation by the double bond must be responsible for a significant rate acceleration and for the retention of configuration in **171**, though cyclic products are not observed. In contrast, the *anti*-7-norbornenyl (**172**) and 7-norbornadienyl (**174**) systems have

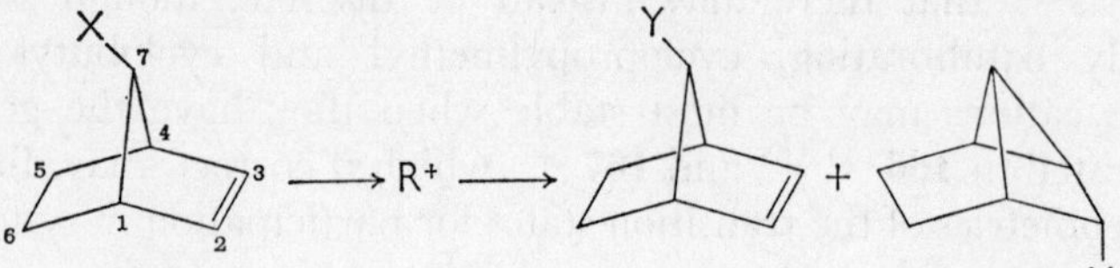

(172) → R⁺ → (173)

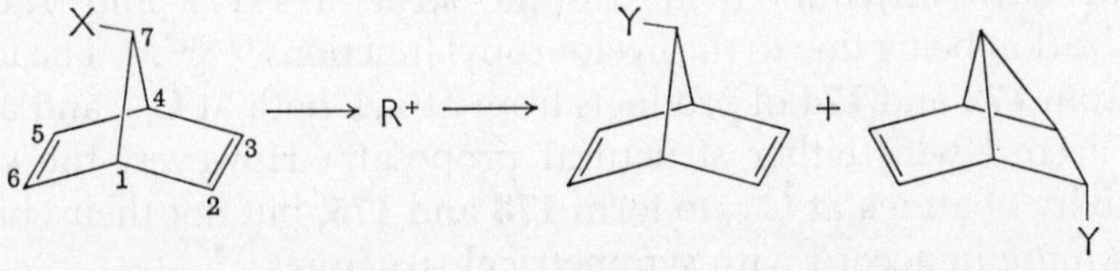

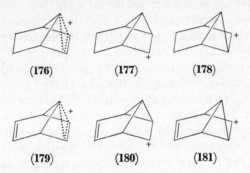

(174) (175)

geometries that particularly favour a symmetrical interaction between a double bond and a reactive carbon. Studies of these systems indicate that participation is unusually important and that symmetrical structures (**176** and **179**) probably are more stable (and certainly are

(176)   (177)   (178)

(179)   (180)   (181)

not much less stable) than asymmetric structures (**177** and **180**) for the resultant cations[194].

Anchimeric assistance to the formation of cations from **172** and **174** is particularly large[194]. Acetolysis of *anti*-7-norbornenyl *p*-toluenesulphonate[199] (**172**, X = *p*-toluenesulphonate) is $\sim 10^{11}$ times faster than that of the corresponding saturated compound, and hydrolysis of 7-norbornadienyl chloride[200] (**174**, X = Cl) in 80% aqueous acetone is $\sim 10^3$ times faster than that even of *anti*-7-norbornenyl chloride. These accelerations of $\sim 10^{11}$ and $\sim 10^{14}$ are much larger than others discussed in this chapter except that observed with **135**, a compound in which the reactive carbon also is placed symmetrically with respect to the double bond, though with a different spatial relationship than in **172** and **174**.

Symmetrical structures **176**[155] and **179**[201] have been proposed for the ions formed in these reactions. Such structures have been termed 'bishomocyclopropenyl' to indicate the resemblance of the conjugated systems to that of the cyclopropenyl cation, though there are two interruptions in the framework connecting the conjugating carbons[155]. Alternatively, it has been proposed that these ions may instead have unsymmetrical tricyclic structures **177** and **180**, the stabilization being due to the cyclopropyl functions[78,202]. The formation from **172** and **174** of products from attack both at $C_{(7)}$ and at $C_{(2)}$ is in accord with either structural proposal. However, the stereoselectivity of attack at $C_{(2)}$ to form **173** and **175**, but not their epimers, seems more in accord with symmetrical structures[194].

Proton magnetic resonance spectra of stable solutions of ions that form from **172**[203,204] and **174**[203,305] have been observed. In the spectra of both fully formed ions, $C_{(2)} = C_{(3)}$, $C_{(5)} = C_{(6)}$ and $C_{(1)} = C_{(4)}$, consistent with symmetrical structures **176** and **179**. This equivalence of carbons is consistent with asymmetric structures only if **177** and **180** are in rapid equilibrium with their isomers, **178** and **181**; even then, symmetrical ions would be midpoints in the rapid equilibrations and therefore not much less stable than the asymmetric ions. The chemical shifts exhibited by the protons of 7-norbornenyl and 7-norbornadienyl cations and by the corresponding 7-methyl substituted cations[203,206] suggest that the charge resides principally at $C_{(2)}$ and $C_{(3)}$ rather than at $C_{(7)}$ in the fully formed cations. The absorption of the hydrogen at $C_{(3)}$ in the 2-methylnorbornadienyl cation is >3 ppm downfield from those of cyclopropyl hydrogens in cyclopropylmethyl cations; therefore, the position of this absorption suggests that **182** may be a more adequate representation of this ion

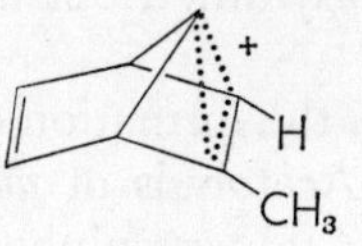

(182)

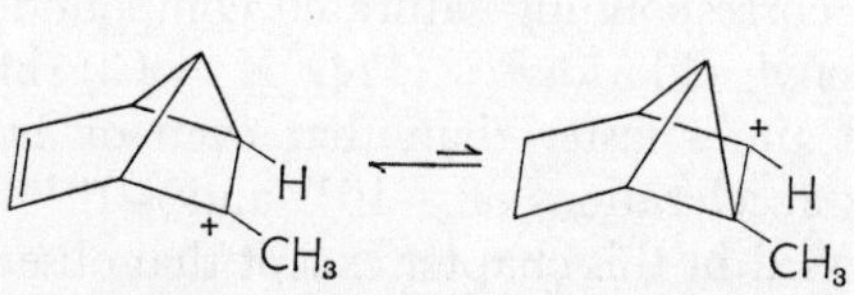

(183)

than **183**, the structure that should predominate if such cations have tricyclic structures[206].

The observation that $C_{(2)} = C_{(3)} \neq C_{(5)} = C_{(6)}$ in the proton magnetic resonance spectrum of the 7-norbornadienyl cation indicates that the system has undergone some distortion to allow $C_{(7)}$ to bond more strongly to one of the original double bonds than to the other. The additional stabilization that is gained is large; achieving a geometry in which the 'double bonds' become equivalent is shown by proton magnetic resonance studies to have $\Delta F^{\ddagger} \geq$ 19·6 kcal/mole[206].

In contrast to the very large anchimeric assistance to solvolysis by the double bond in *anti*-7-norbornenyl derivatives (**172**), hydrolysis of **184**, in which the reactive centre also is placed symmetrically with

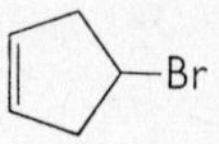

(**184**)

respect to the double bond, proceeds without participation. The hydrolysis rate in 50% aqueous acetone at 60° is ∼0·2 that of cyclopentyl bromide and hydrolysis in 50% aqueous ethanol forms only unrearranged alcohol[207]. The geometry of the 7-norbornenyl system must be much nearer to a particularly favourable geometry for participation than that of the 3-cyclopenten-1-yl system. Distorting the 3-cyclopenten-1-yl cation to a geometry similar to that of the 7-norbornenyl system is estimated to introduce ∼17 kcal/mole of angle strain[207]. This strain energy is greater than the ∼15 kcal/mole of stabilization energy that corresponds at room temperature to the rate acceleration of 10^{11} exhibited by the *anti*-7-norbornenyl system. Therefore, the strain required to establish significant participation in **184** may exceed the stabilization energy that could be gained. The lack of acceleration indicates that the amount of π-overlap between $C_{(1)}$ and $C_{(3)}$ in a relatively unstrained 3-cyclopenten-1-yl cation must be insignificant. The contrasting behaviour of the 7-norbornenyl and 3-cyclopenten-1-yl systems provides a particularly dramatic example of how structural restraints imparted by the starting material can dictate the extent of participation.

C. Anions

No convincing example seems to be known of significant acceleration by a non-conjugated carbon–carbon double bond of the formation of a simple carbanion (or polar organometallic). This deficiency, in contrast to the multitude of examples of participation in cation formation, may indicate a significant difference in the interactions of double bonds with cations and with anions, or may result simply from a lack of appropriate studies.

A carbonyl group should be more effective than a carbon–carbon double bond in stabilizing a transition state for carbanion formation because of the possibility of placing negative charge on oxygen. In fact, carbonyl groups in some fused ring ketones are thought to participate in base-catalysed proton removal at non-conjugated carbons. Nickon and his coworkers have shown that hydrogen isotope exchange in **185** is most rapid at $C_{(6)}$[208,209], and probably involves removal of

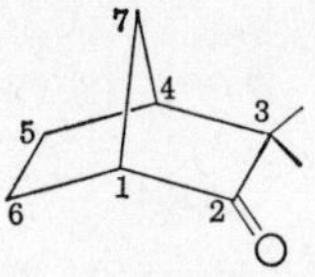

(**185**)

the *exo* rather than the *endo* hydrogen[210]. Racemization of optically active **185** occurs at about the same rate as exchange, suggesting that exchange results in formation of a symmetrical intermediate that can be represented as **186** or as a resonance hybrid of **186**, **187**, and **188**[208].

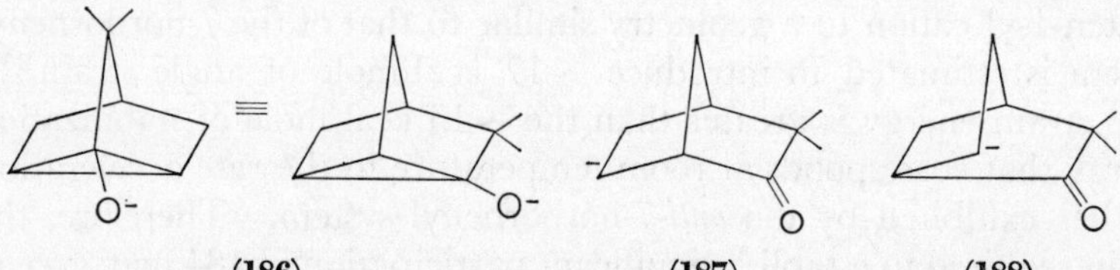

(**186**)   (**187**)   (**188**)

It is attractive to assume that a bonding interaction develops during proton removal, leading to a transition state that resembles this intermediate. If only inductive stabilization by the carbonyl group was involved, the $C_{(1)}$ hydrogens or perhaps the methyl hydrogens should be the most readily removed. Removal of the $C_{(6)}$ *exo* hydrogen permits significant overlap of the developing non-bonded orbital at $C_{(6)}$ with the atomic p orbital at the carbon of the carbonyl group. Removal of hydrogens from other positions of **185** or from any position in compound **189**, which does not undergo exchange under similar

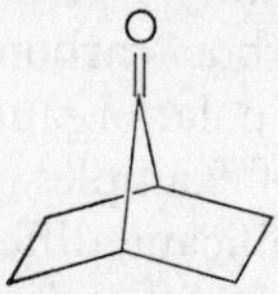

(**189**)

conditions[211], would lead to less favourable overlap. The base-catalysed isomerization of **190** to **191** can also be attributed to activation of proton removal by a non-conjugated carbonyl group[212]. Though the activated proton is one atom farther from the carbonyl group in **190** than in **185**, the functions are held within close proximity of each other.

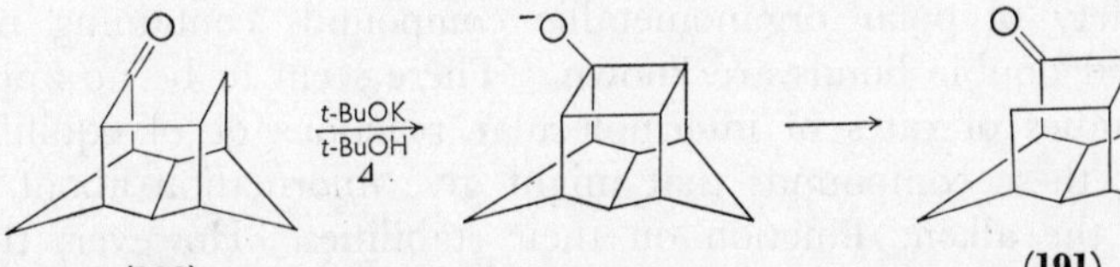

Proton removal from $C_{(4)}$ of **192** does seem to be accelerated significantly by the non-conjugated double bond. Hydrogen isotope exchange catalysed by potassium *t*-butoxide in dimethylsulphoxide at 60° is 3×10^4 faster in **192** than in **194**[213]. This acceleration is too large to attribute solely to the inductive effect of the non-conjugated double bond. It has been suggested that the allyl anion that is being formed has added stability due to the interaction with the other double bond illustrated in **193**. The chemical shifts observed in the proton magnetic resonance spectrum of the anion are consistent with the presence of a significant portion of the negative charge at $C_{(6)}$ and $C_{(7)}$[214]. However, the stabilization of an allyl anion by a non-conjugated double bond in this bicyclic system does not imply a like stabilization of saturated anions by non-conjugated double bonds. Hückel molecular orbital calculations predict that overlap between the allyl anion and a double bond can lead to an appreciable bonding interaction. The delocalization energy is calculated to be $\sim 0{\cdot}2\beta$ if $\beta_{2,7}$ and $\beta_{4,6}$ are assumed to be only $0{\cdot}3\beta$[214]. This anion has a conjugated system similar to that of the aromatic cyclopentadienyl anion, though the ring of conjugated carbons has two interruptions.

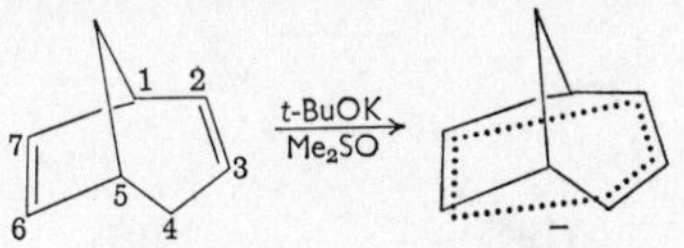

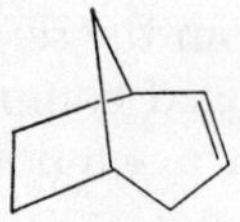

A variety of polar organometallic compounds containing non-conjugated double bonds are known. There seem to be no appropriate studies of rates of intermolecular reactions or of equilibria involving these compounds that might give information about the effect of the alkene function on their stabilities. However, these organometallic compounds do undergo intramolecular cyclizations. To the extent that carbanion-like intermediates are involved, these cyclizations represent examples of participation relevant to the subject of this section.

Cyclizations of Grignard reagents with 3,4-double bonds, such as **195**[215], have been studied by Roberts and his coworkers[215,216]. The

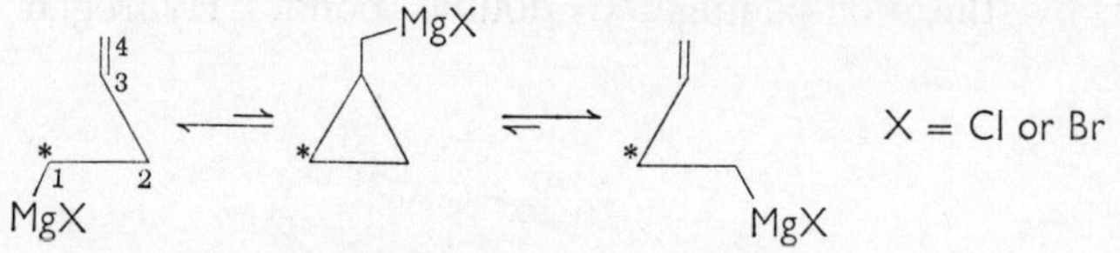

(195)   (196)   (197)   X = Cl or Br

equilibrium concentration of **196** is too small to be detected directly, but its formation is inferred from the scrambling of $C_{(1)}$ and $C_{(2)}$ that is noted when isotopically labeled compounds are used[215]. Isomerization of **198**, a reagent with a 4,5-double bond, to **200** must also

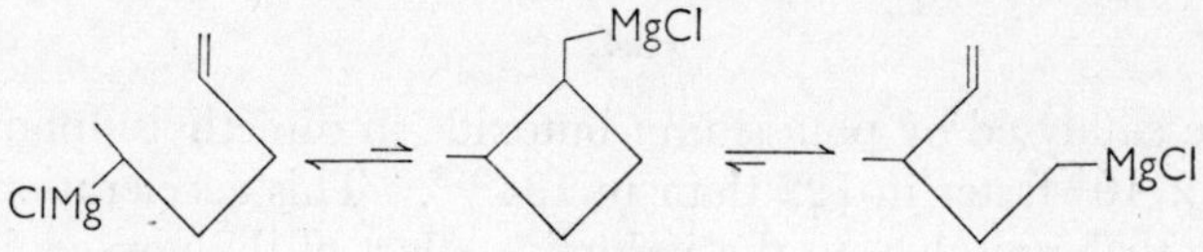

(198)   (199)   (200)

involve cyclization, though the cyclic reagent (**199**) is not actually observed[135,217,218]. Cyclizations of Grignard reagents with 5,6-double bonds[218,219], such as **201**[218], lead to their more stable cyclic

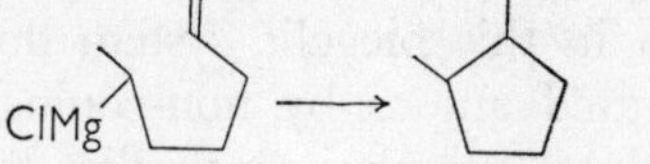

(201)   (202)

isomers, which do not ring open. These cyclizations are remarkably facile. Additions of Grignard reagents to unconjugated double bonds of hydrocarbons, the intermolecular counterparts of the cyclizations, are not known.

Cyclizations occur for organometallics other than Grignard reagents. Maercker and Roberts have observed cyclizations of organolithium and organopotassium compounds with 3,4-double bonds[220]. The cyclization of **198** to **199** also occurs for the corresponding organolithium and organosodium compounds[217]. Additional cyclizations of more complex lithium, sodium, potassium, and caesium derivatives have been reported[221].

The formation of the smaller of the two possible rings is a striking feature of these cyclizations. The preference for the smaller ring size seems in marked contrast to the results of cationic cyclizations, where ring size often can be rationalized by considering the stabilities of the possible cyclic cations. It is clear that ring size in the organometallic cyclizations is not determined solely by the influence on transition states of the stabilities of the cyclized organometallic compounds. For example, the Grignard reagent (**199**) formed in the cyclization of **198** should be more strained than **203**, the other possibility, by ~20

(**203**)   (**204**)

kcal/mole[222]. Similarly, the product (**202**) of the cyclization of **201** should be more strained than **204** by ~7 kcal/mole. Primary organometallics are formed by the observed cyclizations to give smaller rings in these examples, and secondary organometallics would have been formed by cyclizations to give the larger rings. However the difference in stability between primary and secondary Grignard[35,135] or lithium[31,135] reagents probably does not exceed a few kcal/mole.

It has been suggested that a preference for approach of the reactive carbon along the axis of an atomic p orbital at one of the carbons of the double bond may be the significant factor in dictating ring size in the organometallic cyclizations[218,223]. Examination of models suggests that approach with this geometric requirement is less strained to the nearer than to the more distant carbon of 3,4-, 4,5-, and 5,6-double bonds. Preference for a similar approach in radical cyclizations would rationalize such observations[224] as the more rapid cyclization of the 5-hexen-1-yl radical to the cyclopentylmethyl radical than to the more stable cyclohexyl radical.

While the mechanisms of the organometallic cyclizations are not known with certainty, they clearly are not cationic[135,224]. The geometry of approach postulated to rationalize the ring sizes noted in the organometallic cyclizations is consistent with the predictions from molecular orbital calculations that interaction involving a double bond and a carbanion or radical will be most favourable if bonding (and hence approach) is to only one of the carbons of the double bond. In contrast, predictions that maximum stabilization of the approach of a carbonium ion to a double bond will be achieved by bonding to both carbons of the double bond suggest a different transition state geometry for cationic cyclizations, a geometry that perhaps more readily leads to the formation of the larger rings. Even in a system in which initial approach of the cationic carbon is constrained principally to one carbon of a double bond, the possibility in cations (but not in anions or radicals) of 1,2-alkyl migration permits rearrangement to a more stable cation at a stage at which carbon–carbon bond formation has progressed significantly.

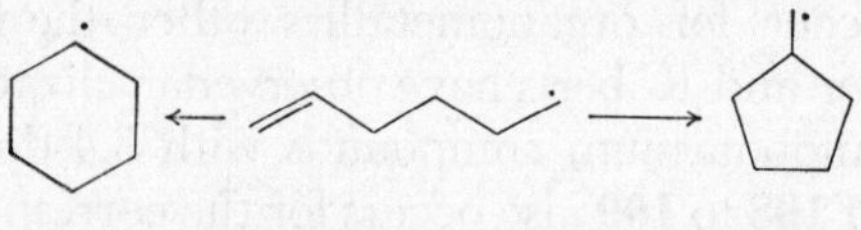

A requirement for concerted formation of the new carbon–metal bond at the same time that the new carbon–carbon bond is being formed might also rationalize the ring sizes observed in organometallic cyclizations[218]. The geometry necessary for a concerted process can be achieved more readily for formation of smaller than larger rings in the examples noted. To the extent that the cyclizations are concerted, the ring sizes formed may not be relevant to the problem of the interaction between double bonds and reactive carbons.

V. ACKNOWLEDGEMENTS

The author wishes to thank N. C. Deno, J. A. Dixon, A. Nickon, P. T. Lansbury, P. S. Skell, A. Streitwieser, Jr., and S. Winstein for helpful comments. Jane M. Richey provided invaluable assistance in the preparation of the manuscript. The National Science Foundation, The Petroleum Research Fund of the American Chemical Society, and the Alfred P. Sloan Foundation have supported the author's research in some of the areas discussed in this chapter. The author wishes to thank the John Simon Guggenheim Memorial Foundation for a fellowship and the members of the Department of Chemistry at the University of California at Berkeley for their hospitality at the time that this chapter was written.

VI. REFERENCES

1. G. S. Hammond, *J. Am. Chem. Soc.*, **77,** 334 (1955).
2. G. A. Olah and P. von R. Schleyer, Eds., *Carbonium Ions*, Vol. I, John Wiley and Sons, Inc., New York, 1968. (Vol. II, III, IV to be published.)
2a. D. Bethell and V. Gold, *Carbonium Ions, An Introduction*, Academic Press, London, 1967.
3. D. J. Cram, *Fundamentals of Carbanion Chemistry*, Academic Press, New York, 1965.
4. C. B. Milne and A. N. Wright in *Rodd's Chemistry of Carbon Compounds*, Vol. 1, Part B, 2nd ed. (Ed. S. Coffey), Elsevier Publishing Co., Amsterdam, 1965, Chap. 7.
5. G. E. Coates and K. Wade, *Organometallic Compounds*, 3rd ed., Vol. 1, Methuen and Co., Ltd., London, 1967, Chap. 1.
6. M. Schlosser, *Angew. Chem. Intern. Ed. Engl.*, **3,** 287, 362 (1964).
7. T. E. Hogen-Esch and J. Smid, *J. Am. Chem. Soc.*, **88,** 307, 318 (1966) and references therein.
8. Ref. 5, Chap. 2.
9. G. A. Balueva and S. T. Ioffe, *Russ. Chem. Rev. (English Transl.)*, 439 (1962).
10. T. L. Brown, *Advan. Organometal. Chem.*, **3,** 365 (1965).
11. E. C. Ashby, *Quart. Rev.* (London), **21,** 259 (1967); B. J. Wakefield, *Organometal. Chem. Rev.*, **1,** 131 (1966).
12. R. Hoffmann, *J. Chem. Phys.*, **40,** 2480 (1964).
13. M. A. Matesich, Ph.D. Thesis, University of California, Berkeley, 1966.
14. H. G. Richey, Jr., in *Carbonium Ions*. (Ed. G. A. Olah and P. von R. Schleyer.) Vol. II, Chap. 21, to be published by John Wiley and Sons, Inc., New York.
15. R. W. Bott, C. Eaborn and D. R. M. Walton, *J. Chem. Soc.*, 384 (1965).
16. D. S. Noyce, M. A. Matesich, M. D. Schiavelli and P. E. Peterson, *J. Am. Chem. Soc.*, **87,** 2295 (1965).
17. T. L. Jacobs and S. Searles, Jr., *J. Am. Chem. Soc.*, **66,** 686 (1944); E. J. Stamhuis and W. Drenth, *Rec. Trav. Chim.*, **80,** 797 (1961); **82,** 385, 394 (1963); G. L. Hekkert and W. Drenth, *Rec. Trav. Chim.*, **82,** 405 (1963).
18. G. L. Hekkert and W. Drenth, *Rec. Trav. Chim.*, **80,** 1285 (1961).
19. W. Drenth and H. Hogeveen, *Rec. Trav. Chim.*, **79,** 1002 (1960); H. Hogeveen and W. Drenth, *Rec. Trav. Chim.*, **82,** 375, 410 (1963).
20. P. E. Peterson and J. E. Duddey, *J. Am. Chem. Soc.*, **85,** 2865 (1963); **88,** 4990 (1966).
21. P. E. Peterson and R. J. Bopp, Abstracts, 152nd National Meeting of the American Chemical Society, New York, September 1966, p. S3; *J. Am. Chem. Soc.*, **89,** 1283 (1967).
22. P. E. Peterson, C. Casey, E. V. P. Tao, A. Agtarap and G. Thompson, *J. Am. Chem. Soc.*, **87,** 5163 (1965).
23. P. E. Peterson and R. J. Kamat, *J. Am. Chem. Soc.*, **88,** 3152 (1966).
24. W. D. Closson and S. A. Roman, *Tetrahedron Letters*, 6015 (1966).
25. C. A. Grob and G. Cseh, *Helv. Chim. Acta*, **47,** 194 (1964).
26. R. W. Fessenden and R. H. Schuler, *J. Chem. Phys.*, **39,** 2147 (1963); E. L. Cochran, F. J. Adrian and V. A. Bowers, *J. Chem. Phys.*, **40,** 213 (1964); P. H. Kasai and E. B. Whipple, *J. Am. Chem. Soc.*, **89,** 1033 (1967).

27. G. D. Sargent and M. W. Browne, *J. Am. Chem. Soc.*, **89**, 2788 (1967); L. A. Singer and N. P. Kong, *J. Am. Chem. Soc.*, **89**, 5251 (1967).
28. H. Fisher in *The Chemistry of Alkenes* (Ed. S. Patai), Interscience Publishers, London, 1964, Chap. 13; A. A. Petrov and A. V. Fedorova, *Russ. Chem. Rev. (English Transl.)*, **33**, 1 (1964); K. Griesbaum, *Angew. Chem. Intern. Ed. Engl.*, **5**, 933 (1966); D. R. Taylor, *Chem. Rev.*, **67**, 317 (1967).
29. W. M. Jones and F. W. Miller, *J. Am. Chem. Soc.*, **89**, 1960 (1967).
30. D. Seyferth, *Progr. Inorg. Chem.*, **3**, 129 (1962).
31. D. E. Applequist and D. F. O'Brien, *J. Am. Chem. Soc.*, **85**, 743 (1963).
32. L. Pauling, *The Nature of the Chemical Bond*, 3rd ed., Cornell University Press, Ithaca, New York, 1960, Chap. 3.
33. C. A. Coulson and W. E. Moffitt, *Phil. Mag.*, **40**, 1 (1949).
34. M. Yu Lukina, *Russ. Chem. Rev. (English Transl.)*, **31**, 419 (1962).
35. R. E. Dessy, W. Kitching, T. Psarras, R. Salinger, A. Chen and T. Chivers, *J. Am. Chem. Soc.*, **88**, 460 (1966).
36. R. Waack and P. E. Stevenson, *J. Am. Chem. Soc.*, **87**, 1183 (1965).
37. R. Waack and P. West, *J. Am. Chem. Soc.*, **86**, 4494 (1964).
38. R. Waack and M. A. Doran, *Chem. Ind.* (London), 496 (1964).
39. A. Streitwieser, Jr. and J. H. Hammons, *Progr. Phys. Org. Chem.*, **3**, 41 (1965).
40. Ref. 3, Chap. 1; A. A. Frost and R. G. Pearson, *Kinetics and Mechanism, A Study of Homogeneous Chemical Reactions*, 2nd ed., John Wiley and Sons, Inc., New York, 1961, Chap. 9; R. P. Bell, *The Proton in Chemistry*, Cornell University Press, Ithaca, New York, 1959, Chap. 10.
41. M. Eigen, *Angew. Chem. Intern. Ed. Engl.*, **3**, 1 (1964).
42. A. Streitwieser, Jr., R. A. Caldwell and M. R. Granger, *J. Am. Chem. Soc.*, **86**, 3578 (1964).
43. R. A. Caldwell, Ph.D. Thesis, University of California, Berkeley, 1964.
44. A. I. Shatenshteĭn, *Advan. Phys. Org. Chem.*, **1**, 155 (1963).
45. Ref. 3, Chap. 3.
46. D. Y. Curtin and J. W. Crump, *J. Am. Chem. Soc.*, **80**, 1922 (1958).
47. D. Seyferth and L. G. Vaughan, *J. Am. Chem. Soc.*, **86**, 883 (1964).
48. N. L. Allinger and R. B. Hermann, *J. Org. Chem.*, **26**, 1040 (1961); A. N. Nesmeyanov, A. E. Borisov, N. V. Novikova and N. A. Chumaevskii, *Dokl. Chem., Proc. Acad. Sci. USSR (English Transl.)*, **148**, 163 (1963).
49. C. S. Johnson, Jr., M. A. Weiner, J. S. Waugh and D. Seyferth, *J. Am. Chem. Soc.*, **83**, 1306 (1961).
50. H. Normant and P. Maitte, *Bull. Soc. Chim. France*, 1439 (1956).
51. G. J. Martin and M. L. Martin, *Bull. Soc. Chim. France*, 1636 (1966).
52. T. Yoshino and Y. Manabe, *J. Am. Chem. Soc.*, **85**, 2860 (1963); T. Yoshino, Y. Manabe and Y. Kikuchi, *J. Am. Chem. Soc.*, **86**, 4670 (1964).
53. G. Fraenkel, D. G. Adams and J. Williams, *Tetrahedron Letters*, 767 (1963).
54. G. J. Martin and M. L. Martin, *J. Organometal. Chem.*, **2**, 380 (1964).
55. R. T. Hobgood, Jr. and J. H. Goldstein, *Spectrochim. Acta*, **18**, 1280 (1962).
56. A. N. Nesmeyanov, A. E. Borisov and N. A. Vol'kenau, *Izv. Akad. Nauk SSSR, Otd. Khim. Nauk* 992 (1954); *Chem. Abstr.*, **49**, 6892 (1955).
57. D. Y. Curtin, H. W. Johnson, Jr. and E. G. Steiner, *J. Am. Chem. Soc.*, **77**, 4566 (1955).
58. D. Y. Curtin and W. J. Koehl, Jr., *J. Am. Chem. Soc.*, **84**, 1967 (1962).

2. The Properties of Alkene Carbonium Ions and Carbanions 109

59. D. H. Hunter and D. J. Cram, *J. Am. Chem. Soc.*, **86,** 5478 (1964); **88,** 5765 (1966).
60. D. J. Cram, C. A. Kingsbury and B. Rickborn, *J. Am. Chem. Soc.*, **83,** 3688 (1961).
61. S. I. Miller and W. G. Lee, *J. Am. Chem. Soc.*, **81,** 6313 (1959).
62. H. Normant, *Advan. Org. Chem.*, **2,** 1 (1960).
63. G. Köbrich, *Angew. Chem. Intern. Ed. Engl.*, **6,** 41 (1967); H. Heaney, *Organometal. Chem. Rev.*, **1,** 27 (1966).
64. G. W. Wheland, *Resonance in Organic Chemistry*, John Wiley and Sons, Inc., New York, 1955.
65. A. Streitwieser, Jr., *Molecular Orbital Theory for Organic Chemists*, John Wiley and Sons, Inc., 1961, Chap. 1.
66. Ref. 65, Chap. 2.
67. Ref. 65, Chap. 9.
68. For example, J. P. Colpa, C. MacLean and E. L. Mackor, *Tetrahedron*, **19,** Suppl. 2, 65 (1963).
69. For example, M. Simonetta and E. Heilbronner, *Theoret. Chim. Acta*, **2,** 228 (1964).
70. D. M. Hirst and J. W. Linnett, *J. Chem. Soc.*, 1035 (1962); 1068 (1963).
71. R. Hoffmann and R. A. Olofson, *J. Am. Chem. Soc.*, **88,** 943 (1966).
72. Ref. 65, Chap. 10; A. W. Krebs, *Angew. Chem. Intern. Ed. Engl.*, **4,** 10 (1965); D. Lloyd, *Carbocyclic Non-Benzenoid Aromatic Compounds*, Elsevier Publishing Co., Amsterdam, 1966, Chap. 2.
73. E. F. Kiefer and J. D. Roberts, *J. Am. Chem. Soc.*, **84,** 784 (1962).
74. T. J. Katz and E. H. Gold, *J. Am. Chem. Soc.*, **86,** 1600 (1964).
74a. R. Breslow, J. Brown, and J. J. Gajewski, *J. Am. Chem. Soc.*, **89,** 4383 (1967) and references therein.
75. N. C. Deno in *Carbonium Ions* (Ed. G. A. Olah and P. Von R. Schleyer), Vol. II, Chap. 18 to be published by John Wiley and Sons, Inc., New York.
76. N. C. Deno, J. Bollinger, N. Friedman, K. Hafer, J. D. Hodge and J. J. Houser, *J. Am. Chem. Soc.*, **85,** 2998 (1963).
77. R. Breslow, H. Höver and H. W. Chang, *J. Am. Chem. Soc.*, **84,** 3168 (1962).
78. N. C. Deno, *Progr. Phys. Org. Chem.*, **2,** 129 (1964).
79. S. Pignataro, A. Cassuto and F. P. Lossing, *J. Am. Chem. Soc.*, **89,** 3693 (1967).
80. C. A. Vernon, *J. Chem. Soc.*, 423, 4462 (1954).
81. W. G. Young, S. Winstein and H. L. Goering, *J. Am. Chem. Soc.*, **73,** 1958 (1951).
82. E. D. Hughes, *J. Chem. Soc.*, 255 (1935).
83. E. Grunwald and S. Winstein, *J. Am. Chem. Soc.*, **70,** 846 (1948).
84. N. C. Deno, H. G. Richey, Jr., N. Friedman, J. D. Hodge, J. J. Houser and C. U. Pittman, Jr., *J. Am. Chem. Soc.*, **85,** 2991 (1963).
85. W. G. Young, S. H. Sharman and S. Winstein, *J. Am. Chem. Soc.*, **82,** 1376 (1960); see also W. G. Young and J. S. Franklin, *J. Am. Chem. Soc.*, **88,** 785 (1966).
86. T. S. Sorenson, *J. Am. Chem. Soc.*, **87,** 5075 (1965).
87. Ref. 65, Chap. 8.

88. N. C. Deno, H. G. Richey, Jr., J. D. Hodge and M. J. Wisotsky, *J. Am. Chem. Soc.*, **84,** 1498 (1962).
89. R. H. DeWolfe and W. G. Young in *The Chemistry of Alkenes* (Ed. S. Patai), Interscience Publishers, London, 1964, Chap. 10.
90. R. H. DeWolfe and W. G. Young, *Chem. Rev.*, **56,** 753 (1956).
91. K. Mackenzie in *The Chemistry of Alkenes* (Ed. S. Patai), Interscience Publishers, London, 1964, Chap. 7.
92. P. B. D. de la Mare in *Molecular Rearrangements* (Ed. P. de Mayo), Part 1, Interscience Publishers, New York, 1963, Chap. 2.
93. M. Cais in *The Chemistry of Alkenes* (Ed. S. Patai), Interscience Publishers, London, 1964, Chap. 12.
94. N. C. Deno, D. B. Boyd, J. D. Hodge, C. U. Pittman, Jr. and J. O. Turner, *J. Am. Chem. Soc.*, **86,** 1745 (1964).
95. N. C. Deno and C. U. Pittman, Jr., *J. Am. Chem. Soc.*, **86,** 1744 (1964).
96. T. S. Sorenson, *Can. J. Chem.*, **42,** 2768 (1964).
97. N. C. Deno, C. U. Pittman, Jr. and J. O. Turner, *J. Am. Chem. Soc.*, **87,** 2153 (1965).
98. T. S. Sorenson, *Can. J. Chem.*, **43,** 2744 (1965).
99. G. A. Olah, C. U. Pittman, Jr. and T. S. Sorenson, *J. Am. Chem. Soc.*, **88,** 2331 (1966).
100. T. S. Sorenson, *J. Am. Chem. Soc.*, **89,** 3782 (1967).
101. T. S. Sorenson, *J. Am. Chem. Soc.*, **89,** 3794 (1967).
102. H. G. Richey, Jr. and A. S. Kushner, unpublished results; A. S. Kushner, Ph.D. Thesis, The Pennsylvania State University, 1966.
103. N. C. Deno, J. S. Liu, J. O. Turner, D. N. Lincoln and R. E. Fruit, Jr. *J. Am. Chem. Soc.*, **87,** 3000 (1965).
104. C. U. Pittman, Jr. and G. A. Olah, *J. Am. Chem. Soc.*, **87,** 5123 (1965).
105. N. C. Deno and J. J. Houser, *J. Am. Chem. Soc.*, **86,** 1741 (1964).
105a. S. Forsén and T. Norin, *Tetrahedron Letters*, 4183 (1966).
106. R. B. Woodward and R. Hoffmann, *J. Am. Chem. Soc.*, **87,** 395 (1965).
107. C. H. DePuy, L. G. Schnack, J. W. Hausser and W. Wiedemann, *J. Am. Chem. Soc.*, **87,** 4006 (1965); S. J. Cristol, R. M. Sequeira and C. H. DePuy, *J. Am. Chem. Soc.*, **87,** 4007 (1965); C. H. DePuy, L. G. Schnack and J. W. Hausser, *J. Am. Chem. Soc.*, **88,** 3343 (1966); C. W. Jefford and R. Medary, *Tetrahedron Letters*, 2069 (1966); G. H. Whitham and M. Wright, *Chem. Commun.*, **6,** 294 (1967); T. Ando, H. Yamanaka, S. Terabe, A. Horike and W. Funasaka, *Tetrahedron Letters*, 1123 (1967); M. S. Baird and C. B. Reese, *Tetrahedron Letters*, 1379 (1967); U. Schöllkopf, K. Fellenberger, M. Patsch, P. von R. Schleyer, T. Su and G. W. van Dine, *Tetrahedron Letters*, 3639 (1967).
108. N. C. Deno, N. Friedman, J. D. Hodge and J. J. Houser, *J. Am. Chem. Soc.*, **85,** 2995 (1963).
109. C. D. Broaddus, T. J. Logan and T. J. Flautt, *J. Org. Chem.*, **28,** 1174 (1963).
110. C. D. Broaddus, *J. Org. Chem.*, **29,** 2689 (1964).
111. A. Streitwieser, Jr., J. I. Brauman, J. H. Hammons and A. H. Pudjaatmaka, *J. Am. Chem. Soc.*, **87,** 384 (1965).
112. R. Kuhn, H. Fischer, F. A. Neugebauer and H. Fischer, *Ann. Chem.*, **654,** 64 (1962).

2. The Properties of Alkene Carbonium Ions and Carbanions

113. Ref. 3, Chap. 5.
114. A. Schriesheim, J. E. Hofmann and C. A. Rowe, Jr., *J. Am. Chem. Soc.*, **83,** 3731 (1961); A. Schriesheim and C. A. Rowe, Jr., *J. Am. Chem. Soc.*, **84,** 3160 (1962).
115. A. Schriesheim, R. J. Muller and C. A. Rowe, Jr., *J. Am. Chem. Soc.*, **84,** 3164 (1962); S. Bank, C. A. Rowe, Jr. and A. Schriesheim, *J. Am. Chem. Soc.*, **85,** 2115 (1963).
116. A. Schriesheim, C. A. Rowe, Jr. and L. Naslund, *J. Am. Chem. Soc.*, **85,** 2111 (1963).
117. A. Schriesheim and C. A. Rowe, Jr., *Tetrahedron Letters*, 405 (1962).
118. S. Bank, A. Schriesheim and C. A. Rowe, Jr., *J. Am. Chem. Soc.*, **87,** 3244 (1965).
119. For example, see ref. 120.
120. S. W. Ela and D. J. Cram, *J. Am. Chem. Soc.*, **88,** 5791 (1966).
121. S. W. Ela and D. J. Cram, *J. Am. Chem. Soc.*, **88,** 5777 (1966).
122. D. H. Hunter and D. J. Cram, *J. Am. Chem. Soc.*, **86,** 5478 (1964).
123. M. D. Carr, J. R. P. Clarke and M. C. Whiting, *Proc. Chem. Soc.*, 333 (1963); W. O. Haag and H. Pines, *J. Am. Chem. Soc.*, **82,** 387 (1960).
124. K. W. Wilson, J. D. Roberts and W. G. Young, *J. Am. Chem. Soc.*, **72,** 215 (1950).
125. S. Bank, *J. Am. Chem. Soc.*, **87,** 3245 (1965).
126. D. J. Cram and R. T. Uyeda, *J. Am. Chem. Soc.*, **86,** 5466 (1964).
127. R. Waack and M. A. Doran, *J. Am. Chem. Soc.*, **85,** 1651 (1963).
128. R. Waack and M. A. Doran, *J. Phys. Chem.*, **68,** 1148 (1964).
129. K. Kuwata, *Bull. Chem. Soc. Japan*, **33,** 1091 (1960).
130. E. A. Rabinovich, I. V. Astaf'ev and A. I. Shatenshtein, *J. Gen. Chem. USSR (Engl. Transl.)*, **32,** 746 (1962).
131. R. H. DeWolfe, D. L. Hagmann and W. G. Young, *J. Am. Chem. Soc.*, **79,** 4795 (1957).
132. E. J. Lanpher, *J. Am. Chem. Soc.*, **79,** 5578 (1957).
133. D. Seyferth and M. A. Weiner, *J. Org. Chem.*, **26,** 4797 (1961).
134. H. G. Richey, Jr. and A. H. Smith, unpublished work; A. H. Smith, M. S. Thesis, The Pennsylvania State University, 1965.
135. H. G. Richey, Jr. and T. C. Rees, unpublished work; T. C. Rees, Ph.D. Thesis, The Pennsylvania State University, 1966.
136. C. Prévost and B. Gross, *Compt. Rend.*, **252,** 1023 (1961).
137. C. S. Johnson, Jr., M. A. Weiner, J. S. Waugh and D. Seyferth, *J. Am. Chem. Soc.*, **83,** 1306 (1961).
138. J. E. Nordlander and J. D. Roberts, *J. Am. Chem. Soc.*, **81,** 1769 (1959).
139. G. M. Whitesides, J. E. Nordlander and J. D. Roberts, *Discussions Faraday Soc.*, **34,** 185 (1962).
140. D. Seyferth and T. F. Jula, *J. Organometal. Chem.*, **8,** P13 (1967).
141. J. E. Nordlander, W. G. Young and J. D. Roberts, *J. Am. Chem. Soc.*, **83,** 494 (1961).
142. R. B. Bates, D. W. Gosselink and J. A. Kaczynski, *Tetrahedron Letters*, 205 (1967).
143. G. M. Whitesides, J. E. Nordlander and J. D. Roberts, *J. Am. Chem. Soc.* **84,** 2010 (1962).
143a. T. J. Katz and P. J. Garratt, *J. Am. Chem. Soc.*, **85,** 2852 (1963); E. A.

LaLancette and R. E. Benson, *J. Am. Chem. Soc.*, **85,** 2853 (1963); G. Wittig, V. Raustenstrauch and F. Wingler, *Tetrahedron,* Suppl. 7, 189 (1966).
144. E. J. Lanpher, L. M. Redman and A. A. Morton, *J. Org. Chem.*, **23,** 1370 (1958).
145. B. Capon, *Quart. Rev.* (London), **18,** 45 (1964).
146. A. Streitwieser, Jr., *Solvolytic Displacement Reactions,* McGraw-Hill Book Co., Inc., New York, 1962.
147. S. Winstein, C. R. Lindegren, H. Marshall and L. I. Ingraham, *J. Am. Chem. Soc.*, **75,** 147 (1953).
148. G. D. Sargent, *Quart. Rev.* (London), **20,** 301 (1966).
149. G. E. Gream, *Rev. Pure Appl. Chem.*, **16,** 25 (1966).
150. J. A. Berson in *Molecular Rearrangements,* Part 1 (Ed. P. de Mayo), Interscience Publishers, New York, 1963, Chap. 3.
151. P. D. Bartlett, *Nonclassical Ions,* W. A. Benjamin, Inc., New York, 1965.
152. M. Simonetta and S. Winstein, *J. Am. Chem. Soc.*, **76,** 18 (1954).
153. M. E. H. Howden and J. D. Roberts, *Tetrahedron,* **19,** Suppl. 2, 403 (1963).
154. R. J. Piccolini and S. Winstein, *Tetrahedron,* **19,** Suppl. 2, 423 (1963).
155. W. G. Woods, R. A. Carboni and J. D. Roberts, *J. Am. Chem. Soc.*, **78,** 5653 (1956).
156. Ref. 154 and references therein.
157. H. E. Zimmerman and A. Zweig, *J. Am. Chem. Soc.*, **83,** 1196 (1961).
158. Ref. 65, Chap. 12.
159. R. Hoffmann, *J. Am. Chem. Soc.*, **86,** 1259 (1964).
160. W. S. Trahanovsky, *J. Org. Chem.*, **30,** 1666 (1965).
161. P. R. Story in *Carbonium Ions* (Ed. G. A. Olah and P. von R. Schleyer), Vol. III, Chap. 23, to be published by John Wiley and Sons, Inc., New York.
162. R. G. Lawton, *J. Am. Chem. Soc.*, **83,** 2399 (1961).
163. P. D. Bartlett and S. Bank, *J. Am. Chem. Soc.*, **83,** 2591 (1961).
164. P. D. Barlett, S. Bank, R. J. Crawford and G. H. Schmid, *J. Am. Chem. Soc.*, **87,** 1288 (1965).
165. P. D. Bartlett and G. D. Sargent, *J. Am. Chem. Soc.*, **87,** 1297 (1965).
166. P. D. Bartlett, W. S. Trahanovsky, D. A. Bolon and G. H. Schmid, *J. Am. Chem. Soc.*, **87,** 1314 (1965).
167. C. C. Lee and E. W. C. Wong, *J. Am. Chem. Soc.*, **86,** 2752 (1964); C. C. Lee and E. W. C. Wong, *Tetrahedron,* **21,** 539 (1965); K. Humski, S. Borcic, and D. E. Sunko, *Croat. Chem. Acta,* **37,** 3 (1965); *Chem. Abstr.*, **63,** 2864 (1965).
168. P. D. Bartlett, W. D. Closson and T. J. Cogdell, *J. Am. Chem. Soc.*, **87,** 1308 (1965).
169. W. S. Trahanovsky, M. P. Doyle and P. D. Bartlett, *J. Org. Chem.*, **32,** 150 (1967).
170. W. S. Johnson, D. M. Bailey, R. Owyang, R. A. Bell, B. Jacques and J. K. Crandall, *J. Am. Chem. Soc.*, **86,** 1959 (1964).
171. W. S. Johnson and R. Owyang, *J. Am. Chem. Soc.*, **86,** 5593 (1964).
172. P. D. Bartlett, E. M. Nicholson and R. Owyang, *Tetrahedron,* Suppl. 8, Part II, 399 (1966).
173. P. S. Skell and J. T. Keating, unpublished manuscript.

174. H. L. Goering and W. D. Closson, *J. Am. Chem. Soc.*, **83,** 3511 (1961).
175. H. L. Goering, A. C. Olson and H. H. Espy, *J. Am. Chem. Soc.*, **78,** 5371 (1956).
176. G. Le Ny, *Compt. Rend.*, **251,** 1526 (1960).
177. M. A. Eakin, J. Martin and W. Parker, *Chem. Commun.*, 955 (1967).
178. S. Winstein and R. L. Hansen, *Tetrahedron Letters*, No. **25,** 4 (1960).
179. G. Leal and R. Pettit, *J. Am. Chem. Soc.*, **81,** 3160 (1959).
180. H. E. Ulery and J. H. Richards, *J. Am. Chem. Soc.*, **86,** 3113 (1964).
181. W. S. Johnson and J. K. Crandall, *J. Org. Chem.*, **30,** 1785 (1965).
182. A. Eschenmoser, D. Felix, M. Gut, J. Meier and P. Stadler in *Ciba Foundation Symposium on the Biosynthesis of Terpenes and Sterols* (Ed. G. E. W. Wostenholme and M. O'Connor), Little Brown and Co., Boston, 1959, p. 217; P. A. Stadler, A. Nechvatal, A. J. Frey and A. Eschenmoser, *Helv. Chim. Acta*, **40,** 1373 (1957).
183. W. S. Johnson and R. B. Kinnel, *J. Am. Chem. Soc.*, **88,** 3861 (1966).
184. J. H. Richards and J. B. Hendrickson, *The Biosynthesis of Steroids, Terpenes, and Acetogenins*, W. A. Benjamin, Inc., 1964, New York, Chap. 6–11; H. J. Nicholas in *Biogenesis of Natural Compounds* (Ed. P. Bernfeld), The MacMillan Co., New York, 1963, Chap. 14; R. B. Clayton, *Quart. Rev.* (London), **19,** 168, 201 (1965).
185. J. A. Marshall and N. H. Anderson, *Tetrahedron Letters*, 1219 (1967).
186. D. J. Goldsmith and B. C. Clark, Jr., *Tetrahedron Letters*, 1215 (1967).
187. T. L. Westman and R. D. Stevens, *Chem. Commun.*, 459 (1965).
188. H. G. Richey, Jr. and M. W. McNeil, unpublished work; M. W. McNeil, Ph.D. Thesis, the Pennsylvania State University, 1968.
189. A. C. Cope and P. E. Peterson, *J. Am. Chem. Soc.*, **81,** 1643 (1959); see also A. C. Cope, J. M. Grisar and P. E. Peterson, *J. Am. Chem. Soc.*, **82,** 4299 (1960).
190. K. L. Servis and J. D. Roberts, *J. Am. Chem. Soc.*, **86,** 3773 (1964).
191. K. L. Servis and J. D. Roberts, *J. Am. Chem. Soc.*, **87,** 1331 (1965).
192. R. S. Bly and R. T. Swindell, *J. Org. Chem.*, **30,** 10 (1965); see also C. F. Wilcox, Jr. and D. L. Nealy, *J. Org. Chem.*, **28,** 3454 (1963).
193. For example, see R. A. Sneen, *J. Am. Chem. Soc.*, **80,** 3977, 3982 (1958).
194. H. G. Richey, Jr., in *Carbonium Ions* (Ed. G. A. Olah and P. von R. Schleyer), Vol. III, Chap. 25, to be published by John Wiley and Sons, Inc., New York.
195. K. B. Wiberg, B. A. Andes, Jr. and A. J. Ashe, *Carbonium Ions* (Ed. G. A. Olah and P. von R. Schleyer), Vol. III, Chap. 26, to be published by John Wiley and Sons, Inc., New York.
196. R. H. Mazur, W. N. White, D. A. Semenow, C. C. Lee, M. S. Silver and J. D. Roberts, *J. Am. Chem. Soc.*, **81,** 4390 (1959).
197. L. F. Fieser and M. Fieser, *Steroids*, Reinhold Publishing Corp., New York, 1959, Chap. 9; N. L. Wendler in *Molecular Rearrangements*, Part 2 (Ed. P. de Mayo), Interscience Publishers, New York, 1963, Chap. 16.
198. C. H. DePuy, I. A. Ogawa and J. C. McDaniel, *J. Am. Chem. Soc.*, **83,** 1668 (1961).
199. S. Winstein, M. Shatavsky, C. Norton and R. B. Woodward, *J. Am. Chem. Soc.*, **77,** 4183 (1955).
200. S. Winstein and C. Ordronneau, *J. Am. Chem. Soc.*, **82,** 2084 (1960).

201. S. Winstein, A. H. Lewin and K. C. Pande, *J. Am. Chem. Soc.*, **85,** 2324 (1963).
202. H. C. Brown and H. M. Bell, *J. Am. Chem. Soc.*, **85,** 2324 (1963).
203. H. G. Richey Jr., and R. K. Lustgarten, *J. Am. Chem. Soc.*, **88,** 3136 (1966).
204. M. Brookhart, A. Diaz and S. Winstein, *J. Am. Chem. Soc.*, **88,** 3135 (1966).
205. P. R. Story, L. C. Snyder, D. C. Douglass, E. W. Anderson and R. L. Kornegay, *J. Am. Chem. Soc.*, **85,** 3630 (1963).
206. M. Brookhart, R. K. Lustgarten and S. Winstein, *J. Am. Chem. Soc.*, **89,** 6352 (1967).
207. P. D. Bartlett and M. R. Rice, *J. Org. Chem.*, **28,** 3351 (1963).
208. A. Nickon and J. L. Lambert, *J. Am. Chem. Soc.*, **88,** 1905 (1966).
209. A. Nickon, J. L. Lambert and J. E. Oliver, *J. Am. Chem. Soc.*, **88,** 2787 (1966).
210. A. Nickon, J. L. Lambert, R. O. Williams and N. H. Werstiuk, *J. Am. Chem. Soc.*, **88,** 3354 (1966).
211. P. G. Gassman and F. V. Zalar, *J. Am. Chem. Soc.*, **88,** 3070 (1966).
212. R. Howe and S. Winstein, *J. Am. Chem. Soc.*, **87,** 915 (1965); T. Fukunaga, *J. Am. Chem. Soc.*, **87,** 916 (1965).
213. J. M. Brown and J. L. Occolowitz, *Chem. Commun.*, 376 (1965).
214. S. Winstein, M. Ogliaruso, M. Sakai and J. M. Nicholson, *J. Am. Chem. Soc.*, **89,** 3656 (1967).
215. M. S. Silver, P. R. Shafer, J. E. Nordlander, C. Rüchardt and J. D. Roberts, *J. Am. Chem. Soc.*, **82,** 2646 (1960).
216. M. E. H. Howden, A. Maercker, J. Burdon and J. D. Roberts, *J. Am. Chem. Soc.*, **88,** 1732 (1966).
217. E. A. Hill, H. G. Richey, Jr. and T. C. Rees, *J. Org. Chem.*, **28,** 2161 (1963).
218. H. G. Richey, Jr. and T. C. Rees, *Tetrahedron Letters*, 4297 (1966).
219. H. G. Richey, Jr. and W. C. Kossa, unpublished work.
220. A. Maercker and J. D. Roberts, *J. Am. Chem. Soc.*, **88,** 1742 (1966).
221. For example, H. Pines, N. C. Sih and E. Lewicki, *J. Org. Chem.*, **30,** 1457 (1965); N. C. Sih and H. Pines, *J. Org. Chem.*, **30,** 1462 (1965); P. T. Lansbury, unpublished work.
222. E. L. Eliel, N. L. Allinger, S. J. Angyal and G. A. Morrison, *Conformational Analysis*, Interscience Publishers, New York, 1965, Chap. 4.
223. H. G. Richey, Jr. and A. M. Rothman, *Tetrahedron Letters*, 1457 (1968).
224. C. Walling, J. H. Cooley, A. A. Ponaras and E. J. Racah, *J. Am. Chem. Soc.*, **88,** 5361 (1966) and references therein.
225. E. A. Hill and J. A. Davidson, *J. Am. Chem. Soc.*, **86,** 4663 (1964).

CHAPTER 3

Alkene Rearrangements

KENNETH MACKENZIE

University of Bristol, England.

I. Cis–Trans Isomerism	115
A. Thermal, Catalysed and Photochemical Stereomutation	115
B. Acid- and Base-Catalysed Stereomutations	121
C. Cis–Trans Isomerism in Polyenes	125
D. Miscellaneous Geometrical Isomerisms	132
II. HYDROGEN TRANSFER AND PROTOTROPIC SHIFTS	132
A. Thermal, Photochemical and Catalysed Isomerization	132
B. Hydrogen Transfer and Prototropy in Polyenes	145
C. Heterogeneous Catalytic Rearrangements of Simple Alkenes	147
III. ANIONOTROPIC REARRANGEMENTS	148
A. Mechanisms of Allylic Rearrangements	148
B. Anionotropy in Polyenes	151
IV. CLAISEN AND COPE REARRANGEMENTS	154
V. MISCELLANEOUS ALKENE REARRANGEMENTS	162
VI. REFERENCES	167

A number of important research papers and reviews have appeared since the publication of 'The Alkenes' and the purpose of this supplement is to highlight significant progress in the field. Full coverage of the literature is not intended; instead a selection of material from the recent literature is given and for this purpose the net has been cast as wide as possible, but subjects which have been expertly reviewed are not given detailed attention unless of considerable novelty and interest. The order of subject matter is similar to 'The Alkenes,' Chapter 7.

I. CIS-TRANS ISOMERISM

A. Thermal, Catalysed and Photochemical Stereomutation

Iodine-catalysed thermal stereomutation of 2-butene has recently been very carefully investigated using more reliable analytical techniques than in earlier work. Thermodynamic parameters evaluated, e.g. ΔH^0 and ΔS^0 agree well with those calculated on the basis of statistical formulae (API tables)[1a] using known vibrational assignments and internal rotational barriers. However, errors in these dynamic quantities constitute a serious source of inaccuracy in the computation of thermodynamic data in general (e.g. for positional isomerism).

It is suggested, however, that it should be possible to measure ΔS^0 from accurate equilibrium studies and use this in conjunction with microwave spectral data to obtain more reliable values for rotational barriers, and thus enable assignment of important low frequency infrared bands ($< 1,000$ cm^{-1})[1b]. Earlier work on the iodine-catalysed stereomutation of diiodoethylene had suggested the formation of an intermediate with double-bond character[2], but the recent work on 2-butene suggests that the mechanism involves formation of the s-butyl radicals **2c** and **2t** by addition of an iodine atom. On the assumption that these radicals are in equilibrium with the rotational transition state, actual and theoretical rates can be compared. Using absolute reaction rate theory and equating the equilibrium constant for **2c** $\rightleftharpoons$ **2t** to the partition function for hindered internal rotation, a barrier for this motion of 3 kcal/mole gives a rate constant k_c, which is somewhat smaller than k_b, leading to an overall rate in steady state theory given by $K_{a,b} \cdot k_c$. $K_{a,b}$ can be computed on the basis of standard entropy and enthalpy values for **I**, *cis* and *trans* olefin, *s*-butyl iodide and ΔC_p^0. The product of $K_{a,b}$ with the calculated k_c gives a rate constant in excellent agreement with the experimental figure, confirming the proposed mechanism and supporting the assumed values in calculation of k_c. Clearly there is considerable scope for the development of this approach[3].

$$\underset{(1)}{\diagup\!=\!\diagdown} + \text{I} \cdot \underset{b}{\overset{a}{\rightleftharpoons}} \underset{(2c)}{\diagdown\!\diagup} \underset{c'}{\overset{c}{\rightleftharpoons}} \diagup\!\diagdown \underset{a'}{\overset{b'}{\rightleftharpoons}} \underset{(2t)}{\diagdown\!=\!\diagup} \underset{(1')}{\diagdown\!=\!\diagup} + \text{I} \cdot$$

A similar mechanism has been proposed for the iodine-catalysed stereomutation of p,p'-disubstituted stilbene derivatives. Electron-donating substituents accelerate the rate, apparently by combined inductive-radical resonance effects of the substituent[4].

An important factor determining the position of equilibrium between stereoisomers is the entropy difference between various geometrical pairs, attributable to differences in barriers to free rotation of end groups in the *cis* isomers. Similar considerations apply in determining the relative stability of positional isomers, where enthalpy differences between isomer pairs ought to be similar, but owing to rotational effects of end groups, entropies for geometrical pairs will differ. Determination of iodine-catalysed equilibrium data for 2- and 3-heptene isomers illustrates this; the *trans* positional isomers have entropy difference of ~0·3 e.u. but in the *cis–trans* isomeric pairs the entropy difference covers the range 0·5–1·5 e.u. depending on, and correlating with the size of the end groups[5].

Nitric oxide-catalysed geometrical isomerism has also been carefully reinvestigated. As pointed out by Cundall[6] the very large decrease in activation energy for stereomutations catalysed by nitric oxide, especially in the case of *trans*-dideuterioethylene[7] and *cis*- and *trans*-2-butenes[6], implies chemical bonding, since the values observed are much lower than the triplet-state energies measured photochemically. As with iodine, addition to the olefin generates a radical which undergoes rotation. In the case of 1,3-pentadiene for the nitric oxide catalysed reaction, rotation in the radical is the rate-determining step but for the iodine-catalysed reaction it is the initial addition step which is rate determining; it is notable that in comparison with the *s*-butyl radical produced from 2-butene, the intermediate produced from 1,3-pentadiene is stabilised by 12·6 kcal/mole—the allylic resonance contribution. For nitric oxide-addition this stabilizing effect is offset by the decreased strength of the C—NO bond as compared to the corresponding C—I bond. Thermodynamic parameters for the 1,3-pentadiene equilibrium differ from those earlier published (API tables). The latter imply that the *cis* isomer is more stable than the *trans* isomer, when in fact the opposite is true.

As a general rule Benson and Egger and their colleagues[3,6] propose that for any radical- or atom-catalysed stereomutation, the mechanism leads to a reaction which is rate controlled by rotation in the intermediate if the bond strength of the *s*-C—A bond and the resonance energy in the radical formed with species A·, do not add up to the π-bond energy being absorbed to dissipate the π-bond (i.e. $\Delta H^0_{a,b}$ should be positive for the reaction of 2-butene depicted above).

Recent work in the field of haloolefin stereochemistry has been concerned with the processes involved in thermal isomerization and the synthesis of isomers of preferred *cis* configuration.

The stereomutation of 1,2-dichloroethylene may occur by a mixed mechanism in which there is an important surface effect neglected in the earliest work, as well as radical chain and true unimolecular pathways. Extrapolation of the results to zero surface to volume ratio in the presence of radical scavengers, suggests typical unimolecular parameters.[7] Very similar parameters are found for the gas-phase stereomutation of *trans*-octafluoro-2-butene (where the activation energy is rather lower than for the hydrocarbon or for dideuterio-ethylene by 6–8 kcal/mole, which is of the same order of magnitude as the relative enthalpy differences for reactions of the fluoro- and hydro-compounds[8]).

Investigations in the field of mixed halogenoolefins continues to uncover interesting phenomena. The *trans* isomer of 1-bromo-2-iodoethylene is the more stable one contrary to earlier reports; its thermal stereomutation is accompanied by decomposition and formation of 1,2-dibromo- and 1,2-diiodoethylenes, the latter compound being obtained pure for the first time[9]. 1-Fluoro-2-iodoethylene, readily obtained from the bromofluoro compound, is predominantly the *cis* isomer; its reaction with ethylene and separation of the products into stereoisomeric mixtures of 1-fluoro-4-iodobutene and 1,4-difluoro-4-iodobutene followed by dehydroiodination, allows isolation by gas-liquid chromatography of *cis* and *trans*-1-fluorobutadiene and *cis*, *cis*-, *cis*, *trans*-, and *trans*, *trans*-1,4-difluorobutadienes respectively. Isomeric 1,4-dichlorobutadienes are also accessible. In each case here the *cis* isomers are the more stable ones, as deduced by iodine-catalysed equilibration studies, and this is explicable in terms of favourable intramolecular hydrogen-halogen Van der Waals forces in the absence of H—H or I—I repulsions, as represented in structure **3**[10]. The *cis*-isomer of 1-*H*-perfluoropropene is also the more stable of the pair and is obtained virtually pure by photolysis of the stereoisomeric mixture[11]. In connexion with these observations the deviation from coplanarity of 2-fluorobiphenyl, being scarcely different from that in biphenyl itself, lends support to this idea[12].

The very high initial *cis*:*trans* isomer ratio observed in the formation of 1-bromopropene by low temperature addition of hydrogen bromide to propyne, indicates that the intermediate vinylic radical does not stereomutate, but retains a bent form; arguments based on the difference in frequency of isomerization derived from the isomer ratio and the difference of in-plane infrared bending vibrations, point to an activation energy for stereomutation of the vinylic radical of ca 17 kcal/mole, which agrees well with spectroscopic data for the bent ground-state to linear excited-state transition for the formyl radical. The vinylic radical can thus be added to the growing number of vinylic and allylic intermediates which retain stereochemical integrity during chemical reactions[13] (see also Chapter 2).

$$\underset{(3)}{\overset{H\quad H}{\underset{H\quad H}{X\diagdown\diagup\diagdown\diagup X}}}$$

Photo- and radiolytically produced thiyl radicals stereomutate 2-butene and ethylene-1,2-d_2; the lack of effect of pressure on the reaction rate in the photo-process with methanethiol, leads to a simple thermal decomposition mechanism for the thiyl radical olefin adduct, the activation energy for stereomutation actually being very small. Rate constants for the important attack step are calculated to be[14] ca 4·5 × 10⁶–4·5 × 10⁸. A similar mechanism involving removal of HS· radical from the intermediate hydrogen sulphide adduct, accounts for the features of the ⁶⁰Co γ-radiation-hydrogen sulphide-catalysed reaction[15].

The resurgence of interest in photoactivation is apparent in all areas of organic chemistry, not least in the field of alkene rearrangements. This is partly due to the increasing use of sensitization processes (see also Chapter 6). For example the gas-phase *cis–trans* isomerization of 2-butene is sensitized by benzene or pyridine, the rate being reduced by triplet quenchers such as oxygen, nitric oxide, ethylene, etc. Radical pathways may be ruled out by the cleanliness of the reaction and the failure to observe typical radical reactions with additives; singlet-state benzene is not involved since high butene pressures fail to quench the characteristic fluorescence associated with this state. The failure to observe tetramethyl-cyclobutane in the reaction products may be due to the formation of the vibrationally excited molecule which instantly fragments. The use of benzene as a gas-phase sensitizer represents an important advance over the use of mercury vapour with its associated multifarious products[16]. Sensitized reactions are much more common in solution however, using aromatic carbonyl compounds and hydrocarbons as well as solvent-sensitizers such as acetone as triplet-energy carriers. Olefin triplets formed in these reactions are probably very short lived ($< 10^{-9}$ sec.) because of overlapping of the ground-state potential curve with that of the twisted triplet-energy profile. This may be one of the reasons why triplet-energy transfer between olefins is rare. It is therefore possible to remove—even selectively—triplet molecules with olefins without effecting excited singlet species or acceptor triplet olefins concurrently produced.[17]

Triplet-energy transfer from carbonyl groups can also be intramolecular. Among other products of irradiation of *trans*-4-hexen-2-one, the *cis* isomer is isolated; however, under identical conditions in the presence of *trans*-2-hexene no isomerization of the latter is observed. Similar stereomutation of *trans*-5-hepten-2-one also occurs. In these reactions the n-π^* excited ketone must intramolecularly transfer triplet energy to the olefin bond by suitable orbital overlap. Filtered light illumination of *trans*-1-phenyl-2-butene (absorption of light by benzene chromophore) gives a single photoproduct by a similar process[18]. Obvious objectives of future work could be determination of steric effects on the efficiency of these intramolecular processes with similar readily accessible compounds and their analogues, including possible transannular effects in medium rings.

The mechanism of photoisomerization of olefins and stilbene in particular, and the nature of the energy-transfer process has recently been examined in some detail. Two approaches have been used[19,20]. First the composition of photostationary mixtures of α-methylstilbene isomers and quantum yield measurements at low conversions have been made in the presence of a range of sensitizing compounds (triplet energies 40–75 kcal/mole)[21]. The second approach to the problem of studying transitory olefin triplets produced in sensitized stereomutations is to observe the behaviour of the sensitizer triplets quenched by the isomerizing olefin. Again stilbenes and diphenylpropenes are useful substrates for these studies. These approaches are described in more detail in Chapter 6.

Triplet-energy transfer is subject to steric effects, as might be expected if close approach of, for example, the carbonyl group and the olefin bond is required for efficient interaction. Introduction of two o-isopropyl groups into benzophenone markedly reduces its sensitization efficiency, the effect being especially noticeable with *cis* olefins[22].

Stereomutation can occur by triplet transfer from the $^3B_{1u}$ state in benzene solutions but the radiation energy required appears to be rather high; high frequency ultraviolet, γ-rays and fast electrons geometrically isomerize 2-butene and other olefins[23], whilst pulses of γ-rays are effective with dimethyl fumarate[24]. The intermediacy of triplet intermediates in the 2-butene case is apparent from the lack of side products expected from radical reactions and the absence of quenching effects by added acceptors. In the fumaric ester case, a triplet intermediate is apparent from the fact that the concurrent yield of triplet anthracene in benzene is reduced, compared to solutions of anthracene alone. The isomerization of *cis*-stilbene in solution in benzene or cyclohexane by ^{60}Co γ-rays appears to involve chain reactions initiated by ejection of solvent electrons which add to the olefin, giving a reactive radical carbanion capable of isomerizing by transferring its electron to further *cis*-stilbene, regenerating itself until chain termination occurs at an ionizable site[25]; genuine cases of stereomutation of stilbene by triplet $^3B_{1u}$ have been described as well[26]. The benzene sensitized stereomutation of dideuterioethylene is accompanied by hydrogen scrambling, but to a considerably lesser extent than in the Hg($^3P_1$)-sensitized reaction, the amount of energy available being somewhat lower (3.6 eV for benzene, 4.9 eV for mercury)[27].

Cases of $S_0 \rightarrow T_1$ ('forbidden') transitions were earlier observed spectroscopically by Evans for photoexcitation in the presence of high pressure oxygen[28]. Isomerization of *cis*- and *trans*-1,2-dichloroethylene by irradiation in the paramagnetically induced singlet-triplet zone similarly produced, appears to involve a triplet state common to both isomers[29].

An interesting case of solid-phase stereomutation has been observed in γ-irradiation of pressed potassium bromide discs of *cis*-α-phenylcinnamic acid. The same phenomenon is observed in the presence of hexabromoethane, but not with hexabromobenzene, fairly clearly implicating bromine atoms and indeed in the presence of efficient bromine scavengers such as allene, no isomerism is observed[30]. The mechanism of solid-state stereomutation of *cis*-dibenzoylethylene cannot involve $(2 + 2)\pi$ photoaddition followed by facile *thermal* scission of the cyclobutane (as required by orbital symmetry arguments), since irradiation of an equimolar mixture of the diketoolefin and the ringdecadeuterated diketoolefin (*trans* isomers) and mass spectroscopic analysis of the *cis* isomer produced, shows that molecules with only five deuterium atoms are absent[31].

Methods of isomerizing non-conjugated polyolefins with iodine and radiation suffer from the disadvantage that considerable double-bond shifts occur; however, the use of filtered light and acetophenone sensitizer allows smooth stereomutations, e.g. with 5-decene which gives a mixture richer in *trans*-5-decene than the thermal equilibrium mixture[32].

B. Acid- and Base-Catalysed Stereomutations

Stereomutation of alkenes by nucleophilic reagents was briefly discussed in 'The Alkenes' by Mackenzie (p. 403) and in more detail by Patai and Rappoport (p. 565). In general such reactions involve reversible addition of the nucleophilic reagent to the olefin, the stability of the resulting carbanion being determined by the electronic nature of the groups present in the olefin. As a result the carbanion can stereomutate and revert to olefin, e.g. with highly reactive α-cyano-β-o-methoxyphenylacrylic esters, or if very unstable the carbanion may protonate, when isomerization requires a further molecule of base to remove the proton [33a]. Fragmentation of the carbanion intermediate is an alternative possibility, the ratio of fragmentation to isomerization varying widely; since the energy requirements for rotation are small, the observed ratio reflects the stability of the fragment carbanion.

The addition-elimination mechanism also applies in the isomerization of 4-nitrochalcone by a variety of nucleophiles, the order of relative reactivity for the reagents being closely similar to that observed in earlier experiments with, for example, maleic ester and α-cyano-β-o-methoxyphenylacrylates [33b], i.e. OH^- > piperidine > N_3^- > $BuNH_2$ > $n\text{-}Pr_2NH$ > $i\text{-}Pr_2NH$ ≫ Et_3N, C_5H_5N, $n\text{-}Bu_3N$ or Br^-. The different degrees of reactivity amongst the strongly basic, secondary amines used in these studies are probably due to steric effects. Their actual positions in the reactivity order reflect the subtle competition between electron availability and steric requirement, which of course may be different with different substrates. Piperidine is presumably only second to hydroxyl ion on account of its basicity and small steric requirement. The much higher reactivity of azide ion compared to bromide ion is however surprising; apart from its greater nucleophilicity (observed only in nucleophilic aromatic substitutions), the steric effect for azide ion might be quite different from that of bromide ion [34].

The base-catalysed stereomutation of stilbene and α,α'-dideuteriostilbene isomers in powerfully basic aprotic media at 50–196° has recently been quantitatively investigated.[35] For stereomutations in t-butanol and in t-butanol with tetrahydrofuran or dimethylsulphoxide as cosolvent, vinylic carbanions are involved (as they are for deuterium exchange), since the relevant addition product of cis-stilbene and t-butanol does not yield trans-stilbene under comparable conditions; however an addition-elimination sequence *does* seem to be involved in methanolic media.

The ratio of exchange to isomerization (k_e^c/k_i^c) for cis-stilbene changes steadily (covering a range of 5×10^3 to 0.8) as the solvent, t-butanol, is enriched with dimethyl sulphoxide, because the latter is a better ion-pair dissociating solvent, and the carbanion therefore has a longer lifetime in the medium, increasing the probability of stereomutation. Interestingly in dimethyl sulphoxide-d_6-potassium perdeuteriodimsyl reagent, isomerization of *cis*-stibene occurs with significant retention of the original isotope, even though an anion is involved; this could be due to the involvement of a bis-solvent liganded potassium ion—carbanion ion pair, in which one of the liganded solvent molecules is that formed by removal of hydrogen from the stilbene by perdeuteriodimsyl anion. Rotation of the solvent liganded cation in the ion pair enables hydrogen or deuterium to be delivered to the carbanion. Protonation from the side remote from the potassium ion occurs at a much slower rate, because the result is a product-separated ion-pair.

Similar results are observed for *p*-nitrostilbene which is of course much more reactive in both stereomutation and exchange.

The nature and geometrical form of the vinylic carbanions involved in these reactions is uncertain. Some delocalization of charge into the aromatic ring is possible, and with *p*-nitrostilbene the nitro group must surely be involved, probably in an essentially allenic type structure.

For the isotopic exchange versus isomerization of stilbenes, two effects are probably operative. More strain is released in going to the transition state for the *cis*-anion than for the *trans* anion if the carbanion rehybridizes to *sp* linear geometry on the one hand; and on the other, in the ground state the *cis* isomer with one aromatic ring necessarily twisted, is in a better conformation for charge delocalization than the *trans* isomer, where to achieve the correct conformation for charge spreading one ring must be taken out of conjugation with the double bond. Both of these effects ought to enhance the rate of formation of the carbanion from the *cis* olefin relative to the *trans* isomer. For the *p*-nitrostilbenes, since charge delocalization is thought to be more complete, a larger value of k_e^c/k_e^t might be expected; but it is not significantly different from that of unsubstituted stilbene. In more polar media k_e^c/k_e^t is higher, clearly because charge-delocalization effects are more important here[35].

The geometrical lability of allylic carbanions has received recent attention in studies of prototropic equilibria of diphenylbutenes and allylbenzene (*vide infra*); in this connection n.m.r. data on solutions of 1,3-diphenylpropenyl lithium shows that the carbanion prefers a *trans,trans* orientation with the magnetically equivalent α-protons and the β-proton constituting an AB_2 spectrum[36]. Calculated bond orders suggest a fairly high activation energy for rotation from the *trans,trans* form **6**.

$$\underset{\substack{(4)\\(5)\ cis\ isomer}}{\overset{Ph}{\underset{H}{>}}C=C\overset{H}{\underset{CH_2Ph}{<}}} \xrightarrow[\text{THF}]{n-C_4H_9Li} \underset{(6)}{\overset{Ph}{\underset{H}{>}}\overset{\overset{H}{|}}{C}-\overset{Ph}{\underset{H}{<}}}$$

Stereomutation of *cis*-stilbene to the *trans* isomer can also be acid catalysed, and is indeed quantitative in 50–60% sulphuric acid. The rate data correlate linearly with H_0 and give a negative value for ρ in Hammett's σ^+-ρ relationship, indicative of the development of positive charge at the benzylic position. The reaction is marked by an induction period suggestive of an intermediate which is possibly 1,2-diphenylethanol; this alcohol dehydrates faster than *cis*-stilbene isomerizes, whilst racemization of the optically pure compound is very much faster still. With 1,2-diphenylethanol-2-*d* the rate of dehydration is reduced by a factor of two, the product stilbene being 78.8% deuterated; consequently the rate limiting process is the loss of a proton from the final carbonium ion, implying that the rate-determining step in the stereomutation of the *cis*-stilbene is the addition of a proton. This is confirmed by the reduction in rate of isomerization in deuteriosulphuric acid, the comparison being actually more marked for p,p'-dimethoxystilbene ($k_{H_2O}/k_{D_2O} = 6.0$ compared to 2.4 for the unsubstituted stilbene). This effect can be understood on the grounds that as the stability of the carbonium ion is increased by p-substituents, the extent of C—H bond formation with the incoming proton is lessened, the transition state resembling that involved in cinnamic acid stereomutation[37] (discussed in Part I).

Comment has been made on the assumption that carbonium ions derived from alcohols in acids are formed directly by loss of water from the conjugate acid, on the grounds that 1-, 1,1-di-, and 1,1,2-triphenylethanols appear to dehydrate completely in 10^{-3}–10^{-4} M perchloric acid in methylene chloride. The olefins thus formed become subsequently protonated. This would appear to suggest a rapid bimolecular elimination from the conjugate acids which is faster than loss of water to form the carbonium ion[38]. This effect, observed by ultraviolet spectroscopy, could be a medium effect, and ought to be further investigated.

Acid-catalysed stereomutation of *cis*-dibenzoylethylene is a first-order process in the alkene, in alcohols and aprotic solvents the rate being dependent on the structure of the acid. The rate dependence on concentration, e.g. with acetic acid in benzene, indicates that molecular acid is active agent, which therefore involves a molecular addition-elimination sequence, rather than attack by dissociated acid[39].

C. Cis—Trans Isomerism in Polyenes

Coupling of vinylacetylene (**7** → **8**) followed by a partial hydrogenation gives mainly cyclooctatriene (**10**) rather than **9** and a small yield of *trans,cis*-1,3,5,7-octatetraene (**11**); the latter is best made from *trans*-1,3,7-octatrien-5-yn by semihydrogenation at the acetylene linkage[40]. The reverse of this type of valence tautomerism has also been observed and is potentially useful for making polyolefins with fixed stereochemistry. The carbinols obtained from cyclooctarienone and Grignard reagents, exist mainly in the bicyclic form (n.m.r. data) and ring-chain tautomerism occurs very easily on warming[41], e.g. **13** → **14** → **15** or **16**. Since cyclooctatrienone is readily available from cyclooctatetraene hydrobromide by hydrolysis and oxidation, this represents a potentially interesting route to *cis* olefins of type **15** and possibly chain substituted derivatives based on simple acetylenes.

The acetylenic route has also been used to confirm assignments for the positional isomers of 3-methyl- and 3-chlorohexa-1,3,5-triene

$$CH_2=CHC\equiv CH \longrightarrow CH_2=CH(C\equiv C)_2CH=CH_2 \longrightarrow$$
$$\quad\quad\quad (7) \quad\quad\quad\quad\quad\quad\quad (8)$$

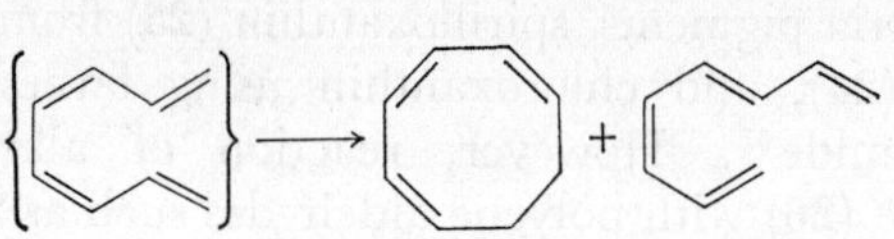

{ (9) } → (10) + (11)

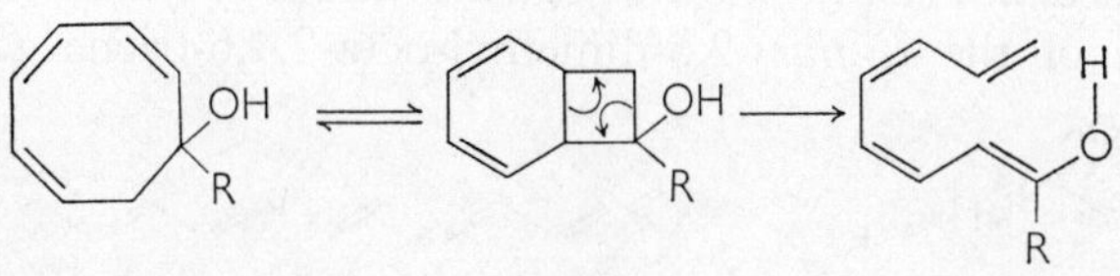

(12) (R = H, Me or Ph) ⇌ (13) → (14)

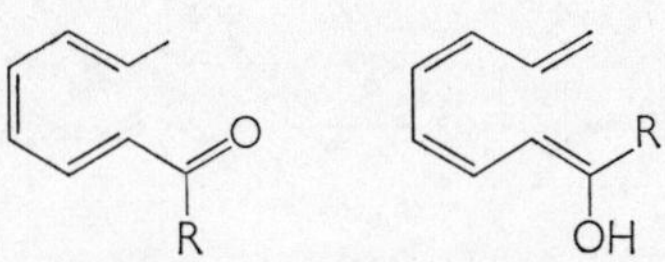

(15)   (16)

made by more conventional routes, e.g. from 2-chlorobut-2-enal[42].

The field of natural product synthesis continues to stimulate experiments towards the development of stereospecific methods of making stereochemically interesting polyenes with *cis* linkages. Alkenylidenetriphenylphosphoranes, e.g. **17** react with aldehydes such as **18** generating mixtures of *cis* and *trans* olefins (**19**), but quite frequently in this type of synthesis one of the isomers—generally the *trans* compound—is formed to the virtual exclusion of the other[43].

$$CH_2=CHCH=PPh_3 + OCHCH=CH(C\equiv C)_2CH=CHMe \longrightarrow$$
$$\quad\quad (17) \quad\quad\quad\quad (18)$$

$$CH_2=CHCH=CH\ CH=CH(C\equiv C)_2CH=CHMe$$
$$\text{cis} \quad\quad\quad (19)$$
$$+$$
$$\text{trans}$$

The reaction has been employed in the vitamin A and carotenoid field; thus **20** reacts with phosphonium bromide **21** in the presence of base, to give the all-*trans* ester **22**—the precursor of vitamin A[44].

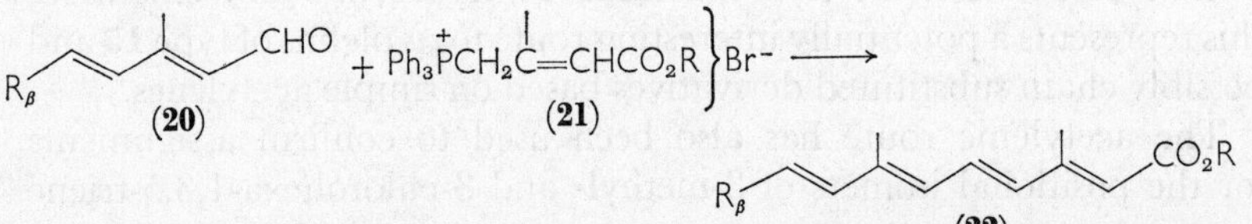

A more recent example of the reaction is the synthesis of a constituent of the purple bacteria pigments, spirilloxanthin (**25**) from the ylid **23** and crocetindial (**24**), and chloroxanthin using farnesyltriphenylphosphonium bromide[45]. However, reaction of alkenylidenetriphenylphosphorane (**26**) with polyene aldehydes such as **27** gives not only the all-*trans* polyenes **28** but most significantly the hindered-4,5 *cis* isomers of **28**; (for n = 1 the *cis:trans* ratio is 1:4). The ultraviolet spectrum of the *cis,trans*-2,3-dimethyl-octa-2,4,6-triene (**28**, n = 1)

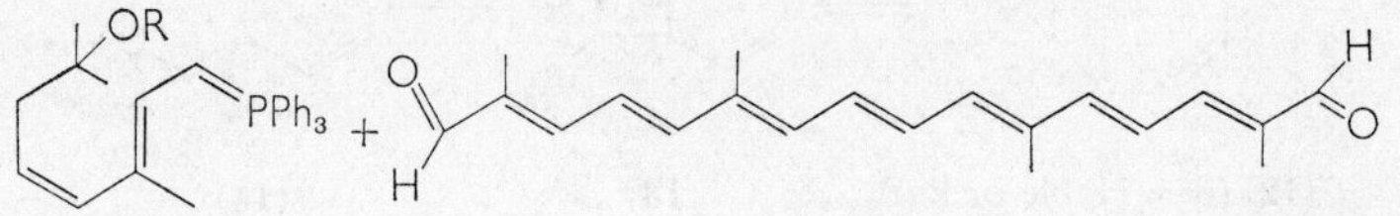

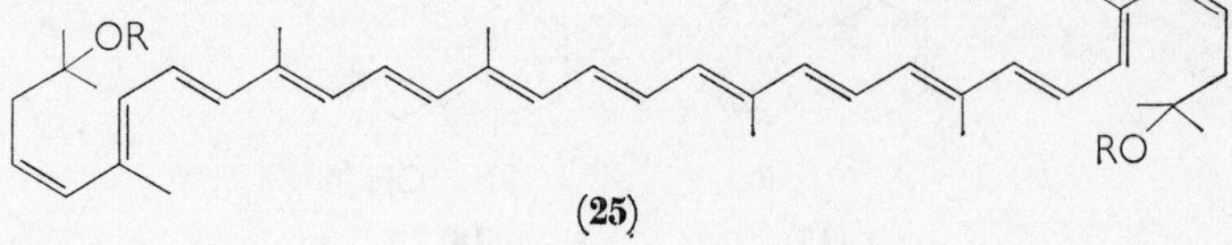

exhibits a typically degraded hindered *cis*-absorption (flat $\varepsilon_{max} \sim 11{,}600$, cf. **28a**) which dramatically converts to the characteristic all-*trans* polyene spectrum on heating with iodine (double ε_{max} 30–39,000)[46]

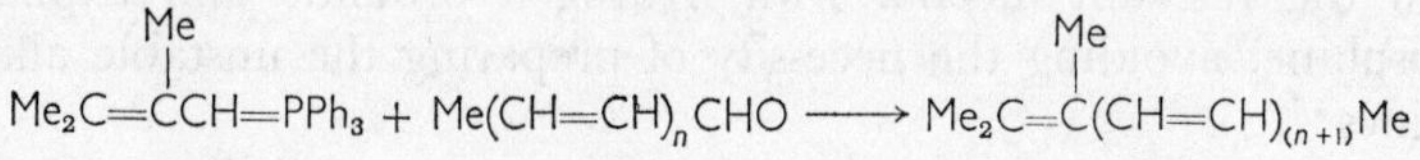

Me₂C=CCH=PPh₃ + Me(CH=CH)ₙCHO ⟶ Me₂C=C(CH=CH)₍ₙ₊₁₎Me
    |Me |Me
   (**26**)   (**27**)   (**28**)

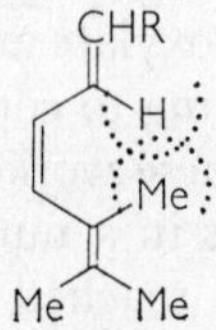

(**28a**)

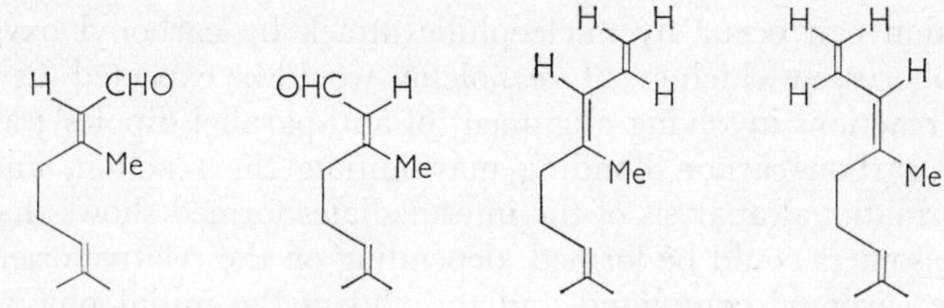

(**29**)   (**30**)   (**31**)   (**32**)

Another more conventional approach to the problem of introducing *cis* linkages by Wittig synthesis, is to start with a suitable α,β-unsaturated aldehyde; for example citral a (**29**) and citral b (**30**) give **31** and **32** with complete retention of initial stereochemistry. More significantly however, hydroxybutenolide (**33**) reacts with various phos-

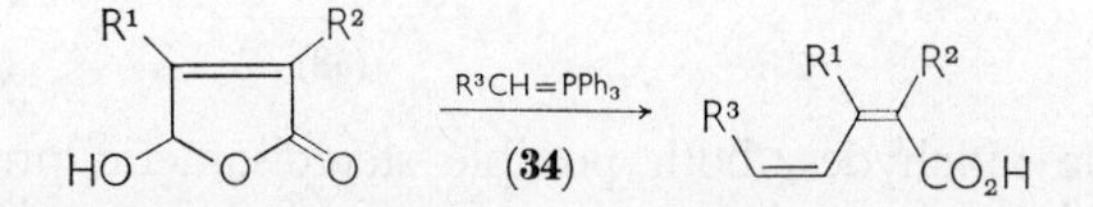

(**33**) (R¹ = Me R² = H)   (**34**)   (**35**) (R¹ = Me R² = H)

R³ =

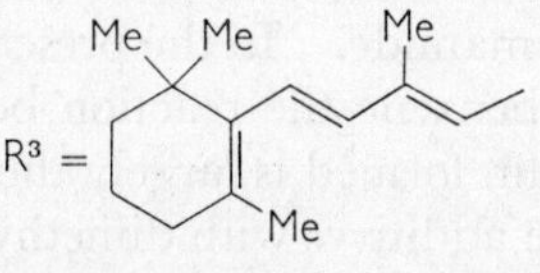

phoranes in ether to give products not only retaining the original *cis* configuration, but also compounds with a hindered *cis* geometry around the newly formed bond—a novel and unexpected result.

Application of this principle to the synthesis of vitamin A analogues gives with phosphorane **34**, the hitherto unknown *cis*-2-*cis*-4 isomer **35** of vitamin A acid[47]. Phosphoranes are advantageously prepared from the relevant alcohol with hydrogen bromide and triphenylphosphine, avoiding the necessity of preparing the unstable alkenyl halides[48].

Bergelson and Shemyakin have described how Wittig reactions in the presence of Lewis bases can give specifically *cis* olefins[49] and in the butenolide reactions the carboxylate anion might well function as an internal Lewis base. In seeking to control the stereochemistry of the Wittig olefination reaction, these workers have explored the effects of various nucleophilic additives in a number of aprotic media. Optimum conditions seem to be reached with iodide ion in dimethylformamide. The explanation of the effect is seen from a consideration of the possible modes of reaction of the ylid with a carbonyl compound. Reaction can occur by nucleophilic attack by carbonyl oxygen at phosphorus, in which event *trans* olefins would be expected (as well as from reactions involving alignment of anti-parallel dipoles); alternatively carbon–carbon bonding may initiate the reaction, and here conformational analysis of the intermediates formed shows that both stereoisomers could be formed, depending on the relative orientation of the carbonyl compound and the ylid in the initial phase of the reaction, e.g. **36** → **37** → *cis* olefin and **38** → **39** → *trans* olefin. In model reactions with benzylidenetriphenylphosphorane in benzene

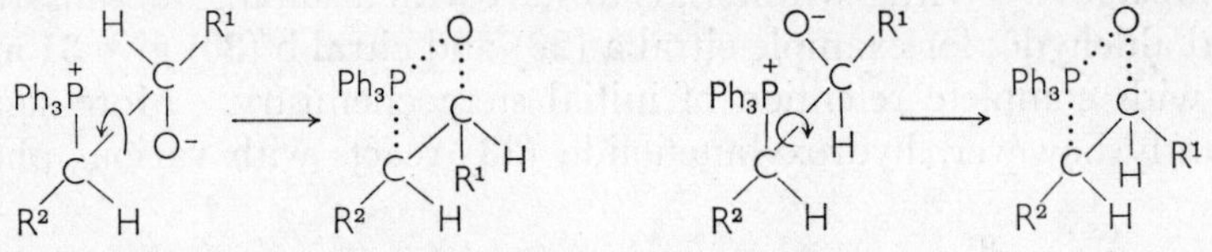

(**36**)   (**37**)   (**38**)   (**39**)

using simple aldehydes, both possible stereoisomeric products are isolated. The *cis* isomeric component increases however in reactions in ether, tetrahydrofuran and alcohol and becomes the predominant product in dimethylformamide. In the presence of amines or lithium bromide or iodide in benzene the reaction becomes selective for the *cis* isomer, and the olefin formed is largely the *cis* isomer for reactions in the presence of these additives with dimethyl formamide as solvent. Stereoselectivity here is interpreted in terms of complex formation between the ylid and the Lewis base additive; the colour of the ylid solutions being discharged on addition of these bases appears to confirm this, for the solutions retain their reactivity towards carbonyl compounds, although the rate of reaction appears to be reduced compared to that of the uncomplexed ylid solutions; *cis* olefins are formed via the sterically more favourable intermediates **40**, where B: could be either Lewis base or merely solvent of suitable polarity.

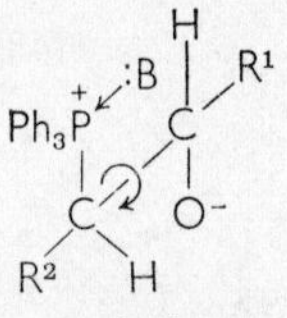

$$
\underset{(40)}{Ph_3\overset{+}{P}\cdots B\cdots \underset{\underset{R^2}{\underset{|}{C}}\diagdown\underset{H}{\overset{|}{C}}\diagup}{\overset{H}{\underset{|}{C}}\diagdown\overset{R^1}{\underset{|}{C}}\diagup}O^-}
$$

Stereoselective synthesis of *trans* olefins can also be achieved by using ylids carrying electronegative substituents on the methylene carbon, e.g. ethoxycarbonylmethylenetriphenylphosphorane. Specificity is the result of reduced electron density at the ylid carbon, making nucleophilic attack by carbonyl oxygen at ylid phosphorus more likely. The final product is determined by steric factors in the intermediate. Application of these stereoselective Wittig reactions is illustrated by the synthesis of various naturally occurring *cis* ethylenic fatty acids and extension of the method leads to diolefins having the characteristic *cis,cis* divinylethane system, e.g. **41** → **43**. α-Eleostearic acid (*cis*-9-*trans*-11-*trans*-13-octadecatrienoic acid) has similarly been made from *trans*-2-*trans*-4-nonadienal by means of (7-ethoxycarbonylheptylmethylene)triphenylphosphorane. With a large excess of the phosphorane the method can be extended to the use of ketones, leading to alkylated olefinic acids[49].

R CH=CHCH$_2$CH$_2$CH=PPh$_3$ + OCH(CH$_2$)$_n$CO$_2$Me $\xrightarrow{\text{DMF/I}^-}$
cis (**41**)   (**42**)

R CH=CHCH$_2$CH$_2$CH=CH(CH$_2$)$_n$CO$_2$Me
    cis   *cis*
(**43**)

Corey and Kwiatkowsky subsequently reported a superior stereoselective olefin synthesis which depends on isolation of diastereomeric β-hydroxyphosphonamides (e.g. **48**) from reaction of lithioalkylphosphonamides (e.g. **47**) with carbonyl compounds. The diastereomers are separately decomposed to the pure stereoisomeric olefins[50].

Further examples of the standard technique for introducing *cis* double bonds into natural polyenes of the vitamin A and carotenoid

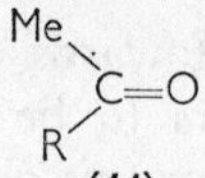

Me\
  C=O   Ph₃P=CHX   Me\
R/   C=CHX
  (44)   (45) (a) X = Br   R/
            (b) X = Cl   (46)

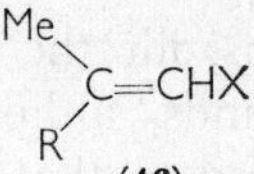

MeCH(Li)PO(NMe₂)₂ —PhCHO→ PhCH(OH)CHMePO(NMe₂)₂   R =
       (47)   −78°   (48)

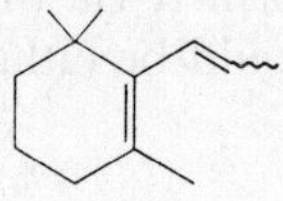

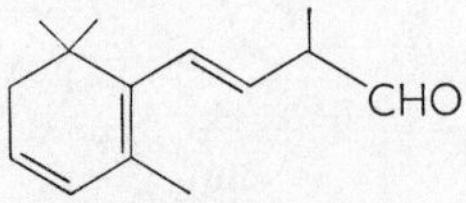

(49)

type using acetylenic intermediates discussed in Part I, have been described. All-*trans* vitamin A_2 and two isomers with Pauling hindrance, e.g. as in **28a**, namely 11,13-*cis,cis*- and 11-*cis*-vitamin A_2 have been made by acetylenic organometallic synthesis starting from *cis*- and *trans*-3-methylpent-2-en-4-ynols and **49**[51]. Spectral data (ultraviolet and n.m.r.) have been tabulated for the 9-, 11-, 13-*cis* and for the 9,13- and 11,13-*cis,cis* isomers, the ultraviolet spectra being characterized by the appearance of further short wavelength bands, with the longest wavelength absorption bands bathochromically shifted by about 25 mμ, compared to the vitamin A isomers[52]. The field of natural all-*trans* carotenoid synthesis has recently been reviewed by Jensen and by Weedon and *cis–trans* isomerism in this field is the subject of a book by Zechmeister[53]. The use of acetylenes in the synthesis of stepped 1,4-poly-ynoic acids and the derived all-*cis* polynoic acids has recently been authoritatively discussed by Osbond[54] in a comprehensive review. The synthesis of arachidonic acid (eicosa-5,8,11,14-tetraenoic acid) in particular has stimulated considerable experimentation in this area. A recent example is the cuprous cyanide Gensler coupling of the Grignard complex of 4-chloro-2-butynol with hexynoic acid **50**; bromination of the product **51** gives **52** which is then coupled with the Grignard reagent **55** of terminal acetylene **54** (prepared from acetylene magnesium bromide and 2-octynyl bromide **53**), to give poly-yne **56** and the polyene **57** is finally achieved by partial hydrogenation by the usual Lindlar catalytic technique[55]. Reactions similar to **50** → **52** and other cases using 1,4-dichloro-2-butyne which 'build in' the required propargylic halide, had previously been described[56].

3. Alkene Rearrangements

HO$_2$C(CH$_2$)$_3$C≡CH + ClCH$_2$C≡CCH$_2$OMgBr ⟶

(50)

HO$_2$C(CH$_2$)$_3$C≡CCH$_2$C≡CCH$_2$R

(51) R = OH
(52) R = Br

Me(CH$_2$)$_4$C≡CCH$_2$Br + HC≡CMgBr ⟶ Me(CH$_2$)$_4$C≡CCH$_2$C≡CX

(53)

(54) X = H
(55) X = MgBr

Me(CH$_2$)$_4$(C≡CCH$_2$)$_4$(CH$_2$)$_2$CO$_2$H

(56)

Me(CH$_2$)$_4$(CH=CHCH$_2$)$_4$(CH$_2$)$_2$CO$_2$H
cis (57)

An instance where Gensler coupling of a propargylic halide with an acetylene breaks down is the reaction of **58** with **59** giving **60**: and an

C$_5$H$_{11}$(C≡CCH$_2$)$_3$Br + BrMgC≡CCH$_2$CO$_2$H ⟶ C$_5$H$_{11}$(C≡CCH$_2$)$_3$CHCO$_2$H
                                                                        |
                                                                        C≡CH

(58)   (59)   (60)

HC≡CCH$_2$C≡CCH$_2$O—⟨THP⟩

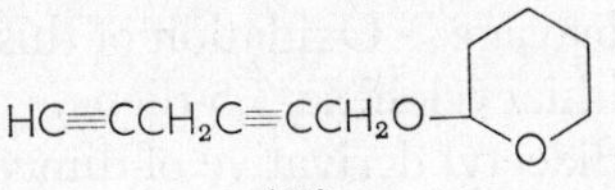

(61)

alternative route to all-*cis*-octadec-3,6,9,12-tetraenoic acid must be employed[57]. A further technique for developing polyene chains is the Gensler coupling of the tetrahydropyranyl ether of hexan-1,4-diyn-6-ol (**61**) with propargylic halides; 4,7,10,13,16,19-docosahexynoic acid is accessible by this route but unfortunately overreduction occurs in the final step to the polyene, the product being a *cis*-pentaenoic acid[58].

The scope of a new coupling reaction of terminal acetylenes and halides using lithiumamide in liquid ammonia has recently been explored; poor yields of acetylenic acids are obtained from 4-bromobutanoic acid but alk-3-,4-, and -5-ynoic acids can be made by condensing alkyl halides with dilithio derivatives of ω-acetylenic alcohols, followed by oxidation of the derived primary alcohol; bromination and incorporation into malonic ester synthesis is an alternative method of introducing carboxyl, and finally partial hydrogenation gives the all-*cis* polyene[59]. Examples of *cis,cis*-1,3-dienoic acids are available by Chodkiewicz coupling of bromoacetylene with terminal acetylenes, e.g. *cis,cis*-hexadec-8,10-dienoic acid[60] and its homologues from **62** and **63** via **64**, and the synthesis of *trans*-octadec-13-en-9,11-diynoic

Me(CH$_2$)$_4$C≡CBr + HC≡C(CH$_2$)$_6$CO$_2$H —Cu$^+$→ Me(CH$_2$)$_4$(C≡C)$_2$(CH$_2$)$_6$CO$_2$H

(62)   (63)   (64)

acid from *trans*-oct-3-en-1-yne and 1-bromodec-1-yne also illustrates the method[61].

D. Miscellaneous Geometrical Isomerisms

Miscellaneous observations reported recently are the preparation of the novel *cis*-pent-2-enal by manganese oxide oxidation of *cis*-pent-2-enol[62]; owing to their extremely easy stereomutation *cis* olefins of this type have not previously been reported, although *cis* diolefinic aldehydes are known[63]. The synthesis of *cis*- and *trans*-dimercaptoethylene and the stereomutation of aqueous solutions of the sodium salts have been investigated; the solutions equilibrate within a few hours. The salts are prepared by cleavage of *cis*-1,2-dibenzylmercaptylethylene by means of sodium in liquid ammonia, followed by acetylation and base hydrolysis[64]. Addition of benzylmercaptan to butadiene gives *cis,cis*-dibenzylmercaptobutadiene, a source of *cis,cis*-dimercaptobutadiene by sodium cleavage. Oxidation of this product with ferric chloride gives 1,2-dithiacyclohexa-3,5-diene. The *cis,cis–cis,trans* photoisomerism of the diacetyl derivative of dimercaptobutadiene has also been investigated; the all-*trans* form is accessible by thermal cleavage from *trans*-1,2-thioacetoxycyclobut-3-ene—obtained from *trans*-7,8-dithioacetoxybicyclo(4,2,0)octan-2,4-diene by Diels-Alder reaction with acetylenedicarboxylic ester and thermal extrusion of dimethylphthalate[65].

II. HYDROGEN TRANSFER AND PROTOTROPIC SHIFTS

A. Thermal, Photochemical and Catalysed Isomerization

The far reaching implications of the conservation of orbital symmetry in concerted reactions manifest themselves in sigmatropic shifts in aliphatic alkenes, an area perhaps not so well explored however as the field of electrocyclic reactions[66]. A number of authors[67,68] have described facile intramolecular dienyl hydrogen transfers represented, for example, by the process **65 → 66**. In the transition state (**67**) the hydrogen is held above the plane of the pentadienyl system, in a concerted suprafacial 1,5-hydrogen migration predicted by calculation[66] and supported by the substantial negative entropy of activation[68,69] ($\Delta S^\ddagger \sim -8$ to -9 e.u. for R = Me). Kinetic analysis of the rearrangement of *cis*-1,1-dideuteriopentadiene gives the rate constants for hydrogen and deuterium migration k_H and k_D; these turn out to be markedly different ($k_H/k_D = 1.15\ e^{1,400/RT}$; 12.2 at 25°) and the entropy of activation change is again substantially negative $(-7\cdot1$ e.u.); mass spectral analysis of the product of rearrangement shows no loss of deuterium[70]. Similar sigmatropic migration is observed for

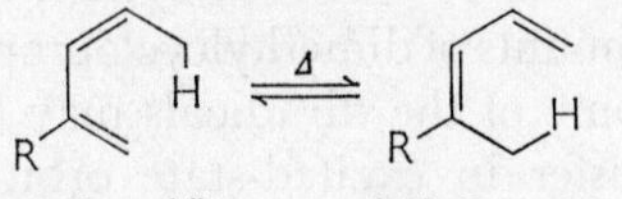

(65) (R = H)   (66) (R = H)
(68) (R = Me)   (69) (R = Me)

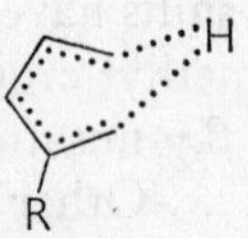

(67) (R = H)
(70) (R = Me)

cis-hexa-1,3-diene which rearranges to cis,trans-hexa-2,4-diene[71]. Interestingly these types of reaction can also occur through ring systems, 'homodienyl' reactions, and may even involve hetero atoms, e.g. **71** → **72** via **73** (followed by prototropy for X = O)[66,67].

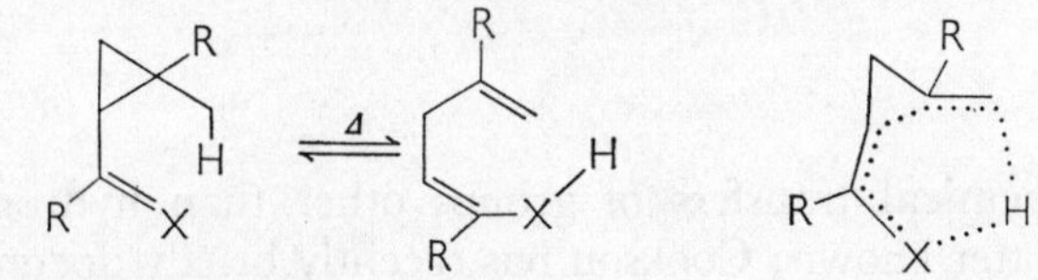

(71) R = Me X = O   (72)   (73)
or R = H X = CH$_2$

Whilst orbital symmetry selection rules allow concerted suprafacial 1,5 hydrogen shifts in these systems to be more easily achieved than other modes of reaction, it has been pointed out that due allowance should be made for relative ground-state energy in determining the activation energy for alternative reaction paths[66].

A further example of this type of reaction is the transformation **74** → **75**, the *trans* isomer thermally formed from **74** being removed by polymerization[72].

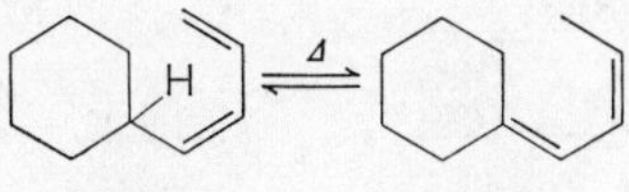

(74)   (75)

Hydrogen transfers in olefins need not necessarily be concerted processes however; the products obtained from the mercury photosensitized isomerization of 1-pentene strongly suggest the intermediacy of triplet diradicals which appear to undergo 1,3- and 1,4 hydrogen shifts giving intermediates which then cyclise, mainly to methylcyclobutane with lesser amounts of dimethylcyclopropane and cyclopentane. It is possible that some of the diradicals may be formed by primary 1,3-sigmatropic transfer in excited-state orbital symmetry from **76** followed by 1,2 hydrogen shifts in the product, pent-2-ene, but the absence of ethylcyclopropane renders this less likely[73,74]. However, photochemical 1,3 hydrogen shifts have been observed in a number of α,β-unsaturated esters, e.g. methyl crotonate and ethyl ββ-dimethylacrylate give their respective β,γ-unsaturated isomers, **81** gives **82** and in methanol **83** gives **84**[75]. Other examples are described in Chapter 6.

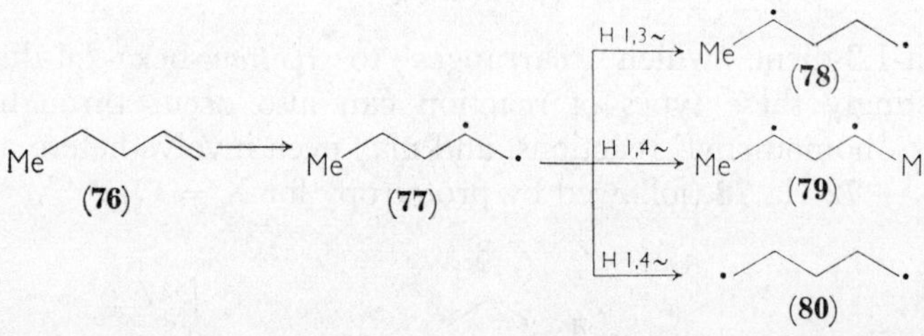

Me—(76)—⟶ Me—(77)• ⟶
  H 1,3~ ⟶ Me—(78)•
  H 1,4~ ⟶ Me—(79)—Me
  H 1,4~ ⟶ •—(80)—•

Photochemical transfers of groups other than hydrogen are becoming better known; Cookson has recently briefly documented relevant information in this area[76a] and described the reactions of some allylic systems. Thus stereomutation of the *trans* isomer of **85** is followed by 1,3 benzyl migration to **86**, and a similar process occurs with geranonitrile giving **87** and subsequently **88** and the stereoisomers of **89**. Cinnamyl ester also rearranges especially under sensitizing conditions, the benzyl ester more rapidly than alkyl esters, perhaps because it has a better absorbing group. Here the 'oxy'

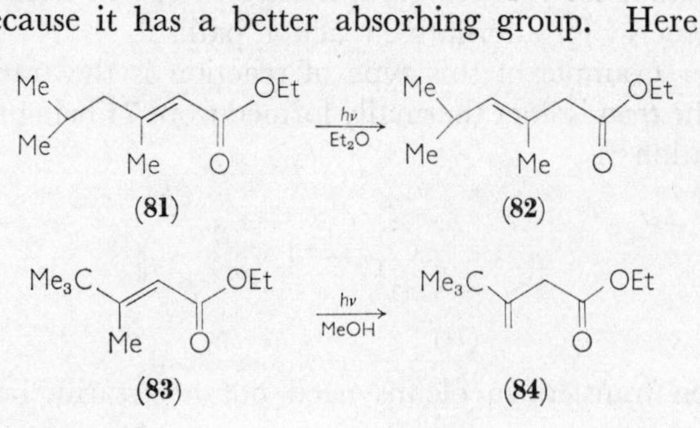

(81) Me₂CH—C(Me)=CH—C(O)—OEt ⟶^{hν}_{Et₂O} (82) Me₂CH—C(Me)=CH—CH₂—C(O)—OEt

(83) Me₃C—C(Me)=CH—C(O)—OEt ⟶^{hν}_{MeOH} (84) Me₃C—CH=CH—C(O)—OEt

analogue of the Claisen rearrangement seems to be symmetry forbidden for passage of the excited intermediate through the preferred six-membered four-centre transition state. All these reactions may be

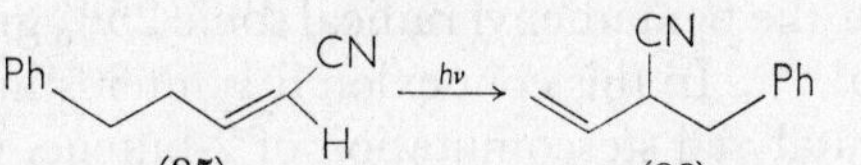

Ph~~~CN(H) —hν→ ~~CN~~Ph
(85)   (86)

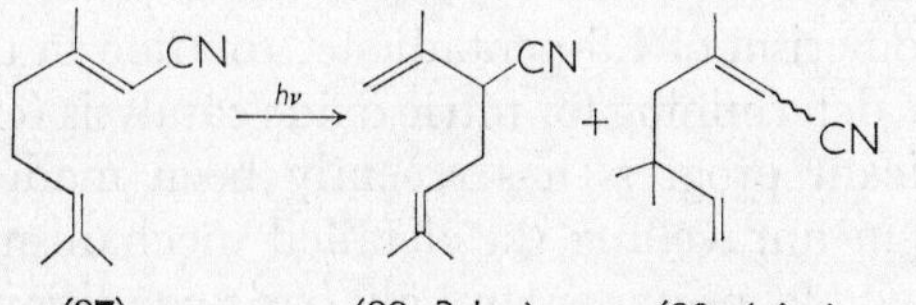

(87) —hν→ (88, 3,1~) + (89, 4,6~)

radical reactions in which no separation of fragments occurs. They could also be examples of concerted 1,3 sigmatropic shifts; examples of sigmatropic rearrangements involving C—C scission are known in cyclic systems, e.g. trimethyltropilidine [77]. Experiments with labelled compounds invite themselves. (These have now been done [76b] and show that the reactions behave as concerted sigmatropic processes.)

Benson and his colleagues in their continuing study of the fundamental thermodynamics of olefin isomerization, have directly measured isomer compositions (gas–liquid chromatography) and shown that the variation of equilibrium composition with temperature for the iodine-catalysed positional isomerism of 1-butene, differs from that predicted by calculation based on statistical thermodynamic formulae, but remains within the limits of error normally quoted for the API tables [78]. The main source of error is the uncertainty regarding low vibrational frequencies which leads to appreciable errors in the calculation of ΔS^0.

The resonance energy of the allyl radical, important in isomerization studies, has also been obtained directly from this work. The activation energy for the positional isomerization of 1-butene, is that for iodine attack on the allylic C—H bond, the critical step in the reaction mechanism which involves formation of the allyl radical; subtraction from the activation energy for the back reaction, i.e. attack by hydrogen iodide on a methyl allyl radical, gives the bond dissociation energy difference for s-C—H and hydrogen iodide. Assuming that the reaction of hydrogen iodide with methyl allyl radicals has the same activation energy as for attack at isopropyl radicals, the difference between activation energies for iodine attack on butene and on propane gives the allylic radical stabilization energy (12·6 ± 0·8 kcal/mole), in good agreement with other work[79]. Similar studies with the 1,4-pentadiene ⇌ 1,3-pentadiene system give a resonance-stabilization energy for the pentadienyl radical some 25% greater than that of the allyl radical[80]. In this connexion it is interesting that in contrast to the iodine-catalysed stereomutation of 2-butene, where rotation in the intermediate radical adduct formed with iodine is rate determining, it is the addition of iodine to the olefin which determines the rate of geometrical isomerism of 1,3-pentadiene; rotation in the intermediate is however rate determining for nitric oxide catalysis (cf Section I.A)[81].

Very significant progress has recently been made by Cram and his colleagues in unravelling the detailed mechanism of prototropic shifts in arylated olefins under strongly basic conditions. A considerable element of intramolecularity attends these rearrangements. For 3-phenyl-1-butene in t-butanol-d/potassium t-butoxide systems, the rate of deuterium incorporation is much slower (by a factor of 10 to 100) than that of rearrangement to the mainly cis isomer of 2-phenyl-2-butene, and similarly for 3-phenyl-1-butene-3-d in unlabelled solvent, although usually the rates of deuterium incorporation are faster than racemization of optically pure substrates, incorporation occurring with a high degree of stereospecificity. The extent of intramolecularity is higher for the protio compounds and the percentage of intramolecularity in the total rearrangement process varies over the range 6–56% with the solvent/base system used.

These rearrangements of 3-phenyl-1-butene can be visualized as passing through an ambident allylic anion bonded at its termini by the hydroxylic compound formed in the initial deprotonation step; *the degree of intramolecularity is then determined by the rate of collapse of the hydrogen-bonded assembly to olefin, and the rate of isotopic exchange of the intermediate with the surrounding solvent.* The stereochemistry is determined by back or front attack at the anion by the proton donor. A possible multistage process for the intermolecular component of the rearrangement could involve a planar carbanion, hydrogen bonded to a protonated (or deuterated) molecule of solvent, where the proton was removed from the benzylic position, and at the opposite terminal of the ion, to solvent of opposite isotopic content. This intermediate would yield the same product for deuterated substrate in unlabelled solvent, as for unlabelled substrate in deuterated solvent. This prediction does not accord with experimental observation, however, and the reactions appear best explained by the processes in Scheme I[81].

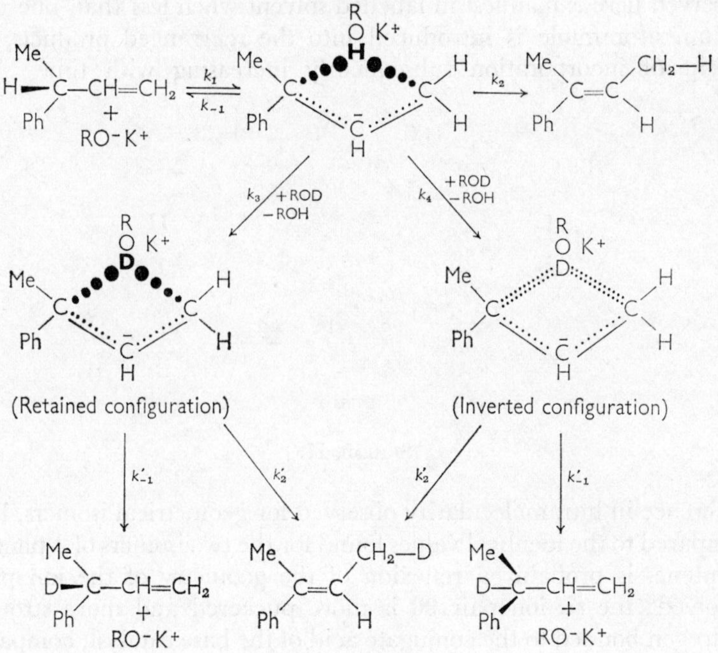

3. Alkene Rearrangements

[Scheme I showing rearrangement pathways with structures]

SCHEME I.

Since for these reactions experimental rate-constant relationships are $k'_2/k'_{-1} \simeq k_2/k_{-1}$, $k_2 \gg k_1$ and $(k_2 + k_3) \gg k_{-1}$, recovery of the originally bonded proton is slow compared to the rate of isomerization, conflicting with Ingolds' statement that 'when a proton is supplied by acids to the mesomeric anion of weakly ionizing tautomers of markedly unequal stability, then the tautomer which is more quickly formed is the thermodynamically least stable'[82].

The isomerization of α-benzylstyrene into cis- and trans-α-methyl stilbene occurs with 55% and 36% intramolecularity respectively, in t-butanol–t-butoxide, and again the rate of rearrangement is faster than either exchange or of isomerization of cis- or trans-stilbene. The kinetics of isomerization and deuterium incorporation are consistent with non-interconverting cis and trans allylic carbanions. Consideration of various models for the interconversion, and comparison of calculated equilibrium constants derived from the experimentally determined rate constants, with the directly determined equilibrium constants based on isomer composition (glc), shows that the reactions are best accounted for as in Scheme II. The true intramolecularity observed here is manifest in labelled solvent when less than one deuterium atom/mole is introduced into the rearranged products, the extent of incorporation subsequently increasing with time. The

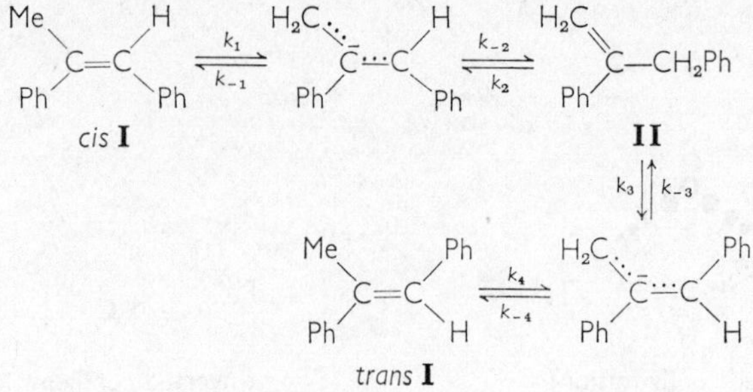

SCHEME II.

difference in intramolecularity observed for geometrical isomers, here compared to the identical values found for the two isomers of 2-phenyl-2-butene, is probably a reflexion of the geometry of the ion pairs involved; the *cis* ion pair **90** is more puckered and more strongly hydrogen bonded to the conjugate acid of the base catalyst, compared to the *trans* ion pair **91**, so that intramolecularity is favoured for the former[83].

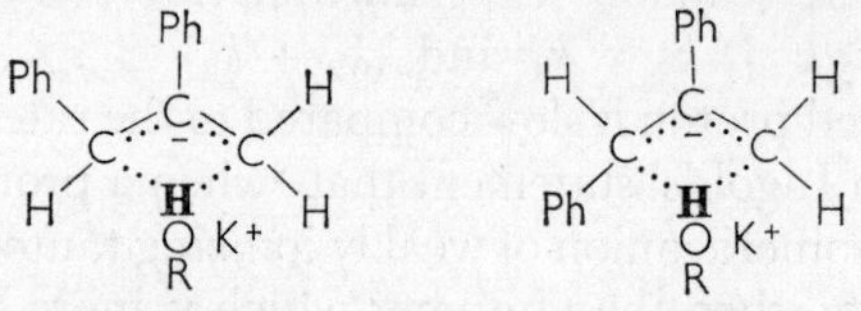

(90)   (91)

Prototropic rearrangement of allylbenzene has also been quantitatively investigated using these techniques; equilibration in dimethyl sulphoxide with *t*-butoxide anion gives 97·75% *trans*-propenylbenzene, 2·2% of the *cis* isomer and only traces of the starting material remain at room temperature. The *trans* isomer is formed nearly 13 times faster than the *cis* isomer, proton capture by the *trans* allylic carbanion being about 3 times faster to give *trans* isomer than return to allylbenzene. Correction for intramolecularity gives collapse ratios for the *cis* and *trans* carbanions as shown in Scheme III, where they are compared to those for related structures studied in this work[82-85].

Combined steric and inductive effects in these carbanions are

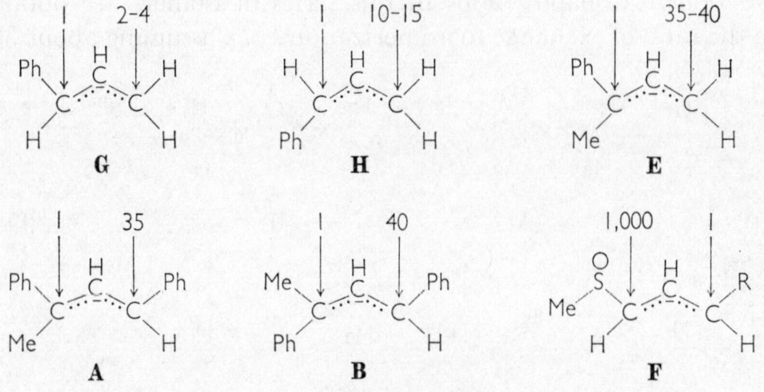

3. Alkene Rearrangements

[Scheme III showing carbanion structures G (2-4), H (10-15), E (35-40), A (35), B (40), F (1,000) with various Ph, Me, H, and S(O)Me substituents]

SCHEME III.

primarily responsible for the unbalanced collapse ratios observed, for whilst the pair of olefins related by carbanions **A** and **B** are about the same in thermodynamic stability, the two olefins connected by carbanion **E** exhibit an equilibrium constant of 500. For the *trans*-propenylbenzene-allylbenzene system the equilibrium constant is about 2,000! Even then this wide difference in stability of the tautomers is not reflected in the rate of protonation of the anion **G**, for which simple molecular-orbital theory calculations confirm the equal division of electron density between the termini; protonation must occur with very little activation energy. The collapse ratios for anion **F** throw the hydrocarbon data into perspective and clearly here the sulphoxide function exhibits its familiar behaviour in stabilizing adjacent charge. The stability of these carbanions in the relevant ion pairs can best be interpreted in terms of internal 1,3 steric interactions; thus **G** is more stable than **H**[84].

The stability relationship for carbanions in ion pairs of the type involved in these prototropic phenomena is also seen from the equilibrium data for isomerization of 1,3-diphenylbutenes in *t*-butanol-butoxide ion at 40° (Scheme IV)[85]. The product composition is 32% *trans*-1,3-diphenyl-1-butene (**I**), 52% *cis*-1,3-diphenyl-2-butene (**II**), 15% *trans*-1,3-diphenyl-2-butene (**III**) and 1% *cis*-1,3-diphenyl-1-butene (**IV**). In the ground state, diphenylbutenes **I** and **II** are the most stable and most easily interconverted, whilst **III** is slightly less stable than the former isomers; **IV** is by far the least reactive. As carbanion intermediates **B** and **D** are directly derivable from **III** and **C** and **D** from **IV**, the order of stability of carbanions is therefore **A** > **B** > **C** ≃ **D**, harmonizing with direct destabilizing 1,3-steric interactions. Collapse ratios in this series of isomers are obtained from the ratio of exchange to isomerization k_e/k_i, assuming about 50%

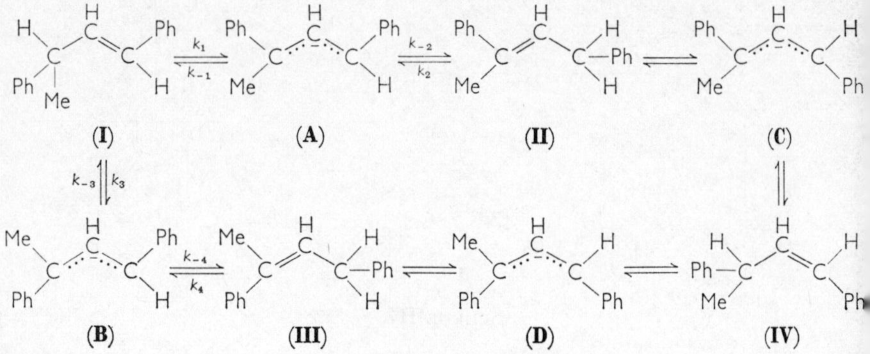

SCHEME IV.

intramolecularity; the difference in the collapse ratios at the termini, is here probably due to the inductive effect of the methyl group directing protonation to the less substituted sites. Again the relative ground-state energies of the interconnected olefins appears to have little effect on the magnitude of the collapse ratios. An elementary observation ruling out equilibrating carbanions is the formation of mainly deuterated *cis*-diphenyl-2-butene from this olefin and the deuterated *trans* olefin from the *trans*-diphenyl-2-butene; in any case the k_e/k_i ratios do not allow for a single intermediate. The barrier to rotation in the isomeric carbanions must be > 3 kcal/mole. The k_e/k_i ratios can be seen to be $(k_e/k_i)_{\text{II}} = k_{-2}/k_{-1}$ and $(k_e/k_i)_{\text{III}} = k_{-4}/k_{-3}$. Making allowance for 50% intramolecularity leads to $k_{-2}/k_{-1} = 2(k_e/k_i)_{\text{II}}$. It is possible to neglect olefin **IV** in considering the interconversions in Scheme IV, and the adjacent anions **C** and **D**. Since **I** can react via **A** or **B** whilst **II** can react via **A** or **C** and **III** via **B** or **D**, the latter two olefins react via carbanions *not* adjacent in the stability order and very likely react *only via the more stable anions*.

The *kinetic acidity* of the olefins may be obtained from the sum $(k_e + k_i)$ using corrected values for the collapse ratios obtained from transition-state free energies evaluated from kinetic data. A knowledge of the relative carbanion stabilities and those of the ground-state olefins then enables an energy profile for the reactions to be set up. The difference in reactivity, i.e. kinetic acidity, is strikingly larger than would be expected on the basis of ground-state free-energy differences, demonstrating that a detailed knowledge of *kinetic and thermodynamic* data is necessary for interpretation of base-catalysed rearrangements of olefins. Effects present in the carbanions themselves determine the difference in energy of transition states.

Cram and his colleagues have stated the results of their extensive work in this area in the following form. From a process involving anion M

$$X \underset{k_{-a}}{\overset{k_a}{\rightleftharpoons}} M \underset{k_b}{\overset{k_{-b}}{\rightleftharpoons}} Y$$

three cases can be envisaged; (i) $Y/X > 1$, $k_{-b}/k_{-a} < 1, k_a/k_b \gg 1$; (ii) $X/Y > 1$, $k_{-b}/k_{-a} > 1$, $k_a/k_b > 1$; (iii) $Y/X > 1$, $k_{-b}/k_{-a} \gg 1$, $k_a/k_b < 1$. Here k_a/k_b is the kinetic acidity ratio and k_{-b}/k_{-a} is the collapse ratio. Case (i) is Ingolds' case, whereas (ii) violates the first part of Ingolds' hypothesis, i.e. anion protonation *favours* the thermodynamically more stable tautomer which is kinetically the weaker acid; (iii) violates Ingolds' rules completely, protonation giving the more stable tautomer which is also the kinetically stronger acid. The systems studied exhibit all three types of behaviour; **I → B → III** is case (i), isomerization of 3-phenylbutene to 2-phenyl-2-butene via ion **E** is case (ii) and **II → A → I** is case (iii).

In considering the role of structure in determining collapse ratios, steric effects of methyl and phenyl groups do not appear to be important, whereas the inductive effect of methyl does appear to direct protonation to the least alkylated position; since the hyperconjugative effects of methyl at a double bond will tend to localize it at the more alkylated site, protonation at benzylic carbon is all the more favoured[85].

Bank, Schriesheim and Rowe[86] have also argued that the prototropic rearrangements of olefins are governed by the intrinsic properties of the intermediate carbanions; they also comment that the rearrangement of 1-butene cannot be controlled by conformational preferences in the olefin as earlier suggested (Part I, p. 427). Such effects would be more important if the rate of conformational change was slower than deprotonation. The rearrangement of 1-methylsulphinyl-dodec-1-ene to the more stable 2-olefin (the reverse of the usual experience with analogues) has also been discussed in terms of faster protonation to the more stable tautomer[87]. The earlier claim that N,N-dialkylallyl-amines rearrange to the *trans*-propenyl isomers is only partly correct; the *cis*-propenyl isomers are the primary products in aprotic media, but these rapidly stereomutate in the presence of alcohols, yet another case of kinetically determined *cis* product[88].

The question of hydrogen acidity in olefins has been approached from other quantitative points of view. The rate of base-catalysed isomerization of 2-ethyl-1-butene (**92**) can be compared to the rate of bromination of 3-pentanone (**93**). Plots of log k fit the same comparative plots used in the alkylidene-cycloalkene/cycloalkanone series[89]. For a series of olefin groups ((i) 3-substituted propenes, (ii) 3-substituted 2-methylpropenes, (iii) 3-substituted-1-butenes and (iv) 2-substituted-1-butenes) the rate of isomerization in each group decreases as the substituent is changed from methyl to t- butyl, whereas phenyl and vinyl groups as expected enhance the rates, e.g. allylbenzene and 1,4-pentadiene are about 10^5 times more reactive than 1-butene. Both inductive and steric effects appear to be important from the Taft-Hammett treatment of the data, and the negative entropy of activation accords with a rigid planar transition state for the reaction, in harmony with Cram's observation of intramolecularity. Steric effects can be analysed into those effecting the approach of the base and those mitigating coplanarity in the carbanion. These steric effects can be seen from the data for the four groups of olefins (i)–(iv). Alkylation has a very similar effect for all members and evaluation of log (k_a/k_0) in each series (k_a = alkyl substituted olefin, k_0 = methyl substituted olefin) shows that it is virtually constant (-0.23 ± 0.04). The average value for each series plotted against Tafts σ^* constant shows a somewhat large deviation from linearity which *increases* with increasing size of the substituent. The effect of spreading the charge in the carbanion can be demonstrated by taking the difference in total calculated π-energy for and without orbital overlap with the carbanion site (E_r^π), and this also correlates well with the increasing rates due to vinyl and phenyl substitution. [1-butene, $E_r^\pi = 0.82\beta$, relative rate 1; 1,4-pentadiene, $E_r^\pi = 1.46\beta$, relative rate 20×10^5; allylbenzene, $E_r^\pi = 1.38\beta$, relative rate 9.4×10^5)[90].

(**92**)   (**93**)

Another approach to the question of anion structure in the context of prototropic rearrangements is to observe the n.m.r. spectra of the anions at low temperatures. Treatment of various hydrocarbon precursors of the 1,4-pentadienyl anion with butyl lithium enables isolation of tetrahydrofuran solutions of the lithium salt. The products of protonation indicate that the rate of attack at the central carbon is not necessarily faster than at the termini, and the n.m.r. chemical shifts support the view that electron density at primary carbon is greater than at secondary carbon, and at odd numbered carbons compared to even numbered carbon atoms. The coupling constants for protons at adjacent positions indicate that for the simplest pentadienyl anion the 'W' shape is preferred to the 'U' shape and indeed models of the latter structure show considerable steric hindrance, whilst bond orders based on the coupling constants indicate 1·33 for inner pairs of bonds and 1·67 for outer pairs. The terminal methylene protons become equivalent above 40°, but below 15° they exhibit different signals indicative of rotation about the outside carbon–carbon bonds (60 times/sec at 30°). The simplest mechanism for these rotational effects is that of a lithium ion forming a covalent bond with the anion for a sufficient length of time to allow rotation about the resulting single bonds. Addition of lithium cation at $C_{(3)}$ gives a symmetrical structure with rotational possibilities for the inside $C_{(2)}$—$C_{(3)}$, $C_{(4)}$—$C_{(5)}$ bonds (**95**) but lithium attached to $C_{(1)}$ allows both inner and outer rotation (**96**); the rotational barrier about the inner bonds is expected to be lower, as observed for the 2-methyl ion where outer bond rotation appears to be preferentially 'frozen out' on cooling the solution (broadening of the n.m.r. peaks of the odd numbered hydrogens in the range 10 to $-20°$). The calculated rate for rotation of the outer round the inner bonds based on n.m.r. data is about 130 times/sec. at $-5°$.[91a]

(**94**)   (**95**)   (**96**)

Similar studies with *cis,cis*- and *trans,cis*-tetraphenylbutadiene involving treatment of the hydrocarbon with metallic sodium, lithium or *n*-butyllithium in tetrahydrofuran and quenching of the resulting blue solutions with alcohols, indicates the equilibrium value for the hydrocarbon isomeric composition to be 70–75% *trans,trans* and 25–30% *cis,cis* which actually represents the equilibrium between the stereomutating species, thought here to be a radical anion. The rearrangement might also occur through the dianion present in small concentration[91b].

Solvent effects on isomerization rates of 3-butenylbenzene show that for polyethyleneglycol ethers the increase in rate is dependent on the chain length; the reaction in hexaethyleneglycol polyether is about

2,000 times as fast as in 1,4-dioxan. Another approach is to study in aprotic solvents, olefins having similar prototropy rates in hydroxylic media. Allylbenzene in hexaethyleneglycol polyether for example rearranges 18×10^4 times as fast as 3-butenylbenzene, whereas they are comparably reactive in methanol. These effects are due to the interaction between the anion base and the cation being greater in less cation-solvating media than the electrostatic interaction in the transition state for proton removal from the substrate, thus enhancing the activation energy and therefore reducing the rate. Considerations of configurational entropy shows that as the number of oxygen atoms in the solvent molecule increases, the entropy change in solvation of cations should reach a limiting upper figure; this is qualitatively confirmed by the decreasing acceleration of reactions as the polyether increases in complexity, reflecting the fact that only some of the oxygen atoms in any one molecule can be employed in cation solvation, and that the number of molecules per unit volume of solution decreases with molecular weight[92].

Further work on the prototropy of allyl alkyl ethers shows that with n-butyllithium at low temperatures in hydrocarbons the protonation of the alkoxy allylic anion to *cis*-propenyl ether (by alkoxy allyl compound as proton source) is more rapid than deprotonation of the propenyl product by n-butyllithium. Generation of the *cis*-propenyl anion from the ether must be *very* slow since quenching of a mixture of the propenyl ether and n-butyllithium in hydrocarbon solvent with deuterium oxide does not introduce deuterium[93]. In this connexion, 3-methoxy-1-phenylbutene is isomerized very largely to 3-methoxy-1-phenyl-2-butene, which must surely reflect electron-density effects in the carbanion rather than lack of conjugative stabilization of the olefin by the phenyl group[94]. Halogenated ambident carbanions have recently been examined; 1-(*p*-nitrophenyl)allyl chloride rearranges to the vinylic chloride isomer in the presence of aliphatic amines[95].

Comparatively little work has been done recently on acid-catalysed prototropic rearrangements in simple olefins. However, homogeneous gas-phase n-butene isomerization catalysed by hydrogen bromide involves molecular acid, and a six-membered transition state is pictured, as a proton developing at a site adjacent to a double bond is replaced by hydrogen abstracted from an allylic position[96]. The formation of carbonium ions from fatty acid olefins which allow double bond migration and finally lactonization, has been discussed. The existence of equilibria between isomeric olefins here is confirmed by reactions in deuteriosulphuric acid with alkan-4-olides; mass spectral examination of the products shows that simpler members (e.g. C_6–C_{11} compounds) are deuterated in all positions, but longer chains are only partially deuterated; octadecan-4-olide is deuterated up to $C_{(11)}$[97].

B. Hydrogen Shifts and Protrotropic Rearrangements in Polyenes

The field of acetylene-allene conjugated olefin rearrangements has recently been reviewed[98]; newer techniques, e.g. the use of aprotic media, alkyl- and aryl-lithiums as bases and preparative gas-liquid chromatographic methods for isolating all constituent products, promises to reveal new details of transformations and equilibria in these reactions, whilst also making previously inaccessible polyenes available for experiment[99]. The mechanism of base-catalysed protropic rearrangement of an en-allene or en-yne systems does not appear to have previously been explored. Two steps are probably involved in the transformation of methyl *cis*-9-octadecen-12-ynoate to *trans,cis,cis*- and *trans,cis,trans*-8,10,12-octadecatrienoic acid. An en-allene intermediate is formed by base catalysis (protonation of the intermediate carbanion to give isolable 9,11,12-trienoic acid), and this is then thermally isomerised, by 1,5 hydrogen transfer (from the $C_{(8)}$-methylene), to *trans,cis,trans*-octadecatrienoic acid[100].

In the field of vitamin A synthesis further techniques for isomerizing the retroionylidene compounds, encountered as unwanted end products in certain reaction sequences, (*cf* Part I, p. 433, 448) are available; hydrolysis of vitamin A aldehyde enol ether gives a substantial yield of the retro isomer **97**, that can be converted to the cyclohexenyl isomer **98** by passage through silica gel columns or by more conventional acid catalysis[101,102]. Reduction of **97** with lithium aluminium hydride gives vitamin A alcohol directly[102].

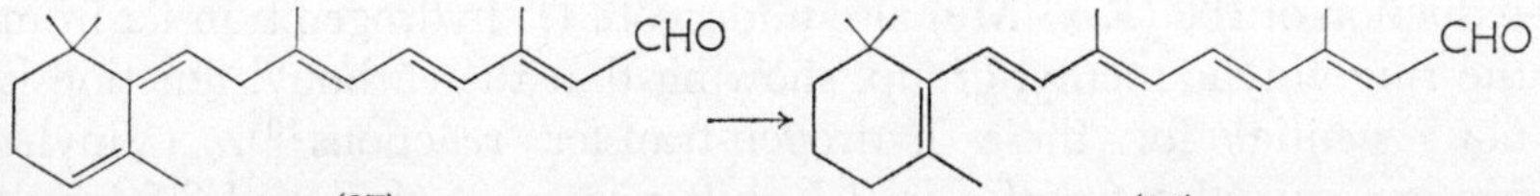

(**97**)   (**98**)

Photochemical reactions of α-ionone and related compounds appear to involve among other reactions such as stereomutation, the type of

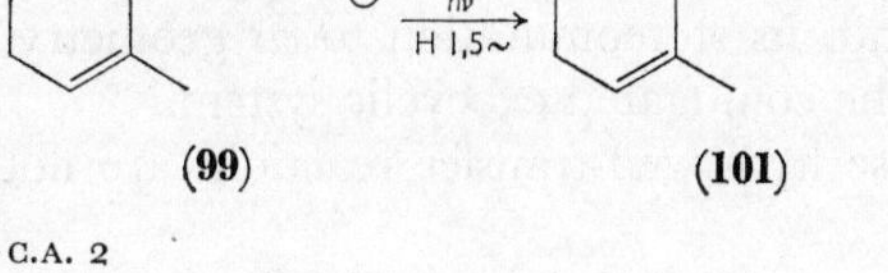

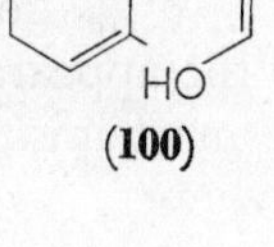

(**99**) $\xrightarrow{h\nu}_{H\,1,5\sim}$ (**101**)   (**100**)

hydrogen transfer discussed in Section II-A; e.g. **99 → 101** via enol **100**. An analogous reaction occurs with *trans*-α-cyclocitrylideneacetic acid which gives the rearranged ester in ethanol because of attack of alcohol on the intermediate enediol. Hydrogen transfer to the excited carbonyl group may be another reaction pathway followed in these reactions, summarized in general in formulae **102–105**[103].

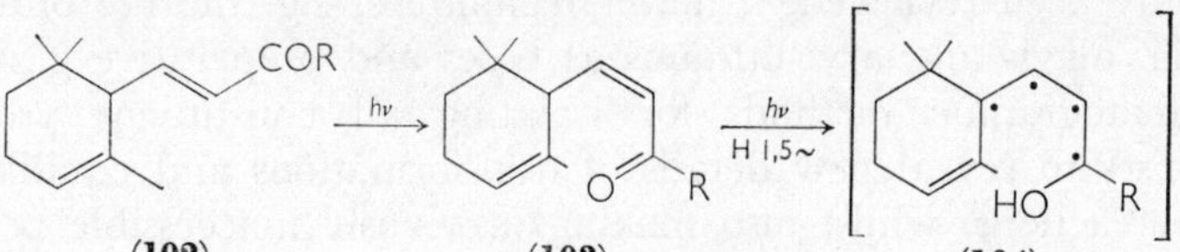

(102)   (103)   (104)

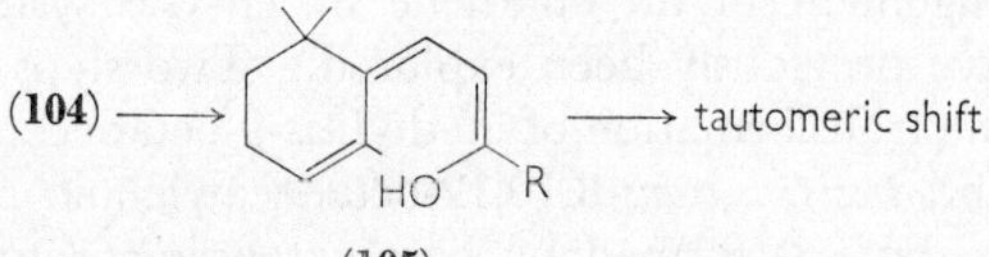

(104) ⟶ (105) ⟶ tautomeric shift

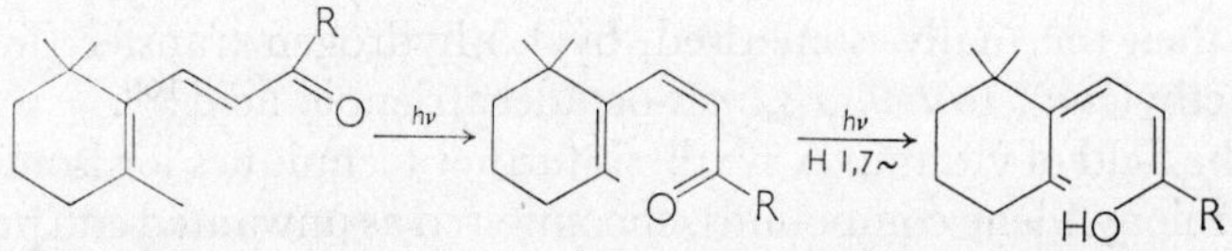

(106)   (107)   (108)

Hydrogen transfer also occurs in the β-ionone series; thus when R = Me, **106 → 108**; a 1,5 shift is observed with *trans*-β-cyclocitrylideneacetic acid (**106**, R = OH), which does not esterify on photolysis in alcohols because no enolic intermediate is involved here unlike the case of the *trans*-α-compounds. The alcohol obtained by carbonyl reduction of **106** (R = Me) also undergoes 1,5 hydrogen transfer from the ring vinylic methyl group, showing that the carbonyl function is not essential for these hydrogen-transfer reactions[104]. (Vinylic protons can also transfer in 1,5 shift processes; thus *cis*-1,3,5-hexatriene gives the allene *cis*-1,2,4-hexatriene on photolysis, and the analogous *cis*-2-6-dimethylocimene behaves similarly[105]). Very similar processes are found in the dehydro-β-cyclocitrylidene series and in the β-ionylideneacetic esters (1,5 shifts) and the reactions can be monitored using n.m.r. techniques to observe the disappearance of the *trans*-$C_{(7)}$—$C_{(8)}$ double bond, its stereomutation to *cis* geometry and finally the appearance of the conjugated exocyclic system.

The mechanisms of these hydrogen-transfer reactions are not entirely clear; it is attractive to picture hydrogen transfer as passing through cyclic hydrogen-bridged transition states in concerted reactions, but orbital symmetry considerations may make such processes unfavourable (whilst not actually precluding them); triplet states and diradicals or zwitterions could also be involved.

There is as yet no evidence for the $C_{(7)}$–$C_{(8)}$ stereomutation to hindered *cis* geometry, as occurs with these photolytic reactions, in the vitamin A series; complex mixtures are isolated from photolysis experiments but it *is* certain that $C_{(13)}$ stereomutation occurs[106].

C. Heterogeneous Catalytic Rearrangements of Simple Alkenes

Advances in technique have stimulated much activity in the field of heterogeneous catalysis. For the isomerization of *n*-butene over silica-alumina and other surfaces, a number of suggestions as to mechanism have been made. In one picture, the Lewis acid (electrophilic oxide sites), formed by dehydration of the catalyst surface, reacts with the olefin by hydride abstraction, to form a carbonium ion which may then react with a further butene molecule by proton donation, giving a further carbonium ion, this acting as the chain-carrying isomerization intermediate. For a mixture of butene and hexadeuteriopropene two types of initial chain-carrying isomerization site can therefore be formed; because of the deuterated site the isomerized butene can contain deuterium. If R and P are reactant and product three possible steps arising for reaction with a proton must be considered:

$$R + H^+ \longrightarrow RH^+$$
$$R + H^+ \longrightarrow P + H^{+\prime}$$
$$RH^+ \longrightarrow P + H^{+\prime}$$

where $H^{+\prime}$ represents a different proton to that accepted by the olefin. If either of the first two possibilities is rate determining, deuterium might be expected to concentrate in the product in the presence of deuterated compounds. This proves to be the case with silica-alumina surfaces and other composite catalysts containing silica; for alumina alone a nearly equal distribution of heavy isotope is found between reactant and product olefins, so that the third possibility may be rate determining here, unless simple exchange subsequently occurs[107]. Approaching the question of mechanism from a different standpoint, prior deuteration of the catalyst surface followed by admission of the olefin shows that the extent of incorporation of the isotope into the rearranged product is too low to allow for a mechanism which includes exchange of hydride as a step involving reaction at the catalyst's surface with the isomerization intermediate. For example, considerable exchange of deuterium occurs between 1-butene and hexadeuteriopropene or silica-alumina, but the extent of exchange with non-labelled olefin on deuteroxylated catalyst is a hundredfold less. The data are however consistent with an allylic carbonium ion mechanism, which involves hydride transfer *from other olefin molecules*[108]. Deuterated ion-exchange resins have also been applied to olefin isomerization and here again hydrogen exchange appears to be associated with the slow step for the isomerization[109]. In many of the isomerizations of 1-butene studied, considerable stereospecificity is observed for the initial reaction products. Since spectroscopic data do however support the idea of hydride abstraction by the catalyst, the stereospecificity observed could be due to the sterically controlled assumption of *cis* geometry at the surface of the catalyst by the cation involved. If kinetic control of the product obtains, the resulting olefin will be expected to be *cis* product; rotation in the intermediate carbonium ions (see Section III.A)[110], as in carbanions, is expected to be relatively slow[111]. A similar explanation of the stereospecificity observed could apply whatever the source of hydride in the final product forming step.

Isomerization by mixed Ziegler-type catalysts has been explored, e.g. by passing 1-butene through boiling solutions in pentane or hexane containing $AlEt_3$-$NiCl_2$-$2C_5H_5N$ complex; chromatographic analysis of the effluent olefins showed the composition to be near the thermal equilibrium value for *cis*-2-butene/1-butene, with some *trans*-2-butene also being formed. Such catalysts rapidly lose their activity and addition of triphenylphosphine quenches catalytic power. Similar results are observed with $AlCl_3$-$NiCl_2$-C_5H_5N, approximate thermodynamic equilibrates appearing for 1-butene, but surprisingly, for methylbutenes the product composition differs widely from the equilibrium figure; 4-methyl-1- and 2-pentenes show little migration of the olefin bond to the tertiary position[112].

III. ANIONOTROPIC REARRANGEMENTS

A. *Mechanisms of Anionotropic Rearrangements in Allylic Systems*

Among the more important recent observations of anionotropic phenomena, is the significant barrier to rotation in the 2-butenyl cations involved in solvolysis of 1-chloro-2-butene and 3-chloro-1-

butene in solvolysis[110]. The high proportion of *trans*-2-butene-1-ol in the hydrolysis products of 3-chloro-1-butene suggests that it gives rise to the same *trans* cation as is formed from *trans*-1-chloro-2-butene, and to very little of the *cis* cation associated with the *cis*-1-chloro-2-butene reaction. Experimentally it is easier to determine small amounts of *trans* alcohols in stereoisomeric mixtures rich in *cis* isomer than vice-versa, and the Winstein group therefore studied the solvolysis of the *cis* chloride.

Analysis of the products of silver ion-catalysed hydrolysis of pure *cis*-1-chloro-2-butene, with allowance for the stereomutation of the alcoholic product, indicates 99% retention of geometric configuration for the *cis* ion involved! The result suggests that in cases where geometrical isomerization appears to involve cationic species, neutral precursors of stereomutated molecules may be involved as intermediates. Allylic carbanions and cations therefore closely resemble each other stereochemically due to the maintenance of considerable double bond character between the original olefin carbons; delocalization is not complete and carbon atoms do not become equivalent. These findings invite comparison with allylic free radicals, but considerably less is known in this area.

The reactions of allylic halides with amines has received comparatively little recent attention. The reaction of 3-chloro-1-butene with diethylamine may involve a cyclic transition state or is possibly an example of the rare S_n2' process. Reactions of this kind are complicated by the possibility of allylic rearrangement preceding substitution, especially in media of low dielectric constant with relatively weak nucleophiles, as apparently occurs for cinnamyl chloride in its reaction with triethylamine[113]. Heterogeneous catalysis of rearrangement of the allylic compound by solid amine salt formed, or by the walls of the reaction vessel, is also a complicating factor[114]. In this connexion it is useful to know free-energy differences between isomers, and for the equilibrium of 1-chloro-2-butene with 3-chloro-1-butene, ΔF is -0.6 kcal/mole favouring the 2-butenyl isomer compared to $1\cdot2-1\cdot9$ kcal/mole for the 1- and 2-butene equilibrium, the decrease in ΔF being ascribed to the steric conflict of the hydrogen *cis* to chlorine in the 2-butenylchloro compound.

Some unusual allylic compounds are being investigated; the rearrangement of 2-butenyl azide is accelerated only slightly by increased solvent polarity and this, with the negative entropy of activation, suggests a cyclic S_ni' process leading to a 3-substituted 1-butene[115]. The reaction of 1,4-dibromo-2-methyl-2-butene with azides is complicated by several rearrangements, for beside the expected product **109, 110** and **111** are also formed; unfortunately these compounds prove difficult to handle and detailed mechanisms for their formation have not been worked out. For the transformations of the simpler

$$N_3CH_2CMe= CHCH_2N_3 \qquad CH_2= CMeCHN_3CH_2N_3$$
$$(109) \qquad\qquad\qquad (110)$$

$$CH_2= CHCMeN_3CH_2N_3$$
$$(111)$$

bis-azide **112** ($R_1 = R^2 = H$) however, a cyclic transition state (**115**) appears to be most consistent with the experimental data. The reduced rate of rearrangement compared to 2-buten-1-yl azide for example, probably reflects the electron withdrawal due to the other azide group. Surprisingly, rearrangement of the tertiary azide **113**

$$N_3CH_2CR^1= CR^2CH_2N_3 \longrightarrow CH_2= CR^1CR^2N_3CH_2N_3 \rightleftharpoons$$
$$cis \qquad\qquad\qquad\qquad\qquad\qquad\qquad N_3CH_2CR^1= CR^2CH_2N_3$$
$$\qquad\qquad\qquad\qquad\qquad\qquad\qquad\qquad\qquad trans$$
$$(112) \qquad\qquad\qquad (113) \qquad\qquad\qquad (114)$$

$$\begin{array}{c} C \\ \delta+ \diagup \quad \diagdown \delta+ \\ C \qquad C \\ \diagup \qquad \diagdown \\ N \qquad N \\ \delta- \diagdown \quad \diagup \delta- \\ N \end{array}$$
$$(115)$$

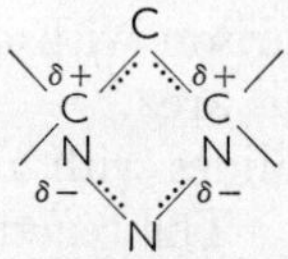

($R^1 = R^2 = Me$) to *trans* **114** is slower than the reverse reaction of the primary azide **114** ($R^1 = R^2 = Me$), but steric compression in the transition state may account for this[116].

The question of the nature of the transition state or intermediate involved in carbonyl addition reactions with allylic organo-metallic compounds, has been examined recently. It has been suggested that cyclic transition states may be involved in some of these reactions[117], e.g. **116**:

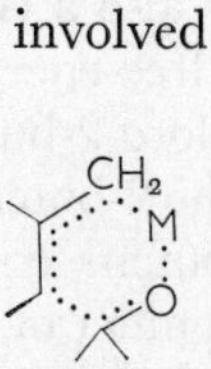

$$\begin{array}{c} CH_2 \\ \diagup \quad \diagdown M \\ \qquad \vdots \\ \diagdown \quad \diagup O \\ \diagup\diagdown \end{array}$$
$$(116)$$

However for acetylenic analogues such a mechanism seems untenable on steric grounds, whilst allylic carbanions which reasonably retain their stereochemistry seem more likely to be the effective reactive species derived from allylic Grignard reagents[118]. Interesting cases of rearrangements can occur with diisobutylaluminium alkenes, obtained by addition of diisobutylaluminium hydride to various olefins, e.g. **118** → **120** (as shown by deuterium oxide quenching and n.m.r. analysis of the products). Unexpectedly **120** is the predominant allylic isomer. These reactions are strongly solvent sensitive and the presence of donor compounds such as amines, stabilizes the phenyl conjugated isomers (**118,119**), possibly because the aluminium group is sterically hindered in isomers such as **120**[119].

The reactions of olefins with *t*-butyl peresters in the presence of copper ion, gives the less stable allylic isomer; it is likely that an acyl free radical and a carboxylate anion are produced by electron transfer from cuprous ion, and subsequent reaction of the olefin with the former by hydrogen abstraction, generates an allylic radical which combines with the carboxylate anion through the agency of cupric ion as electron acceptor; cuprous ion formed in the last step may function as co-ordinating agent for the forming π-bond. Tetramethylethylene gives 1,1,2-trimethylallylbenzoate in this type of reaction, the product being kinetically controlled to that site of highest electron density, much as for a carbanion[120].

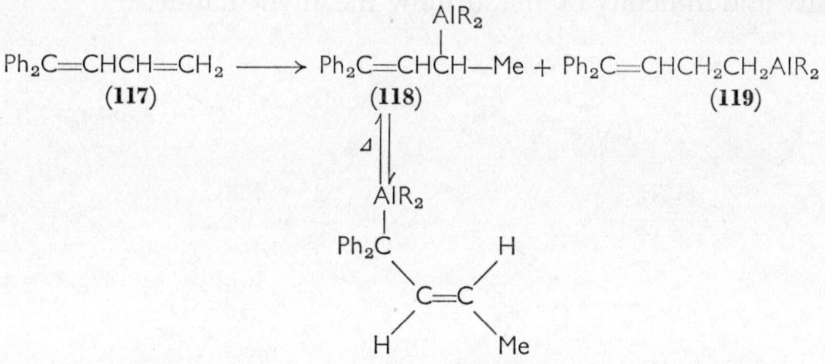

$$Ph_2C\!\!=\!\!CHCH\!\!=\!\!CH_2 \longrightarrow \underset{(118)}{Ph_2C\!\!=\!\!CHCH(AlR_2)\!-\!Me} + \underset{(119)}{Ph_2C\!\!=\!\!CHCH_2CH_2AlR_2}$$

(117)

$$\Delta \updownarrow$$

(120): Ph$_2$C(AlR$_2$)/H C=C H/Me

B. Anionotropic Rearrangements in Polyenes

In their work on extended anionotropic systems, Jones and McOmbie[121] found that 1-ethynylcinnamyl alcohol (**121**) did not rearrange in acids into conjugation in the expected manner, whereas later authors discovered that the compound gave the aldehyde **125** on steam distillation from acidic media[122]. However **121** with *p*-toluenesulphonic acid also gives **126** following Jones-McOmbie rearrangement, in analogy with the 1-ethynylcrotyl alcohol system, so that this type of anionotropy now appears to be perfectly general[123].

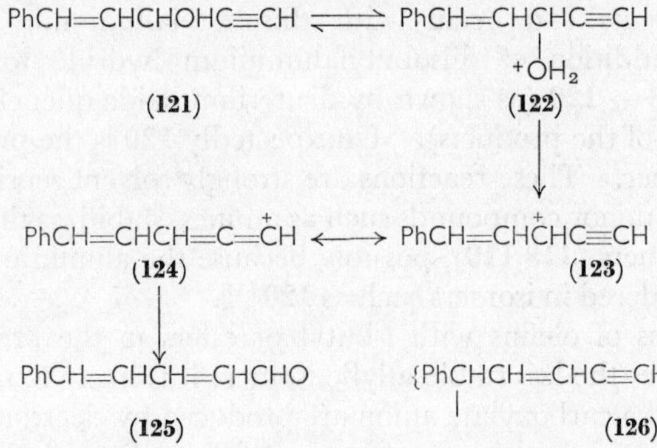

PhCH=CHCHOHC≡CH ⇌ PhCH=CHCHC≡CH
                                  |
                                  +OH₂
(121)   (122)

PhCH=CHCH=C=⁺CH ⟷ PhCH=CH⁺CHC≡CH
(124)   (123)

PhCH=CHCH=CHCHO   {PhCHCH=CHC≡CH}₂O
(125)   (126)

Numerous examples of extended anionotropic shifts have been described in recent synthetic work, mainly of the type discussed in Part I. More interesting perhaps are rearrangements involving formation of triphenylphosphonium salts which have recently become known from work with Wittig reagents. In the synthesis of vitamin A₂ treatment of the tertiary diallylic alcohol **127** with triphenylphosphine and hydrobromic acid usefully gives the salt **128**, directly[124], avoiding the necessity and difficulty of first forming the allylic halide.

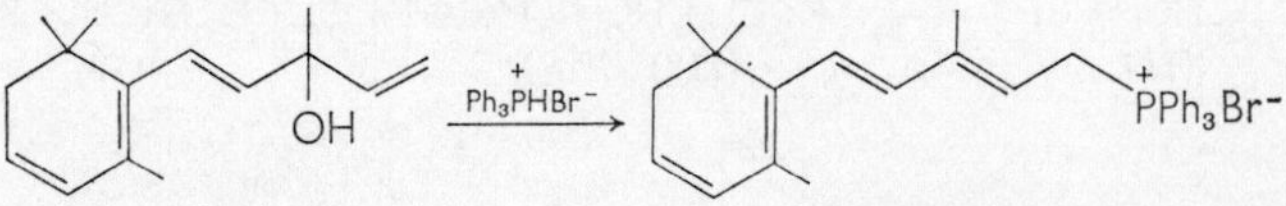

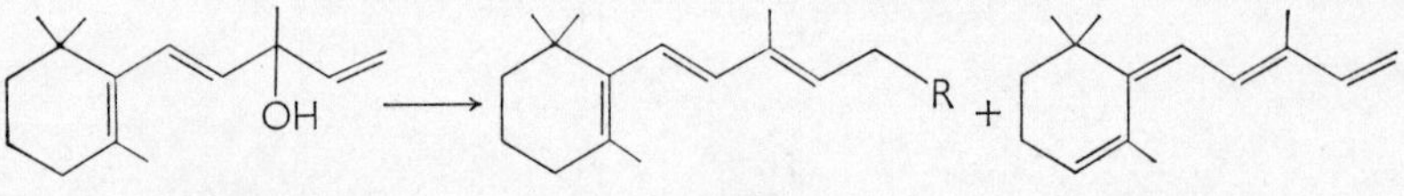

A similar result is observed with the analogous β-ionylidene compounds (cyclohexene ring system); in these reactions the carbonium ion first formed must be very rapidly scavanged, although some retro-hydrocarbon **131** is formed by proton loss in reactions of **129** or its primary allylic isomer. Treatment of hydrocarbons such as **131** with triphenylphosphine and acids results in rearrangement to the normal structure, e.g. **131** gives **130** (R = PPh₃⁺Br⁻), a most useful preparative observation. The reaction of alcohol **129** or its primary allylic isomer with triphenyl phosphite and acids gives a mixture of the tertiary $C_{(9)}$ and primary $C_{(11)}$ phosphonates (reaction of the mesomeric carbonium ion with the phosphite and expulsion of Et$^+$, in effect). Examination of models indicates that other than $C_{(11)}$, $C_{(9)}$ is the only non-hindered potential site for attack in the intermediate mesomeric cation. Again rapid scavanging of the intermediate by phosphite to a kinetically controlled allylic isomeric mixture, largely precludes deprotonation to retrohydrocarbon, although all three components should arise from the less reactive intermediate species.

Anionotropic rearrangement-dehydration to retro compounds in these systems has been further investigated; it is found that **129** reacts with acetic acid to give a substantial proportion of 'forward' allylically rearranged acetate **130** (R = AcO), as well as the retro-dehydro-hydrocarbon **131** earlier observed as the preponderant product. In ethanol–acetic acid mixtures the product ratio of **130** (R = EtO) to hydrocarbon **131**, is similarly due to the stability of the ether function formed by ethanol attack at the carbonium ion[125].

In earlier work it was found that retrodehydrative rearrangements with vitamin A precursors and related compounds, could be avoided by using intermediates with electronegative substituents adjacent to the site of the forming double bond e.g. **132** → **133**[126]. Similarly, the nitrile **134** gives as the major product **135** and small amounts only of the retro compound **136**, whilst catalytic amounts of *N*-haloimides convert **134** exclusively to **135**[127].

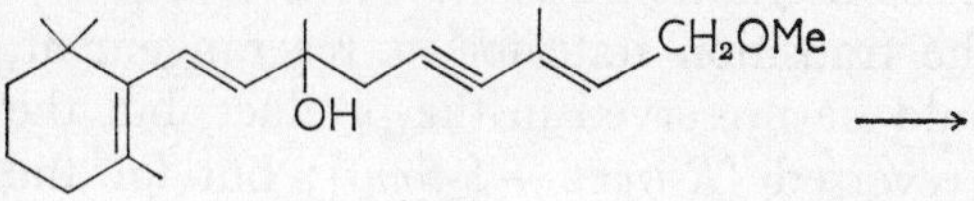

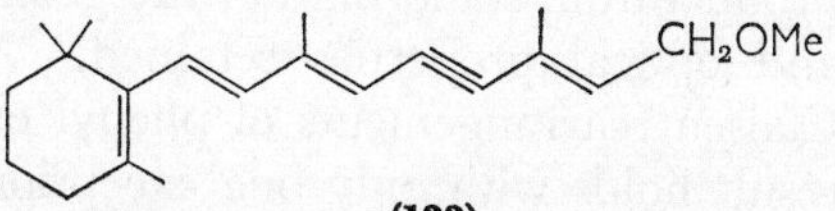

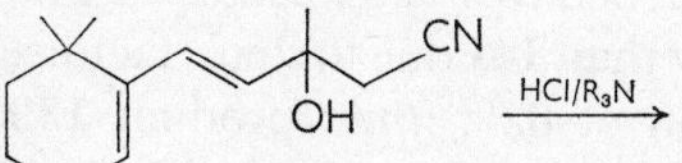

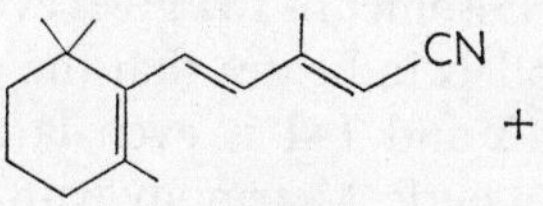

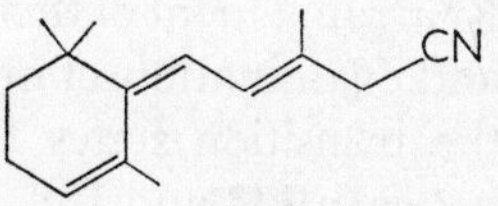

Examples of allylic rearrangement-dehydration in the carotenoid

IV. CLAISEN AND COPE REARRANGEMENTS

Facilitation of accurate measurements of kinetic parameters and product analysis by chromatographic techniques, is establishing the detailed reaction features for Claisen and Cope rearrangements on an even firmer footing. Frey and his colleagues find the isomerization of methylallyl vinyl ether follows as expected, much the same course as allyl vinyl ether and allyl isopropenyl ether, pre-exponential factors A varying from $10^{11.5}$ to $^{11.73}$ with E_a 29,100–30,600 cal/mole. The substantial negative entropy values found reflect the more rigid transition state with respect to reactant, with reduced internal rotational freedom in the proposed six-membered assembly[129]. Similar experiments with allyl 3-allyloxy-2-butenoate and crotyl 3-crotyloxy-2-butenoate show roughly comparable kinetic parameters with reduced activation energies, compared with the simpler allyl vinyl ethers (ΔE_a 4–5 kcal/mole) and heterogeneous catalysis by ammonium chloride[130]. In intramolecular reactions of allylic compounds embracing Claisen rearrangements, the stereo- and optical properties of the product are determined by the conformational features of the transition state[131]; thus if conformation **137** of 1,3-dimethylallyl phenyl ether pertains to the transition state for its rearrangement, geometric relationships should be preserved in the product, but the optical properties will be reversed (*R-trans* → *S-trans*); but for the product from conformation **140** geometrical relationships are reversed and optical properties retained. Application of this principle to Claisen rearrangements of phenyl ethers in general, shows that this result holds with only one exception; repetition of the faulty result shows that it was based on faulty preparation of the optically active reactant[132], and the authentic material behaves as expected. The transition state **138** is actually more likely than **141** due to steric factors, and a reflexion of this is the formation of 82% *trans* product **139** (Z = *o*-HOC$_6$H$_4$—) and only 18% of the analogous *cis* product. Rearrangement of the *cis* isomer of **137** gives even less of the *cis* product, and examination of models indicates that the difference in stability of the transition states **143** and **141** is even larger than the difference between **141** and **138**, in perfect harmony with the result. These ideas have been further elaborated by evaluation of transition-state energy differences[133] from relative reaction rates; as expected the lowest

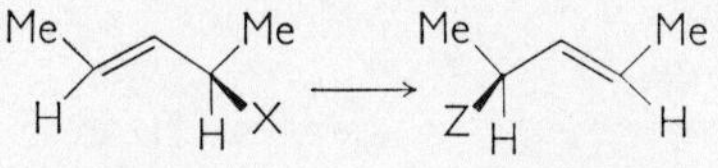

3. Alkene Rearrangements

[Structures 137, 139, 140, 142, 138, 141, 143 shown]

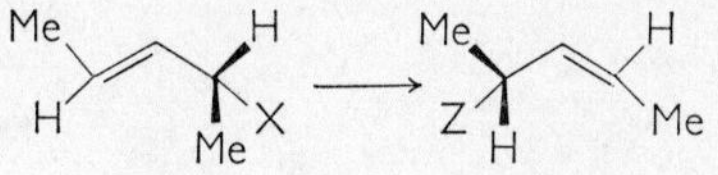

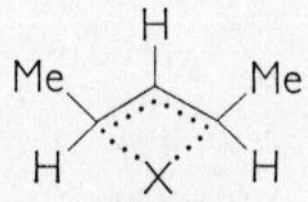

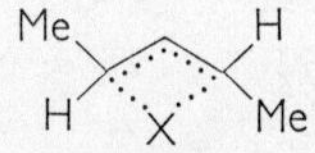

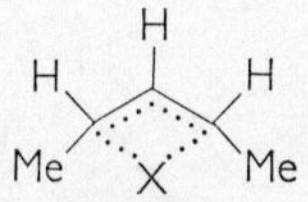

energy transition state is for the *trans* → *trans* rearrangement, that leading to *cis* product being 1·9 kcal/mole higher—of the correct order of magnitude with respect to the explanation. For the *cis* → *trans* and *cis* → *cis* rearrangements the energy differences rise to > 3·6 kcal. If allowance is made for the difference in ground-state free energy of the reactants (ΔF *cis*- and *trans*-4-methyl-2-pentene ~ 0·6 kcal/mole) the transition state for the *cis* → *trans* rearrangement lies about 1 kcal above the *trans* → *trans* activated complex. Hence an energy profile for the two sets of reactions can be constructed.

Consideration of cyclic transition states for these reactions, however, should take account of the conformational effects present, i.e. whether the transition state is chair or boat like, and to what extent substituents interact. Possible transition states for the four-centre rearrangement of **137** are represented in **144–151**.

The simple idea that orbital repulsion may operate in favour of chair-like transition states in Claisen and Cope rearrangements (*vide infra*) seems to be supported by theoretical arguments[134]; for chair structures, equatorial methyls are preferred to axial, and for boat forms, *endo* methyls are less favourable than *exo*. Carrying such arguments further, it seems a reasonable assumption that a pair of mutually repelling substituents in the *exo-cis* conformation will be less stable than in a *trans* relationship. Experimental facts seem a little at variance however with the detailed analysis of stability relationships in these hypothetical transition states; for example one might expect **148** to be *less* stable than **146** due to interaction of the axial methyl group with the aromatic ring, but since this conflicts with the experimental data it is suggested that a non-symmetrical transition state is involved, in which the forming C—C bond is

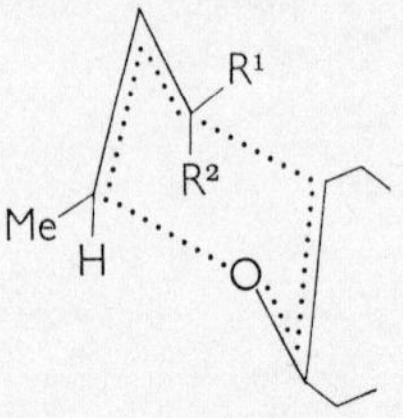

(**144**, R¹ = Me R² = H)
(**148**, R¹ = H R² = Me)

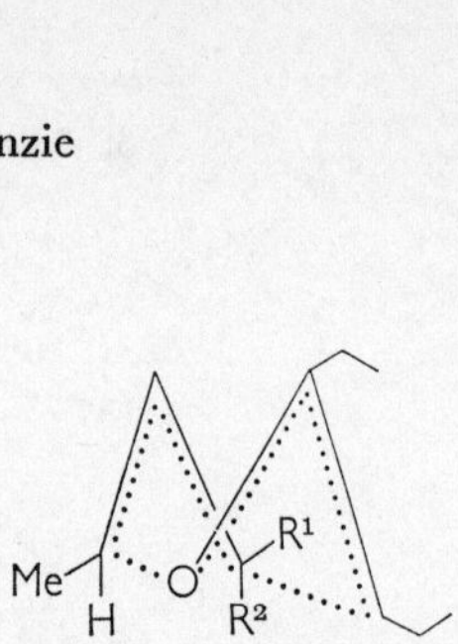

(**145**, R¹ = Me R² = H)
(**149**, R¹ = H R² = Me)

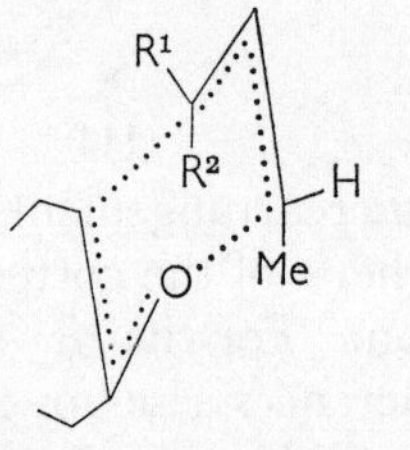

(**146**, R¹ = Me R² = H)
(**150**, R¹ = H R² = Me)

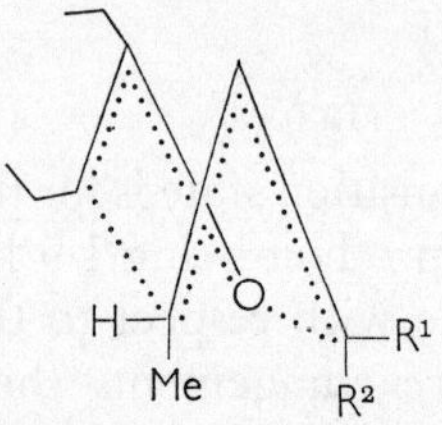

(**147**, R¹ = Me R² = H)
(**151**, R¹ = H R² = Me)

'stretched' compared to the C—O bond, mitigating any steric effect between the substituent attached to the reaction site and the aromatic ring. This accords with the lack of any effect of *o*-substituents in the ring, and is not in conflict with orbital symmetry requirements for a concerted reaction.

Similar stereo-optical effects are seen with cyclopent-2-enyl vinyl ether (**152**), where optical activity is inverted and absolute configuration retained in its Claisen rearrangement to **154**[135]. That this rearrangement resembles the one of enzymatic conversion of chroisinic

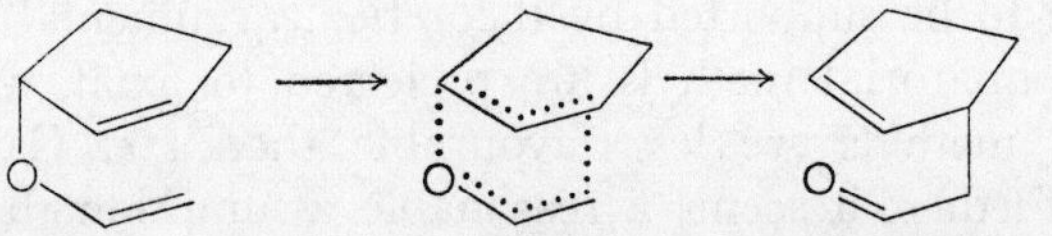

(**152**) [R—(+)]   (**153**)   (**154**) [R—(−)]

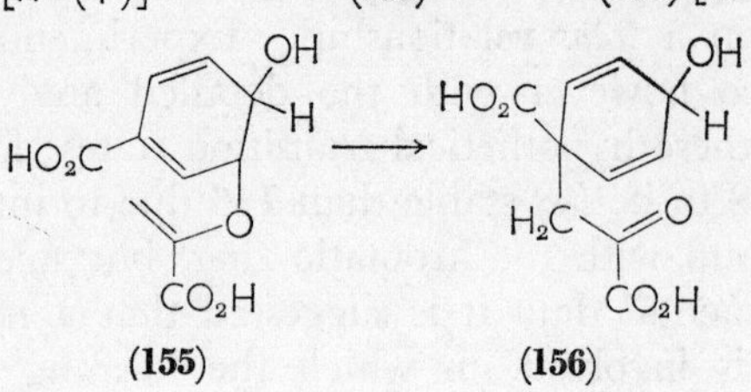

(**155**)   (**156**)

acid to prephenic acid (**155** → **156**) is demonstrative of the geometric relationship of the olefin bonds involved[136a]. An especially facile Claisen rearrangement occurs with 3-(1,4-pentadienyl) vinyl ether below room temperature[136b].

A number of preparatively interesting rearrangements of vinylic esters have recently been described; **157** → **158** and **159** → **160** are novel reactions illustrative of an older principle[137,138,139]. Other related reactions are observed with pyruvyl enol ester (**161**) which

$$\underset{(157)}{\underset{|}{\text{RCO—O}}\atop\underset{|}{\text{CH}_2\!=\!\text{COEt}}} \longrightarrow \left[\underset{}{\underset{|}{\text{RCO}}\atop\underset{}{\text{CH}_2\!-\!\text{C}}}\diagdown_{\text{OEt}}^{\text{O}}\right] \xrightarrow{157} \underset{(158)}{\underset{|}{\text{RC}\!=\!\text{CHCO}_2\text{Et}}\atop\underset{|}{\text{OCOR}}}$$

$$\underset{(159)}{\underset{|}{\text{RCO—O}}\atop\underset{|}{\text{CH}_2\!=\!\text{CMe}}} \longrightarrow \underset{(160)}{\underset{|}{\text{RCO}}\atop\underset{}{\text{CH}_2\!-\!\text{C}}}\diagdown_{\text{Me}}^{\text{O}}$$

gives **162** by decarbonylation under very mild conditions (80°), whilst thermolysis of **163** leads via **164** to **165**[137]. The mechanisms of these reactions have not been investigated in any detail, but extension of the principle to β- and γ-keto acids e.g. acetoacetic acid possibly proceeds by way of a ketene intermediate as in the sequence **166** → **167**[140].

$$\underset{(161)}{\text{RCO—C}\diagdown_{\text{O}}^{\text{O}}\atop\text{CH}_2\!=\!\text{COEt}} \xrightarrow[80°]{-\text{CO}} \underset{(162)}{\underset{|}{\text{RCO}}\atop\text{CH}_2\text{CO}_2\text{Et}}$$

Examples of Claisen transformations with propargyl vinyl ethers have also been described in the recent literature; in general since the achievement of the required transition state is somewhat more difficult, higher temperatures are required. Thus **168** prepared by carboxylation of vinyl propargyl Grignard reagent is rearranged on heating to **170**, presumably via the allene **169** (**170** actually exists in equilibrium with the hydrate and is a powerful antibiotic)[141]. A number of instances where the allenic product is isolable are known[142], e.g. **171** gives **172** and **173**. A similar type of reaction with the propargylic acetal **174** proceeds by way of elimination of methylbutynol and Claisen rearrangement of the ether product **175**; the analogous reaction with 25% optically pure S-(−)-butynol as the source of **174**,

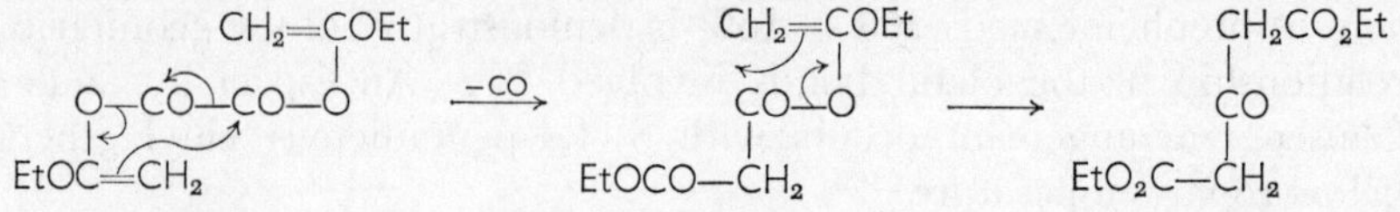

(163), (164), (165), 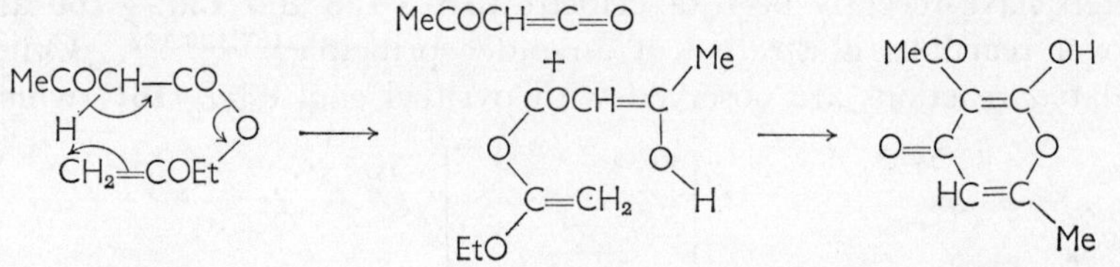 (166), (167)

gives optically active R-($-$) allene **176**[143]. Increasing substitution in these propargyl vinyl ethers allows lower temperatures to be used for rearrangement, suggestive of at least partial radical development in the transition state for the reaction, e.g. **177** → **178**. Propargyl vinyl malonic esters undergo the Cope rearrangement (*vide infra*)[144]. An interesting variant of the Claisen transformation is the sequence **179** → **180** → **181** which uses commercially available dimethyl acetal of dimethylacetamide[145]. Phosphite esters of allyl alcohols also undergo Claisen-type reaction; thus **182** gives **183**, and **185** is formed from **184** together with some **183**, the proportion of the latter rising with

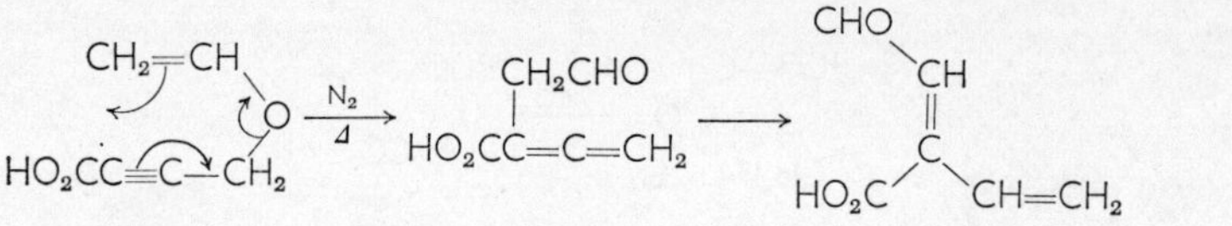

(168), (169), (170)

HC≡CCHMeOCH=CH$_2$ $\xrightarrow{200°}$ MeCH=C=CHCH$_2$CHO + MeCH=CHCH=CHCHO
(**171**)   (**172**) (~60%)   (**173**), (~40%)
                                                                                                     (*cis*: *trans* 2:1)

Me$_2$CH(OCHMeC≡CH)$_2$ ⟶
(**174**)

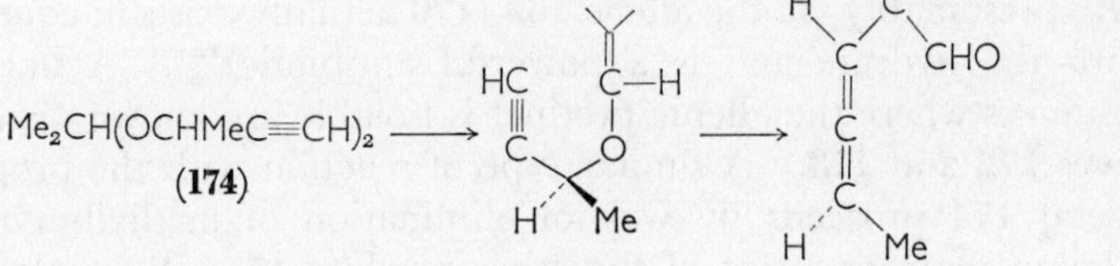

(175), (176)

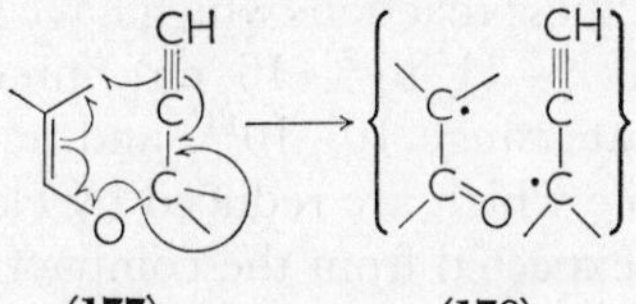

(177) (178)

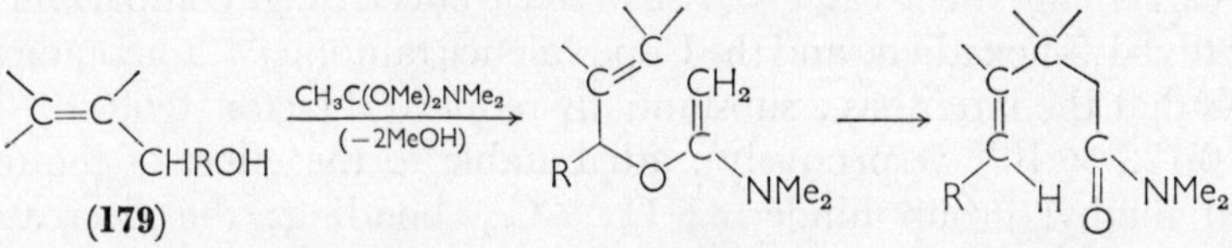

(179) →[CH₃C(OMe)₂NMe₂, (−2MeOH)] (180) → (181)

(EtO)₂P—O—CHMeCH=CH₂ ⟶ (EtO)₂P(=O)CH₂CH=CHMe
(182)   (183)

(EtO)₂P—O—CH₂CH=CHMe ⟶ (EtO)₂P(=O)CHMeCH=CH₂
(184)   (185)

temperature perhaps due to catalysis by decomposition products[146,147].

'Oxy-Cope' rearrangements, e.g. 1,5-hexadien-3-ols to carbonyl products, through enolic intermediates have recently become known in cyclic systems[148]. The parent system **186** has now been prepared and it smoothly rearranges at 300° to the alkenal **187**, giving only traces of dehydration product[149]. Such a result speaks for the favourability of the concerted process involving a six-membered transition state in a four-centre reaction. Whilst the appearance of some acrolein in the product might suggest the intermediacy of free-radical species in the rearrangement, the acrolein could also arise in a separate reaction; further work in hand should clarify this point.

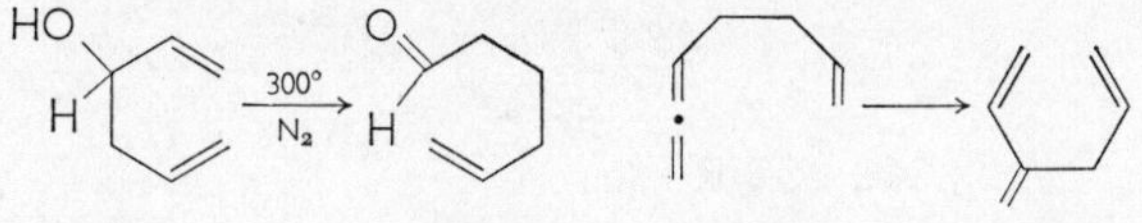

(186) →[300°, N₂] (187)   (188) ⟶ (189)

Kinetic parameters have been measured in a number of cases of gas-phase Cope rearrangements of alka-1,5-dienes; comparison of the results with those observed by Cope and his colleagues[150] for reactions in the condensed phase with allylalkenylmalononitriles, indicates the general similarity of all these reactions which have substantial negative entropies of activation (-11 to -16 eu), preexponential factors covering the approximate range 10^9–10^{11}, and activation energies of from 25 to 32 kcal/mole which are reduced by electron-withdrawing groups. As might be expected from the compact rigid nature of the transition state in these rearrangements, substituents at the termini of the rearranging diene can give rise to steric effects; e.g. comparison of 3-methyl-1,5-hexadiene and the Cope valence tautomer 1,5-heptadiene shows that the latter has a substantially reduced A factor ($0 \cdot 69 \times 10^{10}$ and $0 \cdot 123 \times 10^{10}$ respectively), attributable to the effect of the terminal methyl group hindering $C_{(1)}$–$C_{(6)}$ bonding; the decreased motional freedom in the transition state is reflected in the change in entropy of activation ($-14 \cdot 1$ and -16 eu)[151]. Kinetic parameters for the simplest case of Cope rearrangement, i.e. of 1,1-dideuterio-hexa-1,5-diene, also shows the trend to higher values for E_a in the absence of 'activating' substituents ($E_a 35 \cdot 5$ kcal/mole), and the greater flexibility in the transition state ($\Delta S^{\ddagger} - 9$ eu)[152]. Kinetic results for the homogeneous first-order rearrangement of the allene hepta-1,2,6-triene (**188**) to 3-methylenehexa-1,5-diene (**189**), strongly suggest that this is another type of Cope reaction; the reduced energy of activation (E_a 28·5 kcal/mole) compared to comparatively simple systems, may reflect greater electron availability in the allene (higher polarizability), whilst consideration of possible internal rotations in the reactant indicates that if only the allenyl and vinyl groups can rotate freely—a possibility on account of the internal hindrance to rotation about the $C_{(4)}$–$C_{(5)}$ bond—the observed entropy decrease in the activated complex for the reaction can be accounted for almost completely by the removal of these rotations[153]. However for Cope rearrangement of propargylisobutenylmalonic ester (**190**) the activation energy for a cyclic transition state must be very little different from that for scission of the molecule, since fragmentation products are also isolated. As in

$$\text{(190)} \xrightarrow{250°} \text{(191)}$$

the case of acetylenic Claisen rearrangement the reaction pathway may involve a radical pair[144].

An important recent paper by Doering and Roth[154a] examines the theoretical and experimental possibilities for rearrangement of *meso-*

and *racemic*-3,4-dimethylhexa-1,5-diene; for a boat-like six-membered transition state with six overlapping atoms the *meso* diene will give *trans,trans* and/or *cis,cis*-octa-2,6-diene, but the *racemic* isomer (**192**) will give only the *cis,trans* diene; for a chair-like six-membered transition state involving a four-centre reaction, the result will be precisely the reverse, the *meso* isomer (**193**) giving the *cis,trans* diene. Indeed *cis, trans*-octa-2,6-diene *is* obtained from the *meso* starting material and 90% *trans,trans*-octa-2,6-diene from the racemic compound (with 9% *cis,cis* isomer); in these Cope rearrangements where a free choice of transition-state geometry is possible the four-centre chair-like arrangement is preferred by more than 300:1 ($\Delta\Delta F^\ddagger$ 5·7 kcal/mole at 225°). It is pointed out[154b] that if the transition state is considered to be composed of two allyl radicals, then whilst the single electron in the highest occupied orbital can interact with the other identically placed electron in the other half of the transition state only at the termini of the radicals (since the orbital has a node passing through the central atom in each radical), the mutual repulsion of the paired electrons in the next orbital will be greater for the six-atom overlap situation in the boat-like transition state, than for the four-atom interaction in the chair-like transition complex.

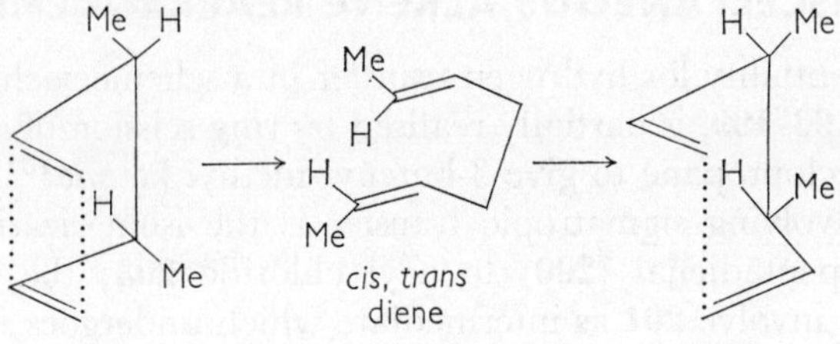

(**192**)
(racemic)

cis, trans diene

(**193**)
(meso)

A theoretical treatment of the Cope rearrangement of 1,5-hexadiene, which takes account of bond angle bending strain, torsional effects, steric repulsion, π-delocalization, compression of σ-bonds in the conjugated part of the molecule and the making and breaking of σ-bonds for various degrees of advancement along the reaction coordinate, shows that the chair-like six-membered transition state is preferred to the boat-like form by about 3 kcal/mole; in view of the approximations inherent in such calculations the rough agreement with the experimental value found in the 3,4-dimethyl-hexadiene system is remarkable[154c].

Interestingly Cope-type rearrangements have been invoked to explain the stereoisomerism of **194** derived by oxidation of the dimer of cyanodithioformic acid, which is converted uniquely to the *trans,trans* isomer (**195**) on warming in carbon tetrachloride. No evidence for the formation of any *cis,trans* isomer is apparent and this isomer, independently prepared, cannot be isomerized to the *cis,cis* or *trans,trans* compounds, in harmony with the Cope-type mechanism, since for the *cis,trans* isomer the rearrangement becomes degenerate [155].

The carbon analogue of the Claisen rearrangement has been realised; with *t*-butoxide anion in *t*-butanol the isomeric 1-phenylbutenes are equilibrated with the isomeric 1-(*o*-tolyl)propenes [156].

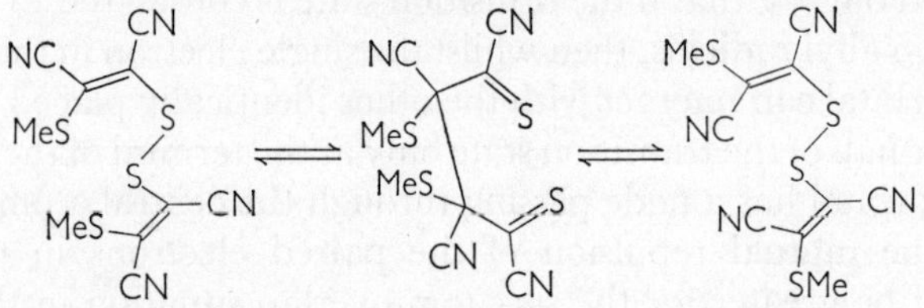

V. MISCELLANEOUS ALKENE REARRANGEMENTS

The potentiality for hydrogen transfer in a scheme such as the one involving **196–199**, is partially realised by ring scission of *cis*-1-acetyl-2-methylcyclopropane to give 3-butenyl methyl ketone [157]. Another reaction involving sigmatropic transfer is the isomerization of perchloro-2,4-pentadienal (**200**) to acyl chloride **203**; the reaction is believed to involve **201** as intermediate which undergoes sigmatropic 1,5 chlorine transfer; ^{14}C labelled carbonyl compound gives acyl chloride with the label at $C_{(5)}$, not in the carbonyl group [158].

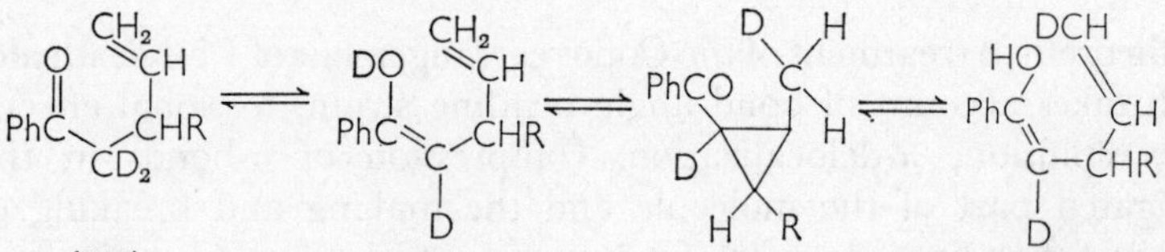

For Wittig rearrangements of 9-lithiofluorenyl methallyl ethers, e.g. **204** (R^2 = Me, R^3 = H) → **205** (R^3 = Me, R^2 = H), radical pair intermediates such as **206** are preferred to ion pairs, since exclusively *trans* olefinic bonds seem to be favoured in the products, whereas in an ion-pair mechanism involving allylic carbanions, the more stable *cis* carbanion would be expected to lead to *cis* olefin preponderantly[159]. It is however possible that rotation of a carbanion in an ion-pair mechanism is too slow to effect the stereochemistry of the product. However, in this connexion crotylmagnesium bromide in ether at room temperature exhibits stereoisomerism, about equal amounts of both isomers being formed. In aprotic solvents such as hexamethylphosphoramide no metal–carbon bond is present and the carbanion exists exclusively in the *cis* configuration[160]. Stereochemistry is however retained in the reactions of *cis*- and *trans*-propenyllithium with germanium tetrachloride, giving tetrapropenyl compounds[161], and largely retained with vinylic Grignard reagents prepared from *cis*- and *trans*-1-bromo-1-propenes. Interestingly, for the latter stereoisomer, retention of configuration is reduced compared to the *cis* compound but this might be expected if the *cis* carbanion is the more stable. Carbonation of these Grignard reagents gives good yields of the respective stereoisomeric carboxylic acids[162]. The preparation of vinylic lithium derivatives has recently been discussed by Seyferth and his collaborators[163]; they are most conveniently made from tetraalkenylmetalloids, e.g. tetrapropenyltin with lithium, and can be estimated by gas–liquid chromatographic analysis of the derived trimethylsilyl compounds formed from trimethylsilyl chloride and the lithio compound. In this connexion tetra-*trans*-propenyltin when partially cleaved by excess lithium shows some isomerization of unchanged organo-tin compound, whilst *cis*- and *trans*-propenyltrimethyltin are isomerized by lithium in ether without cleavage. Propenyltrimethylgermanium (**208**) and the silicon analogue (**207**) are not however isomerized under similar conditions but are with lithium in tetrahydrofuran to **210**. No isomerization occurs with the all-carbon analogues. These isomerizations can best be thought of as involving anion radicals **209** formed by electron donation into the antibonding π-orbital with concomitant stereomutation. The metalloid anion

(**200**) (**201**) (**202**) (**203**)

radicals are probably stabilized by $d\pi$—$p\pi$ interaction compared to the carbon analogues.

The decreasing ratio of *trans* to *cis* isomers in these isomerizations on

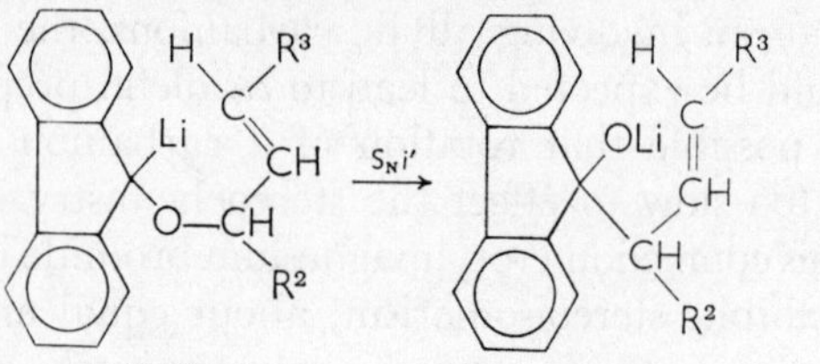

(204)   (205)

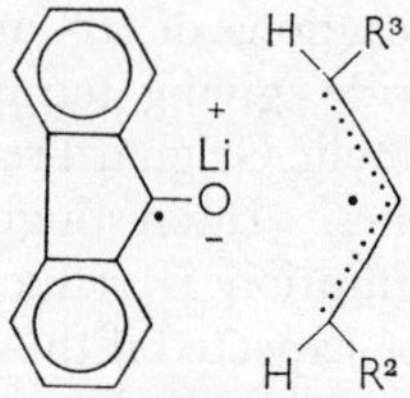

(206)

passing down the series Si, Ge, Sn, is nicely explained by the increasing atomic radius which allows the three methyl groups greater freedom from steric interaction with the *cis* substituent at $C_{(2)}$[163].

Vinylic silver compounds are preparable stereospecifically from *cis*-1-propenyllithium, e.g. *cis*-1-propenyl(tri-*n*-butylphosphine)-silver (**211**); decomposition of this silver compound with iodine gives 96% *cis*-1-iodopropene in high yield, while thermal decomposition in solution yields stereospecifically *cis,cis*-2,4-hexadiene. Similar reactions occur with the corresponding cuprous compounds[164].

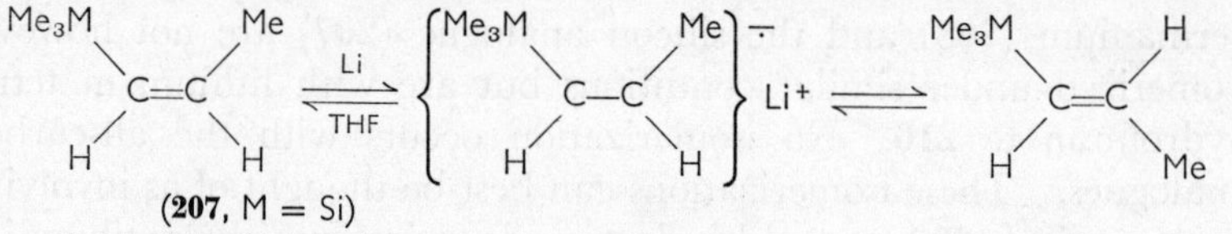

(207, M = Si)
(208, M = Ge)   (209)   (210)

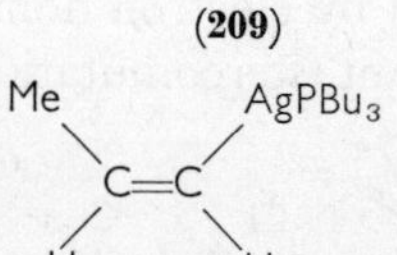

(211)

The proliferation of work with organometallic reagents of the transition metals in connexion with alkene rearrangements, commends itself to the attention of more conventional organic chemists, and justifies a brief account here; a full treatment would clearly require a full review[165]. A number of workers have observed positional isomerism of olefins with derivatives of the Group VII transition elements. In these reactions, whether thermal or photochemical, the isomer ratios found are closely similar to the thermal equilibrium values and parallel those computed on the basis of free-energy differences calculated from heats of formation. In general for n-alkenes and iron carbonyls internal olefins isomerize much more slowly than terminal isomers and *cis* olefins are more reactive than *trans*, but the more hindered the double bond the slower the reaction. The activity of the various iron carbonyls depends to some extent on the olefin used, since reactivity is temperature dependent and isomerizations commonly involve the boiling of mixtures. The use of autogenous pressure in an autoclave is of little value for raising the reaction temperature, since the reactions are inhibited by the overpressure of carbon monoxide. However, additives are often useful and ethanol or acetone have been used; acids and acid chlorides inhibit isomerization, but pyridine does allow use of pentacarbonyl iron with the lower boiling olefins. In some cases catalytic activity is exhibited under very mild conditions if the reaction mixture is irradiated at intermediate wavelengths[166]. Isomer ratios for propenyl ethers, made from allyl ethers by irradiation in the presence of pentacarbonyl iron, also resemble the thermodynamically stable mixtures[167].

The mechanism of double-bond shifts catalysed by transition-metal compounds is the subject of a continuing debate (*vide infra*). However it does seem clear that in some instances complex-metal hydrides are involved transiently or otherwise, and the thermodynamic stability of the product does seem important. Experiments have been conducted using hexane solutions of hydrogen tetracarbonyl cobaltate ($HCo(CO)_4$) which isomerizes olefins under remarkably mild conditions with concomitant hydroformylation. The suggestion that a 1,2 addition, followed by a reverse elimination involving a second olefin as catalyst acceptor, accounts for these isomerizations does not accord with experimental facts, since under conditions where the concentration of alkyl cobalt compound should be optimum no olefin exchange occurs[168], although there is some evidence that an olefin addition compound with the cobaltate may be formed[165]. There is some evidence that allylic exchange takes place; rearrangement of allyl alcohol with deuterium tetracarbonylcobaltate places deuterium exclusively in the methyl group of the propionaldehyde formed[169], consistent with a mechanism involving a transition state such as **212**. For catalysis by iron carbonyls it has been suggested that adventitiously formed iron hydrides are the active species[166] and reactions such as

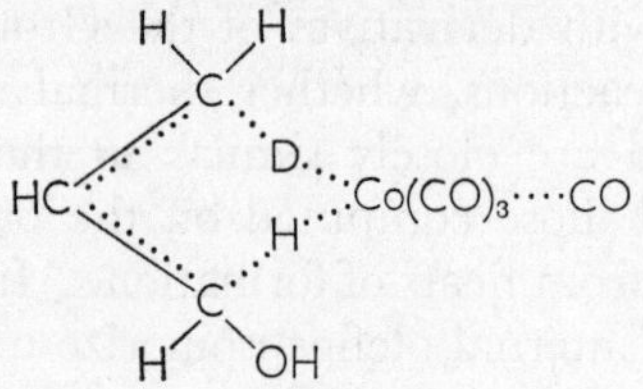

(212)

213 → 214 → 215, or π-allyl iron complexes e.g. **216 → 217 → 218** (where R may be olefinic) may account for the observed results.

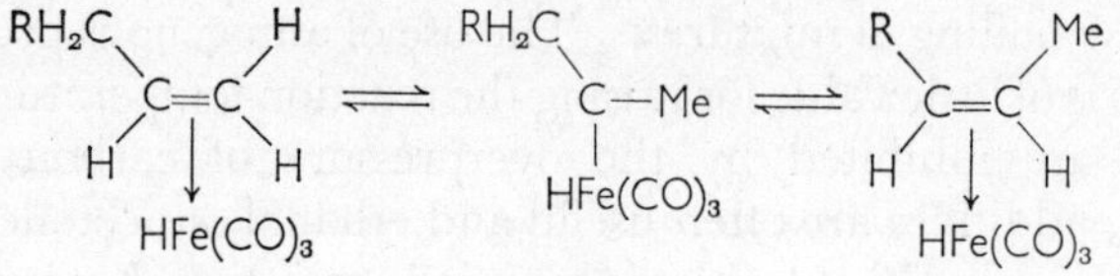

(213) (214) (215)

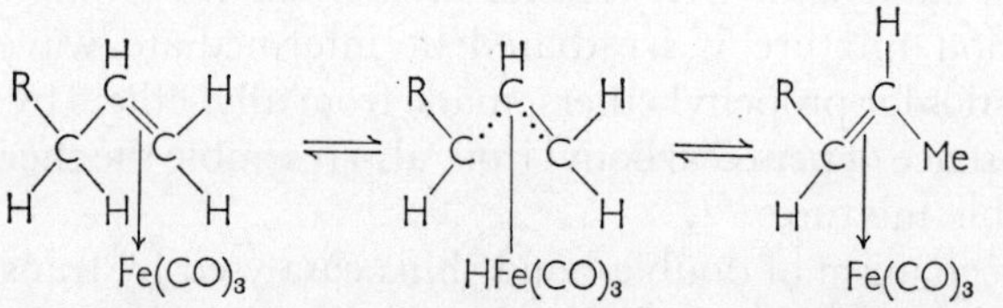

(216) (217) (218)

A number of Group VIII transition–metal catalyst-cocatalyst systems are known; dichlorobis(ethylene)-μ, μ'-dichloroplatinum(II) is active in the presence of alcohols which apparently function as hydride donors, but acids and bases destroy catalytic activity. These catalysts isomerize 1-hexene with a noticeable initial preponderance of *cis*-2 isomer, but eventually the thermodynamic equilibrium composition is achieved. With rhodium trichloride trihydrate there is a *persistent* high preponderance of the *cis* isomeric 3-hexene in the initial rearrangement products. These reactions it is thought[170] involve sequences very similar to those in the transformations **213 → 215**, although this mechanism has been criticized[171] since deuterium is not incorporated into the isomers produced from 1-octene with similar catalysts. It is however possible that this is due to a large isotope effect; in support of this, rearrangement of 1,2-dideuteriohexene occurs somewhat faster than 1-hexene. The explanation is that since the essential step in the metal hydride addition-elimination sequence is addition at $C_{(2)}$, whilst reversion to 1-olefin requires elimination of either MH or MD from the labelled compound compared to elimination of MH for isomerization, the faster rate can be understood on the basis of the unfavourable rate for MD elimination[172]. More recently it has emerged that deuterium tetracarbonylcobaltate isomerizes allylbenzene with only insignificant incorporation of the isotope into isomerized or unchanged starting material, and at the same rate as for the protio cobalt compound; clearly a mechanism involving allylic exchange or addition-elimination in this case is quite incompatible with the results, and it appears that the C—H bond at the active centre and the D—Co bond remain effectively intact, i.e. the transfer of hydrogen appears to be *intramolecular*[173]. Intramolecularity has also been demonstrated in the thermal rearrangement of a mixture of tritiated n-1-octene and unlabelled n-undecene with enneacarbonyldiiron (and in the photochemical rearrangement with pentacarbonyliron); none of the tritium label is transferred between the octene and the undecene isomers. A similar result is observed for n-hexene and allylically tritritiated n-octene with hydrogen tetracarbonyl-cobaltate. It is suggested that the first step is the formation of a π-complex with the tetracarbonylcobalt hydride followed by intramolecular rearrangement[174]. It seems possible that these intramolecular reactions involve metal d-olefin π orbital mixing, leading to symmetry which allows facile suprafacial 1,3-sigmatropic hydrogen transfer; this is perhaps an area rich in possibilities for theoretical investigation.

VI. REFERENCES

1a. *Selected Values of Physical and Thermodynamic Properties of Hydrocarbons and Related Compounds*, American Petroleum Institute, Carnegie Press, Pittsburgh, Pa., 1953.
1b. D. M. Golden, K. W. Egger and S. W. Benson, *J. Am. Chem. Soc.*, **86,** 5416 (1964). *Cf.* A. Macoll and R. A. Ross, *J. Am. Chem. Soc.*, **87,** 1169 (1965).
2. A. R. Olsen, *J. Chem. Phys.*, **1,** 418 (1933).
3. S. W. Benson, K. W. Egger and D. M. Golden, *J. Am. Chem. Soc.*, **87,** 468 (1965); *cf.* K. W. Egger and S. W. Benson, *J. Am. Chem. Soc.*, **87,** 3311, 3314 (1965); J. A. Kerr, R. Spencer and A. F. Trotman-Dickenson, *J. Chem. Soc.*, 6652 (1965).
4. W. J. Muizebelt and R. J. F. Nivard, *Chem. Comm.*, 148, (1965).
5. K. W. Egger, *J. Am. Chem. Soc.*, **89,** 504 (1967); S. W. Benson, K. W. Egger and D. M. Golden, *J. Am. Chem. Soc.*, **87,** 468 (1965).
6. K. W. Egger and S. W. Benson, *J. Am. Chem. Soc.*, **87,** 3314 (1965); *J. Am. Chem. Soc.*, **88,** 236,241 (1966).
7. I. D. Hawton and G. P. Semeluk, *Can. J. Chem.*, **44,** 2143 (1966). *Cf.* C. Steel, *J. Phys. Chem.*, **64,** 1588 (1960).
8. E. W. Schlag and E. W. Kaiser, *J. Am. Chem. Soc.*, **87,** 1171 (1965).

9. H. G. Viehe and E. Franchimont, *Chem. Ber.*, **96**, 3153 (1963).
10. H. G. Viehe and E. Franchimont, *Chem. Ber.*, **97**, 598, 602 (1964); H. G. Viehe, J. Dale and E. Franchimont, *Chem. Ber.*, **97**, 244 (1964).
11. D. Sianesi and R. Fontanelli, *Ann. Chim. (Rome)*, **55**, 850 (1965).
12. O. Bastiansen and L. Smedvik, *Acta Chem. Scand.*, **8**, 1593 (1954).
13. P. S. Skell and R. G. Allen, *J. Am. Chem. Soc.*, **86**, 1559 (1964).
14. D. M. Graham, R. L. Mieville and C. Sivetrz, *Can. J. Chem.*, **42**, 2239 (1964).
15. K. Sugimoto, W. Ando and S. Oae, *Bull. Chem. Soc. Japan*, **38**, 224 (1965).
16. R. B. Cundall, F. J. Fletcher and D. G. Milne, *Trans. Faraday Soc.*, **60**, 1146 (1964). *Cf.* M. Tanaka, T. Terao and S. Sato, *Bull. Chem. Soc. Japan*, **38**, 1645 (1965); W. Ando, K. Sugimoto and S. Oae, *Bull. Chem. Soc. Japan*, **38**, 226 (1965).
17. R. B. Cundall and P. A. Griffiths, *Chem. Comm.*, 194 (1966). *Cf.* R. F. Borkman and D. R. Kearns, *J. Am. Chem. Soc.*, **88**, 3467 (1966).
18. H. Morrison, *J. Am. Chem. Soc.*, **87**, 932 (1965); *Tetrahedron Letters*, 3653, (1964).
19. G. S. Hammond, J. Saltiel, A. A. Lamola, N. J. Turro, J. S. Bradshaw, D. O. Cowan, R. C. Counsell, V. Vogt and C. Dalton, *J. Am. Chem. Soc.*, **86**, 3197 (1964).
20. W. G. Herkstroeter and G. S. Hammond, *J. Am. Chem. Soc.*, **88**, 4769 (1966).
21. J. Saltiel, E. D. Megarity and K. G. Kneipp, *J. Am. Chem. Soc.*, **88**, 2336 (1966).
22. W. G. Herkstroeter, L. B. Jones and G. S. Hammond, *J. Am. Chem. Soc.*, **88**, 4777 (1966).
23. R. B. Cundall and P. A. Griffiths, *Trans. Faraday Soc.*, **61**, 1968 (1965). *Cf.* M. A. Golub, C. L. Stephens and J. L. Brash, *J. Chem. Phys.*, **45**, 1503 (1966); M. A. Golub and C. L. Stephens. *J. Phys. Chem.*, **70**, 3576 (1966).
24. J. Nosworthy, *Trans. Faraday Soc.*, **61**, 1138 (1965).
25. R. R. Hentz, K. Shima and M. Burton, *J. Phys. Chem.*, **71**, 461 (1967).
26. H. P. Lehmann, G. Stein and E. Fischer, *Chem. Comm.*, 583 (1965).
27. T. Terao, S. Hirokami, S. Sato and R. J. Cvetanovic, *Can. J. Chem.*, **44**, 2173 (1966).
28. D. F. Evans, *J. Chem. Soc.*, 1351 (1957); 1735 (1960); 1987, 2566 (1961).
29. Z. R. Grabowski and A. Bylina, *Trans. Faraday Soc.*, **60**, 1131 (1964).
30. W. G. Brown and S. Jankowski, *J. Am. Chem. Soc.*, **88**, 233 (1966).
31. G. W. Griffen, E. J. O'Connell and J. M. Kelliher, *Proc. Chem. Soc.*, 337 (1964).
32. C. Moussebois and J. Dale, *J. Chem. Soc. (C)*, 260 (1966).
33. (a) Z. Rappoport, C. Degani and S. Patai, *J. Chem. Soc.*, 4513 (1963).
    (b) Z. Rappoport and S. Gertler, *J. Chem. Soc.*, 1360 (1964).
34. S. Patai, Z. Gruenbaum and Z. Rappoport, *J. Chem. Soc. (B)*, 1133 (1966).
35. D. H. Hunter and D. J. Cram, *J. Am. Chem. Soc.*, **88**, 5765 (1966). *Cf.* ref. 83.
36. H. H. Freedman, V. R. Sandel and B. P. Thill, *J. Am. Chem. Soc.*, **89**, 1762 (1967).
37. D. S. Noyce, D. R. Harrter and F. B. Miles, *J. Am. Chem. Soc.*, **86**, 3583 (1964).

38. A. Gandini and P. H. Plesch, *Proc. Chem. Soc.*, 113 (1964).
39. M. Campanelli, U. Mazzucato and A. Foffani, *Ann. Chim.* (*Rome*), **54,** 195 (1964).
40. W. Ziegenbein, *Chem. Ber.*, **98,** 1427 (1965).
41. M. Kroner, *Chem. Ber.*, **100,** 3172 (1967).
42. S. W. Spangler and G. F. Woods, *J. Org. Chem.*, **30,** 2218 (1965).
43. F. Bohlmann and H.-J. Mannhardt, *Chem. Ber.*, **88,** 1330 (1955); *cf.* S. Trippett in *Advances in Organic Chemistry: Methods and Results*, **1,** 63 (1960).
44. G. Wittig, *Angew. Chem.*, **68,** 506 (1956); *Experientia*, **12,** 41 (1954).
45. P. S. Manchand and B. C. L. Weedon, *Tetrahedron Letters*, 2603 (1964); M. S. Barber, L. M. Jackman, P. S. Manchand and B. C. L. Weedon, *J. Chem. Soc.* (*C*), 2166 (1966).
46. D. F. Schneider and C. F. Garbers, *J. Chem. Soc.*, 2465 (1964).
47. G. Pattenden, B. C. L. Weedon, C. F. Garbers, D. F. Schneider and J. P. van der Merwe, *Chem. Comm.*, 347 (1965).
48. H. Freyschlag, H. Grassner, A. Nurrenbach, H. Pommer, W. Rief and W. Sarnecki, *Angew. Chem. Intern. Ed.*, **4,** 287 (1965).
49. L. D. Bergelson and M. M. Shemyakin, *Tetrahedron*, **19**, 149 (1963). *Cf.* L. D. Bergelson and M. M. Shemyakin, *Pure and Appl. Chem.*, **9**, 271 (1964). For recent discussions see C. F. Garbers, D. F. Schneider and J. P. van der Merwe, *J. Chem. Soc.* (*C*), 1982 (1968); G. Pattenden and B. C. L. Weedon, *J. Chem. Soc.* (*C*), 1984, 1987 (1968).
50. E. J. Corey and G. T. Kwiatkowski, *J. Am. Chem. Soc.*, **88,** 5652, 5653 and 5654 (1966).
51. U. Schweiter, G. Saucy, M. Montavon, C. von Planta, R. Rüegg and O. Isler, *Helv. Chim. Acta*, **45,** 517 (1962). Cf. M. Julia and J. Bouchaudon, *Compt. Rend.*, **253,** 111 (1961).
52. U. Schweieter, C. von Planta, L. Chopard-dit-Jean, R. Rüegg, M. Kofler and O. Isler, *Helv. Chim. Acta*, **45,** 548 (1962).
53. (a) S. L. Jensen, *Pure and Appl. Chem.*, **14,** 227 (1967); (b) B. C. L. Weedon, *Pure and Appl. Chem.*, **14,** 265 (1967); (c) L. Zechmeister, *Cis-Trans Isomeric Carotenoids, Vitamin A and Aryl Polyenes*, Springer, Verlag Wien, 1962.
54. J. M. Osbond, *Progress in the Chemistry of Fats and Lipids*, **9,** 119 (1966).
55. Yu. B. Pyatnova, V. V. Fedulova, I. K. Sarycheva and N. A. Preobrzhenskii, *Zh. Obshch. Khim.*, **34,** 3317 (1964).
56. S. S. Nigam and B. C. L. Weedon, *J. Chem. Soc.*, 3868 (1957); R. Ya. Lavina, Yu. S. Shabarov and V. V. Ershov, *Zh. Obshch. Khim.*, **23,** 1124 (1953).
57. H. J. J. Pabon, D. van der Steen and D. A. van Dorp, *Rec. Trav. Chim.*, **84,** 1319 (1965).
58. D. van der Steen, H. J. J. Pabon and D. A. van Dorp, *Rec. Trav. Chim.*, **82,** 1015 (1963).
59. D. E. Ames, A. N. Covell and T. G. Goodburn, *J. Chem. Soc.*, 894 (1965).
60. F. D. Gunstone and P. J. Sykes, *J. Chem. Soc.*, 3055 (1962).
61. L. Crombie and J. C. Williams, *J. Chem. Soc.*, 2449 (1962).
62. D. A. Thomas and W. K. Warburton, *J. Chem. Soc.*, 2988 (1965).
63. M. C. Whiting, *J. Chem. Soc.*, 2541 (1963).
64. W. Schroth, *Tetrahedron Letters*, 195 (1965).

65. W. Schroth, *Z. Chem.*, **5,** 353 (1965); W. Schroth, H. Langguth and F. Billig, *Z. Chem.*, **5,** 352 (1965).
66. R. B. Woodward and R. Hoffmann, *J. Am. Chem. Soc.*, **87,** 2511 (1965). S. Winstein and R. S. Boikess and D. S. Glass, *Tetrahedron Letters*, 999 (1966).
67. J. Wolinsky, D. Chollar and M. D. Baird, *J. Am. Chem. Soc.*, **84,** 2775 (1964). D. S. Glass, J. Zirner and S. Winstein, *Proc. Chem. Soc.*, 276, (1963). S. Winstein, *Proc. Chem. Soc.*, 235 (1964); *Angew. Chem.*, **76,** 378 (1964). W. R. Roth and J. König, *Ann. Chem.*, **688,** 28 (1965).
68. H. M. Frey and R. J. Ellis, *J. Chem. Soc.*, 4770 (1965). Cf. H. M. Frey and R. J. Ellis, *J. Chem. Soc., Suppl.*, **1,** 5578 (1964).
69. E. C. Friedrich and D. C. Glass, unpublished work quoted in reference 66; D. C. Glass, Thesis, U.C.L.A., 1965.
70. W. R. Roth and J. König, *Ann. Chem.*, **699,** 24 (1966).
71. H. M. Frey and B. M. Pope. *J. Chem. Soc. (A).*, 1701 (1966).
72. K. J. Crowley, *Tetrahedron*, **21,** 1001 (1965).
73. J. R. Majer, J. F. T. Pinkard and J. C. Robb, *Trans. Faraday Soc.*, **60,** 1247 (1964).
74. D. W. Placzec and B. S. Rabinovitch, *Can. J. Chem.*, **43,** 820 (1965).
75. M. J. Jorgenson, *Chem. Comm.*, 137 (1965).
76a. R. C. Cookson, V. N. Gogte, J. Hudec and N. A. Mirza, *Tetrahedron Letters*, 3955 (1965).
76b. R. C. Cookson, *Chemistry in Britain*, **5,** 6 (1969); *Quart. Rev. (London)*, **22,** 423 (1968).
77. L. B. Jones and V. K. Jones, *J. Am. Chem. Soc.*, **89,** 1880 (1967).
78. D. M. Golden, K. W. Egger and S. W. Benson, *J. Am. Chem. Soc.*, **86,** 5416, 5420 (1964).
79. R. J. Ellis and H. M. Frey, *J. Chem. Soc.*, 595 (1964).
80. K. W. Egger and S. W. Benson, *J. Am. Chem. Soc.*, **88,** 241 (1966). Cf. M. D. Carr, V. V. Kane and M. C. Whiting, *Proc. Chem. Soc.*, 408 (1964). K. W. Egger and S. W. Benson, *J. Am. Chem. Soc.*, **87,** 3314 (1965).
81. D. J. Cram and R. T. Uyeda, *J. Am. Chem. Soc.*, **86,** 5466 (1964).
82. C. K. Ingold, *Structure and Mechanism in Organic Chemistry*, Cornell University Press, New York, 1953, p. 565. (G. Bell & Sons, London.) See also A. G. Catchpole, E. D. Hughes and C. K. Ingold, *J. Chem. Soc.*, 11 (1948).
83. D. H. Hunter and D. J. Cram, *J. Am. Chem. Soc.*, **86,** 5478 (1964).
84. S. W. Ela and D. J. Cram, *J. Am. Chem. Soc.*, **88,** 5791 (1966).
85. S. W. Ela and D. J. Cram, *J. Am. Chem. Soc.*, **88,** 5577 (1966). Note that on p. 5783 chart II some of the arrows appear to be incorrectly oriented.
86. S. Bank, A. Schriesheim and C. A. Rowe, *J. Am. Chem. Soc.*, **87,** 3244 (1965).
87. D. E. O'Conner and C. D. Broaddus, *J. Am. Chem. Soc.*, **86,** 2267 (1964).
88. J. Sauer and H. Prahl, *Tetrahedron Letters*, 2863 (1966).
89. A. Schriesheim and C. A. Rowe, *Tetrahedron Letters*, **10,** 405.
90. A. Schriesheim, C. A. Rowe and L. Nashund, *J. Am. Chem. Soc.*, **85,** 2111 (1963).
91 (a). R. B. Bates, D. W. Gosselink and J. A. Kaczynski, *Tetrahedron Letters*, 199, 205 (1967). Cf. R. B. Bates, R. H. Carnighan and C. E. Staples, *J.*

Am. Chem. Soc., **85,** 3031 (1963). (b) R. Waack and M. A. Doran, *J. Organometal. Chem.*, **3,** 92 (1965). See D. Seyferth, *Record of Chem. Progr.*, **26,** 87 (1965).
92. J. Ugelstad and O. A. Rokstad, *Acta Chem. Scand.*, **18,** 474 (1964).
93. C. D. Broaddus, *J. Org. Chem.*, **30,** 4131 (1965).
94. I. Elphimoff-Felkin and J. Huet, *Tetrahedron Letters*, 1933 (1966).
95. G. Cignarella, C. R. Pasqualuuci, G. G. Gallo and E. Testa, *Tetrahedron*, **20,** 1057 (1964).
96. A. Maccoll and R. A. Ross, *J. Am. Chem. Soc.*, **87,** 4997 (1965).
97. M. F. Ansell, B. E. Grimwood and T. M. Kafka, *J. Chem. Soc. (C)*, 1802 (1967). Cf. M. F. Ansell, *Quart. Rev. (London)*, **18,** 211 (1964).
98. W. Smadja, *Ann. Chim. (Paris)*, **10,** 105 (1965).
99. W. Smadja, *Compt. Rend.*, **258,** 5461 (1964). K. L. Mikolajczak, M. O. Bagby and I. A. Wolff, *J. Am. Oil Chemists' Soc.*, **42,** 243 (1965). Cf. J. C. Craig and R. J. Young, *J. Chem. Soc. (C)*, 578 (1966).
100. K. L. Mikolajczak, M. O. Bagby, R. B. Bates and I. A. Wolff, *J. Org. Chem.*, **30,** 2983 (1965).
101. J. D. Cawley, C. D. Robeson and W. J. Humphlett, *U.S. Patent No.* 2,676,989; *Chem. Abstr.*, **50,** 7125h (1956).
102. D. M. Burness and C. D. Robeson,*U.S. Patent No.* 2,676,988; *Chem.Abstr.*, **50,** 408d (1956).
103. M. Mousseron-Canet, M. Mousseron and P. Legendre, *Bull. Soc. Chim. France*, 1509 (1961); *Compt. Rend.*, **257,** 3782 (1963); *Bull. Soc. Chim. France*, 50 (1964).
104. P. de Mayo, J. B. Stothers and R. W. Yipp, *Can. J. Chem.*, **39,** 2135 (1961). M. Mousseron-Canet, M. Mousseron, P. Legendre and J. Wylde, *Bull. Soc. Chim. France*, 379 (1963).
105. K. J. Crowley, *Proc. Chem. Soc.*, 17 (1964).
106. M. Mousseron, *Advan. Photochem.*, **4,** 195 (1966).
107. A. Ozaki and K. Imura, *J. Catalysis*, **3,** 395 (1964).
108. H. R. Gerberich, J. G. Larson and W. K. Hall, *J. Catalysis*, **4,** 523 (1965).
109. A. Ozaki and S. Tsuchiya, *J. Catalysis*, **5,** 537 (1966).
110. W. G. Young, S. H. Sharman and S. Winstein, *J. Am. Chem. Soc.*, **82,** 1376 (1960).
111. H. P. Leftin and M. C. Hobson, *Advan. Catalysis*, **14,** 188 (1963). Cf. B. J. Joice and J. J. Rooney, *J. Catalysis*, **3,** 565 (1964).
112. Y. Chauvin and G. Lefebvre, *Compt. Rend.*, **259,** 2105 (1964).
113. G. Valkanas and E. S. Waight, *J. Chem. Soc.*, 531 (1964).
114. D. C. Ditmer and A. F. Marcantonio, *J. Org. Chem.*, **29,** 3473 (1964).
115. A. G. Agneux, S. Winstein and W. G. Young, *J. Am. Chem. Soc.*, **82,** 5956 (1960).
116. C. A. V. Werf and V. L. Heasley, *J. Org. Chem.*, **31,** 3534 (1966).
117. L. Miginiac-Groizeleau, P. Miginiac and C. Prévost, *Compt. Rend.*, **260,** 1442 (1965).
118. M. Andrac, *Ann. Chim. (Paris)*, **9,** 287 (1964).
119. J. J. Eisch and G. R. Husk, *J. Organometal. Chem.*, **4,** 415 (1965).
120. D. Z. Denny, A. Appelbaum and D. B. Denny, *J. Am. Chem. Soc.*, **84,** 4969 (1962). *Cf.* J. K. Kochi, *J. Am. Chem. Soc.*, **84,** 774 (1962).
121. E. R. H. Jones and J. T. McOmbie, *J. Chem. Soc.*, 261 (1943).

122. E. T. Clafferton and W. S. MacGregor, *J. Am. Chem. Soc.*, **72,** 2501 (1950).
123. R. C. Cambie and T. D. R. Manning, *Chem. Ind. (London)*, 1884 (1966).
124. U. Schwieter, C. von Planta, R. Rüegg and O. Isler, *Helv. Chim. Acta*, **45,** 541 (1962).
125. H. Freyschlag, H. Grassner, A. Nurrenbach, H. Pommer, W. Rief and N. Sarnecki, *Angew. Chem. Intern. Ed.*, **4,** 287 (1965).
126. W. Oroshnik, G. Karmas and A. D. Mebane, *J. Am. Chem. Soc.*, **74,** 3807 (1952).
127. K. Eiter, E. Truscheit and H. Oediger, *Angew. Chem.*, **72,** 948 (1960).
128. O. Isler and P. Schudel, *Advances in Organic Chemistry, Methods and Results*, **4,** 115 (1963).
129. H. M. Frey and B. M. Pope, *J. Chem. Soc. (B)*, 209 (1966).
130. J. W. Ralls, R. E. Lundin and G. F. Bailey, *J. Org. Chem.*, **28,** 3521 (1963).
131. H. L. Goering and W. I. Kimoto, *J. Am. Chem. Soc.*, **87,** 1748 (1965).
132. E. R. Alexander and R. W. Kluiber, *J. Am. Chem. Soc.*, **73,** 4304 (1951); H. Hart, *J. Am. Chem. Soc.*, **76,** 4033 (1954).
133. E. N. Marvell, J. L. Stephenson and J. Ong, *J. Am. Chem. Soc.*, **87,** 1267 (1965).
134. R. Hoffmann and R. B. Woodward, *J. Am. Chem. Soc.*, **87,** 4389 (1965). K. Fukui and H. Fujimoto, *Tetrahedron Letters*, 251 (1966).
135. R. K. Hill and A. G. Edwards, *Tetrahedron Letters*, 3239 (1964).
136a. F. Gibson and L. M. Jackman, *Nature*, **198,** 388 (1963).
136b. S. F. Reed, *J. Org. Chem.*, **30,** 1663 (1965).
137. G. R. Banks, D. Cohen and H. D. Springall, *Rec. Trav. Chim.*, **83,** 513 (1964).
138. E. S. Rothman, *J. Org. Chem.*, **31,** 628 (1966).
139. R. C. Cambie and T. D. R. Manning, *Chem. Ind. (London)*, 1918 (1964).
140. D. Cohen and G. E. Pattenden, *J. Chem. Soc. (C)*, 2314 (1967).
141. D. K. Black and S. R. Landor, *Proc. Chem. Soc.*, 183 (1963).
142. P. Cresson, *Compt. Rend.*, 261, 1707 (1965).
143. E. R. H. Jones, J. D. Loder and M. C. Whiting, *Proc. Chem. Soc.*, 181 (1960).
144. D. K. Black and S. R. Landor, *J. Chem. Soc.*, 6784 (1965).
145. A. E. Wick, D. Felix, K. Steen and A. Eschenmoser, *Helv. Chim. Acta*, **47,** 2425 (1964).
146. A. N. Pudovik and I. M. Aladzhyeva, *Dokl. Akad. Nauk SSSR*, **151,** 1110 (1963); *Zh. Obshch. Khim.*, **33,** 3096 (1963).
147. A. L. Lemper and H. Tieckelmann, *Tetrahedron Letters*, 3053 (1964).
148. A. Berson and M. Jones, *J. Am. Chem. Soc.*, **86,** 5019 (1964).
149. A. Viola and L. A. Levasseur, *J. Am. Chem. Soc.*, **87,** 1150 (1965).
150. E. G. Foster, A. C. Cope and F. Daniels, *J. Am. Chem. Soc.*, **69,** 1893 (1947).
151. A. Amano and M. Uchiyama, *J. Phys. Chem.*, **69,** 1278 (1965).
152. W. von E. Doering and V. Toscano cited in W. von E. Doering and J. C. Gilbert, *Tetrahedron Supplement*, **7,** 396 (1966).
153. H. M. Frey and D. H. Lister, *J. Chem. Soc. (A)*, 26 (1967). See also L. Skattebol and S. Solomon, *J. Am. Chem. Soc.*, **87,** 4506 (1965).
154a. W. von E. Doering and W. R. Roth, *Angew. Chem. Intern. Ed.*, **2,** 115 (1963).
154b. R. S. Berry quoted in ref. 154a.

154c. M. Simonetta and G. Favini, *Tetrahedron Letters*, 4837 (1966).
155. H. E. Simmons, D. C. Blomstrom and R. D. Vest, *J. Am. Chem. Soc.*, **84,** 4756 (1962).
156. E. von E. Doering and R. A. Bragole, *Tetrahedron*, **22,** 385 (1966).
157. R. M. Roberts, R. N. Greene, R. G. Landolt and E. W. Heyer, *J. Am. Chem. Soc.*, **87,** 2282 (1965).
158. A. Roedig, J. Marks, F. Frank, R. Kohlhaupt and M. Schlosser, *Chem. Ber.*, **100,** 2730 (1967).
159. U. Schollkopf and K. Fellenberger, *Ann. Chem.*, **698,** 80 (1966). *Cf.* P. T. Lansbury, V. A. Pattison, J. D. Siler and J. B. Bieber, *J. Am. Chem. Soc.*, **88,** 78 (1966).
160. C. Agami, M. Andrac-Taussig and C. Prévost, *Compt. Rend.*, **262, C,** 852 (1966).
161. A. N. Nesmeyanov, A. E. Borisov and N. V. Novikova, *Dokl. Akad Nauk. SSSR*, **165,** 333 (1965).
162. G. J. Martin and M. L. Martin, *Bull. Soc. Chim. France*, 1636 (1966).
163. D. Seyferth, R. Suzuki and L. G. Vaughan, *J. Am. Chem. Soc.*, **88,** 286 (1966). *Cf.* G. Köbrich and W. E. Breckoff, *Angew. Chem. Intern. Ed.*, **6,** 45 (1967).
164. G. M. Whitesides and C. P. Casey, *J. Am. Chem. Soc.*, **88,** 4541 (1966).
165. See for example M. Orchin, *Advanc. Catalysis*, **16,** 1 (1966); N. R. Davies, *Rev. Pure and Appl. Chem.*, **17,** 83 (1967), which specifically deal with olefin isomerization.
166. T. A. Manuel, *J. Org. Chem.*, **27,** 3941 (1962). F. Asinger, B. Fell and K. Schrage, *Chem. Ber.*, **98,** 372 (1965).
167. P. W. Jolly, F. G. A. Stone and K. Mackenzie (in part), *J. Chem. Soc.*, 6416 (1965).
168. G. L. Karapinka and M. Orchin, *J. Org. Chem.*, **26,** 4187 (1961).
169. R. W. Goetz and M. Orchin, *J. Am. Chem. Soc.*, **85,** 1549 (1963).
170. J. F. Harrod and A. J. Chalk, *J. Am. Chem. Soc.*, **86,** 1776 (1964); R. Cramer and R. V. Lindsey, *J. Am. Chem. Soc.*, **88,** 3534 (1966).
171. N. R. Davies, *Austral. J. Chem.*, **17,** 212 (1964).
172. J. F. Harrod and A. J. Chalk, *Nature*, **205,** 280 (1965).
173. L. Roos and M. Orchin, *J. Am. Chem. Soc.*, **87,** 5502 (1965).
174. B. Fell, P. Krings and F. Asinger, *Chem. Ber.*, **99,** 3688 (1966).

CHAPTER 4

Hydrogenation of Alkenes

SEKIO MITSUI

Tohoku University, Sendai, Japan

and

AKIRA KASAHARA

Yamagata University, Yamagata, Japan

I. HETEROGENEOUS CATALYTIC HYDROGENATION	. . .	175
A. Mechanism and Stereochemistry	.	175
B. Selective Hydrogenation	.	192
II. HOMOGENEOUS CATALYTIC HYDROGENATION . . .	.	196
A. Ruthenium Chloride	.	196
B. Pentacyanocobaltate	.	197
C. Tris(triphenylphosphine)halogenorhodium	. .	199
D. Irridium Complex	.	200
E. Tris(triphenylphosphine)dichlororuthenium	. .	201
F. Platinum Complexes	.	201
G. Metal–Carbonyl Complexes	.	202
H. Ziegler-Type Catalysts	.	202
III. CHEMICAL REDUCTION	.	203
A. Dissolving Metals	.	203
B. Metal Hydrides	.	206
C. Diimide	.	208
IV. REFERENCES	.	210

I. HETEROGENEOUS CATALYTIC HYDROGENATION

A. Mechanism and Stereochemistry

Two types of adsorption states have been proposed for the substrate in heterogeneous hydrogenation of olefins: Dissociative chemisorption (**1**)[1], including a step in which hydrogen is transferred from carbon to catalyst, and associative chemisorption (**2**)[2] involving a step in which the olefin is adsorbed on the catalyst by opening of the olefin double bond to form σ bonds with the catalyst metal, taking sp^3 hybridization[3].

It was recently professed that olefins may form π-complexes (3) by associating with a simple atomic centre of the catalyst[4,5]. These structures are assumed to be analogous to the π-olefin complexes of the transition elements[6]. It was also professed that π-allylic complexes (4) may form the catalyst surface[7,8].

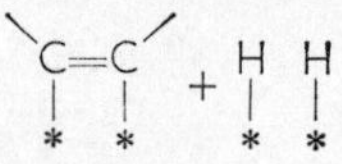

$$\begin{array}{cc} \overset{\diagdown}{\underset{\underset{*\ *}{|\ |}}{C}}{=}\overset{\diagup}{C} + \underset{\underset{*\ *}{|\ |}}{H\ H} & \underset{\underset{*\ *}{|\ |}}{H{-}C{-}C{-}H} \\ (1) & (2) \end{array}$$

$$\begin{array}{cc} \overset{\diagdown}{\underset{*}{C{=}\!\!\!{=}C}}\overset{\diagup}{} & \overset{\diagdown}{\underset{*}{C{\cdots}C{\cdots}C}}\overset{\diagup}{} \\ (3) & (4) \end{array}$$

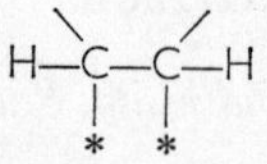

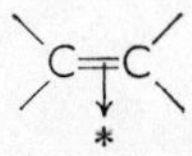

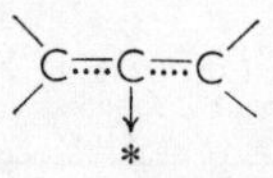

In one of the earliest works on the reaction of deuterium with olefins on nickel catalyst, Farkas, Farkas and Rideal[1] found that deuterium not only added to the double bond of ethylene, but that it also exchanged its hydrogen. Farkas and Farkas[9] proposed a mechanism in which the basic step is the interaction of hydrogen, dissociated on the catalyst surface, with physically adsorbed ethylene. The mechanisms in which chemisorbed ethylene reacts with physically adsorbed hydrogen is called Twigg-Rideal[10] or Rideal-Eley mechanism[11]. These two mechanisms are based on Rideal mechanism:

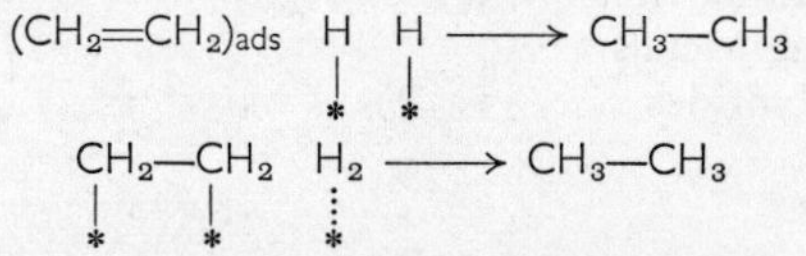

$$(CH_2{=}CH_2)_{ads} \quad \underset{\underset{*\ *}{|\ |}}{H\ H} \longrightarrow CH_3{-}CH_3$$

$$\underset{\underset{*\ *}{|\ |}}{CH_2{-}CH_2} \quad \underset{*}{H_2} \longrightarrow CH_3{-}CH_3$$

On the other hand, the Horiuti–Polanyi mechanism[2] was postulated in terms of the Langmuir–Hinshelwood mechanism (interaction of the chemisorbed species on the adjacent surface sites), and it involves associatively adsorbed ethylene and an intermediate half-hydrogenated state:

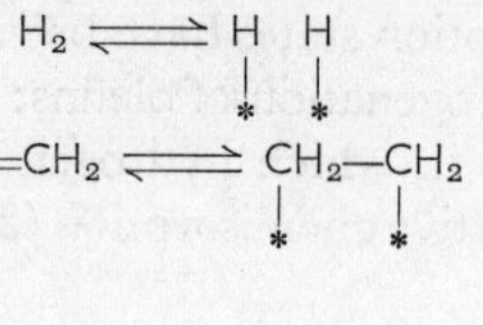

$$H_2 \rightleftharpoons \underset{\underset{*\ *}{|\ |}}{H\ H} \qquad (1)$$

$$CH_2{=}CH_2 \rightleftharpoons \underset{\underset{*\ *}{|\ |}}{CH_2{-}CH_2} \qquad (2)$$

4. Hydrogenation of Alkenes

$$\begin{array}{cc} \underset{*}{CH_2}-\underset{*}{CH_2} \quad \underset{*}{H} & \rightleftharpoons CH_2-CH_3 \quad (3) \\ & \underset{*}{|} \\ \underset{*}{CH_2}-CH_3 \quad \underset{*}{H} & \rightleftharpoons CH_3-CH_3 \quad (4) \end{array}$$

$$\begin{array}{c} R \quad R \\ | \quad | \\ -C=C-H \\ | \\ H \end{array} \xrightarrow{\underset{*}{D}} \begin{array}{c} R \quad R \\ | \quad | \\ -C-C-C-H \\ | \quad | \\ H \quad * \quad D \end{array} \Bigg\{ \begin{array}{l} \xrightarrow{D} \begin{array}{c} R \quad R \\ | \quad | \\ -C-C-C-H \\ | \quad | \\ H \quad D \quad D \end{array} \quad (5) \\[2ex] \xrightarrow{-H} \begin{array}{c} R \quad R \\ | \quad | \\ -C=C-C-H \\ | \\ D \end{array} \quad (6) \\[2ex] \xrightarrow{-D} \begin{array}{c} R \quad H \\ | \quad | \\ -C-C=C-R \\ | \\ H \end{array} \quad (7) \\[2ex] \xrightarrow{-H} \begin{array}{c} R \quad R \\ | \quad | \\ -C-C=C-D \\ | \\ H \end{array} \quad (8) \end{array}$$

The hydrogenation mechanisms proposed above have been reviewed[12]. Other mechanisms have been postulated by Beeck[13], Jenkin and Rideal[14] and Twigg[15].

The mechanism has been widely accepted because it is possible to interpret addition (5), double bond migration (6), *cis–trans* isomerization (7), and hydrogen exchange (8) taking place on the surface of the catalyst through only one intermediate—the half-hydrogenated state.

The above mentioned mechanisms were proposed based on the kinetics of the process. The results of the stereochemical investigations made it possible to postulate more detailed reaction mechanisms.

On assuming that catalytic hydrogenation occurs by the simultaneous addition of two atoms from the same hydrogen molecule, Farkas and Farkas[16] pointed out that the addition to a double bond or triple bond would be *cis*[17]. Greenhalgh and Polanyi[18], however, pointed out that *cis* addition to ethylene derivatives was not a proof of the simultaneous addition of a pair of hydrogen atoms.

On the other hand, it is known that the catalytic hydrogenation of organic compounds proceeds by *cis* addition from the less hindered side. This fact shows that the molecule is preferably adsorbed on the less hindered side, and the simultaneous addition of a pair of hydrogen atoms[19]. For example, the hydrogenation of α-pinene (5) over platinum oxide in acetic acid or ethanol at room temperature, proceeds to give *cis*-pinane (**6**) almost quantitatively[20].

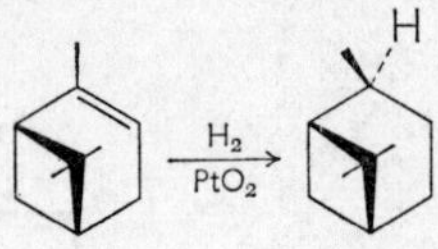

The addition of deuterium[21] or tritium[22] to androst-1-ene-3,7-dione (**7**) over palladium is selectively *cis* from the less hindered α-side.

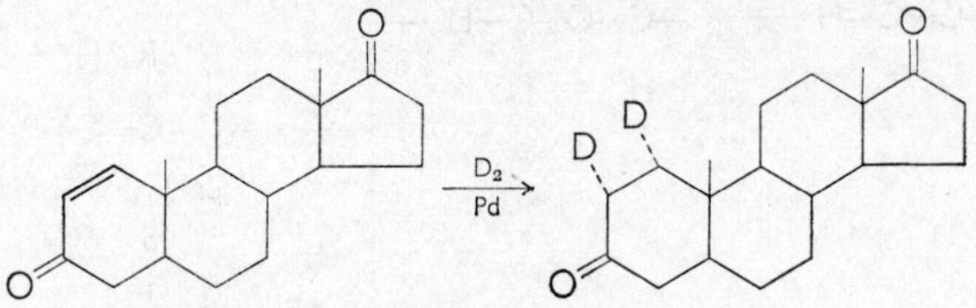

When the Δ^7 (**8**) and Δ^8 (**9**) steroids isomerize to the $\Delta^{8(14)}$-steroid (**10**) in the presence of hydrogen and platinum or palladium, the addition and elimination of hydrogen occur at the α-face of the double bond[23], showing that the isomerization proceeds on the surface of the catalyst.

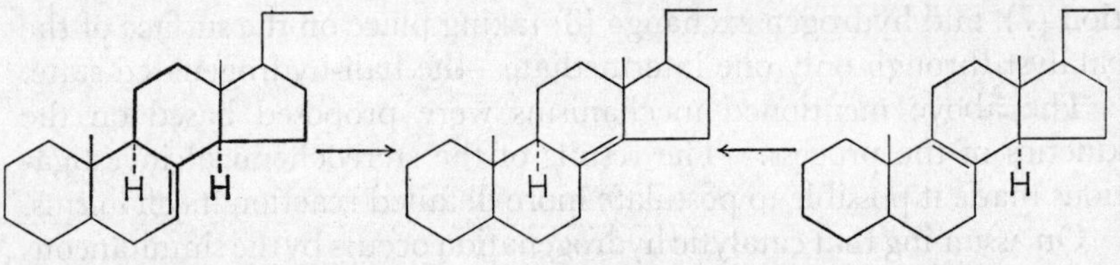

The hydrogenation of *cis*- and *trans*-cyclodecene was performed on platinum oxide by Smith[24]. The 'top-side' of the adsorbed *cis* isomer **11** is less hindered than that of the adsorbed isomer **12** which is blocked by a polymethylene alkyl chain. If the hydrogen attacks the double bond from the opposite side of the catalyst on which cyclodecene is adsorbed, the rate of the hydrogenation of *cis* isomer should be far greater than that of the *trans* isomer. But the experimental data show that *cis* is only slightly faster than *trans*. This result was attributed to hydrogen attacking from the 'bottom side' of the chemisorbed molecules.

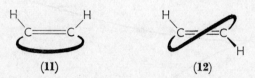

If hydrogenation takes place by *cis* addition, *meso* compounds should be obtained from tetrasubstituted *cis* olefins and racemic mixtures of the optically active compounds from *trans* olefins. Similarly, *cis* disubstituted cyclohexanes would be obtained from 1,2-disubstituted cyclohexenes. Nevertheless, both *trans* and *cis* compounds are obtained as shown in Table 1. With phenyl or carbomethoxy substi-

TABLE 1. Hydrogenation of 1,2-disubstituted cycloalkenes

Substrate	Catalyst	*Cis* Addition Products (%)	Ref.
1,2-Dimethylcyclopentene	PtO_2	42	25
1,2-Dimethylcyclohexene	PtO_2	82	26
1,2-Dimethylcyclohexene	Pd-C	25	27
1,2-Dicarbomethoxycyclohexene	PtO_2	100	28
1,2-Diphenylcyclohexene	PtO_2	95	29
1,2-Diphenylcyclohexene	Pd-C	100	29
1,2-Diphenylcyclohexene	Ni(Raney)	100	29
$\Delta^{9,(10)}$-Octalin	PtO_2	51	30
$\Delta^{9,(10)}$-Octalin	Pd-C	10	31

tuents in the cycloalkene ring, *cis* compounds are almost exclusively obtained. However, certain amounts of *trans* isomer were produced from other compounds, especially when using palladium catalyst.

Siegel[32] found that in the reduction of 1,2-dimethylcyclohexene and 1,2-dimethylcyclopentene over platinum oxide in acetic acid at 25°C[25,26], the proportion of *cis*-1,2-dimethylcycloalkane produced increases on increasing the hydrogen pressure. On the other hand, in the case of the corresponding 2,3-dimethylcycloalkenes the opposite effect was observed, as shown in Figure 1[26].

Siegel considered that the Horiuti–Polanyi mechanism accommodated these results. Assuming that the concentration of hydrogen on the catalyst surface is a function of the hydrogen pressure, the relative rate of the reactions (2–4) should vary with the pressure. When the

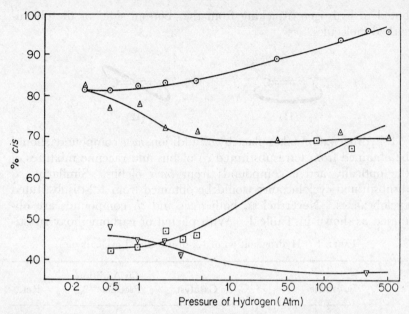

Figure 1. Variation with the pressure of hydrogen of the proportion of *cis*- and *trans*-dimethylcycloalkanes obtained from 1,2-dimethylcyclohexene (○), 2,3-dimethylcyclohexene (△), 1,2-dimethylcyclopentene (□), and 2,3-dimethylcyclopentene (▽), reduced PtO_2 in glacial acetic acid at 25°C. [Reprinted from S. Siegel, *Advances in Catalysis*, Vol 16, Academic Press, New York, 1966, p. 133, by permission of the publisher.]

concentration of hydrogen becomes higher, the rate of reaction (2) may not vary significantly, whereas those of (3) and (4) may increase—especially the rate of reaction (4), which is a function of the concentration of both half-hydrogenated intermediates and hydrogen, increases significantly. Consequently, the rate of the double bond migration (reaction 6) decreases and the population of *cis* product from 1,2-dimethylcycloalkene may increase. At lower hydrogen pressures, 1,2-dimethylcycloalkenes must be converted to the 2,3-dimethylcycloalkene or 2-methyl-methylenecycloalkane from which both *cis* and *trans* saturated products are formed via *cis* addition. This is illustrated for the dimethylcyclopentenes in Scheme I. Siegel and his co-workers[25,26] detached the formation of 2,3-dimethylcycloalkenes from the 1,2-isomers by gas chromatography and showed that these isomers are reduced more easily than their precursors. They also found that *trans*-cycloalkenes were derived mainly from 2,3-isomers[33]. However the initial rate of hydrogenation of 2,3-dimethylcycloalkene

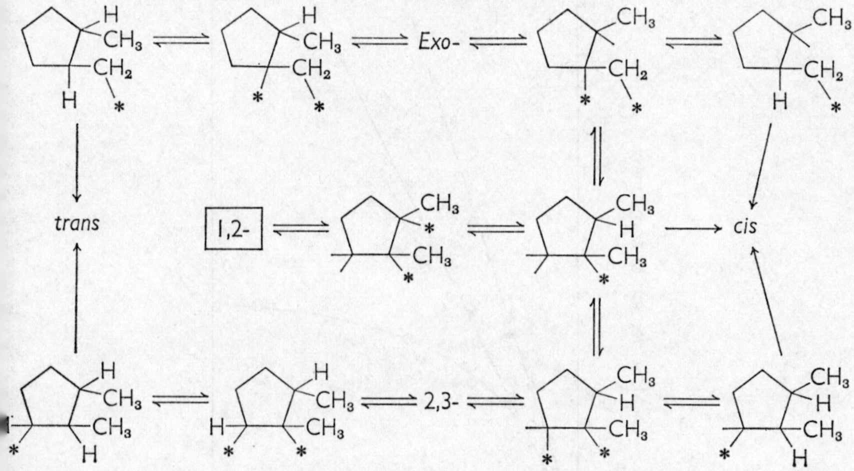

SCHEME I. Reduction and isomerization of 1,2-dimethylcyclopentene according to the Horiuti–Polanyi mechanism. [Reprinted from S. Siegel, *Advances in Catalysis*, Vol. 16, Academic Press, New York, 1966, p. 133. By permission of the publisher.]

is less than the initial rate of reaction, and there is little if any more *cis* isomer formed at the beginning of the reduction than after 2,3-dimethylcycloalkene has reached its steady-state concentration. Apparently, part of the 1,2-dimethylcycloalkene is converted directly to the *trans* saturated product[33]. This is illustrated in Figure 2.

From the investigation of deuterium exchange reaction and the isomerization of polymethylcycloalkanes, Rooney and his coworkers[7] obtained results which could not be explained by simple addition and elimination reactions on the catalyst. They considered the presence of a π-allylic complex intermediate adsorbed on the catalyst and the attack of hydrogen from either top or bottom of the carbons at the ends of the allyl system[8], affording thus *cis* and *trans* isomers from a common intermediate[34,35], as shown in Scheme II. The fact that more *trans* saturated compound is obtained in the hydrogenation over palladium than over platinum catalyst, suggests the easy formation of π-allylic complex over the former catalyst[34,35].

Smith studied the deuterium exchange reaction of cycloalkenes over platinum and palladium catalyst. The extent of hydrogen exchange is approximately the same at both the olefinic and the allylic positions over both catalysts and 70% of the exchanged cyclohexenes are monodeuterated compounds. Since these results are not accommodated by

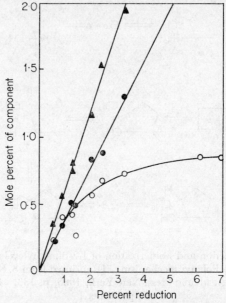

Figure 2. Hydrogenation of 1,2-dimethylcyclopentene (PtO$_2$); 2,3-dimethylcyclopentene (○), cis (●), and trans (▲) 1,2-dimethylcyclopentane. [Reprinted from S. Siegel, P. A. Thomas and J. T. Holt, *J. Catalysis*, **4**, 74 (1965), by permission of the publisher.]

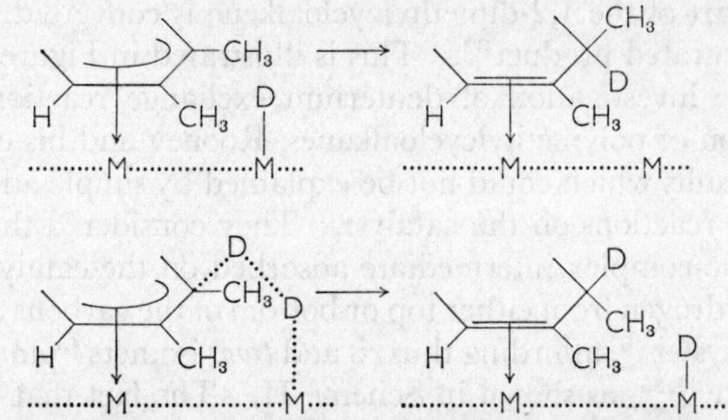

SCHEME II. Allylic intermediate for deuterium exchange and isomerization. [Reprinted from J. J. Rooney, *Chemistry in Britain*, **2**, 245 (1966). By permission of the publisher.]

the classical mechanism postulated by Horiuti and Polanyi or that postulated by Rooney, Smith suggested the presence of either a separate intermediate or the transition state for an intramolecular hydrogen shift [36]:

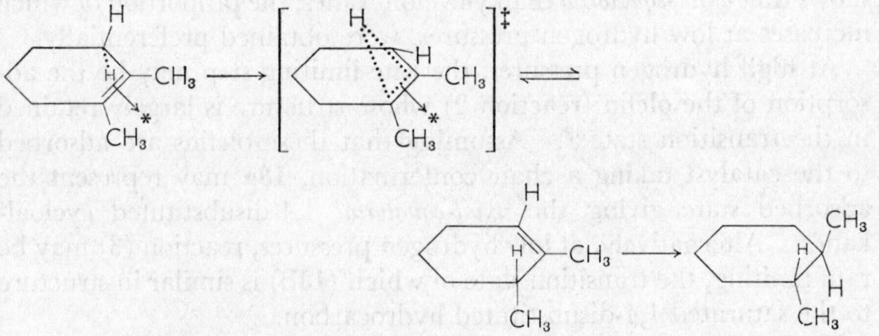

On the other hand, 1,2-diphenyl- and 1,2-dicarbomethoxycyclohexene are hydrogenated exclusively to *cis* products regardless of the catalysts[28,29]. This might be due to a large extent to substituents being strongly adsorbed on the catalyst, thus avoiding the inversion of half-hydrogenated states[37]. The fission of the σ bond between benzylic carbon and metal easily occurs by the attack of hydrogen because the overlap of the σ orbitals of this carbon atom and the metallic orbitals, is less effective owing to the adsorption of the aryl groups[38]. Consequently, the reactions (3) and (4) proceed promptly[29].

Siegel and Smith[26] hydrogenated substituted methylenecyclohexanes over platinum oxide in acetic acid to study the effect of hydrogen pressures on the configuration of the products. Table 2

TABLE 2. Pressure effects on hydrogenation of alkyl-substituted methylene cyclohexanes

Substrate	Mole per cent of *cis* isomer					
	Pressure of Hydrogen (atm)					
	0.25	0.50	1.0	3.0	50	100–200
2-Methyl[a]	70	70	70	69	69	68
3-Methyl	25	25	28	35[b]	43	46
4-Methyl	78	76	73[c]	70[b]	66	67
4-*t*-Butyl[d]	87	86	84	76	62	61

[a] Reference 26.
[b] An earlier report by S. Siegel and M. Dunkel, *Advan. Catalysis*, **9**, 15 (1957), is in error.
[c] Sauvage, Baker and Hussey[30] report 74 *cis* %.
[d] S. Siegel and B. Dmuchovsky, *J. Am. Chem. Soc.*, **84**, 3132 (1962).
[Reprinted from S. Siegel, M. Dunkel, G. V. Smith, W. Halpern and J. Cozort, *J. Org. Chem.*, **31**, 2802 (1966). By permission of the publisher.]

shows that *axial-equatorial* dialkylcyclohexanes, the proportion of which increases at low hydrogen pressures, were obtained preferentially.

At high hydrogen pressures, the rate-limiting step may be the adsorption of the olefin (reaction 2) whose structure is largely retained in the transition state[39]. Assuming that these olefins are adsorbed to the catalyst taking a chair conformation, **13a** may represent the adsorbed state giving the *axial-equatorial* 1,4-disubstituted cycloalkanes. Alternatively, at low hydrogen pressures, reaction (3) may be rate-limiting, the transition state of which (**13b**) is similar in structure to the saturated 1,4-disubstituted hydrocarbon.

The results obtained from alkylmethylcyclohexanes—other than 1,2-disubstituted olefins (Figure 1)—show no systematic relationships with hydrogen pressures (Table 3)[39].

TABLE 3. Pressure effects of disubstituted cyclohexenes

Substrate	Mole per cent of *cis* isomer					
	Pressure of Hydrogen (atm)					
	0·25	0·50	1·0	3·0	50	100-200
2,3-Dimethyl[a]	81	77	77	71	69	70
1,3-Dimethyl	75	75	76[b]	77	80	78
2,4-Dimethyl	—	49	49	51	—	48
1,4-Dimethyl	55	56	56[b]	58	64	65
4-*t*-Butyl-1-methyl[c]	35	35	36	40	47	47

[a] Reference 26.
[b] Sauvage, Baker and Hussey[30], reporting reductions at 1 atm only, give for 1,3-dimethylcyclohexene 74% *cis* and for 1,4-dimethylcyclohexene 57% *cis*.
[c] S. Siegel and B. Dmuchovsky, *J. Am. Chem. Soc.*, **84**, 3132 (1962).

[Reprinted from S. Siegel, M. Dunkel, G. V. Smith, W. Halpern and J. Cozort, *J. Org. Chem.*, **31**, 2802 (1966). By permission of the publisher.]

Sauvage and his coworkers[30] assumed that substituted cyclohexenes were adsorbed in a boat conformation, a very large group such as the *t*-butyl group being forced to take a position *exo* to the boat. Their arguments sought to explain the results obtained at about 1 atm of hydrogen which approximate better the limiting *cis–trans* ratios obtained at low pressures. And indeed, their suggestion can be modified[39] so as to be based on the formation of the half-hydrogenated state, which the previous analysis suggests is the product-determining one at low pressures. Clearly, a large group at $C_{(4)}$ should preferably be *exo-trans* to the $C_{(1)}$ methyl group (**14b**): however the driving force, which causes the cycle to adopt the boat conformation, would also

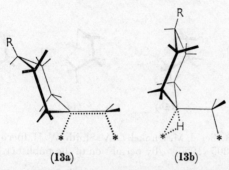

(13a)   (13b)

[Reprinted from S. Siegel, M. Dunkel, G. V. Smith, W. Halpern and J. Cozort, J. Org. Chem., **31,** 2804 (1966). By permission of the publisher.]

cause a small substituent to prefer to be *endo-cis* to the $C_{(1)}$ methyl group (**14a**) and indeed the proportion of *cis* isomers obtained from 4-alkyl-1-methylcyclohexenes increases in the order *t*-butyl < isopropyl < methyl, the percent being 37, 43, and 56 respectively at about 1 atm of hydrogen[30]. The driving force for the adoption of the above conformation in this instance has been assumed to be the spreading surface pressure arising from the saturation of the surface by various adsorbed species[40]. Similar considerations apply in the case of 4-substituted 2-methylcyclohexenes, the half-hydrogenated states of which can be depicted as **14c** and **14d**.

Accordingly, one can rationalize the proportions of epimers obtained from disubstituted cyclohexenes by recognizing this external force which acts to fold the molecule onto itself and increase its extension in a direction away from the surface. The resultant of the interactions produces a boat-like conformation in which substituents at $C_{(3)}$ or $C_{(6)}$ preferably occupy the bowsprit positions, while groups at $C_{(4)}$ and $C_{(5)}$ will be *exo* unless interactions with substituents at $C_{(1)}$ and $C_{(2)}$ overcome this tendency. Thus at low pressure the yields of *cis*-dimethylcyclohexanes are somewhat larger for 2,3- than for 1,3- and for 1,4- than for 2,4-dimethylcyclohexenes (Table 3).

Siegel rationalized that on palladium-hydrogenation of olefins the step from half-hydrogenated state to product (reaction 4) may be rate limiting. If so, a pre-equilibrium may be established for the intermediates appearing before reaction (4). Conformational analysis may be used to predict the relative stabilities of the intermediate and, if the transition state of the postulated rate-limiting step resembles the preceding intermediates, then the configuration of the main saturated isomer may also be predicted.

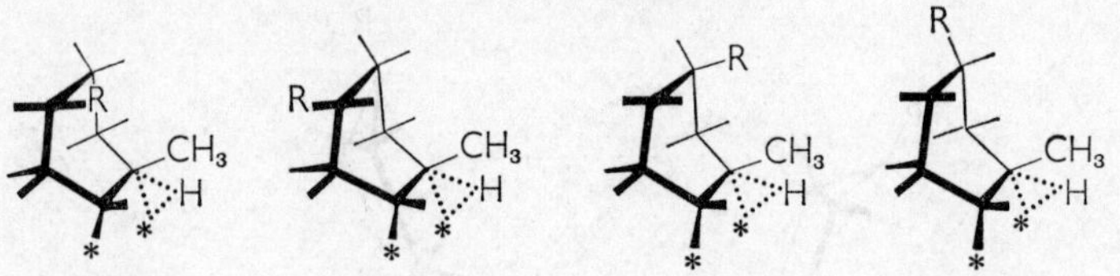

(14a)   (14b)   (14c)   (14d)

[Reprinted from S. Siegel, M. Dunkel, G. V. Smith, W. Halpern and J. Cozort, J. Org. Chem., **31**, 2805 (1966). By permission of the publisher.]

The relative stabilities of the half-hydrogenated states can be thus deduced. The intermediates of 2,3-dialkylcyclohexenes may be displayed as (**15**) and (**16**). The carbon–metal bond seems to be equatorial, considering that the effective size of the catalyst is probably greater than that of other groups in these two half-hydrogenated intermediates. Consequently, the alkyl groups are bonded axially and equatorially in **15**, whereas both are bonded equatorially in **16**.

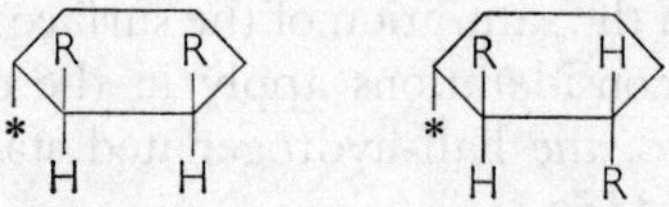

(15)   (16)

The stability of these intermediates is also affected by the interaction between the substituents and catalyst. When the substituent is relatively small, the interaction may be negligible. Thus, in these cases having little catalyst hindrance the stability of the products reflects that of the possible half-hydrogenated intermediates, so that the preferential formation of the more stable isomer usually occurs over palladium catalyst. Thus, the half-hydrogenated intermediate derived from 2,3-dimethylcyclohexene should be more stable in the **16** form, and therefore the *trans* substituted compound is obtained preferably.

On the other hand, (+)-α-pinene (**5**) whose catalyst hindrance should be much greater, yields a large excess of *cis*-pinane (**6**) over palladium catalyst at room temperature. The proportion of the *trans* isomer increases at high temperature[41], as shown in Table 4.

House and his coworkers[42] reported that in the course of the palladium catalysed hydrogenation of lactone **17**, the stereospecificity of

TABLE 4. Effect of temperature on the hydrogenation of (+)-α-pinene in propionic acid using palladium charcoal as catalyst

Temp.	% Pinanes	
	(+)-Cis-	(+)-Trans-
0°	80	20
20	73·5	26·5
52	64·5	35·5
73	62·5	37·5
89	55·5	44·5
111	52·0	48·0
127	47·0	53·0
138	48·5	51·5

[Reprinted from W. Cocker, P. V. R. Shannon and P. A. Staniland, *J. Chem. Soc.*, 42 (1966). By permission of the authors and publisher.]

cis isomer formation increased as the weight of catalyst to substrate decreased, as follows:

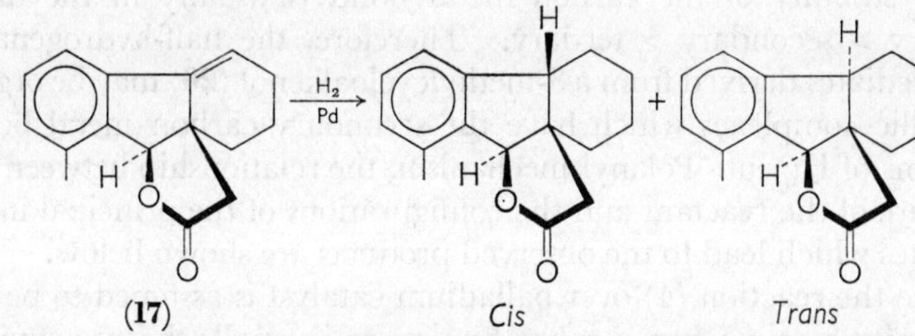

(17) → Cis + Trans

Weight of Catalyst/Weight of Substrate	Cis/Trans
0·025	1·7
0·10	1·4
0·20	0·4

In the presence of large amounts of catalyst or at high temperature where the hydrogenation proceeds relatively rapidly, the hydrogen concentration on the catalyst surface becomes depleted preventing the occurrence of reaction (4) and permitting the equilibrium 18 ⇌ 19 to become important.

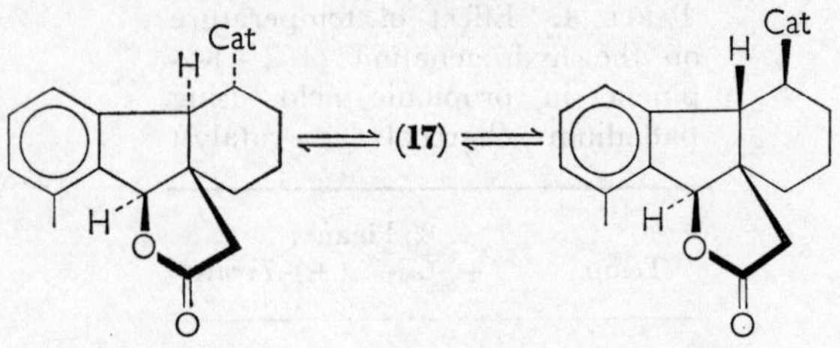

                    (18)   (19)

The hydrogenation of 2-cyclopentylidenecyclopentanol over Raney nickel at high pressure yields 98% of *trans*-cyclopentylcyclopentanol[43]. The fact that a similar result is obtained at atmospheric pressure shows there is no pressure dependence of the isomeric ratio of the products[44]. The hydrogenation of 3-methyl-2-cyclopenten-1-ol and 3-methyl-2-cyclohexen-1-ol over Raney nickel occurs selectively to give *trans* isomers (75% and 87% respectively). On the other hand, over palladium catalyst *cis* isomers are obtained predominantly (60% and 84% respectively)[45]. These results suggest that hydrogen attacks stereoselectively the double bond from the same side as the hydroxyl group over Raney nickel, whereas no directive effect of the hydroxyl group is found on palladium catalyst.

The stability of the carbon–metal bond is usually in the order primary > secondary > tertiary. Therefore, the half-hydrogenated intermediates derived from a 3-methylcycloalkenol (**20**) may be organometallic complexes which have the secondary carbon–metal bond. In terms of Horiuti–Polanyi mechanism, the relationship between the structure of the reactant and the configurations of the principal intermediates which lead to the observed products are shown below.

Since the reaction (4) over palladium catalyst is assumed to be the rate-limiting step when catalyst hindrance is small, the pre-equilibrium of the half-hydrogenated state may be established[26]. Thus **22** is more stable than **25** because of the lesser interaction of the substituents with the catalyst in the former.

Consequently, a *cis*-3-methylcycloalkanol (**23**) is predominantly obtained over palladium catalyst. This occurs even if the *cis* product is the less stable as is the case with a 3-methylcyclopentanol[46].

On the other hand, on hydrogenation of the cyclic allylic alcohols over Raney nickel which has a strong affinity for the oxygen atom[47], the reaction will proceed mainly through intermediates **24** and **25** to afford a *trans* product (**26**).

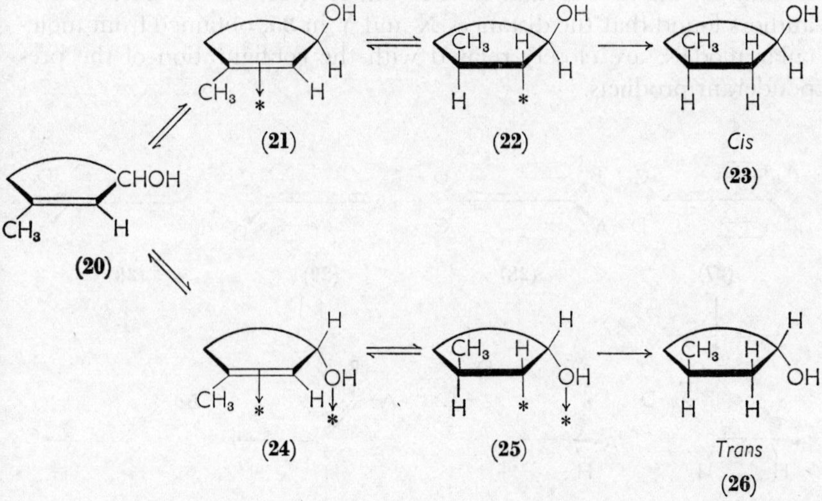

4. Hydrogenation of Alkenes

Crombie and his coworkers[48] partially hydrogenated a series of substituted allenes, and the products were analysed by gas chromatography as shown in Table 5. Properly substituted allene can adopt four distinguishable postures in relation to a solid catalyst (27–30)

TABLE 5. Product analyses at the semi-hydrogenation point of allenes

Allene	Residual allene	Olefins %				Saturated Products%	Selectivity[a]
		(31)	(32)	(33)	(34)		
A = i-Pr; B,C,D = H	4	25	25	0	69	2	0.94
A = CH_2OH; B,C,D = H	2	26	26	8	63	0	0.98
A = CH_2OAc; B,C,D = H	0	24	24	3	69	4	0.96
A = B = i-Bu; C,D = H	0	17	17	80	80	3	0.97
A = Et, B = Me, C,D = H	1	18	18	28	52	0	0.99
A = i-Pr, B = Me; C,D = H	1	29	29	12	58	0	0.99
A = t-Bu, B = Me; C,D = H	12	15	15	9	54	11	0.77
A = Et; B,C = Me, D = H	3	31	2	62	62	1	0.94
A = CO_2H; B,C,D = H	5	0	0	2	89	4	0.91
A = CO_2Me; B,C,D = H	1	0	0	3	95	1	0.98
A = CO_2H; B = Me; C,D = H	4	4	4	6	85	1	0.95
A = CO_2Me, B = Me; C,D = H	0	4	4	7	87	2	0.98

[a] Selectivity is the fraction of olefinic compounds at the end of the reaction.
[Reprinted from L. Crombie, P. A. Jenkins, D. A. Mitchard and J. C. Williams, *Tetrahedron Letters*, 4297 (1967). By permission of the publisher.]

leading by 1,2-*cis* addition to four different olefins (**31–34**). These authors found that the distances X and Y in **35**, obtained from molecular models, are closely related with the configuration of the preponderant products.

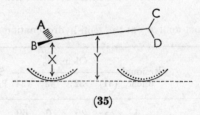

[Reprinted from L. Crombie, P. A. Jenkins, D. A. Mitchard and J. C. Williams, *Tetrahedron Letters*, 4297 (1967). By permission of the publisher.]

[Reprinted from L. Crombie, P. A. Jenkins, D. A. Mitchard and J. C. Williams, *Tetrahedron Letters*, 4297 (1967). By permission of the publisher.]

The gas-phase hydrogenation of 1,3-butadiene has been studied in a static system using alumina-supported ruthenium (0°–25°C), rhodium (15°–80°C), palladium (0°–45°C), osmium (25°–70°C), irridium (minus 20°–75°C), and platinum (0°–15°C). 1-Butene and *cis*- and *trans*-2-butene were first obtained under all conditions[49] (Table 6). Palladium yields olefins exclusively (selectively 1·000), while the other metals afford also n-butane as an initial product and it is the major product of the irridium-catalysed reaction.

1-Butene is formed by 1,2-addition of two hydrogen atoms to the

4. Hydrogenation of Alkenes

TABLE 6. The dependence of butene distribution and selectivity upon catalyst and temperature

Catalyst	Temp. (°C)	1-Butene %	2-Butene % trans	cis	trans/cis	Selectivity[a]
Ru	0	69	19	12	1·6	0·736
	25	64	21	15	1·4	0·820
Rh	16	51	32	17	1·9	0·743
	36	52	31	17	1·8	0·834
Pd	0	64·4	33·2	2·4	13·8	1·000
	21	60·2	37·0	2·8	13·2	1·000
Os	24	65	19	16	1·2	0·431
Ir	24	59	19	22	0·9	0·251
Pt	0	72	18	10	1·8	0·501
	15	65	18	17	1·1	0·563

[a] Selectivity: percent C_4H_8/ (percent C_4H_8 + percent C_4H_{10}).

diene as shown in equation (9); the conformations of the adsorbed diene are irrelevant to this process:

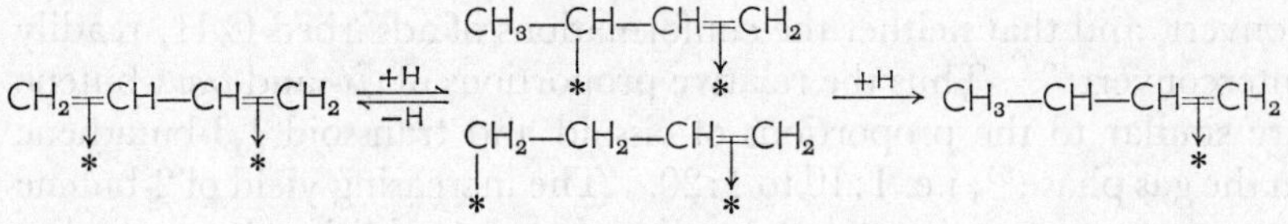

$$CH_2{=}CH{-}CH{=}CH_2 \underset{-H}{\overset{+H}{\rightleftharpoons}} \begin{array}{c} CH_3{-}CH{-}CH{=}CH_2 \\ * \quad * \\ CH_2{-}CH_2{-}CH{=}CH_2 \\ * \quad * \end{array} \overset{+H}{\longrightarrow} CH_3{-}CH{-}CH{=}CH_2 \quad (9)$$

2-Butene may be formed as an initial product (a) by 1,4-addition of two hydrogen atoms to the diene (b) by the isomerization of 1-butene in equation (9) by alkyl reversal (equation 10) or (c) by a combination of both processes (a) and (b)

$$\text{1-Butene (ads)} \underset{-H}{\overset{+H}{\rightleftharpoons}} C_4H_9(\text{ads}) \underset{+H}{\overset{-H}{\rightleftharpoons}} \text{cis and trans 2-Butene (ads)} \quad (10)$$

Two important deductions may be made immediately concerning the mechanisms of 2-butene formation. First, since palladium is an extremely active catalyst for the hydrogenation of butenes, the authors[49] attribute the absence of n-butane formation during 1,3-butadiene hydrogenation to the absence of adsorbed C_4H_9. Consequently, 2-butene must have been formed by 1,4-addition of hydrogen to the diene. This conclusion is in agreement with that of Meyer and Burwell[50]. Secondly, the other metals give n-butane as an initial product from adsorbed C_4H_9. Consequently, butene isomerization by equation (10) will almost certainly complicate the butene distribution.

In the 1,4-addition process the conformation of adsorbed C_4H_6 determines the configuration of the 2-butene produced. Equation (11) shows how the addition of the second hydrogen atom is envisaged: the 1-methyl-π-allylic species **39** may be an intermediate state between **38** and **40**[49].

$$
\begin{array}{ccc}
\mathrm{CH_2{=}CH} & \mathrm{CH_2{=}CH} & \mathrm{CH_2{\cdots}CH} \\
{*}\quad\mathrm{CH{=}CH_2} \rightleftharpoons & {*}\quad\mathrm{CH{-}CH_3} \rightleftharpoons & \quad{*}\,\mathrm{CH{-}CH_3} \\
{*} & {*} & \\
(36) & (38) & (39)
\end{array}
$$

$$\Updownarrow \qquad (11)$$

$$
\begin{array}{ccc}
\mathrm{CH{-}CH} & \mathrm{CH{-}CH} & \mathrm{CH_3{-}CH} \\
\mathrm{H_2C}\diagdown\diagup\mathrm{CH_2} \quad\text{or}\quad \mathrm{H_2C}\cdots\cdots\mathrm{CH_2} & & {*}\,\mathrm{CH{-}CH_3} \\
{*} & {*} & \\
(37) & & (40)
\end{array}
$$

The palladium catalysed reaction always yields a high proportion of *trans*-2-butene (typically *trans/cis* = 10). It is suggested that the conformations of adsorbed 1,3-butadiene, **36** and **37**, do not readily interconvert, and that neither the conformations of adsorbed C_4H_7 readily interconvert[49]. Thus the relative proportions of *cis*- and *trans*-butene are similar to the proportions of cissoid and transsoid 1,3-butadiene in the gas phase[51], i.e. 1 : 10 to 1 : 20. The increasing yield of 2-butene as temperature is increased implies that 1,4-addition has a higher activation energy than 1,2-addition. This may be associated with the transformation of adsorbed C_4H_7 from a σ-π-diadsorbed form (**38**) to the adsorbed 1-methyl-π-allylic species **39**. If 1,4-addition of hydrogen to the diene is also important for catalysts other than palladium in 2-butene formation, then the low *trans/cis* values listed in Table 6 tend to show that the various conformations of adsorbed diene and/or C_4H_7 may be readily interconvertible.

B. Selective Hydrogenation

In general, the rate of hydrogenation of the double bond in monoolefins decreased roughly with an increase in the number and branching of the alkyl substituents.

For instance, the following sequence of diminishing rates found by

$$\mathrm{RCH{=}CH_2} > \mathrm{RCH{=}CHR} > \underset{R}{\overset{R}{\diagdown}}\mathrm{C{=}CH_2} > \underset{R}{\overset{R}{\diagdown}}\mathrm{C{=}CHR}$$

4. Hydrogenation of Alkenes

DuPont[52] in the hydrogenation of ethylene derivatives in the presence of Raney nickel.

However, when the molecule contains highly polar groups (COOH, etc.), or a phenyl group, the situation becomes very complicated and depends on the catalyst, e.g. the rates of platinum-catalysed hydrogenation[53] decrease in the order **41 > 42 > 43 > 44** but in the presence of palladium the order is **43 > 42 > 44 > 41**. On Raney nickel **42** hydrogenates as fast as on palladium, while **43** and **44** hydrogenate faster on the latter catalyst. Compounds **42–44** hydrogenate faster on these two catalysts than on platinum[54].

$$\begin{array}{cc} H_5C_2 \diagdown \quad \diagup H \\ C=C \\ H_5C_2 \diagup \quad \diagdown C_2H_5 \\ (41) \end{array} \qquad \begin{array}{cc} H_3C \diagdown \quad \diagup H \\ C=C \\ H_3C \diagup \quad \diagdown C_6H_5 \\ (42) \end{array}$$

$$\begin{array}{cc} H_5C_6 \diagdown \quad \diagup H \\ C=C \\ H_5C_6 \diagup \quad \diagdown CH_3 \\ (43) \end{array} \qquad \begin{array}{cc} H_5C_6 \diagdown \quad \diagup H \\ C=C \\ H_5C_6 \diagup \quad \diagdown C_6H_5 \\ (44) \end{array}$$

The sequence reactivity obtained when hydrogenating a mixture of these compounds is not always identical with that stemming from the hydrogenation of each compound separately. This is due to the competitive adsorption of the substrates on the catalyst and is a characteristic feature of heterogeneous catalytic reactions. For instance, the rate of the hydrogenation of ethylene with palladium catalyst is far faster than that of acetylene. The mixture of these compounds, however, shows that acetylene is hydrogenated preferably, whereas ethylene is not hydrogenated before most of the acetylene has disappeared, suggesting a stronger adsorption of acetylene which monopolizes the surface of the catalyst[54a].

The rate of the hydrogenation of cyclic olefins over palladium-charcoal in methanol and R_F values of these compounds for a certain solvent-adsorbant chromatographic system, were obtained (Table 7) by Jardine and McQuillin[5]. The results indicate that adsorption is a reflexion of the molecular geometry of the cyclic alkene, whereas the rates of hydrogenation follow an order which is that of the heat of hydrogenation.

The results of hydrogenation of olefin mixtures are shown in Table 8. It is clear that cyclooctene is strongly inhibitory of the hydrogenation

TABLE 7. Hydrogenation of cyclic olefins

	10^{-2} Rate[a] (cc./min./mg. catalyst)	$1/R_E{}^b$	$-\Delta H^c$ (kcal/mole at 25°)
Bicyclo [2,2,1] hepta-2,5-diene	20·9	1·05	35
Cyclohexene	10·8	1·3	27·1
Cycloheptene	9·7	1·4	25·9
Cyclopentene	8·1	1·5	25·6
Cycloocta-1,5-diene	5·7	1·7	25·0
Cyclooctene	1·6	2·2	23·0

[a] Initial rates of hydrogenation for 0·1 M solution in methanol (25 cc.) with Pd-charcoal (60 mg.).

[b] In chloroform–acetic acid (10:1) on silica gel with 2·5% silver nitrate.

[c] Data from R. B. Turner, in *Theoretical Organic Chemistry*, Butterworths, London, 1959, p. 76. The value for cyclooctadiene is from interpolation between the ΔH values for cyclooctene, the triene, and the tetraene, which are linear with the number of double bonds (cf. R. B. Turner and his coworkers, *J. Am. Chem. Soc.*, **79**, 4127 (1957)).

[d] Value for hydrogenation to monoene.

[Reprinted from I. Jardine and F. J. McQuillin, *J. Chem. Soc.*, 459 (1966). By permission of the authors and publisher.]

of cyclohexene, yet sampling during the reaction showed that cyclohexene is in fact preferentially reduced at a relative rate of ca 2:1. These observations establish that cyclooctene is more strongly adsorbed, but more slowly hydrogenated than cyclohexene. The other mixtures show a similar, but smaller effect. From Table 7 and 8 the order: cyclooctene > cyclohexene > bicycloheptadiene appears reasonable as the sequence of strengths of adsorption. The rates of reaction are, however, in the inverse order.

TABLE 8. Hydrogenation of olefin mixtures[a]

Mixture	$10^{-2}M$	Rate (cc./min./mg. catalyst)	
		Observed	Additive
Cyclohexene	4·7	2·0	9·5
Cyclooctene	4·6		
Cyclooctene	1·7	24·1	31·0
Bicycloheptadiene	4·0		
Cyclohexene	5·5	30·7	38·1
Bicycloheptadiene	4·0		

[a] 0·01 M solution in methanol (25cc.) with Pd-charcoal (30 mg.).

[Reprinted from I. Jardine and F. J. McQuillin, *J. Chem. Soc.*, 459 (1966). By permission of the authors and publisher.]

4. Hydrogenation of Alkenes

Cis olefins are usually hydrogenated faster than *trans* olefins, and because of the stronger adsorption of the *cis* olefins[55], these are converted preferentially when mixed with *trans* olefins. A further example is $\Delta^{1(9)}$-octalin which is hydrogenated faster than $\Delta^{9(10)}$-octalin, the difference in the rates becoming larger in the reduction of the mixture over platinum[56] or irridium[57].

The selective hydrogenation of the double bond which coexists with other ones in the same molecule, is almost identical with that of the mixture of monoolefin. The unsaturated bond which can approach more closely to the catalyst surface is hydrogenated predominantly. For example, in the case of limonene (**45**) the isopropenyl group can be hydrogenated over platinum without affecting the olefinic bond in the ring[58].

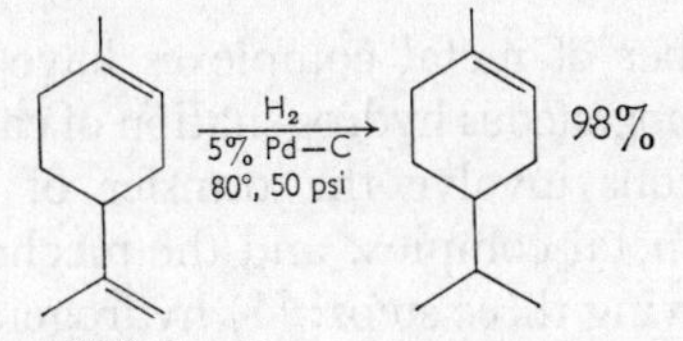

$$\underset{(45)}{\text{limonene}} \xrightarrow[80°, 50\text{ psi}]{\text{H}_2, 5\% \text{ Pd}-\text{C}} \text{product} \quad 98\%$$

In the hydrogenation of conjugated diolefins such as isoprene (**46**), 1,2- and 1,4-addition of hydrogen and isomerization occur to give a mixture of all possible products, since the hydrogenation is less selective. But the result of half-hydrogenation shows that the olefinic bond which has no substituent is preferentially hydrogenated as compared to the olefinic bond carrying some substituents[59].

	Pt	Pd	Ni
$CH_2=\overset{CH_3}{\underset{|}{C}}-CH_2-CH_3$	54	31	41.5
$CH_3-\overset{CH_3}{\underset{|}{CH}}-CH=CH_2$	15	26	17
$CH_3-\overset{CH_3}{\underset{|}{C}}=CH-CH_3$	31	43	41.5

(starting material: $CH_2=\overset{CH_3}{\underset{|}{C}}-CH=CH_2$ (**46**))

The half-hydrogenation of α,β-unsaturated aldehydes or ketones over platinum, rhodium, ruthenium, palladium or nickel give the saturated carbonyl compounds.

It was asserted that methyl isobutyl ketone, which is the hydrogenation product of mesityl oxide, is not hydrogenated at all in the presence of the latter because of much stronger adsorption of mesityl oxide on the catalyst[60].

The selective hydrogenation of the carbonyl group in α,β-unsaturated carbonyl compounds is rather difficult. This is possible, however, over platinum containing iron or zinc salts[61], Raney copper containing cadmium[62] or rhenium catalyst[63].

$$CH_3CH\!=\!CHCHO \xrightarrow[\text{Pt, Pd, Rh, etc.}]{H_2} CH_3CH_2CH_2CHO$$

$$CH_3CH\!=\!CHCHO \xrightarrow[\text{Re, Cu—Cd, etc.}]{H_2} CH_3CH\!=\!CHCH_2OH$$

II. HOMOGENEOUS CATALYTIC HYDROGENATION

Recently a number of metal complexes have been introduced as catalysts for the homogeneous hydrogenation of carbon–carbon double bonds. The reactions involve the transfer of hydrogen from the hydrido-transition metal complex, and the mechanism[64] can be considered as one involving three steps: (1) hydrogen activation, (2) substrate activation, (3) hydrogen transfer.

The activation of molecular hydrogen by transition-metal ions or complexes in homogeneous solution is well-known. The complexes include those of Cu^{II}, Cu^{I}, Co^{II}, Co^{I}, Pd^{II}, Pt^{II}, Rh^{III}, Rh^{I}, Ru^{III}, and Ir^{I}, and the hydride complexes are formed as the intermediates in the hydrogen-activating step. Now three routes of formation of the hydride intermediate have been recognized[65]:

(a) heterolytic splitting of H_2, e.g.
$$[Ru^{III}Cl_6]^{3-} + H_2 \longrightarrow [HRuCl_5]^{3-} + H^+ + Cl^-$$
(b) homolytic splitting of H_2, e.g.
$$2[Co^{II}(CN)_5]^{3-} + H_2 \longrightarrow 2[HCo^{III}(CN)_5]^{3-}$$
(c) insertion of H_2, e.g.
$$Ir^{I}Cl(CO)(PPh_3)_2 + H_2 \longrightarrow Ir^{III}H_2Cl(CO)(PPh_3)_2$$

A. Ruthenium Chloride

Ruthenium(II) chloride in aqueous hydrogen chloride solution is a homogeneous catalyst for the hydrogenation of olefinic compounds such as maleic, fumaric and acrylic acids in which the double bond is activated by an adjacent carbonyl group, while in the case of simple olefins such as ethylene and propylene no hydrogenation was observed. Reduction of fumaric acid with deuterium

4. Hydrogenation of Alkenes

in water yields undeuterated succinic acid, whereas hydrogenation with hydrogen (or deuterium) in heavy-water yields predominantly *dl*-2,3-dideuteriosuccinic acid, indicating stereospecific *cis* addition. Thus the hydrogen adding to the double bond originates predominantly from the solvent. Haplern[66] proposed the mechanism reaction (12), involving the fast formation of a ruthenium(II)-olefin complex. The rate-determining step is the reaction between ruthenium(II)-olefin complex and hydrogen, involving heterolytic splitting of hydrogen with formation of a hydrido-ruthenium(II) complex. Then follows the rearrangement of the hydrido-π-olefin complex to a σ-alkyl complex by 'insertion' of the olefin into the metal-hydride bond which is a well-known reaction in the field of organometallic chemistry[67]. Finally comes an electrophilic attack on the metal-bonded carbon atom by a proton to complete the hydrogenation. The overall rate law of succinic acid production is $k(H_2)(Ru^{II}\text{-olefin})$.

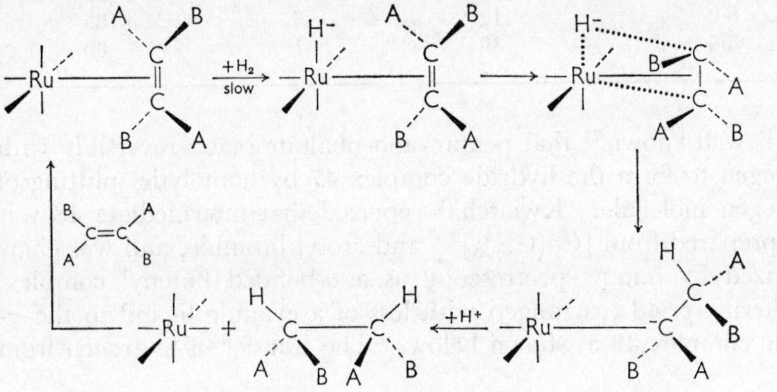

(12)

B. Pentacyanocobaltate

Several workers[68-70] reported that pentacyanocobaltate in aqueous solutions is a catalyst for the homogeneous hydrogenation of conjugated olefins such as butadiene and isoprene at atmospheric pressure, to yield monoolefins. Monoolefins and non-conjugated dienes were not hydrogenated. Although propenylbenzene and stilbene were not reduced, styrene and its derivatives such as α-methylstyrene, cinnamic acid and cinnamyl alcohol readily yielded the corresponding dihydroderivatives. The following mechanism[69] was proposed for the homogeneous hydrogenation of butadiene to butene:

$$2[Co(CN)_5]^{3-} + H_2 \rightleftharpoons 2[HCo(CN)_5]^{3-}$$

(47)

$$[HCo(CN)_5]^{3-} + C_4H_6 \longrightarrow [C_4H_7Co(CN)_5]^{3-}$$
$$(48)$$
$$[C_4H_7Co(CN)_5]^{3-} + [HCo(CN)_5]^{3-} \longrightarrow 2[Co(CN)_5]^{3-} + C_4H_8$$

Table 9 shows that the relative ratios of butene isomer depend on the cyanide–cobalt ratio present in solution. The product at low cyanide concentrations $((CN^-)/(Co^{2+}) < 5.5)$ is predominantly *trans*-2-butene, whereas at higher cyanide concentrations $((CN^-)/(Co^{2+}) > 6)$ it is predominantly 1-butene.

TABLE 9. Hydrogenation of butadiene catalysed by cobalt(II)-cyanide complexes

$[CN^-]/[Co^{2+}]$	*trans*-2-Butene(%)	*cis*-2-Butene (%)	1-Butene(%)
4.5	86	1	13
5.5	70	1	29
6.0	12	3	85
8.5	19	1	80

It is well known[71] that pentacyanocobaltate reacts reversibly with hydrogen to form the hydride complex **47** by homolytic splitting of hydrogen molecule. Kwiatch[72] reported that intermediate **48** was also prepared from $[Co(CN)_5]^{3-}$ and crotyl bromide, and was characterized by n.m.r. spectroscopy as a σ-bonded butenyl complex. Furthermore, **48** rearranged with loss of a cyanide ligand to the π-allylic complex **49** as shown below. The transfer of hydrogen from

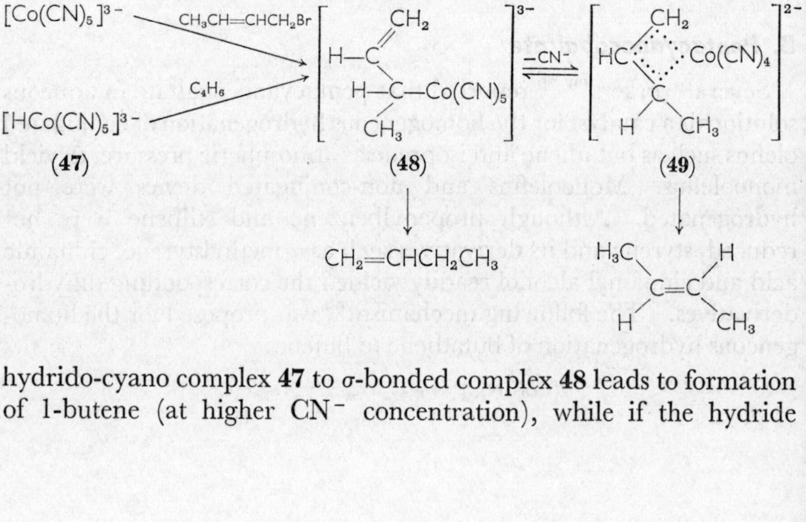

hydrido-cyano complex **47** to σ-bonded complex **48** leads to formation of 1-butene (at higher CN^- concentration), while if the hydride transfer occurs in π-allylic complex **49** it leads to 2-butene (at lower CN^- concentration).

C. Tris(triphenylphosphine)halogenorhodium

Tris(triphenylphosphine)chlororhodium(I), $RhCl(PPh_3)_3{}^{64,73-75}$, prepared by the interaction of an excess of triphenylphosphine with rhodium(III) chloride in ethanol, is a very effective catalyst for hydrogenation of olefinic compounds such as cyclohexene and 1-heptene at ordinary pressure and room temperature. The rates of hydrogenation of cyclohexene by this catalyst in benzene-ethanol solution is faster than by Adams catalyst.

Homogeneous hydrogenation using $RhCl(PPh_3)_3$ has many advantages as compared with other reduction methods:
(1) functional groups such as keto, nitro, chloro, azo or ester are not reduced; (2) terminal olefins are reduced more rapidly than internal olefins, e.g. 1-hexene > 2-hexene; (3) *cis* olefins are reduced faster than *trans* olefins, e.g. *cis*-2-hexene > *trans*-2-hexene; (4) conjugated olefins, such as butadiene, are not reduced at ordinary pressure, though they are reduced at higher pressures (ca. 60 atm); (5) ethylene is not reduced, however, propene, butene and higher olefins are hydrogenated at atmospheric pressure; (6) $RhCl(PPh_3)_3$ is still an active catalyst in the presence of sulphur compounds such as thiophenol or sulphide, which are poisons for the heterogeneous catalyst[76].

Homogeneous hydrogenation with $RhCl(PPh_3)_3$ shows some selectivity when more than one double bond is present. (+)-Carvone (**50**) was rapidly and specifically hydrogenated on the isopropylidene group, to give carvotanacetone (**51**)[74]:

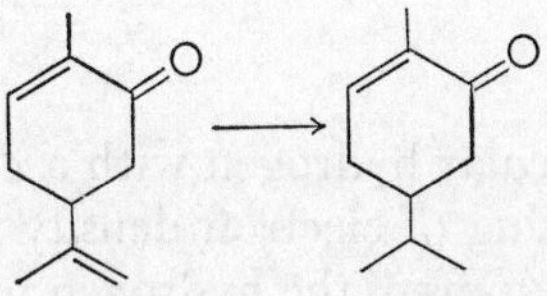

(**50**)   (**51**)

Also $\Delta^{1,4}$-androstadiene-3,17-dione (**52**) and $\Delta^{4,6}$-androstadiene-3, 17-dione (**53**) are converted into the Δ^4-3-ketone **54**[77]. Unhindered olefins such as Δ^1-, Δ^2- and Δ^3-chloestene were reduced to 5α-cholestane, however, more highly substituted olefins such as Δ^4-androstene were not hydrogenated. Deuteration of Δ^1-cholesten-3-one preceded

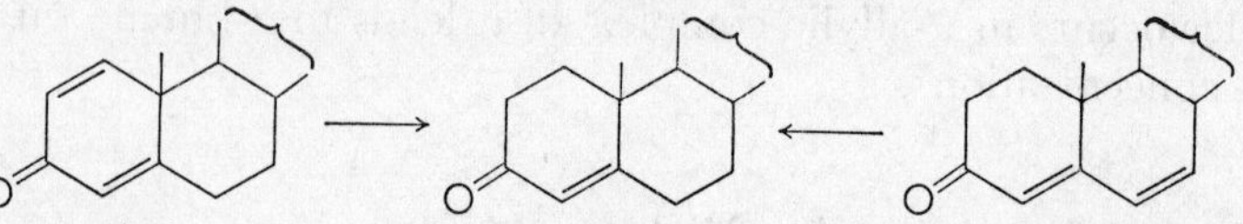

(52) → (54) ← (53)

from the α-face. The reduction[75] with deuterium of 2,2-dihydroergosteryl acetate led to 5α,6-d_2-ergost-7-en-3β-ol, indicating stereospecific *cis* addition.

The complex $RhCl(PPh_3)_3$ dissociates in benzene solution and reacts reversibly with molecular hydrogen according to the process of insertion of hydrogen to form the *cis*-dihydro species $RhCl(H_2)(PPh_3)_2$. The mechanism proposed to explain the above observation is shown in reaction (13)[64], where P and S represent a triphenylphosphine and a solvent ligand respectively:

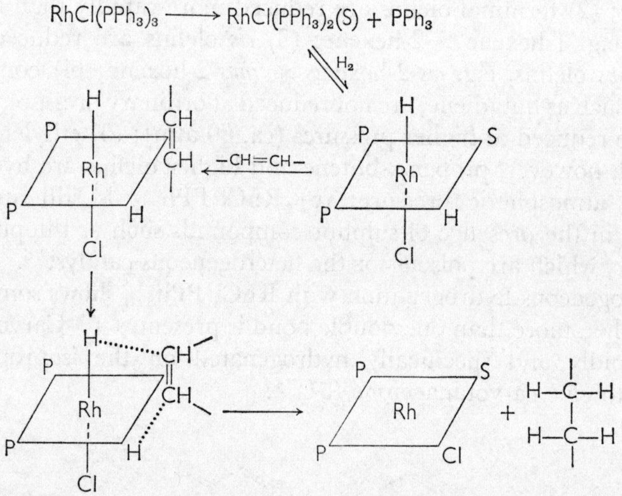

$$RhCl(PPh_3)_3 \longrightarrow RhCl(PPh_3)_2(S) + PPh_3$$

(13)

The reaction of molecular hydrogen with a d^8 complex to give a d^6 species involves the raising of electron density in an acceptor orbital at one of the hydrogen atoms of the hydrogen molecule at the expense of electron density in filled metal orbitals. In this case Nyholm[78] suggested that an anti-bonding orbital of the acceptor could be used.

D. Irridium Complex

The planar irridium complex $Ir(CO)(PPh_3)_2Cl$ is a very reactive one towards hydrogen to give the octahedral $IrH_2Cl(CO)(PPh_3)_2$

(reaction 14) and also forms adducts with alkenes, alkyl and acyl halides, sulphur and hydrogen chloride. This irridium complex is an effective catalyst for the homogeneous hydrogenation of ethylene and propylene in benzene or toluene solutions at sub-atmospheric pressure and at mild temperature (60°C). The rhodium analogue [Rh(CO)-(PPh$_3$)$_2$Cl] of this irridium complex is slightly less effective catalyst[79].

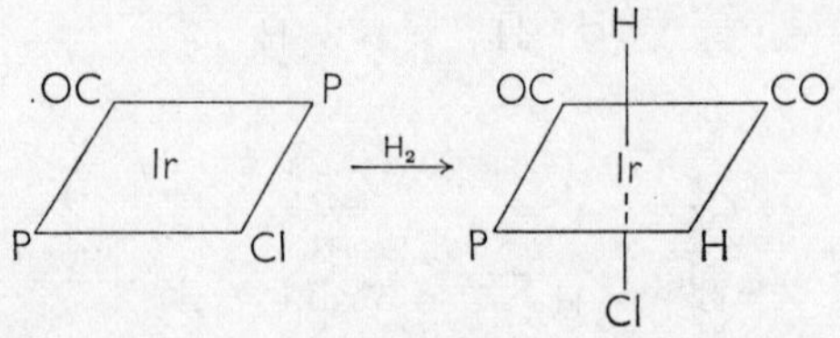

$$\text{OC}\diagdown\text{Ir}\diagup\text{P} \quad \xrightarrow{H_2} \quad \text{OC}\diagdown\text{Ir}(H)\diagup\text{CO} \qquad (14)$$

E. Tris(triphenylphosphine)dichlororuthenium

Tris(triphenylphosphine)dichlororuthenium, $RuCl_2(PPh_3)_3$, is effective for homogeneous hydrogenation in benzene–ethanol solutions[80]. A more active and highly selective catalyst[81], the hydrido-complex $RuCl(H)(PPh_3)_3$ can be readily prepared *in situ* in benzene solution by hydrogenolysis of tris(triphenylphosphine)dichlororuthenium at 25°C and 1 atm using an organic base such as triethyl amine:

$$RuCl_2(PPh_3)_3 + H_2 + Et_3N \rightleftharpoons RuCl(H)(PPh_3)_3 + Et_3HNCl$$

This hydride complex is highly specific for terminal olefins such as 1-pentene, 1-hexene, and 1-decene and is more active than RhCl-(PPh$_3$)$_3$ in benzene, although it is ineffective for the hydrogenation of internal olefins such as 2-hexene and 3-heptene.

F. Platinum Complexes

Polyolefins such as methyl linolenate and cyclooctadiene were selectively hydrogenated at ca 40 atm of hydrogen, at 90°C, in the presence of chloroplatinic acid, hydrido-chlorobis(triphenylphosphine)platinum(II) and dichlorobis(triphenylarsine)platinum(II) each in mixture with stannous chloride in methanol–benzene solution[82-84]. Bidentate diene-Pt-complexes such as dichloro-1,5-cyclooctadiene platinum $PtCl_2(C_8H_{12})$ also behave similarly. The product of hydrogenation is specifically a monoolefin in all cases. Hydrogenation proceeds via metal–hydride (MH) formation (reaction 15), stepwise migration of double bonds to attain conjugation (reaction 16) and conversion of the conjugated diene to monoolefin (reaction 17).

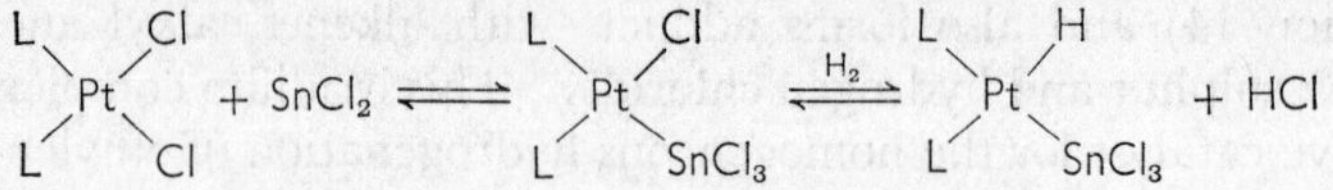

$$L_2Pt(Cl)_2 + SnCl_2 \rightleftharpoons L_2Pt(Cl)(SnCl_3) \underset{}{\overset{H_2}{\rightleftharpoons}} L_2Pt(H)(SnCl_3) + HCl \quad (15)$$

$$\text{>C=C-C-} + MH \longrightarrow \text{>C}\substack{+\\=}\text{C-C-} \rightleftharpoons -\text{C-C-C-} \rightleftharpoons -\text{C-C}\substack{+\\=}\text{C<} \quad (16)$$

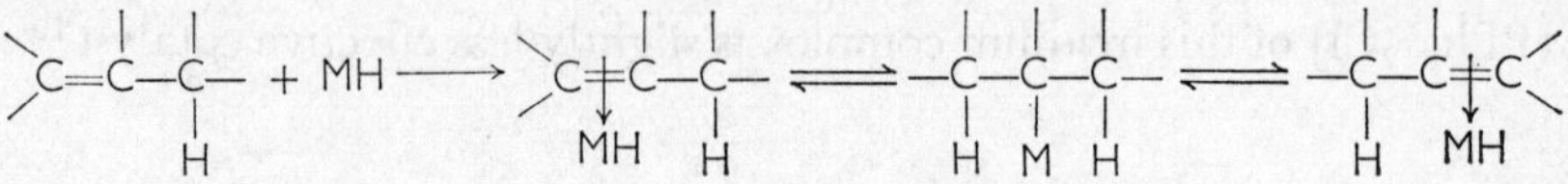

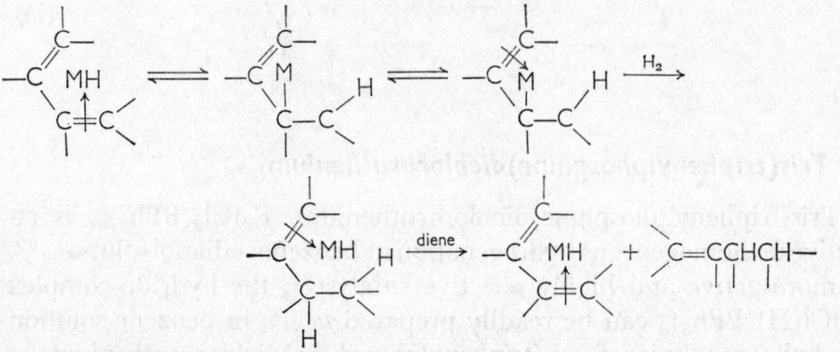

$$(17)$$

G. Metal-Carbonyl Catalysts

Unsaturated fatty esters such as methyl linoleate and methyl linolenate were also selectively hydrogenated at ca 80 atm of hydrogen and 180°C, in the presence of iron carbonyl in cyclohexane to monoolefinic ester[85,86]. Intermediates in the reduction include conjugated dienes-$Fe(CO)_3$ complex formation. Cobalt carbonyl tri(n-butyl)phosphine complex was also an effective and selective catalyst for the homogeneous hydrogenation of cyclic polyenes such as cyclododecatriene and cyclooctadiene, to cyclic monoolefins[87].

H. Ziegler-Type Catalysts

Some Ziegler-type catalyst such as bis(cyclopentadienyl)titanium dichloride-triethylaluminium, triisobutylaluminium–tetraisopropyltitanate and triisobutylaluminium-acetylacetonate of chromium, manganese and cobalt, were found to be reactive for the hydrogenation of olefins[88]. As an example cyclohexene reduced quantitatively with triisobutylaluminium–tetraalkyltitanate in heptane solution at room temperature under an initial hydrogen pressure of 4·4 atm using an aluminium–titanium ratio of 3·3:1. A mechanism involving the addition of a metal hydride complex to the olefin followed by hydrogenolysis of the metal–carbon bond was proposed.

III. CHEMICAL REDUCTION

A. Dissolving Metals

The electron released when a metal is dissolved in acid, neutral or basic solution can reduce carbon–carbon double bonds conjugated with carbonyl groups, aromatic systems or other carbon–carbon multiple bonds. The dissolution of zinc in acetic acid generally reduced α,β-unsaturated carbonyl compounds to saturated carbonyl compounds. The treatment of 7,11-diketolanosteryl acetate (55) with zinc in boiling acetic acid affords diketolanostanyl acetate (56)[89].

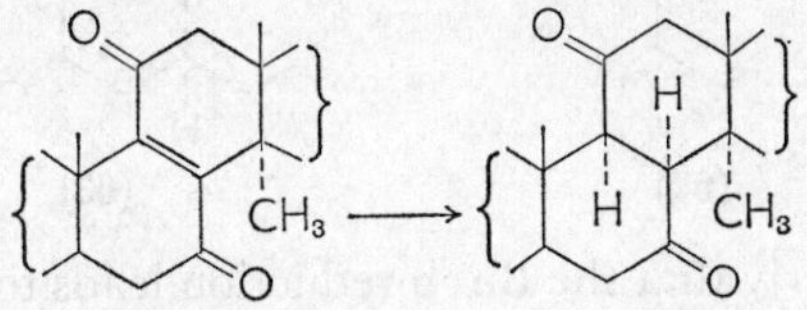

Because of its widespread application, the reduction of conjugated carbon–carbon double bonds is frequently carried out with sodium or lithium in liquid ammonia (Birch reduction). The reactions involve the transfer of electrons with the formation of a carbanionic species. The primary product of the Birch reduction of α,β-unsaturated ketones is enolate anion (58) formed by ammonia protonation of a dianion (or anion radical) intermediate (57). In a subsequent step, the addition of a proton donor stronger than ammonia (e.g. ethanol, t-butanol or ammonium chloride) ketonizes the enolate anion (58) and leads to the saturated ketones (59)[90].

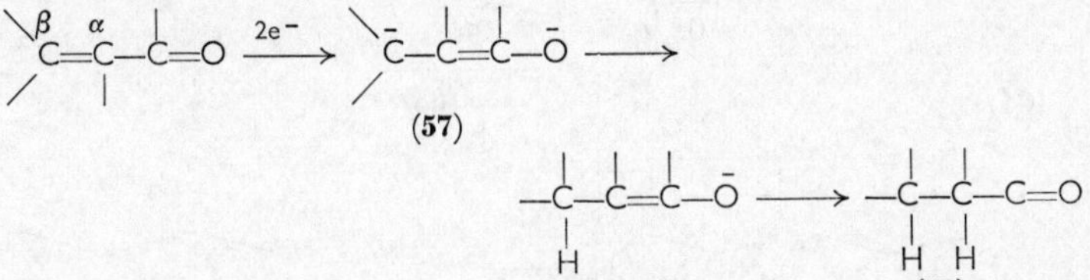

The presence of an enolate anion (58) in the reaction mixture has been proved by methylation to form a C-methylated product[91].

This reduction is stereospecific. Thus 9-methyl-$\Delta^{5(10)}$-octalin-1,6-dione (**60**) leads to a *trans*-diketone (**61**) which is different from the major product obtained on catalytic hydrogenation[92]. Cholest-5-en-7-one (**62**) forms cholestan-7-one (**63**)[93].

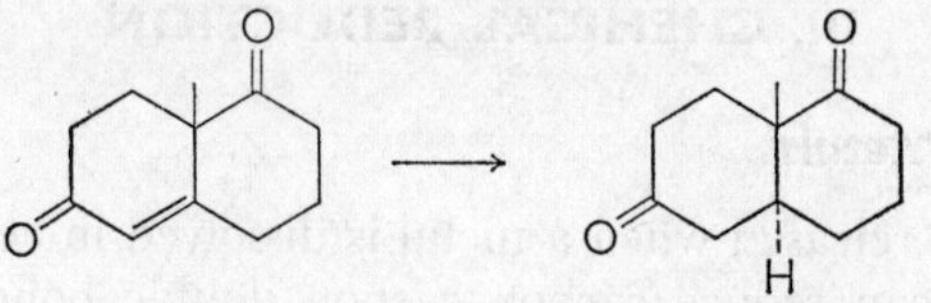

(**60**)   (**61**)

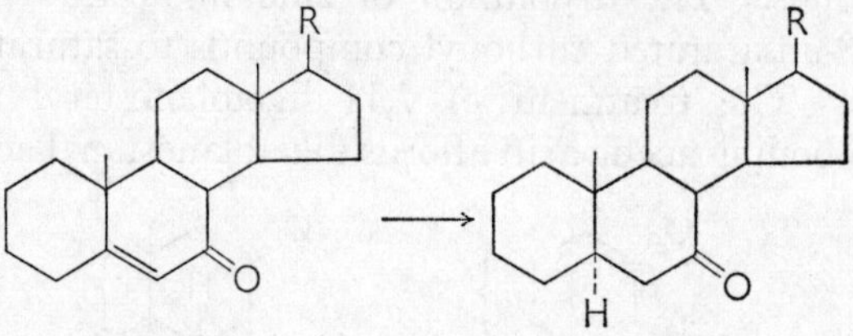

(**62**)   (**63**)

It was noted early that the Birch reduction leads to the formation of the thermodynamically more stable isomer[93]. However, the stereochemistry of the product at the β-carbon is governed by the stereochemistry of the protonation of the anion radical **57**. In the transition state for the protonation of **57**, the developing *p* orbital at the β-carbon overlaps continuously with the carbon–carbon double bond of the enolate system and must remain perpendicular to the O=C—C=C plane (i.e. **65**). In an octalone system, the proton will be introduced at the β-carbon from the axial direction with respect to ring A (con-

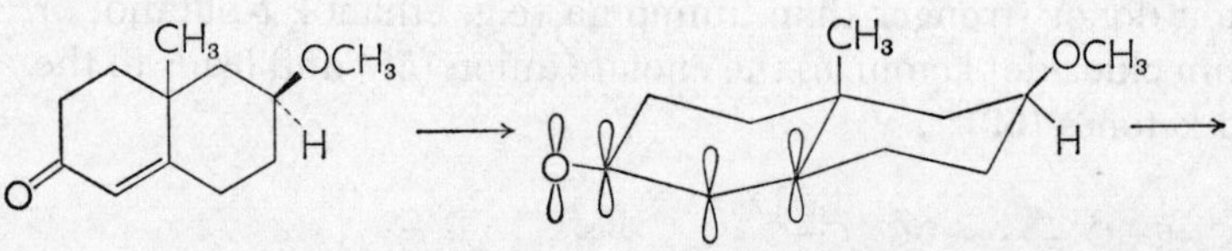

(**64**)   (**65**)

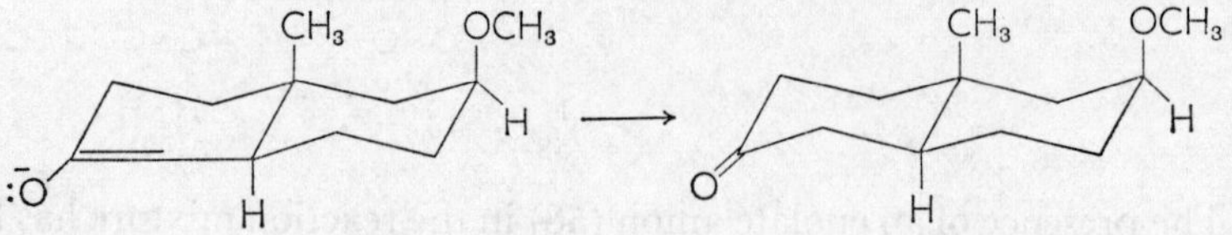

(**66**)

taining the carbonyl group) and the reduction of a $\Delta^{1(9)}$-octal-2-one derivative such as **64** affords a trans-decalone system **66**[94].

The configuration at the α-carbon atom of the reduction product is determined in the kinetically controlled ketonization step. The enolic system is still sp^2 hybridized in the transition state (i.e. **68**), and the proton will attack the trigonal α-carbon atom from less hindered side of the enolic system. Thus the lithium–liquid ammonia reduction[95] of 2-methyl-3-phenylindone (**67**) leads to the formation of *cis*-2-methyl-3-phenylindanone (**69**).

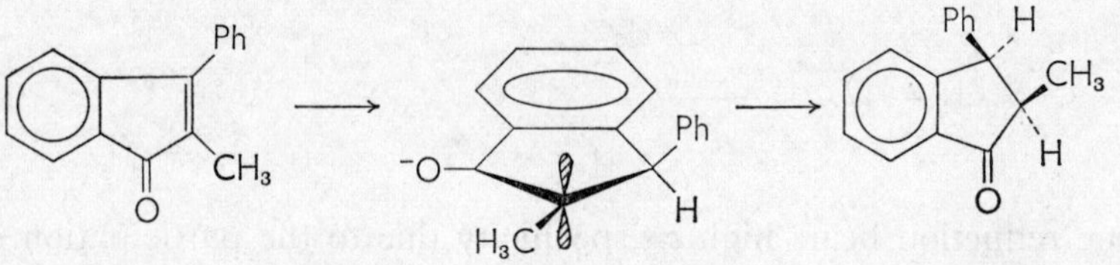

(**67**)   (**68**)   (**69**)

The reduction[96] of the phenanthrene derivative **70** by lithium and ethanol in ammonia leads initially to the *cis* isomer **71** which is convertible into the more stable *trans* isomer **72**. Although the ketoniza-

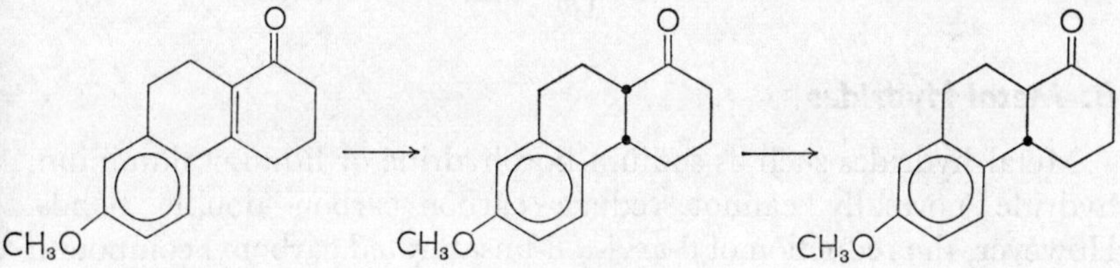

(**70**)   (**71**)   (**72**)

tion intermediate **73** is the more stable anion, the path of proton approach is hindered due to its being axial to both fused rings, while in **74** it is axial to one ring and equatorial to the other. Lithium–liquid

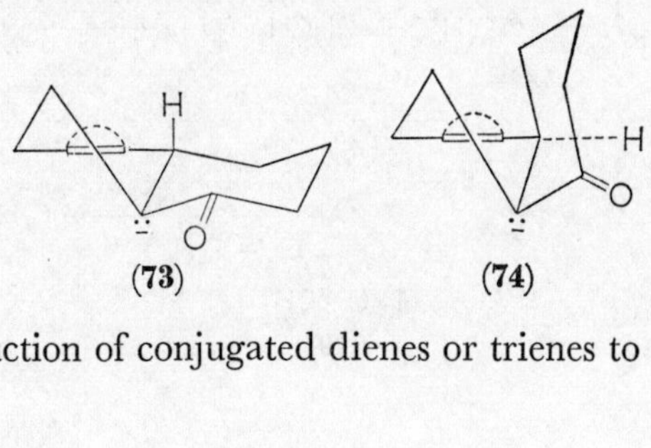

(**73**)   (**74**)

ammonia reduction of conjugated dienes or trienes to olefins, usually affords the thermodynamically more stable product. The process can be represented as a carbanion protonation[93]:

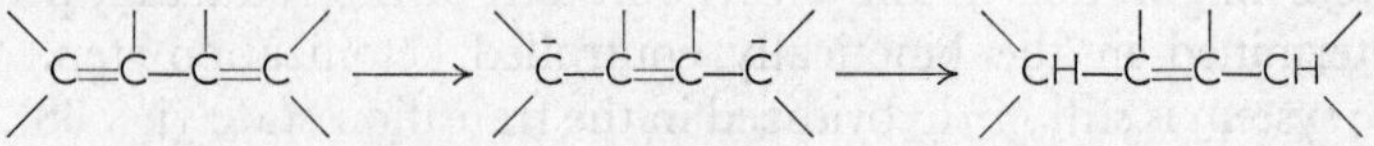

$$\text{C=C-C=C} \longrightarrow \text{C}^-\text{-C=C-C}^- \longrightarrow \text{CH-C=C-CH}$$

In the sodium–liquid ammonia reduction of butadiene (reaction 18) at the boiling point of ammonia ($-33°$C) the yield of *cis*-2-butene was 13% while at $-78°$C it was 50%. The reduction of 1,3-pentadiene was more specific (68% *cis*)[97]. It was assumed therefore that this

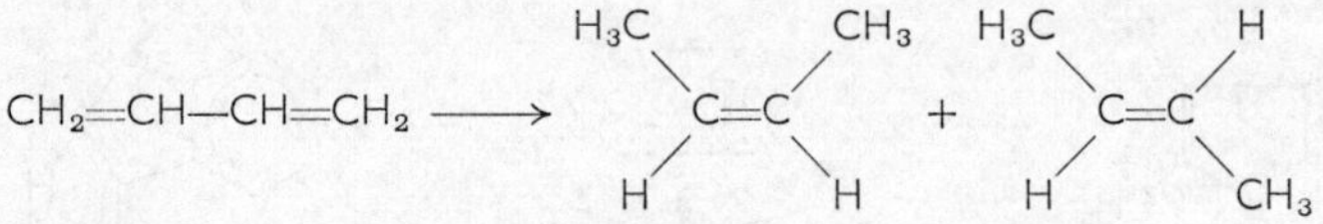

$$CH_2\!=\!CH\!-\!CH\!=\!CH_2 \longrightarrow \begin{matrix} H_3C \\ \\ H \end{matrix}\!\!C\!=\!C\!\!\begin{matrix} CH_3 \\ \\ H \end{matrix} + \begin{matrix} H_3C \\ \\ H \end{matrix}\!\!C\!=\!C\!\!\begin{matrix} H \\ \\ CH_3 \end{matrix} \quad (18)$$

diene reduction bears high *cis* specificity due to the participation of intermediates such as butenyl sodium (**75**).

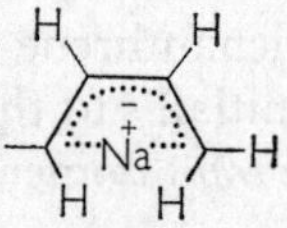

(structure **75**: butenyl sodium complex)

B. Metal Hydrides

Metal hydrides such as sodium borohydride or lithium aluminium hydride normally cannot reduce carbon–carbon double bonds. However, the reduction of β-aryl-α,β-unsaturated carbonyl compound such as cinnamaldehyde (**76**) to the saturated compound **77** was reported[98]. Furthermore, the reduction of cinnamyl alcohol (**78**)

$$C_6H_5CH\!=\!CHCHO \xrightarrow{\text{LiAlH}_4} C_6H_5CH_2CH_2CH_2OH$$
$$(76)(77)$$

$$C_6H_5CH\!=\!CHCH_2OH \longrightarrow (C_6H_5CH\!=\!CHCH_2O)_2Al\text{-}H \longrightarrow$$
$$(78)$$

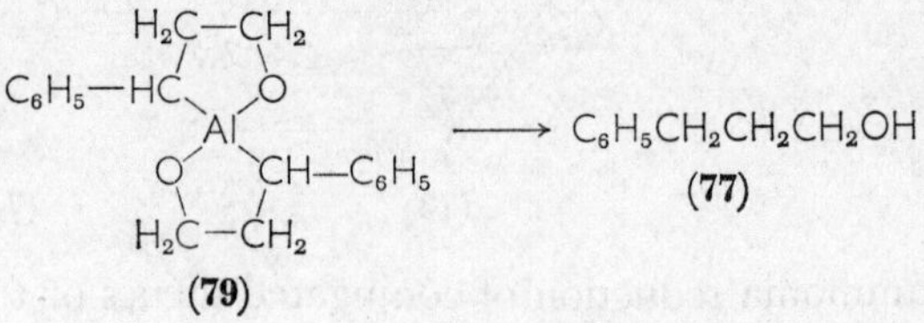

$$\text{(79)} \longrightarrow C_6H_5CH_2CH_2CH_2OH \quad (77)$$

with lithium aluminium hydride affords hydrocinnamyl alcohol (**77**), involving an intermediate organo-aluminium compound such as **79**. *Syn*-7-hydroxynorbornene (**80**) was reduced with lithium aluminium hydride to form 7-norbornanol (**81**), whereas *anti*-7-hydroxynorbornene (**82**) under more severe conditions, did not undergo double-bond reduction[99]. Double-bond reduction with lithium aluminium hydride occurred with 7-hydroxynorbornadiene (**83**) to form **82**. Results from deuterium-labelling experiments indicated the occurrence of *exo* deuteride addition followed by formation of an *exo* aluminium-carbon bond. Deuterolysis of this intermediate (**84**) resulted in *exo*, *exo*-dideuterio-*anti*-7-hydroxynorbornene (**85**)[99].

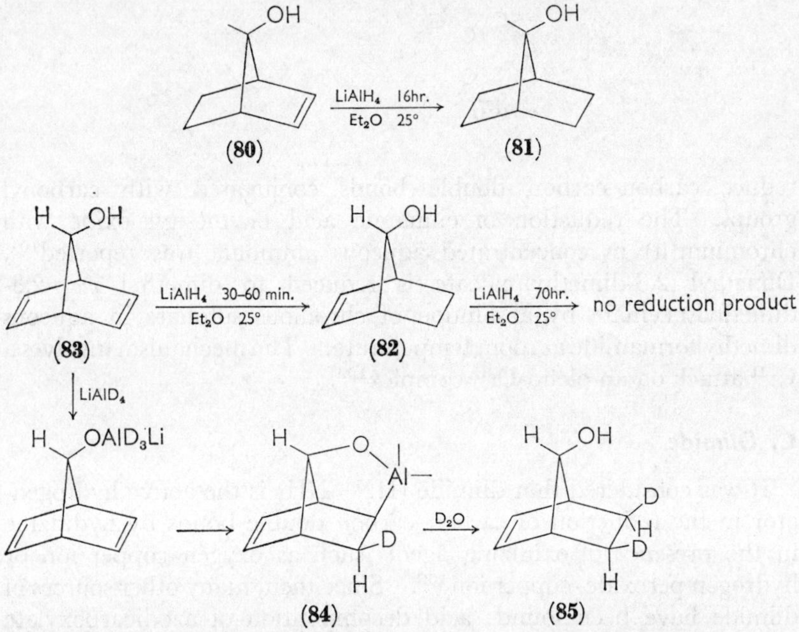

α,β-Unsaturated aldehydes and ketones (e.g. **86**) were reduced to the saturated carbonyl compounds by cobalt hydrocarbonyl through a π-oxa-propenyl tricarbonyl intermediate complex (**87**)[100].

Triphenylstannane, prepared from triphenyltin chloride and lithium aluminium hydride, can reduce the double bond of mesityl oxide[101].

$(CH_3)_2C=CHCOCH_3 + 2Ph_3SnH \longrightarrow (CH_3)_2CHCH_2COCH_3 + Ph_3SnSnPh_3$

Low-valent transition-metal species such as Cr^{II}[102] or Ti^{III}[103] can

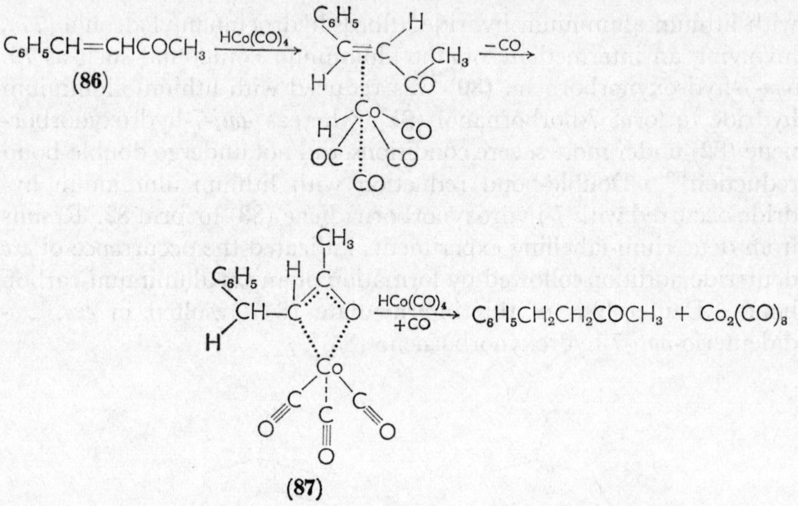

$$C_6H_5CH=CHCOCH_3 \xrightarrow{HCo(CO)_4} \text{(86)} \xrightarrow{-CO}$$

(86)

(87) $\xrightarrow{HCo(CO)_4}{+CO}$ $C_6H_5CH_2CH_2COCH_3 + Co_2(CO)_8$

reduce carbon–carbon double bonds conjugated with carbonyl groups. The reduction of cinnamic acid or mesityl oxide with chromium(II) in concentrated aqueous ammonia was reported[104]. Dimethyl 2,3-dimethylmaleate is reduced to dimethyl *meso*-2,3-dimethylsuccinate by a solution of chromous sulphate in aqueous dimethylformamide at room temperature. The mechanism involves a Cr^{II} attack on an olefin-Cr^{II} complex[105].

C. Diimide

It was considered that diimide (HN=NH) is the active hydrogenator in the reduction of carbon–carbon double bonds by hydrazine in the presence of oxidizing agents such as oxygen–copper ion or hydrogen peroxide–copper ion[106]. Since then, many other sources of diimide have been found: acid decomposition of azodicarboxylate salts[107–110], alkali decomposition of certain sulphonyl and acrylhydrazides[111], thermal decomposition of anthracene-9,10-diimide, alkali decomposition of hydroxyamino-*O*-sulphonic acid ($NH_2^-OSO_3H$)[112] and alkali decomposition of chloroamine[113].

Reduction with diimide is a stereospecific *cis* addition of hydrogen, giving no migration or *cis–trans* isomerization of double bonds, which is advantageous when compared with catalytic reduction. The reduction[114] of maleic acid (88) and fumaric acid (90) either by deuteriohydrazine-oxidizing agent or by potassium azodiformate-deuterium oxide, affords stereospecifically *meso-* (**89**) and *dl-*2,3-dideuteriosuccinic acid (**91**), respectively, indicating *cis* addition of hydrogen.

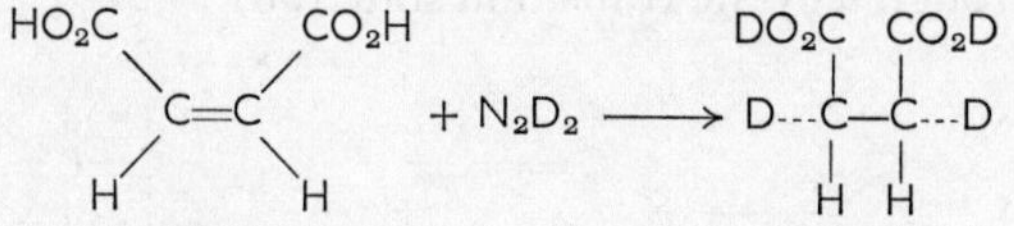

$$\underset{(88)}{\overset{HO_2C}{\underset{H}{>}}C=C\overset{CO_2H}{\underset{H}{<}}} + N_2D_2 \longrightarrow \underset{(89)}{DO_2C-\underset{H}{\overset{|}{C}}-\underset{H}{\overset{|}{C}}-CO_2D}$$

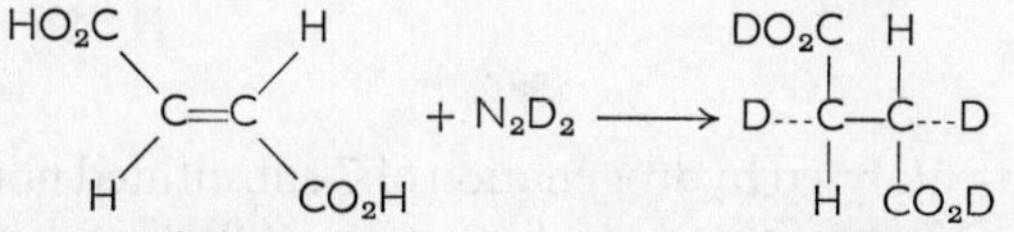

$$\underset{(90)}{\overset{HO_2C}{\underset{H}{>}}C=C\overset{H}{\underset{CO_2H}{<}}} + N_2D_2 \longrightarrow \underset{(91)}{DO_2C-\underset{H}{\overset{|}{C}}-\underset{CO_2D}{\overset{|}{C}}-D}$$

Cis hydrogenation was also observed in the reaction of dimethylmaleic acid (**92**) which afforded only *meso-*2,3-dimethylsuccinic acid (**93**)[114].

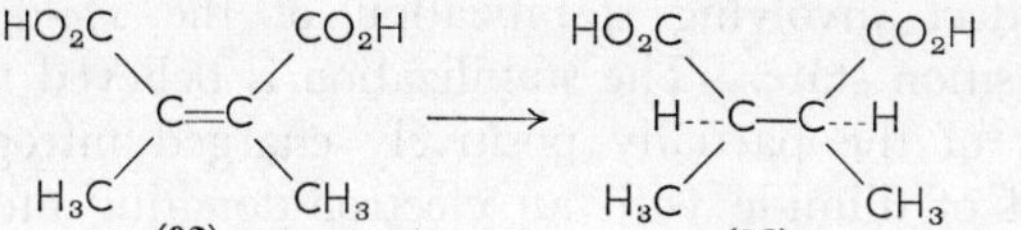

$$\underset{(92)}{\overset{HO_2C}{\underset{H_3C}{>}}C=C\overset{CO_2H}{\underset{CH_3}{<}}} \longrightarrow \underset{(93)}{\overset{HO_2C}{\underset{H_3C}{>}}H-\overset{|}{C}-\overset{|}{C}-H\overset{CO_2H}{\underset{CH_3}{<}}}$$

Attack by diimide takes place more easily at *trans* double bonds. Thus, fumaric acid is reduced more rapidly than maleic acid[115]. The *trans* double bonds of *cis,trans,trans-*cyclododeca-1,5,9-triene (**94**) can be selectively reduced yielding *cis-*cyclooecene (**95**)[116].

The hydrogen *cis* addition in general occurs from the less hindered side of the molecule. Thus, 2-norbornene-2,3-dicarboxylic acid (**96**) leads to the *endo-cis* product **97**[107].

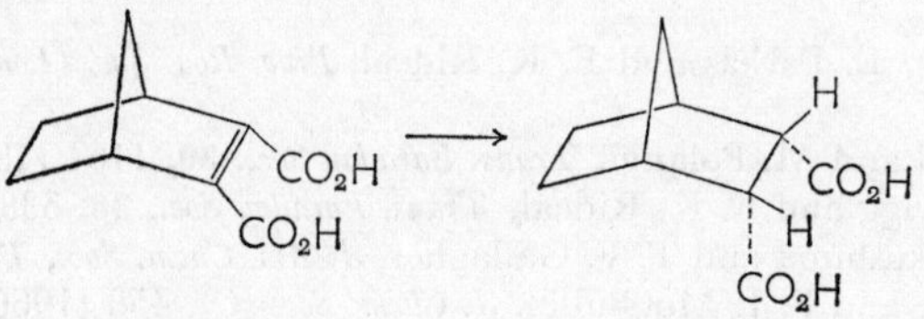

The hydrogen *cis* addition and the observed stereoselectivity both support the mechanism involving a synchronous transfer of a pair of hydrogens through a cyclic transition state (**98**)[107,110b,114].

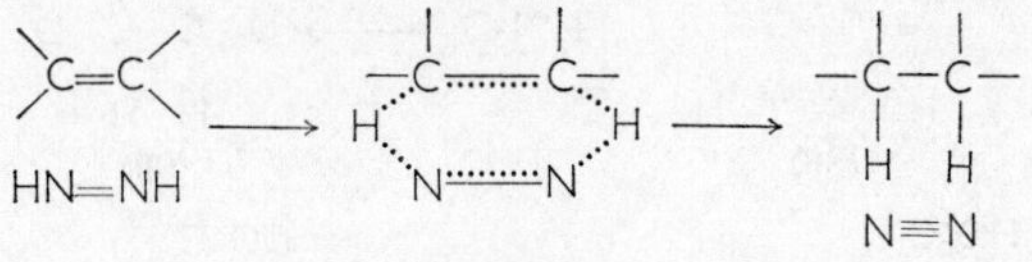

(**98**)

The platinum-catalysed hydrogenation of 7-substituted norbornadienes (**99**) yields *syn*-7-substituted norbornenes (**100**)[117], whereas diimide reduction of these dienes gives *anti*-7-substituted norbornenes (**101**) as the exclusive products[118]. This anomalous result has been tentatively rationalized by the suppression of adverse steric factors by a potent electronic effect involving stabilization of the *syn*-double bond-diimide transition state. The stabilization is believed to arise from coordination of the partially positively charged nitrogen–nitrogen double bond of diimide with an electron-donating atom (oxygen) in the 7-position[118].

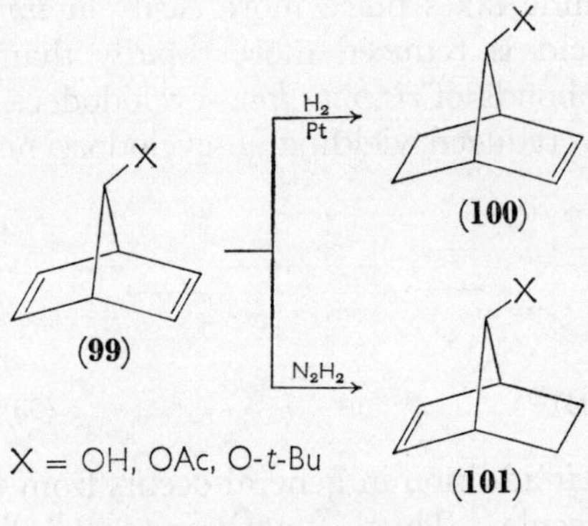

X = OH, OAc, O-*t*-Bu

IV. REFERENCES

1. A. Farkas, L. Farkas and E. K. Rideal, *Proc. Roy. Soc. (London)*, **146**, 630 (1934).
2. J. Horiuti and M. Polanyi, *Trans. Faraday Soc.*, **30**, 1164 (1934).
3. G. H. Twigg and E. K. Rideal, *Trans. Faraday Soc.*, **36**, 533 (1940).
4. D. K. Fukushima and T. F. Gallagher, *J. Am. Chem. Soc.*, **77**, 139 (1955).
5. I. Jardine and F. J. McQuillin, *J. Chem. Soc. (C)*, 458 (1966).
6. J. Chatt and L. A. Dancanson, *J. Chem. Soc.*, 2934 (1953).

7. J. J. Rooney, F. G. Gault and C. Kemball, *Proc. Chem. Soc.*, 407 (1960); F. G. Gault, J. J. Rooney and C. Kemball, *J. Catalysis*, **1,** 255 (1962).
8. J. J. Rooney, *J. Catalysis*, **2,** 53 (1963).
9. A. Farkas and L. Farkas, *J. Am. Chem. Soc.*, **60,** 22 (1938), *Trans. Faraday Soc.*, **35,** 715 (1939).
10. G. H. Twigg and E. K. Rideal, *Proc. Roy. Soc. (London)*, **171,** 55 (1939).
11. D. E. Eley, *Proc. Roy. Soc. (London)*, **178,** 452 (1941).
12. T. I. Taylor, *Catalysis*, Vol. V, Reinhold Publishing Corp., New York, 1957, p. 257.
13. O. Beeck, *Rev. Mod. Phys.*, **17,** 61 (1945), **20,** 127 (1948).
14. G. I. Jenkin and E. K. Rideal, *J. Chem. Soc.*, 2490 (1955).
15. G. T. Twigg, *Discuss. Faraday Soc.*, **8,** 90 (1950).
16. A. Farkas and L. Farkas, *Trans. Faraday Soc.*, **33,** 837 (1937).
17. K. N. Campbell and B. K. Campbell, *Chem. Rev.*, **31,** 77 (1942).
18. R. K. Greenhalgh and M. Polanyi, *Trans. Faraday Soc.*, **35,** 520 (1939).
19. R. P. Linstead, W. E. Doering, S. B. Davis, P. Levine and R. R. Whetstone, *J. Am. Chem. Soc.*, **64,** 1985 (1942).
20. C. L. Arcus, L. A. Cort, T. J. Howard and Le. BuLoc, *J. Chem. Soc.*, 1195 (1960).
21. H. J. Ringold, M. Gut, M. Hayano and A. Turner, *Tetrahedron Letters*, 835 (1962).
22. H. J. Brodee, M. Hayano and M. Gut, *J. Am. Chem. Soc.*, **84,** 3766 (1962).
23. J. B. Bream, D. C. Eaton and H. B. Hembest, *J. Chem. Soc.*, 1974 (1957).
24. G. V. Smith, *J. Catalysis*, **5,** 152 (1966).
25. S. Siegel and B. Dmuchovsky, *J. Am. Chem. Soc.*, **86,** 2192 (1964).
26. S. Siegel and G. V. Smith, *J. Am. Chem. Soc.*, **82,** 6082 (1960).
27. S. Siegel and G. V. Smith, *J. Am. Chem. Soc.*, **82,** 6087 (1960).
28. S. Siegel and G. V. McCalel, *J. Am. Chem. Soc.*, **81,** 3655 (1959).
29. S. Mitsui, Y. Nagahisa, G. Tomomura and M. Shionoya, *Shokubai (catalyst)* **13,** 74 (1968).
30. J.-F. Sauvage, R. H. Baker and A. S. Hussey, *J. Am. Chem. Soc.*, **82,** 6090 (1960).
31. J.-F. Sauvage, R. H. Baker and A. S. Hussey, *J. Am. Chem. Soc.*, **83,** 3874 (1961).
32. S. Siegel, *Advances in Catalysis*, Vol. 16, Academic Press, New York, 1966, p. 123.
33. S. Siegel, P. A. Thomas and J. T. Holt, *J. Catalysis*, **4,** 73 (1965).
34. J. J. Rooney and G. Webb, *J. Catalysis*, **3,** 488 (1964).
35. J. J. Rooney, *Chem. Brit.*, **2,** 242 (1966).
36. G. V. Smith and J. R. Swoap, *J. Org. Chem.*, **31,** 3904 (1966).
37. R. D. Schuetz and L. R. Caswell, *J. Org. Chem.*, **27,** 486 (1962).
38. G. V. Smith and J. A. Roth, *J. Am. Chem. Soc.*, **88,** 3879 (1966).
39. S. Siegel, M. Dunkel, G. V. Smith, W. Halpern and C. Cozort, *J. Org. Chem.*, **31,** 2802 (1966).
40. C. Kemball and E. K. Rideal, *Proc. Roy. Soc. (London)*, **187,** 53 (1946).
41. W. Cocker, P. V. R. Shannon and P. A. Staniland, *J. Chem. Soc.(C)*, 41 (1966).
42. H. O. House, R. G. Carlson, H. Muller, A. W. Noltes and C. D. Slates, *J. Am. Chem. Soc.*, **84,** 2614 (1962).

43. T. J. Howard, *Chem. Ind. (London)*, 1899 (1963).
44. S. Mitsui, Y. Senda and H. Saito, *Bull Chem. Soc. Japan*, **39,** 694 (1966).
45. S. Mitsui, K. Hebiguchi and H. Saito, *Chem. Ind. (London)*, 1746 (1967).
46. B. Fucks and R. G. Haber, *Tetrahedron Letters*, 1447 (1966).
47. S. Mitsui, K. Iijima and T. Masuko, *J. Chem. Soc. Japan (pure chemical section)*, **84,** 833 (1963); S. Mitsui, Y. Senda and K. Konno, *Chem. Ind. (London)*, 1354 (1963).
48. L. Crombie, P. A. Jenkins, D. A. Mitchard and J. C. Williams, *Tetrahedron Letters*, 4297 (1967).
49. G. C. Bond, G. Webb, P. B. Wells and J. M. Winterbottom, *J. Chem. Soc.*, 3218 (1965).
50. E. F. Meyer and R. L. Burwell, Jr., *J. Am. Chem. Soc.*, **85,** 2881 (1963).
51. W. B. Smith and J. L. Massingill, *J. Am. Chem. Soc.*, **83,** 4301 (1961).
52. Y. DuPont, *Bull. Soc. Chim.*, **3**(5), 1021 (1936).
53. L. W. Kern, R. L. Schunes and R. Adams. *J. Am. Chem. Soc.*, **47,** 1147 (1925).
54. B. A. Kazanskii and I. E. Grushko, *Dokl. Akad. Nauk. SSSR*, **87,** 767 (1952).
54a. G. C. Bond, *Catalysis by Metals*, Academic Press, London, 1962, p. 291.
55. N. A. Dobson, G. Eglinton, K. A. Raphael and R. G. Wils, *Tetrahedron*, **16,** 16 (1961).
56. G. V. Smith and R. L. Burwell, Jr., *J. Am. Chem. Soc.*, **84,** 925 (1960).
57. A. W. Weitkemp, *J. Catalysis*, **6,** 431 (1966).
58. W. F. Newhall, *J. Org. Chem.*, **23,** 1274 (1958).
59. B. A. Kazanskii, I. V. Gostunskaya and H. M. Granat, *Izv. Akad. Nauk. SSSR, Otd. Khim. Nauk.*, **4,** 670 (1953).
60. E. Breitner, E. Roginskii and P. N. Rylander, *J. Org. Chem.*, **24,** 1855 (1959).
61. W. F. Tuley and R. Adams, *J. Am. Chem. Soc.*, **47,** 3061 (1925); P. N. Rylander, N. Himelstein and M. Kilroy, *Engelhard Ind. Tech. Bull.*, **4,** 49, 131 (1963).
62. S. Yada, K. Yamauchi and S. Kudo, *Catalyst (Japan)*, **5,** 2 (1963); K. Yamagishi, S. Hamada, H. Arai and H. Yokoo, *J. Syn. Org. Chem. (Japan)*, **24,** 54 (1966).
63. H. S. Broadbent, G. C. Campbell, W. J. Bartley and J. H. Johnson, *J. Org. Chem.*, **24,** 1847 (1959).
64. J. A. Osborn, F. H. Jardine, J. F. Young and G. Wilkinson, *J. Chem. Soc. (A)*, 1711 (1966).
65. J. Halpern, *Ann. Rev. Phy. Chem.*, **16,** 103 (1965); *Chem. Eng. News Oct. 31.* 68 (1966).
66. J. Halpern, J. F. Harrod and B. R. James, *J. Am. Chem. Soc.*, **83,** 753 (1961); **88,** 5150 (1966).
67. R. F. Heck, *Mechanism of Inorganic Reactions*, Advances in Chemistry Series, No. 49, American Chemical Society, Washington D.C., 1965, p. 181.
68. J. Kwiatch, I. L. Mador and J. K. Seyler, *J. Am. Chem. Soc.*, **84,** 304 (1962).
69. J. Kwiatch, I. L. Mador and J. K. Seyler, *Reactions of Coordinated Ligands and Homogeneous Catalysis*, Advances in Chemistry Series, No. 37, American Chemical Society, Washington D.C., 1963, p. 201.
70. M. Murakami and J. W. Kang, *Bull. Chem. Soc. Japan*, **34,** 1243 (1962).

71. N. K. King and M. E. Winfield, *J. Am. Chem. Soc.*, **83,** 3366 (1961).
72. J. Kwiatch and J. K. Seyler, *J. Organometal. Chem.*, **3,** 421 (1965).
73. J. F. Young, J. A. Osvorn, F. H. Jardine and G. Wilkinson, *Chem. Comm.*, 131 (1965).
74. A. J. Birch and K. A. M. Walker, *J. Chem. Soc. (C)*, 1894 (1966).
75. A. J. Birch and K. A. M. Walker, *Tetrahedron Letters*, 4939 (1966).
76. A. J. Birch and K. A. M. Walker, *Tetrahedron Letters*, 1935 (1967).
77. C. Djerassi and J. Gutzwiller, *J. Am. Chem. Soc.*, **88,** 4537 (1966).
78. R. S. Nyholm, *Proc. 3rd. Internat. Cong. Catal.* (Ed. W. M. H. Sachter, G. C. A. Schutt and P. Zwietering), Vol. 1, North Holland Pub. Co., Amsterdam, 1965, p. 25.
79. L. Vaska and R. E. Rhodes, *J. Am. Chem. Soc.*, **87,** 4970 (1965).
80. D. Evans, J. A. Osborn, F. H. Jardine and G. Wilkinson, *Nature*, **208,** 1203 (1965).
81. P. S. Hallman, D. Evans, J. A. Osborn and G. Wilkinson, *Chem. Comm.*, 305 (1967).
82. E. N. Frankel, E. A. Emken, H. Itatani and J. C. Bailer, Jr., *J. Org. Chem.*, **32,** 1447 (1967).
83. J. C. Bailer, Jr. and H. Itatani, *J. Am. Chem. Soc.*, **89,** 1592, 1600 (1967).
84. H. A. Tayim and J. C. Bailer, Jr., *J. Am. Chem. Soc.*, **89,** 4330 (1967).
85. E. N. Frankel, E. A. Emken, H. M. Peters, V. L. Davison and R. O. Butterfield, *J. Org. Chem.*, **29,** 3292 (1964).
86. E. N. Frankel, E. A. Emken and V. L. Davison, *J. Org. Chem.*, **30,** 2739 (1965).
87. A. Misono and I. Ogata, *Bull. Chem. Soc. Japan*, **40,** 2718 (1967).
88. M. F. Sloan, A. S. Matlack and D. S. Breslow, *J. Am. Chem. Soc.*, **85,** 4014 (1963).
89. C. Doree, J. F. McGhie and F. Kurzer, *J. Chem. Soc.*, 988 (1948).
90. A. J. Birch and H. Smith, *Quart. Rev.*, **12,** 17 (1958).
91. G. Stork, P. Rosen and N. L. Goldman, *J. Am. Chem. Soc.*, **83,** 2965 (1961); M. J. Weiss, R. F. Schaub, G. R. Allen, Jr., J. F. Poletto, C. Pidacks, R. B. Conrow and C. J. Cuscia, *Tetrahedron*, **20,** 357 (1964).
92. C. B. C. Boyce and J. S. Whitehurst, *J. Chem. Soc.*, 2680 (1960).
93. D. H. R. Burton and C. H. Robinson, *J. Chem. Soc.*, 3045 (1954).
94. G. Stork and S. D. Darling, *J. Am. Chem. Soc.*, **82,** 1512 (1960); 86, 1761 (1964).
95. H. E. Zimmerman, *J. Am. Chem. Soc.*, **78,** 1168 (1956).
96. A. J. Birch, H. Smith and R. E. Thorton, *J. Chem. Soc.*, 1339 (1957).
97. N. L. Bauld, *J. Am. Chem. Soc.*, **84,** 4347 (1962).
98. F. A. Hochstein and W. G. Brown, *J. Am. Chem. Soc.*, **70,** 3484 (1948).
99. B. Franzus and E. I. Snyder, *J. Am. Chem. Soc.*, **87,** 3423 (1965).
100. R. W. Goetz and M. Orchin, *J. Am. Chem. Soc.*, **85,** 2782 (1963).
101. M. Pereyre and J. Valade, *Compt. Rend.*, **260,** 581 (1965).
102. W. Traube and W. Passarge, *Chem. Ber.*, **49,** 1692 (1961).
103. P. Karrer, Y. Yen and I. Reichstein, *Helv. Chim. Acta*, **13,** 1308 (1930).
104. K. D. Kopple, *J. Am. Chem. Soc.*, **84,** 1586 (1962).
105. C. E. Castro, R. D. Stephens and S. Moje, *J. Am. Chem. Soc.*, **88,** 4964 (1966).
106. C. E. Miller, *J. Chem. Education*, **42,** 254 (1965).

107. S. Hünig, R. H. Müller and W. Shier, *Tetrahedron Letters*, 353 (1961).
108. E. J. Corey, W. L. Mock and D. J. Pasto, (a) *Tetrahedron Letters*, 347 (1961). (b) *J. Am. Chem. Soc.*, **83,** 2957 (1961).
109. E. E. Van Tamelan and R. J. Timmons, *J. Am. Chem. Soc.*, **84,** 1067 (1962).
110. (a) E. E. Van Tamelen and R. S. Dewey, *J. Am. Chem. Soc.*, **83,** 3725 (1961).
    (b) E. E. Van Tamelen, R. S. Dewey, M. F. Leass and W. H. Pirkles, *J. Am. Chem. Soc.*, **83,** 4302 (1961).
111. R. S. Dewey and E. E. Van Tamelen, *J. Am. Chem. Soc.*, **83,** 3729 (1961).
112. R. Buyle, A. Van Overstractem and F. Floy, *Chem. Ind. (London)*, 839 (1964).
113. E. J. Corey and W. L. Mock, *J. Am. Chem. Soc.*, **84,** 685 (1962).
114. E. J. Corey, D. J. Pasto and W. L. Mock, *J. Am. Chem. Soc.*, **83,** 2957 (1961).
115. S. Hünig and H. R. Müller, *Angew. Chem.*, **74,** 215 (1962).
116. M. Ohno, M. Okamoto and S. Torimitsu, *Bull. Chem. Soc. Japan*, **39,** 316 (1966).
117. B. Franzus, W. C. Baird, Jr., E. I. Snyder and J. H. Surridge, *J. Org. Chem.*, **32,** 2845 (1967).
118. W. C. Baird, Jr., B. Franzus and J. H. Surridge, *J. Am. Chem. Soc.*, **89,** 410 (1967).

CHAPTER 5

Alkene Complexes of Transition Metals as Reactive Intermediates

JEAN-FRANÇOIS BIELLMANN

Faculte des Sciences, Strasbourg, France

HENRI HEMMER

Société Nationale des Pétroles d'Aquitaine, Lacq, France

and

JACQUES LEVISALLES

Faculte des Sciences, Nancy, France

I. INTRODUCTION	216
II. STRUCTURE OF THE COMPLEXES	216
A. Alkene Complexes	216
1. Dewar's MO Picture	216
2. An Organic Chemist's Picture	218
3. Stability	219
B. π-Allyl Complexes	221
III. DYNAMIC STEREOCHEMISTRY OF THE COMPLEXES	222
A. The Kinetic *Trans* Effect	222
B. The *Cis* Migration Reaction	223
IV. DOUBLE BOND MIGRATIONS	224
A. The Metal Hydride Addition-Elimination Mechanism	225
B. The π Allyl Complex Mechanism	228
1. The Iron-Carbonyl Catalysed Reaction	228
2. The Palladium Catalysed Reaction	230
C. Other Reactions	230
V. HYDROGENATION OF ALKENES	230
A. The Pentacyanocobaltate Catalysed Reaction	231
B. The Rhodium(I) Catalysed Reaction	232
C. The Ruthenium(II) Catalysed Reaction	235
1. Ruthenium(II) Chloride	235
2. Tris(triphenylphosphine)hydridochlororuthenium(II)	236
D. Other Catalysts	236

VI.	NUCLEOPHILIC REACTIONS OF THE COMPLEXES	236
	A. Oxidation of Alkenes	236
	1. Experimental Facts	237
	2. Reaction Mechanism	238
	3. Reaction in Acetic Acid	240
	B. Other Reactions	240
VII.	CARBONYLATION REACTIONS	242
	A. The Reaction of Carbon Monoxide within the Alkene Complex	243
	B. Reactions of π Allyl Complexes	244
	C. Reactions of Alkene Complexes	245
VIII.	OLIGOMERIZATION AND MIXED OLIGOMERIZATION	247
	A. Oligomerization of Ethylene	247
	B. Oligomerization of Higher Alkenes	248
	C. Oligomerization of Conjugated Dienes	249
	1. Linear Oligomerization	249
	2. Cyclic Oligomerization	250
	D. Condensations (Mixed Oligomerizations)	253
	1. Linear Condensations	253
	2. Cyclic Condensations	255
	E. Dimerization of Alkenes (Cyclobutanation)	255
IX.	REFERENCES	256

I. INTRODUCTION

Alkene complexes of transition metals were known at a very early stage of chemistry[1], but their significance as reactive intermediates is comparatively recent and can be traced down to the famous work by Reppe and his coworkers, on the chemistry of the related acetylene[2,3] and olefin complexes[4]. In the last ten years the growth of the field has been exponential, and there are no signs at the moment of any slackening in this expansion. Therefore it is perhaps timely to survey the present state of the chemistry of alkene complexes of transition metals. As there are many excellent reviews on various aspects of this field[5-18], the present chapter will be more critical than exhaustive.

II. STRUCTURE OF THE COMPLEXES

A. Alkene Complexes

1. Dewar's MO Picture

Although many complexes have been postulated as intermediates in the reaction of alkenes, comparatively few complexes have been

5. Alkene Complexes of Transition Metals as Reactive Intermediates 217

isolated. The structures of some of them are reviewed in Chapter 6 of Volume 1. Light was first shed on their electronic structure, when Dewar[19] suggested in 1951, a picture for the silver complexes, which has subsequently been largely confirmed. Dewar's picture, which was soon extended by Chatt and Duncanson[20] to platinum complexes, is as follows:

(a) The metal has an empty orbital (s or sp in the case of Ag^+, dsp^2 in the case Pt^{2+}), which can share an electron pair donated by a ligand. In the case of a ligand alkene, the electron pair is the π-electron pair, with the appropriate energy and symmetry to overlap with the vacant orbital of the metal (Figure 1A).

The situation is thus very similar to what happens in a sulphoxide. A formal positive charge is placed on the alkene, a formal negative charge on the metal.

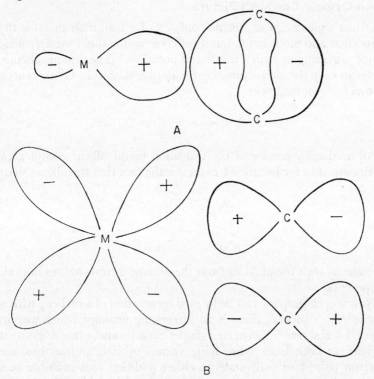

Figure 1. Orbital overlap in metal–olefin complexes. A, hybrid sp or dsp^2 orbital of metal with π orbital of olefin; B, d orbital of metal with π^* orbital of olefin.

(*b*) The electric balance can be restored, when the metal has *filled d* orbitals, by back donation of a *d* electron pair from the metal to the π^* vacant orbital of the alkene, which has the appropriate energy and symmetry (Figure 1B).

The situation is thus similar, but opposite to what happens in a sulphoxide, when the oxygen can donate back electrons to an *empty d* orbital on sulphur. It is also similar to what happens with a triphenyl phosphine complex of a transition metal, when the phosphorus gives to the metal a sp^3 electron pair, and receives back a *d* electron pair from the metal, into one of its vacant *d* orbitals.

This molecular-orbital picture thus distinguishes a σ bond between the metal and the alkene (cf. Figure 1A), which is a $sp\sigma\text{-}p\sigma$ bond, and a π bond (cf. figure 1B), which is a $d\pi\text{-}\pi^*\pi$ bond.

2. An Organic Chemist's Picture

Dewar's picture of an alkene complex of a transition metal is thus quite clear and most useful, but it suffers from the slight inconvenience of not providing us with a shorthand notation. The organic chemists prefer to keep the old picture, i.e. the double dash (C=C) and curved arrows (⌢), for instance:

$$CH_3-CH\overset{\frown}{=}CH_2 \; H^+ \longrightarrow CH_3-\overset{+}{C}H-CH_3$$

An analogous picture of the transition metal–alkene complex can be drawn, thus for instance **1** expresses the fact that the alkene shares

(1) 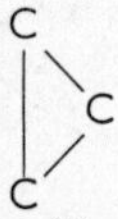   (2)

4 electrons with metal M (2 from the alkene, 2 from M), as in cyclopropane (**2**).

Now a cyclopropane can be formed by reaction of a carbene with an alkene[21,22]. Indeed, there is an interesting analogy, which was first drawn by Halpern[23], between a singlet carbene and a transition-metal derivative: both have a low-lying vacant orbital, and an unshared electron pair; both will undergo either addition to a multiple bond (e.g. an alkene), or insertion into a single bond (e.g. H—Cl or H—H). There is however a significant difference: the bonds formed by the carbene–alkene interaction are usually much stronger than the bonds formed by the metal–alkene interaction, as shown by the enthalpy differences.*

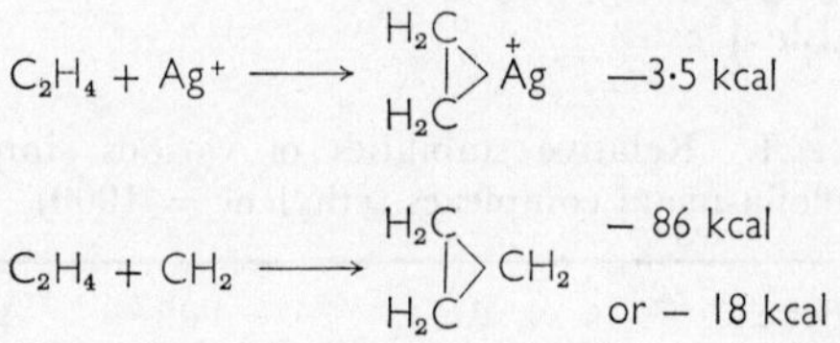

$$C_2H_4 + Ag^+ \longrightarrow \begin{array}{c} H_2C \\ H_2C \end{array}\!\!\!\!\triangleright Ag^+ \quad -3\cdot5 \text{ kcal}$$

$$C_2H_4 + CH_2 \longrightarrow \begin{array}{c} H_2C \\ H_2C \end{array}\!\!\!\!\triangleright CH_2 \quad \begin{array}{l} -86 \text{ kcal} \\ \text{or} -18 \text{ kcal} \end{array}$$

It must be stressed here that the 'organic chemist's picture' has one draw-back, because it may lead one to think that the alkene occupies two coordination sites, one site for each carbon atom. This is not true: each alkene molecule occupies only one coordination site, exactly in the same way as the oxygen atom occupies only one coordination site of the sulphur in a sulphoxide.

3. Stability

The stability of the transition metal–alkene complexes depends both on the metal and on the alkene, and certain generalizations can be made:

(a) For complexes of the same general composition the complex is stronger the lower the metal is in the periodic table. Thus in the series $[(\text{alkene})MX_3]^-$, Zeise's salt $[Pt(C_2H_4)Cl_3]^-$, the first olefin complex ever recorded in the chemical literature,[1] is quite stable and fairly unreactive. The corresponding palladium complex has been detected kinetically, and, under some circumstances, will be transformed into $[(C_2H_4)PdCl_2]_2$, a bimetallic complex which can be isolated[26-28]. The corresponding nickel complex has never been detected, and no nickel–alkene complex is known, unless the alkene is chelating (e.g. 1,5-cyclooctadiene or norbornadiene)[29];

(b) For metals of the same periodic row, stability differences for various olefin complexes are the more pronounced, the lower the oxidation state (Table 1);

(c) The more electron donating the alkene, the less stable is the complex, at least for negatively substituted or even slightly positively substituted ligands, fluorine being more effective than chlorine (Table 1 and rhodium(I) complexes of Table 2);

* The -86 kcal value is based on the estimated enthalpy of formation of cyclopropane[24] and the -18 kcal value is based on the additivity of bond energies[25].

(d) Symmetrical withdrawal of negative charge by the alkene stabilizes the complex better than dissymetrical withdrawal (Table 2);
(e) Steric strain (e.g. *cis* as compared to *trans*) in the alkene stabilizes the complex (Table 1);

TABLE I. Relative stabilities of various simple olefin-metal complexes (ethylene = 1000)

Olefin	$Rh^{I\ a}$	$Pd^{II\ b}$	$Ag^{I\ c}$
Ethylene	1000	1000	1000
Propylene	78	830	430
1-Butene	92	640	500
cis-2-Butene	4·1	490	280
trans-2-Butene	2·0	260	91
iso-Butene	0·35	very small	

[a] At 25°C[30].
[b] At 25°C[31].
[c] At 25°C[32].

(f) For a diene, the chelating power is at maximum efficiency when the double bonds are separated by two sp^3 carbon atoms (Table 2). The mutual disposition of the two double bonds may be important (Table 2);

Indeed these are the facts which make the chemistry of transition metal–alkene complexes both fascinating and frustrating: fascinating because of the wide scope, which is left to the chemist, but frustrating

TABLE 2. Relative stabilities of various substituted olefin-metal complexes

Olefin	Complex[a] with Rh^I	Olefin	Complex[b] with Ag^I
Ethylene	1	Ethylene	1
Fluoroethylene	0·32	1,3-Butadiene	0·19
Chloroethylene	0·17	1,4-Pentadiene	0·45
Methoxyethylene	0·018	1,5-Hexadiene	1·29
Phenylethylene	0·08	1,6-Heptadiene	0·66
Cyanoethylene	> 50	1,5-Cyclooctadiene	3·36
cis-1-2-Dichloroethylene	0·07	1,4-Cyclohexadiene	0·22
cis-1-2-Difluoroethylene	1·59	Norbornadiene	1·51
Tetrafluoroethylene	59		
Tetrachloroethylene	No reaction		

[a] At 25°C[30].
[b] At 40°C[33].

because it has not yet been very easy to work with known complexes toward definite products.

B. π-Allyl Complexes

The chemistry of transition metal–alkene complexes is complicated by the fact that alkenes with allylic C—H or C—X (X = halogen usually) can exhibit further transformations to give π-allyl complexes **3** and **4**[7] according to equations (1) and (2).

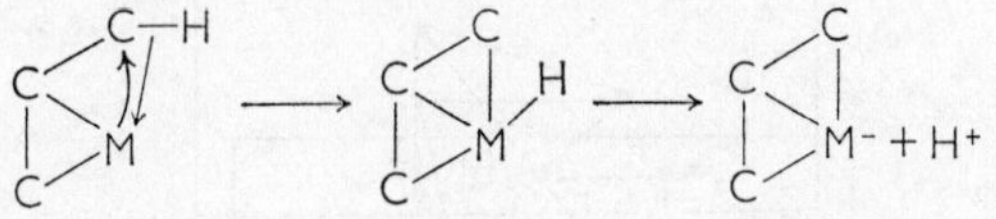

$$\begin{array}{c}\text{(equation 1, structure 3)} \longrightarrow \longrightarrow M^- + H^+ \end{array} \quad (1)$$

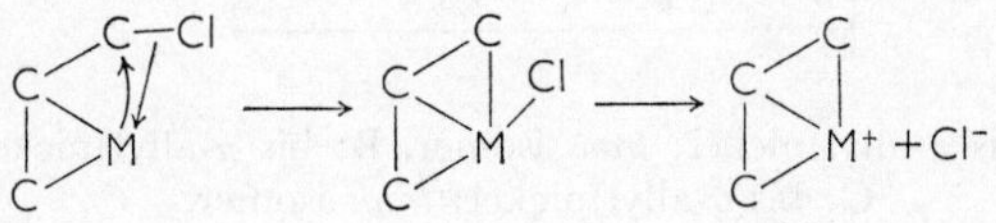

$$\begin{array}{c}\text{(equation 2, structure 4)} \longrightarrow \longrightarrow M^+ + Cl^- \end{array} \quad (2)$$

In fact, the structure of a π-allyl complex will depend on the metal (and its attached ligands), and on the physical state. At low temperature,[34] in crystalline bis (π-allyl) nickel complexes, the structure is symmetrical with regard to the allyl moiety: i.e., for a propene derivative, both C—C bonds are of equal length (1.376 Å) and hydrogen atoms in symmetrical positions cannot be distinguished (by n.m.r for instance). X-ray studies[35] have shown that the allyl moiety occupies *two* coordination sites, and, in bis-(π-allyl) nickel complexes, the two allyl fragments are symmetrically situated, with regard to the nickel atom, which is a centre of symmetry. This is the so-called *trans* complex (Figure 2A).

Such a centrosymmetric structure is of course only compatible with a dsp^2 (square-planar) hybridization of nickel, and not with sp^3 (tetrahedral) hybridization. Contrary to other statements[6] this must mean that nickel is at the +2 oxidation level, and not at the zero oxidation level (as it is in tetrahedral nickel tetracarbonyl). The change in oxidation level is adequately expressed in equation (1).* It has been shown that in solution, bis (π-allyl) nickel is an

---

* It should be stressed, however, that **3** and **4** do not depict quite accurately the structure of a π-allyl complex, in that only two, and not three, coordination sites are occupied by the allyl moiety, but they are very useful for the understanding of reaction processes.

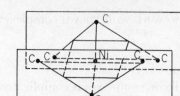
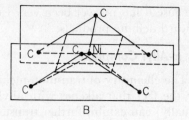
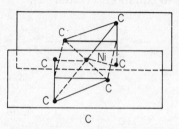

Figure 2. A: bis(π-allyl)nickel: *trans* isomer. B: bis(π-allyl)nickel: *cis* isomer. C: bis(σ-allyl)nickel: *trans* isomer.

equilibrium mixture of two species,[34,36,37] the *trans* and the *cis* complexes (Figure 2B).

Another structure of π-allyl complexes is suggested by n.m.r. studies: the so-called σ-allyl structure (Figure 2C) which contains an unsymmetrical allyl group, one of the terminal carbon atoms being bonded to the metal by an ordinary, although probably highly ionic, σ bond, the other two being probably bonded to the metal by the type of double bond described in Section A. In some cases at least the σ-allyl structure is more stable at elevated temperatures than the π-allyl structure: the loss in heat content (resonance energy) of the π-allyl grouping is then compensated by the higher entropy of the σ-allyl grouping.

For the understanding of chemical reactions, it does not seem at present that a π-allyl complex behaves very differently from a σ-allyl complex. Therefore the simple representations **3** and **4** are satisfactory and will be used in the following sections.

III. DYNAMIC STEREOCHEMISTRY OF THE COMPLEXES

A. The Kinetic Trans Effect

Detailed kinetic and stereochemical studies on substitution reactions with square-planar and octahedral transition-metal complexes

5. Alkene Complexes of Transition Metals as Reactive Intermediates

have shown that some ligands L on metal M powerfully accelerate, in a predictable way, the departure of another ligand X, provided that X and L are located in *trans* positions with respect to M, i.e. X-M and M-L bonds are colinear.[40] Some idea of this effect can be gained from examination of Table 3, in which a parallel behaviour is seen between the kinetic *trans* effect and the substituent influence in the unimolecular decomposition of *p*-substituted benzenediazonium ions.

TABLE 3. *Trans* effect in dsp^2 complexes and substituent effect in *p*-substituted benzenediazonium ions

Ligand or substituent	*trans* displacement in dsp^2 complexes		Relative rates unimolecular N_2 loss in *p*-substituted benzenediazonium ions[b]
	Relative rates[a]	Similar ligands	
H	10^5	CO, CN^-	530
CH_3	10^2	I^-, C_6H_5	26
Cl	1	Br^-	1
OH	10^{-1}	C_5H_5N, NH_3	0·66

[a] From reference 40.
[b] From reference 41.

Since the origin of this effect is not yet completely understood, a full discussion will be avoided here. The following sequence of the *trans* effect should be kept in mind:

$$\text{C=C} > Cl^- > OH^-$$

It should be emphasized that *cis* ligands usually have no predictable effect.

B. The Cis Migration Reaction

This has been termed frequently the *cis* ligand insertion reaction [42,43] and can be represented, for instance, by equation (3) for the case of a doubly bonded ligand (here an alkene). The alkene is inserted into the M—R bond, yielding a metal complex with carbenoid properties. For such an electronic reorganization to take place readily (i.e.

$$\underset{M}{\overset{R}{\underset{|}{\rightleftharpoons}}}\underset{C}{\overset{C}{\bigg|}} \longrightarrow \underset{M:}{\overset{R}{\underset{|}{}}}\underset{C}{\overset{C}{\bigg|}} \quad (3)$$

without large nuclear displacements), both ligands, R and the alkene, must be *cis* and not *trans* with respect to M. Of course equation (3) is just a special case of the formally simpler reaction equation (4a), e.g. the last step of the decarbonylation of aldehydes by $RhCl(PPh_3)_2$ equation (4b)[44,45].

$$\underset{M}{\overset{R}{\underset{|}{\rightleftharpoons}}} L \longrightarrow \overset{\square}{M:} + \underset{L}{\overset{R}{\bigg|}} \quad (4a)$$

$$\underset{LCl}{\overset{L}{\underset{|}{Rh}}} \xrightarrow{R-CH=O} \underset{LCl}{\overset{LH}{\underset{|}{Rh}}}\overset{\diamond}{\underset{C}{\underset{\|}{O}}}R \longrightarrow \underset{LCl}{\overset{LH(R}{\underset{|}{Rh}}}\overset{}{\underset{C}{\underset{\|}{O}}} \longrightarrow \underset{L}{\overset{H-R}{\underset{Rh}{\overset{L+Cl}{\underset{|}{}}}}}\overset{}{\underset{C}{\underset{\|}{O}}} \quad (4b)$$

$L = Ph_3P$

Now it is obvious that equation (4a) is just the reverse of equation (5), which an organic chemist would call an insertion reaction (in carbene chemistry). It seems therefore more appropriate to discard

$$\underset{M:}{\overset{R}{\underset{\square}{\rightleftharpoons}}} L \longrightarrow \underset{M-L}{\overset{R}{\underset{|}{}}} \quad (5)$$

the term 'ligand insertion reaction' for equation (3) and to keep the term 'ligand migration reaction,' which encompasses similar reactions such as the Pummerer reaction[46–48] (equation 6) or the Hillman-Brown reaction[49,50] (equation 7).

$$\underset{H_3C}{\overset{H_3C}{\underset{\diagup}{}}}\overset{+}{S}-OR \longrightarrow \underset{H_3C}{\overset{H_2C}{\underset{\diagup}{}}}\overset{+}{S}-OR \longrightarrow \underset{H_3C}{\overset{H_2C}{\underset{\diagup}{}}}S-O-R \quad (6)$$

$$R_2B-\overset{R}{\underset{|}{C}}\overset{+}{\equiv}\overset{\cdot\cdot}{O} \longrightarrow R_2B-\overset{R}{\underset{|}{C}}=O \quad (7)$$

IV. DOUBLE BOND MIGRATIONS

Transition-metal compounds are able to transform olefins into double-bond isomers. This interesting reaction must always be borne in mind when working with transition metal–alkene complexes, because the product of a reaction may be very different from the one expected from the structure of the substrate alkene.

Two mechanisms have been proposed for these double-bond migrations, the metal hydride addition-elimination and the π-allyl complex intermediate.

A. The Metal Hydride Addition-Elimination Mechanism

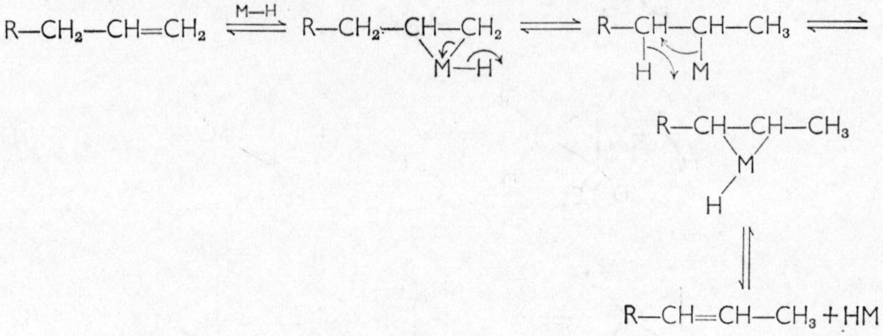

$$R-CH_2-CH=CH_2 \xrightleftharpoons{M-H} R-CH_2-CH-CH_2 \rightleftharpoons R-CH-CH-CH_3 \rightleftharpoons$$
$$\phantom{R-CH_2-CH=CH_2 \xrightleftharpoons{M-H} R-CH_2-}M-HHM$$

$$R-CH-CH-CH_3$$
$$M$$
$$H$$

$$R-CH=CH-CH_3 + HM$$

The rhodium(I) catalysed isomerization of butene is the reaction which is best understood[51]. Cramer first showed that the reaction is reversible and that the rate equation has the form

$$\text{rate} = k \,(\text{alkene})\,(\text{Rh}^\text{I})(\text{H}^+)(\text{Cl}^-)$$

The reaction is strongly inhibited by ethylene and other strongly coordinating ligands. Moreover, u.v. spectral data show that a hexacoordinated rhodium species is involved. The following schemes can explain these facts:

$$\text{Alkene} + \text{Rh}^\text{I} \rightleftharpoons [(\text{alkene})\,\text{Rh}^\text{I}] \qquad (8)$$
$$[(\text{Alkene})\text{Rh}^\text{I}] + \text{H}^+ + \text{Cl}^- \rightleftharpoons [(\text{alkene})\,\text{Rh}^\text{III}\,\text{HCl}]$$

or

$$\text{Rh}^\text{I} + \text{H}^+ + \text{Cl}^- \rightleftharpoons (\text{Rh}^\text{III}\,\text{HCl}) \qquad (9)$$
$$\text{Alkene} + (\text{Rh}^\text{III}\,\text{HCl}) \rightleftharpoons [(\text{Alkene})\,\text{Rh}^\text{III}\,\text{HCl}]$$

It is not known whether equations (8) or (9), or both, are involved, but this is not very important. The intermediacy of a rhodium hydride complex was not definitely established, but other works with rhodium[52-55], cobalt[56-58]*, iridium[60-62] or platinum[63-64] compounds and u.v. spectral data support the assumption of the presence of a rhodium hydride. Inhibition by ethylene is easily understood, since ethylene can compete very efficiently for the metal with the alkene.

* For a conclusion to the contrary, see Reference 59.

The reaction is intermolecular, as shown by the work of Harrod and Chalk[52-53]: deuterated 1-heptene and undeuterated 1-pentene, exchange deuterium. In deuterated solvents, 1-butene is deuterated (at $C_{(1)}$) and isomerized (to undeuterated 2-butenes) at about the same rates. This can be explained in terms of equations (10–13).

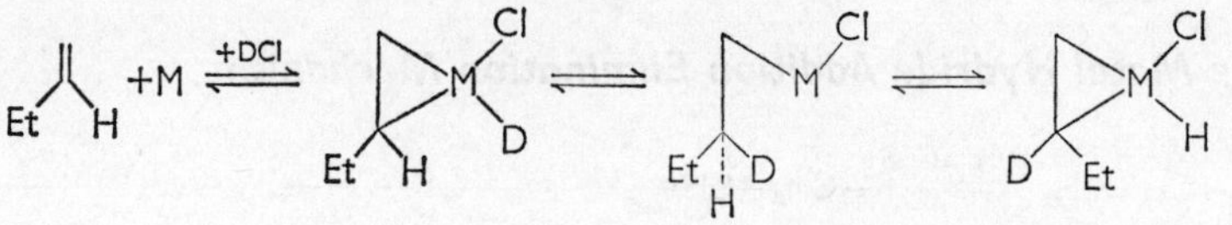

$$\text{Et-CH=CH}_2 + M \xrightarrow{+DCl} \text{[complex with Et, H, D, Cl, M]} \rightleftharpoons \text{[complex]} \rightleftharpoons \text{[complex]} \quad (10)$$

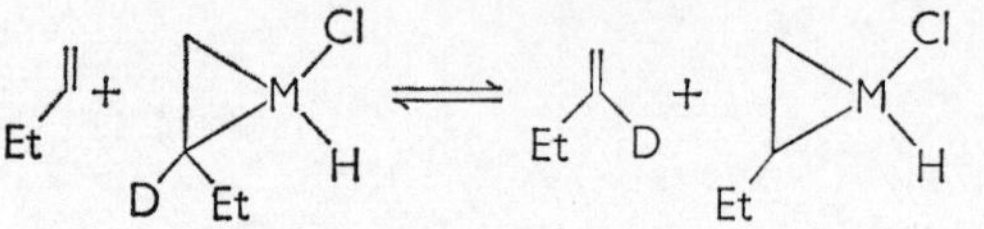

$$\quad (11)$$

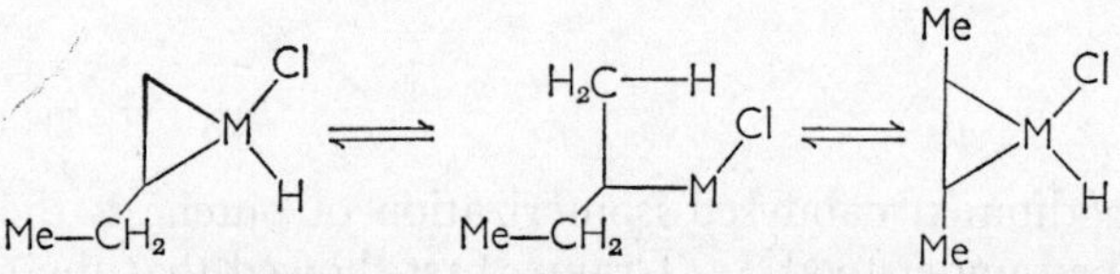

$$\quad (12)$$

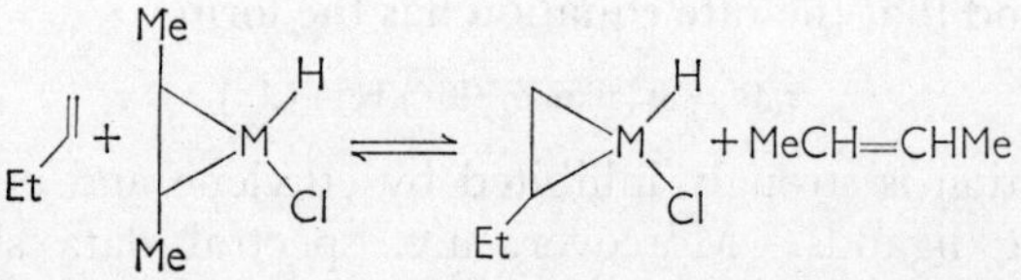

$$\quad (13) + \text{MeCH=CHMe}$$

The transition metal–alkene complex can be isomerized to an alkyl rhodium compound: primary carbanions being more stable than secondary ones, the preferred step should be the one shown in equation (10). If the $C_{(1)}-C_{(2)}$ bond is rotated by 120°, the complex which can then be formed will have protium bonded to rhodium, instead of deuterium (equation 10). Ligand exchange (equation 11) will thus produce specifically deuterated 1-butene and a hydride complex.

As the reaction proceeds, the concentration of this hydride complex will increase. At this stage the alternative isomerization of the hydride complex can make itself felt (equation 12). Now 1-alkene forming more stable complexes than 2-alkenes (table 1), the 2-alkene will be replaced by more 1-alkene in the complex. In a deuterated solvent, *cis*-2-butene initially gives *trans*-2-butene, which is partly deuterated at $C_{(2)}$, and 1-butene which is fully deuterated (at $C_{(3)}$). This is depicted in equations (14–19).

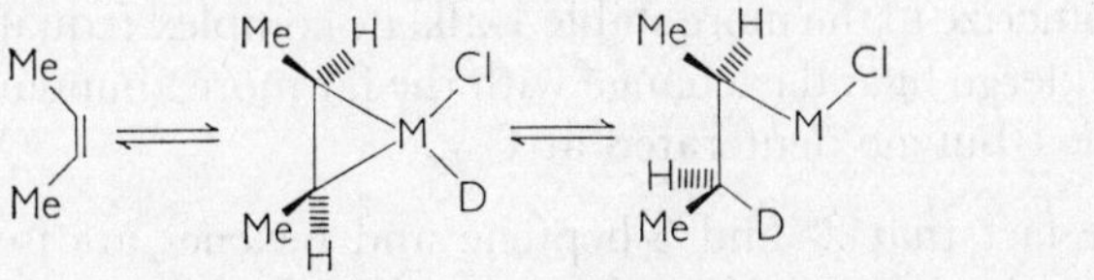

5. Alkene Complexes of Transition Metals as Reactive Intermediates

$$\text{(14)}$$

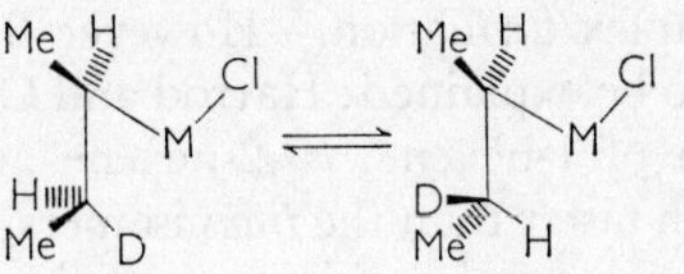

$$\text{(15)}$$

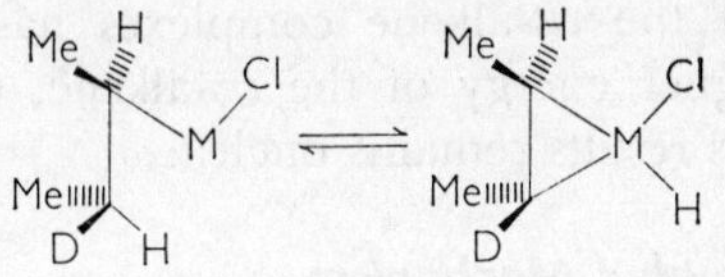

$$\text{(16)}$$

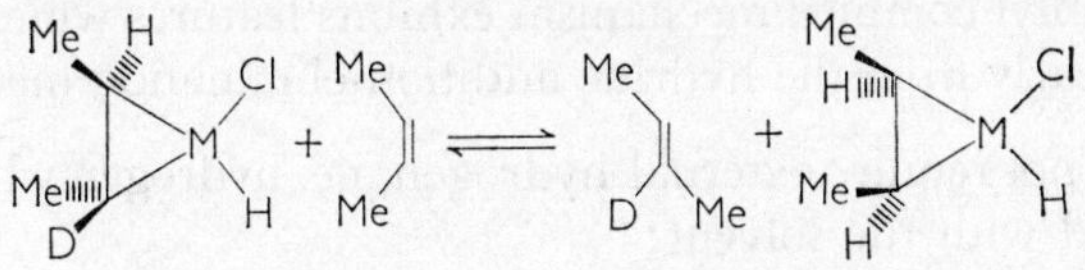

$$\text{(17)}$$

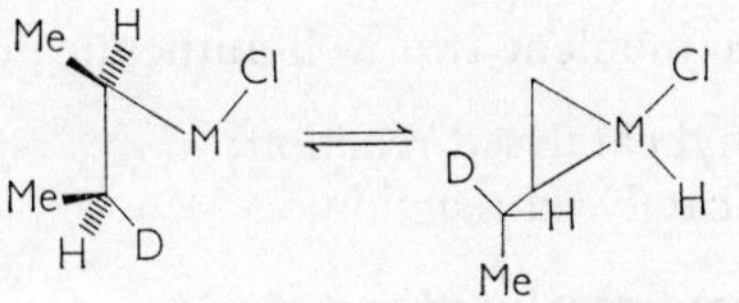

$$\text{(18)}$$

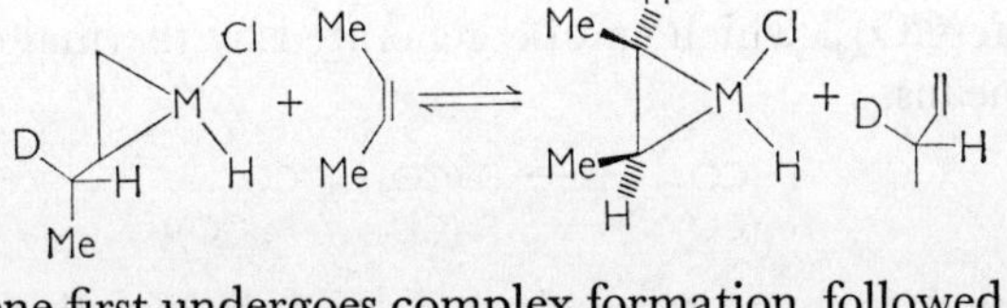

$$\text{(19)}$$

Cis-2-butene first undergoes complex formation, followed by hydride migration to give the alkyl rhodium (equation 14), which may react by three paths:

(*a*) Give back the *cis*-2-butene complex;
(*b*) Undergo rotation around the $C_{(2)}$–$C_{(3)}$ bond (equation 15) to give another conformer, which can revert now to the *trans*-2-butene complex (equation 16); the latter will undergo ligand exchange to give *trans*-2-butene deuterated at $C_{(2)}$ (equation 17);

(c) Isomerize to the more stable 1-alkene complex (equation 18), which can undergo ligand exchange with the far more abundant *cis*-2-butene, to give 1-butene deuterated at $C_{(3)}$.

The fact that 2- and 3-heptene and hexenes are isomerized more slowly than 1-hexene is easily accounted for by an unfavourable pre-equilibrium for complex formation. However there is one reaction which still remains to be explained: Harrod and Chalk[52] noticed that, in the isomerization of 1-hexene, *cis*-2-hexene and *cis*-3-hexene are initially formed much faster than the *trans* isomers, although they form more stable complexes (this is again true of the heptenes). As the greater stability of the *cis*-alkene complexes has reasonably been assigned to the higher energy of the *cis*-alkene, the reason behind Harrod and Chalk's results remains unclear.

B. The π-Allyl Complex Mechanism

The π-allyl complex mechanism exhibits features which distinguish it very clearly from the hydride addition–elimination mechanism:

1. It does not require external hydrogen, i.e. hydrogen which could be exchanged with the solvent;
2. It involves 1,3 shifts and not a 1,2 shift.

There are at the moment two well-authenticated such reactions:

1. The iron carbonyl catalysed reaction;
2. The palladium catalysed reaction.

I. The iron-carbonyl catalysed reaction

This reaction[65-69] requires a coordinatively unsaturated iron carbonyl, $Fe(CO)_4$, which can be generated by thermal or by photochemical means.

$$Fe(CO)_5 \rightleftharpoons Fe(CO)_4 + CO$$
$$Fe_2(CO) \rightleftharpoons Fe(CO)_4 + Fe(CO)_5$$

As expected, carbon monoxide and strongly coordinating ligands, like amines or phosphines, inhibit the reaction by preventing the alkene to coordinate with the metal complex. Acetone and isopropanol also inhibit the reaction by competing with the alkene[67].

The second condition (1,3 shifts, no 1,2 shift) has not been conclusively demonstrated, but the first condition (no exchange with external hydrogen) has been shown to be fulfilled[69]. Thus 3,3-ditritio-1-octene is isomerized in the presence of unlabelled 1-hexene

5. Alkene Complexes of Transition Metals as Reactive Intermediates

or undecene, without exchanging tritium. The reaction can be represented by equation (20). The complex of the isomerized alkene can react further according to equations (21) or (22). The latter will

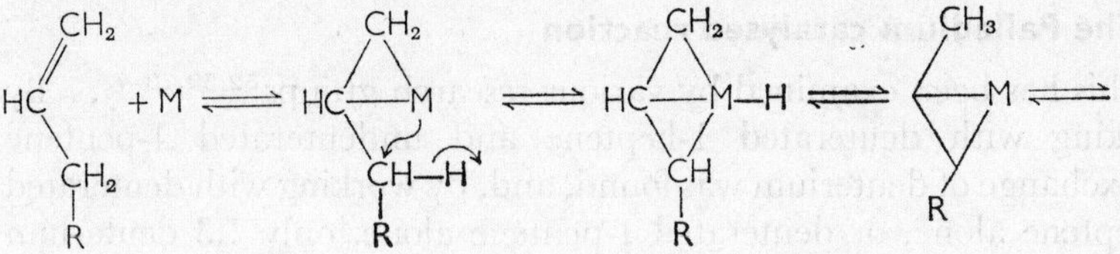

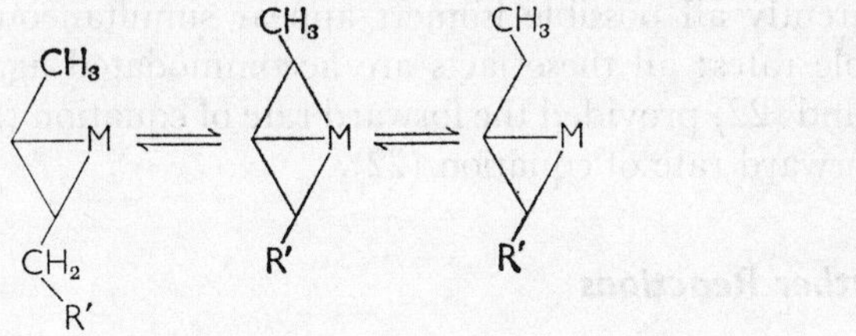

$$\underset{R}{\overset{CH_2}{\underset{CH_2}{HC}}} + M \rightleftharpoons \underset{R}{\overset{CH_2}{\underset{CH-H}{HC-M}}} \rightleftharpoons \underset{R}{\overset{CH_2}{\underset{CH}{HC-M-H}}} \rightleftharpoons \underset{R}{\overset{CH_3}{\underset{}{\triangleleft-M}}} \rightleftharpoons \text{etc.} \quad (20)$$

$$\underset{\underset{R'}{CH_2}}{\overset{CH_3}{\triangleleft-M}} \rightleftharpoons \underset{R'}{\overset{CH_3}{\triangleleft-M}} \rightleftharpoons \underset{R'}{\overset{CH_3}{\triangleleft-M}} \quad (21)$$

$$\underset{\underset{R}{CH_2}}{\overset{CH_3}{\triangleleft-M}} \rightleftharpoons \underset{CH_2R}{\overset{CH_3}{\|}} + M \quad (22)$$

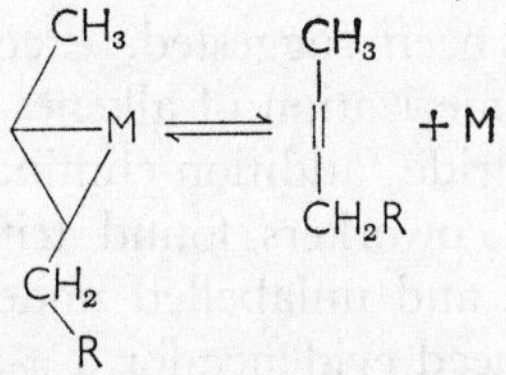

lead to a 2-alkene while equation (21) leads to olefins isomerized further down the chain. From the result of Asinger and coworkers[69], it seems that the forward rate of reaction (21) is faster than the forward rate of reaction (22), so that the equilibrium is established very fast.

The rates of the whole series of reactions increase with the concentration of the catalyst when this is low, but at high concentrations there seems to be a strong inhibition of the reaction, which should probably be ascribed to recombination of the unsaturated iron-carbonyl fragments, with saturated iron carbonyls which compete efficiently with the alkene molecules.

This type of isomerization has been used to convert unsaturated alcohols to ketones[70-71], for instance

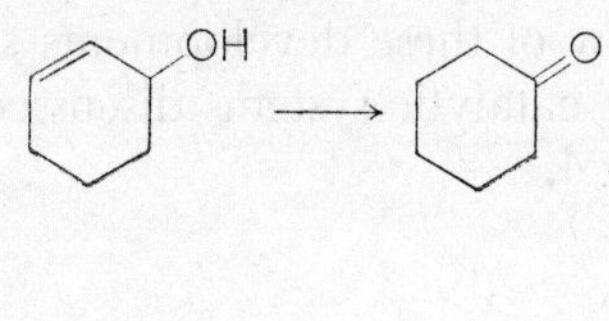

In this case the reaction is made irreversible by the much larger stability of the ketone compared with that of the unsaturated alcohol. Cobalt carbonyl is also an efficient catalyst[72].

2. The Palladium catalysed reaction

This has been examined by various research groups[52,53,73-75]. By working with deuterated 1-heptene and undeuterated 1-pentene no exchange of deuterium was found, and, by working with deuterated 1-heptene alone, or deuterated 1-pentene alone, only 1,3 deuterium shifts and no 1,2 shift[53] were found (the catalyst was $(PhCN)_2PdCl_2$). Apparently all possible isomers appear simultaneously and at comparable rates: all these facts are accommodated again by equations (21) and (22) provided the forward rate of equation (21) is larger than the forward rate of equation (22).

C. Other Reactions

Many other transition-metal catalysts[56-64,76] have been examined, but no mechanism has been suggested, except in one case. The $CoH(CO)_4$ catalysed isomerization of alkenes is usually regarded as proceeding by the hydride addition-elimination mechanism: for instance Asinger[69] and coworkers found tritium partial exchange between tritiated octene and unlabelled undecene. However Roos and Orchin[59] have adduced evidence for a π-allyl complex mechanism: $CoD(CO)_4$ isomerizes allylbenzene to propenylbenzene with no detectable kinetic isotope effect, and only very slight introduction of deuterium into propenylbenzene:

$$C_6H_5CH_2-CH=CH_2 \xrightarrow{CoD(CO)_4} C_6H_5-CH=CH-CH_3$$

The very much enhanced acidity of the allylic protons in allylbenzene, compared to those of octene, might well be the cause of a different mechanism for each of these two cases.

V. HYDROGENATION OF ALKENES

The use of transition-metal complexes as homogeneous catalysts for olefin hydrogenation has been developed quite recently with good success. For a review of these developments see reference 77. For further details on the catalytic systems discussed here and additional ones, see also chapter 4.

A. The Pentacyanocobaltate Catalysed Reaction

A solution of cobalt(II) chloride and potassium cyanide has been shown to absorb hydrogen[78]. Systematic studies of the reduction of different unsaturated groups by this solution have been undertaken[79].

Simple olefins with non-conjugated double bonds are not hydrogenerated[79a,c]. Conjugated dienes are reduced to olefins mainly by 1,4 addition: isoprene gives 2-methyl-2-butene (84%) and a mixture of 2-methyl-1-butene and 3-methyl-1-butene (15%). Styrene is reduced, but not stilbene[79c]. A careful study of hydrogenation of unsaturated acids has been made[79d]. Sorbic acid gives a mixture of 82% 2-, 17% 3-, 1% 4-hexenoic acids. Acrylic acid is not reduced, but a higher temperature gives 2-methylglutaric acid when hydrogen is present, and 4-methylglutaconic acid in the absence of hydrogen[79c]. Several α,β-unsaturated acids, aldehydes, esters and acetylenic acids have been reduced to the saturated derivatives[79b,c]. Because of the strongly basic medium the reduction of aldehydes is of limited use.

The precise nature of the catalyst is beginning to be known. The first observation of a hydride complex in a solution of pentacyanocobaltate, which had been reduced by sodium borohydride, was made by n.m.r. measurements[80]. Later it was shown that the same hydride occurs even in the absence of hydrogen or borohydride[81].

A recent careful study of the hydride solution has led to the isolation of several crystalline complexes from the hydrogenated pentacyanocobaltate solution: the hydride $[Co(CN)_5H]^{5-}$ whose existence has been proposed without convincing support, and two other complexes whose structures have been postulated to be of $[Co(CN)_4H_2]^{3-}$ and $[Co(CN)_6]^{5-}$ [82,83]. The catalytic activity of the three isolated species is yet unknown. The formation of hydride may occur in the presence of hydrogen[81,84,85,86] either by direct splitting of the hydrogen molecule and a dimeric cyanocobaltate complex:

$$2\,[Co(CN)_5, H_2O]^{3-} \rightleftharpoons [Co_2(CN)_{10}]^{6-} + 2\,H_2O$$
$$[Co_2(CN)_{10}]^{6-} + H_2 \rightleftharpoons 2\,[Co(CN)_5H]^{3-}$$

or by previous formation of a dihydride complex

$$[Co(CN)_5, H_2O]^{3-} + H_2 \rightleftharpoons [Co(CN)_5H_2]^{3-} + H_2O$$
$$[Co(CN)_5H_2]^{3-} + [Co(CN)_5, H_2O]^{3-} \rightleftharpoons 2\,[Co(CN)_5H]^{3-} + H_2O$$

the reaction:

$$[Co_2(CN)_{10}]^{6-} + H_2O \longrightarrow [Co(CN)_5H]^{3-} + [Co(CN)_5OH]^{3-}$$

would explain the slow appearance of the cobalt hydride in the pentacyanocobaltate solution in the absence of hydrogen or borohydride.

In solution, the hydride $[Co(CN)_5H]^{3-}$ adds, for example, to butadiene, forming first a σ-allyl cobalt derivative. This organo-cobalt intermediate has been prepared in two other ways: addition of the hydride to butadiene, and alkylation of pentacyanocobaltate ion with crotyl bromide. In the presence of an excess of cyanide this σ-allyl complex is stable and gives on protonation or by reaction with $[Co(CN)_5H]^{3-}$ mainly 1-butene. If no excess of cyanide is present, the complex dissociates to a π-allyl complex according to:

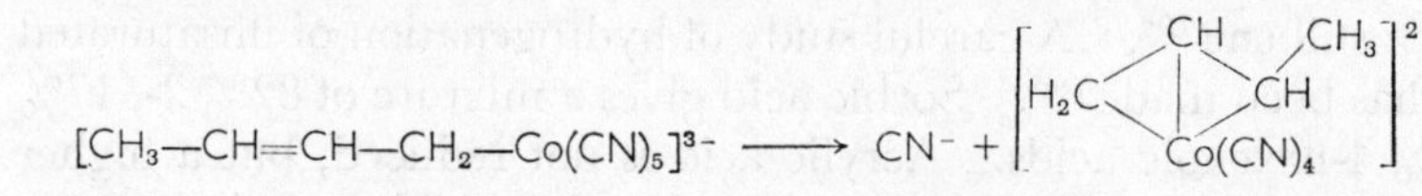

$$[CH_3-CH=CH-CH_2-Co(CN)_5]^{3-} \longrightarrow CN^- + \begin{bmatrix} H_2C\underset{\underset{Co(CN)_4}{|}}{\overset{CH}{\diagup}}\overset{CH_3}{\underset{CH}{\diagdown}} \end{bmatrix}^{2-}$$

which on protonation gives mainly *trans*-2-butene. This explains the influence of the ratio Co/CN on the nature of the hydrogenation products. The reaction with unsaturated ketones seems to be more complex[87]. The kinetics of the addition of the pentacyanocobalt-hydride $Co(CN)_5H^{-3}$ to substituted olefins have recently been studied[242] and the use of this catalytic system in solvents other than water has been made possible by the use of the lithium cyanide[243].

More definite studies on the mechanism of hydrogenation by the pentacyanocobaltate system are needed for a better understanding of the process.

B. The Rhodium(I) Catalysed Reaction

The rhodium complex $RhCl(Ph_3P)_3$ forms with hydrogen a dihyd-ride, which hydrogenates olefins[88]. The reaction is in general fast with mono- and disubstituted olefins, extremely slow with trisubstituted double bonds and no hydrogenation is observed for tetrasubstituted ones[89,90]. For instance, olefins such as 1-octene, 1-decene, cyclo-hexene and 5α-2-cholestene are easily reduced. 3-Methyl-5α-2-cholestene is not reduced under normal conditions. The very useful possibility of reducing selectively in high yield one double bond in the presence of another which remains unchanged, has already been taken advantage of[91,92]. For example linalool (5) has been reduced to

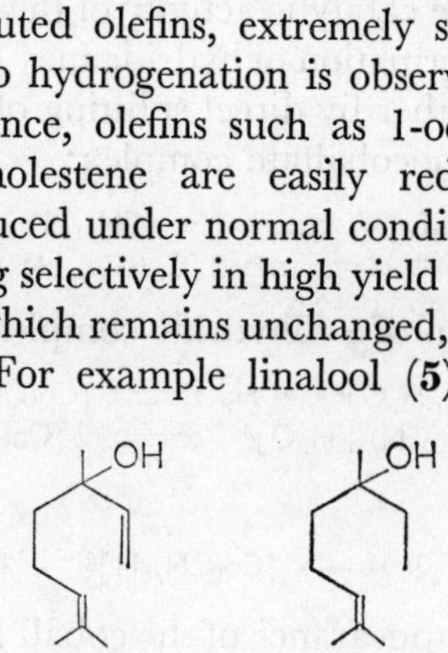

(5)   (6)

dihydrolinalool **6** in high yield[90]. Another example is the sesquiterpene γ-gurjunene (**7**): on hydrogenation by heterogeneous catalysts, after absorption of one equivalent of hydrogen, a mixture of about equal amounts of the starting material, the tetrahydro and both dihydrogurjunenes, is obtained while using the rhodium complex the dihydro derivative (**8**) is quantitatively produced[91].

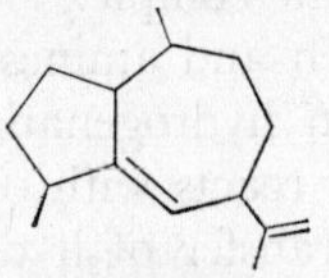

(7) 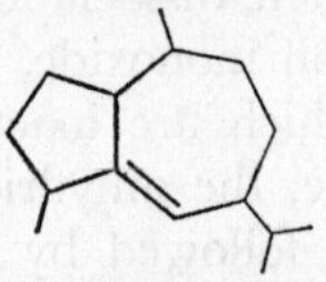 (8)

The selective hydrogenation of the steroidal 3-keto, 1,4-dienes has been questioned[250]. Under certain conditions, in the presence of deuterium, there is a *cis* addition[88,92] of two deuterium atoms, almost without participation of protons from the solvent and without exchange of the hydrogens of the molecule being hydrogenated[88,91-93]. The experimental conditions for good deuteration with this catalyst are critical[251].

Being a molecular species, this catalyst is more specific in its reactions than the surface of a heterogeneous catalyst: for instance, derivatives of 1,4-cyclohexadiene and naphthoquinone may be reduced without disproportionation for the first, and aromatization for the second[90,94]. Hydrogenolysis is also strongly repressed: for instance benzyl cinnamate is reduced without the formation of dihydrocinnamic acid[89] and olefinic thioethers are reduced without cleavage of the carbon–sulphur bond[95]. The hydrogenation of thiophene derivatives by this catalyst has been described, the thiophene ring remaining unaltered during the process[244].

Many functional groups remain untouched during hydrogenation with this catalyst. Ketones, esters, acids, amides, nitro groups and aromatic systems are not reduced[88b,89,90]. Aldehydes are decarbonylated[96,97], but the decarbonylation reaction may be avoided by working with a small concentration of catalyst and with a high pressure of hydrogen[98].

The mechanism of hydrogenation with $RhCl(Ph_3P)_3$ has been studied[88b,99]. The complex $RhCl(Ph_3P)_3$ (**9**), prepared by reaction of hydrated rhodium trichloride with triphenylphosphine, dissociates in solution:

$$RhCl(Ph_3P)_3 \rightleftharpoons RhCl(Ph_3P)_2 + Ph_3P$$
$$\quad\quad (9) \quad\quad\quad\quad\quad\quad (10)$$

Species **10** reacts reversibly with hydrogen to give a dihydride which may be isolated: $RhH_2Cl(Ph_3P)_2$, whose structure in solution has been shown by n.m.r. to be **11**. However, the dissociation of tris(triphenylphosphine)rhodium chloride in solution has been seriously questioned[245]. On the other hand, activation of the catalyst by a trace of oxygen may result from the oxidation of the phosphine, which leads to the complete dissociation of the complex[246].

Ethylene, carbon monoxide, oxygen and amines react with **10** to give complexes which are inactive in hydrogenation[98]. According to kinetic evidence, the dihydride **11** reacts with olefins in the rate-determining step, followed by the transfer of hydrogen. A simultaneous transfer of both hydrogens has been proposed[98], but one example where the isomerization is the main reaction path[100] and the fact that besides dideuterated species, trideuterated molecules are obtained[100], are in favour of the formation of an alkyl rhodium hydride as an intermediate. The isotope effect on the rate agrees with formation of a complex between **11** and the olefin as the rate-determining

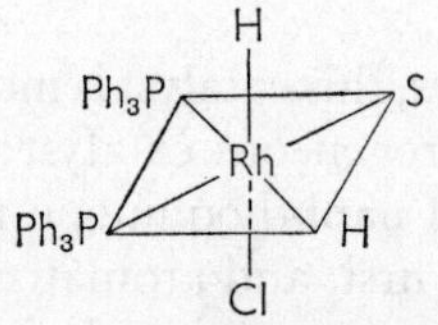

$$Ph_3P\diagdown\underset{\underset{Cl}{|}}{\overset{\overset{H}{|}}{Rh}}\diagup S$$
$$Ph_3P\diagup\diagdown H$$

(**11**)

step. Other examples of double bond isomerization by this catalyst alone[247] and during the hydrogenation[248] have been described. However, the hydrogenation of ethylene is possible. This fact and competition experiments are consistent with the simultaneous existence of two hydrogenation processes: reaction of the dihydride with the olefin and reaction of the catalyst first with the olefin, and then with hydrogen[249].

The reactivity of $RhI(Ph_3P)_2$ is higher than that of $RhBr(Ph_3P)_2$ which is higher than that of the chloro complex. The rate is increased in more polar solvents, but solvents which can act as ligands may prevent reduction, e.g. pyridine, thiophene and acetonitrile[88b]. The change of the substituent at phosphorus is of little value, phenyl seeming to be best at the present time[252].

Asymmetric induction during the hydrogenation by asymmetric phosphines has been demonstrated[253].

C. The Ruthenium(II) Catalysed Reaction

I. Ruthenium(II) chloride[101]

Ruthenium(II) chloride is prepared by reduction of aquochlororuthenite in aqueous hydrochloric acid solution by titanium(III) chloride. In the presence of hydrogen, reduction of fumaric, maleic, acrylic and crotonic acids is observed. If dimethylacetamide or dimethylformamide are used as solvents, ethylene and cyclohexene can be reduced. But it is not known if the same mechanism is operating in both groups of substrates.

The use of ruthenium(II) chloride in aqueous solution gives rise to an interesting observation: the formation of the hydride occurs after the formation of a 1:1 complex between ruthenium and the olefin. This was detected spectroscopically in solutions of the above mentioned acids, dimethylmaleic anhydride, 5-norbornene-2,3-dicarboxylic anhydride, ethylene and propylene. The last four olefins are not reduced, but their complexes catalyse the exchange of D_2 with H_2O.

In the reduction experiments of fumaric and maleic acids the following observations were made: starting with fumaric acid, using D_2 and H_2O, the succinic acid obtained contained no deuterium atoms but with D_2 and D_2O or H_2 and D_2O, DL-2,3-dideuteriosuccinic acid was obtained. This shows that the addition is *cis*. The experiments with maleic acid indicate competition between isomerization to fumaric acid and reduction. In this case no experiments were undertaken where hydrogenation was interrupted before completion.

The following mechanism of hydrogenation has been proposed: in a first step, formation of Ru^{II}-olefin complex followed by a rate-determining reaction between this complex and hydrogen to give a hydridoruthenium complex. The formation of this hydride is reversible in the case of olefins which do not undergo hydrogenation, explaining the exchange between D_2 and H_2O. The next step would be hydride migration onto the olefin to give an alkylruthenium, which in a further step reacts with a proton to give Ru^{II} and the saturated compound:

$$Ru^{II} + \overset{|}{C}=\overset{|}{C} \longrightarrow \begin{array}{c} \diagdown C \diagup \\ | | \\ \diagup Ru \diagdown \\ -C- \\ | \end{array} \xrightarrow[-H^{\oplus}]{+H_2} \begin{array}{c} -\overset{H}{\overset{|}{C}}- \\ \diagdown | \\ Ru \\ | \\ -C- \\ | \end{array} \longrightarrow \begin{array}{c} -\overset{H}{\overset{|}{C}}\diagup \\ | \\ Ru \\ | \\ -C- \\ | \end{array} \xrightarrow{+H^{\oplus}}$$

$$Ru^{II} + H\overset{|}{\underset{|}{C}}-\overset{|}{\underset{|}{C}}H$$

It is interesting to note than an electron-attracting group on the double bond is required for the addition of hydrogen to take place in the system Ru^{II}-water.

2. Tris(triphenylphosphine)hydridochlororuthenium(II)

This complex has only recently been prepared in pure form [102] and it is quite likely that catalysis observed with $RuCl_3$ in the presence of triphenylphosphine [103] and with $RuCl_2(Ph_3P)_n$ ($n = 3$ and 4) [104] is due to this hydride.

Monosubstituted olefins: 1-pentene, 1-hexene, etc., are reduced at high rates. Disubstituted olefins are very slowly reduced with rates about 10^{-3} times lower than those of monosubstituted olefins [102,103]. The case of acetylenic compounds has been studied with $RuCl_3$ in the presence of triphenylphosphine: diphenylacetylene gives cis-stilbene [103]. The deuteride $RuCl(Ph_3P)_3D$ has been described as exchanging deuterium with the hydrogen of mono- as well of disubstituted olefins. Surprisingly no isomerization of cis-2-hexene to trans-2-hexene was detectable in 24 hours [102]. The usefulness of this catalyst for selective hydrogenation is evident. The full paper has been published [254].

D. Other Catalysts

Several derivatives of titanium [105], chromium [106], manganese [107], iron [108–110], osmium [111], cobalt [109b,109d,112–117], rhodium [118–120], iridium [62,89,95,120–123], nickel [110,124–127], palladium [128], platinum [64,89,129–131] and copper [132], and Ziegler–Natta complexes [134–136], have also been used as catalysts for the hydrogenation of olefins. Selective hydrogenation of conjugated olefins by arene chromium tricarbonyl has been reported [255]. Nickel [256] and cobalt [257] derivatives are active hydrogenation catalysts. The hydridocarbonyltris(triphenylphosphine)rhodium is a very effective hydrogenation catalyst for terminal olefins [258].

VI. NUCLEOPHILIC REACTIONS OF THE COMPLEXES

A. Oxidation of Alkenes

Most reactions of transition-metal complexes are nucleophilic, as evidenced by the nature of the reagents: halides, water, alcohols, alkene, alkynes, dienes, and even carbon monoxide. This is readily understandable since the transition metal, like a singlet carbene, ex-

5. Alkene Complexes of Transition Metals as Reactive Intermediates

hibits electrophilic properties. From the point of view of the alkene these reactions are electrophilic attacks at carbon, with retention of configuration, as evidenced by the isomerization and hydrogenation reaction described in the previous sections.

Among these, the reaction which was perhaps best studied is the palladium-catalysed oxidation of ethylene to acetaldehyde, which acquired industrial importance in recent times[136].

$$C_2H_4 + PdCl_4^= + OH_2 \longrightarrow CH_3CHO + 2\,Cl^- + 2\,HCl + Pd$$

There are excellent reviews on this reaction[10,136], and the following discussion will be a summary of the most important aspects.

I. Experimental facts

(a) The rate equation has the following form[137]

$$\text{rate} = K_1 k'[Pd^{2+}][C_2H_4]/[Cl^-]^2[H^+]$$

K_1 and k' are given in Table 4 for various simple olefins[31];

TABLE 4. Complex formation constants, K_1, and rate constants k' for the palladium-catalysed oxidation of simple alkenes

Olefin	K_1	k'
Ethylene	17·4	20·3
Propylene	14·5	6·5
cis-2-Butene	8·7	3·5
trans-2-Butene	4·5	7·5
1-Butene	11·2	3·5

(b) The kinetic isotope effect in D_2O is $k_H/k_D = 4\cdot05$[138];
(c) The kinetic isotope effect for C_2D_4 in D_2O is $k_H/k_D = 1\cdot07$[137];
(d) 1-alkenes are oxidized mainly to methyl ketones, the isomeric aldehydes being formed in minor amounts[28,136a,139,140]; (Markownikov-type reaction)
(e) The reaction temperature for lower alkenes can be kept to 20–50°C[136a];

(f) When the reaction is carried out with ethylene in D_2O, no deuterium is introduced[141].

2. Reaction mechanism

Equations (23–26) lead to the observed rate equation (fact a), with $k' = K_2K_3k$. As K_2 and K_3 should not be very much affected by the

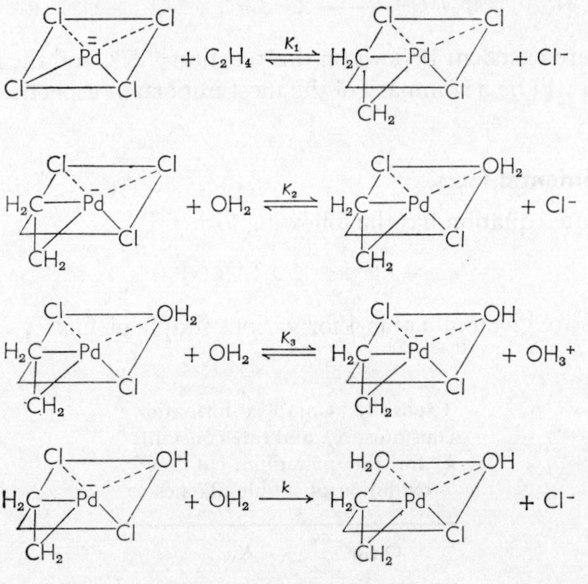

$$[Pd_2Cl_4]^{2-} + C_2H_4 \underset{}{\overset{K_1}{\rightleftharpoons}} [Pd_2Cl_3(C_2H_4)]^- + Cl^- \qquad (23)$$

$$[Pd_2Cl_3(C_2H_4)]^- + OH_2 \overset{K_2}{\rightleftharpoons} [Pd_2Cl_2(C_2H_4)(OH_2)] + Cl^- \qquad (24)$$

$$[Pd_2Cl_2(C_2H_4)(OH_2)] + OH_2 \overset{K_3}{\rightleftharpoons} [Pd_2Cl_2(C_2H_4)(OH)]^- + OH_3^+ \qquad (25)$$

$$[Pd_2Cl_2(C_2H_4)(OH)]^- + OH_2 \overset{k}{\rightarrow} [Pd_2Cl(C_2H_4)(OH)(OH_2)] + Cl^- \qquad (26)$$

olefin structure, the slight variation in k' probably expresses a similar variation in k for the slow step, indicating some kind of steric hindrance for the entry of the second water molecule (compare k' for cis and trans 2-butene in Table 4).

The kinetic isotope effect in D_2O (fact b) is of the expected size for equation (25)[142]. The very small isotope effect with C_2D_4 (fact c) indicates that no C—H bond is broken before the slow step. Because of the small differences in reactivity between ethylene and propylene, the preferred Markownikov orientation (fact d) must mean that, in the product-determining step (equation 27), the Pd—C bond which is *not*

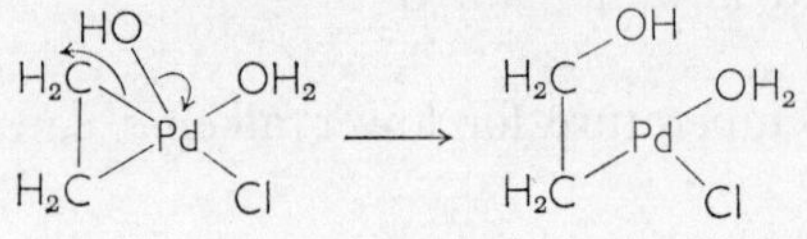

$$\text{(Pd complex with HO, OH}_2\text{, Cl)} \longrightarrow \text{(Pd complex with OH, OH}_2\text{, Cl)} \qquad (27)$$

5. Alkene Complexes of Transition Metals as Reactive Intermediates 239

broken has considerable carbanion character. In agreement with this, α,β unsaturated esters give methyl ketones[136a] (equation 28).

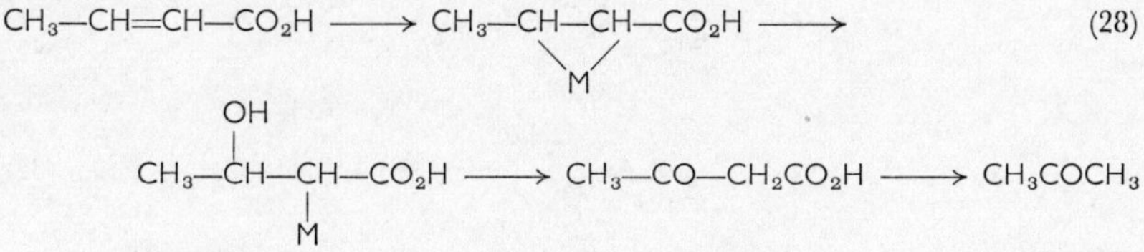

$$CH_3-CH=CH-CO_2H \longrightarrow CH_3-\underset{M}{CH-CH}-CO_2H \longrightarrow \qquad (28)$$

$$CH_3-\underset{}{\overset{OH}{CH}}-\underset{M}{CH}-CO_2H \longrightarrow CH_3-CO-CH_2CO_2H \longrightarrow CH_3COCH_3$$

Interestingly vinyl halides are hydrolyzed to acetaldehyde, and 1-halo-1-alkenes give *methyl ketones*[136a]. This can again be explained by the carbanion character of the unbroken Pd—C bond, which is enhanced by halogen substituent (equation 29).

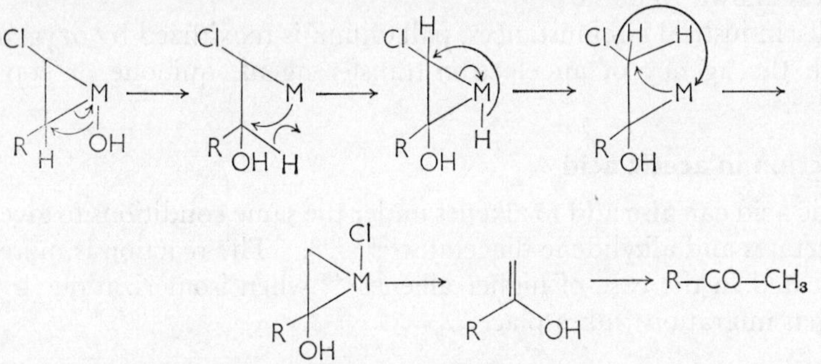

$$\longrightarrow R-CO-CH_3 \qquad (29)$$

1-Halo-1-alkenes having no hydrogen at $C_{(2)}$ give unsaturated aldehydes[136a] (equation 30).

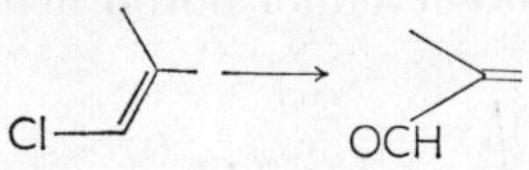

$$(30)$$

The relatively low temperature of the reaction (fact e) probably precludes any π-allylic intermediate for higher alkenes, and explains the product specificity[147-148]. With conjugated dienes, when π-allylic complexes are very probable, only aldehydes are formed[143]:

$$CH_3CH=CH-CH=CH_2 \longrightarrow CH_3CH_2-CH=CH-CH=O$$

Finally the absence of deuterium in acetaldehyde, when the reaction is run in D_2O (fact f) makes it compulsory that a proton from ethylene should be transferred intramolecularly from one end of the double bond to the other (equations 31 and 32).

This scheme also agrees with the fact that the strongest *trans* labilizing ligand around palladium is ethylene: ethylene will make it very

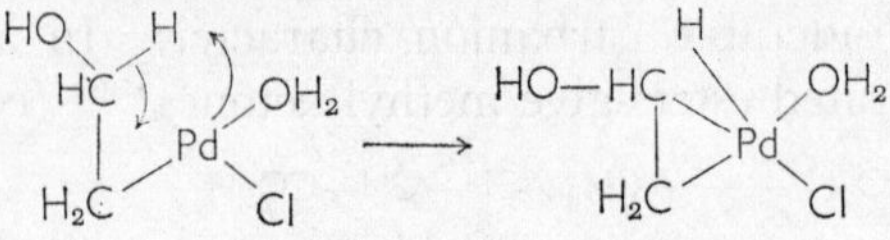

$$\text{(31)}$$

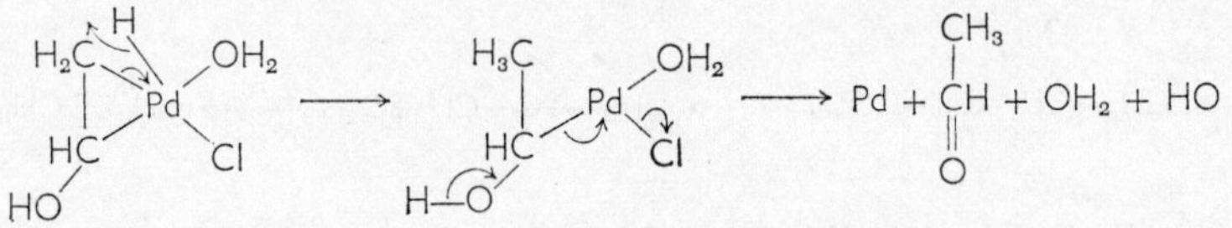

$$\text{(32)}$$

easy for the first molecule of water to be linked to palladium[144]. However, this water molecule, being *trans* to ethylene is unable to react unless a second molecule of water is introduced into a *cis* position, a process known to be slow.

Under industrial circumstances, palladium, is reoxidized by oxygen through the agency of an electron-transfer agent, quinone or copper(II)[136].

3. Reaction in acetic acid

Acetic acid can also add to alkenes under the same conditions to give enol acetates and alkylidene diacetates[136a,145]. The reaction is more complicated in the case of higher alkenes[146] when isomerization (by hydrogen migration) takes place.

B. Other Reactions

As for reactions discussed in the previous sections, *cis* additions onto the double bond, were observed for norbornene[149] (equation 33).

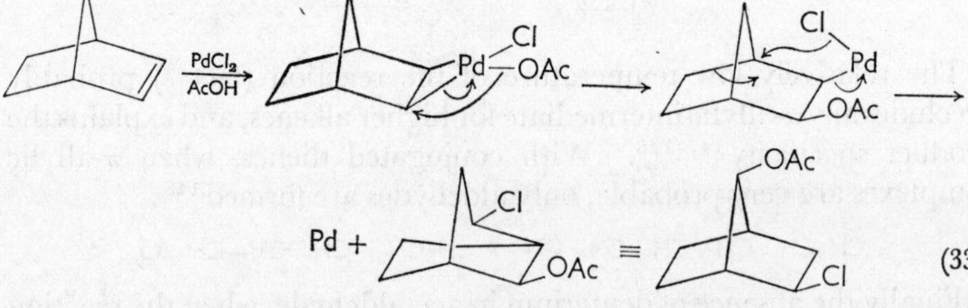

$$\text{(33)}$$

However, with chelating dienes (1,5 dienes) there are instances where addition *trans* to the metal has been found to be operative, as was shown from n.m.r. studies and X-ray data[150-154].

Sodium acetate and amines can act in the same way[155] and this inversion of configuration at carbon seems to be linked with the

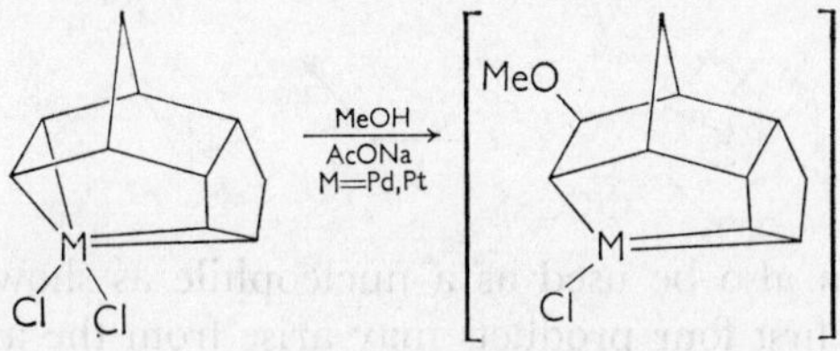

5. Alkene Complexes of Transition Metals as Reactive Intermediates 241

$$\text{(34)}$$

electron-accepting power of the second double bond and the cage structures of the chelating dienes. The situation, however, is by no means clear, because Anderson and Burreson[156] have reported the following reactions

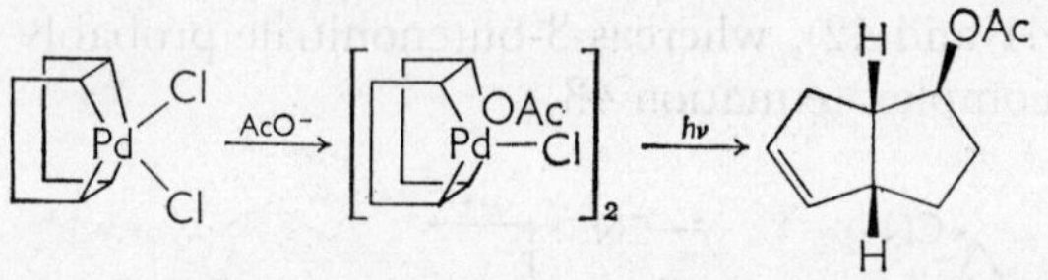

$$\text{(35)}$$

Other nucleophiles like acetylacetone[157,158], enamines[159], sodio malonic esters[159] or dienes[157] react with dienes, complexes, π-allyl complexes and alkene complexes. Equations (36–38) illustrate the behaviour of a β-diketone.

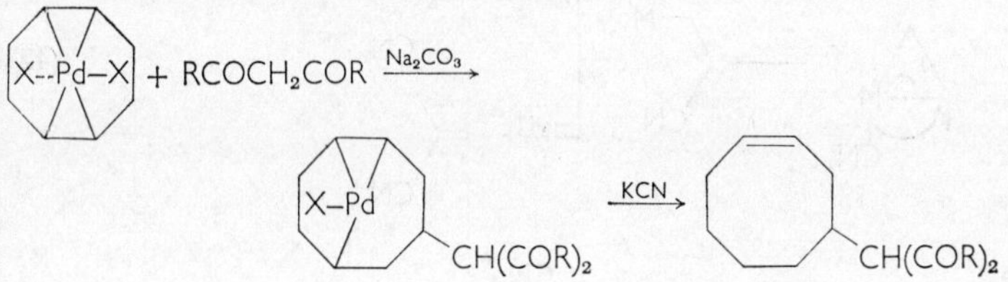

$$\text{(36)}$$

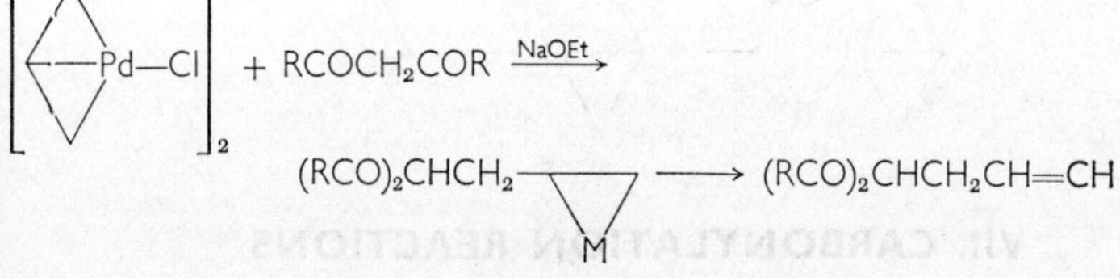

$$\text{(37)}$$

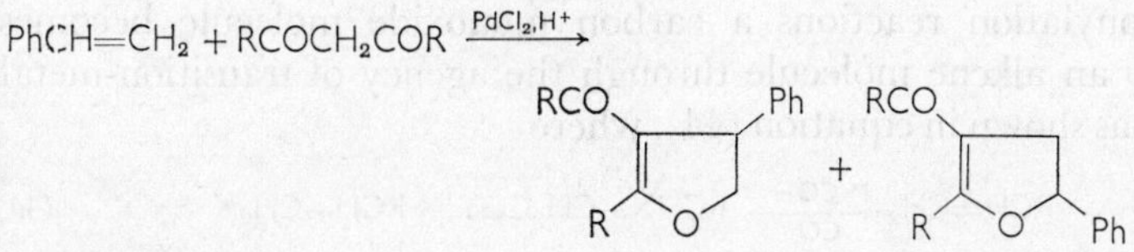

$$PhCH=CH_2 + RCOCH_2COR \xrightarrow{PdCl_2, H^+} \quad \text{(38)}$$

Under alkaline conditions, the reaction may go further as exemplified in equation (39)[160].

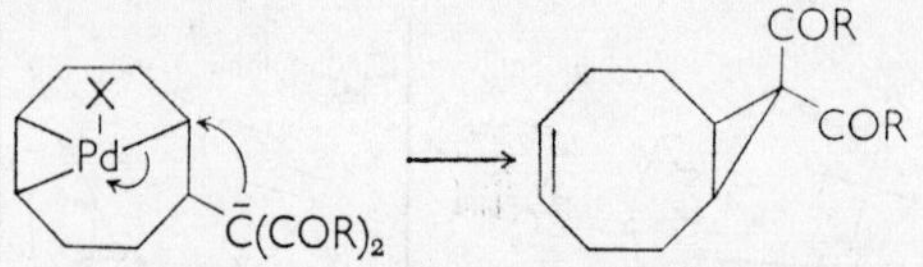

Cyanide can also be used as a nucleophile as shown in equation (40)[161]. The first four products may arise from the alkene complex (equations 41 and 42), whereas 3-butenonitrile probably comes from the π-allyl complex (equation 43).

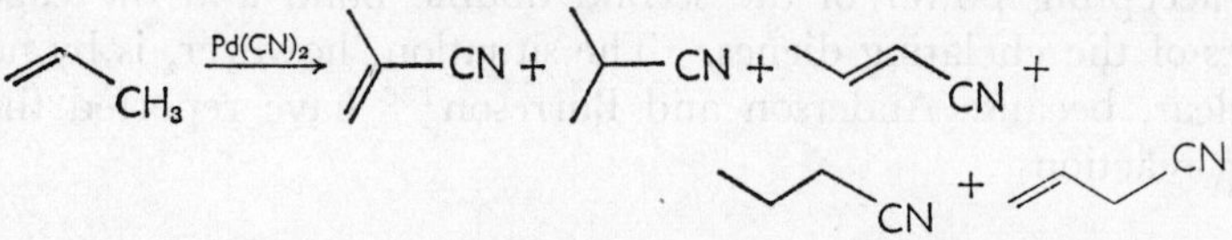

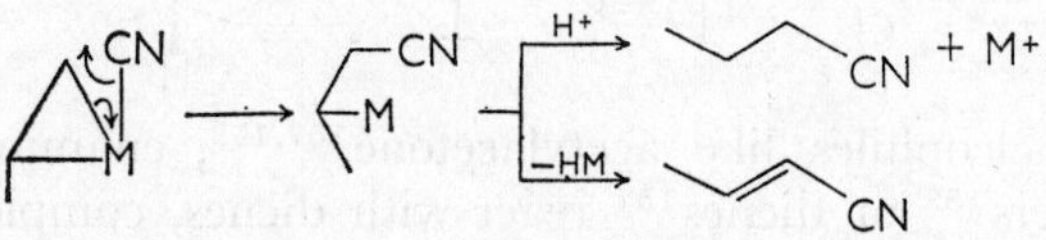

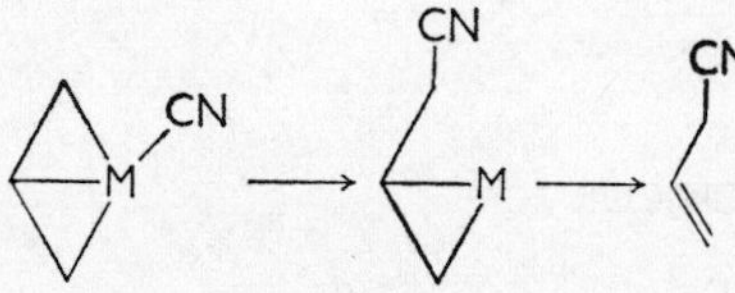

VII. CARBONYLATION REACTIONS

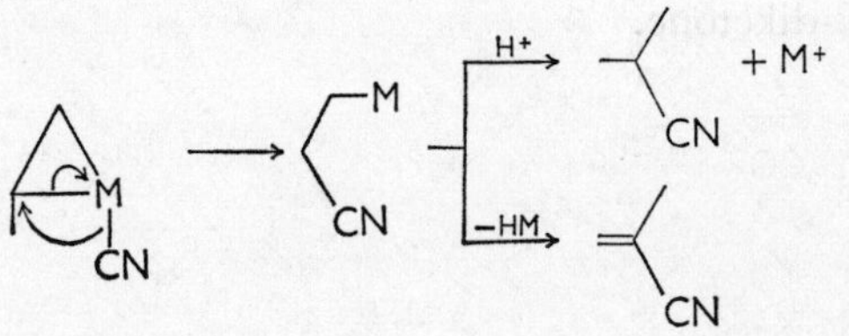

In carbonylation reactions a carbon monoxide molecule becomes linked to an alkene molecule through the agency of transition-metal catalyst, as shown in equation (44), where

$$RCH\!=\!CH_2 \xrightarrow[CO]{M(CO)_n} RCHX\!-\!CH_2COY + RCH\!-\!CH_2X \qquad (44)$$
$$\phantom{RCH\!=\!CH_2 \xrightarrow[CO]{M(CO)_n} RCHX\!-\!CH_2COY + RCH\!-\!CH} \underset{COY}{|}$$

X and Y are usually H, halogen, OR′.

In some cases, it has been possible to isolate an intermediate transition metal–alkene complex[162].

A. The Reaction of Carbon Monoxide within the Alkene Complex

In most cases carbon monoxide is already linked to the metal and the alkene has been transformed to an alkyl–metal derivative. It has been shown conclusively, with stable alkyl metal–carbonyl complexes[163–164], that the first reaction involves a migration of the alkyl group to a bonded carbonyl group, as in equation (45)

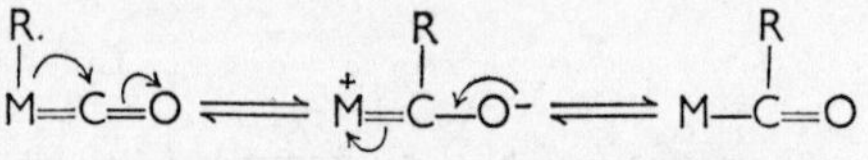

$$\underset{M=C=O}{\overset{R}{|}} \rightleftharpoons \underset{M=C-O^-}{\overset{R}{|}} \rightleftharpoons \underset{M-C=O}{\overset{R}{|}} \quad (45)$$

The arguments in support of this are the following

(a) The reactions follow first-order kinetics[164–166], the metal–carbonyl complex only being involved in the rate equation;
(b) Entering ligands (CO, PR$_3$) are *cis* to the newly formed acyl group[163–167];
(c) ^{13}C labelled CO, as entering ligand, will not end up in the newly formed acyl group[163].

There is no reason to believe that the reaction should be different with alkyl–metal carbonyls which are too unstable to be isolated at room temperature.

The reaction is usually reversible, in the absence of some substance which can strongly coordinate with the metal at the site which has become vacant. It can be made irreversible in the presence of compounds such as phosphines[169,170], phosphites[167,170], amines[170], alcohols[171], water[171], hydrogen[172] and carbon monoxide[173].

After the alkyl migration has taken place, a second reaction can happen, which detaches the acyl grouping RCO from the metal. Under the usual **oxo** synthesis conditions, hydrogen is present and can coordinate with the metal to give an intermediate hydride, which will undergo a hydride migration to release aldehyde and a metal hydride[174]. Similarly, the cleavage may be attained with water, alcohols or hydrogen halides, as shown in equation (46)[14].

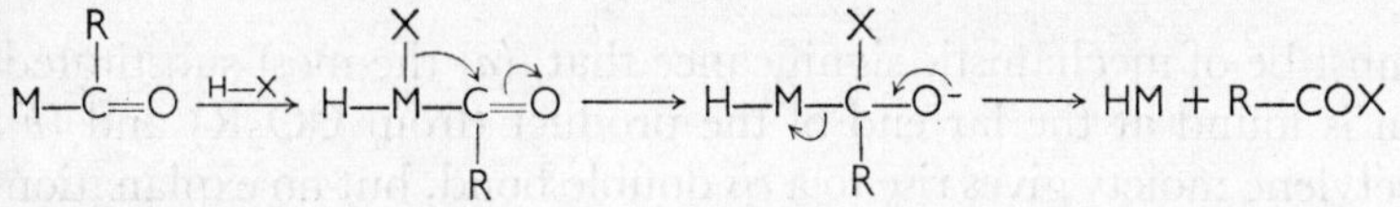

$$\underset{M-C=O}{\overset{R}{|}} \xrightarrow{H-X} \underset{\underset{R}{|}}{\overset{X}{|}} H-M-C=O \longrightarrow \underset{\underset{R}{|}}{\overset{X}{|}} H-M-C-O^- \longrightarrow HM + R-COX \quad (46)$$

X = H, halogen, OH, OR

Although these metal hydrides have only been isolated in some cases (Fe[175,176], Co[56-58], Rh[52-55], Ir[60-62], Pt[63,64]), it seems reasonable to assume their intermediacy.

B. Reactions of π-Allyl Complexes

These reactions have been mostly examined with nickel and palladium derivatives. With the former, whereby π-allyl complexes are formed rapidly, the reaction is straightforward and gives β,γ-unsaturated acid derivatives[13,14,177] (equation 47).

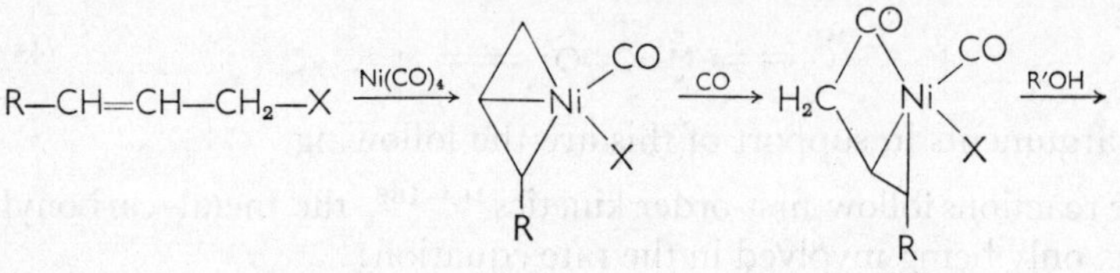

$$R-CH=CH-CH_2-X \xrightarrow{Ni(CO)_4} \text{[complex]} \xrightarrow{CO} \text{[complex]} \xrightarrow{R'OH}$$

(12)

$$R-CH=CH-CH_2-CO_2R' \quad (47)$$

The intermediacy of **12** has been established spectroscopically[178]. In the presence of acetylene, there is an additional reaction, coordinated acetylene being apparently more reactive towards the π-allyl moiety than carbon monoxide[14,178] (equation 48).

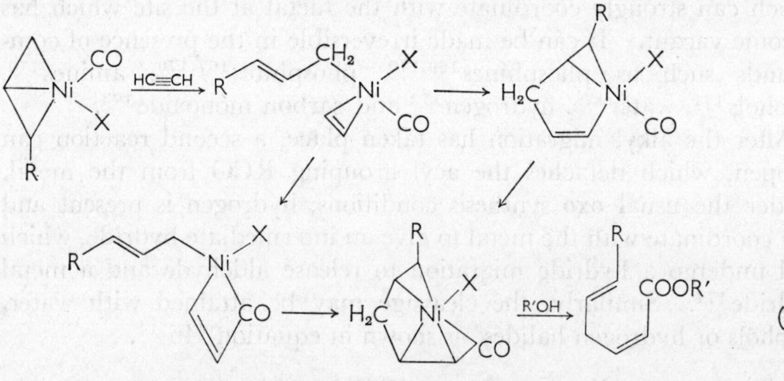

$$\text{[scheme]} \quad (48)$$

It must be of mechanistic significance that, (a) the most substituted carbon is found at the far end of the product (from CO_2R) and (b) the acetylene moiety gives rise to a *cis* double bond, but no explanation has yet been given to this fact.

The same is true for palladium complexes[178], although in some cases, a complicating feature interferes: at low temperature (i.e. room temperature) the palladium complexes from allylic chlorides are rather of σ type[147-148] and yield different products, i.e.

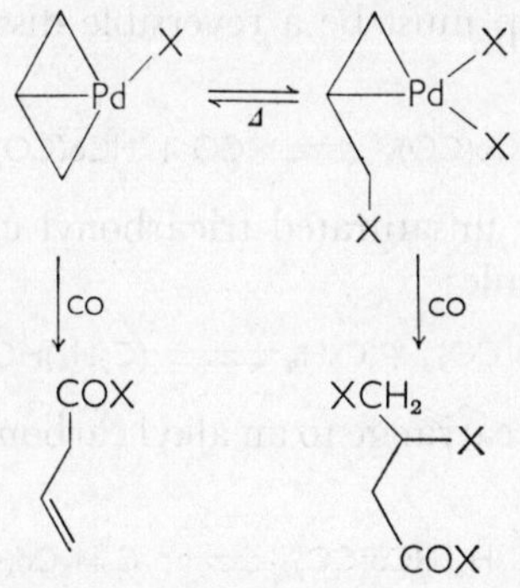

[structure diagrams showing Pd-allyl complexes with X ligands, reacting with CO to give COX and XCH₂...COX products]

The high temperature reaction implies a reduction of allylchloride by CO, the mechanism of which has been established by Shaw and coworkers[180], see also [181].

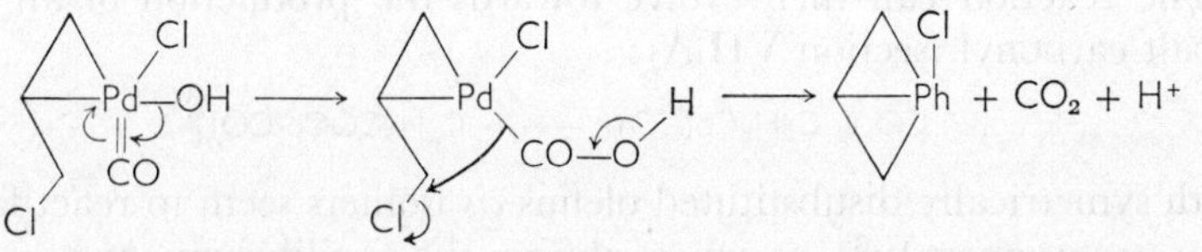

[mechanism scheme showing Pd-allyl chloride with OH and CO, proceeding through CO-O intermediate to give Ph-allyl + CO_2 + H^+]

It should also be borne in mind that dimerization can occur if CO pressure is too low[182].

$$2\ CH_2=CHCH_2Cl \xrightarrow{Ni(CO)_4} CH_2=CHCH_2CH_2CH=CH_2$$

This reaction has indeed been used synthetically[14].

C. Reactions of Alkene Complexes

In the case of simple alkenes, double bond shifts take place with most catalysts, and even with the most active ones, e.g. $HCo(CO)_4$, two products can be expected and are indeed obtained. Although the reaction was first discovered with iron carbonyl[4], the reaction has been most extensively studied with cobalt hydrocarbonyl and the latter will be examined here with more details.

Heck and Breslow[183] have shown with ethylene, that the first steps involve the reversible formation of an ethylcobalt carbonyl (see also Karapinka and Orchin[174]):

$$C_2H_4 + HCo(CO)_4 \rightleftharpoons C_2H_5Co(CO)_4$$

This reaction is inhibited by carbon monoxide, a fact which indicates that the very first step must be a reversible dissociation of the tetracarbonyl:

$$HCo(CO)_4 \rightleftharpoons CO + HCo(CO)_3$$

The coordinatively unsaturated tricarbonyl can then be saturated by an ethylene molecule:

$$HCo(CO)_3 + C_2H_4 \rightleftharpoons (C_2H_4)HCo(CO)_3$$

which can reversibly rearrange to an alkyl carbonyl, by hydride migration:

$$(C_2H_4)HCo(CO)_3 \rightleftharpoons C_2H_5Co(CO)_3$$

The reaction can be made irreversible if an appropriate molecule is available as a ligand (e.g. Ph_3P or CO)

$$C_2H_5Co(CO)_3 + CO \longrightarrow C_2H_5Co(CO)_4$$

The reaction can then evolve towards the production of an acyl cobalt carbonyl (section VII.A):

$$CO + C_2H_5Co(CO)_4 \longrightarrow C_2H_5COCo(CO)_4$$

With symetrically disubstituted olefins *cis* isomers seem to react faster than *trans* isomers[57,68], as expected from the equilibrium constants for the formation of transition metal–alkene complexes.

With unsymmetrical olefins reports are contradictory: at low CO pressure (1 atm) isobutene and methyl acrylate are carbonylated mostly on the more substituted carbon at 0°[183], but mostly at the less substituted carbon at 120°[184-187]. At high CO pressure (over 100 atm) and 120°, 1-alkenes are carbonylated mostly at the least substituted carbon, although there is some double bond shift[57,172]. These results cannot be explained presently in a satisfactory way.

The reaction with $Fe(CO)_5$ has been studied mostly by Wender and coworkers[184,188], and Manuel[65], and the results are similar to those described above.

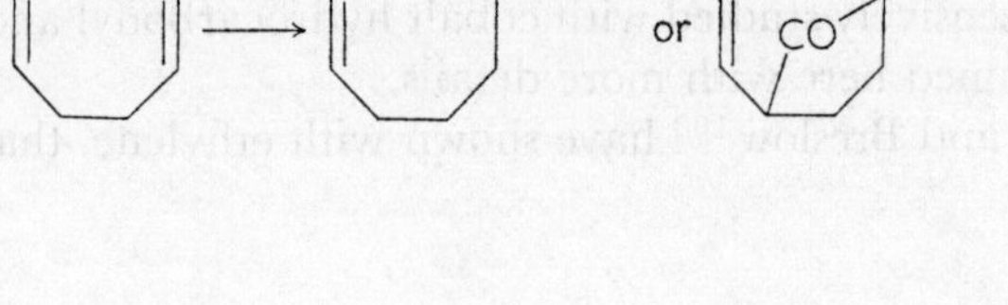

(49)

The stereochemistry of the reaction has been examined by Cookson and his colleagues[189] who could show, with norbornadiene, that the reaction is a *cis-exo* addition. Other 1,4 and 1,5 dienes, have been used in these reactions[190-191], for instance 1,5-cyclooctadiene (equation 49) and 1,4-pentadiene (equation 50). With dicobalt octacarbonyl, allylic alcohols can be transformed into γ-butyrolactones[192] (equation 51).

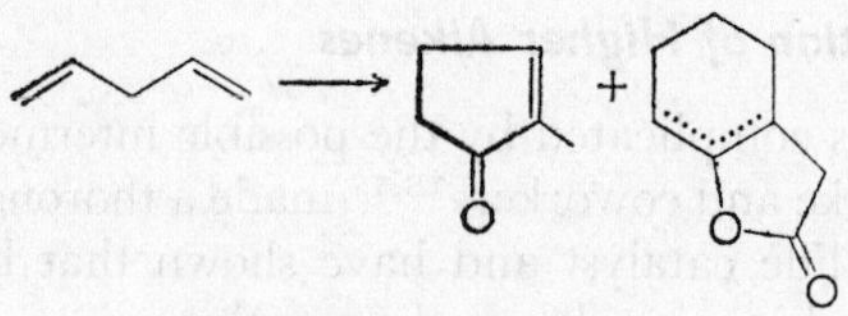

$$\text{1,4-pentadiene} \longrightarrow \text{cyclopentenone} + \text{bicyclic lactone} \quad (50)$$

$$\text{allyl alcohol} \longrightarrow \text{γ-butyrolactone} \quad (51)$$

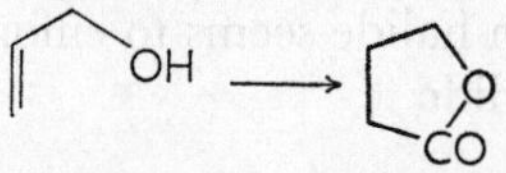

Many other analogous reactions have been described. Industrial application has also been described[193].

VIII. OLIGOMERIZATION AND MIXED OLIGOMERIZATION

A. Oligomerization of Ethylene

In this case there is no complicating factor arising from the possible formation of a π-allyl complex. The reaction which has been most thoroughly studied is that of rhodium trichloride[194].

Cramer[195] could show that:

1. Rhodium(III) must be reduced to rhodium(I) to be active. This can be brought about by ethylene itself in the presence of water:

$$6\,C_2H_4 + RhCl_3 + 2\,H_2O \longrightarrow 2\,MeCHO + 4\,HCl + [(C_2H_4)_2RhCl]_2$$

It is also possible to use a preformed rhodium(I) catalyst such as $Rh(C_2H_4)_2MeCOCH_2COMe$. Acetylacetone is instantly displaced on HCl addition, as shown by n.m.r. experiments;

2. The rate equation can be described as

$$-d[C_2H_4]/dt = k[C_2H_4][H^+][Cl^-][Rh]$$

3. The intermediate $[C_2H_5RhCl_3(C_2H_4)]^-$, with a trivalent rhodium, can be detected spectroscopically and is formed reversibly (experiments with DCl);

4. The rearrangement of this intermediate is rate determining.

The reaction can thus be represented as follows, where s is a molecule of solvent:

$$[Cl_2Rh^I(C_2H_4)_2]^- + H^+ + Cl^- \rightleftharpoons [C_2H_5Rh^{III}Cl_3(C_2H_4)s]^-$$
$$s + [C_2H_5Rh^{III}Cl_3(C_2H_4)s]^- \rightleftharpoons C_2H_4 + [C_2H_5Rh^{III}Cl_3s_2]^-$$
$$s + [C_2H_5Rh^{III}Cl_3(C_2H_4)s]^- \xrightarrow{slow} [C_4H_9Rh^{III}Cl_3s_2]^-$$
$$[C_4H_9Rh^{III}Cl_3s_2]^- \longrightarrow [(C_4H_8)HRh^{III}Cl_3s] + s$$

B. Oligomerization of Higher Alkenes

The reaction is complicated by the possible intermediacy of π-allyl complexes. Wilke and coworkers[16,17] made a thorough study of their π-allyl nickel halide catalyst and have shown that both aluminium halides and phosphines accelerate the reaction.

The role of aluminium halide seems to enforce dissociation onto the dimeric π-allyl nickel halide

$$\begin{array}{c}\text{(}\pi\text{-allyl)}_2Ni_2X_2\end{array} + Al_2X_6 \rightleftharpoons 2\left[\begin{array}{c}\text{(}\pi\text{-allyl)}Ni\end{array}\right]^+ [AlX_4]^-$$

The π-allyl nickel ion thus formed retains two vacant sites for coordination (dsp^2 hybridization) and is active for dimerization. It is made inactive by CO or excess phosphine. Nevertheless if only one mole of phosphine per nickel atom is added, the catalyst has a higher activity. The mechanism might be depicted as the sequence of steps in chart 1.

The function of the phosphine can thus be understood: the strong *trans* labilizing effect of the phosphine should accelerate the migration of the hydride and the alkyl groups, and also the release of the olefins. In agreement with this view, Wilke and coworkers could show that the isomerization of the initially formed olefins decreases with increasing electron-donating power of the phosphine[17,196].

The nature of the phosphine does not seem to influence to any large extent the hydride migration to the α or the β carbon, but it does so for the migration of the alkyl group. The better electron donating is the phosphine, the greater is the proportion of alkyl migration to the β carbon atoms, suggesting that this migration is probably rate determining. This was found by Cramer[195] for the dimerization of ethylene with a rhodium catalyst.

Other nickel catalysts have been employed[197–199] and ruthenium[200], rhodium[194,201,202], palladium[201–202] have also been tried as catalysts.

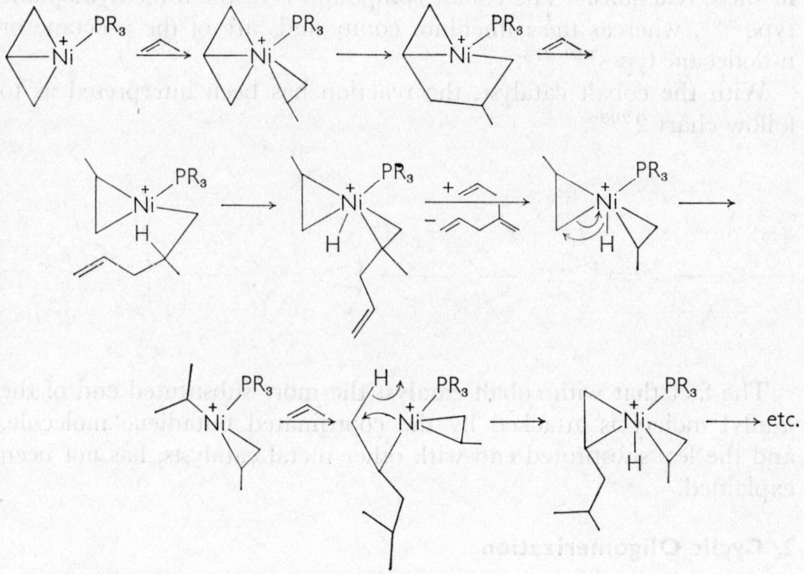

C. Oligomerization of Conjugated Dienes

Oligomerization of conjugated dienes is complicated by the occurrence of interconversion between dimers, and uncertainties about products arising from kinetic or thermodynamic control of the reaction. A wide variety of catalysts have been used, but only a few are well defined.

I. Linear Oligomerization

Linear dimers can be formed from 1,3 dienes with complexes of iron [203-205], cobalt [206-208], rhodium [194-209] and palladium [210-211]. For butadiene, three types of carbon skeleton could arise:

n-octane   from 1,4 + 1,4 addition
3-methylheptane   from 1,4 + 1,2 addition
3,4-dimethylhexane   from 1,2 + 1,2 addition

The last type seems to have never been observed. The 3-methylheptane type can be formed with iron or cobalt catalysts, and the n-octane type is formed nearly exclusively with palladium or rhodium catalysts. These facts agree with the X-ray structures determined for one cobalt and several ruthenium complexes which have been isolated in these reactions. The cobalt compound is of the 3-methylheptane type[206b], whereas the ruthenium compounds are of the n-octane or n-dodecane types[212-213].

With the cobalt catalyst, the reaction has been interpreted as to follow chart 2[206b].

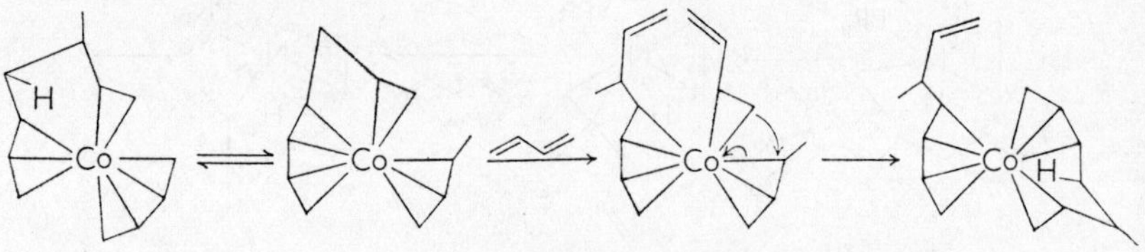

The fact that with cobalt catalyst the more substituted end of the π-allyl moiety is attacked by the coordinated butadiene molecule, and the less substituted end with other metal catalysts, has not been explained.

2. Cyclic Oligomerization

This reaction has been extensively studied with zero valent nickel catalysts[214-215], mostly by Wilke and coworkers[16-18]. Other metals, like titanium[216-220] and iron[221], also give cyclic dimers and trimers. Actually the reaction is a special case of a linear polymerization, where the ends of the chain are linked together in the last step.

Derivatives of zero valent nickel or bivalent nickel[16-17] (plus an appropriate Lewis acid) will give with dienes an active catalyst (equation 52).

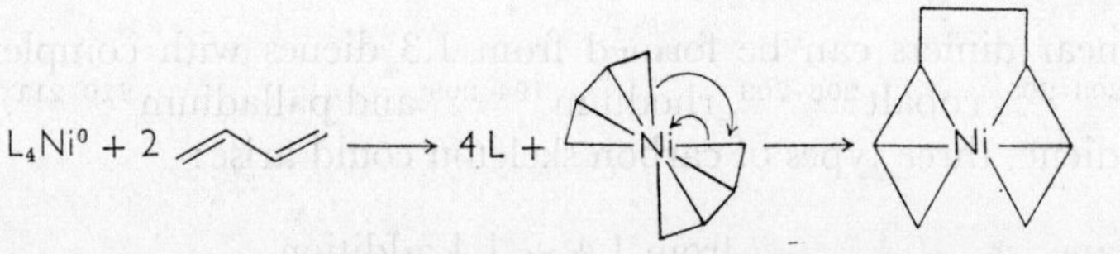

$$L_4Ni^0 + 2 \diagup\!\!\diagdown \longrightarrow 4L + [Ni] \longrightarrow [Ni] \qquad (52)$$

The presence of only one strongly coordinating ligand P (phosphine, phosphite), as in equation (53) should not be disturbing because nickel still retains a low lying $4p$ orbital available for bonding, as in equation (54) leading to **13**.

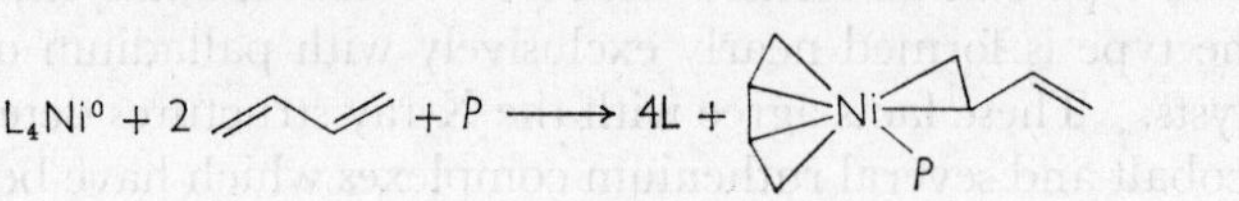

$$L_4Ni^0 + 2 \diagup\!\!\diagdown + P \longrightarrow 4L + [Ni\text{-}P] \qquad (53)$$

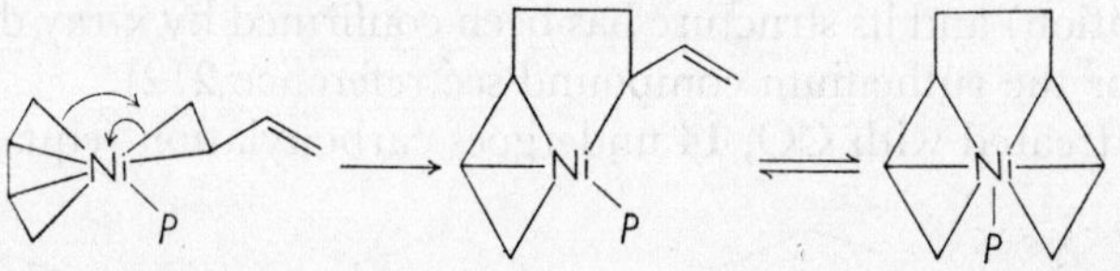

5. Alkene Complexes of Transition Metals as Reactive Intermediates

$$\text{[Ni complex]} \longrightarrow \text{[Ni complex]} \rightleftharpoons \text{[Ni complex]} \quad (54)$$

(13)

However a second such ligand would prevent the formation of the bis-π-allyl complex, and partially destroys the catalytic activity.

In the presence of excess butadiene at high pressure, a change from π-allyl to σ-allyl structure takes place in **13**, making a new coordination site available on nickel, and the reaction goes on to incorporate a new molecule of butadiene, to give **14**, as in the sequence (55).

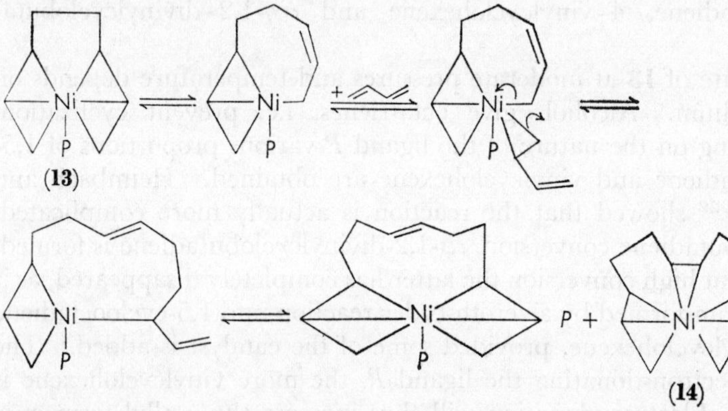

$$\text{(13)} \rightleftharpoons \text{[Ni]} \xrightleftharpoons[]{+\text{butadiene}} \text{[Ni]} \rightleftharpoons$$

$$\rightleftharpoons \text{[Ni]} \rightleftharpoons \text{[Ni complex]} \quad P + \text{[Ni]} \quad (55)$$

(14)

In principle the last equilibrium of equation (55) should be the more displaced to the right, the less strongly coordinated the ligand P is. Thus with AsR_3 or $PhC \equiv CH$ the equilibrium is in favour of the bis π-allyl complex and trimerization prevails. This must mean that ring closure (equation 56) of the bis π-allyl complex must be rate determining.

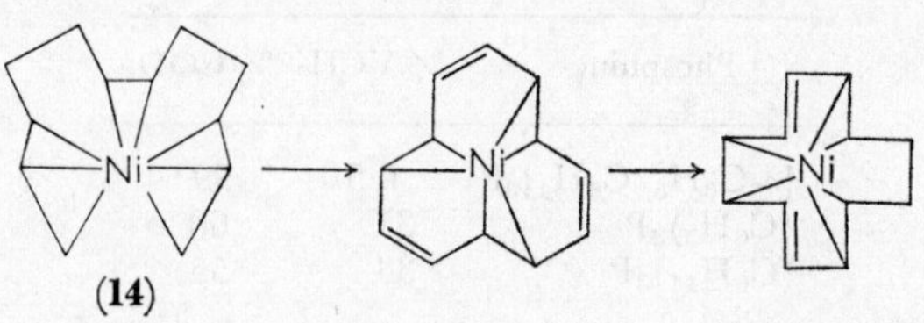

$$\text{(14)} \longrightarrow \text{[Ni complex]} \longrightarrow \text{[Ni complex]} \quad (56)$$

The cyclododecatriene complex can be isolated under the *trans, trans,cis* configuration (and not as one would expect the *trans,trans,trans* configuration) and its structure has been confirmed by x-ray determinations (for the ruthenium compound see reference 212).

When treated with CO, **14** undergoes carbonylation (equation 57).

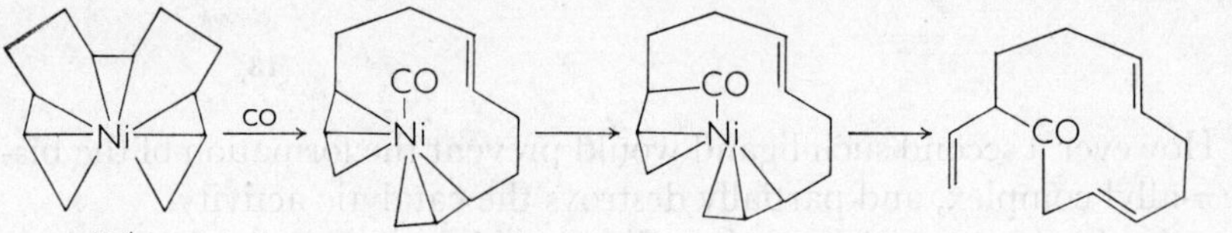

(14) (57)

Of much interest is the fact that the analogous Pd compound gives only the linear dodecatetraene[17].

With lower butadiene pressure **13** will undergo ring closure to 1,5-cyclooctadiene, 4-vinylcyclohexene and cis-1,2-divinylcyclobutadiene[18].

The fate of **13** at moderate pressures and temperature depends on the medium. Alcohols give octatrienes, i.e. prevent cyclization. Depending on the nature of the ligand P various proportions of 1,5-cyclooctadiene and vinylcyclohexene are obtained. Heimbach and Brenner[222] showed that the reaction is actually more complicated. At low butadiene conversion, cis-1,2-divinylcyclobutadiene is formed, whereas at high conversion the latter has completely disappeared, as it can be transformed by a zeroth-order reaction into 1,5-cyclooctadiene and vinylcyclohexene, provided some of the catalyst is added. The better electron-donating the ligand P, the more vinylcyclohexene is formed. Electron donation will thus increase the σ-allyl form over the π-allyl and favour the formation of vinylcyclohexene, as shown by Table 5 for various phosphines.

TABLE 5. Influence of electron donation on the proportions of vinylcyclohexene (VCH) and 1,5-cyclooctadiene (COD) formed

Phosphine	% VCH	% COD
$[o\text{-}C_6H_5\text{-}C_6H_4]_3P$	0.5	99
$(C_6H_5)_3P$	31	60
$(C_6H_{11})_3P$	33	55

Analogous isomerization can also be brought about by other metals such as palladium[223].

D. Condensations (Mixed Oligomerizations)

Mixtures of alkenes or of dienes can be oligomerized to mixtures of oligomers. Thus mixtures of ethylene and propene will afford mixtures of butenes, pentenes and hexenes. More interesting perhaps are the reactions of dienes with alkenes. Again linear condensation and cyclic condensation can take place.

I. Linear Condensations

Linear condensation of dienes and alkenes can be brought about by iron[224–225], cobalt[226–227], nickel[228] and rhodium[194,229] catalysts.

With rhodium catalysts, the complex **15** derived from butadiene has

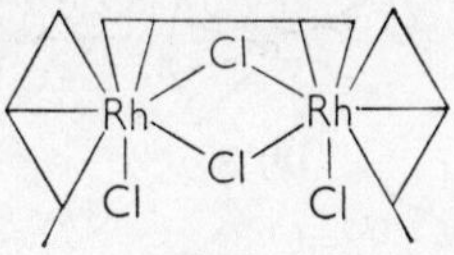

(15)

been isolated[230,231]. In the presence of ethylene, which is a better ligand than butadiene, this could be expected to give **16**, which could

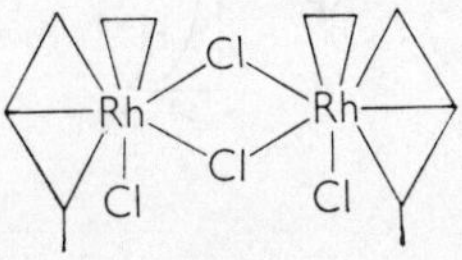

(16)

undergo isomerization to **17**. On further reaction with butadiene, the hexenyl moiety would be displaced to give an alkyl rhodium (**18**) capable of transferring a hydride to butadiene, thus releasing 1,4-hexadiene and an unsaturated π-allyl complex. This would in turn be able to add up ethylene to carry the reaction further.

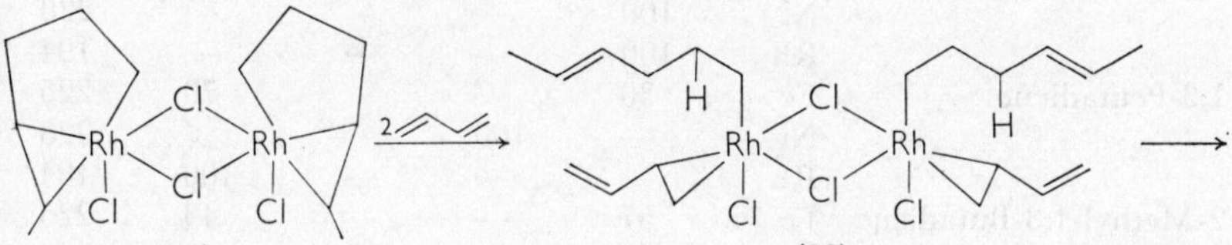

(17) $\xrightarrow{2}$ (18) $\longrightarrow$

$2\left[\right]_2 \longrightarrow 2\left[\right]_2 \xrightarrow{2C_2H_4}$ (16)

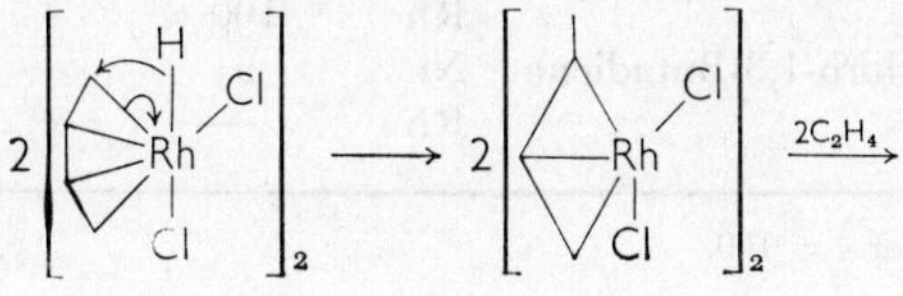

The rate equation can be written [229]:

$$-d[C_2H_4]/dt = k[C_2H_4][Cl^-][H^+][Rh][C_4H_6]$$

at about 50°. This agrees with what is known for ethylene dimerization. It means that the rate-determining step is the release of hexadiene, as the hydride transfer has been shown to be a fast reaction. With rhodium some stereomutation to *trans* isomer and some isomerization to conjugated dienes takes place. It is not so with appropriate cobalt[227] or iron[225] derivatives (with bidentate phosphine ligands, P) as shown in equation (58).

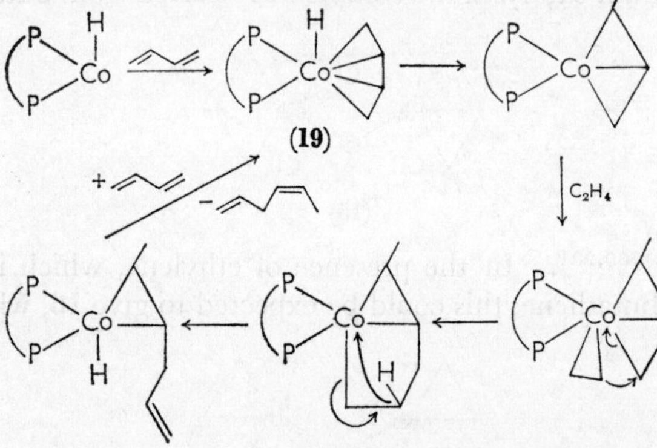

(19)   (58)

TABLE 6. Condensation of Ethylene with 1,3-dienes

Diene	Metal	Ratio (%) of attacked site in diene				Ref.
		$C_{(1)}$	$C_{(2)}$	$C_{(3)}$	$C_{(4)}$	
1,3-Butadiene	Fe	100	—	—	—	225
	Ni	100	—	—	—	228
	Rh	100	—	—	—	194
1,3-Pentadiene	Fe	30	—	—	70	225
	Ni	—	100			228
	Rh	—	—	—	100	194
2-Methyl-1,3-Butadiene	Fe	57	—	—	43	225
	Ni	69	—	31	—	228
	Rh	100	—	—	—	194
2-Chloro-1,3-Butadiene	Ni	91	—	9	—	228
	Rh	—	x^a	y^a	—	194

[a] x or $y = 100$.

The various condensation products of ethylene with conjugated dienes are given in Table 6. It can be seen that usually the carbon atom in the diene, which is substituted is the one at which the electron density is greatest.

2. Cyclic Condensations

These have been less thoroughly studied. But Wilke and coworkers[16] could show that complex **13** will add to ethylene to give cyclodecadiene.

Acetylenes with **13** will give cyclododecatrienes[232].

E. Dimerization of Alkenes (Cyclobutanation)

Various alkenes have been dimerized to cyclobutanes in the presence of rhodium[233], iron[234] or cobalt[234b,235] catalyst. Mango and Schachtschneider[236] have suggested that the participation of the metal d orbitals would make possible a concerted cyclization, which would otherwise be forbidden by the Woodward–Hoffmann rules[237].

As pointed out by Heimbach[222c], this is not necessarily true, because divinylcyclobutane seems to be formed from butadiene by two steps mechanisms, and not a one step (concerted) mechanism. In this respect, one should also notice the rhodium catalysed dimerization of norbornadiene[233], which, *inter alia*, gives a cyclobutane type dimer and also dimer **19** as shown in chart 3.

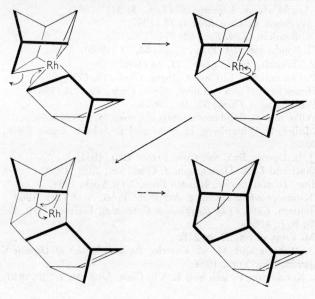

This can be explained by a multistep mechanism, so that the actual status of the reaction still remains to be clarified.

Finally it should be pointed out that the reversed reaction (cyclobutane giving alkenes) is also catalysed by metals[238-240]. It is quite conceivable that the remarkable *metathesis* depicted by equation (59)

$$CH_3CH{=}CHCH_3 + C_2H_5CH{=}CHC_2H_5 \rightleftharpoons 2\ CH_3CH{=}CHC_2H_5 \qquad (59)$$
$$25\% \qquad\qquad 25\% \qquad\qquad\qquad 50\%$$

which readily takes place in ethanol solution at room temperature in the presence of WCl_6 and $AlCl_2Et$[241], is actually a cyclobutanation followed by the reverse reaction.

IX. REFERENCES

1. W. E. Zeise, *Poggendorf Annalen der Physik.*, **9**, 632 (1827).
2. W. Reppe, O. Schlichting, K. Klager and T. Toepel, *Ann. Chem.*, **560**, 1 (1948); W. Reppe, O. Schlichting and H. Meister, *Ann. Chem.*, **560**, 93 (1948); W. Reppe and W. J. Scheckendieck, *Ann. Chem.*, **560**, 104 (1948).
3. W. Reppe, *Ann. Chem.*, **582**, 1 (1953).
4. W. Reppe and H. Kröper, *Ann. Chem.*, **582**, 38 (1953).
5. R. Pettit and G. F. Emerson, *Advan. Organometallic Chem.*, **1**, 1 (1964).
6. G. N. Schrauzer, *Advan. Organometal. Chem.*, **2**, 1 (1964).
7. M. L. H. Green and P. L. I. Nagy, *Advan. Organometal. Chem.*, **2**, 325 (1964).
8. P. M. Maitlis, *Advan. Organometal. Chem.*, **4**, 95 (1966).
9. R. F. Heck, *Advan. Organometal. Chem.*, **4**, 243 (1966).
10a. A. Aguiló, *Advan. Organometal. Chem.*, **5**, 321 (1967).
10b. E. W. Stern, *Catalysis Rev.*, **1**, 73 (1967).
11. M. A. Bennett, *Chem. Rev.*, **62**, 611 (1962).
12. G. C. Bond, *Ann. Rep. Progr. Chem.*, **53**, 27 (1966).
13. G. P. Chiusoli, *Angew. Chem.*, **72**, 74 (1960).
14. G. P. Chiusoli and L. Cassar, *Angew. Chem.*, **79**, 177 (1967).
15. G. Henrici-Olivé and S. Olivé, *Angew. Chem.*, **79**, 764 (1967).
16. G. Wilke, *Angew. Chem.*, **75**, 10 (1963).
17. G. Wilke and B. Bogdanović, *Angew. Chem.*, **78**, 170 (1966).
18. H. Muller, D. Wittenberg, H. Seibt and E. Scharf, *Angew. Chem.*, **77**, 318 (1965).
19. M. J. S. Dewar, *Bull. Soc. Chim. France*, **C79** (1951).
20. J. Chatt and L. A. Duncanson, *J. Chem. Soc.*, 2939 (1953).
21. J. Hine, *Divalent Carbon*, Ronald Press, New York, 1964.
22. W. Kirmse, *Carbene Chemistry*, Academic Press, New York, 1964.
23. J. Halpern, *Catalysis by Coordination Compounds*, Prentice Hall, Englewood Cliffs, N.J., 1964.
24. H. M. Frey, in ref. 22, p. 218.
25. J. D. Roberts and M. C. Caserio, *Basic Principles of Organic Chemistry*, Benjamin, New York, 1964.
26. J. L. Kondakow, F. Bais and L. Vit, *Chem. Listy*, **24**, 1, 26 (1930).

27. M. S. Kharasch, R. C. Seyler and F. R. Mayo, *J. Am. Chem. Soc.*, **60**, 882 (1938).
28. R. Huttel, J. Kratzer and M. Bechter, *Chem. Ber.*, **94**, 766 (1961).
29. B. Bogdanović, M. Kröner and G. Wilke, *Ann. Chem.*, **699**, 1 (1966).
30. R. Cramer, *J. Am. Chem. Soc.*, **89**, 4621 (1967).
31. P. M. Henry, *J. Am. Chem. Soc.*, **88**, 1595 (1966).
32. R. J. Cvetanović, F. J. Duncan, W. E. Falconer and R. S. Irwin, *J. Am. Chem. Soc.*, **87**, 1827 (1965).
33. M. A. Muhs and F. T. Weiss, *J. Am. Chem. Soc.*, **84**, 4697 (1962).
34. G. Wilke and B. Bogdanović, *Angew. Chem.*, **73**, 756 (1961).
35. H. Dietrich and R. Uttech, *Z. Krist*, **122**, 112 (1965).
36. J. K. Becconsall, B. E. Job and S. O'Brien, *J. Chem. Soc.*, A, 423 (1967).
37. H. Bönnemann, B. Bogdanović and G. Wilke, *Angew. Chem.*, **79,** 817 (1967).
38. J. C. W. Chien and H. C. Dehm, *Chem. Ind.*, 745 (1961).
39. R. G. Schultz, *Tetrahedron Letters*, 301 (1964).
40. C. H. Langford and H. B. Gray, *Ligand Substitution Processes*, Benjamin, New York, 1965.
41. J. F. Bunnett and R. E. Zahler, *Chem. Rev.*, **49**, 273 (1956).
42. P. Cossee, *J. Catalysis*, **3,** 80 (1964).
43. P. Cossee, *Rec. Trav. chim.*, **85,** 1151 (1966).
44. M. C. Baird, C. J. Nyman and G. Wilkinson, *J. Chem. Soc.*, A, 348 (1968).
45. J. Tsuji and K. Ohno, *Tetrahedron Letters*, 3669 (1965).
46. J. A. Smythe, *J. Chem. Soc.*, **95**, 349 (1909).
47. R. Pummerer, *Chem. Ber.*, **43,** 1401 (1910).
48. C. R. Johnson, J. C. Sharp and W. G. Phillips, *Tetrahedron Letters*, 5299 (1967).
49. M. E. D. Hillman, *J. Am. Chem. Soc.*, **84,** 4715 (1962).
50. H. C. Brown and M. W. Rathke, *J. Am. Chem. Soc.*, **89,** 2737 (1967).
51. R. Cramer, *J. Am. Chem. Soc.*, **88,** 2272 (1966).
52. J. F. Harrod and A. J. Chalk, *J. Am. Chem. Soc.*, **86,** 1776 (1964).
53. J. F. Harrod and A. J. Chalk, *J. Am. Chem. Soc.*, **88,** 3491 (1966).
54. F. Asinger, B. Fell and P. Krings, *Tetrahedron Letters*, 633 (1966).
55. J. C. Trebellas, J. R. Olechowski, H. B. Jonassen, and D. W. Moore, *J. Organometal. Chem.*, **9,** 153 (1967).
56. R. F. Heck and D. Breslow, *J. Am. Chem. Soc.*, **83,** 4023 (1961).
57. M. Johnson, *J. Chem. Soc.*, 4859 (1963).
58. B. Fell, P. Krings and F. Asinger, *Chem. Ber.*, **99,** 3688 (1966).
59. L. Roos and M. Orchin, *J. Am. Chem. Soc.*, **87,** 5502 (1965).
60. J. K. Nicholson and B. L. Shaw, *Tetrahedron Letters*, 3533 (1965).
61. R. S. Coffey, *Tetrahedron Letters*, 3809 (1965).
62. G. G. Eberhardt and L. Vaska, *J. Catalysis*, **8,** 183 (1967).
63. G. C. Bond and M. Hellier, *J. Catalysis*, **7,** 217 (1967).
64. H. A. Tayim and J. C. Baylar, *J. Am. Chem. Soc.*, **89,** 3420 (1967)
65. T. A. Manuel, *J. Org. Chem.*, **27,** 3941 (1962).
66. F. Asinger, B. Fell and G. Collin, *Chem. Ber.*, **96,** 716 (1963).
67. F. Asinger, B. Fell and K. Schrage, *Chem. Ber.*, **98,** 372 (1965).
68. F. Asinger, B. Fell and K. Schrage, *Chem. Ber.*, **98,** 381 (1965).
69. B. Fell, P. Krings and F. Asinger, *Chem. Ber.*, **99,** 3688 (1966).
70. P. W. Jolly, F. G. A. Stone and K. Mackenzie, *J. Chem. Soc.*, 6416 (1965).

71a. R. Damico and T. J. Logan, *J. Org. Chem.*, **32,** 2536 (1967).
71b. W. T. Hendrix, F. G. Cowherd and J. L. von Rosenberg, *Chem. Comm.*, 897 (1968).
72. V. Macho, M. Polievka and L. Komora, *Chem. Zvesti*, **21,** 170 (1967).
73. N. R. Davies, *Austral. J. Chem.*, **17,** 212 (1964).
74. G. C. Bond and M. Hellier, *J. Catal.*, **4,** 1 (1965).
75. G. Pregaglia, M. Donati and F. Conti, *Chim. Ind. (Milan)*, **49,** 1277 (1967).
76a. Y. Chauvin, N. H. Phung, N. Guichard-Loudet and G. Lefebvre, *Bull. Soc. chim. France*, 3223 (1966).
76b. N. H. Phung, Y. Chauvin and G. Lefebvre, *Bull. Soc. chim. France*, 3618 (1967).
77. J. Halpern, *Developments in Homogeneous Catalysis, Proc. Intern. Congr. Catalysis*, 3rd Amsterdam 64, North-Holland, Amsterdam, 146 (1965).
78. M. Iguchi, *J. Chem. Soc. Japan*, **63,** 634 (1942).
79a. J. Kwiatek, I. L. Mador and J. K. Seyler, *J. Am. Chem. Soc.*, **84,** 304 (1962).
79b. M. Murakami, K. Suzuki and J. W. Kang, *Chem. Abstr.*, **59,** 13 868a (1963).
79c. J. Kwiatek, I. L. Mador and J. K. Seyler, *Reactions of coordinated ligands. Advan. Chem. Ser.*, **37,** 201 (1963).
79d. A. F. Mabrouk, H. J. Dutton and J. C. Cowan, *J. Am. Oil Chemists' Soc.*, **41,** 153 (1964).
80. W. P. Griffith and G. Wilkinson, *J. Chem. Soc.*, 2757 (1959).
81. B de Vries, *J. Catalysis*, **1,** 489 (1962).
82. M. G. Burnett, P. J. Connolly and C. Kemball, *J. Chem. Soc. (A)*, 800 (1967).
83. R. G. S. Banks and J. M. Pratt, *Chem. Comm.*, 776 (1967).
84. N. N. King and M. E. Winfield, *J. Am. Chem. Soc.*, **83,** 3366 (1961).
85. L. Simandi and F. Nagy, *Chem. Abst.*, **44,** 15037f (1966).
86. J. M. Pratt and R. J. P. Williams, *J. Chem. Soc. (A)*, 1291 (1967).
87. J. Kwiatek and J. K. Seyler, *J. Organometal. Chem.*, **3,** 421 (1965).
88. J. F. Young, J. A. Osborn, F. H. Jardine and G. Wilkinson; (a) *Chem. Comm.*, 131 (1965); (b) *J. Chem. Soc. (A)*, 1711 (1966).
89. J. F. Biellmann and H. Liesenfelt, *C.R. Acad. Sci., Paris, Ser. C.*, **263**, 251 (1966).
90. A. J. Birch and K. A. M. Walker, *J. Chem. Soc. (C)*, 1894 (1966).
91. J. F. Biellmann and H. Liesenfelt, *Bull. Soc. Chim.*, 4029 (1966).
92. A. J. Birch and K. A. M. Walker, *Tetrahedron Letters*, 4939 (1966).
93. W. Voelter and C. Djerassi, *Chem. Ber.*, **101,** 58 (1968).
94. A. J. Birch and K. A. M. Walker, *Tetrahedron Letters*, 3457 (1967).
95. A. J. Birch and K. A. M. Walker, *Tetrahedron Letters*, 1935 (1967).
96. J. Tsuji and K. Ohno, *Tetrahedron Letters*, 3969 (1965); *Tetrahedron Letters*, 2173 (1967); *J. Am. Chem. Soc.*, **88,** 3452 (1966); *Tetrahedron Letters*, 4713 (1966).
97. J. Blum, *Tetrahedron Letters*, 1605 (1966); J. Blum, E. Oppenheimer and E. D. Bergmann, *J. Am. Chem. Soc.*, **89,** 2338 (1967).
98. F. H. Jardine and G. Wilkinson, *J. Chem. Soc. (C)*, 270 (1967).
99. F. H. Jardine, J. A. Osborn and G. Wilkinson, *J. Chem. Soc. (A)*, 1574 1967).

100. J. F. Biellmann and M. J. Jung, to be published.
101. J. Halpern, J. F. Harrod and B. R. James, *J. Am. Chem. Soc.*, **83,** 753 (1961); **88,** 5150 (1966).
102. P. S. Hallman, D. Evans, J. A. Osborn and G. Wilkinson, *Chem. Comm.*, 305 (1966).
103. I. Jardine and F. J. McQuillin, *Tetrahedron Letters*, 4871 (1966).
104. D. Evans, J. A. Osborn, F. H. Jardine and G. Wilkinson, *Nature*, **208,** 1203 (1965).
105. K. Sonogashira and N. Hagihara, *Bull. Chem. Soc. Japan*, **39,** 1178 (1966). K. Shikata, K. Nishino, K. Azuma and Y. Takegami, *Chem. Abstr.* **65,** 10 452b (1966); K. Shikata, K. Nishino and K. Azuma, *Chem. Abstr.*, **63,** 71119a (1965).
106. V. A. Tulupov, *Chem. Abstr.*, **57,** 14471b (1962).
107. V. A. Tulupov, *Chem. Abstr.*, **52,** 14302c (1958).
108a. E. N. Frankel, H. M. Peters, E. P. Jones and H. J. Dutton, *J. Am. Oil Chemists' Soc.*, **41,** 186 (1964).
108b. E. N. Frankel, E. A. Emken, H. M. Peters, V. L. Davison and R. O. Butterfield, *J. Org. Chem.*, **29,** 3292 (1964).
108c. E. N. Frankel, E. A. Emken and V. L. Davison, *J. Org. Chem.*, **30,** 2739 (1965).
109a. Y. Takegami, T. Ueno and T. Fujii, *Chem. Abstr.*, **61,** 13931g (1964).
109b. Y. Takegami, T. Ueno and T. Fujii, *Bull. Chem. Soc. Japan*, **38,** 1279 (1965).
109c. Y. Takegami and T. Fujimaki, *Chem. Abstr.*, **57,** 4271f (1962).
109d. Y. Takegami, T. Ueno, and K. Kawajiri, *Chem. Abstr.*, **62,** 7661b (1965).
109e. Y. Takegami, T. Ueno, and K. Kawajiri, *Bull. Soc. Chem. Japan*, **39,** 1 (1966).
110. V. A. Tulupov, *Chem. Abstr.*, **51,** 17 776i (1957).
111. P. Fotis, Jr. and J. D. McCollum, *Chem. Abstr.*, **67,** 53 616v (1967).
112a. L. Marko, *Chem. Ind.*, 260 (1962).
112b. F. Ungvary, B. Babos, and L. Marko, *J. Organometal. Chem.*, **8,** 329 (1967).
113. R. D. Mullineaux, *Chem. Abstr.*, **60,** 7504a (1964).
114a. R. W. Goetz and M. Orchin, *J. Org. Chem.*, **27,** 3698 (1962).
114b. R. W. Goetz and M. Orchin, *J. Am. Chem. Soc.*, **85,** 2782 (1963).
115a. E. N. Frankel, E. J. Jones, V. L. Davison, E. A. Emken, and H. J. Dutton, *J. Am. Oil Chemists' Soc.*, **42,** 130 (1965).
115b. E. A. Emken, E. N. Frankel and R. O. Butterfield, *J. Am. Oil Chemists' Soc.*, **43,** 14 (1966).
116a. A. Misono, Y. Uchida, T. Saito and K. M. Song, *Chem. Comm.*, 419 (1967).
116b. A. Misono and I. Ogata, *Bull. Chem. Soc. Japan*, **40,** 2718 (1967).
117. R. F. Heck, *Chem. Abstr.*, **65,** 16 857d (1966).
118. R. D. Gillard, J. A. Osborn, P. B. Stockwell and G. Wilkinson, *Proc. Chem. Soc.*, 284 (1964).
119. B. R. James and G. L. Rempel, *Canad. J. Chem.*, **44,** 233 (1966).
120a. L. Vaska and R. E. Rhodes, *J. Am. Chem. Soc.*, **87,** 4970 (1965).
120b. L. Vaska, *Inorg. Nucl. Chem. Letters*, **1,** 89 (1965).

121a. Y. M. Y. Haddad, H. B. Henbest, J. Husbands and T. R. B. Mitchell, *Proc. Chem. Soc.*, 361 (1964).
121b. J. Trocha-Grimshaw and H. B. Henbest, *Chem. Comm.*, 544 (1967).
122. M. A. Bennett and D. L. Milner, *Chem. Comm.*, 581 (1967).
123. R. S. Coffey, *Chem. Comm.*, 923 (1967).
124. H. Itatani and J. C. Bailar, Jr., *J. Am. Chem. Soc.*, **89,** 1600 (1967).
125. M. G. Burnett, *Chem. Comm.*, 507 (1965).
126. S. J. Lapporte and W. R. Schuett, *J. Org. Chem.*, **28,** 1947 (1963).
127. *Patent: Neth. Appl.*, 296 137 (*Chem. Abst.*, **63,** 9878b (1965)).
128. H. Itatani and J. C. Bailar, Jr., *J. Am. Oil. Chem. Soc.*, **44,** 147 (1967).
129. R. D. Cramer, E. L. Jenner, R. V. Lindsey and U. G. Stolberg, *J. Am. Chem. Soc.*, **85,** 1961 (1963).
130. J. C. Bailar, Jr. and H. Itatani, *Inorg. Chem.*, **4,** 1618 (1965).
131a. H. A. Tayim and J. C. Bailar, Jr., *J. Am. Chem. Soc.*, **89,** 4330 (1967).
131b. J. C. Bailar, Jr. and H. Itatani, *J. Am. Oil Chemists' Soc.*, **43,** 337 (1966).
131c. J. C. Bailar, Jr., Adams and Batley, Private Communication.
131d. J. C. Bailar, Jr. and H. Itatani, *J. Am. Chem. Soc.*, **89,** 1592 (1967).
132. V. A. Tupulov and M. I. Gagarina, *Chem. Abstr.*, **61,** 11371g (1964).
133. W. R. Kroll, *Chem. Abstr.*, **67,** 53621t (1967).
134. J. V. Kalechits, V. G. Lipovich and F. K. Shmidt, *Chem. Abstr.*, **66,** 94 632v (1967).
135. M. F. Sloan, A. S. Matlack and D. S. Breslow, *J. Am. Chem. Soc.*, **85,** 4014 (1963).
136a. J. Smidt, W. Hafner, M. Jira, R. Sieber, R. Ruttinger and H. Kojer, *Angew. Chem.*, **71,** 176 (1965).
136b. J. Smidt, *Chem. Ind.*, 54 (1962).
137. P. M. Henry, *J. Am. Chem. Soc.*, **86,** 3246 (1964).
138. I. I. Moiseev, M. N. Vargaftik and Ya. K. Syrkin, *Izv. Akad. Nauk. SSSR, Otd. Khim. Nauk.*, 1144 (1963).
139. W. H. Clement and C. M. Selwitz, *J. Org. Chem.*, **29,** 241 (1964).
140. W. Hafner, R. Jira, J. Sedlmeier and J. Smidt, *Chem. Ber.*, **95,** 1575 (1962).
141. J. Smidt, W. Hafner, R. Jira, R. Sieber, J. Sedlmeier and A. Sabel, *Angew. Chem.*, **74,** 93 (1962).
142. K. B. Wiberg, *Chem. Rev.*, **55,** 713 (1955).
143. R. Huttel and H. Christ, *Chem. Ber.*, **97,** 1439 (1964).
144. R. Jira, J. Sedlmeier and J. Smidt, *Ann. Chem.*, **693,** 99 (1966).
145a. I. I. Moiseev, M. N. Vargaftik and Ya. K. Syrkin, *Dokl. Akad. Nauk. SSSR*, **133,** 377 (1960).
145b. I. I. Moiseev and M. N. Vargaftik, *Izv. Akad. Nauk. SSSR, Ser. Khim.*, 759 (1965).
146. W. Kitching, Z. Rapoport, S. Winstein and W. G. Young, *J. Am. Chem. Soc.*, **88,** 2054 (1966).
147a. G. L. Statton and K. C. Ramey, *J. Am. Chem. Soc.*, **88,** 1327 (1966).
147b. K. C. Ramey and G. L. Statton, *J. Am. Chem. Soc.*, **88,** 4387 (1966).
147c. W. B. Wise, D. C. Lini and K. C. Ramey, *Chem. Comm.*, 463 (1967).
148a. K. Vrieze, C. MacLean, P. Cossee and C. W. Hilbers, *Rec. Trav. Chim.*, **85,** 1077 (1966).
148b. K. Vrieze, P. Cossee, C. W. Hilbers, and A. P. Praat, *Rec. Trav. Chim.*, **86,** 769 (1967).

5. Alkene Complexes of Transition Metals as Reactive Intermediates 261

149. W. C. Baird, *J. Org. Chem.*, **31,** 2411 (1966).
150a. N. C. Baenziger, J. R. Doyle and C. Carpenter, *Acta Cryst.*, **14,** 303 (1961).
150b. N. C. Baenziger, C. F. Richards and J. R. Doyle, *Acta Cryst.*, **18,** 924 (1965).
151a. J. K. Stille, R. A. Morgan, D. D. Whitehurst and J. R. Doyle, *J. Am. Chem. Soc.*, **87,** 3282 (1965).
151b. J. K. Stille and R. A. Morgan, *J. Am. Chem. Soc.*, **88,** 5135 (1966).
152. W. A. Whitla, H. M. Powell and L. M. Venanzi, *Chem. Comm.*, 310 (1966).
153a. R. G. Schultz, *J. Organometal. Chem.*, **6,** 435 (1966).
153b. M. Green and R. I. Hancock, *J. Chem. Soc. A.*, 2054 (1967).
154. J. Chatt, L. M. Vallarino and L. M. Venanzi, *J. Chem. Soc.*, 3413 (1957).
155. J. Paiaro, A. de Renzi and R. Palumbo, *Chem. Comm.*, 1150 (1967).
156. C. B. Anderson and B. J. Burreson, *Chem. Ind.*, 620 (1967).
157. Y. Takahashi, S. Sakai and Y. Ishii, *Chem. Comm.*, 1092 (1967).
158. S. Uemura and K. Ichikawa, *Bull. Chem. Soc. Japan*, **40,** 1016 (1967).
159. J. Tsuji, H. Takahashi and M. Morikawa, *Tetrahedron Letters*, 4387 (1965); *Kogyo Kagaku Zasshi*, **69,** 920 (1966).
160. J. Tsuji and H. Takahashi, *J. Am. Chem. Soc.*, **87,** 3275 (1965).
161. Y. Odaira, T. Oishi, T. Yukawa and S. Tutsumi, *J. Am. Chem. Soc.*, **88,** 4105 (1966).
162. A. J. Chalk, *Tetrahedron Letters*, 2627 (1964).
163. K. Noack and F. Calderazzo, *J. Organometal. Chem.*, **10,** 101 (1967).
164. P. J. Craig and M. Green, *Chem. Comm.*, 1246 (1967).
165. R. J. Mawby, F. Basolo and R. G. Pearson, *J. Am. Chem. Soc.*, **86,** 5043 (1964).
166. W. D. Bannister, M. Green and R. H. Haszeldine, *Chem. Comm.*, 54 (1965).
167. M. Green and D. C. Wood, *J. Am. Chem. Soc.*, **88,** 4106 (1966).
168. C. S. Kraihanzel and P. K. Maples, *J. Am. Chem. Soc.*, **87,** 5267 (1965).
169. F. Calderazzo and F. A. Cotton, *Chim. Ind. (Milan)*, **46,** 1165 (1964).
170. R. J. Mawby, F. Basolo and R. G. Pearson, *J. Am. Chem. Soc.*, **86,** 3994 (1964).
171. J. Tsuji and K. Ohno, *J. Am. Chem. Soc.*, **90,** 94 (1968).
172. F. Piacenti, R. Pino, R. Lazzaroni and M. Bianchi, *J. Chem. Soc., C.*, 488 (1966).
173. F. Calderazzo and F. A. Cotton, *Inorg. Chem.*, **1,** 30 (1962).
174. G. L. Karapinka and M. Orchin, *J. Org. Chem.*, **26,** 4187 (1961).
175. P. Krumholz and H. M. A. Stettiner, *J. Am. Chem. Soc.*, **71,** 3035 (1949).
176. H. W. Sternberg, R. Markby and I. Wender, *J. Am. Chem. Soc.*, **78,** 5704 (1956); **79,** 6116 (1957).
177. E. O. Fischer and G. Burger, *Z. Naturforsch*, **17b,** 484 (1962).
178. R. F. Heck, *J. Am. Chem. Soc.*, **85,** 2013 (1963).
179a. J. Tsuji, S. Imamura, and J. Kiji, *J. Am. Chem. Soc.*, **86,** 4491 (1964).
179b. J. Tsuji and S. Imamura, *Bull. Chem. Soc. Japan*, **40,** 197 (1967).
180. J. K. Nicholson, J. Powell, and B. L. Shaw, *Chem. Comm.*, 174 (1966).
181. J. Tsuji and N. Iwamoto, *Chem. Comm.*, 828 (1966).
182. G. P. Chiusoli and G. Cometti, *Chim. Ind. (Milan)*, **45,** 404 (1963).
183. R. F. Heck and D. S. Breslow, *J. Am. Chem. Soc.*, **83,** 4023 (1961).

184. I. Wender, J. Feldman, S. Metlin, B. H. Gwynn and M. Orchin, *J. Am. Chem. Soc.*, **77,** 5760 (1955).
185. H. Adkins and G. Krsek, *J. Am. Chem. Soc.*, **71,** 3051 (1949).
186. Y. Takegami, C. Yokokawa and Y. Watanabe, *Bull. Chem. Soc. Japan*, **39,** 2430 (1966).
187. A. Matsuda, *Bull. Chem. Soc. Japan*, **40,** 135 (1967).
188. I. Wender, S. Metlin, S. Ergun, H. W. Sternberg and H. Greenfield, *J. Am. Chem. Soc.*, **78,** 540 (1958).
189. C. W. Bird, R. C. Cookson, J. Hudec and R. D. Williams, *J. Chem. Soc.*, 410 (1963).
190a. J. Tsuji, S. Hosaka, J. Kiji and T. Susuki, *Bull. Chem. Soc. Japan*, **39,** 141 (1966).
190b. J. Tsuji, S. Hosaka, J. Kiji and T. Nogi, *Bull. Chem. Soc. Japan.*, **39,** 146 (1966).
191a. S. Brewis and P. R. Hughes, *Chem. Comm.*, 489 (1965); 6 (1966).
191b. S. Brewis and P. R. Hughes, *Chem. Comm.*, 71 (1967).
192. S. Falbe, *Angew. Chem.*, **78,** 532 (1966).
193. N. von Kutepow and H. Kindler, *Angew. Chem.*, **72,** 802 (1960).
194. T. Alderson, E. L. Venner and R. V. Lindsey, *J. Am. Chem. Soc.*, **87,** 5638 (1965).
195. R. Cramer, *J. Am. Chem. Soc.*, **87,** 4717 (1965).
196. B. Bogdanović and G. Wilke, World Petrol. Cong., Proc., 7th, N.Y., 351 (1967).
197. M. Uchino, Y. Chauvin and G. Lefevre, *C.R. Acad. Sci. Paris, Ser. C*, **265,** 104 (1967).
198. G. Hata and A. Mikaye, *Chem. Ind.*, 921 (1967).
199. N. S. H. Feldblyum, N. Vobeshchalova, A. I. Leshcheva and T. I. Baranova, Neftekhimiya, **7,** 379 (1967).
200. J. K. Nicholson and B. L. Shaw, *J. Chem. Soc.*, A, 807 (1966).
201. A. D. Ketley, L. P. Fischer, C. R. Morgan, E. H. Gorman and T. R. Steadman, *Inorg. Chem.*, **6,** 657 (1967).
202a. N. H. Phung, Y. Chauvin and G. Lefebvre, *Bull. Soc. Chim. France*, 3618 (1967).
202b. N. H. Phung and G. Lefebvre, *C.R. Acad. Sci., Paris, Ser. C*, **265,** 519 (1967).
203a. H. Takahashi, T. Kimata and M. Yamaguchi, *Tetrahedron Letters*, 3173 (1964).
203b. H. Takahashi, S. Tai and M. Yamaguchi, *J. Org. Chem.*, **30,** 1661 (1965).
204a. M. Hidai, Y. Uchida and A. Misono, *Bull. Chem. Soc. Japan*, **38,** 1243 (1965).
204b. M. Hidai, K. Tamai, Y. Uchida and A. Misono, *Bull. Chem. Soc. Japan*, **39,** 1357 (1966).
205. A. Carbonaro, A. Greco and G. Dall'Agata, *Tetrahedron Letters*, 2037 (1967).
206a. G. Natta, U. Giannini, P. Pino and A. Cassata, *Chim. Ind. (Milan)*, **47,** 524 (1965).
206b. G. Allegra, F. Logiudice, G. Natta, U. Giannini, G. Fagherazzi and P. Pino, *Chem. Comm.*, 1263 (1967).
207a. S. Otsuka, T. Taketomi and T. Kikuchi, *J. Am. Chem. Soc.*, **85,** 3709 (1963); *Kogyo Kagaku Zasshi*, **66,** 1094 (1963).

207b. T. Saito, T. Ono, T. Uchida and A. Misoni, *Kogyo Kagaku Zasshi*, **66**, 1099 (1963).
208. S. Tanaka, K. Mabuchi and N. Shimazaki, *J. Org. Chem.*, **29**, 1626 (1964).
209. K. C. Dewhurst, *J. Org. Chem.*, **32**, 1297 (1967).
210. S. Takahashi, T. Shibano and N. Hagihara, *Tetrahedron Letters*, **89**, 2451 (1967).
211. E. J. Smutny, *J. Am. Chem. Soc.*, **89**, 6793 (1967).
212a. J. E. Lydon, J. K. Nicholson, B. L. Shaw and M. R. Truter, *Proc. Chem. Soc.*, 421 (1964).
212b. J. K. Nicholson and B. L. Shaw, *J. Chem. Soc. A*, 807 (1966).
212c. J. E. Lydon and M. R. Truter, *J. Chem. Soc. A*, 362 (1968).
213. L. Porri, M. C. Gallazzi, A. Colombo and G. Allegra, *Tetrahedron Letters*, 4187 (1965).
214. H. W. B. Reed, *J. Chem. Soc.*, 1931 (1954).
215. L. I. Zakharkin and G. G. Zhigareva, *Izv. Akad. Nauk. SSSR, Ser. Khim.*, 386 (1963).
216. G. Wilke, *Angew. Chem.*, **69**, 397 (1957).
217. H. Takahashi and M. Yamaguchi, *J. Org. Chem.*, **28**, 1409 (1963).
218. M. Weber, W. Ring, U. Hochmuth and W. Franke, *Ann. Chem.*, **681**, 10 (1965).
219. L. I. Zakharkin and V. M. Akhmedov, *Zh. Org. Khim.*, **2**, 1557 (1966).
220. G. Longiave, R. Castelli and A. Andreetta, *Chim. Ind. (Milan)*, **49**, 497 (1967).
221a. A. Yamamoto, K. Morifuji, S. Ikeda, T. Saito, Y. Uchida and A. Misono, *J. Am. Chem. Soc.*, **87**, 4652 (1965).
221b. A. Misono, Y. Uchida, M. Hidri and Y. Ohsawa, *Bull. Chem. Soc. Japan*, **39**, 2425 (1966).
222a. P. Heimbach and W. Brenner, *Angew. Chem.*, **79**, 813 (1967).
222b. P. Heimbach and W. Brenner, *Angew. Chem.*, **79**, 814 (1967).
222c. P. Heimbach, Private communication.
223. H. Frye, E. Kuljian and J. Viebrock, *Inorg. Nucl. Chem. Letters*, **2**, 119 (1966).
224a. M. Iwamoto and S. Yuguchi, *Bull. Chem. Soc. Japan.*, **39**, 2001 (1966).
224b. M. Iwamoto and S. Yuguchi, *J. Org. Chem.*, **31**, 4290 (1966).
225a. G. Hata, *J. Am. Chem. Soc.*, **86**, 3903 (1964).
225b. G. Hata and D. Aoki, *J. Org. Chem.*, **32**, 3754 (1967).
226. A. Misono, Y. Uchida, T. Saito and K. Uchida, *Bull. Chem. Soc. Japan*, **40**, 1889 (1967).
227a. M. Iwamoto, K. Tani, H. Igaki and S. Yuguchi, *J. Org. Chem.*, **32**, 4148 (1967).
227b. M. Iwamoto and S. Yuguchi, *Chem. Comm.*, 28 (1968).
228a. R. G. Miller, *J. Am. Chem. Soc.*, **89**, 2785 (1967).
228b. R. G. Miller, T. J. Kealy and A. L. Barney, *J. Am. Chem. Soc.*, **89**, 3756 (1967).
229. R. Cramer, *J. Am. Chem. Soc.*, **89**, 1633 (1967).
230. J. Powell and B. L. Shaw, *Chem. Comm.*, 236 and 323 (1966).
231. G. Paiaro, A. Musco and G. Diana, *J. Organometal. Chem.*, **4**, 466 (1965).
232a. P. Heimbach, *Angew. Chem.*, **78**, 983 (1966).

232b. P. Heimbach and W. Brenner, *Angew. Chem.*, **78,** 983 (1966).
233a. J. J. Mrowca and T. J. Katz, *J. Am. Chem. Soc.*, **88,** 4012 (1966).
233b. T. J. Katz, J. C. Carnahan and R. Boecke, *J. Org. Chem.*, **62,** 1301 (1967).
233c. T. J. Karz and N. Acton, *Tetrahedron Letters*, 2601 (1967).
234a. D. Lemal and K. S. Shim, *Tetrahedron Letters*, 369 (1961).
234b. C. W. Bird, D. L. Colinese, R. C. Cookson, J. Hudec, and R. R. Williams, *Tetrahedron Letters*, 373 (1961).
234c. P. W. Jolly, F. G. A. Stone and K. Mackenzie, *J. Chem. Soc.*, 6416 (1965).
235a. D. R. Arnold, D. J. Trecker and E. B. Whipple, *J. Am. Chem. Soc.*, **87,** 2596 (1965).
235b. G. N. Schrauzer, B. N. Bastian and G. A. Fosselius, *J. Am. Chem. Soc.*, **88,** 4890 (1966).
236. F. D. Mango and J. H. Schachtschneider, *J. Am. Chem. Soc.*, **89,** 2484 (1967).
237. R. B. Woodward and R. Hoffmann, *J. Am. Chem. Soc.*, **87,** 395, 2046, 2511, 4388 and 4389 (1965).
238. W. Merck and R. Pettit, *J. Am. Chem. Soc.*, **89,** 4788 (1967).
239a. H. Diehl and P. M. Maitlis, *Chem. Comm.*, 759 (1967).
239b. B. I. Booth, R. N. Haszeldine and M. Hill, *Chem. Comm.*, 1118 (1967).
239c. H. Hogeveen and H. C. Volger, *Chem. Comm.*, 1133 (1967).
240. H. Hogeveen and H. C. Volger, *J. Am. Chem. Soc.*, **89,** 2486 (1967).
241. N. Calderon, H. Y. Chen and K. W. Scott, *Tetrahedron Letters*, 3327 (1967).
242. J. Halpern and L. Y. Wong, *J. Am. Chem. Soc.*, **90,** 6665 (1968).
243. G. Pregaglia, D. Morelli, F. Conti, G. Gregorio and R. Ugo, *Trans. Faraday Soc.*, 1969, to be published.
244. A. B. Hörnfelt, J. S. Gronowitz and S. Gronowitz, *Acta. chim. Scand.*, **22,** 2725 (1968).
245. D. R. Eaton and S. R. Suart, *J. Am. Chem. Soc.*, **90,** 4170 (1968).
246. H. N. Bekkum, F. V. Rantwijk and T. V. de Putte, *Tetrahedron Letters*, 1, (1969).
247. A. J. Birch and G. S. R. Subba Rao, *Tetrahedron Letters*, 3797 (1968).
248. J. J. Sims, V. K. Honwad and L. H. Selman, *Tetrahedron Letters*, 87 (1969). R. Rüesch and T. J. Mabry, *Tetrahedron Letters*, **25,** 805 (1969); A. S. Hussey and Y. Takeuchi, *J. Am. Chem. Soc.*, **91,** 672 (1969); G. C. Bond and R. A. Hillyard, *Trans. Faraday Soc.*, 1969, to be published.
249. J. P. Candlin and A. R. Oldham, *Trans. Faraday Soc.*, 1969, to be published.
250. P. Wieland and G. Anner, *Helv. chim. Acta*, **51,** 1698 (1968).
251. W. Voelter and C. Djerassi, *Chem. Ber.*, **101,** 58 (1968).
252. S. Montelatici, A. Van der Ent, J. A. Osborn and G. Wilkinson, *J. Chem. Soc. (A)*, 1054 (1968).
253. W. S. Knowles and M. J. Sabacky, *Chem. Comm.*, 1445 (1968); L. Horner, H. Siegel and H. Büthe, *Ang. Chem.*, **24,** 1034 (1968).
254. P. S. Hallman, B. R. McGarvey and G. Wilkinson, *J. Chem. Soc. (A)*, 3143 (1968).
255. M. Cais, E. N. Frankel and A. Rejoan, *Tetrahedron Letters*, 1919 (1968); E. N. Frankel, E. Selke and C. A. Glass, *J. Am. Chem. Soc.*, **90,** 2446 (1968); A. Miyake and H. Kondo, *Ang. Chem.*, **80,** 663 (1968).

256. E. W. Duck, M. M. Locke and C. J. Mallinson, *Ann.*, **719**, 69 (1968).
257. R. Stern and L. Sajus, *Tetrahedron Letters*, 6313 (1968).
258. C. O'Connor, G. Yagupsky, D. Evans and G. Wilkinson, *Chem. Comm.*, 420 (1968); C. O'Connor and G. Wilkinson, *J. Chem. Soc. (A)*, 2665 (1968).

CHAPTER **6**

Photochemistry of Olefins

K. J. CROWLEY
Trinity College, Dublin, Ireland
and
P. H. MAZZOCCHI
University of Maryland, College Park, Maryland, U.S.A.

I. INTRODUCTION	268
II. EXCITED STATES	269
A. The Absorption Process	269
B. The Immediate Fate of the Initial Excited State	271
C. Product formation	273
1. On direct irradiation	273
2. On sensitized irradiation	274
III. ISOMERIZATIONS	275
A. *Cis–Trans* Isomerizations	275
1. Photosensitized stereomutation	275
2. Unsensitized stereomutation	277
3. Stereomutation and bond migration	277
B. Stereochemical Predictions (Woodward-Hoffman Rules)	279
C. Sigmatropic Isomerizations	281
1. 1,3-Shifts	281
2. 1,5-Shifts	282
3. 1,7-Shifts	284
D. Electrocyclic Reactions	284
1. Cyclobutene formation	285
2. Bicyclobutane formation	288
3. Photochemical ring opening and 1,3-cyclohexadiene formation	289
E. Complex Electrocyclic Transformations	291
1. Formal analogues of cyclobutene formation	291
2. Bicyclo[3.1.0]hexene formation	292
3. Bullvalene and Semibullvalene	295
4. Cyclooctatriene photoisomerizations	296
IV. INTRAMOLECULAR ADDITIONS	297
A. Carbocyclic Photoproducts	297
1. From simple olefins	297

    2. From more complex hydrocarbons 298
    3. From α,β-unsaturated carbonyl compounds . . 299
    4. Cage products 300
  B. Heterocyclic Photoproducts 302
V. INTERMOLECULAR PHOTOADDITION REACTIONS . . . 303
  A. Simple Addition 303
  B. Cycloadditions to Conjugated Olefinic Bonds . . . 304
  C. Paterno-Buchi and Related Reactions 308
  D. Additions to Aromatic Compounds 311
VI. DIMERIZATIONS 312
  A. Cyclobutane Photodimers 312
    1. From nonconjugated olefins 312
    2. From conjugated olefins 313
  B. Other Photodimers 315
VII. REFERENCES 316

I. INTRODUCTION

The study of the photochemical reactions of olefinic compounds has now become one of the principal and most rapidly developing regions of olefin chemistry. An attempt is made here to indicate the range of photochemical reactions undergone by olefins, with particular attention being given to those relatively few cases which have been subjected to mechanistic study. Many aspects of this topic have been included in several recent publications[1] and some features have been covered to considerable depth in reviews describing the photochemistry of olefins[2a], conjugated dienes[2b], ionones[2c], conjugated carbonyl compounds[2d], naturally occurring compounds[2e], cyclohexadienones[2f], stilbenes[2g], quinones[2h], photocycloaddition reactions[2i], and intramolecular additions[2j]. Mechanistic aspects are dealt with in several recent texts[3], and other reviews[4]. Those aspects of olefin photochemistry which are adequately covered in recent articles have been deemphasized and, in some cases, omitted.

Topics such as the reactions of photochemically generated carbenes with olefins, and the photoisomerizations of cyclohexadienones, which seem generally to proceed via $n \to \pi^*$ states, are beyond the scope of this chapter.

Although a clear distinction should be made between direct and sensitized photochemical transformations, since a comparison between the two provides a very important means of examining the mechanisms involved, it has not been always possible to do so. Isomerizations and intramolecular additions are treated separately although the division between them is blurred. Reactions involving valency isomerization of conjugated groups are regarded as isomerizations, while those involving only groups having little or no conjugative interaction are regarded as intramolecular additions.

II. EXCITED STATES

A. The Absorption Process

The light absorption process is governed by two well-established laws: that of Grotthus and Draper, which states that only light which is absorbed by a molecule is effective in producing a photochemical reaction, and the more recent (1908–1913) Stark-Einstein law, which states that each quantum or photon absorbed excites one molecule to the primary excitation state, so that the summation of quantum yields of the primary processes must be unity. This may not be strictly true with very high intensity sources when a single species can absorb two quanta in rapid succession, as in flash photolysis and laser photochemistry. The quantum yield is a quantitative measure of the efficiency of any photochemical process, and may be defined as the ratio of the rate at which molecules are undergoing the process, to the rate at which ground-state molecules are absorbing light. A simple alkene absorbs ultraviolet light in the 180–200 mμ region. This absorption is due to a $\pi \to \pi^*$ transition and corresponds to the promotion of an electron from a π bonding molecular orbital to a π^* antibonding molecular orbital. By applying the equation

$$E \text{ (kcal/einstein)} = 28591/\lambda \text{ (in m}\mu\text{)}$$

(1 einstein is 1 mole of quanta) the energy corresponding to the absorption in the above wavelength range can be calculated at 158·8–142·9 kcal/mole. For a conjugated diene the absorption maximum shifts to 217 mμ or more, and the energy required for the electron excitation is less (ca 130 kcal/mole).

If the alkene contains a carbonyl group, or if a carbonyl sensitizer is used, then the lower-energy promotion of a non-bonding electron to a π^* antibonding orbital becomes a possibility. Such $n \to \pi^*$ transitions are symmetry forbidden, and thus give rise to lower intensity absorption bands than the permitted $\pi \to \pi^*$ transitions.

The ultraviolet absorption spectroscopy familiar to organic chemists is due almost exclusively to transitions in the singlet manifold. The other relevant manifold, the triplet, is equally important but is not accessible on direct irradiation, since a singlet-triplet transition involves spin inversion, which is highly forbidden by the selection rules and occurs particularly inefficiently in olefins[5]. In the ground state the electrons of most organic molecules are all spin paired, and the molecules are said to be in a singlet state (S_0). Absorption of light results in the promotion of one of the two electrons in the highest-energy filled orbital to a vacant orbital. This normally occurs with spin conservation, giving an electronically excited singlet, but the highly forbidden singlet-triplet transitions can be detected in some cases as very weak absorptions. Theoretically this is explained by the borrowing of transition probability from an allowed transition, giving rise to the mixing of singlet and triplet states known as *spin-orbital* coupling. This effect becomes much more pronounced when the energy difference between the triplet and singlet states involved is minimal and when there is a heavy atom present in either the molecule or the solvent[6,7].

It should be noted, however, that the Pauli principle does not require that the spins remain paired in most excited configurations, and spin inversion (intersystem crossing) may occur, giving rise to another excited state, a triplet (T_1). Unlike the singlet, a triplet does not have a net spin angular momentum of zero, and it is actually triply degenerate. In the singlet state a molecule is diamagnetic, but the parallel electron spins of the triplet state give rise to paramagnetism, and in a magnetic field separation into three energy levels occurs. These changes can be represented for the case of ethylene as in Figure 1, where each arrow represents the spin of one electron.

According to the Franck-Condon principle (1925, 1929) nuclear motion is so slow that, for the duration of an electronic transition (ca 10^{-15} sec), there will be little change in nuclear position. Since the distributions of electrons differ in the ground and excited states, the equilibrium nuclear configurations are usually also different. Thus

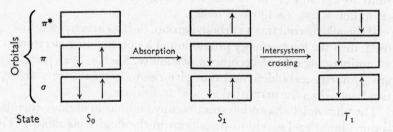

FIGURE 1.

the initial vertical (Franck-Condon) excitation of a vibrationally unexcited ground-state (S_0^0) molecule usually gives rise to a molecule which is vibrationally, as well as electronically, excited. At ordinary temperatures most molecules are in the S_0^0 state. Therefore, if the lowest-energy (0–0) band of a spectrum is strong relative to the higher energy transitions, it is probable that there is little difference in nuclear configuration between the ground and excited states. If one of the higher-energy bands (0–1, 0–2) is stronger, a change in nuclear configuration is to be expected. For example Srinivasan[8] has suggested that the 0–0 band for butadiene may be in the region of 256 mμ, where the compound shows no detectable absorption. This suggests that the first excited state has a very different geometry from the ground state.

B. The Immediate Fate of the Initial Excited State

A typical relationship between the low-lying excited states of an organic molecule is indicated in the modified Jablonski diagram (Figure 2) which gives the processes available after initial excitation. Fluorescence is a radiative transition between states of like multiplicity. It does not generally occur from higher singlets since these rapidly undergo internal conversion to lower excited singlets, but, owing to the relatively larger energy separation between S_1 and S_0, internal conversion is slower in this step. Phosphorescence, a radiative transition between states of different multiplicity, is generally observed only from T_1 to S_0. In this case too, internal conversion from higher triplets is more rapid than emission, which here involves a (forbidden) change in spin multiplicity. Because of rapid bimolecular diffusion quenching of triplets by impurities, very few molecules phosphoresce at room temperature, but at liquid nitrogen temperatures lifetimes of several seconds have been measured, and spectra recorded. A simple colour test for compounds having short-lived triplet states has recently been reported[9].

Non-radiative paths for energy redistribution of the initially excited molecule are internal conversion, intersystem crossing, transfer to other molecules and new bond formation. Internal conversion is the isoenergetic radiationless transition between states of like multiplicity, ($S_1 \leadsto S_0$; $T_2 \leadsto T_1$) while intersystem crossing is the related process involving spin inversion ($S_1 \leadsto T_1$; $T_1 \leadsto S_0$). Internal conversion from upper electronic levels to S_0 dominates fluorescence, so that it occurs in 10^{-11} to 10^{-13} sec. It may produce ground-state molecules with much more vibrational energy than is available through normal thermal means. This has given rise to interesting

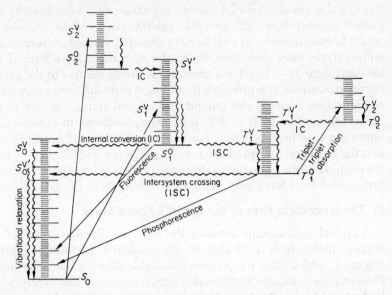

FIGURE 2. Manifold of excited states and intramolecular transitions between these states in a 'typical' organic molecule. Radiative and nonradiative transitions are solid and wavy lines respectively. IC = internal conversion between states of like multiplicity; ISC = intersystem crossing between states of unlike multiplicity; other wavy lines indicate vibrational relaxation processes. Vibrational and rotational levels are shown approximately equally spaced only for ease in presentation; in reality the levels become closer as their quantum numbers increase. (Reproduced by kind permission from Reference 3a).

ground-state reactions of 'hot' molecules produced photochemically; in fact an appreciable proportion of the photochemical reactions of alkenes, especially in the vapour phase, may occur in this manner, but in the great majority of cases the mechanism involved has not been investigated. Energy transfer in the vapour state from such 'hot' molecules[10] to added quenchers seems to be normally less than about 30 kcal/mole per collision, although the internally converted molecule may have as much as 115 kcal/mole of vibrational energy[11,12] The high probability of collisional deactivation with solvent molecules should make such 'hot ground-state' reactions relatively unimportant in solution photochemistry.

Lamola and Hammond[13] have developed a simple chemical method for measuring the quantum yield of intersystem crossing (Φ_{ISC}), and obtained values in the range 0·15 to 1·0 for a variety of sensitizers. In the case of olefins ISC does not seem to occur to any appreciable extent, and access to the triplet from the singlet formed on direct excitation does not seem possible.

The directly formed singlet has another path for energy dissipation —direct energy transfer to another molecule. Singlet-singlet energy transfers in solution occur over distances of 50 to 150 Å, which are much greater than the collision diameters of the molecules involved. Such long-distance transfers occur by a resonance mechanism which involves energetic coupling between the donor and acceptor molecules. This coupling depends on a certain overlap of the emission spectrum of the former with the absorption spectrum of the latter, but the phenomenon is quite distinct from the energy transfer which occurs when emission of the donor is absorbed by the acceptor.

C. Product Formation

I. On direct irradiation

After a $\pi \rightarrow \pi^*$ transition, the excited molecule has several pathways open to it for return to an S_0 state. Of especial interest to organic chemists are the pathways leading to new bond formation (i.e., product formation). Of the various factors which may influence product formation [14], one of the most important is the electronic distribution in the excited state. Consideration of the simple MO picture of butadiene shown below in Figure 3, shows, for the first two occupied (ground-state) molecular orbitals, that the electrons have a bonding contribution between all four carbons in ψ_1 and within the pairs $C_{(1)}-C_{(2)}$ and $C_{(3)}-C_{(4)}$ in ψ_2. A $\pi \rightarrow \pi^*$ ($S_0 \rightarrow S_1$) excitation of the molecules corresponds to the promotion of one electron from the highest occupied molecular orbital (ψ_2) to the lowest unoccupied one

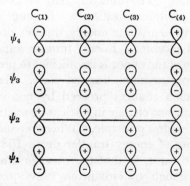

FIGURE 3.

(ψ_3). The form of ψ_3 is such that there is now bonding between $C_{(2)}$ and $C_{(3)}$ and potentially between $C_{(1)}$ and $C_{(4)}$. Thus if $C_{(1)}$ and $C_{(4)}$ should come in close enough proximity to each other bonding takes place between two lobes of like sign, and there is a decrease in energy along the reaction pathway making bond formation in this manner a favoured reaction. Using the top (+) and bottom (−) lobes of ψ_3, 1,4 bond formation results in a σ^* state and a large rise in energy along the reaction pathway.

Steric factors can obviously be decisive when the two reacting atoms are sterically prohibited from coming within bonding distance of each other. When the lifetime of the excited singlet approaches that of molecular motions (10^{-11} to 10^{-12} sec), the ground-state geometry may become crucial in determining the reaction path. In such a case the initial excited molecule (S_1) may either undergo internal conversion back to S_0 (starting material) or, given the correct geometry, yield S_0 (i.e. product in ground state). Dauben and Wipke[15] have pointed out that the product may possess more potential energy than the starting material, but that the reisomerization to the S_0 state will be barred at moderate temperatures if the product does not possess a chromophore.

2. On sensitized irradiation

Because of its relatively long lifetime the triplet state of a molecule is *a priori* more likely than the singlet to be involved in energy transfer. Exchange-energy transfer, the usual mechanism of sensitized organic photochemistry, requires that the donor and acceptor molecules be within collision diameters of one another. For efficient transfer relatively high concentrations of sensitizer are thus necessary, especially in solid solution. This difficulty may be avoided in some cases by heterogeneous photosensitization[16]. Kinetic arguments indicate that for high-energy sensitizers energy transfer occurs on every colision of donor and acceptor. Energy transfer sometimes occurs even if the triplet energy of the donor is insufficient to promote the acceptor to any known (spectroscopic) excited state. Hammond has concluded[4a,17] that such changes proceed by non-vertical transitions, which involve significant change in nuclear geometry, and thus do not obey the Franck-Condon principle. Whenever such transitions must be invoked the rate of energy transfer drops 10^3 to 10^5 below the diffusion-controlled limit, so that the average time required for the change is long enough to circumvent any quantum mechanical restriction to nuclear motion.

The difference in product formation from the first excited singlet and triplet states (compare, for example, the direct and sensitized dimerizations of butadiene, Section VI) is governed by the difference in electronic distribution in those states. The distinguishing factor between the two states is the spin of one electron. When electrons are paired they tend to occupy the same region in space. However when they have parallel spins they tend to be as far away from each other as possible, giving rise to the 'diradical' character of triplets, demonstrated by their reactions.

III. ISOMERIZATIONS

A. Cis-Trans *Isomerizations*

Photochemical *cis–trans* isomerization, which is also discussed in chapter (3), when sterically allowed is probably a universal photoreaction of alkenes. It is a key process in the chemistry of vision, and often affords the most convenient means of preparing a *cis* isomer. In direct isomerizations the photostationary state can often be adjusted to give a preponderance of the *cis* isomer, since the photostationary state is dependent on the quantum yields of the two reactions, on the relative extinction coefficients of the isomers, and on the wavelength of the light used (the two isomers normally have their maxima at different wavelengths). In some photosensitized isomerizations the *cis–trans* ratio can be varied by the appropriate choice of sensitizer. Thus Nozaki and colleagues[18] report that *cis*, *trans*, *trans*-1,5,9-cyclododecatriene is isomerized to both the all-*trans* isomer and the *cis*, *cis*, *trans* isomer, in proportions which can be varied greatly by the choice of a suitable sensitizer.

Quantum mechanical calculations indicate that both the S_1 and T_1 states of ethylene have energy minima when the molecule is twisted 90° about the carbon–carbon bond, where the overlap of the π and π^* orbitals is minimal[19,20]. In fact, for this twisted configuration, T_1 is of lower energy than S_0, and is actually the ground state[21]. Thus for an ethylene-like molecule the minimum energy pathway for *cis–trans* isomerization involves intersystem crossing to the triplet at about 60° and back to the other singlet at 120°.

1. Photosensitized stereomutation

Both intermolecular and intramolecular photosensitized *cis–trans* isomerizations are known. The former have been subjected to detailed study, particularly by Hammond and coworkers[17]. They consider the isomerization in terms of the following general mechanism:

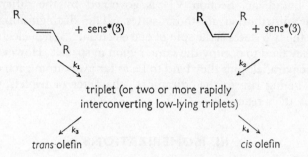

The rapidly interconverting triplets could be cisoid, transoid and non-planar (twisted, 'phantom'). The stilbene (PhCH=CHPh) system[17] has been studied most intensively, and the available evidence suggests that both transoid and phantom triplets are involved in the isomerization, but that the cisoid triplet has, at most, only a very transient existence. On the basis of the above mechanism it can be predicted that the photostationary *cis–trans* ratios established in the presence of high-energy sensitizers will all be the same. As the available energy is reduced so that it falls below that of one of the isomers, the k_1–k_2 ratio should alter, and thus also the *cis–trans* ratio. In the stilbene case the $S_0 \rightarrow T_1$ excitation energy of the *trans* isomer is 49 kcal/mole, and that of the *cis* isomer is 57 kcal/mole. Thus as the available excitation energy goes below 57 kcal/mole the value of k_2 will decrease, while that of k_1 will remain unchanged. As the sensitizer energy is decreased below 49 kcal/mole the value of k_1 should also start to decrease, and both k_1 and k_2 should then fall continuously as the sensitizer energy is reduced. Their ratio should again become constant however, so that with sensitizers of energy less than 49 kcal/mole, the *cis–trans* ratio should again become constant, although much higher than with high-energy sensitizers. As is shown in Fig. 4 these predictions are nicely borne out above 57, and in the 49–57 kcal/mole range, but not with low-energy sensitizers, where the value of k_2 falls off much more slowly than predicted. This result, which is paralleled in all other cases examined, supports the postulate that energy can be transferred from sensitizer to acceptor by non-Franck–Condon excitation, giving rise directly to a twisted triplet. Subsequent work[22] involving flash photolysis of stilbene and of 1,2-diphenylpropene is in agreement, showing that the efficiency of energy transfer is still high even when the

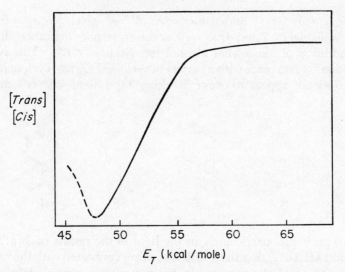

FIGURE 4. Stationary state composition on irradiation of stilbene with various sensitizers.

triplet excitation energy of the sensitizer is too low, by over 10 kcal/mole, to produce any known optical transition in the olefin.

The occurrence of intramolecularly sensitized photochemical cis–trans isomerization has been observed spectroscopically[23] and directly[24]. Morrison[24] has reported instances of intramolecular cis–trans isomerization involving light absorption by a carbonyl and an aromatic moiety, which results in isomerization at a non-conjugated olefinic bond in the same molecule.

2. Unsensitized stereomutation

Mechanistic examination of direct cis–trans photoisomerization has also centred on the stilbenes. It seems that the initially obtained S_1 state is not directly responsible for the isomerization. Planar and twisted triplets[17,25(a)], twisted excited singlets[25(b)], and a highly energetic freely rotating ground state resulting on $S_1 \rightsquigarrow S_0$ conversion[26] have been proposed as intermediates by various groups. The current situation has been recently discussed[27].

3. Stereomutation and bond migration

When cis–trans isomerization is sterically impeded, sensitized irradiation can give rise to positional rearrangement of the double bond.

Thus carene (1)[28] and 1-methene (3)[28,29] give, irreversibly, the exocyclic olefins 2 and 4, as well as the corresponding ethers, formed by addition of the solvent alcohol (see Section V.A). This type of reaction[30] takes place with a variety of cyclohexenes and cycloheptenes, but does not appear to occur in either rigid or acyclic systems. It

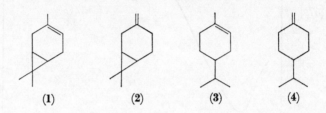

seems probable, particularly in the light of the results of Liu[31] (see Section III.D.1), that these differences are connected with the relative ease with which the systems can accommodate a twisted olefin triplet and, perhaps, an unstable *trans* olefin. Relief of strain could be achieved via protonation followed by either elimination yielding starting material or exocyclic product, or ether formation[28].

In the acyclic series double-bond migration and stereomutation can be brought about photochemically using a tungsten filament lamp in the presence of iodine or iron carbonyl derivatives. Iodine is generally used to effect *cis–trans* isomerization[32] while the iron derivatives bring about positional rearrangement more effectively[33,34], but complex mixtures frequently result with both. Thus irradiation[34] with visible light of pentane solutions of 1-octene in the presence of iron dodecacarbonyl, gives *cis*-(2%) and *trans*-2-octene (19%), *cis*-(6%) and *trans*-3-octene (33%) and *cis*-(8%) and *trans*-4-octene (32%). The thermodynamic *cis–trans* equilibrium in non-conjugated olefins can be approached without double-bond migration by irradiation in the presence of diphenyl sulphide[35].

A variety of other, more specialized, cases of olefinic bond migration are known, such as the unsensitized photoisomerization of 5 to 7. This has been shown[36] to take place via the cyclopropane 6 formed by a novel 1,2-methyl migration.

B. Stereochemical Predictions (Woodward-Hoffmann Rules)

Woodward and Hoffmann [37] have proposed that the courses of concerted thermal and concerted photochemical valence isomerizations are governed, respectively, by the form of the highest occupied ground-state orbital and that of the first excited-state orbital. One of the simplest examples available is the photochemical cyclization of butadienes to cyclobutenes. Thus on cyclization of *trans,trans*-2,4-hexadiene, there are two possible products: *cis*- and *trans*-3,4-dimethylcyclobutene. The two possible* modes of ring closure of the first excited state are shown in equations (1) and (2). Overlap of orbitals

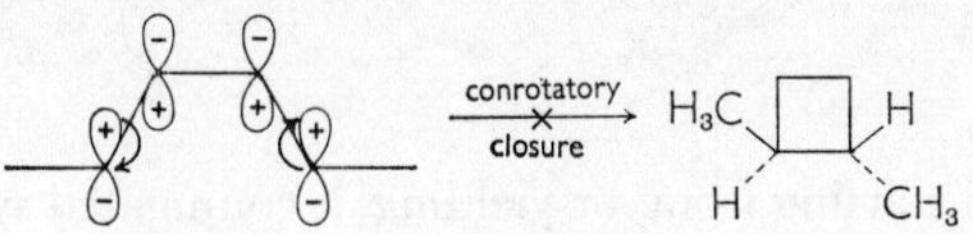

$$\text{butadiene orbitals} \xrightarrow[\text{closure}]{\text{conrotatory} \times} H_3C \square H \atop H \quad CH_3 \qquad (1)$$

$$\text{butadiene orbitals} \xrightarrow[\text{closure}]{\text{disrotatory}} H_3C \square CH_3 \atop H \quad H \qquad (2)$$

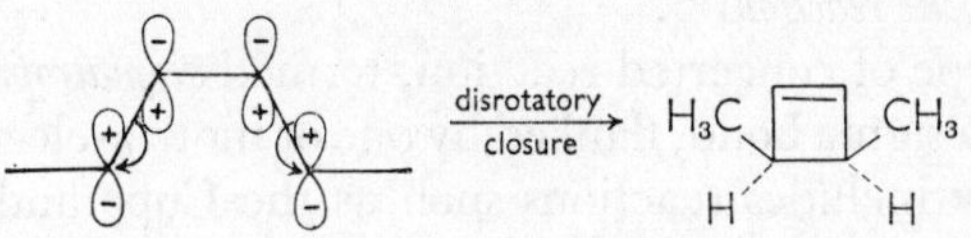

of like sign will have a bonding character, while overlap of orbitals of unlike sign will have an antibonding character (see Section II.C.1). It is apparent that the conrotatory closure will result in a $+,-$ overlap between $C_{(2)}$ and $C_{(5)}$ and the reaction is 'disallowed'. Conversely disrotatory closure will take place with overlap of like signs between $C_{(2)}$ and $C_{(5)}$ resulting in net bonding interaction, and the process is 'allowed'. The Woodward-Hoffmann rules have furnished numerous predictions on the 'allowedness' and stereochemical course of concerted reactions, and several of these predictions have since been vindicated. Thus the photochemical hexatriene-cyclohexadiene ring closure and opening should be, and is, conrotatory (equation 3)[93,97]. The somewhat more involved rearrangement of hexatriene to bicyclo-[3.1.0]hexene can also be deduced from the form of the triene π^* molecular orbital, as illustrated (equation 4) for a *trans,cis,trans*-1,6-disubstituted triene [38,117].

The following reactions (equations 1-4) involve concerted electron

---

* In fact there are two pairs of possible modes of ring closure, but both conrotatory modes give a single product, and both disrotatory modes give another single product; in the reverse of reaction (1), however, the two possible conrotatory ring openings would yield different products.

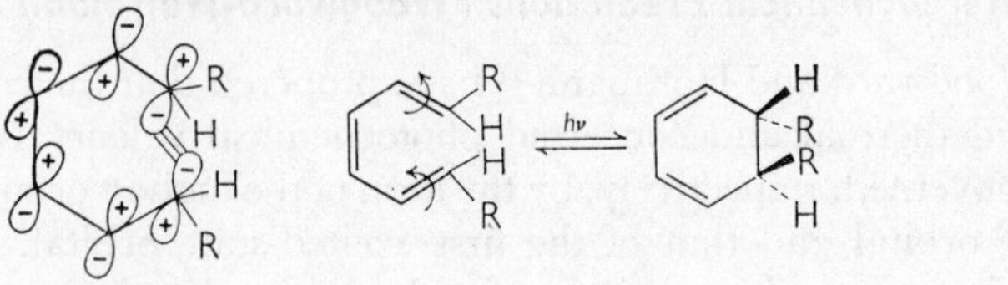

redistribution starting from, or yielding, a conjugated system, and are called *electrocyclic reactions*[37].

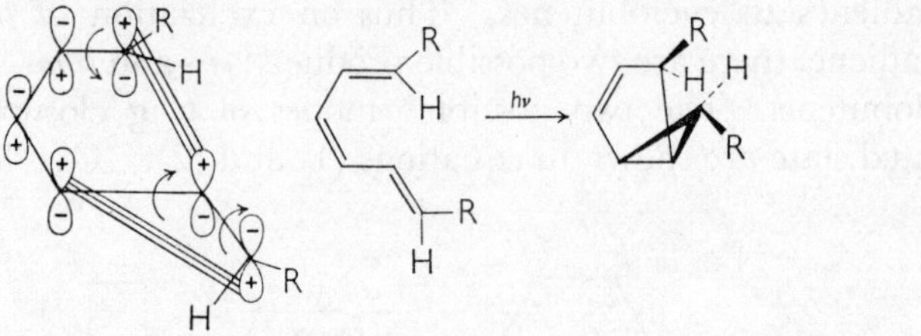

Another type of concerted reaction, termed *sigmatropic*, involves the migration of a sigma bond, flanked by one or more π-electron systems[39]. This category includes reactions such as the Cope and Claisen rearrangements, but only those aspects relevant to olefin photochemistry need be illustrated here.

A concerted hydrogen migration in a terminally substituted cisoid *cis*-diene can, *a priori*, take place in two ways. The migration can proceed *suprafacially* with the migrating group remaining on the same face (side of the plane) of the molecule (equation 5), or *antarafacially* with the group migrating from the top to the bottom face of the molecule (equation 6). The predictions for the course of some thermal and photochemical sigmatropic isomerizations are listed in Table 1.

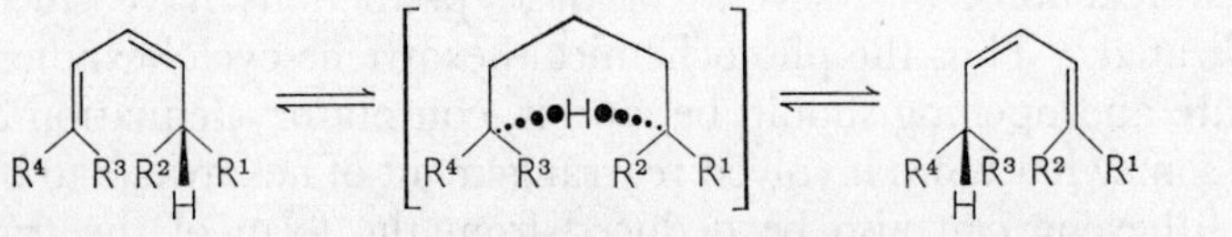

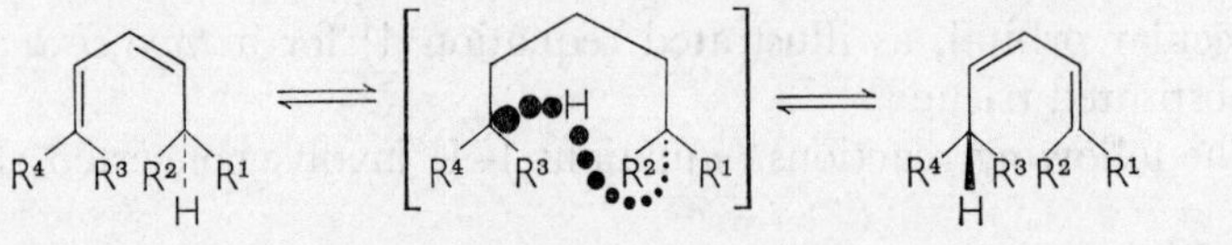

It should be pointed out that steric factors may thwart some of these processes. Thus it will be impossible for a reaction to occur antarafacially in a normal cyclic system. The fact that in many cases

TABLE 1. Allowed sigmatropic reactions

Hydrogen migration	Thermal	Photochemical
1,3	Antarafacial	Suprafacial
1,5	Suprafacial	Antarafacial
1,7	Antarafacial	Suprafacial

'disallowed' processes are observed is not an argument against the validity of these concepts, but merely suggests that the reactions are multistep or non-concerted (radical intermediate) processes.

There has been some disagreement about the validity of the above approaches, and several groups have offered alternatives [4b, 40, 41].

C. Sigmatropic Isomerizations

I. 1,3-Shifts

As described above, a concerted 1,3-shift in the first excited state is normally suprafacial. Only a small number of such shifts is known, possibly because higher-order changes are preferred when permissible, since they give a greater degree of linear conjugation in the transition state [39]. The photoisomerization of the α,β-unsaturated ketone, verbenone, to the cyclobutanone, chrysanthenone [42], afforded an early example, and a related isomerization (equation 7, $n = 1,2$ and 3) now gives a general synthesis of cyclobutanones [43]. Among reports of reactions which may involve concerted photochemical 1,3-hydrogen shifts are the transformation of various phenylcyclopropanes to 4-phenylbut-1-enes [44], the isomerization of 1,3- to 1,2-cyclononadiene [45], and the formation of 2-phenyl-1,3,5-cycloheptatriene from the 7-phenyl isomer [46]. The latter may proceed by successive 1,7-shifts, but this should give rise to the 1-phenyl isomer as an intermediate and this was not detected. The overall result of the photoisomerization of ergostadiene (8) to 9 [15] is a 1,3-shift of hydrogen. This may formally proceed through the corresponding bicyclobutane, but this species, if formed, should be in such a high vibrationally excited ground state that it may be meaningless to ascribe to it a single structure before it cascades to a lower vibrational energy level (see also Section III.D.1) [15].

Cookson has proposed the generalized 1,3-allyl migration (8) in the course of reporting a number of photoisomerizations such as equation

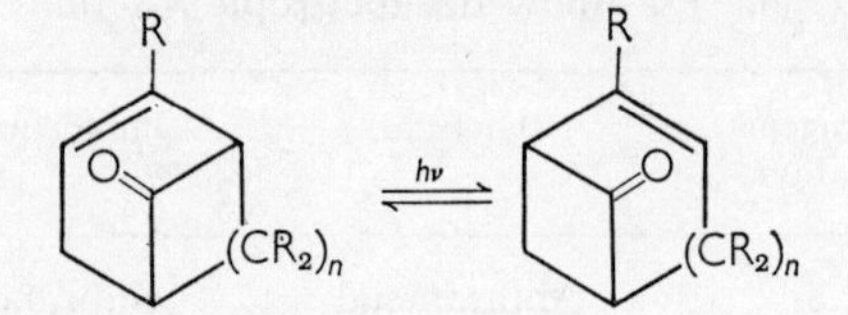

$$\text{(diagram: bicyclic ketone with R and } (CR_2)_n \rightleftharpoons \text{isomer, } h\nu) \qquad (7)$$

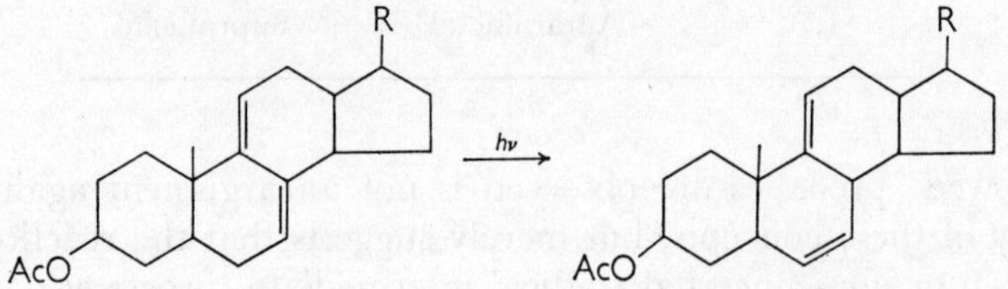

$$\text{(8)} \xrightarrow{h\nu} \text{(9)}$$

(9)[47]. Deuterium labelling of **10** suggests that *cis–trans* equilibration of the migrating allyl group is an integral part of the rearrangement[48].

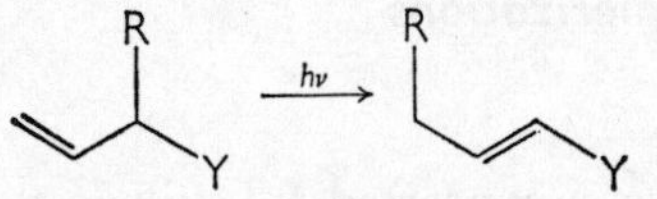

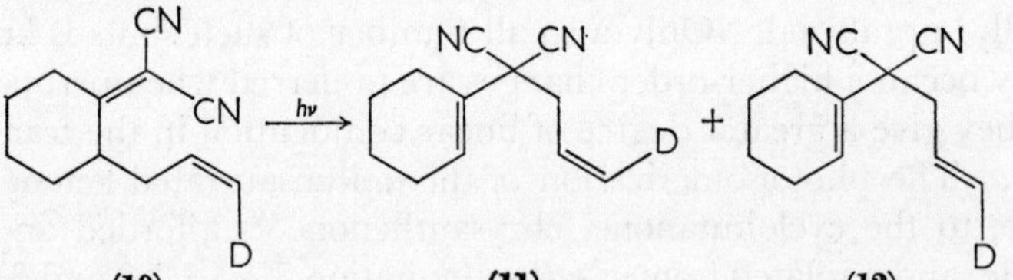

$$\text{allyl-Y } \xrightarrow{h\nu} \text{rearranged allyl-Y} \qquad (8)$$

$$\text{(10)} \xrightarrow{h\nu} \text{(11)} + \text{(12)} \qquad (9)$$

The hydrogen shift which occurs in the photoisomerization of α,β-unsaturated acids to β-lactones, (equation 10), is probably not photochemical but merely due to ketolization. This isomerization apparently involves the intermediacy of a hydroxyoxetene, in a reaction analogous to the photoisomerization of 1,3-dienes to cyclobutenes.

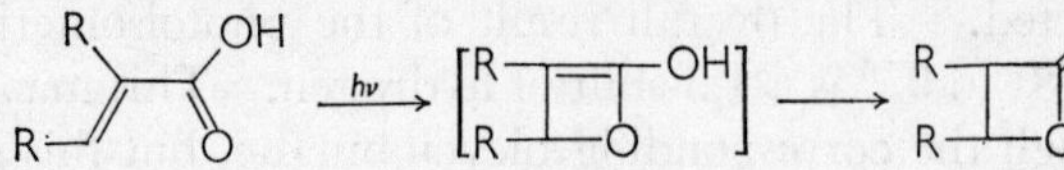

$$\underset{R}{\text{R}}\!\!=\!\!\text{C(COOH)} \xrightarrow{h\nu} [\text{hydroxyoxetene}] \longrightarrow \beta\text{-lactone} \qquad (10)$$

2. 1,5-Shifts

Thermal 1,5-hydrogen shifts in conjugated carbon chains are widely recognized, occurring in cyclopentadienes[50], in cycloheptatrienes[51], and in acyclic systems[52]. In contrast with the thermal reaction,

6. Photochemistry of Olefins

photochemical 1,5-hydrogen shifts should occur antarafacially, and are to be expected only in acyclic or large ring systems. Thus photochemical equilibria such as equation (11) are known to occur in several acyclic instances[53], but not in cyclic systems.

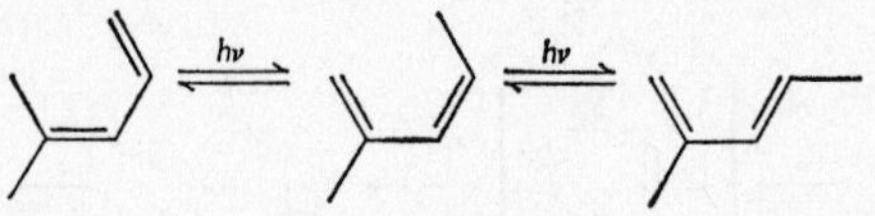

$$\diagdown\diagup\diagdown\diagup \xrightleftharpoons{h\nu} \diagdown\diagup\diagdown\diagup \xrightleftharpoons{h\nu} \diagdown\diagup\diagdown\diagup \qquad (11)$$

These 1,5-hydrogen shifts may become photochemically non-reversible if the migrating diene system is conjugated with another unsaturated group in the starting material, but not in the product. Thus, in all-carbon conjugated systems several examples[54–56] of allene formation have been reported (**14** from **13**; **17** from **16**) and the photoisomerization of 2,4-dienoic to 3,4-dienoic acids[57] may proceed similarly, by way of the enolic form of the allenic product.

1,5-Shifts yielding the enolic form of the product have been shown to be involved in the photoisomerization of 5-methylhex-3-en-2-one to 5-methylhex-4-en-2-one[58] and the same mechanism is probably applicable to analogous α,β- to β,γ-unsaturated ketone photoisomerizations[59–62].

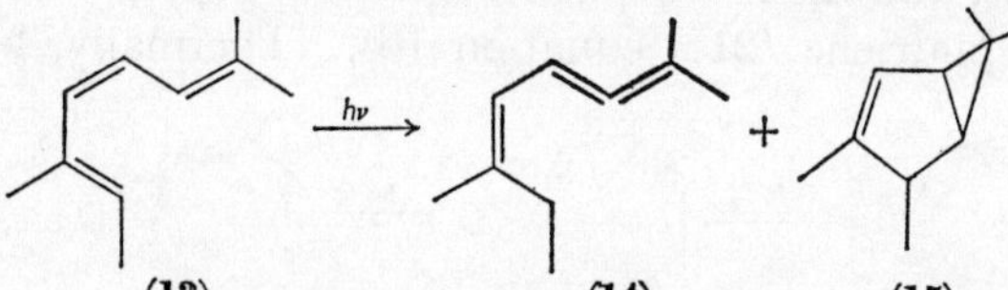

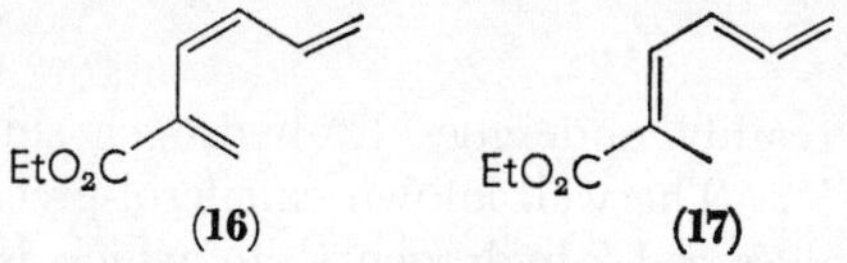

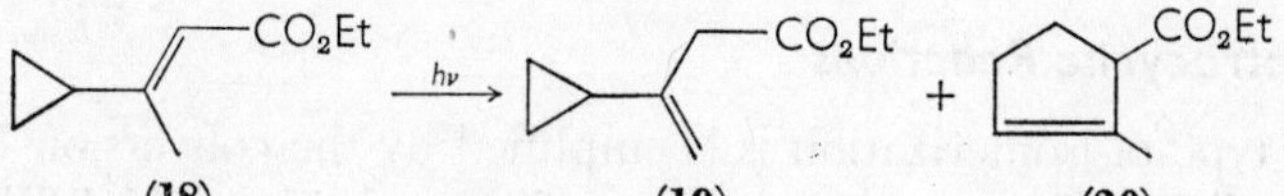

$$\text{(13)} \xrightarrow{h\nu} \text{(14)} + \text{(15)} \qquad (12)$$

(13)   (14)   (15)

EtO$_2$C (16)   EtO$_2$C (17)

$$\text{(18)} \xrightarrow{h\nu} \text{(19)} + \text{(20)} \qquad (13)$$

(18)   (19)   (20)

The conversion of the ester **18** to its isomer **19**[63] may well occur by way of a 1,5-shift, giving the enolic form of the product, but a 1,3-shift cannot be ruled out. A 1,5-phenyl migration from carbon to oxygen has been shown to occur[64] in the phototransformation (in ethanol) of dibenzoylethylene to ethyl 4-phenyl-4-phenoxy-3-butenoate.

A photochemical 1,5-homodienyl hydrogen shift may be involved in the isomerization shown in equation (14) (R = H or CH_3)[65], and an example of what is formally a 1,5-homodienyl shift in a small ring system has also been reported (equation 15)[66]. The authors suggest that an intermediate diradical is involved.

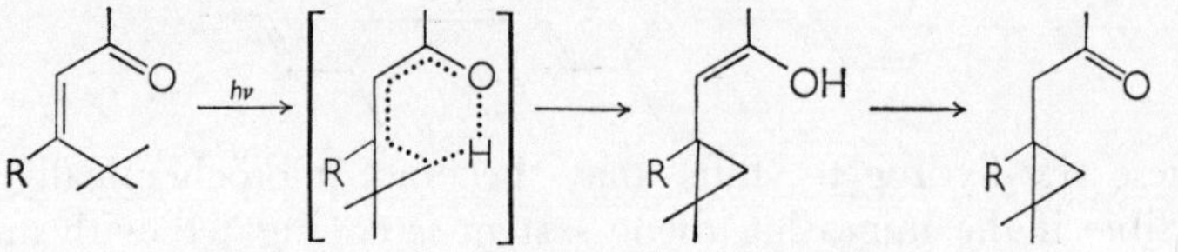

$$ (14) $$

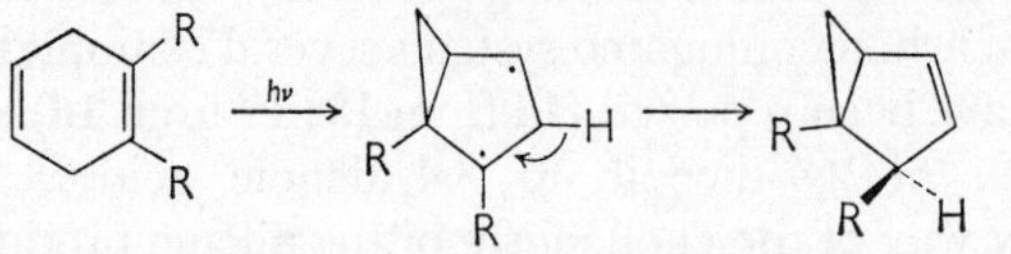

$$ (15) $$

3. 1,7-Shifts

Photochemical 1,7-hydrogen shifts, being suprafacial, can occur in cyclic systems, and the first reported instance[67] dealt with deuterated 1,3,5-cycloheptatriene (**21**) (equation 16). Thermally, by contrast,

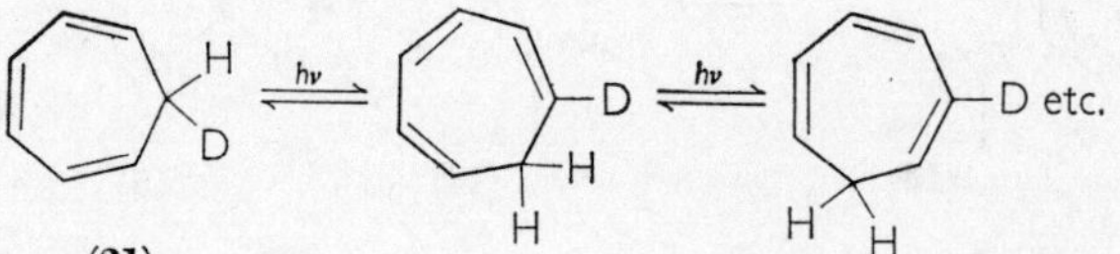

$$ (16) $$

(**21**)

this compound readily undergoes 1,5-hydrogen shifts, but not 1,7-hydrogen shifts[51a]. The well known calciferol-precalciferol thermal equilibrium involves a 1,7-hydrogen shift, which is possible in this acyclic system antarafacially. Other photochemical 1,7-shifts have been reported more recently[68–70].

D. Electrocyclic Reactions

This type of isomerization is exemplified by the conversion (17) of 1,3-butadiene (**22**) to cyclobutene (**23**) and bicyclo[1.1.0]butane (**24**)[71]. Cyclobutene formation is a common reaction in organic

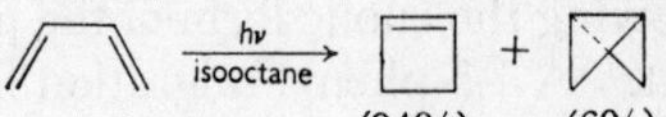

$$ (17) $$

(**22**)   (94%) (6%)
           (**23**) (**24**)

photochemistry, but few examples of bicyclobutane formation are known, partly because the high reactivity of such species complicates their isolation.

I. Cyclobutene formation

The first proven instance of cyclization of a 1,3-diene to a cyclobutene was given by Dauben and Fonken in 1959[72]. On irradiation of pyrocalciferol (**25**), the pentacyclic valence tautomer **26** was formed; the other *syn* isomer of this structure, isopyrocalciferol, also gives the

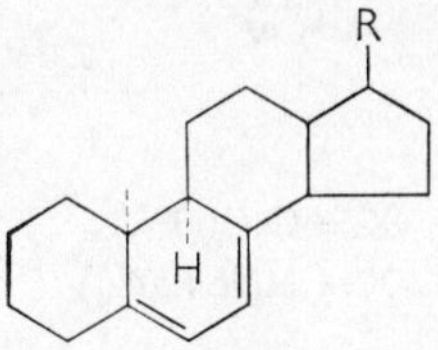

(**25**) 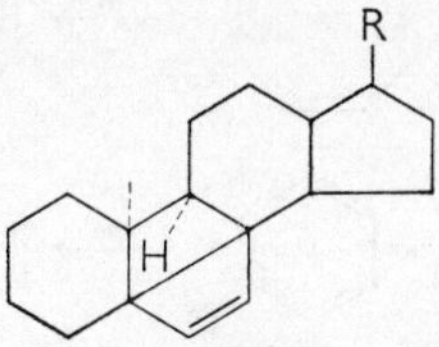   (**26**)

corresponding cyclobutene[72]. Since then numerous cases of such isomerizations have been reported. It should be noted here that ring opening (Section III.D.3) competes with cyclobutene formation when even-membered (e.g. cyclohexadiene, cyclooctatriene) rings are involved, and the two *anti* isomers of **25** open to the hexatriene, precalciferol. This difference is understandable in that photochemical (conrotatory) ring opening of either *syn* isomer would result in a highly labile *trans*-cyclohexene, so that cyclobutenes result, while photochemical $C_{(5)}$–$C_{(8)}$ bonding in the *anti* isomers would give rise to a strained system with a *trans* fused cyclohexane-cyclobutane ring junction. Ring opening reactions on the whole seem to be less common than ring forming reactions[73]. This is to be expected in the nature of photochemistry, since the ring-formed products are, in general, more transparent to ultraviolet light than the opened products with more extended chromophoric systems.

Some representative examples of cyclobutene formation are given in equations (18)[74], (19)[75a], (20)[76], (21)[77], (22)[78], and (23)[79]. Tropolones such as α-tropolone methylether (**27**) give initially the cyclobutenes expected, but these further rearrange as shown in reaction (24), and in some cases open up to a cyclopentenone[75b].

Acyclic dienes also yield cyclobutenes on irradiation. The yield is greatly influenced by the substitution pattern, 2- and 3-substitution raising, and 1- and 4-substitution generally lowering the yield. Thus

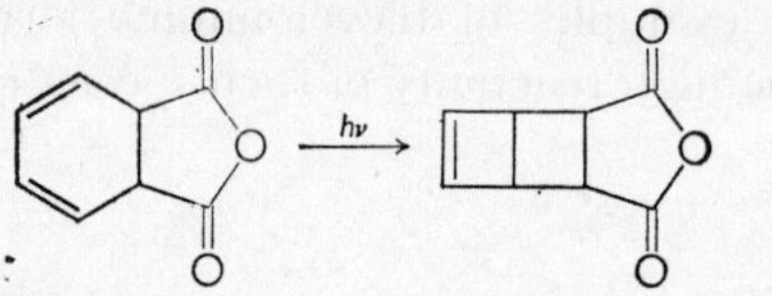
(18)

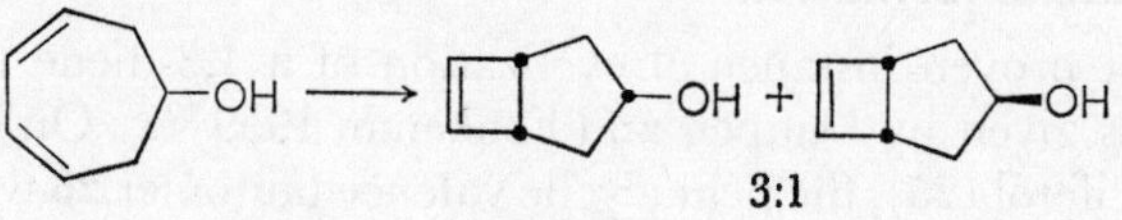
(19)

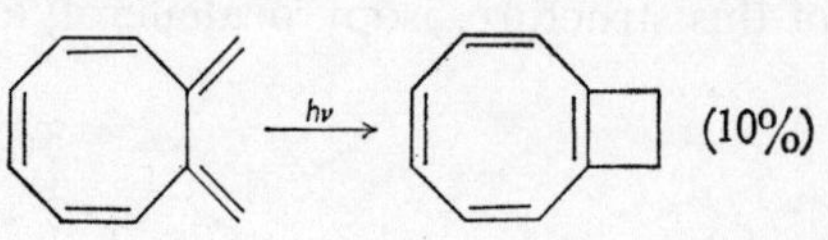
(20)

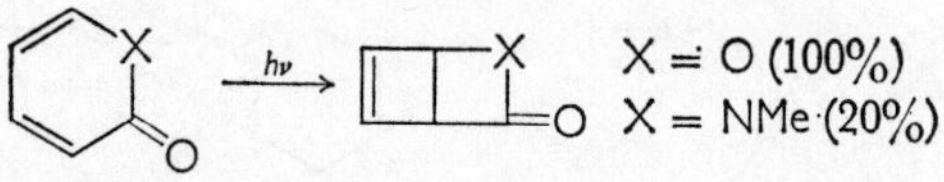
(21)

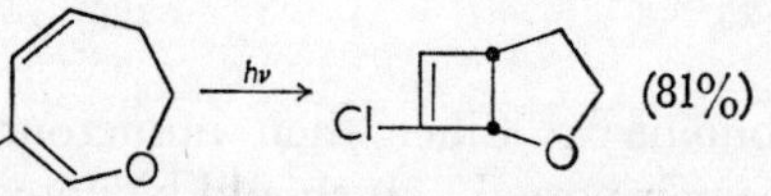
(22)

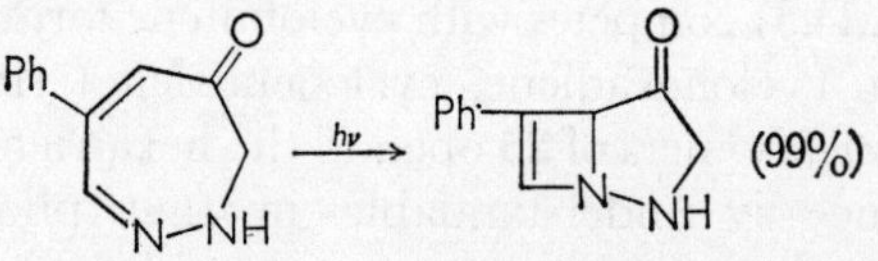
(23)

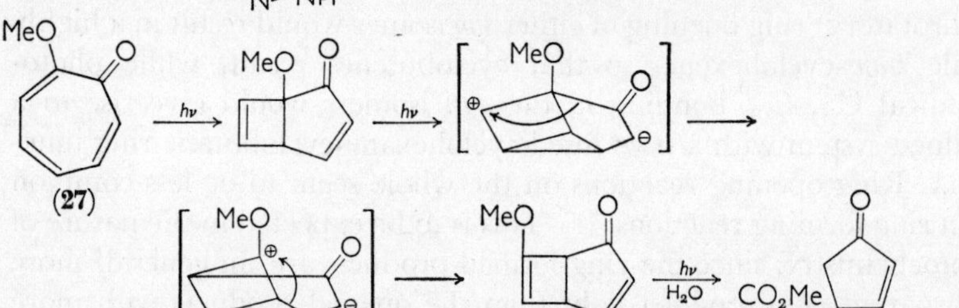
(24)

2,3-dimethylbutadiene cyclizes in 71% yield, myrcene (Section IV. A.2) in 68%, 1-cyclohexylbutadiene in 43% and hexa-2,4-dien-1-ol in 13%, while 2,4-dimethyl-1,3-pentadiene gives no detectable yield of cyclobutene[53]. The known quantum yields for the cyclization range from 0·12 to zero[80].

The mechanism of the transformation is still unclear in most cases. On sensitized irradiation in the liquid phase no monomeric products are reported[81], and in general the vapour-phase reaction cannot be quenched[80]. Thus the triplet state does not appear to be involved.

Exceptionally, *cis,cis*-1,3-cyclooctadiene gives bicyclo[4.2.0]oct-7-ene (85%) under photosensitizing conditions[81a]. However, Liu has recently shown[31] that the function of the sensitizer is to convert the *cis,cis*- to the *cis,trans*-isomer, which then readily undergoes thermal (conrotatory) cyclization to the *cis*-fused cyclobutene. The same explanation may[31] be applicable to the only other known exception, the photosensitized isomerization of bicyclohexenyl (**30**) to **31**[82]; *cis,trans*-bicyclohexenyl should be very unstable, and would be expected to undergo facile thermal cyclization. The cyclobutene **31** is also formed on direct irradiation of **30**[53], and its stereochemistry has been determined by Dauben and coworkers[82], who also identified one of the minor products of direct irradiation as the non-conjugated diene **32**. In the direct irradiation the ratio **31**–**32** decreases with shorter wavelength of incident light (6·5 with Vycor filter, 1·7 with quartz). The *s-cis* form of **30** should absorb maximally at longer wavelength

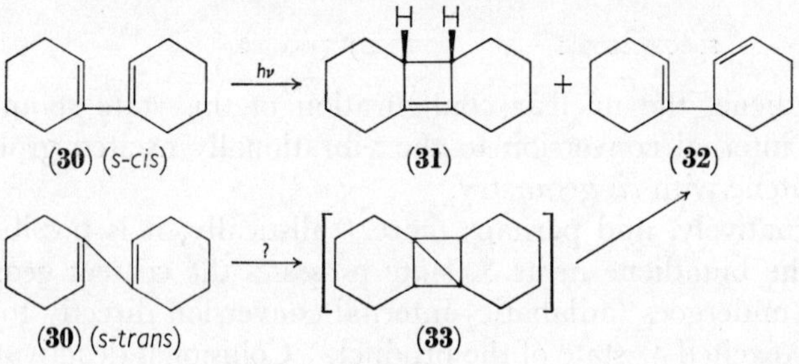

[Scheme I structures showing (30)(s-cis) → (31) + (32), and (30)(s-trans) → [33]]

SCHEME I

than the *s-trans*, so that the *s-cis* form apparently gives rise to the cyclobutene, and *s-trans* to **32**, perhaps via the bicyclobutane **33** (cf. **8 → 9**, Section III. C.1). However it is not necessary to invoke the intermediacy of **33**, and Dauben[82] favours instead, two non-interconvertible excited states **34** and **35**, the populations of which are wavelength dependent.

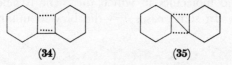

(**34**)   (**35**)

In all known cases the cyclobutene formed by irradiation of a cyclic conjugated diene is part of a *cis*-fused ring system. According to the Woodward-Hoffmann electrocyclization rules this system should

arise, in a concerted reaction, from either the first excited state of the
cis,cis diene or the ground state of the *cis,trans* diene. Since the above
findings of Liu cannot be extrapolated to the extent of proposing
cis,trans intermediates on the cyclization of other cyclic conjugated
dienes to cyclobutenes, only the first alternative needs be considered.
However, consideration of the energetics led Dauben[82] to conclude
that the first excited state of cyclobutene is probably not involved.
He suggests that since the butadiene S_1 state will be 20–50 kcal/mole
less energetic than the corresponding cyclobutene S_1 state, the latter
will be formed only very inefficiently, and its intermediacy need not
be invoked. By assuming a scheme (25) whereby partial 1,4-bonding
occurs (see Section II.C.1) to stabilize a non-spectroscopic excited state

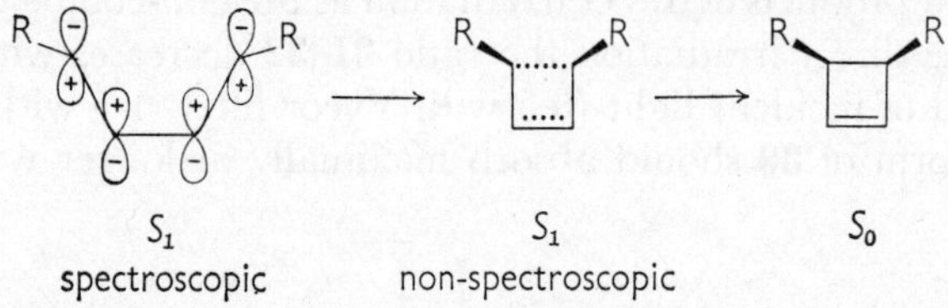

$$\begin{array}{ccc} S_1 & S_1 & S_0 \\ \text{spectroscopic} & \text{non-spectroscopic} & \end{array} \tag{25}$$

of the diene, the nuclear configuration of this state should persist
during internal conversion to the vibrationally excited ground-state
cyclobutene with *cis* geometry.

Alternatively, and perhaps more realistically, it is possible[14] that
when the butadiene in its S_1 state possesses the correct geometry, it
merely undergoes (adiabatic) internal conversion directly to a vibrationally excited S_0 state of the product. Collisional deactivation with
solvent molecules would then afford the product in its S_0^0 state.

2. Bicyclobutane formation

Appurtenant to this topic, the formation of the parent hydrocarbon
has already been cited, as has the possible intermediacy of bicyclobutanes in the isomerizations **8 → 9** and **30 → 32**. Dauben's
group[15,83] has given several instances, such as (26), of the reaction occurring in rigid molecules in which the diene is held in the *s-trans*
conformation. In some cases[15,84] the bicyclobutane could not be

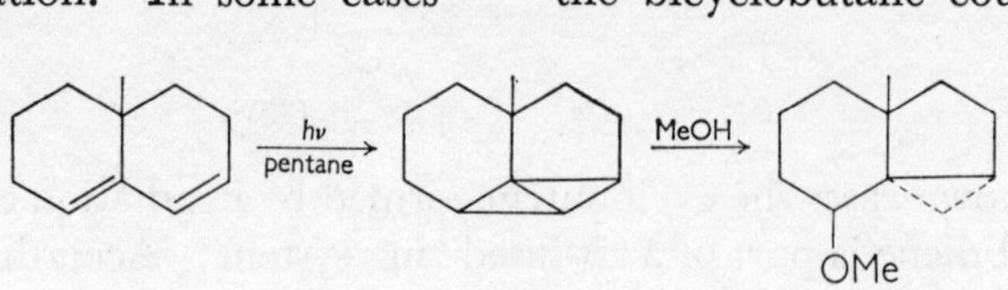

$$\text{(36)} \xrightarrow[\text{pentane}]{h\nu} \text{(37)} \xrightarrow{\text{MeOH}} \text{(38)} \tag{26}$$

isolated owing to its high reactivity, but further reaction products analogous to **38** were identified.

Bicyclobutanes can also be formed on irradiation of homoannular conjugated systems, at least in the aromatic series. Thus the benzvalene **39** was obtained, together with the prismane **40** and the Dewar benzene **41**, in the course of the elegant studies of Wilzbach and Kaplan[85] on the photochemical interconversion of 1,2,4-tri-*t*-butylbenzene (**42**) and 1,3,5-tri-*t*-butylbenzene (**43**), (Scheme II). The

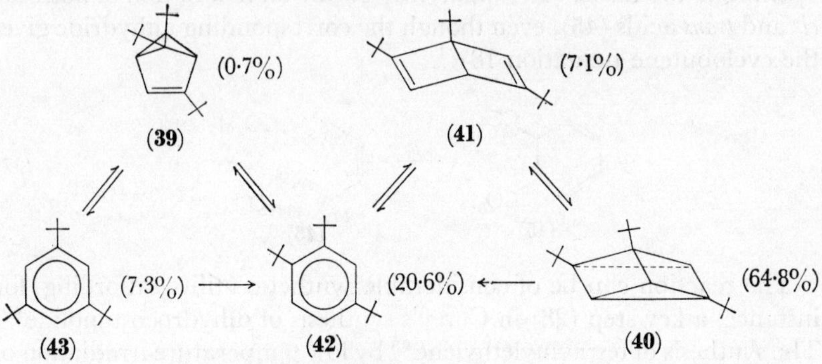

SCHEME II

composition of the photostationary state mixture on irradiation with 2537 Å light is given in parentheses; quantum yields for these reactions range from 0.43 (**39** → **42**) to 0.008 (**43** → **39**). More recently the parent benzvalene has been obtained[86]. It is formed together with fulvene[86,87], on photolysis of liquid benzene with 2537 Å light; with 1849 Å light and in the vapour phase, fulvene and *cis*-1,3-hexadien-5-yne[87a] are obtained. A common diradical precursor, 'prefulvene' (**44**) has been suggested[88] for these isomerizations and for 1,3-cycloaddition[89]. The direct photoisomerization of benzene itself to prismane and to Dewar benzene has not been reported, and may[88] require the intermediacy of a different biradical. (See also equation 31.)

(**44**)

3. Photochemical ring opening and 1,3-cyclohexadiene formation

Photochemical ring opening of cyclobutenes to 1,3-dienes has been inferred[53], but normally its extent is only slight or negligible. Irradiation of 1,3-cyclohexadienes frequently results in ring opening, although since the reaction is reversible, and since both the diene and the ring-opened product may undergo further photochemical reaction, this may not always be evident. Many instances of such ring opening have been reported[1-4,90-96], and only a few examples need be given here.

Although ring-forming reactions seem to be, in practice, more common (Section III.D.1), the preferred reaction would appear to be ring opening when no constraint is present in the molecule. Thus, opening to the triene (**46**) apparently occurs on irradiation of both the *cis* and *trans* acids (**45**), even though the corresponding anhydride gives the cyclobutene (equation 18)[74].

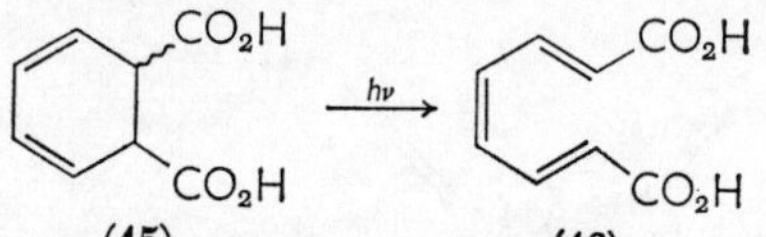

$$\text{(45)} \xrightarrow{h\nu} \text{(46)} \tag{27}$$

The reaction can be of considerable synthetic utility, affording, for instance, a key step (28) in Corey's synthesis of dihydrocostunolide[97]. The synthesis of tetravinylethylene[98] by low temperature irradiation of 2,3-divinyl-1,3-cyclohexadiene illustrates the strength of the method in the preparation of highly reactive products. Like most other ring-opening reactions of similar type, this transformation cannot be effected under sensitizing conditions, so that the triplet is probably not involved as an intermediate.

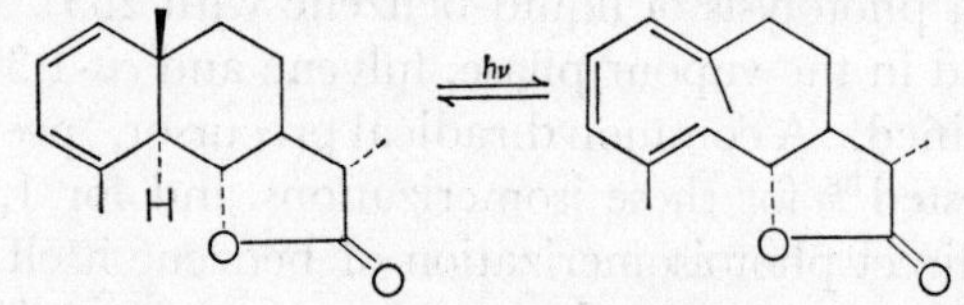

$$\xrightarrow{h\nu} \tag{28}$$

$$\text{(47)} \xrightarrow{h\nu} [\text{(48)}] \longrightarrow \text{products} \tag{29}$$

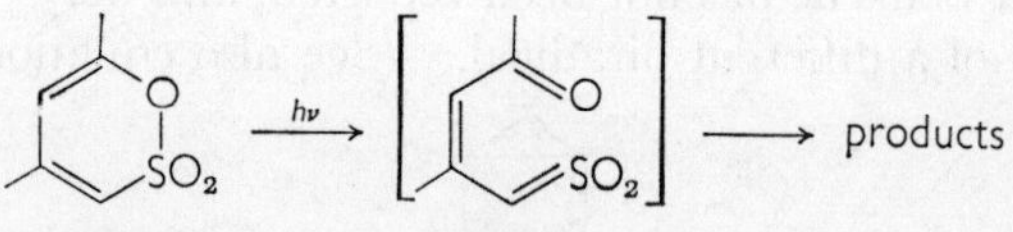

Only a few cases of ring opening of heterocyclic analogues are known as yet[1d,99]. In one of these[99a] the unsaturated sultone **47** yields, in appropriate solvent systems, an ester and an amide which are apparently derived from the sulphen **48**, although such a species has not been isolated.

The ring opening of 1,3,5-cyclooctatriene is mentioned in Section III.E.4, and the stilbene-dihydrophenanthrene cyclization has been discussed in detail elsewhere[2,100].

E. Complex Electrocyclic Transformations

The mechanisms of some of the reactions described in this Section are not known, and may well not be electrocyclic.

I. Formal analogues of cyclobutene formation

Norbornadiene[101] and its derivatives[103], such as the dicarboxylic acid **49**[102], and their oxabicyclic analogues[104] undergo the same type

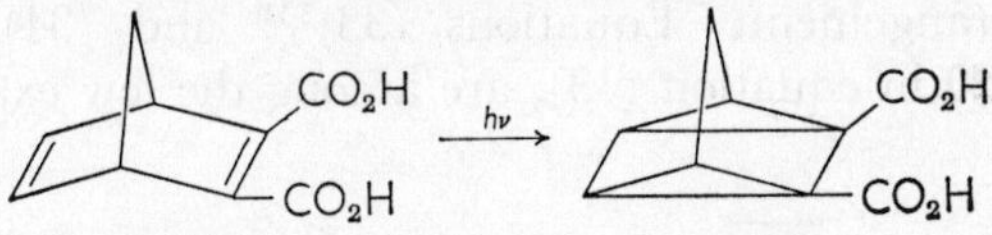

$$\text{(49)} \xrightarrow{h\nu} \text{(50)} \quad (30)$$

of photoisomerization (equation 30) to yield quadricyclanes. The isomerization of the parent hydrocarbon has been examined by several groups. It can be effected in solution by direct irradiation, or under sensitizing conditions with a range of organic sensitizers[101b] and with cuprous chloride[105]. Norbornadiene shows strong ultraviolet absorption in the conjugated diene region (ca. 220 mμ), due to conjugative interaction of the two double bonds. This interaction has been discussed in molecular-orbital terms, and for the lowest excited state **51**, of the diene, a bond order of 1·50 has been calculated for the

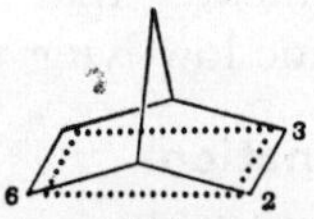

(51)

$C_{(2)}$—$C_{(3)}$ bond and either 0·12 or 0·50, depending on the model used, for the $C_{(2)}$—$C_{(6)}$ bond[106]. Direct irradiation of the diene in the vapour state with 2537 Å light gives a complex mixture, which seems to contain no quadricyclene[107], and the diene triplet does not appear to be involved in the reaction.

The preparation[108] of hexamethyl prismane affords a related example (equation 31) of this type of photoisomerization, while the

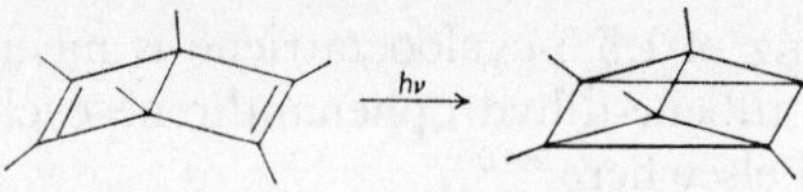

$$\text{(31)}$$

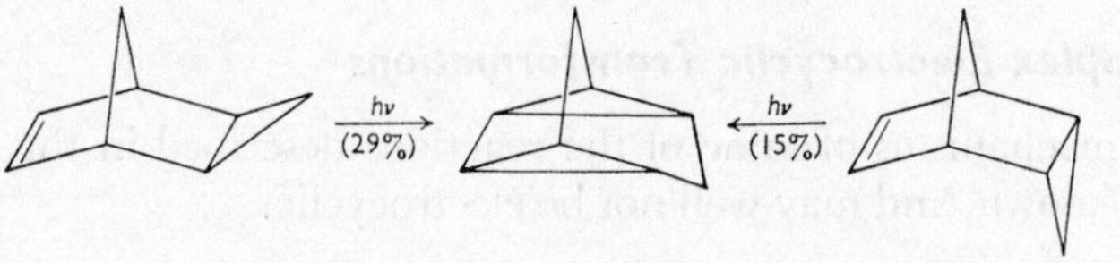

$$\xrightarrow{h\nu} \text{(29\%)} \xleftarrow{h\nu} \text{(15\%)} \quad \text{(32)}$$

findings[109] shown in equation (32), where both *endo* and *exo* isomers give the same polycyclic product, affords another variation.

Another reaction related at least formally to the butadiene-cyclobutene isomerization is the photochemical vinylcyclopropane-cyclopentene rearrangement. Equations (33)[110] and (34)[111] and the formation of **20** in equation (13), are among the few examples so far

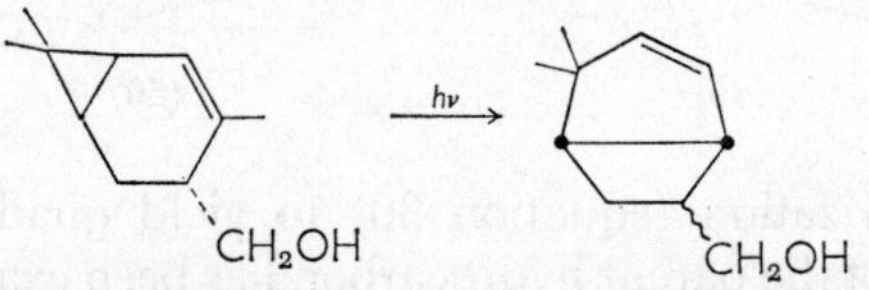

$$\underset{CH_2OH}{\diagup} \xrightarrow{h\nu} \underset{CH_2OH}{\diagup} \quad \text{(33)}$$

$$\underset{RO_2C \quad CO_2R}{\diagup} \underset{\Delta}{\overset{h\nu}{\rightleftarrows}} \underset{CO_2R}{\overset{CO_2R}{\diagup}} \quad \text{(34)}$$

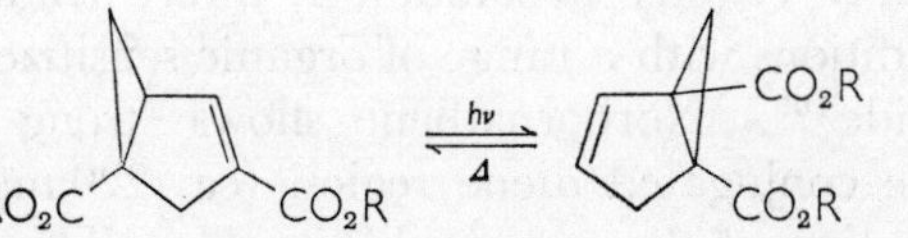

reported. While in some instances[110] the reaction proceeds under sensitizing conditions, in others[112] it does not, so that a single mechanism cannot be proposed to cover all cases. In (33) evidence has been adduced[110] which indicates that both direct and sensitized reactions proceed via the same low-lying triplet.

2. Bicyclo[3.1.0]hexene formation

The formation of bicyclo[3.1.0]hexenes on irradiation of cyclohexadienes and/or acyclic 1,3,5-trienes represents another general reaction of this system. With this reaction Scheme III, giving the known photoisomerizations of the system, can be completed. All except B → G are electrocyclic.

The first example of bicyclo[3.1.0]hexene formation, namely the conversion of calciferol (**52**) to suprasterol II (**53**), was reported by Dauben and coworkers in 1958[113], and since then many others have been published[96,114-117]. Suprasterol I (**54**), a concomitant pro-

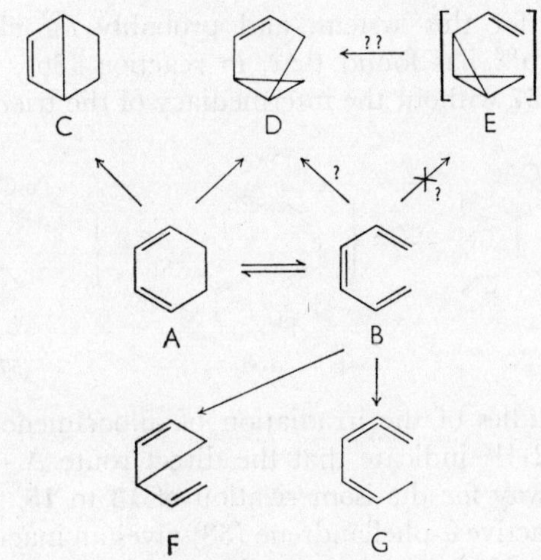

6. Photochemistry of Olefins

SCHEME III

duct, is now known[118] to be a stereoisomer, and the steric course of these cyclizations has been explained in terms of the bonding properties of the excited orbital[118].

The immediate precursor of the bicyclo[3.1.0]hexene is not known in most instances, and there seem to be at least two routes in different isomerizations. Meinwald and Mazzocchi[114] have shown, by deuteration experiments, that 1,1-dimethylhexatriene is photoisomerized to 6,6-dimethylbicyclo[3.1.0]hexene without passing through a vinyl-bicyclobutane, so that the pathway B → E → D in Scheme III can

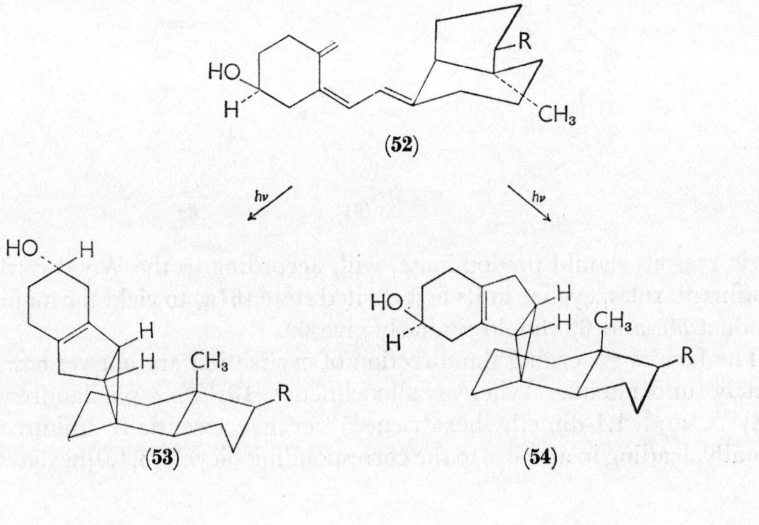

(35)

be discounted for this system, and probably for close analogues. Ullman's group[96] has found that, in reaction (36), the cyclohexadiene **55** gives **57** without the intermediacy of the triene **56**. On the

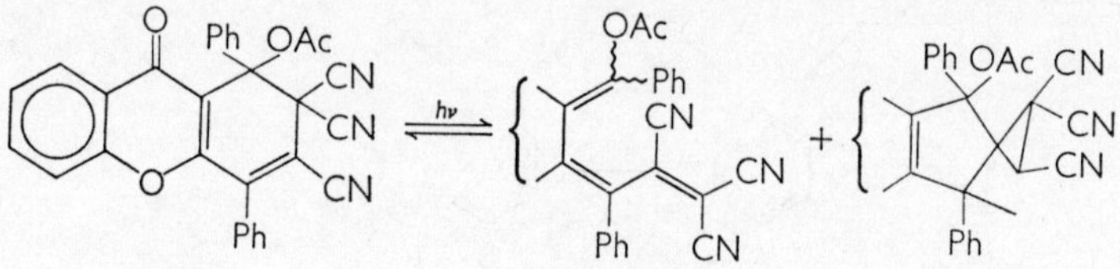

(36)

(55)   (56)   (57)

other hand, studies of the irradiation of alloocimene (equation 12, Section III.C.2)[116] indicate that the direct route A → D is not the principal pathway for the isomerization of **13** to **15**. Furthermore, since optically active α-phellandrene (**58**) gives an inactive[119] product (**59**)[120] (equation 37), some intermediate in which chirality is lost must be invoked. The results in equation (37) are in accord with the intermediacy of the trienes **61** and **62**[117]. The former, which for

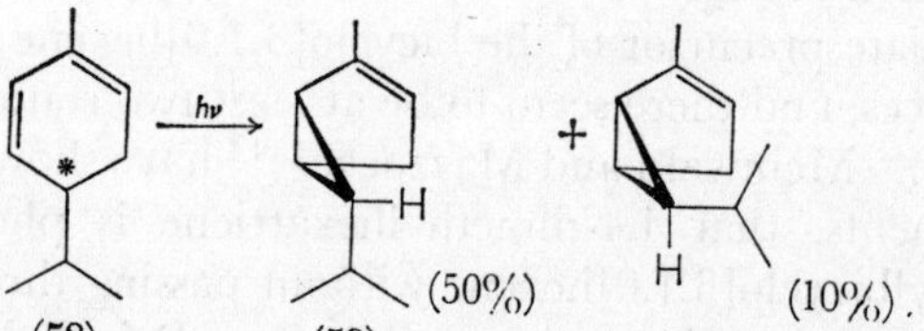

(37)

(58)   (59) (50%)   (60) (10%)

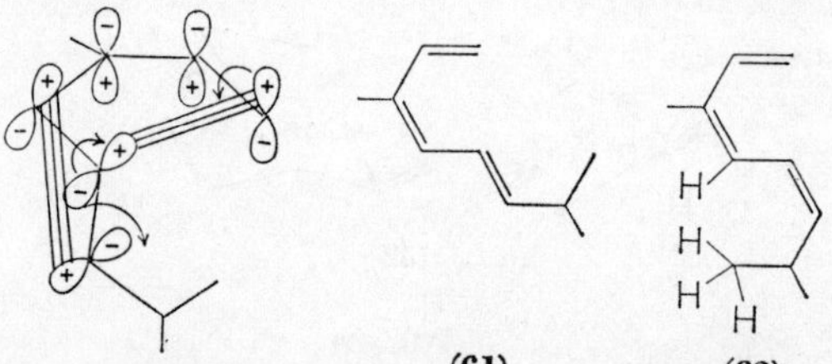

(61a)   (61)   (62)

steric reasons should predominate, will, according to the Woodward-Hoffmann rules, cyclize in its first excited state (**61a**) to yield the major product **59**, and **62** should similarly give **60**.

The factors governing the direction of cyclization are not yet completely understood. Whereas alloocimene (**13**)[116], α-phellandrene (**58**)[117], and 1,1-dimethylhexatriene[114] cyclize essentially unidirectionally, leading in all cases to the corresponding bicyclo[3.1.0]hexenes, the parent 1,3-cyclohexadiene affords equal amounts of bicyclo-[3.1.0]hexene and 3-vinylcyclobutene[121]. However in the case of

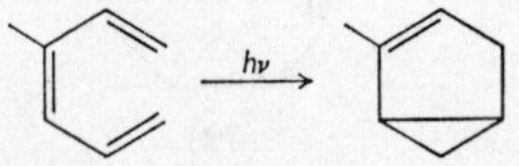

$$\text{structure} \xrightarrow{h\nu} \text{structure} \quad (38)$$

$$\text{structure} \xrightarrow{h\nu} \text{structure} \quad (39)$$

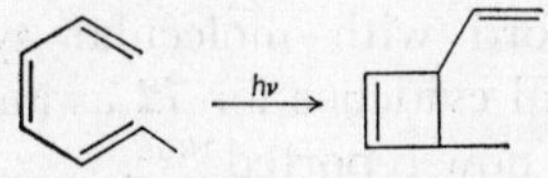

methylhexatrienes the 3-methyl isomer affords mainly the bicyclo-[3.1.0]hexene (equation 38), while the 1-methyl isomer gives the 3-vinylcyclobutene (equation 39) as the predominant product[122].

3. Bullvalene and Semibullvalene

The interconversions of bullvalene (65) and its valence isomers affords what is currently one of the most intriguing aspects of olefin photochemistry. Several groups[123–126,130] are working in the area, and the probable relationships so far reported are shown in Scheme IV.

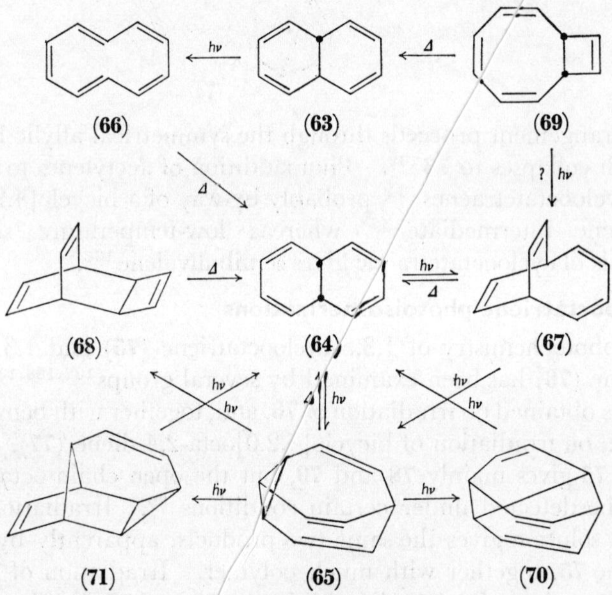

SCHEME IV

With the exception of 70, 71 and cyclodecapentaene (66), all the compounds in this scheme have also been obtained by non-photochemical methods[124,125,127–129]. A plausible intermediate is 72, the formation

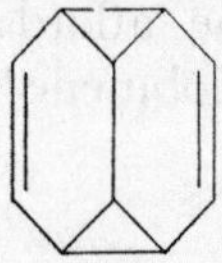

(72)

of which is in accord with molecular symmetry considerations. Physical and chemical evidence for **72** as an intermediate in the isomerization scheme is now reported [130a].

Related to Scheme IV is the preparation (equation 40) of 'semibullvalene' (**73**) [131]. On further irradiation this product yields cyclooctatetraene [133]. Irradiation of hexadeuteriobarrelene indicates that

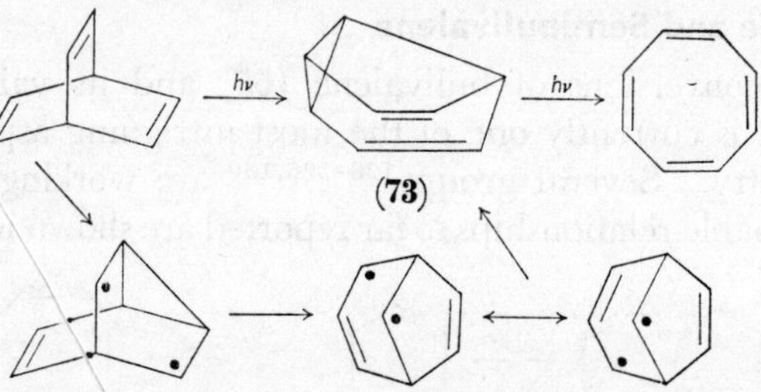

(73)   (40)

(74)

the rearrangement proceeds through the symmetrical allylic biradical **74** which collapses to **73** [132]. Photoaddition of acetylenes to benzene yields cyclooctatetraenes [134], probably by way of a bicyclo[4.2.0]octa-2,4,7-triene intermediate [135], whereas low-temperature sensitized photolysis of cyclooctatetraene gives semibullvalene [136].

4. Cyclooctatriene photoisomerizations

The photochemistry of 1,3,5-cyclooctatriene (**75**) and 1,3,6-cyclooctatriene (**76**) has been examined by several groups [137,138,139]. The former is obtained on irradiation of **76**, and, together with benzene and ethylene, on irradiation of bicyclo[4.2.0]octa-2,4-diene (**77**). Irradiation of **75** gives mainly **78** and **79**, but the open chain octatetraene **80** can be detected under certain conditions [139]. Irradiation of the latter in solution gives the same two products, apparently by way of the triene **75**, together with much polymer. Irradiation of **76** gives five new products (**81–85**), in addition to **78** and **79** [137,139].

Other examples of similarly complex photoisomerizations are known [45,93,140], but the mechanisms involved are, in general, uncertain.

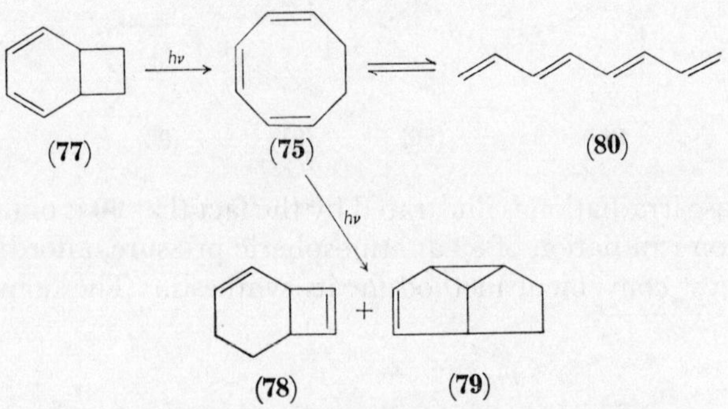

(77) → hν → (75) ⇌ (80)   (41)

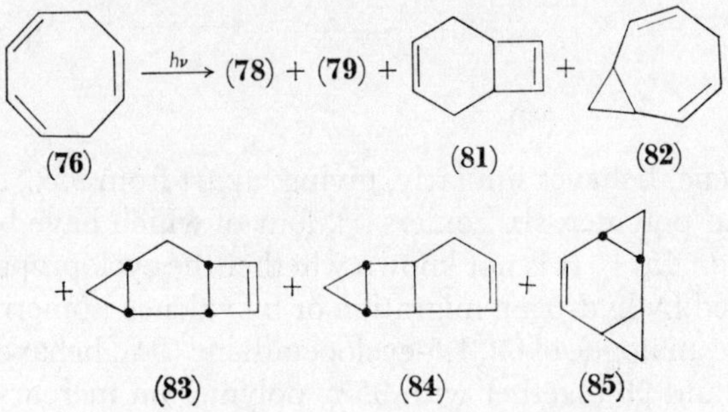

↓ hν

(78) + (79)

(76) → hν → (78) + (79) + (81) + (82)   (42)

+ (83) + (84) + (85)

IV. INTRAMOLECULAR ADDITIONS

This field has been recently reviewed[2j].

A. Carbocyclic Photoproducts

I. From simple olefins

The mercury sensitized vapour-phase photoisomerizations of a variety of non-conjugated dienes have been examined. Irradiation of 1,4-pentadiene (equation 43) and several methylated derivatives in the vapour state, afforded **86** and **87**[141,142] and in one case **88**[141]. 1,5-Hexadiene (**89**) gives mainly **90** and **91** together with a little **92**, in

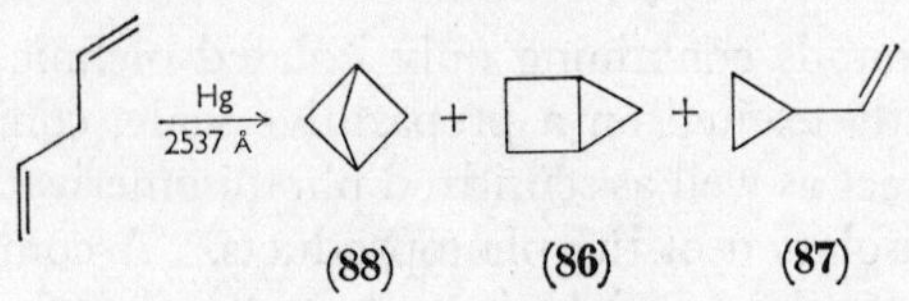

$$\xrightarrow{\text{Hg}, 2537\,\text{Å}}$$ (88) + (86) + (87)   (43)

proportions varying with the pressure[143,144]. The synthetic utility of

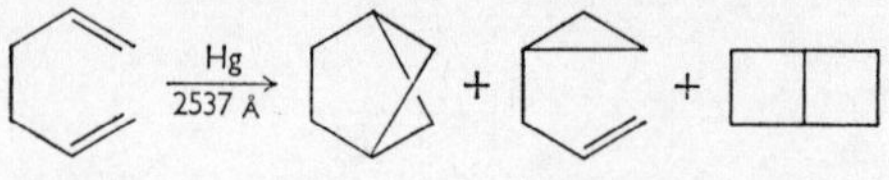

$$\text{(89)} \xrightarrow[2537\text{ Å}]{\text{Hg}} \text{(90)} + \text{(91)} + \text{(92)} \qquad (44)$$

vapour-phase irradiation is illustrated by the fact that **90** is obtained in 18% yield on irradiation of **89** at atmospheric pressure, affording perhaps the most convenient method for its synthesis. The homologue,

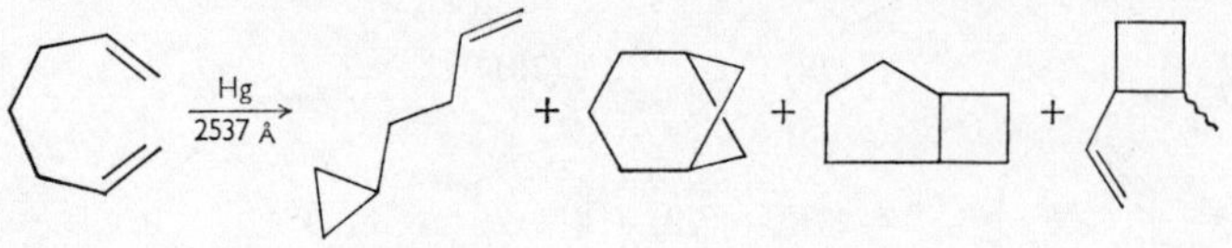

$$\xrightarrow[2537\text{ Å}]{\text{Hg}} \text{(93)} + \cdots + \cdots + \cdots \qquad (45)$$

1,6-heptadiene, behaves similarly, giving, apart from 95% free radical products and polymer, six isomers[145], four of which have been identified (equation 45). It is not known whether the cyclopropanes **91** and **93** are formed by hydrogen migration or by valence isomerization.

The cyclic analogue of **89**, 1,5-cyclooctadiene (**94**), behaves similarly, yielding **95** and **96** together with 95% polymer on mercury sensitized photolysis in the vapour phase. In ethereal solution, with cuprous

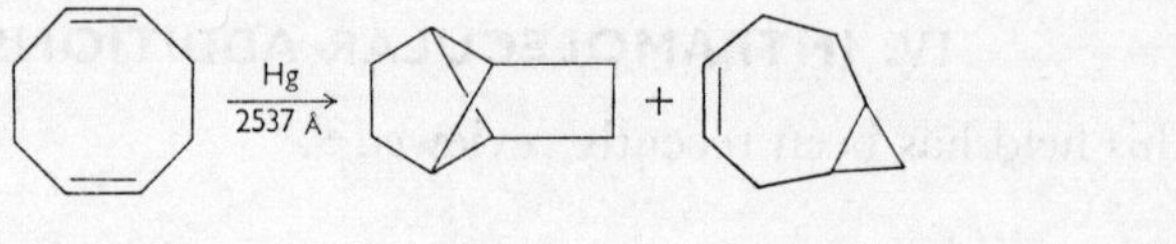

$$\text{(94)} \xrightarrow[2537\text{ Å}]{\text{Hg}} \text{(95)} + \text{(96)} \qquad (46)$$

chloride as catalyst, **95** is obtained in 30% yield[146]. After some dispute[147] it is now clear that the reaction proceeds intramolecularly, and is not a free radical process[148]. Irradiation of the 1,5-cyclo-octadiene-rhodium chloride complex in ether gives 1,3- and 1,4-cyclo-octadiene and bicyclo[4.2.0]oct-7-ene[146].

2. From more complex hydrocarbons

Unlike compounds containing only isolated olefinic bonds, which cannot be directly excited on a preparative scale, conjugated dienes can undergo direct as well as sensitized photoisomerizations on a scale permitting the isolation of the photoproducts. A comparison of the products may afford an insight into the nature of the intermediate excited states. Myrcene (**98**), on sensitized irradiation, gives exclusively **97**[149], while on direct excitation it yields mainly **99**, together with small amounts of **100, 97** and unidentified products[150,151].

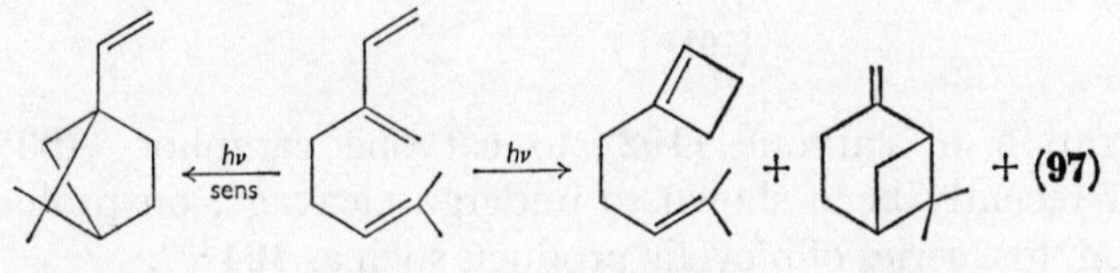

$$\underset{(97)}{\vcenter{\hbox{[structure]}}} \xleftarrow{h\nu\ \text{sens}} \underset{(98)}{\vcenter{\hbox{[structure]}}} \xrightarrow{h\nu} \underset{(99)}{\vcenter{\hbox{[structure]}}} + \underset{(100)}{\vcenter{\hbox{[structure]}}} + (\mathbf{97}) \quad (47)$$

It is virtually certain that the triplet state of the diene is involved in the sensitized reaction. If we assume that there is only one pathway for the photochemical formation of **97** from myrcene, then the fact that **97** is only a very minor product in the direct irradiation shows that intersystem crossing from the excited singlet to the triplet does not take place efficiently.

Dauben has suggested[15] that since intersystem crossing from the excited singlet of a diene to the triplet has not been observed[5,152], the convergence point of the pathways $\mathbf{98}(S_1) \rightarrow \mathbf{97}$ and $\mathbf{98}(T_1) \rightarrow \mathbf{97}$ is at a vibrationally excited ground state which can descend the vibrational energy cascade to either starting material or product.

Several analogous instances (equation 48) of this type of reaction have been reported[153,154], and it appears to be a fairly general reaction.

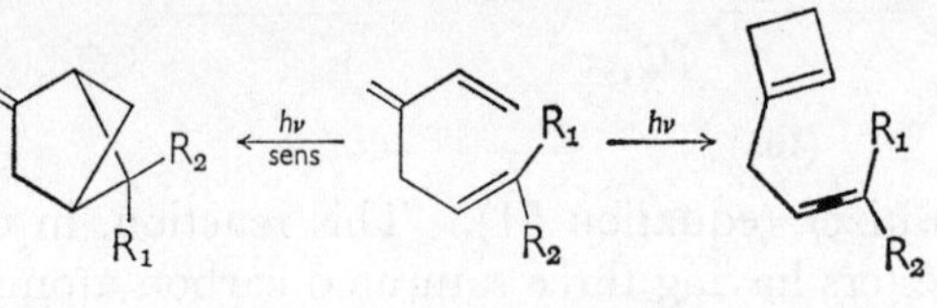

$$\vcenter{\hbox{[structure with R_2, R_1]}} \xleftarrow{h\nu\ \text{sens}} \vcenter{\hbox{[structure with R_1, R_2]}} \xrightarrow{h\nu} \vcenter{\hbox{[structure with R_1, R_2]}} \quad (48)$$

These bicyclic photoproducts, as well as those from the above non-conjugated dienes, are apparently formed in a two-step process from biradical intermediates. In the formation of the latter a strong preference for five-membered ring formation is apparent[141,149].

3. From α,β-unsaturated carbonyl compounds

When conjugation is achieved with a carbonyl group rather than an olefinic bond, loss of carbon monoxide becomes another possible photoreaction, although cyclobutane formation is still the usual reaction on direct irradiation in solution[155-157], and **101** gives the corresponding bicyclo[2.1.1]hexanone (49) (30%)[155].

The first instance of photochemical cyclobutane formation was the

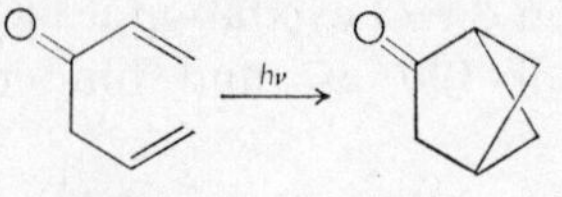

$$\text{(101)} \xrightarrow{h\nu} \text{product} \quad (49)$$

isomerization of carvone (**102**) to carvone camphor (**103**)[157,158]; this has recently been shown to undergo cleavage, on prolonged irradiation, to a series of bicyclic products such as **104**[159].

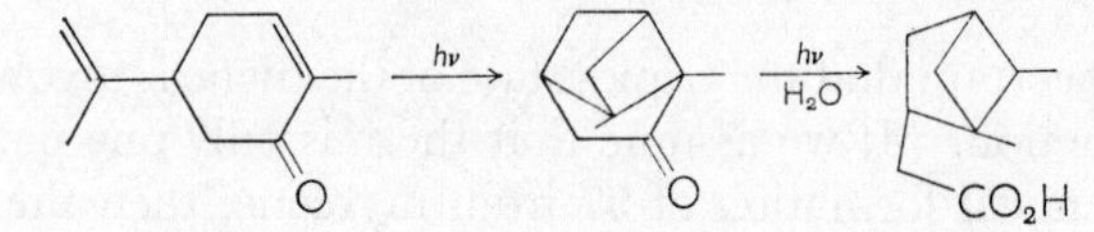

$$\text{(102)} \xrightarrow{h\nu} \text{(103)} \xrightarrow[H_2O]{h\nu} \text{(104)} \quad (50)$$

Several instances are known of cyclobutane formation on irradiation of α,β-unsaturated esters containing another olefinic bond[160,161], while the α,β-unsaturated aldehyde, citral, yields both an analogous cyclobutane and an isopropylcyclopentene aldehyde[162]. Exceptionally, one of these esters (**105**)[160] cyclizes equally well (55% yield) with or

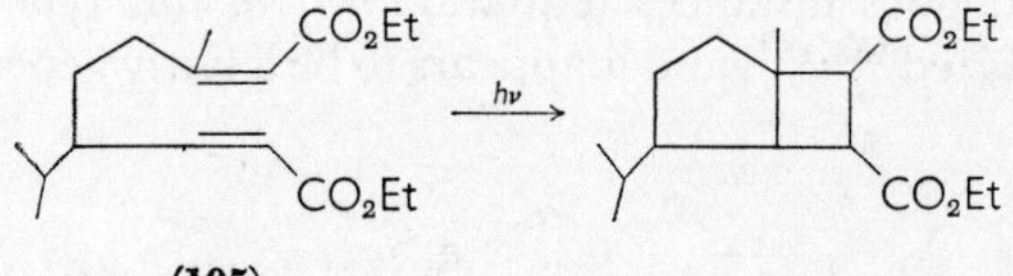

$$\text{(105)} \xrightarrow{h\nu} \text{product} \quad (51)$$

without a sensitizer (equation 51). This reaction, in common with that of other esters having three saturated carbon atoms between the two olefinic bonds[161], proceeds only by head-to-head addition, while only head-to-tail products have been reported in other such intramolecular cycloadditions[160,161].

4. Cage products

A classic example of this type of reaction occurs in Eaton's cubane synthesis[163]. This is the isomerization, in 95% yield, of the α,β-unsaturated ketone **106** to **107**. Several analogous α,β-unsaturated ketones have since been cyclized similarly[164-167]. In one case[167] (equation 52) a competing reaction (**108**→**109**) was observed when $\pi \to \pi^*$ excitation (high pressure lamp, quartz apparatus) was achieved, but with $n \to \pi^*$ excitation (pyrex filter and potassium

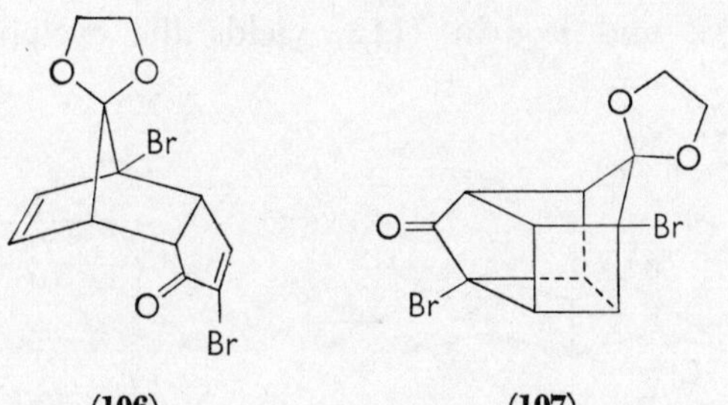

6. Photochemistry of Olefins

(106)   (107)

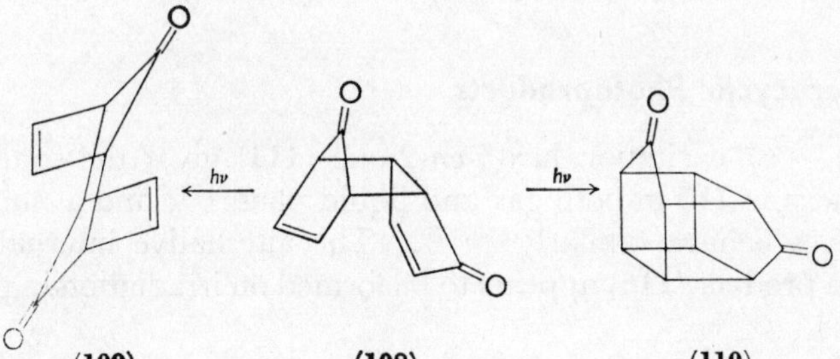

(109) ← hv — (108) — hv → (110)   (52)

hydrogen phthalate filter solution) the normal cycloaddition process (108→110) predominated. This cycloaddition can also be effected by triple-energy transfer from added acetophenone sensitizer. The rearrangement 108→109 was shown, by deuteration experiments, not to involve dissociation into two cyclopentadienone molecules, but to be intramolecular[167].

Acetone sensitized photocycloaddition[168,169] has been used recently by three groups[170-172] to obtain the 1,1'-bishomocubane system. Dauben[171] has extended the process (equation 53) to obtain the

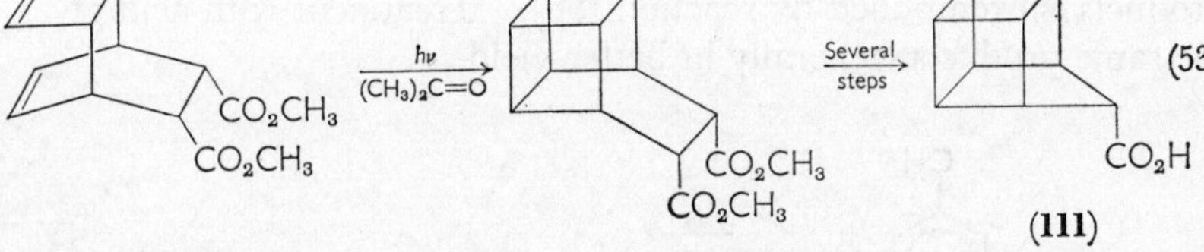

CO_2CH_3 CO_2CH_3 —hv, $(CH_3)_2C=O$→ CO_2CH_3 CO_2CH_3 —Several steps→ CO_2H (111)   (53)

homocubanecarboxylic acid 111, which is readily converted into the remarkably stable parent hydrocarbon homocubane.

Non-conjugated dienes can also undergo this type of cycloaddition on direct irradiation if substituents are present which shift the absorption region to longer wavelength. Halogen substitution is most common[173-176], and isodrin (**112**) yields the cyclobutane photoproduct **113**[173].

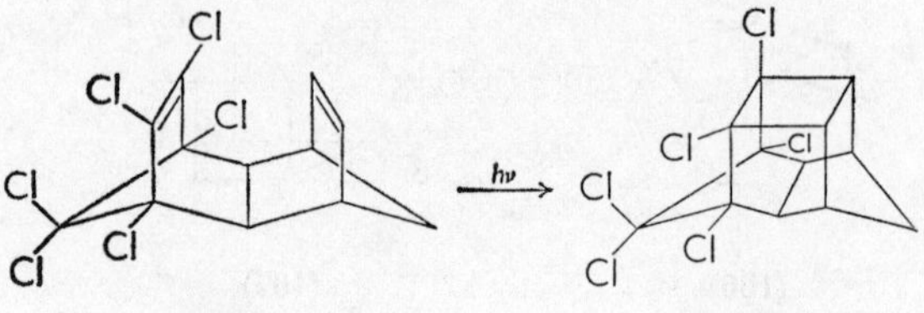

$$(112) \xrightarrow{h\nu} (113) \qquad (54)$$

B. Heterocyclic Photoproducts

On $n \rightarrow \pi^*$ excitation hex-5-en-2-one (**114**) gives the oxabicyclo-[2.1.1]hexane **115** in both gas and liquid phase[177], and a number of derivatives behave similarly[178,179]. The alternative internal cycloaddition product (**116**) appears to be formed on irradiation in pentane

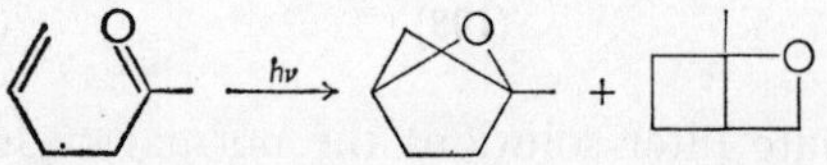

$$(114) \xrightarrow{h\nu} (115) + (116) \qquad (55)$$

solution, but is unstable, although a dimethyl homologue has been isolated[179]. The quantum yields for such oxetane formation are low (<0.01) in all the cases where these were determined[177,178]. These γ,δ-unsaturated ketones do not undergo the normal 'type II' fission[180], to any appreciable extent. This fission occurs universally in the corresponding saturated ketones with γ-hydrogen atoms.

Another type[181,182] of intramolecular reaction yielding heterocyclic products is exemplified by reaction (56). Treatment with acid gives the same products, generally in better yield.

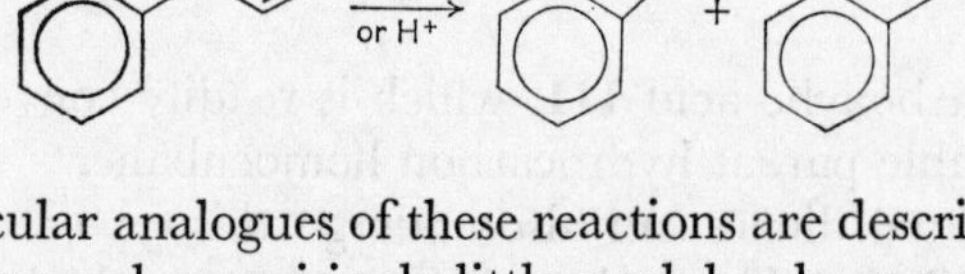

$$\xrightarrow{h\nu \text{ or } H^+} \qquad (56)$$

Intermolecular analogues of these reactions are described in Section V.C, but in general, surprisingly little work has been published in this field.

V. INTERMOLECULAR PHOTOADDITION REACTIONS

A. Simple Addition

Photoaddition of various compounds (H_2S, HBr, ROH, O_2, etc.) to simple olefinic molecules has long been recognized, and has been reviewed.[2a] (For recent leading references see refs. 183–187.) Such additions usually proceed by radical intermediates, and may involve chain reactions.

Acetone sometimes adds across a double bond as a side reaction when it is used as a sensitizer[188,189]. Two modes of addition are illustrated in equation (57)[190]; the more common mode, exemplified by **117 → 119**, probably occurs by a radical mechanism, and there is

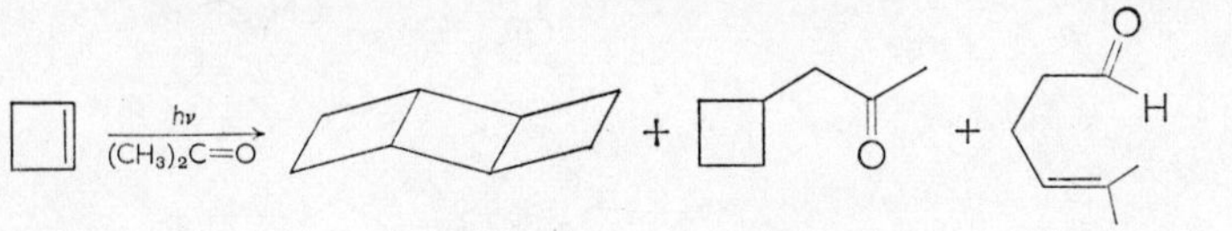

(117)   (118)   (119)   (120)   (57)

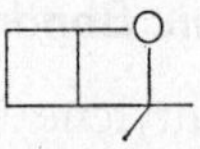

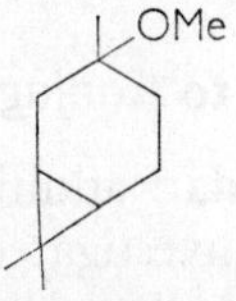

(121)   (122)

evidence that the aldehyde **120** is formed via the unisolated oxabicyclohexane **121**. The analogous oxabicycloheptene has been obtained on irradiation of cyclopentadiene in acetone[191].

Formamide adds photochemically to the olefinic bond in various environments, to give good yields of amides[192,193]. Photoaddition of N-nitrosodialkylamines affords a method of cleaving unhindered carbon–carbon double bonds[194,195], and photoaddition of bromoacetic acid offers a convenient method for carboxymethylation of olefins[196].

Irradiation of several 3,5- and 4,6-steroidal dienes and related compounds[15,197,198] in the presence of alcohols (or water) results in ether (or alcohol) formation as in equation (58)[15]. In some of these additions an excited polar species abstracts a proton from the alcohol to give an intermediate carbonium ion which reacts subsequently with the olefin[197], and in others, not strictly photochemical, the alcohol reacts with an unstable bicyclobutane formed photochemically[15]. (See **36** → **38**, Section III.D.2.) Alcohols also add photochemically to flexible cyclic *cis* olefins[28,29,198] to yield ethers such as **122** [from carene (**1**)][28].

α,β-Unsaturated ketones may react with alcohols on irradiation according to equation (59)[199]. This is thought [200] to proceed from a π–π* triplet, but it is not clear whether an intermediate ground-state species is also involved.

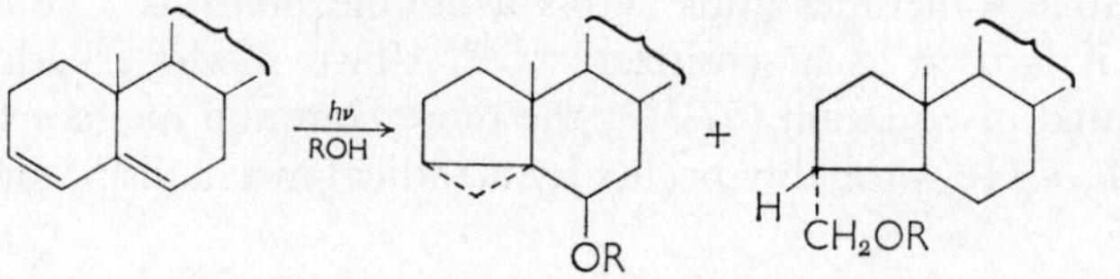

$$\xrightarrow{h\nu}_{\text{ROH}} \qquad (58)$$

$$\xrightarrow{h\nu}_{\text{ROH}} \qquad (59)$$

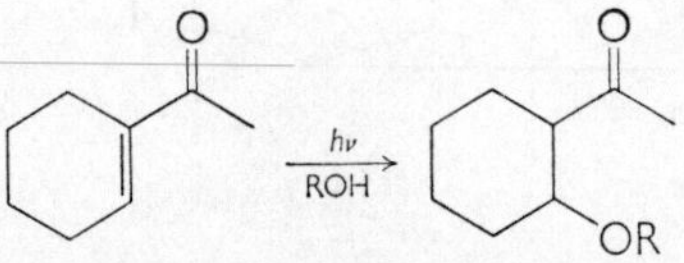

B. Cycloadditions to Conjugated Olefinic Bonds

The intermolecular addition of isobutylene to 2-cyclohexenone (equation 61) was investigated by Corey and used as a key step in his synthesis of caryophyllene[201]. Further investigations indicated that

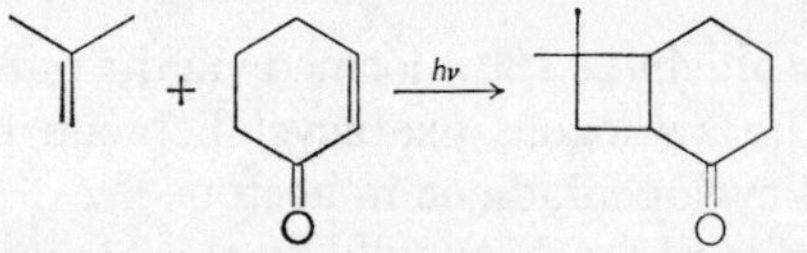

$$\xrightarrow{h\nu} \qquad (61)$$

the reaction is quite general[203] for cyclopentenone and cyclohexenone, affording in many cases a mixture of *cis*- and *trans*-fused bicyclo-[*n*.2.0]alkanones[202]. On the basis of these results Corey[202] has suggested the preliminary formation of an excited state π-complex such as **123**. Complex formation would then be followed by stepwise bond formation to give either one of a pair of diradicals **124** and **125** which could then collapse to the observed products (Scheme V). This mechanism clearly explains the orientational specificity of addition on the basis of the known polarity of the excited states[4b], of α,β-unsaturated ketones. It is also consistent with the fact that *cis*- and *trans*-2-butene afford identical product mixtures (Scheme VI).

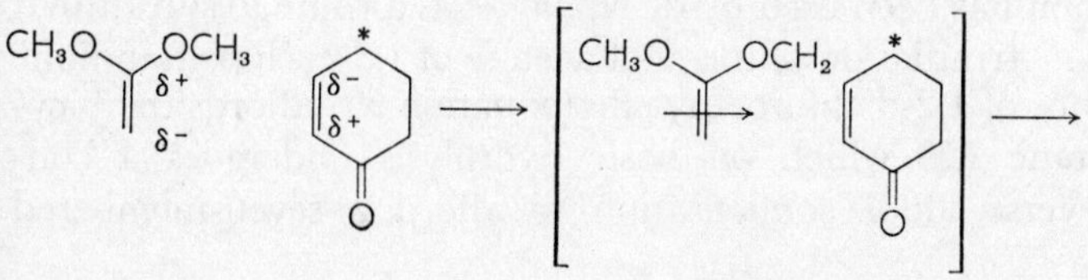

6. Photochemistry of Olefins

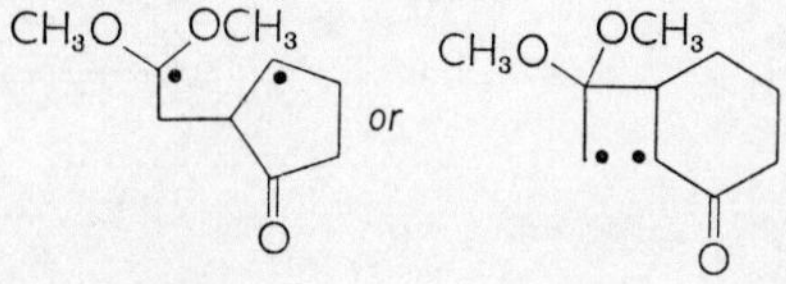

SCHEME V

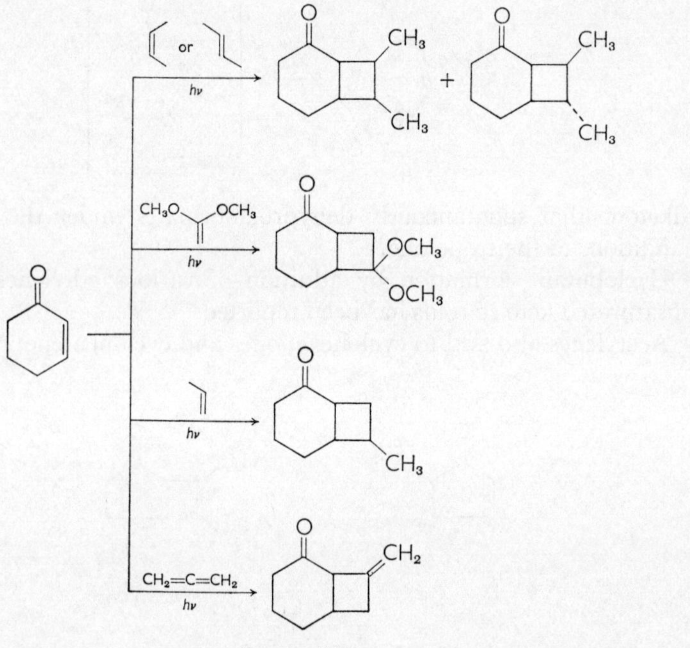

SCHEME VI

The intermediacy of a *trans*-cyclohexenone is unlikely in view of the fact that 2-cyclooctenone and 2-cycloheptenone, which are known to photoisomerize to their *trans* isomers[204-206], do not undergo an analogous reaction.

The reaction has been used by de Mayo[207] as a route to substituted γ-tropolones. Irradiation of the enol acetate of 1,3-cyclopentandione in the presence of 1,2-dichloroethylene (equation 62) affords the bicyclo[3.2.0]octane **126** which on basic hydrolysis undergoes a concomitant reverse aldol condensation to afford a seven-membered

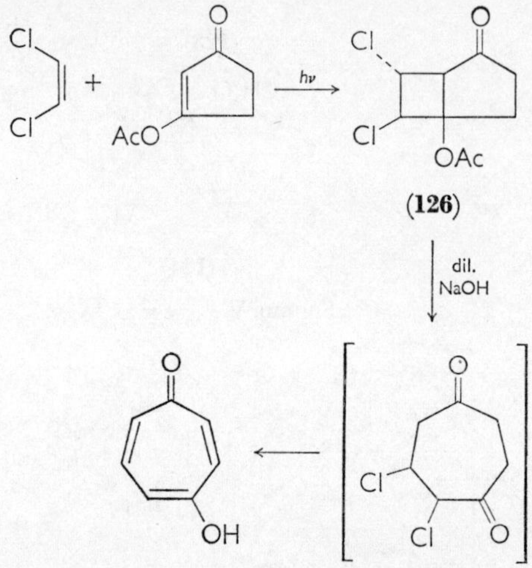

(62)

(**126**)

diketone that spontaneously dehydrochlorinates under the reaction conditions to the tropolone.

Cyclobutane formation by addition of various ethylenes to α,β-unsaturated keto steroids has been reported[208-210].

Acetylenes also add to cyclohexenones and cyclopentenones to give

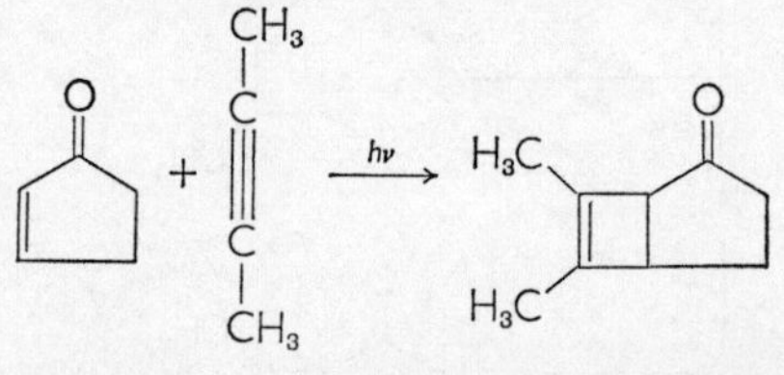

(63)

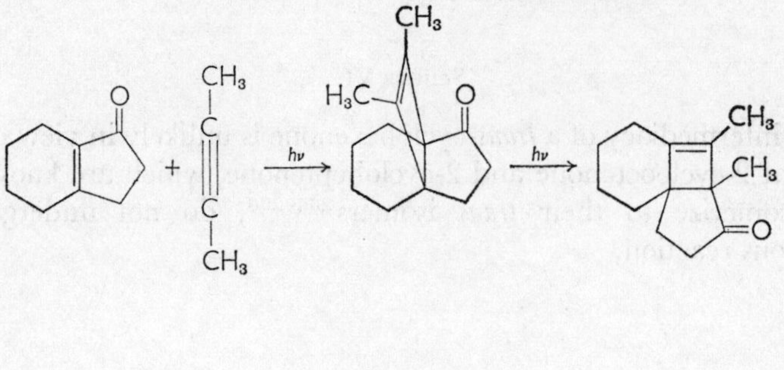

(64)

the corresponding bicyclic[211] (equation 63) or tricyclic[212] (equation 64) olefins in good yield. The tricyclic compounds are mixtures of rearranged products.

A formally similar reaction is the photoaddition of maleates or maleic anhydride to olefins. Addition of dimethyl maleate[213] and

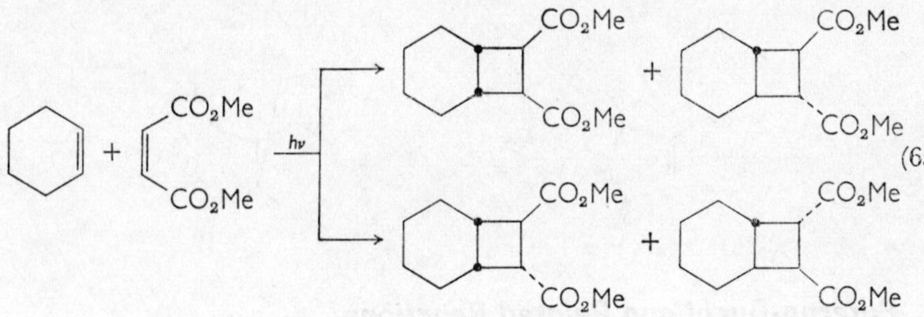

$$\text{(65)}$$

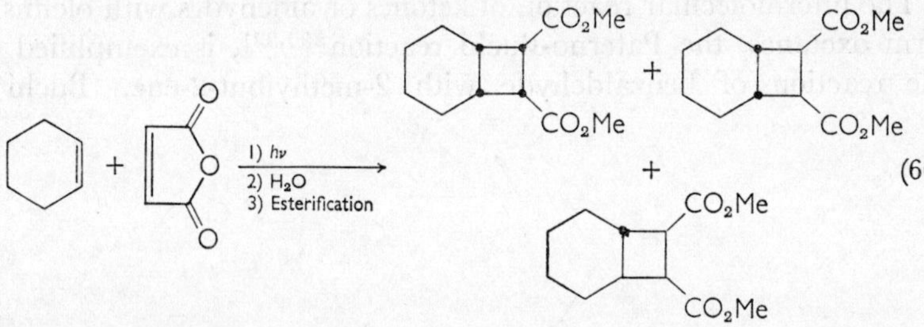

$$\text{(66)}$$

maleic anhydride[214] to cyclohexene gives mixtures of *cis-* and *trans-*fused bicyclo[4.2.0]octanes.

An investigation of this reaction by de Mayo[215] has led to the conclusion that the sensitized reaction takes place via a triplet intermediate (equation 67), in agreement with the fact that maleate and fumarate

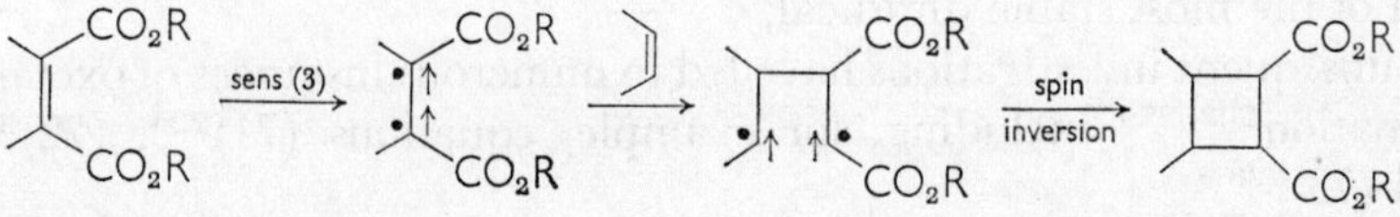

$$\text{(67)}$$

afford the same product mixtures. It is concluded, however, that for direct irradiation, a duality of mechanisms including both singlet and triplet intermediates is in effect.

The reaction has been extended to acetylenes in a number of cases[216-219] (equations 68 and 69) affording bi- and tricyclic products in good yield.

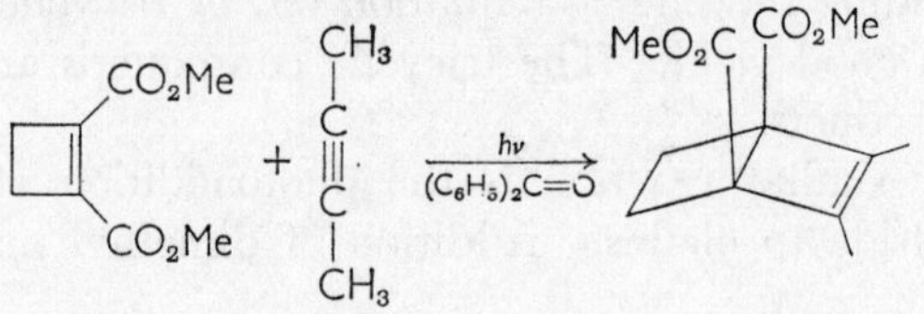

(68)

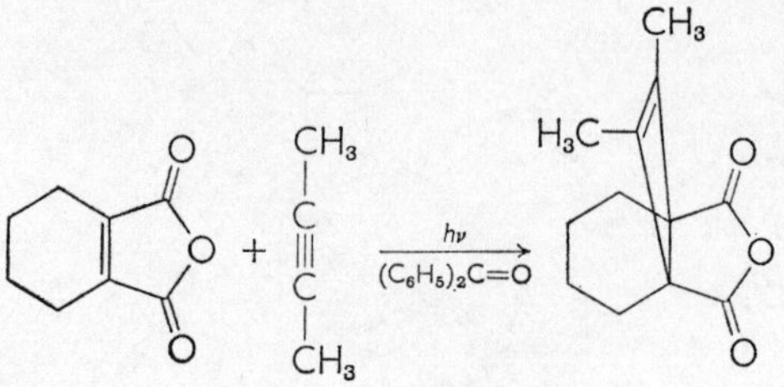

(69)

C. Paterno-Buchi and Related Reactions

The intermolecular reaction of ketones or aldehydes with olefins to form oxetanes, the Paterno-Buchi reaction[220,221], is exemplified by the reaction of benzaldehyde with 2-methylbut-2-ene. Buchi[221]

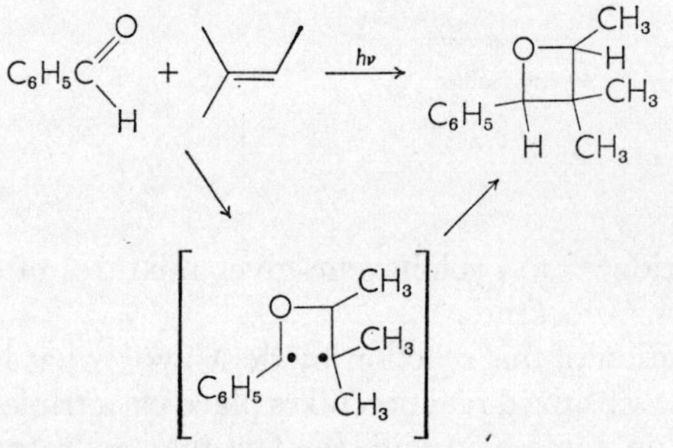

(70)

has rationalized the orientational selectivity on the basis of the formation of the most stable diradical.

Subsequent investigations have led to numerous instances of oxetane formation[223-229] including, for example, equations (71)[224], (72)[225] and (73)[226].

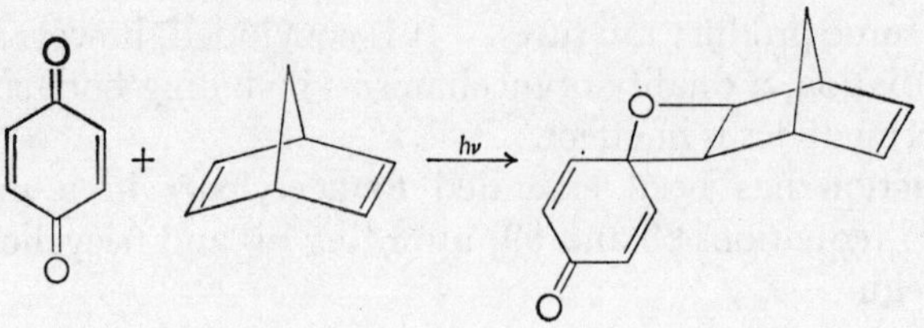

(71)

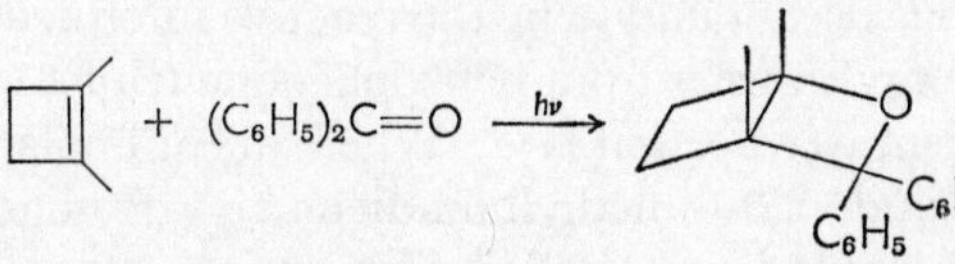

(72)

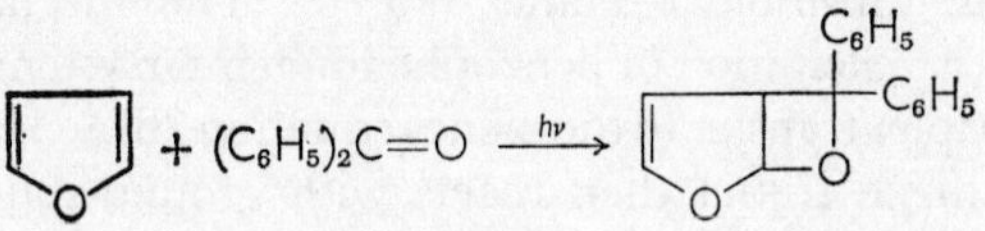

(73)

When an acetylene is substituted for the alkene (equations 74 and 75), the expected oxetenes (**127**) cannot be isolated as they rearrange to α,β-unsaturated ketones (**128**)[230-232].

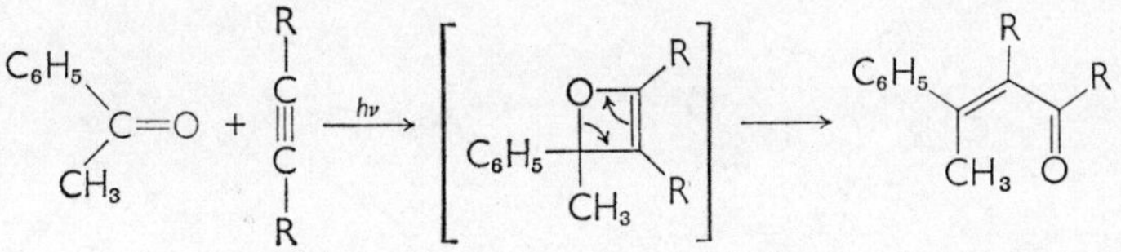

(74)

(**127a**)   (**128a**)

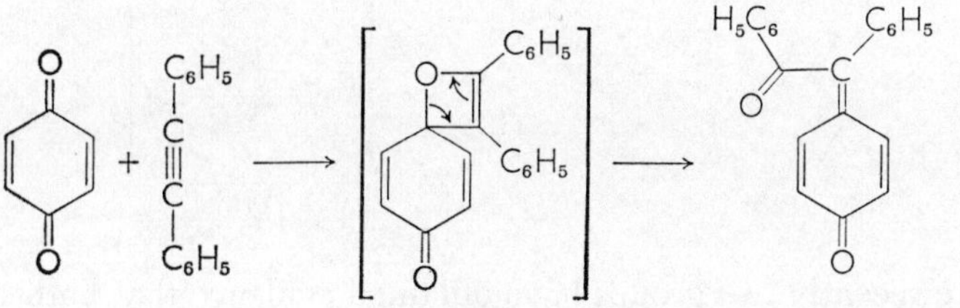

(75)

(**127b**)   (**128b**)

The Paterno-Buchi reaction has been the subject of several mechanistic studies. Arnold and coworkers[222] came to the conclusion that the $n \rightarrow \pi^*$ excited state of the ketone is the reactive one, since only those ketones which undergo photoreduction in isopropanol ($n \rightarrow \pi^*$) are reactive. They pointed out, however, that in those cases where the olefin $\pi \rightarrow \pi^*$ state lies below the ketone $n \rightarrow \pi^*$ state, only ketone to olefin energy transfer takes place. Yang[233] came to the same conclusion as to $n \rightarrow \pi^*$ being the reactive excited state, but also suggested that the orientational specificity was a function of the relative electron deficiency of carbonyl oxygen in the ketone $n \rightarrow \pi^*$ state.

An olefin triplet has been found to be the reactive intermediate in at least one case. Energy transfer from benzophenone triplet to a diene should be very efficient (see Section II). When an equimolar mixture of benzophenone and 2,3-dimethylbutadiene is irradiated, 1,1-diphenyl-2,3-dimethylbutadiene (**130**) is formed in a clean reaction, apparently via the unstable oxetane **129**[225]. The mechanism is formulated as $n \rightarrow \pi^*$ excitation of benzophenone, intersystem crossing to the triplet and triplet transfer to the diene (equation 76). Dauben[225] suggests that the diene triplet then reacts with ground-state benzophenone to form the oxetane (equation 77).

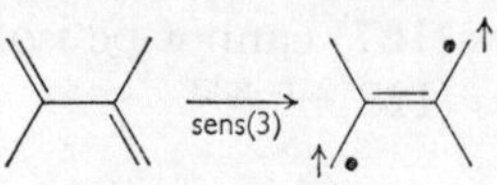

$$\text{diene} \xrightarrow{\text{sens}(3)} \text{diene triplet} \qquad (76)$$

$$\text{(C}_6\text{H}_5\text{)}_2\text{CO} + \text{diene triplet} \longrightarrow \text{biradical} \qquad (77)$$

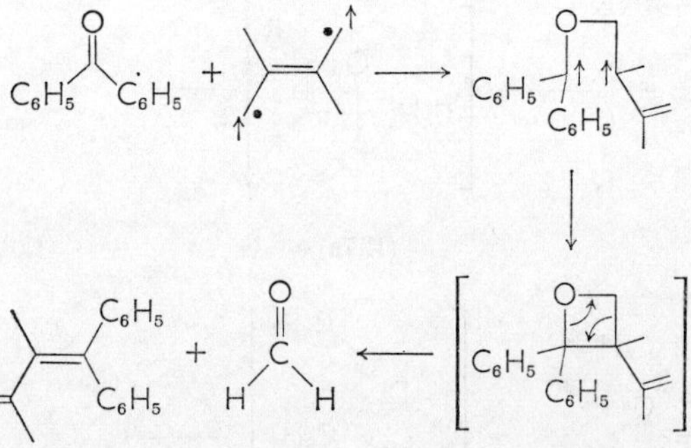

$$\downarrow$$

(**130**) + HCHO ⟵ [**129**]

More recently two groups have obtained evidence that both singlet and triplet states are involved in some cases. Yang[234] has suggested that a singlet state might be in part responsible for the Paterno–Buchi reaction on the basis that 9-anthraldehyde addition to 2,3-dimethyl-but-2-ene is only partially quenched by di-t-butylnitroxide. Turro[235] has also used the absence of both quenching and sensitization in the reaction of *cis*- or *trans*-dicyanoethylene with alkyl ketones as evidence for a singlet intermediate. He further found that reaction (78) is stereospecific, in agreement with predictions for a singlet state.

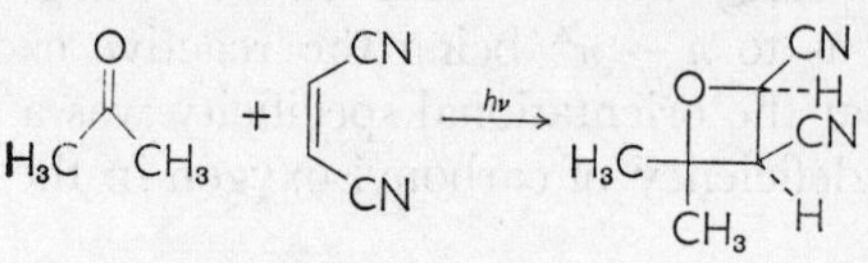

$$\text{H}_3\text{C-CO-CH}_3 + \text{(CN)CH=CH(CN)} \xrightarrow{h\nu} \text{oxetane} \qquad (78)$$

D. Additions to Aromatic Compounds

Olefins have been found to undergo addition to aromatic compounds on irradiation. Maleic anhydride and benzene form a 2:1 adduct **131**[236-238] presumably through the initially formed bicyclo[4.2.0]octane intermediate (equation 79). It has been shown[236] that

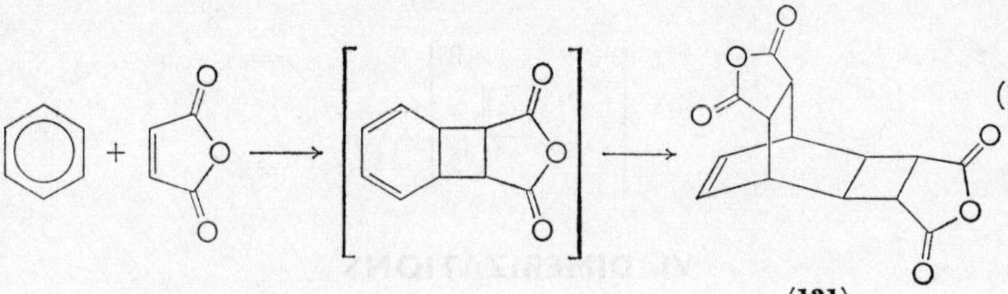

$$(79)$$

(**131**)

the first step is the result of excitation of a benzene–maleic anhydride charge-transfer complex. It is suggested[239] that addition of the second molecule of maleic anhydride is a simple Diels–Alder reaction, but no intermediate 1:1 adduct could be characterized. Several groups have obtained related 2:1 adducts using a variety of benzenes[240,241], anhydrides[242] and maleimides[243,244].

Addition of 2-methyl-2-butene to benzonitrile affords only a single bicyclo[4.2.0]octadiene (**132**)[135] (equation 80), while with a range of chloroolefins, several benzenes have given the corresponding open-chain tetraenes[245]. Photoaddition of cyclobutene and benzene has been reported to give tetracyclo[4.4.0.0^{2,5}.0^{7,10}]dec-3-ene[246], but this is now in doubt[247].

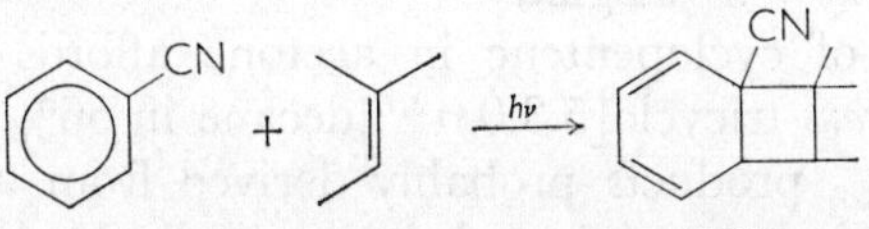

$$(80)$$

(**132**)

While the 1,2-modes of addition of olefinic bonds to aromatic groups are the most common[89,248,249], 1,3- and 1,4-photoadditions[89] are also known. The 1,3-photoadduct **133** is formed in good yield[249] from

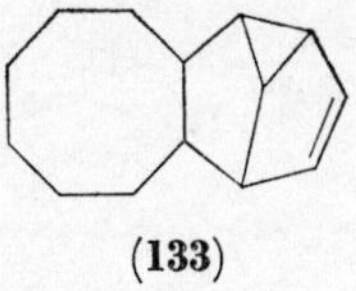

(**133**)

cyclooctene and benzene on direct irradiation, and a variety of 1,4-adducts result on irradiation of a mixture of benzene and 1,3-dienes[89].

Addition of alkyl acetylenes[250] (equation 81), dimethyl acetylenedicarboxylate[251,252] (equation 81, R = CO_2Me), phenyl acetylene[252] and methyl propiolate[252] lead to the corresponding substituted cyclooctatetraenes.

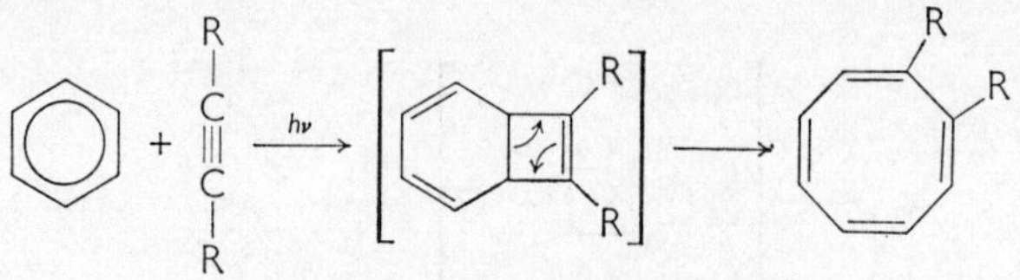

$$\text{(equation 81)} \tag{81}$$

VI. DIMERIZATIONS

A. Cyclobutane Photodimers

I. From nonconjugated olefins

Photodimerizations of olefins to cyclobutanes have been carried out in the vapour, liquid, and crystalline phases. Well over one hundred instances are known but the great majority involve conjugated olefins. Of the remainder, most are dimerizations of small-ring olefins and only a very few examples of the dimerization of simple alkenes are known.

Irradiation of ethylene in the presence of mercury vapour gives cyclobutane together with several other products[253], while tetramethylethylene gives octamethylcyclobutane on direct excitation.[254] The latter, which occurs only under non-sensitizing conditions, seems to involve the $\pi \rightarrow \pi^*$ singlet.

Irradiation of cyclopentene in acetone affords the cyclobutane photodimer, *trans*-tricyclo[5.3.0.0²,⁶]decane in 56% yield, along with four other $C_{(10)}$ products probably derived from a diradical intermediate[255]. Cyclopentadiene behaves similarly, in giving the corresponding *trans*-tricyclic diene[256,257].

A closely related reaction is the dimerization of norbornene to the *exo-trans-exo* dimer **134**, via the cuprous bromide complex. Sensitiza-

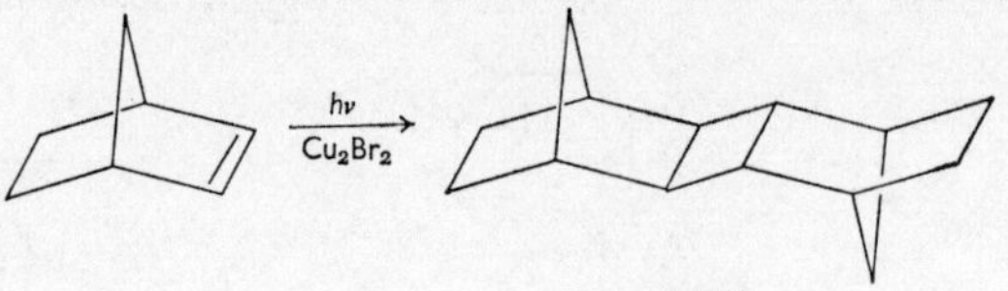

$$\tag{83}$$

(**134**)

tion through cuprous halide complexation was found to be quite specific and is reminiscent of the 'criss-cross' reaction of 1,5-cyclo-octadiene (Section IV.A.1).

A few cyclopropenes have been shown to dimerize to tricyclo-[3.1.0.0^{2,4}]hexanes. In the case of the 1,3,3-trimethylcyclopropene

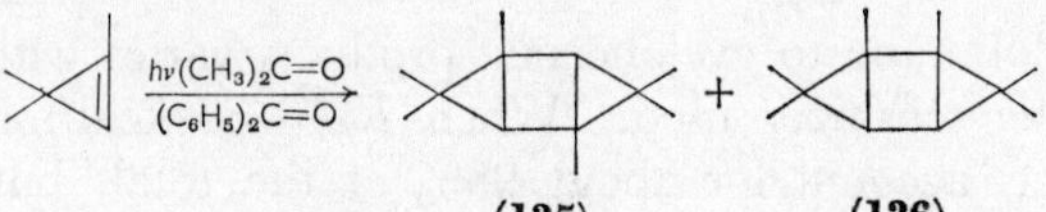

$$\text{[structure]} \xrightarrow[\text{(C}_6\text{H}_5)_2\text{C=O}]{h\nu(\text{CH}_3)_2\text{C=O}} \text{(135)} + \text{(136)} \qquad (84)$$

a 4:1 ratio of **135** to **136** is obtained[259]. It is not clear whether benzophenone or acetone sensitizes the reaction.

3-Acetyl-1,2-diphenylcyclopropene affords an analogous tricyclic product, which further photolyses to 1,2,4,5-tetraphenyl benzene[260].

2. From conjugated olefins

The photodimerization of thymine is one of the most important reactions in photobiology[261]. In the course of a study on the photochemistry of 2,3-diphenylbutadiene, White and Anhalt[262] have shown that the tricyclo[4.2.0.0^{2,5}]octane **138** is a secondary product formed by photodimerization of the originally formed cyclobutene **137**.

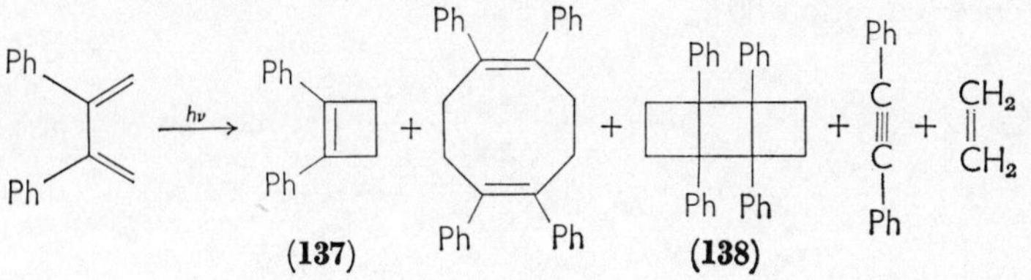

$$\text{(137)} \xrightarrow{h\nu} \text{[structures]} + \text{(138)} + \text{[structures]} \qquad (85)$$

Cyclopent-2-enone, dimerizes to give a 1:1 ratio (solvent dependent) of *syn* and *anti* dimers in yields up to 90% (equation 86). A triplet was indicated as the reactive intermediate[263].

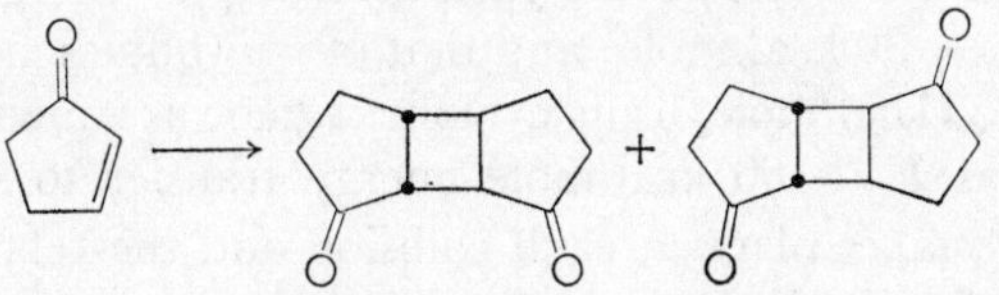

$$\text{[structures]} \qquad (86)$$

The most complete study of a photochemical dimerization reaction was carried out by Hammond and coworkers[264,265] on the dimerization of butadiene and isoprene. Irradiation of concentrated solutions of butadiene in the presence of triplet sensitizers gives a mixture of products containing *cis*- and *trans*-divinylcyclobutane and 4-vinyl-cyclohexene (equation 87). They found that, with both dienes, the

$$\text{butadiene} \xrightarrow[\text{sens}]{h\nu} \text{divinylcyclobutane (cis)} + \text{divinylcyclobutane (trans)} + \text{4-vinylcyclohexene} \qquad (87)$$

ratio of cyclobutane to cyclohexene products varies with the triplet energy of the sensitizer used. When $E_T = 60$ kcal/mole cyclobutane products account for about 95% of the total, but this value drops steadily to about 55% with $E_T = 53$ kcal/mole sensitizers. These results are interpreted on the basis of stereoisomeric triplet states[264,265,266]. Butadiene is an equilibrium mixture of *s-cis* and the predominant *s-trans* conformers. Simple MO theory predicts that the first excited state of a diene should be bonding between $C_{(2)}$ and $C_{(3)}$ and that there should be a distinct barrier to rotation about the $C_{(2)}$—$C_{(3)}$ bond. Franck-Condon excitation of *s-cis*- and *s-trans*-butadiene should then give two stereoisomeric non-interconvertible excited states.

Reaction of the *s-trans*-triplet with ground state *s-trans*-diene would be reasonably expected to form a predominance of cyclobutanes (equation 88), while the *s-cis*-triplet would be more prone to form cyclohexenes (equation 89).

$$\text{s-trans-diene} + \text{s-trans-triplet} \longrightarrow \text{biradical} \longrightarrow \text{cyclobutanes.} \qquad (88)$$

$$\text{s-cis-diene} + \text{s-cis-triplet} \longrightarrow \text{biradical} \longrightarrow \text{cyclohexenes.} \qquad (89)$$

The dependence of product ratios on E_T of the sensitizers, can be rationalized on the basis of the measured triplet energy of *s-trans*-butadiene ($E_T = 60$ kcal/mole) and that of *s-cis*-butadiene (estimated to have $E_T = 53$ kcal/mole, using cyclohexadiene as a model). When a sensitizer has $E_T > 60$ kcal/mole energy transfer to *s-cis*- and *s-trans*-butadiene takes place at each collision and the relative amount of each triplet formed is dependent only on the concentration of each rotomer. As E_T of the sensitizer drops below 60 kcal/mole energy transfer to *s-trans*-butadiene becomes increasingly inefficient, the efficiency of excitation of *s-cis*-butadiene remains unchanged, and the ratio of *cis* triplet to *trans* triplet increases with a concomitant increase in the cyclohexene-cyclobutane product ratio. At $E_T < 53$ kcal/mole energy transfer takes place via non-Franck-Condon excitation (see Section III.A.1).

In contrast to the sensitized dimerization, the unsensitized dimerization of butadiene affords 2-vinylbicyclo[3.1.0]hexene as the major product, in addition to the 'normal' vinylcyclohexene and butane dimers of equation (87)[267].

Cyclohexa-1,3-diene, like 1,3-butadiene, gives cyclobutane and 'Diels-Alder' dimers on sensitized irradiation[268]. This is also a multistep process, but it does not show the same variation in product composition with sensitizer energy, since transoid forms are not possible.

B. Other Photodimers

One of the few examples of the 'allowed' $(4 + 4\pi$ electron) photochemical cycloaddition is the dimerization of 2-pyridones[269,270]

4,6-Dimethyl-2-pyrone (**139**) yields the dimer affording **140–142**,

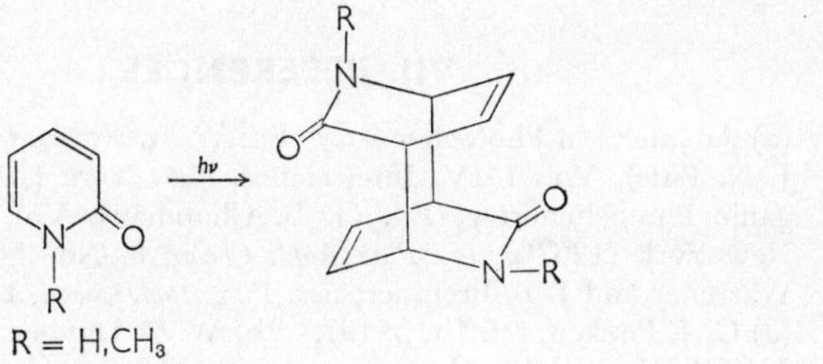

(90)

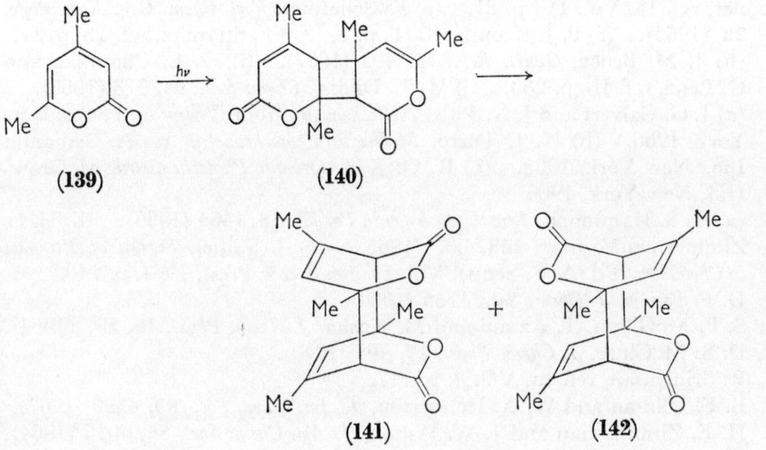

R = H, CH$_3$

(139) (140)

(91)

(141) (142)

but de Mayo[271] has shown that dimer **141** is the product of a Cope rearrangement of **140** and not a direct product of **139**.

Tropone dimerization has been studied by two groups and has been found to form $(6 + 2\ \pi$ electron) (**143**), $(4 + 2\ \pi$ electron), and $(6 + 4\ \pi$ electron) dimers on[272,273] direct irradiation in acetonitrile

(**143**) $\xleftarrow{h\nu}{CH_3CN}$ tropone $\xrightarrow{h\nu}{H_2SO_4}$ (**144**) \hfill (92)

solution. In 2N sulphuric acid only a $(6 + 6\ \pi$ electron) dimer (**144**) is formed[274]. The formation of the $(6 + 4\ \pi$ electron) dimer, which should be photochemically disallowed, has recently been shown to involve triplet intermediates[275].

VII. REFERENCES

1. (a) Advances in Photochemistry (Ed. W. A. Noyes, G. S. Hammond and J. N. Pitts), Vols I–IV, Interscience, New York (1963–1966). (b) Organic Photochemistry (Ed. O. L. Chapman), Vol. I, Marcel Dekker, New York (1967). (c) *Pure Appl. Chem.*, **9**, No. 4 (1964). (d) R. N. Warrener and J. B. Bremner, *Rev. Pure Appl. Chem.*, **16**, 117 (1966).
2. (a) G. J. Fonken, ref. 1b, p. 197. (b) W. G. Dauben, *Chem. Weekblad*, **60**, 381 (1964). (c) M. Mousseron, ref. 1a, Vol. IV, p. 195. (d) K. Schaffner, ref. 1a, Vol. IV, p. 81. (e) K. Schaffner, *Fort. Chem. Org. Naturstoffe*, **22**, (1964). (f) P. J. Kropp, ref. 1b, p. 1. (g) F. Stermitz, ref. 1b, p. 247. (h) J. M. Bruce, *Quart. Rev.*, **21**, 405 (1967). (i) O. L. Chapman and G. Lenz, ref. 1b, p. 283. (j) W. L. Dilling, *Chem. Rev.*, **66**, 373 (1966).
3. (a) J. G. Calvert and J. N. Pitts, *Photochemistry*, John Wiley and Sons, New York, 1966. (b) N. J. Turro, *Molecular Photochemistry*, W. A. Benjamin Inc., New York, 1965. (c) R. O. Kan, *Organic Photochemistry*, McGraw-Hill, New York, 1966.
4. (a) G. S. Hammond, *Kagaku to Kogyo* (*Tokyo*), **18**, 1464 (1965). (b) H. E. Zimmerman, *Science*, **153**, 837 (1966). (c) J. Saltiel, *Survey of Progress in Chemistry* (Ed. A. F. Scott), Vol. II, Academic Press, 1964, p. 240.
5. D. F. Evans, *J. Chem. Soc.*, 1735 (1960).
6. S. P. McGlynn, T. Azumi and M. Kasha, *J. Chem. Phys.*, **40**, 507 (1964).
7. D. S. McClure, *J. Chem. Phys.*, **17**, 905 (1949).
8. R. Srinivasan, ref. 1a, Vol. 4, p. 117.
9. E. F. Ullman and W. A. Henderson, *J. Am. Chem. Soc.*, **89**, 4390 (1967).
10. H. E. Zimmerman and J. W. Wilson, *J. Am. Chem. Soc.*, **86**, 4036 (1964).

11. D. W. Setser, B. S. Rabinovitch, J. W. Simons, *J. Chem. Phys.*, **40**, 1751 (1964).
12. I. Haller and R. Srinivasan, *J. Chem. Phys.*, **42**, 2977 (1965).
13. A. A. Lamola and G. S. Hammond, *J. Chem. Phys.*, **43**, 2129 (1965).
14. J. P. Malrieu, *Photochem. Photobiol.*, **5**, 301 (1966).
15. W. G. Dauben and W. T. Wipke, ref. 1c, p. 539.
16. P. A. Leermakers and F. C. James, *J. Org. Chem.*, **32**, 2898 (1967).
17. G. S. Hammond, J. Saltiel, A. A. Lamola, N. J. Turro, J. S. Bradshaw, D. O. Cowan, R. C. Counsell, V. Vogt and C. Dalton, *J. Am. Chem. Soc.*, **86**, 3197 (1964).
18. (a) H. Nozaki, Y. Nisikawa, M. Kawanisi and R. Noyori, *Tetrahedron*, **23**, 2173 (1967). (b) See also J. K. Crandall and C. F. Mayer, *J. Am. Chem. Soc.*, **89**, 4374 (1967).
19. R. S. Mulliken and C. C. J. Roothaan, *Chem. Rev.*, **41**, 219 (1947).
20. See also R. Hoffmann, *Tetrahedron*, **22**, 521 (1966).
21. C. A. Coulson and E. T. Stewart, *The Chemistry of Alkenes* (Ed. S. Patai) Vol. I, Interscience, New York, 1964, p. 138.
22. W. G. Herkstroeter and G. S. Hammond, *J. Am. Chem. Soc.*, **88**, 4769 (1966).
23. P. A. Leermakers, G. W. Byers, A. A. Lamola and G. S. Hammond, *J. Am. Chem. Soc.*, **85**, 2670 (1963). R. A. Keller and L. J. Dolby, *J. Am. Chem. Soc.*, **89**, 2768 (1967).
24. H. Morrison, *J. Am. Chem. Soc.*, **87**, 932 (1965). See also J. K. Crandall, J. P. Arrington and R. J. Watkins, *Chem. Comm.*, 1052 (1967); Z. J. Barneis, D. M. S. Wheeler and T. H. Kinstle, *Tetrahedron Letters*, 275 (1965).
25. (a) D. Schulte-Frohlinde, H. Blume and H. Güsten, *J. Phys. Chem.*, **66**, 2486 (1962). (b) J. Saltiel, O. C. Zafiriou, E. D. Megarity and A. A. Lamola, *J. Am. Chem. Soc.*, **90**, 4759 (1968).
26. G. N. Lewis, T. T. Magel, D. Lipkin, *J. Am. Chem. Soc.*, **62**, 2973 (1940).
27. L. D. Weis, T. R. Evans and P. A. Leermakers, *J. Am. Chem. Soc.*, **90**, 6109 (1968).
28. P. J. Kropp and H. J. Krauss, *J. Am. Chem. Soc.*, **89**, 5199 (1967); P. J. Kropp, *J. Am. Chem. Soc.*, **89**, 3650 (1967).
29. J. A. Marshall and R. D. Carroll, *J. Am. Chem. Soc.*, **88**, 4092 (1966).
30. See also G. Camaggi and F. Gozzo, *Chem. Comm.*, 236 (1967); J. A. Marshall and A. R. Hochstetler, *Chem. Comm.*, 732 (1967).
31. R. S. H. Liu, *J. Am. Chem. Soc.*, **89**, 112 (1967).
32. K. Mackenzie. *The Chemistry of Alkenes* (Ed. S. Patai), Interscience, New York, 1964, p. 396.
33. F. Asinger, B. Fell and K. Schrage, *Chem. Ber.*, **98**, 372, 381 (1965). But see P. Heimbach, *Angew. Chem. Int. Ed. Engl.*, **5**, 595 (1966).
34. M. D. Carr, V. V. Kane and M. C. Whiting, *Proc. Chem. Soc.*, 408 (1964).
35. C. Moussebois and J. Dale, *J. Chem. Soc.* [C], 260 (1966); see also E. J. Corey and E. Hamanaka, *J. Am. Chem. Soc.*, **89**, 2758 (1967).
36. H. Kristinsson and G. W. Griffin, *J. Am. Chem. Soc.*, **88**, 378 (1966).
37. R. B. Woodward and R. Hoffmann, *J. Am. Chem. Soc.*, **87**, 395 (1965).
38. R. Hoffmann, personal communication.
39. R. B. Woodward and R. Hoffmann, *J. Am. Chem. Soc.*, **87**, 2511 (1965).
40. M. J. S. Dewar, *Tetrahedron*, Supplement No. 8, 75 (1966).

41. (a) K. Fukui and M. Fujimoto, *Tetrahedron Letters*, 251 (1966). (b) H. C. Longuet-Higgins and E. W. Abrahamson, *J. Am. Chem. Soc.*, **87**, 2045 (1965).
42. J. J. Hurst and G. H. Whitham, *J. Chem. Soc.*, 2864 (1960). See also W. F. Erman, *J. Am. Chem. Soc.*, **89**, 3828 (1967).
43. W. F. Erman and H. C. Kretschmar, *J. Am. Chem. Soc.*, **89**, 3842 (1967).
44. H. Kristinsson and G. W. Griffin, *Tetrahedron Letters*, 3259 (1966).
45. K. M. Shumate and G. J. Fonken, *J. Am. Chem. Soc.*, **88**, 1073 (1966).
46. A. P. ter Borg and H. Kloosterziel, *Rec. trav. chim.*, **84**, 241 (1965).
47. R. C. Cookson, V. N. Gogte, J. Hudec and N. A. Mirza, *Tetrahedron Letters*, 3955 (1965).
48. R. F. C. Brown, R. C. Cookson and J. Hudec, *Chem. Comm.*, 823 (1967).
49. O. L. Chapman and W. R. Adams, *J. Am. Chem. Soc.*, **89**, 4243 (1967).
50. W. R. Roth, *Tetrahedron Letters*, 1009 (1964); S. McLean and P. Haynes, *Tetrahedron Letters*, 2385 (1964).
51. (a) A. P. ter Borg, H. Kloosterziel and N. van Meurs, *Rec. trav. chim.*, **82**, 717 (1963). (b) E. Weth and A. S. Dreiding, *Proc. Chem. Soc.*, 59 (1964). (c) D. S. Glass, R. S. Boikess and S. Winstein, *Tetrahedron Letters*, 999 (1966). (d) K. W. Egger, *J. Am. Chem. Soc.*, **89**, 3688 (1967).
52. J. Wolinsky, B. Chollar and M. D. Baird, *J. Am. Chem. Soc.*, **84**, 2775 (1962).
53. K. J. Crowley, *Tetrahedron*, **21**, 1001 (1965).
54. R. Srinivasan, *J. Am. Chem. Soc.*, **83**, 2806 (1961).
55. H. Prinzbach and E. Druckrey, *Tetrahedron Letters*, 2959 (1965).
56. K. J. Crowley, *Proc. Chem. Soc.*, 17 (1964).
57. K. J. Crowley, *J. Am. Chem. Soc.*, **85**, 1210 (1963).
58. N. C. Yang and M. J. Jorgenson, *Tetrahedron Letters*, 1203 (1964).
59. J. N. Pitts and J. K. S. Wan, *The Chemistry of the Carbonyl Group* (Ed. S. Patai) Interscience, New York, 1966, p. 846.
60. K. J. Crowley, R. A. Schneider and J. Meinwald, *J. Chem. Soc.* [C], 571 (1966) and references cited therein.
61. H. D. Munro and O. D. Musgrave, *J. Chem. Soc.* [C], 702 (1967).
62. H. Nozaki, T. Mori and R. Noyori, *Tetrahedron*, **22**, 1207 (1966).
63. M. J. Jorgenson and C. H. Heathcock, *J. Am. Chem. Soc.*, **87**, 5264 (1965); M. J. Jorgenson, *Chem. Comm.*, 137 (1965).
64. H. E. Zimmerman, H. G. Dürr, R. S. Givens and R. G. Lewis, *J. Am. Chem. Soc.*, **89**, 1863 (1967).
65. M. J. Jorgenson and N. C. Yang, *J. Am. Chem. Soc.*, **85**, 1698 (1963).
66. E. Druckrey, M. Arguelles and H. Prinzbach, *Chimia (Aarau)*, **20**, 432 (1966).
67. (a) W. R. Roth, *Angew. Chem.*, **75**, 921 (1963); (b) W. von E. Doering and P. P. Jaspar, *J. Am. Chem. Soc.*, **85**, 3043 (1963); (c) A. P. ter Borg and H. Kloosterziel, *Rec. trav. chim.*, **84**, 241 (1965).
68. R. W. Murray and M. L. Kaplan, *J. Am. Chem. Soc.*, **88**, 3527 (1966).
69. G. W. Borden, O. L. Chapman, R. Swindell and T. Tezuka, *J. Am. Chem. Soc.*, **89**, 2979 (1967); L. B. Jones and V. K. Jones, *J. Am. Chem. Soc.*, **89**, 1880 (1967).
70. A. P. ter Borg, H. Kloosterziel and Y. L. Westphal, *Rec. trav. chim.*, **86**, 474 (1967).

71. R. Srinivasan and F. I. Sonntag, *J. Am. Chem. Soc.*, **87,** 3778 (1965).
72. W. G. Dauben and G. J. Fonken, *J. Am. Chem. Soc.*, **81,** 4060 (1959).
73. G. S. Hammond and N. J. Turro, *Science*, **142,** 1541 (1963).
74. E. E. van Tamelen and S. P. Pappas, *J. Am. Chem. Soc.*, **85,** 3297 (1963).
75. (a) O. L. Chapman, D. J. Pasto, A. A. Griswold and G. W. Borden, *J. Am. Chem. Soc.*, **84,** 1220 (1962). (b) W. G. Dauben, K. Koch, O. L. Chapman and S. L. Smith, *J. Am. Chem. Soc.*, **83,** 1768 (1961); O. L. Chapman and J. D. Lassila, *J. Am. Chem. Soc.*, **90,** 2449 (1968). For variations and other references see also T. Mukai and T. Miyashi, *Tetrahedron*, **23,** 1613 (1967).
76. J. A. Elix, M. V. Sargent, F. Sondheimer, *J. Am. Chem. Soc.*, **89,** 180 (1967).
77. E. J. Corey and J. Streith, *J. Am. Chem. Soc.*, **86,** 950 (1964).
78. L. A. Paquette, J. H. Barrett, R. P. Spitz and R. Pitcher, *J. Am. Chem. Soc.*, **87,** 3417 (1965).
79. W. J. Theuer and J. A. Moore, *Chem. Comm.*, 468 (1965).
80. R. Srinivasan, *J. Am. Chem. Soc.*, **84,** 4141 (1962).
81. G. S. Hammond, N. J. Turro and R. S. H. Liu, *J. Org. Chem.*, **28,** 3297 (1963).
81a. Mentioned in G. O. Schenck and R. Steinmetz, *Bull. Soc. Chim. Belges*, **71,** 785 (1962).
82. W. G. Dauben, R. L. Cargill, R. M. Coates and J. Saltiel, *J. Am. Chem. Soc.*, **88,** 2742 (1966); W. G. Dauben, *Reactivity of the Photoexcited Organic Molecule*, Solvay Institute 13th Chemistry Conference (Ed. R. Defay), Interscience, New York, 1967, p. 171.
83. W. G. Dauben and C. D. Poulter, *Tetrahedron Letters*, 3021 (1967).
84. P. G. Gassman and W. E. Hymans, *Chem. Comm.*, 795 (1967); P. G. Gassman, *Chem. Comm.*, 793 (1967).
85. K. E. Wilzbach and L. Kaplan, *J. Am. Chem. Soc.*, **87,** 4004 (1965).
86. K. E. Wilzbach, J. S. Ritscher and L. Kaplan, *J. Am. Chem. Soc.*, **89,** 1031 (1967).
87. H. J. F. Angus, J. M. Blair and D. Bryce-Smith, *J. Chem. Soc.*, 2003 (1960).
87a. L. Kaplan, S. P. Walch and K. E. Wilzbach, *J. Am. Chem. Soc.*, **90,** 5646 (1968).
88. D. Bryce-Smith, A. Gilbert and H. C. Longuet-Higgins, *Chem. Comm.*, 240 (1967).
89. L. Kaplan, J. S. Ritscher and K. E. Wilzbach, *J. Am. Chem. Soc.*, **88,** 2881 (1966).
90. R. J. de Kock, N. G. Minaard and E. Havinga, *Rec. trav. chim.*, **79,** 922 (1960).
91. R. Srinivasan, *J. Am. Chem. Soc.*, **82,** 5063 (1960).
92. W. G. Dauben and R. M. Coates, *J. Org. Chem.*, **29,** 2761 (1964).
93. E. Vogel, W. Grimme and E. Dinne, *Tetrahedron Letters*, 391 (1965).
94. P. Courtot and J.-M. Robert, *Bull. Soc. Chim. France*, 3362 (1965).
95. G. Schroder, W. Martin and J. F. M. Oth, *Angew. Chem. Int. Ed. Engl.*, **6,** 870 (1967).
96. K. R. Huffman, M. Loy, W. A. Henderson and E. F. Ullman, *Tetrahedron Letters*, 931 (1967).
97. E. J. Corey and A. G. Hortmann, *J. Am. Chem. Soc.*, **87,** 5736 (1965).

98. L. Skattebøl, J. L. Charlton and P. de Mayo, *Tetrahedron Letters*, 2257 (1966).
99. (a) E. Henmo, P. de Mayo, A.B.M.A. Sattar and A. Stoessl, *Proc. Chem. Soc.*, 238 (1961). (b) G. R. Lenz and N. C. Yang, *Chem. Comm.*, 1136 (1967). (c) R. S. Becker and J. Michl, *J. Am. Chem. Soc.*, **88**, 5931 (1966).
100. For more recent leading references see K. A. Muszkat and E. Fischer, *J. Chem. Soc.* (B), 662 (1967).
101. (a) B. C. Roquitte, *J. Am. Chem. Soc.*, **85**, 3700 (1963). (b) G. S. Hammond, P. Wyatt, C. D. DeBoer and N. J. Turro, *J. Am. Chem. Soc.*, **86**, 2532 (1964).
102. S. J. Cristol and R. L. Snell, *J. Am. Chem. Soc.*, **76**, 5000 (1954).
103. D. Seyferth and A. B. Evnin, *J. Am. Chem. Soc.*, **89**, 1468 (1967); and leading references in J. R. Edman, *J. Am. Chem. Soc.*, **88**, 3454 (1966) and ref. 2j.
104. E. Payo, L. Cortes, J. Mantecon, C. Rivas and G. de Pinto, *Tetrahedron Letters*, 2415 (1967). H. Prinzbach, M. Arguëlles and E. Druckrey, *Angew. Chem. Int. Ed. Engl.*, **5**, 1039 (1966).
105. R. Srinivasan, quoted in reference 2j.
106. C. F. Wilcox, S. Winstein, and W. G. McMillan, *J. Am. Chem. Soc.*, **82**, 5450 (1960); R. B. Hermann, *J. Org. Chem.*, **27**, 441 (1962).
107. B. C. Roquitte, *J. Phys. Chem.*, **69**, 2475 (1965).
108. D. M. Lemal and J. P. Lokensgard, *J. Am. Chem. Soc.*, **88**, 5934 (1966); W. Schäfer, R. Criegee, R. Askani and H. Grüner, *Angew. Chem. Intern. Ed.*, **6**, 78 (1967).
109. P. K. Freeman, D. G. Kuper, V. N. M. Rao, *Tetrahedron Letters*, 3301 (1965); cf. H. Prinzbach, W. Eberbach and G. von Veh, *Angew. Chem. Int. Ed. Engl.*, **4**, 436 (1965).
110. P. J. Kropp. *J. Am. Chem. Soc.*, **89**, 1126 (1967).
111. H. Prinzbach, H. Hagemann, J. H. Hartenstein and R. Kitzing. *Chem. Ber.*, **98**, 2201 (1965).
112. J. Weimann, N. Thoai and F. Weisbuch, *Bull. Soc. Chim. France*, 575 (1966).
113. W. G. Dauben, I. Bell, T. W. Hutton, G. F. Laws, A. Rheiner and H. Urscheler, *J. Am. Chem. Soc.*, **80**, 4116 (1958).
114. J. Meinwald and P. H. Mazzocchi, *J. Am. Chem. Soc.*, **89**, 1755 (1967).
115. (a) M. Pomerantz, *J. Am. Chem. Soc.*, **89**, 694 (1967). (b) J. Meinwald and P. H. Mazzocchi, *J. Am. Chem. Soc.*, **89**, 696 (1967).
116. K. J. Crowley, *Tetrahedron Letters*, 2863 (1965).
117. K. J. Crowley, K. Erickson, A. Eckell, and J. Meinwald, manuscript in preparation.
118. W. G. Dauben, personal communication.
119. K. J. Crowley, *J. Am. Chem. Soc.*, **86**, 5692 (1964).
120. J. Meinwald, A. Eckell and K. L. Erickson, *J. Am. Chem. Soc.*, **87**, 3532 (1965).
121. J. Meinwald and P. H. Mazzocchi, *J. Am. Chem. Soc.*, **88**, 2850 (1966).
122. J. Meinwald and P. H. Mazzocchi, unpublished results.
123. (a) W. von E. Doering and J. W. Rosenthal, *J. Am. Chem. Soc.*, **88**, 2078 (1966); (b) W. von E. Doering and J. W. Rosenthal, *Tetrahedron Letters*, 349 (1967).

124. M. Jones and L. T. Scott, *J. Am. Chem. Soc.*, **89,** 150 (1967); M. Jones, *J. Am. Chem. Soc.*, **89,** 4236 (1967).
125. E. E. van Tamelen and T. L. Burkoth, *J. Am. Chem. Soc.*, **89,** 151 (1967).
126. P. Radlick, W. Fenical, cited in footnote 4, ref. 123 (b).
127. E. E. van Tamelen and B. Pappas, *J. Am. Chem. Soc.*, **85,** 3296 (1963).
128. G. Schroder, *Angew. Chem.*, **75,** 722 (1963).
129. M. Avram, E. Sliam and C. D. Nenitzescu, *Ann. Chem.*, **636,** 184 (1960).
130. S. Masamune, C. G. Chin, K. Hojo and R. T. Seidner, *J. Am. Chem. Soc.*, **89,** 4804 (1967).
130a. S. Masamune, R. T. Seidner, H. Zenda, M. Wiesel, N. Nakatsuka and G. Bigam, *J. Am. Chem. Soc.*, **90,** 5286 (1968).
131. H. E. Zimmerman and G. L. Grunewald, *J. Am. Chem. Soc.*, **88,** 183 (1966).
132. H. E. Zimmerman, R. W. Binkley, R. S. Givens and M. A. Sherwin, *J. Am. Chem. Soc.*, **89,** 3932 (1967); H. E. Zimmerman, R. S. Givens and R. M. Pagni, *J. Am. Chem. Soc.*, **90,** 6096 (1968).
133. J. P. N. Brewer and H. Heaney, *Chem. Comm.*, 811 (1967).
134. D. Bryce-Smith and J. E. Lodge, *J. Chem. Soc.*, 695 (1963); E. Grovenstein and D. V. Rao, *Tetrahedron Letters*, 148 (1961).
135. J. G. Atkinson, D. E. Ayer, G. Buchi and E. W. Robb, *J. Am. Chem. Soc.*, **85,** 2257 (1963).
136. H. E. Zimmerman and H. Iwamura, *J. Am. Chem. Soc.*, **90,** 4763 (1968).
137. J. Zirner and S. Winstein, *Proc. Chem. Soc.*, 235 (1964).
138. O. L. Chapman, G. W. Borden, R. W. King and B. Winkler, *J. Am. Chem. Soc.*, **86,** 2660 (1964).
139. W. R. Roth and B. Peltzer, *Ann. Chem.*, **685,** 56 (1965).
140. E. Wiskott and P. von R. Schleyer, *Angew. Chem. Int. Ed. Engl.*, **6,** 694 (1967).
141. R. Srinivasan and K. H. Carlough, *J. Am. Chem. Soc.*, **89,** 4932 (1967).
142. J. Meinwald and G. W. Smith, *J. Am. Chem. Soc.*, **89,** 4923 (1967).
143. R. Srinivasan, *J. Phys. Chem.*, **67,** 1367 (1963).
144. R. Srinivasan and F. I. Sonntag, *J. Am. Chem. Soc.*, **89,** 407 (1967).
145. R. Srinivasan and K. A. Hill, *J. Am. Chem. Soc.*, **87,** 4988 (1965).
146. R. Srinivasan, *J. Am. Chem. Soc.*, **86,** 3318 (1964).
147. J. E. Baldwin and R. H. Greeley, *J. Am. Chem. Soc.*, **87,** 4514 (1965).
148. I. Haller and R. Srinivasan, *J. Am. Chem. Soc.*, **88,** 5084 (1966).
149. R. S. H. Liu and G. S. Hammond, *J. Am. Chem. Soc.*, **89,** 4936 (1967).
150. K. J. Crowley, *Proc. Chem. Soc.*, 245, 334 (1962).
151. W. G. Dauben and R. L. Cargill, unpublished results quoted in reference 15.
152. E. Havinga and J. L. M. A. Schlatmann, *Tetrahedron*, **16,** 146 (1961).
153. J. L. Charlton, P. de Mayo and L. Skattebøl, *Tetrahedron Letters*, 4679 (1965).
154. R. S. H. Liu, *Tetrahedron Letters*, 2159 (1966).
155. F. T. Bond, H. L. Jones and L. Scerbo, *Tetrahedron Letters*, 4685 (1965).
156. R. T. LaLonde and R. I. Aksentijevich, *Tetrahedron Letters*, 23 (1965).
157. G. Ciamician and P. Silber, *Chem. Ber.*, **41,** 1928 (1908).
158. G. Buchi and I. M. Goldman, *J. Am. Chem. Soc.*, **79,** 4741 (1957).
159. J. Meinwald and R. A. Schneider, *J. Am. Chem. Soc.*, **87,** 5218 (1965).
160. M. Brown, *Chem. Comm.*, 340 (1965); *J. Org. Chem.*, **33,** 162 (1968).

161. E. J. Corey and K. Sestanj, as cited in footnote 28, cf. reference 2(j); for a brief discussion of factors governing ring formation see reference 149.
162. R. C. Cookson, J. Hudec, S. A. Knight and B. R. D. Whitear, *Tetrahedron* **19**, 1995 (1963).
163. P. E. Eaton and T. W. Cole, *J. Am. Chem. Soc.*, **86**, 3157 (1964).
164. J. C. Barborak, L. Watts and R. Pettit, *J. Am. Chem. Soc.*, **88**, 1328 (1966).
165. G. L. Dunn, V. J. DiPasquo and J. R. E. Hoover, *Tetrahedron Letters*, 3737 (1966).
166. R. C. Cookson, E. Crundwell, R. R. Hill and J. Hudec, *J. Chem. Soc.*, 3062 (1964).
167. E. Baggiolini, E. G. Herzog, S. Iwasaki, R. Schorta and K. Schaffner, *Helv. Chim. Acta*, **50**, 297 (1967).
168. G. O. Schenck, J. Kuhls and C. H. Krauch, *Bull. Soc. Chim. Belges*, **71**, 781 (1962).
169. G. O. Schenck and R. Steinmetz, *Chem. Ber.*, **96**, 520 (1963).
170. S. Masamune, H. Cuts and M. G. Hogben, *Tetrahedron Letters*, 1017 (1966).
171. W. G. Dauben and D. L. Whalen, *Tetrahedron Letters*, 3743 (1966).
172. R. Furstoss and J.-M. Lehn, *Bull. Soc. Chim. France*, 2497 (1966).
173. C. W. Bird, R. C. Cookson and E. Crundwell, *J. Chem. Soc.*, 4809 (1961).
174. J. D. Rosen, *Chem. Comm.*, 189 (1967).
175. R. J. Stedman, L. S. Miller and J. R. E. Hoover, *Tetrahedron Letters*, 2721 (1966).
176. R. J. Stedman and L. S. Miller, *J. Org. Chem.*, **32**, 35 (1967).
177. R. Srinivasan, *J. Am. Chem. Soc.*, **82**, 775 (1960).
178. H. Morrison, *J. Am. Chem. Soc.*, **87**, 932 (1965).
179. N. C. Yang, M. Nussim and D. R. Coulson, *Tetrahedron Letters*, 1525 (1965).
180. See ref. 59.
181. W. M. Horspool and P. L. Pauson, *Chem. Comm.*, 195 (1967).
182. G. Frater and H. Schmid, *Helv. Chim. Acta*, **50**, 255 (1967).
183. K. Gollnick, S. Schroeter, G. Ohloff, G. Schade and G. O. Schenck, *Ann. Chem.*, **687**, 14 (1965).
184. L. H. Gale, *J. Am. Chem. Soc.*, **88**, 4661 (1966).
185. H.-G. Viehe, *Chem. Ber.*, **97**, 598 (1964).
186. P. S. Skell and R. R. Pavlis, *J. Am. Chem. Soc.*, **86**, 2956 (1964).
187. R. H. Young and H. Hart, *Chem. Comm.*, 827 (1967).
188. H.-D. Scharf and F. Korte, *Chem. Ber.*, **97**, 2425 (1964).
189. H.-D. Scharf and G. Weisgerber, *Tetrahedron Letters*, 1567 (1967).
190. R. Srinivasan and K. A. Hill, *J. Am. Chem. Soc.*, **88**, 3765 (1966).
191. E. H. Gold and D. Ginsberg, *Angew. Chem. Int. Ed. Engl.*, **5**, 246 (1966).
192. J. Rokach and D. Elad, *J. Org. Chem.*, **31**, 4210 (1966) and earlier papers.
193. M. Fisch and G. Ourisson, *Bull. Soc. Chim. France*, 1325 (1966).
194. Y. L. Chow, *J. Am. Chem. Soc.*, **87**, 4642 (1965).
195. Y. L. Chow, C. Colon and S. C. Chen, *J. Org. Chem.*, **32**, 2109 (1967).
196. N. Kharasch, P. Lewis and R. K. Sharma, *Chem. Comm.*, 435 (1967).
197. G. Bauslaugh, G. Just and E. Lee-Ruff. *Can. J. Chem.*, **44**, 2837 (1966).
198. J. Pusset and R. Beugelmans, *Tetrahedron Letters*, 3249 (1967).
199. B. J. Ramey and P. D. Gardner, *J. Am. Chem. Soc.*, **89**, 3949 (1967).
200. P. Bladon and I. A. Williams, *J. Chem. Soc.* (C), 2032 (1967).
201. E. J. Corey, R. B. Mitra and H. Uda, *J. Am. Chem. Soc.*, **86**, 485 (1964).

202. E. J. Corey, J. D. Bass, R. La Mahieu and R. B. Mitra, *J. Am. Chem. Soc.*, **86**, 5570 (1964).
203. For other examples of additions to α,β-unsaturated ketones see: (a) R. L. Cargill, J. R. Damewood and M. M. Cooper, *J. Am. Chem. Soc.*, **88**, 1330 (1966); (b) R. L. Cargill and J. W. Crawford, *Tetrahedron Letters*, 169 (1967); (c) E. J. Corey and S. Nozoe, *J. Am. Chem. Soc.*, **87**, 5733 (1965); (d) Y. Yamada, H. Uda and K. Nakanishi, *Chem. Comm.*, 423 (1966); (e) P. E. Eaton, *Tetrahedron Letters*, 3695 (1964); (f) K.Wiesner, I. Jirkovsky, M. Fishman and C. A. J. Williams, *Tetrahedron Letters*, 1523 (1967) and earlier papers.
204. P. E. Eaton and K. Lin, *J. Am. Chem. Soc.*, **86**, 2087 (1964).
205. E. J. Corey, M. Tada, R. La Mahieu, and L. Libit, *J. Am. Chem. Soc.*, **87**, 2051 (1965).
206. P. E. Eaton and K. Lin, *J. Am. Chem. Soc.*, **87**, 2052 (1965).
207. H. Hikino and P. de Mayo, *J. Am. Chem. Soc.*, **86**, 3582 (1964).
208. P. Sunder-Plassmann, J. Zderic and J. H. Fried, *Tetrahedron Letters*, 3451 (1966).
209. P. H. Nelson, J. W. Murphy, J. A. Edwards and J. H. Fried, *J. Am. Chem. Soc.*, **90**, 1307 (1968).
210. P. Sunder-Plassmann, P. H. Nelson, L. Durham, J. A. Edwards, and J. H. Fried, *Tetrahedron Letters*, 653 (1967).
211. R. Criegee and H. Furrer, *Chem. Ber.*, **97**, 2949 (1964).
212. R. L. Cargill, M. E. Beckham, A. E. Siebert and J. Dorn, *J. Org. Chem.*, **30**, 3647 (1965).
213. P. de Mayo, S. T. Reid and R. W. Yip. *Can. J. Chem.*, **42**, 2828 (1964).
214. J. A. Barltrop and R. Robson, *Tetrahedron Letters*, 597 (1963).
215. A. Cox, P. de Mayo and R. W. Yip, *J. Am. Chem. Soc.*, **88**, 1043 (1966).
216. R. Criegee, U. Zirngibl, H. Furrer, D. Seebach and G. Freund, *Chem. Ber.*, **97**, 2942 (1964).
217. D. Seebach, *Chem. Ber.*, **97**, 2953 (1964).
218. R. Askani, *Chem. Ber.*, **98**, 2322, 3618 (1965).
219. G. Koltzenburg, P. G. Fuss and J. Leitich, *Tetrahedron Letters*, 3409 (1966).
220. E. Paterno and G. Chieffi, *Gazz. Chim. Ital.*, **39**, 341 (1909).
221. G. Buchi, C. G. Inman and E. S. Lipinsky, *J. Am. Chem. Soc.*, **76**, 4327 (1954).
222. D. R. Arnold, R. L. Hinman and A. H. Glick, *Tetrahedron Letters*, 1425 (1964).
223. Y. Shigemitsu, Y. Odaira and S. Tsutsumi, *Tetrahedron Letters*, 55 (1967).
224. D. Bryce-Smith and A. Gilbert, *Proc. Chem. Soc.*, 87 (1964).
225. J. Saltiel, R. M. Coates and W. G. Dauben, *J. Am. Chem. Soc.*, **88**, 2745 (1966).
226. G. O. Schenck, W. Hartmann and R. Steinmetz, *Chem. Ber.*, **96**, 498 (1963).
227. J. J. Beereboom and M. S. von Wittenau, *J. Org. Chem.*, **30**, 1231 (1965).
228. C. Rivas and E. Payo, *J. Org. Chem.*, **32**, 2918 (1967).
229. J. Leitich, *Tetrahedron Letters*, 1937 (1967).
230. G. Buchi, J. T. Kofron, E. Koller and D. Rosenthal, *J. Am. Chem. Soc.*, **78**, 876 (1956).
231. D. Bryce-Smith, G. I. Fray and A. Gilbert, *Tetrahedron Letters*, 2137 (1964).

232. H. E. Zimmerman and L. Craft, *Tetrahedron Letters*, 2131 (1964).
233. N. C. Yang, M. Nussim, M. J. Jorgenson and S. Murov, *Tetrahedron Letters*, 3657 (1964).
234. N. C. Yang, R. Loeschen and D. Mitchell, *J. Am. Chem. Soc.*, **89**, 5465 (1967).
235. N. J. Turro, P. Wriede, J. C. Dalton, D. R. Arnold and A. H. Glick, *J. Am. Chem. Soc.*, **89**, 3950 (1967).
236. D. Bryce-Smith and J. E. Lodge, *J. Chem. Soc.*, 2675 (1962).
237. E. Grovenstein, D. V. Rao and J. W. Taylor, *J. Am. Chem. Soc.*, **83**, 1705 (1961).
238. G. O. Schenck and R. Steinmetz, *Tetrahedron Letters*, No. 21, 1 (1960).
239. W. M. Hardhan and G. S. Hammond, *J. Am. Chem. Soc.*, **89**, 3200 (1967).
240. D. Bryce-Smith and A. Gilbert, *J. Chem. Soc.*, 918 (1965).
241. J. S. Bradshaw, *J. Org. Chem.*, **31**, 3974 (1966).
242. G. B. Vermont, P. X. Riccobono and J. Blake, *J. Am. Chem. Soc.*, **87**, 4024 (1965).
243. J. S. Bradshaw, *Tetrahedron Letters*, 2039 (1966).
244. D. Bryce-Smith and M. A. Hems, *Tetrahedron Letters*, 1895 (1966).
245. N. C. Perrins and J. P. Simons, *Chem. Comm.*, 999 (1967).
246. R. Srinivasan and K. A. Hill, *J. Am. Chem. Soc.*, **87**, 4653 (1965).
247. D. Bryce-Smith, *Pure and Appl. Chem.*, **16**, 47 (1968).
248. K. E. Wilzbach and L. Kaplan, *J. Am. Chem. Soc.*, **88**, 2066 (1966).
249. D. Bryce-Smith, A. Gilbert and B. H. Orger, *Chem. Comm.*, 512 (1966).
250. J. G. Atkinson, D. E. Ayer, G. Buchi and E. W. Robb, *J. Am. Chem. Soc.* **85**, 2257 (1963).
251. E. Grovenstein, Jr. and D. V. Rao, *Tetrahedron Letters*, 148 (1961).
252. D. Bryce-Smith and J. E. Lodge, *J. Chem. Soc.*, 695 (1963).
253. J. P. Chesick, *J. Am. Chem. Soc.*, **85**, 3718 (1963).
254. D. R. Arnold and V. Y. Abraitys, *Chem. Comm.*, 1053 (1967).
255. H.-D. Scharf and F. Korte, *Chem. Ber.*, **97**, 2425 (1964).
256. G. S. Hammond, N. J. Turro and R. S. H. Liu, *J. Org. Chem.*, **28**, 3297 (1963).
257. E. H. Gold and D. Ginsberg, *Angew. Chem. Int. Ed. Engl.*, **5**, 246 (1965).
258. D. J. Trecker, R. S. Foote, J. P. Henry and J. E. McKeon, *J. Am. Chem. Soc.*, **88**, 3021 (1966).
259. H. Stechl, *Chem. Ber.*, **97**, 2681 (1964).
260. N. Obata and I. Moritani, *Tetrahedron Letters*, 1503 (1966).
261. Several articles reviewing this field are published in *Photochem. Photobiol.*, **7**, 511–835 (1968) (No. 6, June 1968).
262. E. H. White and J. P. Anhalt, *Tetrahedron Letters*, 3937 (1965).
263. P. E. Eaton and W. S. Hurt, *J. Am. Chem. Soc.*, **88**, 5038 (1966); J. L. Ruhlen and P. A. Leermakers, *J. Am. Chem. Soc.*, **89**, 4944 (1967).
264. G. S. Hammond, N. J. Turro, A. Fischer, *J. Am. Chem. Soc.*, **83**, 4674 (1961).
265. G. S. Hammond and R. S. H. Liu, *J. Am. Chem. Soc.*, **85**, 477 (1963).
266. R. S. H. Liu, N. J. Turro and G. S. Hammond, *J. Am. Chem. Soc.*, **87**, 3406 (1965).
267. R. Srinivasan and F. I. Sonntag, *J. Am. Chem. Soc.*, **87**, 3778 (1965).

268. D. Valentine, N. J. Turro and G. S. Hammond, *J. Am. Chem. Soc.*, **86,** 5202 (1964).
269. L. A. Paquette and G. Slomp, *J. Am. Chem. Soc.*, **85,** 765 (1963).
270. E. C. Taylor and R. O. Kan, *J. Am. Chem. Soc.*, **85,** 776 (1963).
271. P. de Mayo and R. W. Yip, *Proc. Chem. Soc.*, 84 (1964).
272. A. S. Kende, *J. Am. Chem. Soc.*, **88,** 5026 (1966).
273. T. Tezuka, Y. Akasaki and T. Mukai, *Tetrahedron Letters*, 1397 (1967).
274. T. Mukai, T. Tezuka and Y. Akasaki, *J. Am. Chem. Soc.*, **88,** 5025 (1966).
275. A. S. Kende and J. E. Lancaster, *J. Am. Chem. Soc.*, **89,** 5283 (1967).

CHAPTER 7

The Mass Spectrometry of the Double Bond

A. G. LOUDON and ALLAN MACCOLL

University College, London, England

I. Introduction	.	327
II. Ionization and Appearance Potentials	.	337
III. Location of Double Bonds by Mass Spectrometry	.	342
IV. The Retro-Diels-Alder Reaction	.	347
V. The Expulsion of Alkyl Radicals from Cyclic Olefins	.	353
VI. References	.	357

I. INTRODUCTION

When an electron of energy between 5 and 100 eV and greater than the ionization potential collides with a molecule in the source of a mass spectrometer, the molecular ion either in its ground state or one of its excited electronic states is produced. Since an electron has been lost, this process may be represented as

$$e + M \longrightarrow M^{+\cdot} + 2e$$

where the · signifies the odd electron in the ion. Evidence is now accumulating to the effect that the charge may be regarded as localized in the molecule[1-6], especially where a heteroatom is present. Thus in n-propylamine, Shannon[3] explains the formation of m/e 30 by process (1).

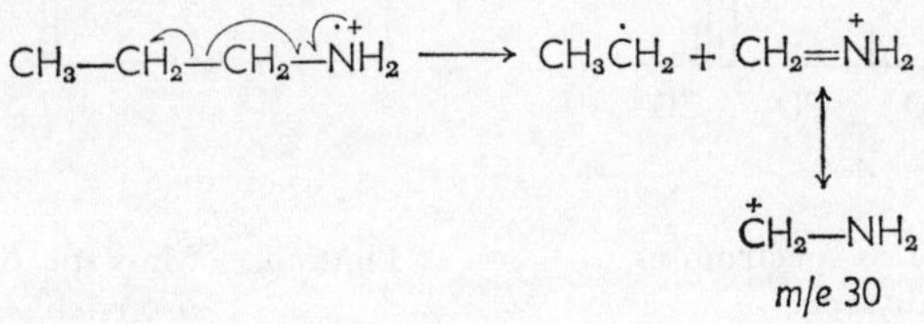

$$CH_3-CH_2-CH_2-\overset{+}{N}H_2 \longrightarrow CH_3\dot{C}H_2 + CH_2=\overset{+}{N}H_2 \quad (1)$$

$$\downarrow$$

$$\overset{+}{C}H_2-NH_2$$
$$m/e\ 30$$

The fish-hook ($\curvearrowright$) is used[2,3] to represent one-electron transfer, the curved arrow ($\curvearrowright$) represents a two-electron transfer as used in ground-state physical organic chemistry[7]. McLafferty[8] has quoted the production of m/e 15 and 45 in the mass spectrum of methyl ethyl ether as examples of the two types of processes (2,3). In the case of the double

$$\text{CH}_3\text{—CH}_2\text{—}\overset{+}{\text{O}}\text{—CH}_3 \longrightarrow \text{CH}_3\cdot + \text{CH}_2\text{=}\overset{+}{\text{O}}\text{—CH}_3 \qquad (2)$$
$$m/e\ 45$$

$$\text{CH}_3\text{—CH}_2\text{—}\overset{\cdot+}{\text{O}}\text{—CH}_3 \longrightarrow \text{CH}_3\text{—CH}_2\text{—O}\cdot + \text{CH}_3^+ \qquad (3)$$
$$m/e\ 15$$

bond, the representations of the molecular ion are either **a** or **b**. The latter will be seen to be important when we come to consider double-bond migration in molecular ions.

$$[\text{CH}_3\text{—CH=CH—CH}_3]^{\cdot+} \qquad \text{CH}_3\text{—}\overset{+}{\text{CH}}\text{—}\overset{\cdot}{\text{CH}}\text{—CH}_3$$
$$\textbf{(a)} \qquad\qquad\qquad \textbf{(b)}$$

Figures 1–4 show the mass spectra of the straight-chain terminal olefins up to 1-pentene. Figures 5 and 6 show the effect of branching at the 4-position[9]. In all the molecules past propylene, fission β to the double bond plays an important role (Table 1). The modes of fragmentation (4) and (5) are thus of considerable importance.

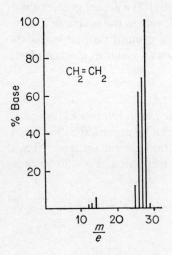

Figure 1. Mass spectrum of ethylene.

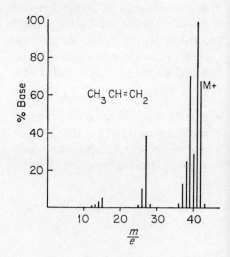

Figure 2. Mass spectrum of propylene.

7. The Mass Spectrometry of the Double Bond

TABLE 1. Fragmentation β to the double bond[a]

Molecule	F_u		F_s	
	m/e	% base	m/e	% base
1-Butene	41	100	15	4
1-Pentene	41	45	29	28
4-Methyl-1-pentene	41	72	43	100
4,4-Dimethyl-1-pentene	41	53	57	100
2-Hexene	55	100	29	27
2-Heptene	55	100	43	22

[a] F_u and F_s represent the unsaturated and saturated ions produced by β-fission.

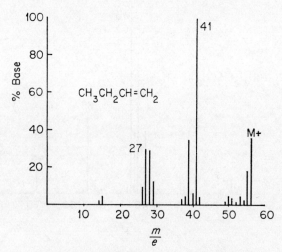

Figure 3. Mass spectrum of 1-butene.

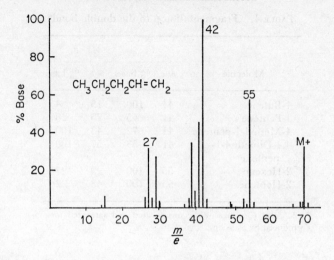

Figure 4. Mass spectrum of 1-pentene.

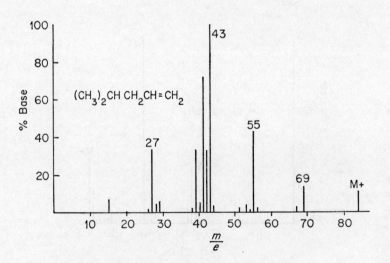

Figure 5. Mass spectrum of 4-methyl-1-pentene.

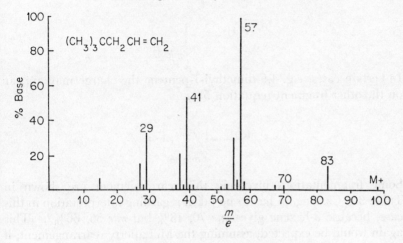

Figure 6. Mass spectrum of 4,4-dimethyl-1-pentene.

While the picture presented works reasonably well for the terminal olefins, difficulties arise when the other isomers are considered. Table 2 shows the partial mass spectra of the n-hexene isomers. It would appear that 3-hexene possibly undergoes extensive isomerization prior to fragmentation. The others must as well, as evidenced by m/e 55 in 1-hexene. This assumes that allylic fission is the main mode of fragmentation.

The case of 1-pentene (Figure 4) is of great interest, since the base peak now moves to the odd electron ion, m/e 42. This may be formulated as a McLafferty rearrangement[10] (equation 5a).

TABLE 2. Abundance (% base) of various ions in the mass spectra of n-hexene isomers[9,a]

Molecule m/e:	69	15	55	29	41	43
1-Hexene	19	7	59	28	100	59
2-Hexene	19	5	100	27	41	12
3-Hexene	26	10	100	25	77	16

[a] The peaks underlined are those that would be expected on the basis of β-fission.

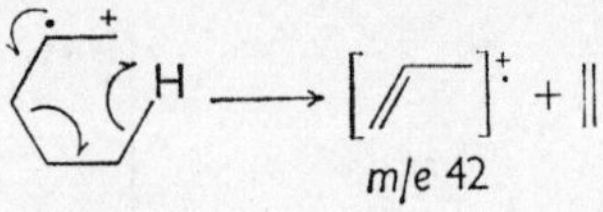

$$\text{(cyclohexyl-H)}^{+\cdot} \longrightarrow [\!/\!/\,]^{+\cdot}_{m/e\ 42} + \|\qquad(5a)$$

In certain cases, e.g. 4,5-dimethyl-1-pentene the charge may appear on the other fragment (equation 5b).

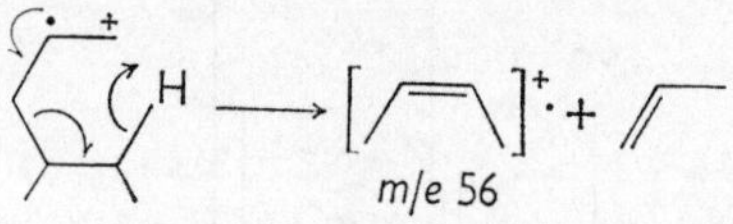

$$\text{(cyclohexyl-H)}^{+\cdot} \longrightarrow [\!/\!\diagdown\!]^{+\cdot}_{m/e\ 56} + /\!/\qquad(5b)$$

Some fragmentations involving this rearrangement are shown in Table 3. Care again has to be taken regarding interpretation in this case, because 3-hexene gives m/e 70, 13% but m/e 56, 66%[9]. This again would be expected, assuming the McLafferty rearrangement, if there was isomerization prior to fragmentation.

The last route to be discussed is observed in cyclic olefins, the retro-Diels-Alder fragmentation[11] (equation 6). This occurs to the extent

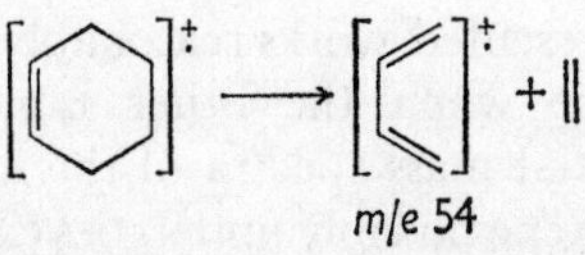

$$[\bigcirc]^{+\cdot} \longrightarrow [\square]^{+\cdot}_{m/e\ 54} + \|\qquad(6)$$

of 76% in cyclohexene—it is the second most abundant peak[9]. In 1-methylcyclohexene the abundance of m/e 68 is 45%, but there is also a peak at m/e 54 (20%). In the case of 1,2-dimethylcyclohexene the abundances of m/e 82, 68 and 54 are 25%, 40% and 14% respectively[9]. Again it would appear that assuming the retro-Diels-Alder reaction, isomerization of the double bond occurs prior to fragmentation.

TABLE 3. The McLafferty Rearrangement

Molecule	m/e	% base	Process
1-Pentene	42	100	5a
1-Hexene	42	75	5a
5-Methyl-1-hexene	56	100	5b
4-Methyl-1-hexene	56	82	5b
3-Methyl-1-hexene	56	64	5a
2-Methyl-1-hexene	56	100	5a
2-Heptene	56	94	5a

7. The Mass Spectrometry of the Double Bond

We have seen that a fundamental problem behind the interpretation of the mass spectra of the unconjugated olefinic bond is the possibility that an electron-impact induced migration occurs before fragmentation takes place (equation 7a). The first stage of electron impact is considered to be the removal of a π-electron and subsequent migrations are accomplished by 1,3 hydrogen migration[12] (equation 7a). However, there is also evidence that 1,2 hydrogen shifts occur resulting in the separation, in the case of a straight-chain olefin, of the two carbon atoms carrying one hydrogen only[12,13,14] (equation 7b). The location of the charge is then a matter of conjecture.

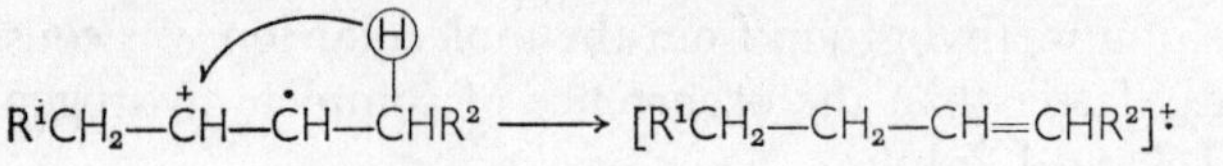

$$R^1CH_2-\overset{+}{C}H-\overset{\cdot}{C}H-CHR^2 \longrightarrow [R^1CH_2-CH_2-CH=CHR^2]^{\ddagger} \qquad (7a)$$

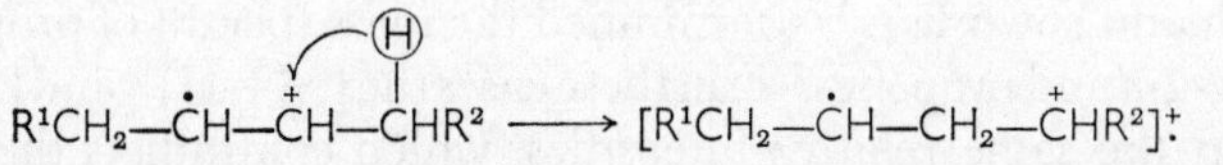

$$R^1CH_2-\overset{\cdot}{C}H-\overset{+}{C}H-CHR^2 \longrightarrow [R^1CH_2-\overset{\cdot}{C}H-CH_2-\overset{+}{C}HR^2]^{\ddagger} \qquad (7b)$$

The implications of the existence of these processes, as far as the interpretation of the mass spectra fragmentation patterns and physical data such as ionization and appearance potential measurements are concerned, are considerable. So before discussing these points in detail, the nature and extent of these rearrangement processes requires examination.

Such migrations were postulated by many early workers[15,16], based mainly on the observations of many similarities between the mass spectra of double-bond isomers. McLafferty[15], in discussing the mass spectra of 4-nonene, attributed the formation of the large peak at m/e 84 as being due to a McLafferty rearrangement to give the ion **c** (equation 8). The presence of a metastable at m^*/e 37·3 (calc. for $84^+ \rightarrow 56^+ + 28$, 37·3) showed the further loss of ethylene from this

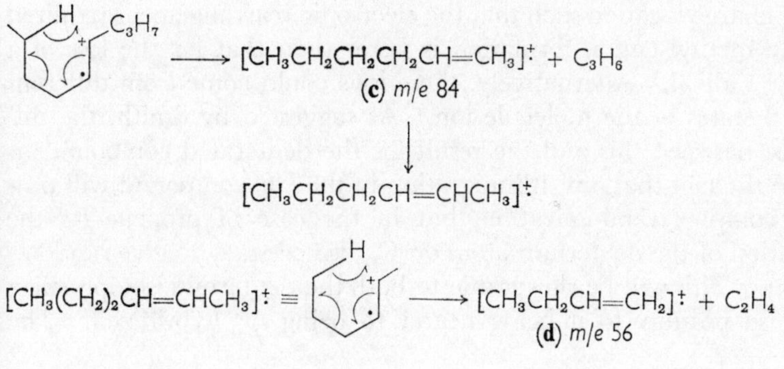

$$\longrightarrow [CH_3CH_2CH_2CH_2CH=CH_3]^{\ddagger} + C_3H_6$$
(**c**) m/e 84

$$\downarrow$$

$$[CH_3CH_2CH_2CH=CHCH_3]^{\ddagger} \qquad (8)$$

$$[CH_3(CH_2)_2CH=CHCH_3]^{\ddagger} \equiv \longrightarrow [CH_3CH_2CH=CH_2]^{\ddagger} + C_2H_4$$
(**d**) m/e 56

ion to give the ion **d** at m/e 56. It was suggested that this involved a hydrogen migration followed by a second McLafferty rearrangement. The results of Millard and Shaw[14] suggest that, owing to the extensive hydrogen migration in simple olefin systems[14], the presence of a McLafferty rearrangement cannot be proved by labelling, so this evidence is not valid.

The concrete evidence for such hydrogen rearrangement has been obtained by the study of labelled compounds. The work of McFadden[12] on the mass spectra of propene-2-d, propene-1,1-d_2 and propene-3,3,3-d_3 was considered to show that only 1,3 shifts were important as far as the loss of a hydrogen or a deuterium were concerned. The loss of a methyl group from the molecular ion was considered to be slightly faster than the occurrence of complete hydrogen scrambling by 1,3 and 1,2 shifts.

Smith and coworkers[13] reexamined the mass spectra of propene-1-d, propene-2-d and propene-3-d; in these cases the $[M—H]^+$ and $[M—D]^+$ ions had the same relative intensities, which contradicts the work of McFadden, and would be best explained by complete proton scrambling before fragmentation. In agreement with McFadden they suggest a lower rate of randomization between $C_{(1)}$ and $C_{(2)}$ than between $C_{(2)}$ and $C_{(3)}$ to explain the slower amount of scrambling before the loss of a methyl group in the 1,1-d_2 compound. The randomization, although still not complete, is more extensive in the case of the other two compounds.

The work of Voge and coworkers[17] on 3-^{13}C-propene indicates that the equilibrium represented by (equation 9) is reached before the

$$[CH_2{=}CH{-}^{13}CH_3]^{+\cdot} \rightleftharpoons [CH_3{-}CH{=}^{13}CH_2]^{+\cdot} \qquad (9)$$

methyl radical is lost, but that the rate of formation of the methyl cation is a little faster than this equilibration. This would seem to imply charge location such that the electronic rearrangement required for the methyl cation formation is faster than that for the loss of a methyl radical. Alternatively, these ions could come from different excited states of the molecule ion. As suggested by Smith, the difference between this and the result for the deuterated compounds is due to the fact that any 1,3 migration in the ^{13}C compound will produce complex randomization, but in the case of propene-2-d the migration of the deuterium atom on $C_{(2)}$ is necessary to give randomization. This will be slower due to both the deuterium isotope effect and also possibly to it being nearer to being sp^2 hybridized. The situation of propene-3-d is a little better off, since only deuterium transfer will not produce complete randomization. In propene-1-d it is not quite clear why the randomization is slower than for the ^{13}C compound, save that the first hydrogen transfer leads to propene-3-d and only hydrogen transfer leads back to propene-1-d.

The work of Bryce and Kebarle[18] on 1-butene-4,4,4-d_3 shows that in this case as well, the [M—Me]$^+$ ion is formed after considerable hydrogen scrambling. Interestingly it appears that the amount of scrambling decreases on increasing the electron voltage. The explanation put forward is that with more energetic electrons the loss of methyl occurs from more highly excited states of the molecular ion and is thus faster*. The rate of hydrogen rearrangement as judged by the ratios of the abundances of the [$C_3H_nD_{5-n}$]$^+$ ions seems much less dependent on the electron voltage†.

Recently the work of McFadden[12] has been reinterpreted by Millard and Shaw[14]. On the basis of his data they suggest that the formation of [$C_2H_nD_{3-n}$]$^+$ ions involve both 1,2 and 1,3 shifts, and that the former are faster than the latter in contradiction to McFadden's ideas. The mass spectrum of $trans$-2-butene-1,4-d_2 determined by these authors, shows a significant [M—CHD$_2$]$^+$ ion approximately consistent with a complete random distribution of hydrogen and deuterium atoms. This is to be compared with the incomplete randomization of 1-butene reported by Bryce[18]. The latter result may be due to the fact that more than one deuterium transfer from one carbon atom ($C_{(4)}$) has to happen for randomization, or may be due to the slower loss of a methyl group from the 2-butene, since vinylic cleavage would be required in the case of the original molecular ion. Millard and Shaw[14] have also examined the mass spectra of several deutero-1-pentenes and have come to the conclusion that the [M−28]$^{+}_{.}$ ion in the spectrum of 1-pentene (Figure 4) cannot be shown to arise by a McLafferty rearrangement or by any other one exclusive rearrangement. They suggest that due to the existence of 1,3 hydrogen shifts, not only $C_{(1)}$ and $C_{(2)}$ but also $C_{(4)}$ and $C_{(5)}$ can be lost as ethylene from the molecular ion. In these compounds the loss of ethyl and methyl radicals occurs at about the same rate as randomization of the hydrogen atoms.

The situation seems to be the same in the case of cyclic olefins. The work of Weinberg and Djerassi[19] on $\Delta^{4(8)}$ menthene (1) was the first

---

* For an alternative explanation see Reference 23.
† For more recent work see Reference 44.

to show that hydrogen migration under electron impact occurred in simple monocyclic olefins. The evidence for the fragmentation of this menthene shown below, was supplemented by the spectra of the labelled compounds **2** and **3**.

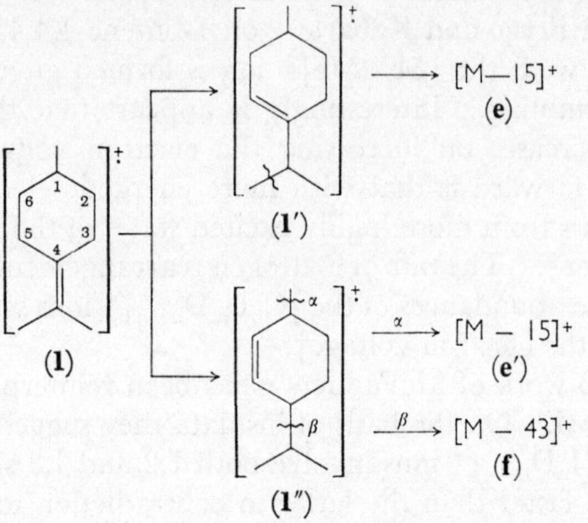

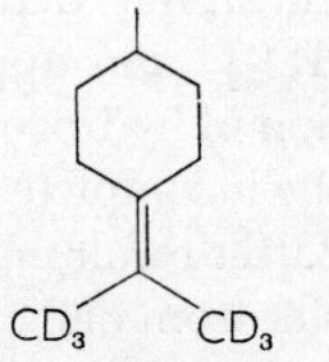

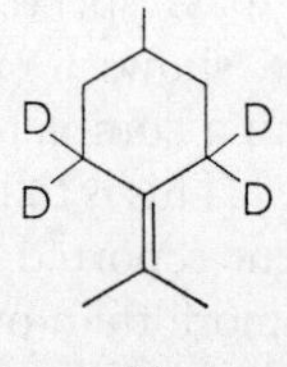

Djerassi considers that the 4:1 ratio for the loss of an isopropyl methyl to yield **e**, compared with the loss of the $C_{(1)}$ methyl group to yield **e'**, shows that the hydrogen migrations are relatively slow. This appears to be based on the idea that the ratio would be about 1:1 if the migrations were fast relative to fragmentation. However, the ion **1''** can undergo two different allylic fragmentations either losing the $C_{(1)}$ methyl group or the isopropyl group. So the second fragmentation would be expected to be preferred, since it is a well-known rule in mass spectrometry that the larger radical is lost preferentially. So the proper comparison to show whether fragmentation is fast or slow compared with isomerization, is to compare the ion current due to the loss of the $C_{(1)}$ methyl and the isopropyl group, with that due to the loss of a methyl from the isopropyl group. These are in the ratio of 7:1 approximately. This would suggest that double-bond migration is fast compared with fragmentation. The ratio would not be expected to be 1:1 since the rate of loss of an isopropyl group may be faster than that for a methyl group.

More recently the work of Kinstle and Stark[20] on some deuterated 1-methylcyclohexenes has confirmed the existence of such hydrogen migrations in cyclic olefins. This work will be considered in more detail in the discussion of the retro-Diels-Alder reaction. With compound 4 the absence of any peaks corresponding to the loss of CH_2D and CHD_2 suggests that the rate of transfer between the ring and the methyl group is very slow. This is perhaps as might be expected since primary hydrogen transfer is in general known to be slower than secondary hydrogen transfer[21]. In contrast, in the case of the *exo*-methylene compound 5, the spectra of deuterated analogues 6 and particularly 7 show that hydrogen transfer to the *exo*-methylene group is fast, since there is a large $[M-CH_2D]^+$ peak in the latter compound.

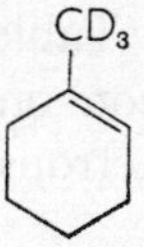

(4) 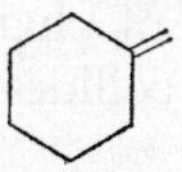   (5) 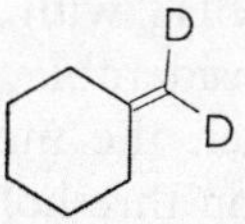   (6) 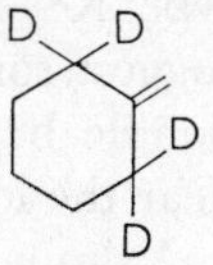   (7)

The explanation of most of the experimental results pertaining to double-bond migration has been made on the basis of ground-state ions and radicals; the possibilities of excited states and unusual valency states has been ignored. This is rather an arbitrary restriction but what is clear is that extensive hydrogen rearrangements occur prior to the fragmentation of the molecular ion in the case of many olefins. This has many implications in the mass spectrometry of the double bond.

II. IONIZATION AND APPEARANCE POTENTIALS[22]

When the energy of the electron beam is greater than the ionization potential of the substrate, the following processes may occur:

$$e + M \longrightarrow M^{+\cdot} + 2e \qquad (10)$$
$$e + M \longrightarrow A^+ + B\cdot + 2e \qquad (11)$$
$$e + M \longrightarrow M_1^{+\cdot} + M_2 + 2e \qquad (12)$$

The lowest energy for which reaction (10) occurs is the ionization potential of M, $I(M)$, while the lowest energy for which (11) or (12) occur is the appearance potential of A^+ or M_1^+, $A(A^+)$ or $A(M_1^+)$. These quantities are usually determined by plotting the ionization efficiency curves of the unknown and of a standard, that is the ion current as a function of electron-beam energy. The plot may either be linear or semi-logarithmic. Such a semi-log plot is shown in Figure 7[5]. It will be seen that the two curves are parallel. If the horizontal distance between them is ΔI and if the ionization potentials of the standard (S) and the unknown (X) are $I(S)$ and $I(X)$, then

$$I(X) = I(S) + \Delta I$$

The same technique may be used for determining appearance potentials of fragments, but the position is sometimes complicated by the ionization efficiency curves not being parallel.

Electron-impact ionization potential data for many olefins have been determined and collected[22,24,25]. The gradual decrease in the difference between the ionization potentials of homologues of the type R^1—CH=CH_2 with increasing size of R^1 (Table 4), would suggest some considerable degree of charge location on the initial double bond when the molecule collides with electrons of energies near the ionization threshold.

A similar effect is noticeable for olefins of the type Me—CH=CHR[2] (Table 5).

The point made by Stevenson[26] that the difference between the ionization potentials of propene and 1-butene (~ 0.1 eV) is small compared with that between butane and propane (~ 0.4 eV), suggests a different type of ion in the two cases, which is in accord with the idea of charge localization on the double bond.

1,1-Disubstitution of the double bond seems slightly less effective than 1,2-disubstitution in reducing the ionization potential of the

TABLE 4. Variation in the ionization potentials of olefins of the type R^1—CH=CH_2

R^1	$I(M)(eV)$	ΔI
H	10·6	—
CH_3	9·84	0·76
C_2H_5	9·72	0·12
n-C_3H_7	9·66	0·06
n-C_4H_9	9·59	0·06
n-C_8H_{17}	9·51	0·08

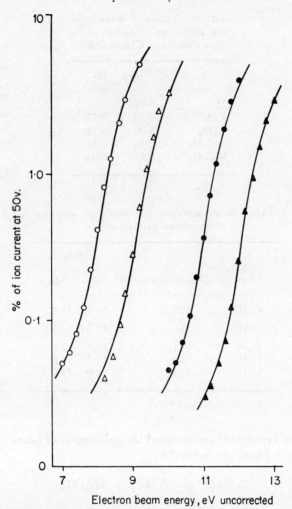

Figure 7. The ionization efficiency and electron beam energy plots for thiourea. ○,●, Thiourea runs 1 and 2; △,▲, methyl iodide runs 1 and 2. For the pair of duplicate runs at the right-hand side of the diagram the scale has been shifted by +3 eV. (Reproduced by permission from *Chemistry and Industry*.)

double bond. Thus the ionization potentials of isobutene and *trans*-2-butene are 9·23 and 9·13 eV respectively.

Relatively little appearance potential data is available for olefins. However propylene has been fully investigated (Table 6). The calculated values are derived as described below.

TABLE 5. Variation in the ionization potentials of olefins of the type trans-Me—CH=CHR^2

R^2	I(M)(eV)	ΔI
H	9·84	—
CH$_3$	9·27	0·57
C$_2$H$_5$	9·06	0·21
n-C$_3$H$_7$	9·16	−0·10
n-C$_5$H$_{11}$	9·11	0·05

TABLE 6. Appearance potentials of fragment ions from propylene[a]

Ion	Neutral species	A(eV) obs.	A(eV) calc.
C$_3$H$_5^+$	H·	11·9	11·8
C$_3$H$_4^{+\cdot}$	H$_2$	12·5	12·1
C$_2$H$_4^{+\cdot}$	CH$_2$:·	12·9	12·8
C$_2$H$_3^+$	CH$_3$	13·7	13·2
CH$_3^+$	C$_2$H$_3$	14·9	14·8

[a] Ionization potential 9·7 eV.

Field and Franklin[22] recommend the tabulation of heats of formation of ions. These are defined by

$$\Delta H_f(M^+) = I(M^+) + \Delta H_f(M) \tag{13}$$

$$\Delta H_f(A^+) = A(A^+) - \Delta H_f(\dot{B}) + \Delta H_f(AB) \tag{14}$$

$$\Delta H_f(M_1^+) = A(M_1^+) - \Delta H_f(M_2) + \Delta H_f(M_1M_2) \tag{15}$$

for the processes (10), (11) and (12) respectively. Thus tabulated values of $\Delta H_f(M^+)$, $\Delta H_f(A^+)$ and $\Delta H_f(M_1^+)$, together with corresponding values for $\Delta H_f(\dot{B})$ and the normal thermodynamic heat of formation of molecules, enable the calculation of appearance potentials. Some values are given in Table 7. It is of course assumed that fragments are formed with no kinetic energy release. Stevenson[27],

7. The Mass Spectrometry of the Double Bond

TABLE 7. Heats of formation of ions and radicals (kcal/mole)[22]

Ion (A$^+$)	CH_2^+	CH_3^+	$C_2H_3^+$	$C_2H_5^+$	$C_2H_4^{\cdot+}$	$C_3H_4^{\cdot+}$	$C_3H_5^+$
$\Delta H_f(A^+)$	333	262	280	224	225	279^a, 280^b	225

Radical(Ḃ)	Ḣ	ĊH$_2$	ĊH$_3$	Ċ$_2$H$_3$	Ċ$_2$H$_5$		Ċ$_3$H$_5$
$\Delta H_f(\dot{B})$	52	59	33	83	24		33

^{a} [CH$_2$=C=CH$_2$]$^{\cdot+}$.
^{b} [CH$_3$C≡CH]$^{\cdot+}$.

from a study of the alkanes, has proposed that for the alternative process

$$e + AB \begin{cases} \rightarrow A^+ + \dot{B} + 2e & (11a) \\ \rightarrow B^+ + \dot{A} + 2e & (11b) \end{cases}$$

A^+ will be produced in its lowest-energy state only if $I(A) < I(B)$. This implies that

$$\Delta H_f(A^+) = A(A^+) - \Delta H_f(\dot{B}) + \Delta H_f(AB)$$

but

$$\Delta H_f(B^+) \leq A(B^+)^* - \Delta H_f(\dot{A}) + \Delta H_f(AB)$$

where now $A(B^+)^*$ is the appearance potential for an excited state†. Care has thus to be taken in using the above type of argument.

A difficulty underlying the energetics of fragmentation, which has already been discussed when considering mechanisms of fragmentation, is the possibility of isomerization of the molecular ion and the structure of the fragment ions. Some evidence is provided by the work of Bryce and Kebarle[18] who examined the mass spectrum of CD$_3$CH$_2$CH=CH$_2$ at varying electron energies (Table 8). A major route of fragmentation is the splitting off of ĊH$_3$ to give m/e 41 in the hydrogen compound.

It is apparent that the C$_3$ fragment contains a considerable amount of deuterium resulting from the transfer from the CD$_3$ group to the rest of the molecule. What is more surprising at first sight is that the amount of deuterium transfer increases as the electron-beam energy is reduced.

† This can also be due to the fact that B$^+$ is not produced from the molecular ion.

TABLE 8. Variation of C_3 ion intensity with electron-beam energy

Electron beam energy (eV)	Relative Abundance				
	m/e 41	42	43	44	(42 + 43 + 44)/41
13·0	1·0	3·0	5·0	2·0	10·0
15·0	4·0	8·0	14·0	6·0	7·0
17·0	12·0	20·5	32·0	16·0	5·7
50·0	74·3	67·0	99·3	59·0	3·0

Bryce and Kebarle explain this in terms of the more highly energized ions having shorter lifetimes and hence less time for rearrangement*. This work confirms the prediction of Stevenson and Wagner[29] that the rearrangement of H and D atoms would be more extensive in the alkenes than in the alkanes.

More work will be required, especially with labelled (D or ^{13}C) compounds, before a detailed understanding of the energetics of fragmentation of alkenes is achieved.

III. LOCATION OF DOUBLE BONDS BY MASS SPECTROMETRY

The direct location of double bonds by mass spectrometry is made difficult by migration of the double bond under electron impact[30]. This problem is particularly important in the field of lipid chemistry. To overcome this problem simple chemical transformations involving a minimal amount of material have been developed. Reduction of the double bond with deuterium and examination of the spectrum of the corresponding saturated compound was the first method investigated. It was hoped that the two deuterium atoms would go on to the carbon atoms of the original double bond (equation 16) and that fragmentation as in equation (17) would lead to the location of the double bond. This means that the deuterium must be introduced specifically and that there must be no hydrogen or deuterium migrations under the experimental conditions. However, these conditions were not fulfilled, since double bonds migrate under catalytic hydrogenation conditions[31] (see Chapter 4). Secondly, catalytic exchange of hydrogen for deuterium also occurs, thus introducing more than two deuterium atoms per double bond. Even without the first two com-

* For an alternative explanation see Reference 23, see also footnote p. 335.

plications the hydrogens and deuterium atoms of the saturated compound will migrate under electron-impact conditions[14,32]. Finally the fragmentation of long-chain paraffins is known not to be as simple as first described. Previously it was considered to be due to straight cleavage of the carbon–carbon bonds giving simple hydrocarbon ions of the type $[C_nH_{2n+1}]^+$. However, for $CD_3(CH_2)_{26}CD_3$ the mass spectrum[32] shows a base peak at m/e 57 $[C_4H_9]^+$, although on the basis of direct cleavage the largest four carbon containing peak would be expected at m/e 60 $[C_4H_6D_3]^+$. The latter is indeed large but the presence of the former implies the expulsion of polymethylene units with a hydrogen capture, from the middle of the molecule. Indeed the $[C_{17}H_{35}]^+$ ion is still noticeable in the spectrum of this compound. This expulsion reaction is imagined to proceed through a coiling mechanism.

To overcome the problems introduced by catalytic hydrogenation the use of perdeuterohydrazine in dioxan as a reducing agent, for the double bond in unsaturated fatty acid esters (reaction 16) was developed by Ryhage and coworkers[30]. However, the fragmentation patterns of saturated fatty acid esters were shown by the same workers to involve not only the simple cleavages (reaction 17) but also the expulsion of alkyl radicals from the centre of the chain, as in the case of

$$R^1CH=CH(CH_2)_nCOOMe \longrightarrow R^1CHDCHD(CH_2)_nCOOMe \quad (16)$$

$$[R\overset{\alpha}{-}CHD\overset{\beta}{-}CHD(CH_2)_nCOOMe]^{\ddot{+}} \begin{array}{c} \xrightarrow{\alpha} [(CHD)_2(CH_2)_nCOOMe]^+ \\ \xrightarrow{\beta} [CHD(CH_2)_nCOOMe]^+ \end{array} \quad (17)$$

the saturated paraffins. Again, for $CD_3(CH_2)_{17}COOCH_3$ the mass spectrum shows there is an ion at m/e 143, $[(CH_2)_6COOCH_3]^+$, but there is also an ion at m/e 146 corresponding to $[C_8H_{12}D_3O_2]^+$ and the intensity of the ion at m/e 199, $[(CH_2)_{10}COOCH_3]^+$, is the same as that at m/e 202 (the corresponding trideuterated ion). Other labelled esters produced the same results under electron impact, so other methods were sought for[32].

The first successful method was developed by Fetizon and his coworkers[33] for simple olefins. The olefin was converted to the epoxide, which on heating with dimethylamine in a sealed tube gave a mixture of two compounds (reaction 18).

$$R^1CH=CHR^2 \begin{array}{c} \longrightarrow R^1CH(OH)CH(NMe_2)R^2 \\ \longrightarrow R^1CH(NMe_2)CH(OH)R^2 \end{array} \quad (18)$$

These compounds fragment as expected under electron impact to give the ions R—CH=$\overset{+}{N}$H$_2$. These ions further fragment as shown in reaction (19), so any substituent on the carbon atoms α or β to the double bond can be detected. This method can also be applied to

$$[R^1CH(NMe_2)CH(OH)R^2]^{+\cdot} \longrightarrow R^1CH=\overset{+}{N}Me_2 \longrightarrow CH_2=CH-CH=\overset{+}{N}Me_2$$
$$[R^1CH(OH)CH(NMe_2)R^2]^{+\cdot} \longrightarrow R_2CH=\overset{+}{N}Me_2 \longrightarrow CH_2=CH-CH=\overset{+}{N}Me_2$$
(19)

cyclic olefins such as cholestene. Often in this type of olefin only one of the two possible compounds is formed for steric reasons (neglecting conformational isomers), but the fragmentation patterns are more complicated. However, the fragmentation patterns of steroidal amines have been worked out[33] and can be applied here with one simplification, the hydroxyl function favours cleavage of the bond between the two carbon atoms carrying the functional groups. An example is outlined for the hydroxy amino derivative (8) of cholest-5-ene.

This method gives large ions corresponding to the cleavages previously discussed, as well as a molecular ion. There is no evidence as to the geometry of the double bond. With esters or amides the chemical transformation produces the substituted amide.

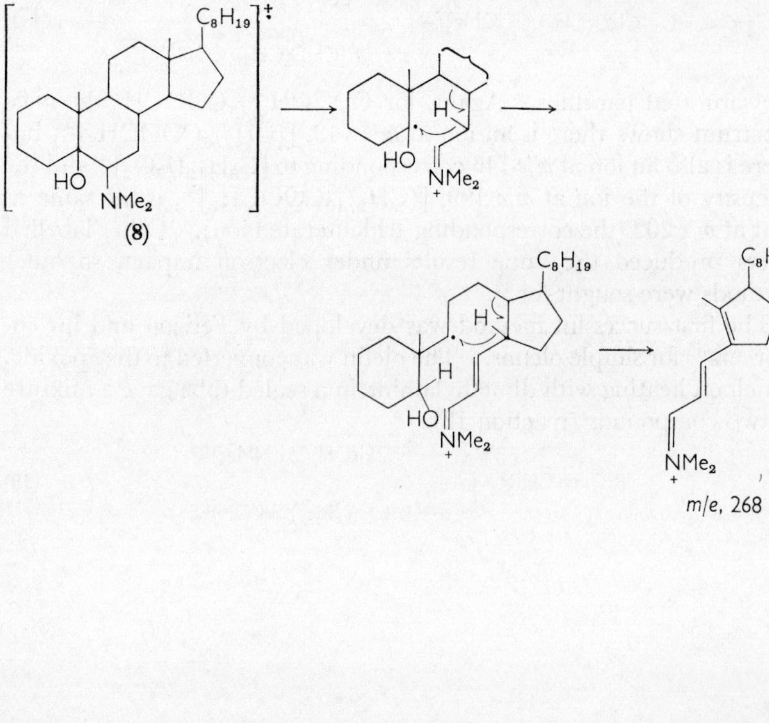

Another method has been developed by McCloskey and his co-workers[35,36]. The method was first used for locating the double bond in unsaturated fatty acid esters[35]. The material (1 mg) was treated with osmium tetroxide which gave the *cis*-diol. Subsequent reaction with acetone gave the isopropylidene derivative, which is then introduced into the mass spectrometer via a chromatographic column. The mass spectrum shows a large $[M-15]^+$ ion thus allowing the determination of the molecular weight. The double bond is located by the small ions formed by the cleavages shown in reaction (20). Use of hexadeuteroacetone or ^{18}O enriched osmium tetroxide allows identification of these peaks by the appropriate shifts in the m/e ratios of these ions (**g,h**). Also the relationship between the mass of **g** (m_g)

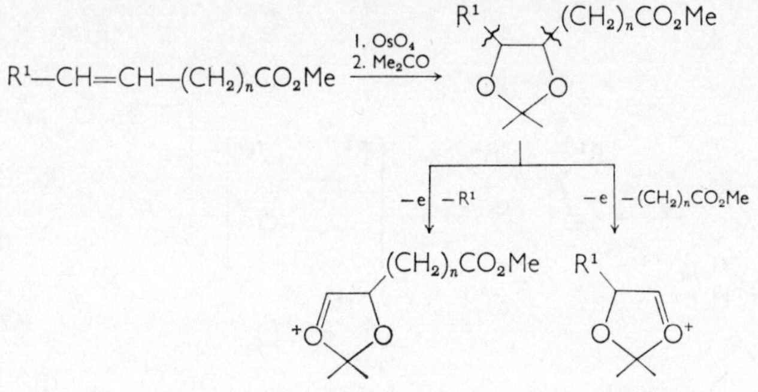

$$R^1\text{—CH=CH—}(CH_2)_n CO_2 Me \xrightarrow[2.\ Me_2CO]{1.\ OsO_4} \cdots \tag{20}$$

(g) (h)

and of **h** (m_h) is given by equation (21). The ion **g** containing the ester

$$m_g + m_h = 100 + \text{molecular weight} \tag{21}$$

grouping is always the more intense. The added advantage of this method is that comparison of the spectra of the *cis* and *trans* isomers allows geometrical assignment. In all these cases the ion **j**, formed by successive loss of ketene and methanol from the $[M-15]^+$ ion, is about twice as intense for the *erythro* derivative formed from the *cis* olefin than for the *threo* derivative formed from the corresponding *trans* olefin. Reaction (22) was substantiated by deuterium labelling.

The proton originally belonging to acetone is lost in the last stage whatever the chain length. This probably reflects the ability of these types of chains to coil round themselves as postulated in the mechanism for the expulsion of alkyl radicals from the middle of the hydrocarbon chains.

The spectra[36] of the same derivatives of straight-chain olefins are much simpler, showing $[M-15]^+$ ions and the two prominent **k** and **l** ions (reaction 23). From these three ions the molecular weight can

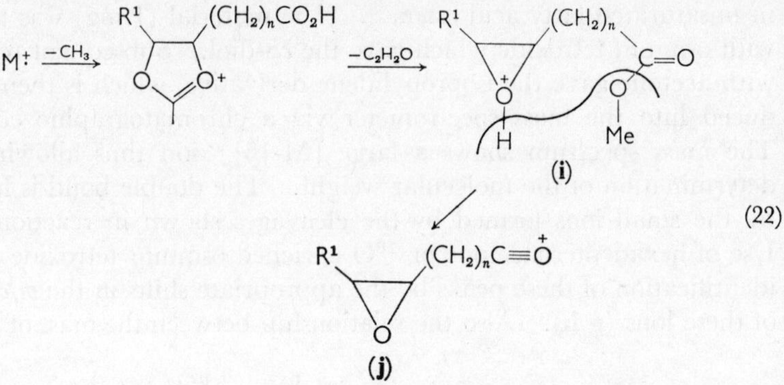

$$\text{M}^+ \xrightarrow{-CH_3} \begin{array}{c} R^1 \quad (CH_2)_nCO_2H \\ \diagup \\ O \quad O^+ \\ \end{array} \xrightarrow{-C_2H_2O} \begin{array}{c} R^1 \quad (CH_2)_n \\ \triangle \\ O^+ \\ H \end{array} \begin{array}{c} C=O \\ | \\ Me \end{array} \quad (i)$$

$$\downarrow$$

$$\begin{array}{c} R^1 \quad (CH_2)_nC\equiv\overset{+}{O} \\ \triangle \\ O \end{array} \quad (j) \tag{22}$$

$$\begin{array}{c} R^1 \diagdown \diagup R^2 \\ R^1 = CH_3(CH_2)_m \\ R^2 = CH_3(CH_2)_n \end{array} \longrightarrow \begin{array}{c} R^1 \quad R^2 \\ O \quad O \\ \diagdown \end{array} \longrightarrow \left[\begin{array}{c} R^1 \quad R^2 \\ O \quad O \\ \diagdown \end{array} \right]^{\cdot +} \begin{array}{c} \nearrow \\ \searrow \end{array} \begin{array}{c} R^1 \\ O \quad O^+ \\ \text{(k)} \\ \\ R^2 \\ ^+O \quad O \\ \text{(l)} \end{array} \tag{23}$$

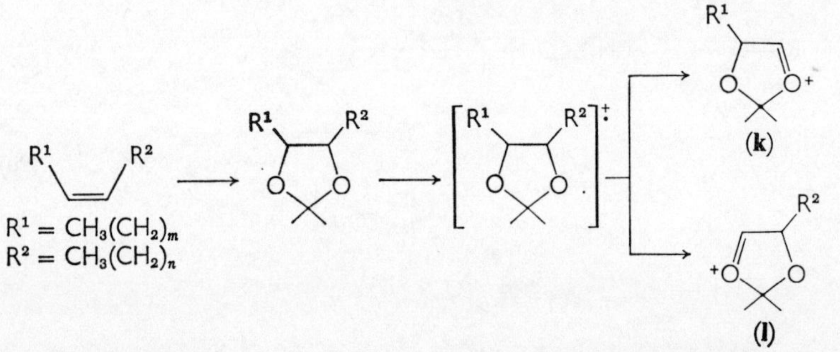

easily be determined by use of equation (24). Confirmation of the

$$m_k + m_l = 100 + \text{molecular weight of } R^1CH=CHR^2 \tag{24}$$

identity of the ions **k** and **l** comes from the relationships given in equations (25) and (26). The routine use of hexadeuteroacetone and

$$m_k = 101 + 14\,(n+1) \tag{25}$$
$$m_l = 101 + 14\,(m+1) \tag{26}$$

if necessary high resolution mass measurements, can also be used to identify these ions. As a general rule if $R^1 > R^2$ then the intensity of the ion **k** is greater than that of the ion **l**.

The intensity of **k** or **l** can be expressed as a fraction of the total ion current $\sum_{40}^{M}$ above m/e 40. For the *cis* and *trans* forms of the same olefin the intensity ratio is governed approximately by equation (27). Also

$$\left(\text{Intensity of } \mathbf{k} \Big/ \sum_{40}^{M} \text{Intensities} \right)_{cis} \Big/ \left(\text{Intensity of } \mathbf{k} \Big/ \sum_{40}^{M} \text{Intensities} \right)_{trans}$$
$$\sim 2 \cdot 6 \tag{27}$$

the ion **i** formed by the loss of ketene from the $[M-15]^+$ ion, has a different value expressed as a percentage of the total ion current for the *cis* and *trans* isomers, as given by equation (28).

$$\left(\text{Intensity of } \mathbf{i}/\sum_{40}^{M} \text{Intensities}\right)_{cis} \Big/ \left(\text{Intensity of } \mathbf{i}/\sum_{40}^{M} \text{Intensities}\right)_{trans}$$
$$\sim 4 \cdot 0 \quad (28)$$

The use of hexadeuteroacetone in the case of the 2-alkenes shows that there is no contribution to the ion current at $[M-15]^+$ due to the ion $\mathbf{k}(R^2 = Me)$. The problem of location of the double bond when **k** or **l** is small can be overcome by identification of the ion **m** of moderate intensity when R^1 or $R^2 \le C_4H_9$. The ion **m** always has an m/e value 29 units above that expected for the ion **l**.

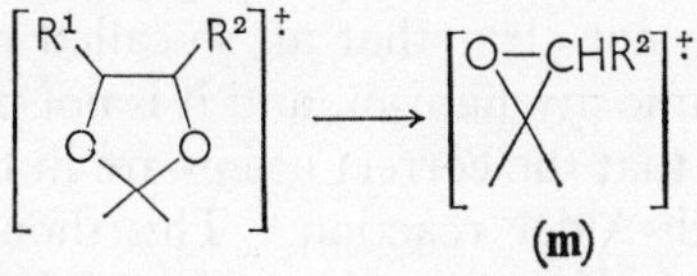

$$\begin{bmatrix} R^1 & R^2 \\ & \\ O & O \\ \end{bmatrix}^{\ddagger} \longrightarrow \begin{bmatrix} O-CHR^2 \\ \\ \end{bmatrix}^{\ddagger} \quad (29)$$
$$(\mathbf{m})$$

Mention is also made[36] of compounds containing two double bonds. The derivative **9** shows only ions corresponding to fragmentations (1) and (2), the ions corresponding to fragmentations (3) and (4) are not

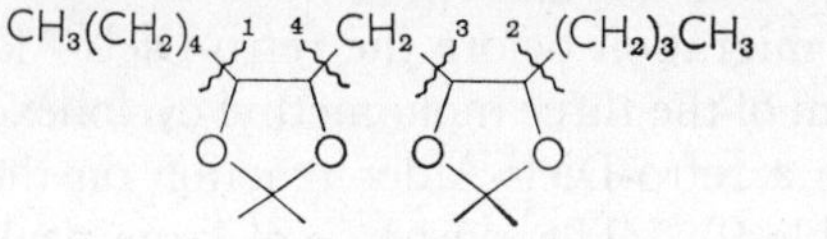

$$CH_3(CH_2)_4 \overset{1}{\underset{OO}{\times}} \overset{4}{} CH_2 \overset{3}{\underset{OO}{\times}} \overset{2}{} (CH_2)_3CH_3$$
$$(9)$$

seen. This may be due to the nearness of the two isopropylidene derivatives, but this has yet to be investigated and more work is required before these methods can be applied to polyene systems.

IV. THE RETRO-DIELS-ALDER REACTION

In general as discussed in the previous section, the migration of double bonds under electron impact causes difficulties in the interpretation of the mass spectra of olefins, but this is not always so. In some cases this migration does not occur. Alternatively if it occurs, either it does not affect the fragmentation pattern or it affects it in a recognizable manner.

The retro-Diels-Alder reaction, first so-called by Biemann[37], exemplifies these points. The general structural requirement for this reaction is a six-membered ring containing one double bond. Formal diallylic cleavage as shown for cyclohexene (reaction 30) is the characteristic of this reaction. In molecules of unknown structure this can often be recognized by the fragmentation pattern involving the expulsion of a neutral molecule, leaving an ion corresponding to a molecule, i.e. all the classical valencies are satisfied. However, in other cases further bond fragmentation may be required to give a fragment ion at all, making identification of this reaction more

$$\left[\bigcirc\right]^{\ddagger} \longrightarrow \left[\diagdown\diagup\right]^{\ddagger} + C_2H_4 \qquad (30)$$

difficult. Also it is not clear that all so-called retro-Diels-Alder reactions go by the same mechanism, and it is not possible to check this save by confirming that the correct atoms are in the ions formed by a suspected retro-Diels-Alder reaction. This should perhaps be contrasted with the McLafferty reaction, where labelling shows whether it is the γ hydrogen which is transferred. In neither case are the details of the electronic rearrangement completely known*.

Examination of the mass spectra of several alkyl derivatives of cyclohexene[9] reveals that the mass spectra can be explained, in part, on the basis of bond migration before the retro-Diels-Alder reaction occurs. In the spectrum of the three monomethyl cyclohexenes the major peak corresponds to a retro-Diels-Alder reaction on the rearranged molecular ion (Table 9). The existence of large peaks corresponding to this reaction on rearranged molecular ions (Table 9), would suggest that here, migration of the double bond and this reaction proceed at about the same rate. Notice that the large difference between the 4-methyl compound and the other two rules out complete equilibration before fragmentation.

Djerassi[11] postulates that the first stage in the reaction is an allylic fission (reaction 31). In the case of cyclohexene he estimates the appearance potential for such a reaction by considering the bond energy of an allylic carbon–carbon bond in cyclohexene as ca. 70 kcal. (3·0 eV) and the ionization potential for the diradical **n** as being 7·7 eV. This leads to the appearance potential for such a mechanism to be

---

* See references 42 and 43.

TABLE 9. Contribution of retro-Diels-Alder reactions to the spectrum of methylcyclohexenes

Compound	retro-Diels-Alder reaction on molecular ion		retro-Diels-Alder reaction on rearranged molecular ion	
	m/e	$\sum_{41}\%$ base	m/e	$\sum_{41}\%$ base
1-Me	68[a]	10	54[c]	5·2
3-Me	68[b]	10	54[d]	6·4
4-Me	54[b]	16·7	68[d]	6·6

[a] Can be formed via 1,3-migration giving the molecular ion of the 3-methyl compound.
[b] Can also still be formed after a 1,3 migration.
[c] This requires a minimum of two 1,3-hydrogen shifts.
[d] Requires one 1,3-hydrogen shift.

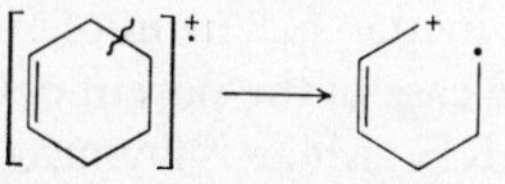

$$[\text{cyclohexene}]^{\ddagger} \longrightarrow \text{(n')} \quad \text{(n)} \tag{31}$$

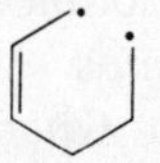

> 10·7 eV. The experimental value of 11·2 eV is thus consistent with the postulated mechanism. However, these estimates are quite crude. The thermal retro-Diels-Alder reaction of cyclohexene is regarded as a one-step process and the energy required for this is known (0·78 eV). Coupled with the ionization potential for butadiene this would require an appearance potential of only 10·1 eV. So the intermediate is regarded as being straight-chained **n′** rather than a ring-expanded transition state, where the two bonds to be broken are equally stretched, since the former calculated value is better than the latter*.

The relatively high intensity of the retro-Diels-Alder reaction of the unrearranged molecular ion in the 4-methyl compound, may on this basis represent the extra stabilization of the radical ion **o**, lowering the activation energy for this reaction.

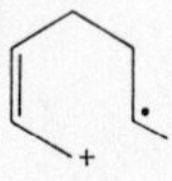

(**o**)

* Recently[28], evidence has been produced for a concerted retro-Diels-Alder reaction.

A similar argument may account for the fact that in the spectra of 1,2-dimethylcyclohexene the largest peak corresponding to a retro-Diels-Alder reaction is at m/e 68. If the double-bond migration is relatively fast, then the important factor will be the energy of the various possible transition states. Of all the possibilities **p** will be expected to be the most stable and would give rise to an ion at m/e 68.

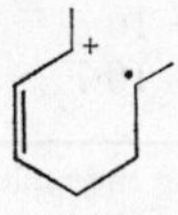

(p)

Djerassi advances a similar argument to account for the appearance of a large $[M-54]^{+\cdot}$ ion in the spectrum of 9-methyl-*trans*-Δ^2-octalin which is very small in the case of the parent octalin. He suggests that of the ions **q** and **r** when R = Me, **r** is favoured but when R = H then **q** is favoured[11]. Evidence for double-bond migration can be found in the spectra of the two compounds. Rearrangement to the cation **s** would require methyl migration when R = Me. When R = H but not when R = Me, an appreciable $[M-28]^+$ ion is seen. Notice in the case of the α-methyl compound the ion **s'** which could give an $[M-28]^+$ ion and would be formed without methyl migration is not formed, perhaps due to a preferential migration of the double bond towards the methyl group.

The retro-Diels-Alder reaction of 1-methylcyclohexene has recently been studied by the means of deuterium compounds **10** and **11**[20].

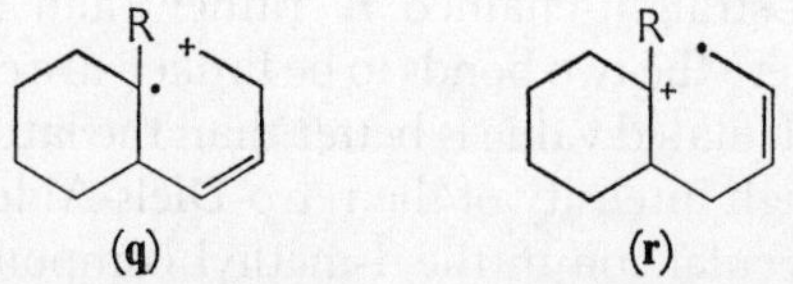

(q)   (r)

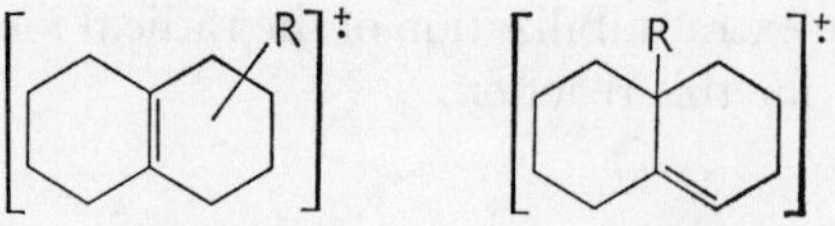

(s)   (s')

On the assumption that the ion at m/e 66 $[M—C_2H_5]^+$ in the spectrum of 1-methylcyclohexene does not come by a loss of the hydrogen from the $[M—C_2H_4]^{+\cdot}$ ion (the retro-Diels-Alder ion), the minimum degree

7. The Mass Spectrometry of the Double Bond 351

of rearrangement before production of the ion at m/e 68 can be calculated. This assumption is probably reasonable since cyclohexene shows a large [M—Me]$^+$ ion which cannot come from the retro-Diels-Alder reaction in that compound. A detailed study of the distribution of the labelling in the two compounds also indicates that at least a considerable portion of the [M—C$_2$H$_5$]$^+$ ion does not come from the corresponding retro-Diels-Alder ion.

From the ratio of [M−28]$^+$/[M−29]$^+$ in the parent compound and the labelled compound **10**, it was calculated that the percentage rearrangement before retro-Diels-Alder fragmentation was 53%. This

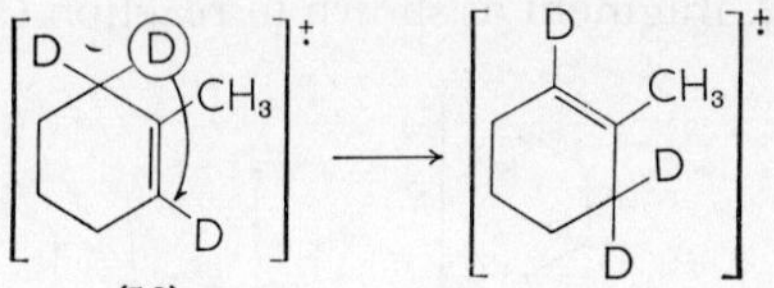

$$\left[\begin{array}{c} \text{D} \cdot \text{D} \\ \text{CH}_3 \\ \text{D} \end{array}\right]^{\ddagger} \longrightarrow \left[\begin{array}{c} \text{D} \\ \text{CH}_3 \\ \text{D} \\ \text{D} \end{array}\right]^{\ddagger} \quad (31)$$

(10)

is the minimum amount due to a retro-Diels-Alder fragmentation before hydrogen migration, since the isomerization shown in reaction (31), although a hydrogen migration, gives the same molecular ion and therefore will lead to the ion at m/e 71, and therefore part of the ion current at m/e 71 is also due to fragmentation after rearrangement.

Similar calculations on the deuterated compound **11** give again a minimum figure of 28% due to the isomerization shown in reaction (32)

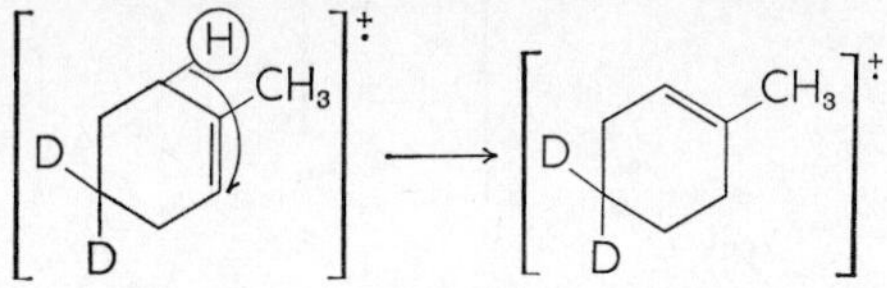

$$\left[\begin{array}{c} \text{(H)} \\ \text{CH}_3 \\ \text{D} \\ \text{D} \end{array}\right]^{\ddagger} \longrightarrow \left[\begin{array}{c} \text{CH}_3 \\ \text{D} \\ \text{D} \end{array}\right]^{\ddagger} \quad (32)$$

Djerassi in his paper on the menthenes[19] suggests that part of the ion at m/e 81 is due to a retro-Diels-Alder reaction on the ion **t**, followed by the loss of a methyl group (reaction 33).

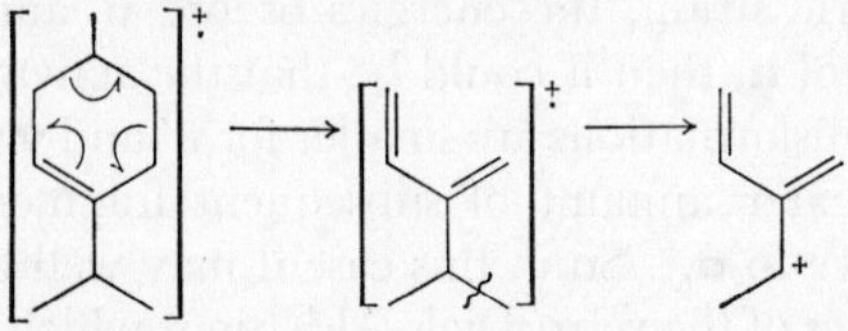

$$\left[\begin{array}{c} \end{array}\right]^{\ddagger} \longrightarrow \left[\begin{array}{c} \end{array}\right]^{\ddagger} \longrightarrow \quad (33)$$

(t)

In general the retro-Diels-Alder reaction seems a fairly high-energy process and if other simple fragmentations are available, these often take precedence. Thus in the mass spectra of 1- and 4-acetyl cyclohexenes the only large fragment ion is the $[M-15]^+$ ion.

Most of the above arguments have neglected the possibility of further fragmentations affecting the relative abundance of these processes. This rests on the stability of conjugated olefins to fragmentation under electron impact, as well as the fact that the simple explanations seem adequate as far as a qualitative picture is concerned.

The retro-Diels-Alder reaction is important for other reasons, particularly for structure determination by mass spectrometry. This is particularly so in the field of triterpene chemistry. The Δ^{12} compounds of general structure **12**, fragment as shown in reaction (34) giving the ion

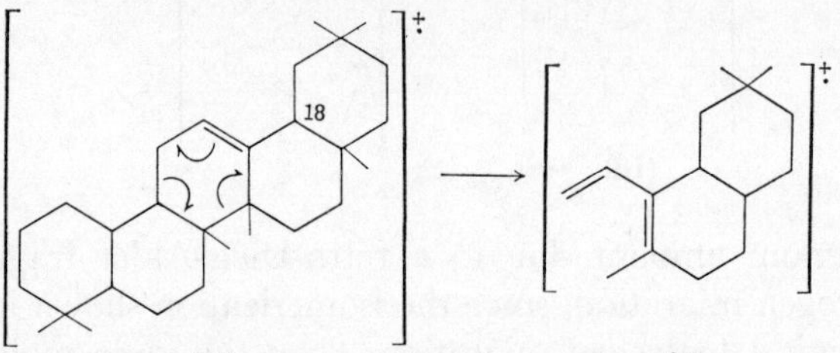

(34)

(12)   (u)

(35)

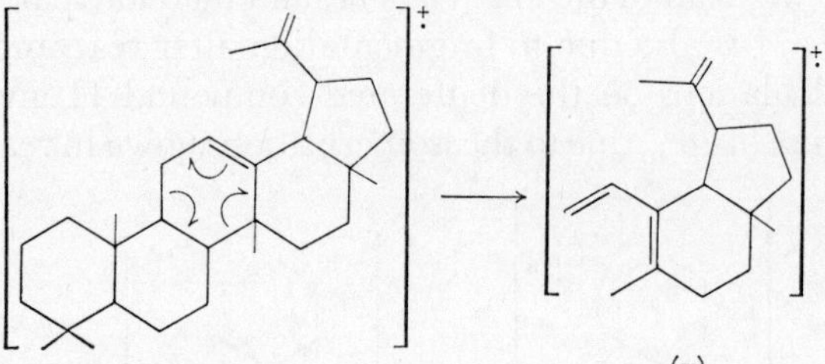

(13)   (v)

u. However, if the ring junction changes to 18α, the corresponding ion **u'**, is less abundant. The corresponding retro-Diels-Alder peak (**v**) shown in reaction (35) is small in the case of the Δ^{12}-lupene series (**13**). If, because of steric strain, the energies of ions **u'** and **v** are slightly greater than that of **u**, then it could be that the activation energies of the subsequent transformations are smaller for **u'** and **v** than for **u**, thus resulting in a greater amount of subsequent fragmentations of the former ions relative to **u**. So in this case it may well be that it is the decomposition rates of the retro-Diels-Alder ions which are important.

In other respects modification of the structure **12** still leaves the retro-Diels-Alder ion, or a simple recognizable decomposition product of it, as the largest peak in the spectrum. Thus the main peaks in the mass spectrum of a triterpene of molecular formula $C_{30}H_{50}O_2$ were at m/e 234, 216 and 201. This suggested the partial structure **14** fragmenting as shown in reaction (36). This leads to the identification of this compound as manila-diol[39]. Many other examples have also been described.

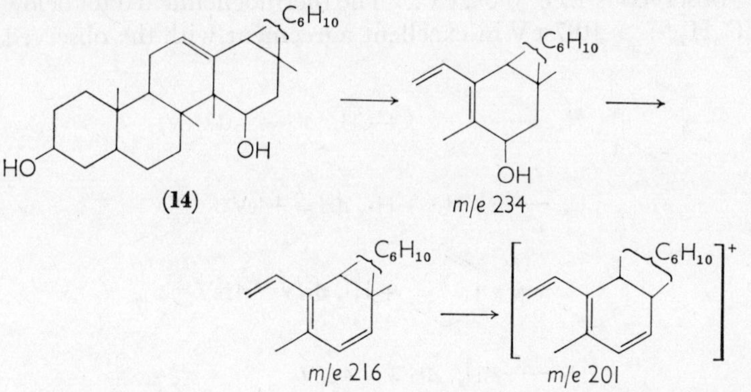

The retro-Diels-Alder reaction has also been invoked to explain the mass spectra of many alkaloids. One example in this context, that of the production of the base peak of vincadiformine (**15**)[40] will suffice. It should be pointed out that a similar reaction is also shown by the dihydrocompound.

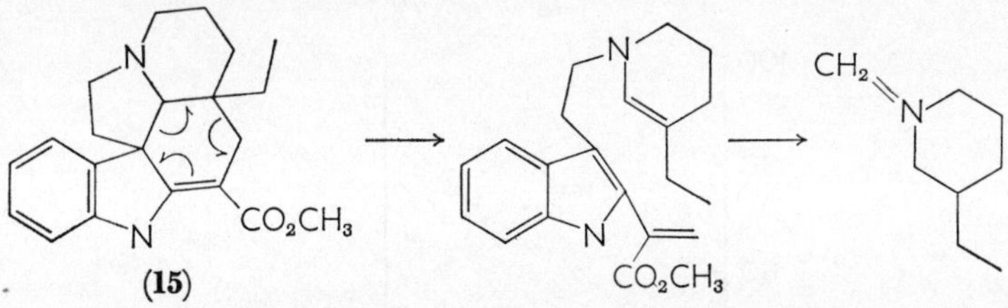

V. THE EXPULSION OF ALKYL RADICALS FROM CYCLIC OLEFINS

In the case of straight-chain olefins, the loss of alkyl radicals by allylic fission has already been discussed. Mention has also been made of the fact that the base peak in cyclohexene is due to the ion $[M-15]^+$. Djerassi and coworkers[11] have studied this particular fragmentation and on the basis of thermochemical data have come up with the mechanism depicted in reaction (38). The appearance potential of

$[\bigcirc]^{\pm} \longrightarrow \text{(cyclohexene cation with H)} \longrightarrow \text{(pentadienyl cation)}$ (38)

$[C_5H_7]^+$ observed is $10\cdot 8 \pm 0\cdot 2$ eV. The thermochemical data below yield $A(C_5H_7{}^+) \geqslant 10\cdot 7$ eV in excellent agreement with the observed

$\bigcirc + H_2 \longrightarrow \diagup\!\!\diagdown + CH_4, \Delta H = -0\cdot 27\text{eV},$

$CH_4 \longrightarrow CH_3\cdot + H\cdot, \Delta H = 4\cdot 4\text{eV}$

$\diagup\!\!\diagdown \longrightarrow \diagup\!\!\diagdown \cdot + H\cdot, \Delta H = 3\cdot 45\text{eV}$

$2H\cdot \longrightarrow H_2, \Delta H = -4\cdot 5\text{eV}$

$\diagup\!\!\diagdown \cdot \longrightarrow \diagup\!\!\diagdown_+, \Delta H = 7\cdot 7\text{eV}$

value. The alternative mechanism (39) is ruled out by the experiments of Kinstle and Stark[20].

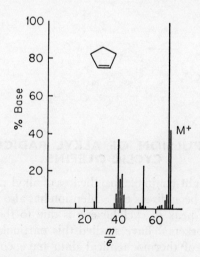

Figure 8. Mass spectrum of cyclopentene.

7. The Mass Spectrometry of the Double Bond

$$\overset{+}{\underset{\cdot}{\bigcirc}} \longrightarrow \underset{+}{\cdot\overset{\cdot}{\bigcirc}} \longrightarrow \overset{+}{\bigcirc} \tag{39}$$

The mass spectra of cyclopentene[9], cyclohexene[9] and cyclooctene[41] are shown in Figures 8–10. Notable is the fact that the base peak in all three cases is at m/e 67. In other words, the three compounds lose H·, CH_3· and C_3H_7· respectively, with ease. This presumably reflects the stability of the system $[C_5H_7]^+$.

$$CH_2{=}CH{-}CH{=}CH{-}\overset{+}{C}H_2 \longleftrightarrow \overset{+}{C}H_2{-}CH{=}CH{-}CH{=}CH_2$$

The retro-Diels-Alder peak is outstanding in cyclohexene at m/e 54 (76%). In cyclooctene at m/e 82 and 54 appear peaks that could be

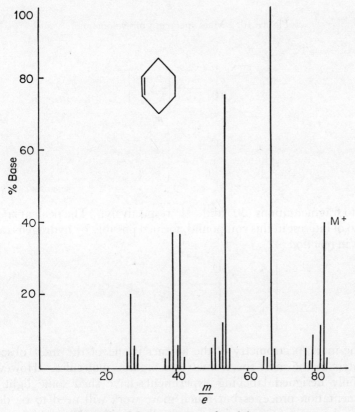

Figure 9. Mass spectrum of cyclohexene.

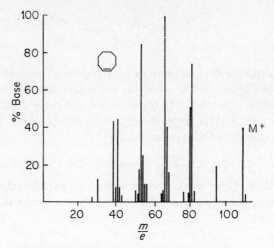

Figure 10. Mass spectrum of cyclooctene.

due to fragmentations (40) and (41) respectively. The peak at m/e 67 is also of interest in this compound, formed possibly by hydrogen transfer as in reaction (42).

The mass spectrometry of the alkenes is one of the most obscure chapters in the mass spectrometry of organic molecules. However, carefully designed labelling experiments have shed some light on fragmentation processes but much more work will need to be done before the various mechanisms are finally elucidated.

VI. REFERENCES

1. F. W. McLafferty in *Determination of Organic Structures by Physical Methods* (Ed. F. C. Nachbod and W. D. Philips), Academic Press, New York, 1962, p. 93.
2. F. W. McLafferty, *Mass Spectrometry of Organic Ions*, Academic Press, New York, 1963, p. 315.
3. J. S. Shannon, *Tetrahedron Letters*, 801 (1963); J. S. Shannon, *Proc. Roy. Aust. Chem. Inst.*, 232 (1962).
4. H. Budzikiewicz, C. Djerassi and D. H. Williams, *Interpretation of Mass Spectra of Organic Compounds*, Holden-Day Inc., San Francisco, 1964.
5. M. Baldwin, A. Kirkien-Konasiewicz, A. Maccoll and B. Saville, *Chem. Ind.*, 286 (1966).
6. M. Baldwin, A. Kirkien-Konasiewicz, A. G. Loudon, A. Maccoll and D. Smith, *Chem. Comm.*, 574 (1966).
7. C. K. Ingold, *Structure and Mechanism in Organic Chemistry*, Bell, New York, 1953.
8. F. W. McLafferty, *Chem. Comm.*, 78 (1966).
9. *Catalogue of Mass Spectral Data A.P.I. Project 44*, National Bureau of Standards, Washington D.C., 1949.
10. F. W. McLafferty, *Anal. Chem.*, **31**, 82 (1959).
11. H. Budzikiewicz, J. I. Brauman and C. Djerassi, *Tetrahedron*, **21**, 1855, (1965).
12. W. H. McFadden, *J. Phys. Chem.*, **67**, 1074 (1963).
13. S. R. Smith, R. Schor and W. P. Norris, *J. Phys. Chem.*, **69**, 1615 (1965).
14. B. J. Millard and D. F. Shaw, *J. Chem. Soc.* (B), 664 (1966).
15. F. W. McLafferty, *Anal. Chem.*, **31**, 2072 (1959).
16. E. Honkanen, T. Moiso, M. Ohno and A. Hatanaka, *Acta Chem. Scand.*, **17**, 2051 (1963).
17. H. H. Voge, C. D. Wagner and D. P. Stevenson, *J. Catalysis*, **2**, 58 (1963).
18. W. A. Bryce and P. Kebarle, *Can. J. Chem.*, **34**, 1249 (1956).
19. D. S. Weinberg and C. Djerassi, *J. Org. Chem.*, **31**, 115 (1966).
20. T. H. Kinstle and R. E. Stark, *J. Org. Chem.*, **32**, 1318 (1967).
21. H. Budzikiewicz, C. Feneslau and C. Djerassi, *Tetrahedron*, **22**, 1391 (1966).
22. F. H. Field and J. L. Franklin, *Electron Impact Phenomena*, Academic Press, New York, 1957.
23. D. H. Williams and R. G. Cooks, *Chem. Comm.*, 663 (1968).
24. R. W. Kiser, *Tables of Ionization Potentials* (TDI-6142), University of Kansas, 1960 and *Supplement*, 1962.
25. T. L. Cottrell, *The Strength of Chemical Bonds*, Butterworths, London, 1958; V. I. Vedeneyev, L. V. Garvick, V. N. Kondrateiv, V. A. Medvedev and Ye. L. Frankevich, *Bond Energies, Ionization Potentials and Electron Affinities*, Edward Arnold, London, 1966.
26. D. P. Stevenson, *J. Am. Chem. Soc.*, **65**, 209 (1943).
27. D. P. Stevenson, *Faraday Soc. Discussions*, **10**, 35 (1951).
28. A. G. Loudon, A. Maccoll and S. K. Wong, unpublished work.
29. D. P. Stevenson and C. D. Wagner, *J. Chem. Phys.*, **19**, 11 (1951).

30. N. Dinh-Nguyen, R. Ryhage and S. Ställberg-Stenhagen, *Arkiv. Kemi*, **15,** 433 (1960).
31. R. O. Tenge, Vol. 3, p. 431 and T. I. Taylor, Vol. 5, Chap. 5, *Catalysis* (Ed. Emmet), Reinhold Publishing Corporation, 1955, 1957.
32. N. Dinh-Nguyen, R. Ryhage, S. Ställberg-Stenhagen and D. E. Stenhagen, *Arkiv Kemi*, **18,** 393 (1961).
33. H. Audier, S. Bory, M. Fetizon, P. Longevialle and R. Toubiana, *Bull. Soc. Chem. France*, 3034 (1964).
34. H. Budzikiewicz, C. Djerassi and D. H. Williams, *Interpretation of the Mass Spectra of Organic Compounds*, Holden-Day Inc., San Francisco, 1964, Ch. 4.
35. J. A. McCluskey and M. J. McClelland, *J. Am. Chem. Soc.*, **87,** 5090 (1965).
36. R. E. Wolff, G. Wolff and J. A. McCluskey, *Tetrahedron*, **22,** 3093 (1966).
37. K. Biemann, *Angew. Chemie*, **74,** 102 (1962).
38. H. Budzikiewicz, C. Djerassi and D. H. Williams, *Structural Elucidation of Natural Products*, Vol. 2, Holden-Day Inc., San Francisco, 1964, Ch. 23.
39. T. G. Halsall and A. W. Oxford, private communication.
40. H. Budzikiewicz, C. Djerassi and D. H. Williams, *Structural Elucidation of Natural Products*, Vol. 1, Holden-Day Inc., San Francisco, 1964, p. 110.
41. A. G. Loudon and P. C. Cardnell, unpublished work.
42. R. C. Dougherty, *J. Am. Chem. Soc.*, **90,** 5780, 5785 (1968).
43. F. P. Boer, T. W. Shannon and F. W. McLafferty, *J. Am. Chem. Soc.*, **90,** 7239 (1968).
44. G. G. Meisels, J. Y. Park and B. G. Giessner, *J. Am. Chem. Soc.*, **91,** 1555 (1969).

CHAPTER **8**

The Radiolysis of Olefins

G. G. MEISELS

Department of Chemistry, University of Houston, Texas, U.S.A.

I.	INTRODUCTION	360
II.	EXPERIMENTAL TECHNIQUES	360
	A. General Aspects	360
	B. Electrons	362
	C. Electromagnetic Radiation	366
	D. Dosimetry	366
III.	RADIOLYSIS MECHANISMS AND REACTIVE INTERMEDIATES	368
	A. General Aspects	368
	B. Historical Aspects	369
	C. Generalized Mechanism	370
	1. Initial event and initial species	371
	2. Classification of unimolecular events according to energy deposited	373
	a. Theoretical aspects	373
	b. Energies below the least endoergic dissociation	373
	c. Energies above the least endoergic dissociation process, but below the ionization potential	374
	d. Energies above the ionization potential	375
	e. Sub-excitation electrons	380
	f. Summary and relative abundance of primary species	381
	g. Division of primary energy absorption in mixtures	382
	3. Spatial distribution of initial events	384
	4. Classification of bimolecular events	388
	a. Primary products and original products	388
	b. Physical effects of collision	388
	c. Chemically reactive collisions of excited states	389
	d. Ion-molecule reactions	389
	e. Neutralization	396
	f. Electron attachment reactions	397
	g. Radical reactions	397
	5. Mixtures	401
	a. Sub-excitation electrons	401
	b. Energy transfer	401

IV. Yields of Products and Intermediates	403
V. Acknowledgement	405
VI. References	405

I. INTRODUCTION

The elucidation of the mechanism of product formation in radiolysis is complicated by the simultaneous contribution of reactions of ionic intermediates, excited intermediates, and free radicals to the totality of observed end products. In the earliest experiments only ionic species were thought to be important participants in radiolysis mechanisms[1,2]. Later, a theoretical analysis suggested that free-radical intermediates were also highly important[3], and for many years thereafter, product formation in olefin radiolysis was interpreted in terms of free-radical reactions alone[4,5]. Mass spectrometric studies of the reactions of ions with neutral molecules called new attention[6-8] to the participation of ionic species in product formation, and extensive work in all phases[9-17] soon revealed that the contributions of ionic and neutral precursors to product formation in olefin radiolysis were approximately equal.

The irradiation of olefins by any form of ionizing radiation leads to the formation of fragmentation products of molecular weight less than that of the starting compound, and to the formation of condensation products of molecular weight exceeding that of the starting compound. Polymer formation is also frequently observed, but constitutes the predominant course of product formation only under particularly chosen conditions.

II. EXPERIMENTAL TECHNIQUES

A. General Aspects

High-energy radiation is frequently also referred to as ionizing radiation, or as penetrating radiation. As such it encompasses both corpuscular and electromagnetic radiation, capable of producing ionization and excitation in molecules. Since energies in excess of 24 eV (1 eV/molecule = 23·06 kcal/mole) can ionize all stable atoms and molecules, such a value might be set as the lower limit for the range of high-energy radiation. On the other hand, most organic compounds and certainly olefins can be ionized at much lower energies, even as low as 8·5–10 eV (Table 1)[18]. Such energies are

TABLE 1. Ionization potentials of some simple olefins*

Compound	Ionization potential (ev)	Reference
Ethylene	10.515	a
Propylene	9.73	a
1-Butene	9.58	a
2-Methylpropene	9.23	a
trans-2-Butene	9.13	a
cis-2-Butene	9.13	a
1-Pentene	9.50	a
2-Methyl-1-butene	9.12	a
3-Methyl-1-butene	9.51	a
3-Methyl-2-butene	8.67	a
cis-2-Pentene	9.11	b
trans-2-Pentene	9.06	b
1-Hexene	9.46	a
trans-2-Hexene	9.16	c
trans-3-Hexene	9.12	c
2,3-Dimethyl-2-butene	8.30	d
3-Ethyl-1-butene	9.21	c
1-Heptene	9.54	c
1-Octene	9.52	c
2-Octene	9.11	c
1-Decene	9.51	c
Isoprene	8.85	a
Cyclopentene	9.01	a
Cyclohexene	8.95	a
4-Methylcyclohexene	8.91	a
4-Vinylcyclohexene	8.93	a
Cyclooctatetraene	7.99	a
1-Methylcyclohexene	8.67	e
3-Methylcyclohexene	8.94	e
Methylenecyclohexane	9.04	e

* The determination of ionization potentials of complex molecules is complicated by the possible presence of vibrational states, etc. Most values must be accepted with reservations.
[a] Reference 18, probable error ±0.02 eV or less, by photoionization.
[b] J. Collin and F. P. Lossing, *J. Am. Chem. Soc.*, **81,** 2064 (1959); probable error ±0.2 eV, by electron impact.
[c] R. E. Honig, *J. Chem. Phys.*, **16,** 105 (1948); probable error ±0.02 eV, by electron impact.
[d] J. P. Teegenand and A. D. Walsh, *Trans. Faraday Soc.*, **47,** 1 (1951); probable error ±0.01 eV, by spectroscopy.
[e] R. E. Winters and J. H. Collins, *J. Am. Chem. Soc.*, **90,** 1235 (1968); probable error ±0.1 eV, by electron impact.

accessible with far vacuum ultraviolet (600–1400 A°) photolysis equipment[19-22] and such techniques are therefore used to complement information obtained by irradiation at higher energies, where the complexity of intermediate formation is greater. Perhaps a good operational definition of high-energy radiation is one which relates the lower-energy limits to the mechanism of energy deposition leading to product formation. For both corpuscular and electromagnetic radiation in excess of approximately 100 or 200 eV, the predominant mode of energy loss is the ejection of an electron from the target molecules or atoms. The secondary electrons then have sufficient energy to cause further ionization and excitation, and are the predominant particles leading to the formation of reactive species. From a practical point of view, however, both electromagnetic and corpuscular energy in the region between 10 eV and several 10^4 eV's is accessible only with difficulty because of the limited penetration of these types of radiation through vessel walls. Optical devices (such as lithium fluoride and calcium fluoride windows) can be employed at energies up to 11·5 eV[21,22] and in the energy range beyond 40,000–50,000 eV it is possible to use thin metal foils, such as those made of aluminium. In the intermediate range, windows only a few hundred nanometers thick or windowless arrangements are required[20,21], which are experimentally very difficult to achieve. We shall therefore restrict our discussions to devices capable of producing radiation of at least 50 keV energy.

B. Electrons

Electron accelerators range from simple high voltage power supplies, suitable for operation up to several hundred kilovolts, through electrostatic machines of the moving belt (VandeGraff) type, cascaded rectifier (Dynamitron) and resonant transformers, to the very high-energy machines such as Betatrons, Synchrotrons, Linear Accelerators, etc. (Table 2). From the radiation chemist's point of view, the most important of these are those which cover the energy range of approximately 0·6 to 10 meV (VandeGraff, Dynamitron, Febetron, Linac, Resonant Transformer). The electron beams typically emerge through thin aluminium or titanium foils (0·001–0·005 inch thick). Beam diameters are usually of the order of 1 cm or less. Electrostatic accelerators and cascaded rectifiers provide a continuous (steady) flow of electrons. Resonant transformers and linear accelerators inherently provide bursts of radiation[24]. The differences in the time characteristics have no appreciable effect on the penetration and

TABLE 2. Characteristics of some electron accelerators

Accelerator type	Energy range	Maximum beam current	Beam characteristics
VandeGraaff	0·6–3·0 meV	10^{-3}A	Continuous (steady) can be modified for pulsed beams.
Dynamitron	1·5–5·0 meV	10^{-2}A	Continuous (steady) can be modified for pulsed beams
Resonant Transformer	1·0–2·0 meV	10^{-2}A	Sinusoidal
Febetron	0·6–2·0 meV	4×10^3A (per pulse)	Pulsed, pulse duration 3–20 nsec, single pulses only.
Linear Electron Accelerator (Linac)	3·0–15·0 meV	10^3A (per pulse)	Pulsed, up to 500 p.p.s., pulse duration 3–30 nsec, fine structure pulses to ·01 nsec.
Betatron	10–300 meV	10^2A (per pulse)	Pulsed, 300–500 p.p.s., pulse duration ca. 1 nsec.

absorption characteristics of high-energy radiation. However, they require a different kinetic examination of reactions occurring in the vessels, as in one case one deals with steady-state kinetics, and in the other with the equivalent of a rotating-sector experiment, or flash photolysis if single pulses of radiation are employed[25]. Electrons may also be provided from radioactive sources such as tritium, nickel 63, cesium 137, etc. These are primarily used for special purposes such as mass spectrometry,[26] hot atom reaction studies[27] and physical measurements[28].

In order to minimize the attenuation of the electron beam through the gas, which will lose approximately 6 keV/inch at 1 atm., the sample vessels employed for gas-phase irradiation should not provide a path length for the incident electrons of more than approximately one inch. While the energy loss per unit path length is small, the scattering of the beam outside of the radiation chamber is considerable if such chambers are long and narrow[29]. On the other hand, the total amount of radiation energy deposited varies approximately linearly with path length of such systems, and in order to obtain amounts of product sufficient for quantitative analysis, it is occasionally necessary to extend the length to a few inches. In such cases allowance must be made for the difficulty in assessing the total energy deposition. One type of irradiation vessel extensively used for gas-phase radiolysis of olefins is shown in Figure 1 [29].

Irradiation vessels for the radiolysis of liquids may simply be open beakers if the accelerator beam is vertical, and if the presence of air in the liquid can be tolerated. Since dissolved oxygen will inhibit free-radical reactions, vessels incorporating thin aluminum walls or other metallic foil windows are usually employed. Radiolysis usually leads to the sizeable formation of gases such as hydrogen, methane, etc., so that an expansion volume must be provided to forestall bursting of the vessel after prolonged irradiation. Moreover, cooling or complete temperature control must be provided, since even at relatively

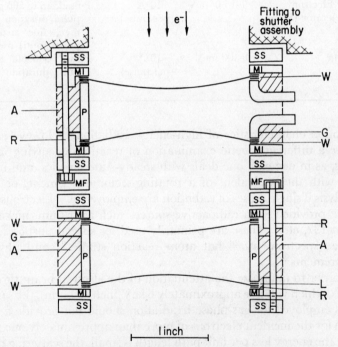

Figure 1. Irradiation cell assembly for gas-phase radiolyses with electrons in the meV range. A—Apiezon wax potting, G—guard rings, L—lead gaskets, MI—Micarta insulator, MF—Micarta fitting, P—Pyrex cylinder, R—retaining bolts, SS—steel pressure rings, W—aluminum window (Ref. 29). The upper portion is filled with the sample while the lower serves as a reference ionization chamber.

low dose rates the rate of heating in liquids is considerable. A typical irradiation vessel used for liquid radiolysis under controlled temperatures is shown in Figure 2[30]. It is noteworthy that the irradiation should not proceed through the gas phase into the liquid, as energy absorption in the gas will lead to products which will be analysed together with those formed in the liquid, possibly leading to confusion.

Actually, radiolysis investigations in the condensed phases are seldom carried out with particle irradiation because of their limited penetration. Exceptions are of course pulse radiolysis studies[25], one of the methods of kinetic spectroscopy, where the intense radiation and limited penetration are desirable because a large amount of energy

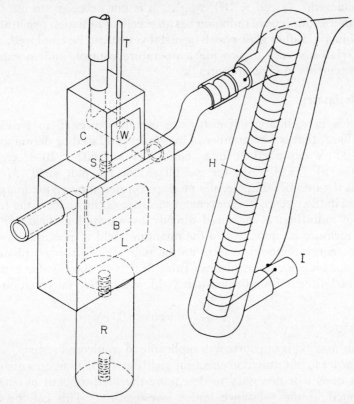

Figure 2. Irradiation cell and temperature regulating device for radiolysis of liquids at sub-ambient temperatures. C—copper cell, 0·1 ml internal volume; T—thermocouple, W—aluminum, brass or titanium foil window. B—cooling block with junction to irradiation cell. R—support rod; I—precooled nitrogen passing through heaters H to cooling loop L in cooling block (Ref. 14).

is deposited in a small volume. This permits the detection of intermediates directly by spectroscopic absorption measurements. For most bulk studies, however, X-rays or γ-rays are preferred.

C. Electromagnetic Radiation

Electromagnetic radiation can be produced either by the impact of highly energetic particles on solid targets, such as tungsten or gold, or by the γ-decay of radioactive isotopes. Cobalt 60 sources, up to several tens of thousands curies (1 C = 3·70 × 10^{10} disintegrations/sec) are most frequently employed. This isotope emits γ-rays of 1·173 and 1·332 meV energy, and dose rates up to several megarads per hour can be achieved. (1 rad = 100 erg/g). It is not necessary to use thin windows as this type of radiation has an excellent radiation penetration capability. Either glass vessels or metal vessels can be employed, and it is relatively easy to provide temperature control, and to extend measurements to high pressures[31].

D. Dosimetry

Yields in radiation chemistry are usually expressed in terms of 'G values', that is, the number of molecules produced or decomposed per 100 eV deposited in the system (1 eV = 1·602 × 10^{-12} erg = 3·838 × 10^{-20} cal.). In the gas phase, the ion pair yield (M/N), that is the number of molecules produced or decomposed per ion pair formed in the system, is a convenient measure of the yield, if the total rate of ionization is measured directly by an applied field[29,32-34]. This measure is equivalent to the quantum yield in photochemistry. If the energy required to form an ion pair (W) in the gas phase is known, as it is for many olefins (Table 3)[29,35-39], the G-value can be obtained directly from the ion-pair yield, and vice-versa (equation 1).

$$G = \frac{100}{W} \cdot \frac{M}{N} \text{ (molecules/100 eV)} \qquad (I)$$

This method is of course not applicable to condensed systems.

In order to put radiation chemical-yield determinations on a quantitative basis it is necessary to determine the total amount of energy deposited in the substance under investigation. This can be accomplished by several means. The first is an absolute method, and relies on the rise of temperature of a sample under investigation. Aqueous systems are normally chosen since here one can, under appropriate conditions, convert all the energy deposited to heat[40].

TABLE 3. Electron energies required to form an ion pair in some olefins

Compound	W, eV/ion pair*	Reference
Ethylene	26·1 ± 0·2	a,b,c,d
Propylene	24·8 ± 0·2	b,d
1-Butene	24·4 ± 0·2	b
	23·8 ± 0·4	d
2-Methylpropene	24·4 ± 0·2	b
	23·8 ± 0·4	d
cis-2-Butene	23·6 ± 0·4	d
trans-2-Butene	23·9 ± 2	b
	23·6 ± 0·4	d
1-Hexene	23·9	c
2-Hexene	23·4	c
Allene	25·2 ± 0·4	d
1,3-Butadiene	25·0 ± 0·4	d
Fluoroethylene	28·3 ± 0·4	d
Chloroethylene	24·7 ± 0·4	d
Bromoethylene	23·7 ± 0·4	d
Iodoethylene	20·9 ± 0·6	d

* Errors are those indicated by the authors, except for ethylene where the average deviation of the values reported by all investigators was taken.

[a] Work prior to 1963, chiefly that of Jesse, summarized by White[36].

[b] G. G. Meisels[29], 1 meV electrons.

[c] P. Adler and H. K. Bothe[37], ^{63}Ni β-rays.

[d] M. Leblanc and J. A. Herman[38], Compton electrons from ^{60}Co γ-rays.

This is a difficult and inconvenient method, and is employed only to establish a secondary standard, or dosimeter. Popular dosimeters for γ-rays have been the oxidation of ferrous ion in aerated solutions of ferrous sulphate in 0·8N sulphuric acid [$G(Fe^{+++}) = 15·6$ ions/100eV][40] and the hydrogen yield from cyclohexane [$G(H_2) = 5·6$ molecules/100 eV][41,42]. In the gas phase one can measure the number of molecules produced per ion pair in the system (M/N) using an applied electrostatic field to collect the total ionization[29,32-34]. The nitrogen yield from nitrous oxide has been employed as a secondary standard, but there appears to be considerable controversy over the exact yield in this instance, the most probable value being $G(N_2) = 10·0$ molecules/100 eV[43]. At present the use of hydrogen formation from ethylene is preferred, with $G(H_2) = 1·3$ molecules/100 eV, as independently evaluated by several investigators[12,13,29,44,45].

When the substance in the radiation vessel is changed, it is necessary to make allowance for differences in stopping power of the medium under irradiation from that of the dosimeter. For γ-rays and highly energetic X-rays this is readily accomplished by comparing the relative electron densities [(number of moles/unit volume) × (number of electrons/molecule)]. The absorption of electromagnetic irradiation at energies of 1–2 meV is primarily by the Compton effect, and a simple consideration of electron densities is therefore sufficient. For particle irradiation the matter is more complicated. When isotopic sources are employed where the total energy is expended in the system, stopping power considerations will not enter, although back scattering into the source may be important for weak β-rays [46]. When beams of high-energy radiation are employed, stopping powers must be calculated by the Bethe equation [47,48], and values substantially different from electron densities may be obtained. Moreover, one must allow for attenuation of the beam by both absorption and scattering. These problems are readily overcome for electrons, but are not as easily accounted for when heavy ions are employed, and these are therefore usually permitted to expend all their energy in the sample.

III. RADIOLYSIS MECHANISMS AND REACTIVE INTERMEDIATES

A. General Aspects

The interaction of high-energy radiation with chemicals in all phases leads to the formation of neutral excited states, and of ionic species in their ground and electronically excited states. All these may fragment, leading to the formation of free radicals and fragment ions. An elucidation of the nature and the reactions of these species is one of the chief goals of the radiation chemist.

Approximately one-half of the product formation occurs via ionic intermediates in the gaseous olefins [49]. In the liquid phase, reaction of ionic species competes with neutralization since the electron does not fully escape the influence of the parent ion [50-55]. The chemistry is therefore more complex in the latter case, but is still dominated by reactions of ionic species, since the lifetime of the ion before neutralization is long compared to collision frequency. Even though physical measurements may not be able to determine the total number of ion pairs, chemical measurements [56-59] indicate clearly that the yield of ionic species is nearly that in the vapour phase and can characterize much of the chemistry observed.

Radiolysis mechanisms involve the same initial energy absorption processes, and the same types of intermediates in all phases. Only the relative importance of these, the time scale, and the spatial distribution of these events are affected by a change in density or phase. We shall therefore examine, primarily, energy absorption processes and the types of intermediates produced in the radiolysis of gases, and indicate differences for condensed-phase systems as required. It is necessary to exercise considerable caution when attempting to apply gas-phase mechanisms to liquids or solids. Major differences arise from reduced ion-pair lifetimes, from non-uniform energy deposition in spurs and from subsequent diffusion-controlled reactions. The effect of solvation on ion-stability, ionization potential, and ion-electron recombination, and collective excitation and ionization, increased probability of intersystem crossing and vibrational deexcitation, and solvent effects on the rate constant are also important factors.

Our presentation shall attempt principally to develop a generalized mechanism for the sequence of events leading from the first steps of energy absorption to the final, stable products.

B. Historical Aspects

The early mechanisms of olefin radiolysis, based on the formation of ionic clusters, did not take account of the participation of excited or radical species at all. A major advance was made in the mid-1930's when Eyring, Hirschfelder and Taylor[3] suggested that the initial formation and reactions of ionic species might lead to free radicals, and that excitation processes lead to the same species with approximately equal efficiency. Further chemical changes could then be ascribed to the radicals produced by either mechanism. This view was eagerly accepted, since it allowed the reduction of radiation chemistry to the much better known areas of photochemistry and free-radical chemistry. Although ionic reactions leading to intermediates or products were considered in detail by these authors, this possibility was almost totally neglected for the next twenty years. The rediscovery of ion-molecule reactions by Tal'rose[6], by Stevenson[7], and by Franklin, Field and Lampe[8] demanded allowance for ionic reaction steps in chemical product formation, and demonstrated their importance in olefins[60-68]. Although the relative importance of the two types of reaction initiation varies with the nature of the compound, no systems are known where one of them fails to make any contribution at all.

The recent progress in an understanding of radiation chemistry can perhaps be traced by following the meaning associated with the 'wiggly arrow' ($\leadsto$), a symbol which is normally read as 'on irradiation yields', and stands, in effect, for our ignorance and our uncertainty. Decades ago, the 'wiggly arrow' could have been used to connect the starting materials to the final end products. Later, it led to the listing of intermediate radicals, and then to specific intermediate ions and excited species. Thus, some of the steps which the arrow had covered were removed and put into separate categories. The frontiers today are largely attempts to separate additional steps, such as unimolecular dissociations of excited and ionized species, detailed descriptions of precursors, and analysis of non-homogeneities in reactive species formation, which will further reduce the significance of the 'wiggly arrow'.

C. Generalized Mechanism

There are several ways of discussing the mechanism of radiolysis generally, but we shall follow the one based on the time scale of events as an approximate guide line. The first stage, which we shall term the initial stage, and which is often called the physical stage[69], refers to all events which occur within a time period short in comparison to a molecular vibration, that is, within less than about 10^{-14} sec. The species produced within that time will be called the initial species. In the gas phase we can conveniently define a second stage in which only reactions of isolated species may occur, that is, a period of unimolecular processes before the energized initial entity experiences a collision. The duration of this 'primary unimolecular' stage will depend on the pressure in the system, but is typically of the order of 10^{-9} to 10^{-11} sec. In liquids, a period of the order of 10^{-12} to 10^{-13} sec would be involved. Lastly, there will be a period in which bimolecular events can occur, and contribute to ultimate product formation. The duration of this 'bimolecular' period will depend on the pressure and the dose rate to which the radiolytic system is subjected, as well as on the kind of the reacting species. At dose rates of approximately 10^{15} eV cc^{-1} sec^{-1}, roughly equivalent to high-energy electrons from an accelerator, ions will have lifetimes on the order of microseconds, and free-radical species will survive for as long as milliseconds. At the much reduced dose rates when Co^{60} γ-radiation is employed, these lifetimes will normally be increased by an order of magnitude or more. The present definitions differ slightly from the physicochemical and chemical stage definitions of Platzman[69], but lend themselves more readily to a systematic discussion in the gas phase. It should be noted that unimolecular dissociation or rearrangement processes can occur during the bimolecular period as well, but can involve 'primary' species (those undergoing their first collision) only when the collision process does not affect them at all.

I. Initial event and initial species

The initial event occurs within a time period very short with respect to a change in interatomic distances, and in this stage therefore, no ultimate product formation can occur. The physical event of depositing energy in a molecule can lead to the excitation of a species to levels below and above the ionization limit ('super-excited states')[70,71], and to immediate ionization.

The optical approximation[47,70] suggests that the initial event induces primarily allowed transitions from the ground state of the molecule. The cross-section for excitation of spin or otherwise forbidden levels is large only when the incident electron energy is at near-resonance with the level to be attained. Only a small portion of the initial events should fall into this category. Although one would thus expect essentially no contribution of forbidden levels to radiolytic action, there is virtually no verification of this from chemical measurements. The reason for this is several-fold. First, there are few systems in the gas phase where the formation of triplet states can be unequivocally established at the exclusion of the participation of singlet states. Secondly, the chemical identification of triplet states is generally based on dissociation or rearrangement processes which are frequently quenched at the pressures convenient for radiation chemical investigation. Lastly, and most importantly, intersystem crossing may occur in complex molecules leading to the formation of triplet states, not in the initial event, but in subsequent intramolecular processes.

There have been a few chemical examples where it was relatively certain that super-excited species were formed and dissociated into neutral products. In all those instances the dissociation process invariably led to the same products as those arising from excitation of singlet states below the onset of ionization[72,73,74]. Although chemical measurements can be helpful in assigning the proper choice of dissociation path[75], they can yield little additional information on the competition between ionization and dissociation to that obtained by physical measurements.

The formation of ions and secondary electrons by first-order processes can occur as a consequence of direct or immediate initial ionization, including autoionization within 10^{-14} sec.,

$$C_2H_4 \longrightarrow C_2H_4^+ + e \tag{1}$$

or as a result of preionization processes from highly excited states

$$C_2H_4 \longrightarrow C_2H_4^* \tag{2}$$
$$C_2H_4^* \longrightarrow C_2H_4^+ + e^- \tag{3}$$

or by ion-pair formation

$$C_2H_4 \longrightarrow C_2H_4^* \tag{4}$$
$$C_2H_4^* \longrightarrow C_2H_3^+ + H^- \tag{5}$$

The last type of process is unimportant in all olefins not containing heteroatoms. Autoionizing levels exist even in the rare gases. Their lifetimes in H_2 are probably on the order of 10^{-13} to nearly 10^{-7} sec[76]. Therefore, they clearly span both the initial unimolecular and the bimolecular reaction stages, in which chemical product formation may occur.

A question of considerable importance concerns the distribution of excess electronic energies with which the parent ions are formed. This knowledge is required because it permits an estimate of the lifetime of the ion before unimolecular dissociation, and a better extrapolation of mass spectrometric data to the higher pressures typical of radiation chemistry experiments. It is usually assumed that energy transfer from 70 eV electrons is representative of energy transfer from high-energy radiation[49]. Calculations of an average excess energy can then be based on the relative abundance of fragment ions in a mass spectrometer[77]. Another approach is the evaluation of the electron-energy deposition distribution function. Chupka and Kaminsky[78] have carefully analysed the excited state and fragment ion-production in butane and propane, and arrived at an approximate distribution function. In order to do so they have assumed that the probability of ionization is linearly proportional to energy above the threshold, at least for a limited range. There is some uncertainty associated with this assumption[79-80], but the main features are probably correct. The excess energy is usually rather small, peaks at 1–2 eV above threshold, and the distribution falls off rapidly at transferred energies in excess of 4 or 5 eV. Moreover, the energy distribution is thought to be essentially independent of incident electron energy at sufficiently high energies. This is supported by the classical observation that fragmentation patterns, even of relatively complex molecules, are fairly insensitive to electron energies for several hundred volts above 50 eV[82,83]. No allowance was made for super-excited states (Section III.C.2.d) in either method. No analogous attempt has been made for olefins.

2. Classification of unimolecular events according to energy deposited

a. Theoretical aspects. Excitation to an allowed state may lead to essentially immediate dissociation ($\sim 10^{-12}$ sec.). Delayed fragmentation may result if the breaking of the bond required predissociation or a series of complex processes to degrade electronic to vibrational excitation energy by radiationless transitions, including intersystem crossing and internal conversions. When dissociation occurs from highly vibrationally excited states, one can estimate rate constants from unimolecular rate theory. This assumes that the vibrationally excited molecules can be regarded as a collection of loosely coupled oscillators, dissociation occurring when sufficient energy is localized in the dissociating bond[84]. Although the physical picture is almost certainly correct, quantitative mathematical treatments are complicated by the energy dependence of the effective number of oscillators and anharmonicity effects, to name but two[85,86,87]. The same considerations and the same difficulties apply to the fragmentation of ions[88,89].

Our chief concern here is a qualitative discussion of the possible events leading to chemical product formation. This can be done conveniently by separating the initial events according to the amount of energy deposited in the molecule. Appropriate dividing points for the discussion are the lowest energies required for fragmentation, and the ionization potential of the molecule.

b. Energies below the least endoergic dissociation. The interaction of the degradation spectrum electrons with molecules does not lead to appreciable vibrational and rotational excitation[90]. The small extent to which such processes might occur should not lead to deposition of enough energy to cause chemical changes. Therefore we need concern ourselves only with electronic excitation. The lowest excited states of olefins are spin-forbidden triplet states, and range from somewhat less than 4·1 eV for ethylene, to less than 3·5 eV in the butenes[91]. Although such states are not produced abundantly in the initial process, they may be reached by intersystem crossing, possibly followed by internal conversion and degradation to the lowest state of the system, and can therefore contribute to the chemistry. If the level is below the least endoergic fragmentation process, chemical changes may still occur if the triplet state has a molecular configuration different from the ground state. The triplet state of olefins has a minimum energy configuration at right angles[92], so that *cis-trans* isomerization can result readily[93]. If the molecule has insufficient energy to dissociate it must eventually lose its energy by a low probability radiative process, or by intersystem crossing and eventual vibrational degradation.

c. Energies above the least endoergic dissociation process, but below the ionization potential. Most excitation processes are well above the lowest fragmentation energy, and the isomerization, degradation, and rearrangement processes mentioned above will therefore be in competition with dissociation. This may occur either by the breaking of a simple bond, producing two radical species, or by the elimination of a stable entity such as molecular hydrogen. Such processes then produce either two stable entities, or a stable species and a diradical.

In ethylene radiolysis, the neutral excited precursors dissociate by one of the three paths:

$$C_2H_4^* \longrightarrow \begin{cases} C_2H_2 + 2H & (6a) \\ (CHCH, CH_2C) + H_2 & (6b) \\ C_2H_3 + H & (6c) \end{cases}$$

with relative probabilities of $0.87:0.10:0.03$ [16]. A sizeable fraction of the C_2H_2 is probably initially formed as CH_2C:, but rearranges rapidly to acetylene as suggested from the isotopic constitution of

$$CH_2C \longrightarrow CHCH \qquad (7)$$

hydrogens produced in the radiolysis of CH_2CD_2, and $CHDCHD$ [13]. The radiolysis of other olefins has not been investigated in sufficient detail to permit such an assignment, although photolytic studies in the far vacuum u.v. have suggested the following initial steps for propylene [19,22,94]:

$$C_3H_6^* \longrightarrow \begin{cases} C_3H_4 + H_2 & (8a) \\ C_2H_2 + CH_4 & (8b) \\ C_2H_3 + CH_3 & (8c) \\ C_3H_5 + H & (8d) \end{cases}$$

The relative probabilities of isomerization-degradation processes and dissociation depend on the state from which the two processes originate. Excitation to a repulsive state clearly will produce immediate dissociation and no rearrangement. Non-dissociative states, where radiationless transitions are required, can be expected to have lifetimes sufficiently long to allow competition between radiationless transitions, radiative return to the ground state, and collisional processes. An excellent summary of the possible photophysical processes is given by Calvert and Pitts[95], and the reader is referred to their general chapters for a very lucid and detailed description.

The dissociation of molecules can yield radicals with an epithermal energy distribution, and such species are referred to as 'hot' radicals or 'hot' atoms. If the dissociation occurred from a highly vibrationally excited ground state, appreciable excess kinetic energy is not to be expected in the products, on the basis of unimolecular dissociation rate theory[96]. However, any fragmentation resulting from a dissociative state may easily lead to retention of excess translational energy in the dissociating partners. Evidence for the participation of 'hot' species, particularly hydrogen atoms, has frequently been sought[97,98], but no definitive evidence for olefinic systems has been reported. Hydrogen atoms add readily to olefins[99-101], for example:

$$H + CH_3-CH=CH_2 \longrightarrow CH_3-CH-CH_3 \qquad (9)$$

but can also abstract a hydrogen atom from the hydrocarbon molecule:

$$H + CH_3-CH=CH_2 \longrightarrow H_2 + CH_2CHCH_2 \qquad (10)$$

The latter process is preferred for 'hot' atoms, and inconsistencies in the ratios of abstraction to addition are sometimes ascribed to the participation of 'hot' atoms in mechanisms. The reactivity of radiolytically produced hydrogen atoms has been used to estimate rate constants for processes such as reactions (9) and (10) for various olefins[102], but the method is suspect because of the complications of energy and charge transfer[53,103].

d. Energies above the ionization potential. Platzman[69,104,105] has repeatedly emphasized an important consequence of the optical approximation[45]: the most abundant radiation interactions are those with the greatest values for $f(E)/E$, the ratio of the oscillator strength to the energy of the transition. Therefore, the major portion of excitation processes involves outer shell and bonding electrons. The bulk of the oscillator strength lies very high, between 10 and 30 eV above the ground state. Even when low states, resulting from double bonds and conjugation, exist in a molecule, the oscillator strength is larger at energies above the ionization potential. While the optical approximation thus leads to the conclusion that perhaps 60–70 percent of the total oscillator strength resides at energies where ionization is possible, experimental values of W can be analysed to show that ionization and excitation in olefins are frequently equally important. In addition to this discrepancy, it has been observed that the increase in ionization when gases of low ionization potential are added to rare gases under irradiation (Jesse Effect)[106] depends on molecular parameters other than the ionization potential. This has been attributed to the formation of neutral molecules excited beyond the ionization potential. This class of states has been termed 'super-excited states', and includes inner-shell excitation, two-electron excitation, and outer-electron vibronic excitation. The latter may result from large vibronic overlap[76]. The Franck-Condon envelope should lead to excitation relatively high on the repulsive portion of the energy surface, since highly excited states of bonding electrons have a looser configuration than in the ground state of the molecule. Since the mode of excitation appears to be immaterial[106], ionization efficiencies can be assessed by studying photoionization efficiency as a function of photon energy[107-109], or by studying the Jesse effect with a series of rare gases of different, discrete long-lived excited states[106]. The latter method employs energy transfer from metastable states and trapped resonance radiation, but suffers from the simultaneous population of several energy levels[110] and limited energy values.

Platzman[104] suggested that competition between ionization and dissociation occurred and depended on the energy of the super-excited state. Ionization of the super-excited state leads presumably only to the molecular ion. The ionization efficiencies, defined as the number of ion pairs produced per quantum absorbed, show an isotope effect. This supports the competitive model since the rate constant k_i for ionization should be unaffected by the difference in zero-point energies of the species involved, while the rate constant for homolytic dissociation (k_d) should show a typical isotope effect[84], $k_d = k_d'/\sqrt{m^*}$, where k_d' is a collection of terms independent of isotopic substitution and m^* is the reduced mass of the dissociated portions A and B, $m^* = m_A m_B/(m_A + m_B)$.

If a portion of δ of the energy deposition processes leads to direct ionization without competitive dissociation, the overall ionization efficiency at a particular energy should be given by

$$\eta(E) = \delta + (1 - \delta)k_i (k_i + k_d'/\sqrt{m^*})^{-1} \qquad \text{(II)}$$

Platzman suggested that all ionization is competitive with dissociation ($\delta \ll 1$) and it seemed reasonable that the modes of decomposition of a molecule could be predicted from the magnitude of the isotope effect on the ionization efficiency. For example, in the case of ethylene activated by collision with excited argon, the measured value for η_D/η_H was 1·085, suggesting that the predominant mode of dissociation of super-excited ethylene was a symmetrical split into two CH_2 (or CD_2) radicals[111]. The value for η_D/η_H predicted for this process was 1·05, while a ratio of 1·27 would be expected if the reaction were H- or D-atom elimination.

The species CH_2 is not, in fact, an important intermediate in the argon-sensitized radiolysis of ethylene[73]. Analysis of radiolysis products showed that the super-excited ethylene molecule dissociates predominantly through the reactions

$$C_2H_4^* \longrightarrow [H + C_2H_3] \text{ or } [H_2 + C_2H_2] \tag{11}$$

More recently, Person and Nicole[114] have measured the total absorption cross-section (σ) and photoionization quantum yields of ethylene as a function of energy (Figure 3). They observed absorption peaks above the ionization threshold for C_2H_4 and C_2D_4. The ionization cross-sections do not increase in the region of these peaks, indicating that the excited state resulting from the photon absorption does not contribute to ion formation. These authors point out that in this energy region, the isotope effect on the ionization efficiency can be interpreted solely by the difference in energy thresholds to reach the ionizing excited states, that for C_2D_4 being 0·1 eV higher than that for C_2H_4. Indeed, the ionization efficiency of the deuterated molecule is lower ($\eta_D/\eta_H < 1$) for ethylene and other hydrocarbons at energies close to the threshold for ionization[23]. This cannot be predicted on the basis of the competitive model. It appears that one can relate the reversal of the isotope effect on the ionization efficiencies at energies near threshold to the fact that the ionization potentials of some deuterated compounds are higher than those of their perprotonated analogues.

If η could indeed be described simply in terms of competition between ionization and dissociation, one would expect that the ratio η_D/η_H would, generally speaking, decrease continuously with increasing energy (i.e. as both η_D and η_H approach unity). Such a trend in η_D/η_H was observed by Jesse[106] for ethylene. However, the precise reverse should be observed near the onset of ionization if differences in ionization potential contribute to the isotope effect, and such a reversal has indeed been reported[23,106]. Ionization efficiencies are thought to approach unity as the photon energy of excitation becomes very high,

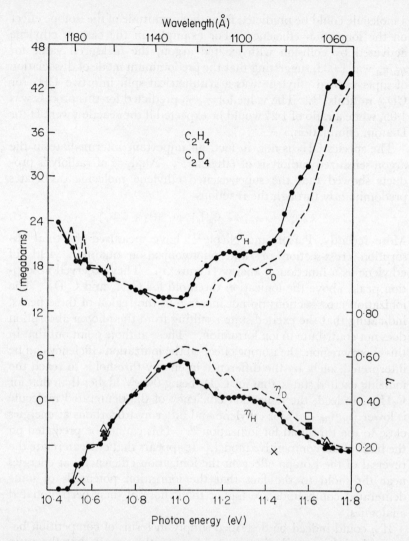

Figure 3. Photon absorption cross-section (σ) and ionization efficiencies (η) for ethylene (subscript H) and deuteroethylene (subscript D) (Ref. 114).

and this appears to be correct for many molecules[107,108]. However, the ionization efficiencies of olefins apparently do not exceed 0·8 even at 21 eV[115]. At higher ionization efficiencies the isotope effect should diminish, and one would therefore expect a maximum in η_D/η_H for most molecules.

It is clear that the interpretation of values η_D/η_H is of great complexity. Besides the effects caused by the existence of nonionizing super-excited states in some systems, and the possible crossover of energy due to differences in ionization potential between the deuterated and non-deuterated analogues, both autoionization[116] and direct ionization[117] may exhibit slight isotope effects which are difficult to evaluate.

Ionization of complex organic molecules by 50–100 eV electrons leads to a distribution of excitation energies which has a maximum at the ionization potential and then essentially decreases with increasing energy[104]. After the electron is ejected and if there is sufficient excitation energy, the ion may dissociate into fragment ions and neutral products, which then may dissociate further. Important ions in the mass spectra of simple olefins are summarized in Table 4. This area

TABLE 4. Most important ions in the mass spectra of some simple olefins[a]

Ion	percent of total ion current					
	Ethylene	Propylene	1-Butene	cis-2-butene	Isobutylene	1-Pentene
CH_2^+	2.4	0.9	0.5	0.3	0.6	0.3
CH_3^+	0.2	1.4	1.4	1.2	1.5	1.4
C_2H^+	4.5	0.6	0.3	0.5	0.3	0.1
$C_2H_2^+$	23.8	2.7	3.1	3.5	1.7	1.4
$C_2H_3^+$	24.2	10.0	9.3	9.3	6.4	8.3
$C_2H_4^+$	38.9	0.4	9.1	8.8	6.7	1.3
$C_2H_5^+$			4.1	4.3	3.4	6.9
$C_3H_3^+$		18.6	10.7	9.9	13.6	8.8
$C_3H_5^+$		26.1	30.9	27.7	31.2	11.5
$C_3H_6^+$		17.7	1.0	0.1	0.1	25.2
$C_4H_7^+$			5.4	6.1	5.0	14.8
$C_4H_8^+$			11.4	13.3	13.3	0.1
$C_5H_{10}^+$						8.1

[a] From the catalog of Mass Spectral data, API Project 44, Texas A & M, College Station, Texas, 70 eV electrons.

has been one of considerable interest to mass spectrometrists, and is the general subject of the quasi-equilibrium theory of mass spectral fragmentation[85,88,89]. Application of this theory to real systems suffers somewhat from a lack of characterization of the ionic species involved, and from problems common to all unimolecular dissociation rate theory. It has, however, proved highly successful in explaining ionic fragmentation processes.

An important aspect of mass spectral theory is its ability to suggest the rate constant of the fragmentation process. In the mass spectrometer, ion lifetimes are usually on the order of 10^{-6} sec or 10^{-5} sec, while typical collision-free periods in radiation chemistry are on the order of 10^{-9} to 10^{-11} sec, and perhaps 10^{-13} sec in condensed phases. The frequently employed practice of using mass spectral fragmentation patterns and W values to calculate the distribution of primary ions in radiolysis is therefore subject to considerable uncertainty[77,118,119].

Application of the statistical theory of mass spectra shows that the fragmentation of the parent ion is rapid and does not change materially between 10^{-5} and 10^{-10} sec[119]. This indicates that the parent-ion abundance should show little dependence on pressure, the chief factor determining the collision-free period. This is reasonably well borne out by studies of the radiolysis of hydrocarbon systems, where the yield of parent ions appeared to be close to that predicted from mass spectrometry[17,120,121]. The situation regarding the abundance of fragment ions, however, is considerably more complex, since in this case one must also consider the possibility of further fragmentation. Secondary fragmentation appears to be considerably slower, and should therefore show a sizable pressure dependence, which finds experimental support in the work of Ausloos[120,121]. One may assume, therefore, that mass spectral fragmentation patterns are applicable to radiation chemistry with the following provisions. The yield of fragment ions arising from two or more successive fragmentation processes may be expected to be drastically dependent on and chiefly decreased under radiation chemical conditions. The yield of secondary ions arising from only one fragmentation process may be enhanced if they undergo considerable further fragmentation in the mass spectrometer, but will remain relatively unchanged if they do not. Superimposed on this will be a general but slow trend towards stabilization of the parent ion at elevated pressures. In liquids and solids extensive parent-ion stabilization may be expected.

e. Sub-excitation electrons. When the energy of the electron falls below that required to excite the lowest electronic level of a molecule, further energy loss can only be by excitation of vibrational and rotational modes, slow processes[122] unless intermediate negative ion states are involved[90,123]. Even then, theoretical considerations for hydrogen[90] have shown that the cross-section peaks at 7·5 eV and drops to zero at 4·5 eV. Apparently electrons typically lose the last few eV of energy inefficiently, as originally pointed out by Platzman[124].

The role of sub-excitation electrons is particularly important if attachment processes can occur at resonant energies only, and in mixtures. Allowance for their contribution requires a knowledge of their yield and energy distribution function. While the former must be less than the ion-pair yield, the distribution function is not directly accessible. El Komoss and Magee[125] have suggested the function

$$F(\varepsilon) = \frac{1}{[1 + (\varepsilon/I)]^3} \frac{[1 + (E_0/I)]^2}{E_0[1 + (E_0/2I)]} \qquad \text{(III)}$$

With the condition $0 < \varepsilon < E_0$, for the number of sub-excitation electrons formed per unit energy interval at energy ε. E_0 and I are the lowest electronic excitation and ionization potentials, respectively. This relationship is consistent with the empirical analysis of Platzman and Jesse[104,106] based on enhancement of ionization in helium and neon by the addition of other rare gases. These authors suggested that the total population of electrons in the energy range ε to E_0 was proportional to $(1 - \varepsilon/E_0)$. This corresponds to a linear decrease with energy, and may be compared with the El Komoss and Magee function integrated between ε and E_0 or $\{[(1 + E_0/I)/(1 + \varepsilon/I)^2 - 1]/(E_0/2I + E_0^2/4I^2)$.

Electrons in the sub-excitation energy range may either be unreactive towards the substrate molecule, attach themselves with the formation of stable negative molecule ions, or undergo a dissociative attachment process[123,126]. Little is known about such processes for simple olefins, and these are probably unreactive (see Section III.C.4.f). In complex systems also containing heteroatoms attachment processes become dominant[126].

f. Summary and relative abundance of primary species. To recapitulate, primary species are all those which have not experienced a collision after their formation by the initial radiation impact. These consist of excited states, both at forbidden and allowed levels, super-excited states, parent ions, fragment ions, and the molecular and radical species resulting from the primary unimolecular dissociation. Moreover, the free electrons at energies below the lowest excited level must be regarded as primary species. An *a priori* assessment of their relative probabilities of formation could be made if one knew the total degradation spectrum of electrons, the energy transfer function to the molecule at each energy, and lastly, the detailed chemical fate of the molecules excited at a particular energy[70,104,127]. Our knowledge in the first area is satisfactory down to energies of a few hundred electron volts, however, we know little about the most important range of even lower energies[48]. Information in the second area is available but incomplete. Lastly, information on the exact chemical consequence of energy absorption at any particular energy is only very slowly becoming available. An alternate and more successful approach has been that of Platzman[104,105], who bases his consideration essentially on the optical approximation, that is, he assumes that high-energy radiation is 'white light' of all frequencies, with equal intensities at all wavelengths, corresponding to the distribution $n(\nu)$ of the number of photons per energy (frequency) interval

$$n(\nu) \simeq \text{constant}/\nu \tag{IV}$$

Since each frequency is absorbed in proportion to its generalized oscillator strength, this approach also requires an estimate of the oscillator strengths for the possible transitions of the outer shell and valence electrons. This avoids the complications in the first two areas above.

A priori estimates are helpful in guiding our approach to an understanding of the detailed events, but at present lead only to the crudest estimates of product yields. Hatano, Shida and Inokuti[127] have attempted to correlate hydrogen yields in butene radiolysis with estimates based on the optical approximation with reasonable success.

g. Division of primary energy absorption in mixtures. The division of initial energy absorption between the components of a mixture has often been estimated on the basis of 'electron fractions'. These are calculated by multiplying the number of moles of each component in the mixture by the number of electrons of the corresponding molecule, and dividing by the total, for example

$$F_A = \frac{m_A n_A}{m_A n_A + m_B n_B} \tag{V}$$

where F_A is the electron fraction of component A in a mixture of m_A and m_B of species A and B, containing n_A and n_B electrons per molecule, respectively. This approximation arose from the well-known attenuation coefficient for ^{60}Co γ-rays, where the Compton effect is the chief absorption process and is proportional to the electron density. Total electron density is therefore a reasonable approximation to total energy deposition in the cell under irradiation. The use of electron fractions to estimate energy division between mixture components is seldom satisfactory, since this would be a special case of the optical approximation [128]. The partition will to some extent depend on the portion of the oscillator strength residing at lower energies, and this method will lead to *a priori* estimates of the energy division which must be considered as rough guidelines only. Experimental evidence suggests that electron fractions are adequate only when comparison is made between compounds of a homologous series [129].

The energy division is actually determined by the degradation spectrum of electrons in the mixture, and the cross sections for individual processes under these specific conditions [104,105,130]. As a compromise one can use cross-sections (σ) calculated for electron-molecule interaction using polarizabilities [130],

$$\sigma \leq \pi e \sqrt{\alpha/2 I} \qquad (VI)$$

where α is the angle-averaged polarizability, e the charge of the electron and I the ionization potential. Alternately, one may calculate stopping power on the basis of the Bethe stopping-power equation for electrons [29,27]. Stopping power for other molecules can be calculated on the basis of Bragg's rule of summing stopping powers of the indivi-

TABLE 5. Mean stopping-power ratios relative to air for electronic equilibrium spectra generated by monoenergetic initial electrons [131]

Initial energy (meV)	Hydrogen		Carbon		Nitrogen			Oxygen	
	at saturated site	at unsaturated site	saturated	unsaturated	highly chlorinated	amines, nitrates	heterocyclic	—O—	=O
0·1	2·52	2·59	1·016	1·021	1·047	0·976	1·018	0·978	0·994
0·2	2·52	2·59	1·015	1·019	1·043	0·978	1·016	0·979	0·995
0·3	2·48	2·55	1·014	1·018	1·040	0·979	1·016	0·981	0·995
0·4	2·46	2·53	1·014	1·018	1·038	0·980	1·015	0·981	0·996
0·5	2·44	2·51	1·013	1·017	1·037	0·980	1·015	0·982	0·996
0·6	2·44	2·50	1·012	1·016	1·035	0·980	1·013	0·981	0·995
0·7	2·42	2·48	1·010	1·013	1·033	0·978	1·011	0·980	0·993
0·8	2·40	2·46	1·009	1·012	1·031	0·978	1·010	0·979	0·992
1·0	2·39	2·44	1·004	1·008	1·026	0·975	1·005	0·977	0·988
1·2	2·37	2·42	1·001	1·004	1·022	0·973	1·002	0·974	0·985
1·3	2·36	2·42	0·999	1·002	1·019	0·971	1·000	0·972	0·983
1·5	2·35	2·39	0·995	0·998	1·015	0·967	0·996	0·969	0·980

dual atoms. This, of course, neglects the influence of binding on the stopping power which is known to be considerable at lower energies. Allowance for this difficulty can be made if different stopping powers are assumed for the atoms in saturated and unsaturated compounds (Tables 5 and 6)[131]. When mixtures consist of olefins only, most

TABLE 6. Mean mass stopping-power ratios relative to air for electronic equilibrium spectra generated by Compton recoil electrons from monoenergetic γ-rays[131]

γ-ray energy (MEV)	Hydrogen		Carbon			Nitrogen		Oxygen	
	at unsaturated site	at saturated site	saturated	unsaturated	highly chlorinated	amines, nitrates	heterocyclic	—O—	=O
0.15	2.73	2.85	1.020	1.027	1.058	0.970	1.022	0.972	0.992
0.25	2.62	2.72	1.017	1.022	1.050	0.974	1.019	0.976	0.994
0.4	2.55	2.63	1.016	1.020	1.045	0.977	1.017	0.978	0.995
0.6	2.50	2.57	1.014	1.018	1.040	0.979	1.016	0.980	0.995
1.0	2.44	2.50	1.008	1.012	1.032	0.977	1.009	0.978	0.991
1.5	2.39	2.45	1.001	1.005	1.023	0.972	1.003	0.973	0.985
2.0	2.36	2.42	0.994	0.997	1.014	0.966	0.995	0.967	0.978
2.5	2.32	2.37	0.987	0.990	1.007	0.960	0.988	0.962	0.973

approximations to the calculation of energy division will be adequate. When mixtures contain alkanes or compounds with other functional groups as well, the simple approaches will not suffice because the greater oscillator strength for the lower-lying allowed states of the π bond should lead to a preferential excitation of such states in mixtures with alkanes. There is at present no reliable way to estimate energy division between dissimilar components in a mixture.

3. Spatial distribution of initial events

In the gas phase all but the heavy-particle irradiations by α-particles and fission recoils essentially lead to homogeneous distributions of energy deposition processes and initial species. At high pressures and with heavy particles, energy deposition occurs in the columns of dense ionization accompanying the track of the particle[132], but such complications have received little attention, as fission-recoil studies of olefins have barely been initiated[133], and α-particles have not been used widely at pressures where non-homogeneity of energy deposition could be an important factor in the subsequent chemical fate of the reactants. It is important, however, that energies required to form an ion pair (W values) differ substantially for α-particles and electrons, and indiscriminate use of W values is therefore not permissible[36]. The difference arises almost certainly from effects along the end of the α-particle track, where some of the energy losses will not be by excitation and ionization in the usual fashion, but by head-on collision and electron acceptance. While such phenomena are not too well understood, it is clear that for fission recoils, which originate with up to 20 positive charges and energies of ca 80 meV, partial neutralization will occur along the track. Even modifications of the simple stopping-power equations do not give satisfactory agreement with experiment[134].

In the condensed phases even γ-rays and electrons will lead to an inhomogeneous distribution of initial species. This arises from considerations of the energy-deposition characteristics of the secondary electrons. The degradation spectrum of electrons (Figure 4)[135,136],

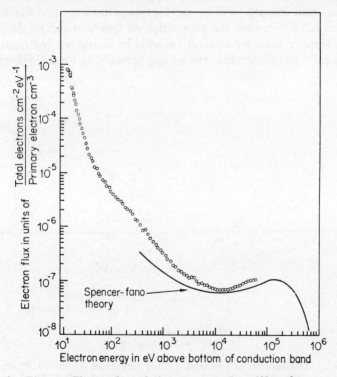

Figure 4. Electron Flux or degradation spectrum from ^{198}Au β-rays absorbed in aluminum (logarithmic scales) (Ref. 135).

defined as the number of electrons passing through unit area per unit time, contains a large fraction with energies low enough so that their further energy loss will occur over a small region of a few angstroms only, as originally pointed out by Samuel and Magee[137]. These regions of energy depositions are called 'spurs', and contain initial species in a relatively high local concentration, and in varying numbers depending on the energy of the electron which has caused the formation of a particular spur. Thus spur sizes vary, and their distributions have been estimated for various media[138].

The general picture of initial events may thus be one of a distribution of regions of high densities of small numbers of initial species (Figures 5 and 6). If these species are ions and electrons, their recombination will compete with diffusion into the bulk of the liquid (or solid). A great majority of the ions and electrons will undergo recombination after a very short period, and this recapture of the electron by the parent ion is termed geminate neutralization. However, a small fraction of these species escape their mutual Coulombic attraction. Of course, the proportion of ions that escape should be much smaller than for neutral radicals in irradiated hydrocarbons. Onsager[139] has shown that the escape probability for a single pair of

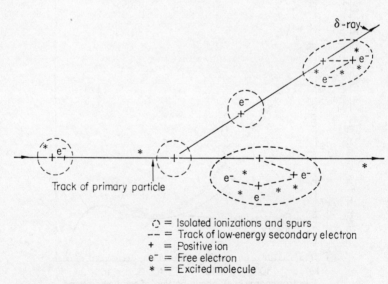

◯ = Isolated ionizations and spurs
-- = Track of low-energy secondary electron
+ = Positive ion
e⁻ = Free electron
* = Excited molecule

Figure 5. Distribution of ions and excited molecules within 10^{-14} sec, after passage of a fast particle through a liquid (not to scale) (Ref. 24). δ-Rays are energetic electrons (>200eV) ejected by passage of the primary radiation.

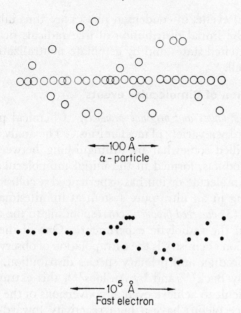

Figure 6. Distribution of spurs in particle tracks. Typical spur diameter would be of the order of 20 Å (not to scale) (Ref. 24).

ions undergoing Brownian motion and separated by a distance r varies as $\exp(-r_c/r)$, where r_c is a 'critical' radius, so that the 'free' or separated ions are not derived merely from those few ions and electrons which have achieved rather large initial charge separations.

The theoretical justification for the application of such a diffusion model to the problem of ion-electron separation in radiolysis depends on the existence of trapped and escaped electrons during the initial stage of the process. The most direct evidence for 'free' ions comes from conductivity studies on hydrocarbons during irradiation[54-55]; these results indicate that the G value for separated (escaped) ions is about 0·1, and therefore considerably less than the total yield (ca 2·6) of ion pairs as determined by chemical methods at high scavenger concentration[56-58,140,141]. It should be added that it is most unlikely that the escaped electron itself survives as such into the steady state, because of the relatively long time available during which it can be scavenged by trace impurities such as oxygen, and it is thus probably a negative ion[53-55]. Strong chemical evidence for free ions also comes from the results of ionic polymerization[142] which show that kinetic chain lengths can be as high as 10^6 in well-purified systems.

Since initial events in condensed phases are thus inhomogeneously distributed, the initial distribution of free radicals produced by dissociation of excited states and by geminate neutralization will not be uniform as well.

4. Classification of Bimolecular events

a. Primary products and original products. Chemical product formation can occur by a variety of mechanisms. The analysis of the events can be simplified somewhat by distinguishing between: (1) *primary species* and products, formed in the initial unimolecular stage before the activated molecule or ion has experienced a collision; (2) *original products*, arising in an ultra-pure system at infinitesimally small conversions; and (3) *observed products*, corresponding to the actual product distribution in the radiolytic experiment. One of the very difficult tasks of radiation chemistry is the extrapolation of observed products to the original product and primary species distribution. As has been pointed out by Back[143] and by Ausloos[144], this extrapolation is exceedingly difficult to achieve even at conversions of the order of 0·001 percent. Since olefins have a high reactivity towards intermediate radical and ionic species this difficulty is a less severe complication here than it is for many other types of compounds.

b. Physical effects of collision. When the primary species is an excited state, the collision process may assist the spectroscopic processes of internal conversion, intersystem crossing, vibrational degradation, etc. The relative enhancement of a given process such as induced predissociation or intersystem crossing will depend on the particular system involved. A non-chemically reactive collision process will, however, invariably affect translational, rotational, and vibrational excitation.

When the excited primary species is an ion its capabilities of further dissociation may be quenched by the collision. This process will frequently be accompanied by an ion-molecule reaction and one normally assumes that the excited parent ion will undergo the same reactions as parent ions in their ground state. It is, however, quite possible that a molecular ion, sufficiently excited to give a characteristic fragment ion, may yield those ion-molecule reaction products one would have expected from the fragment ion formed previously. It is well known that large amounts of translational energy in the ion-molecule collision lead to a different product distribution[145,146] often including reactions less exothermic or even endothermic for thermal ions and molecules. This is an important complication, since the relative abundance of parent ions and fragment ions in radiation chemistry is generally assessed from the neutral products resulting from their ion-molecule reactions. These products can often be assigned to a particular primary ion by comparison with mass spectrometric studies of ion-molecule reactions. This conclusion, however, will be invalidated if excited primary ions and their derived fragment ions give the same ion-molecule reaction products. An example where mass spectrometric information is probably inapplicable to radiolysis systems is the charge transfer

$$C_2H_2^+ + C_2H_4 \longrightarrow C_2H_2 + C_2H_4^+ \tag{12}$$

which has been reported repeatedly[65,66], but probably requires epithermal energies[147] and thus may not be important in the radiolysis.

c. Chemically reactive collisions of excited states. Atom-abstraction reactions and excimer additions in condensed-phase photochemistry are well known[95], but little if any evidence has been presented that such processes are important contributors to radiation chemistry. It is, however, quite conceivable that such reactions, and particularly atom abstractions, contribute to radiolysis mechanisms.

Highly excited molecular species may participate in associative ionization. Such processes have been reported in acetylene[148], (equation 13), but the veracity of this report has been questioned seriously[149].

$$C_2H_2^* + C_2H_2 \longrightarrow C_4H_4^+ + e \tag{13}$$

Prerequisite for associative ionization of olefins are the existence of high-lying metastable or resonance states (excited states). Since lifetimes of states which may return to the ground state by allowed transitions may be even on the order of 10^{-9} sec, such processes might be more important in radiation chemistry than one would anticipate from the absence of mass spectrometric evidence. This method requires a lifetime of ca 10^{-6} sec for the detection of reaction.

d. Ion-molecule reactions. The reactions between ions and neutral molecules frequently occur with collision efficiency. Because of ion-induced dipole forces, rate constants for the collision process are about one order of magnitude larger than those involving only neutral species. They are recognized as being important contributors to radiation chemical mechanisms in olefins and evidence for their participation has been presented in a number of gaseous and condensed-phase systems. Most of our knowledge derives from mass spectrometric evidence and most simple olefins have been investigated by this technique[150].

The application of mass spectrometric data to radiation chemistry is again complicated by the change in time scale. The studies of Wexler[151], Field and Munson[152-154], and Kebarle[65,66] indicate clearly that at the higher pressures typical of radiation chemistry, gas-phase solvation of ions may occur. Moreover, the lifetime of the intermediate complex in condensation reactions appears to be considerably greater than the collision-free period at one atmosphere[17,67,155], suggesting the possibility of collisional deactivation of the intermediate ionic complex and further reactions of the complex itself.

Most ion-molecule reaction pairs yield more than one product ion. Studies with tandem mass spectrometers and ion cyclotron double resonance have shown that competition between the various processes depends on ion energy[145,146,147]. Most experimental investigations have been carried out in the presence of repeller or draw-out fields in the ion source. Since the median energy at which ion-molecule reactions resulting from ion-induced dipole forces occurs is 1/4 of the ion energy at the exit slit, average energies for which such reactions are reported are typically on the order of a few tenths of an eV or considerably above thermal. Although most reactions are relatively insensitive to energy in this range, major differences between mass spectrometry and radiation chemistry can be expected if the alternate reaction paths are nearly thermoneutral.

Ion-molecule reactions can be divided into two general classes: atom or ion transfer and condensation reactions. The first class is quite common and because of the ease of investigation has been extensively studied. Condensation reactions are of dominant importance in olefinic systems, and are important contributors to product formation and to polymerization.

Reactions of ethylene ion have been studied most extensively[15,17,61-68,74,155,156,157] and proceed via a relatively long-lived intermediate

$$C_2H_4^+ + C_2H_4 \longrightarrow [C_4H_8^+] \tag{14}$$

$$[C_4H_8^+] \longrightarrow C_3H_5^+ + CH_3 \tag{15}$$

$$[C_4H_8^+] \longrightarrow C_4H_7^+ + H \tag{16}$$

complex with $k_{15}/k_{16} = 10$ and $k_{15} + k_{16} = 3 \cdot 2 \times 10^7 \text{ sec}^{-1}$ [17,72]. The square brackets denote that the species it encloses is metastable towards dissociation. The intermediate complex ion or the secondary fragment ion produced in reaction (15) may react further with ethylene:

$$[C_4H_8^+] + C_2H_4 \longrightarrow C_4H_8^+ + C_2H_4 \qquad (17)$$

$$[C_4H_8^+] + C_2H_4 \longrightarrow [C_6H_{12}^+] \qquad (18)$$

$$C_3H_5^+ + C_2H_4 \longrightarrow [C_5H_9^+] \qquad (19)$$

$C_4H_7^+$ appears highly unreactive towards ethylene (collision efficiency $< 10^{-3}$)[147,158]. The importance of the direct process of intermediate hexene ion formation by step (18) is in serious question, and it may not occur at all [63,64,68,147,158]. The ion produced in reaction (19) is long lived (10^{-4} sec)[147] and survives as an entity under all conditions of radiation chemical interest.

The stabilized intermediate butene ion may undergo further ion-molecule reactions, but the successive addition steps of higher olefin ions to ethylene are inefficient [17,65-67] and butene ion survives more than 100 collisions before forming a hexene ion reaction complex,

$$C_4H_8^+ + C_2H_4 \longrightarrow (C_6H_{12}^+) \qquad (20)$$

which may dissociate

$$(C_6H_{12}^+) \longrightarrow C_5H_9^+ + CH_3 \qquad (21)$$

$$(C_6H_{12}^+) \longrightarrow \text{other products} \qquad (22)$$

where $(C_6H_{12}^+)$ indicates a metastable entity whose internal energy differs from that of $[C_6H_{12}^+]$ produced by reaction (18). The rate constant for this dissociation has been estimated on the assumption that the competitive deactivation by ethylene occurs with collision efficiency and is probably on the order of 4×10^8 sec^{-1} [67,135]. Further addition reactions of stabilized $C_6H_{12}^+$ to ethylene occur with a collision efficiency of well below 1%[67], and the intermediate product octene ions probably dissociate rapidly with a rate constant of 5×10^9 sec^{-1} or greater[67,155].

Mass spectrometric experiments cannot readily distinguish whether the reaction sequences (14), (17), (19) and (21) or (14), (15) and (19) are dominant under radiation chemical conditions. Not even the recent studies by ion cyclotron double resonance (ICDR) are helpful. At very low pressures in the ion source, $C_4H_8^+$ cannot survive before collisions[159]. However, under all radiation chemical conditions of interest the intermediate formed in reaction (14) must survive, and is presumably deactivated by collision, as shown by the ability of charge acceptors (molecules of low ionization potential) to enhance the yield of stable butenes (Figure 7), presumably by an electron transfer (see also section III.C.5.b)

$$C_4H_8^+ + CA \longrightarrow C_4H_8 + CA^+ \qquad (23)$$

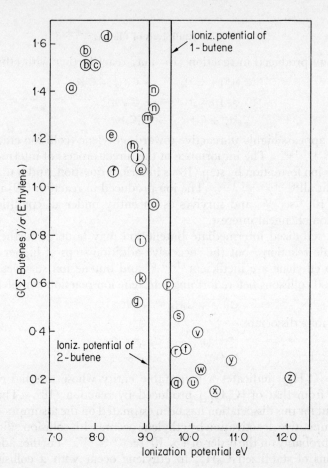

Figure 7. Enhancement of total butene yields in radiolysis of ethylene at 100 torr. Additives present in 10% concentration and oxygen (3%) added when necessary to inhibit free radical reactions. The ionization potentials of the linear butenes are indicated as vertical lines for comparison (Ref. 17). Additives in order of increasing ionization potential:

- (a) Triethylamine (7·50)
- (b) Trimethylamine (7·82)
- (c) Diethylamine (8·01)
- (d) Dimethylamine (8·24)
- (e) Dihydropyran (8·34)
- (f) 2-Methylfuran (8·39)
- (g) Isopropylamine (8·72)
- (h) Toluene (8·82)
- (i) Ethylamine (8·86)
- (j) Furan (8·89)
- (k) Methylamine (8·97)
- (l) Cyclopentene (9·01)
- (m) 2-Methyl-1-butene (9·12)
- (n) Nitric oxide (9·25)
- (o) Benzene (9·25)
- (p) Diethylether (9·53)
- (q) Dichloroethylene (9·66)
- (r) Acetone (9·69)
- (s) Propylene (9·73)
- (t) Cyclohexane (9·88)
- (u) Cyclopropane (10·06)
- (v) Ammonia (10·15)
- (w) Acetaldehyde (10·21)
- (x) Ethylene (10·52)
- (y) Methanol (10·85)
- (z) Oxygen (12·08)

The complete reaction sequence arising from vinyl ion is unknown. The only important first-order process is the proton transfer (reactions 24 and 25).

$$C_2H_3^+ + C_2H_4 \longrightarrow [C_4H_7^+] \tag{24}$$

$$[C_4H_7^+] \longrightarrow C_2H_2 + C_2H_5^+ \tag{25}$$

The lifetime of intermediate $C_4H_7^+$ is unknown, and it has been suggested that the formation of ethyl ion proceeds in part by a stripping type mechanism which would not involve an intermediate ionic complex whose dissociation may be quenched by collision[67]. There is relatively little convincing evidence on the mechanism by which the higher-order reactions of vinyl ions occur. One of the difficulties in studying these reactions is the inability to produce vinyl ions exclusively in an atmosphere of ethylene. While an attempt to do so has been made with the charge-exchange technique, the results, even for the parent-ion reactions, are at such variance with established mechanisms that little weight can be placed on the findings[160]. ICDR experiments[147] have shown that at low pressures the reaction

$$C_2H_5^+ + C_2H_4 \longrightarrow C_3H_5^+ + CH_4 \tag{26}$$

contributes, but this technique does not see complexes of lifetimes less than ca 10^{-5} sec, well below the range of radiation chemical interest. Evidence in the radiolysis of ethylene suggested that methane formation occurs by ionic processes[16,155], and this has been ascribed to reaction (27),

$$[C_4H_7^+] + C_2H_4 \longrightarrow C_5H_7^+ + CH_4 \tag{27}$$

first suggested by Field[63], based on the observation of $C_5H_7^+$ in the high-pressure mass spectrometry of ethylene. In view of the doubt cast on this step[62,147] this detailed assignment of a reaction step in the radiolysis must be accepted with reservations.

Acetylene ion can exchange charge with ethylene

$$C_2H_2^+ + C_2H_4 \longrightarrow C_2H_2 + C_2H_4^+ \tag{28}$$

since the ionization potentials are favourable. This reaction has been reported repeatedly[63,64,67,147] but it is not known whether it involves a long-lived complex. Moreover, it probably is not very important at thermal energies[147]. The condensation-type ion-molecule reactions, however, involve more than the transfer of a simple species and therefore almost certainly a long-lived complex

$$C_2H_2^+ + C_2H_4 \longrightarrow [C_4H_6^+] \tag{29}$$

$$[C_4H_6^+] \longrightarrow C_3H_3^+ + CH_3 \tag{30}$$

$$[C_4H_6^+] \longrightarrow C_4H_5^+ + H \tag{31}$$

Once again, higher-order ion-molecule reactions may involve either the complex ion formed in reaction (29) or the products of its dissociation in a series analogous to that for ethylene ion. It should be noted that addition of $C_3H_3^+$ to ethylene and collisional quenching of the product leads to the same ion ($C_5H_7^+$) as does reaction (27), and some investigators ascribe the formation of this ion to the sequence originating with acetylene ion[62,66].

Mass spectrometric studies have also revealed higher-order ion-molecule reactions in this system with molecular weights corresponding up to 10 or more ethylene units in the ion[63-65]. However, there is no convincing evidence for the participation of long-chain ion-molecule reactions in the radiation-induced polymerization of this compound[116]. In fact, at least over a limited range of up to ca 25 atm, an excellent fit of the data can be obtained by application of simple homogeneous competition kinetics, with the insertion of rate constants known to be applicable to free-radical reactions[15]. However, when studies are extended to pressures up to 70–400 kg/cm², the high pressure radiation-induced polymerization cannot be described by such a simple model. Ionic reactions may be responsible for this deviation[161].

In the liquid phase at -168 to $-110°$C ionic reactions appear to be less important[14] and much of the chemistry can be explained in terms of free-radical reactions; only dimerization and trimerization appear to proceed significantly via ionic intermediates. In solid ethylene[9,10] at $-196°$C neutral products of the type $C_{2n}H_{4n}$ with n up to 5 are most important, and this has been ascribed to a mechanism of ionic condensation competing with geminate neutralization, i.e. return of the parent electron to the growing ion:

$$C_2H_4^+ \xrightarrow{+C_2H_4} C_4H_8^+ \xrightarrow{+C_2H_4} C_6H_{12}^+ \longrightarrow \text{etc.}$$
$$\downarrow e^- \qquad\qquad \downarrow e^- \qquad\qquad \downarrow e^-$$
$$C_2H_4 \qquad\qquad C_4H_8 \qquad\qquad C_6H_{12} \qquad\qquad\qquad (32)$$

While this appears to be an adequate explanation for species containing more than 2 ethylene entities, Ausloos[164] has recently suggested that butenes may also result by a free-radical reaction, with a neighbouring ethylene molecule adding the hydrogen atom produced simultaneously with a vinyl radical and then combining with the vinyl radical. In mixtures of ethylene and perdeuteroethylene, the isotopic composition of the adducts would be C_4H_8, $C_4H_4D_4$ and C_4D_8 and hence indistinguishable from those produced by ionic dimerization.

The vacuum ultraviolet photochemistry of propylene has been investigated[19,94] and the studies have been extended into the region where the parent ion is formed and its reactions are amenable to elucidation[22]. Ion-molecule reactions in this compound have been investigated by Fuchs[60] and Harrison[165], and lead exclusively to condensation ions with four- and five-carbon atom skeletons. The gas-phase radiolysis[49,97] has shown that hydrogen, ethylene, and 2,3-dimethylbutane (or 2-methylpentane) are formed with yields in excess of 1·0 molecules/100 eV, with ethane, propane, allene, n-hexane, 1-pentene, propyne, and several other products formed with G values less than one.

The radiolysis of liquid propylene has been studied chiefly by Wagner[11]. Detailed analysis of the dimers support an ionic reaction mechanism, and 4-methyl-1-pentene, 1-hexene and 2-hexene are formed in appreciable fractions by an ionic mechanism similar to equation (32). However, Abramson and Futrell[166] have seriously questioned the validity of this experiment for 1-hexene, and this is also not consistent with results of the vacuum ultraviolet photochemistry[22].

The gas-phase radiolysis of the butenes has been investigated as an adjunct to liquid-phase studies[167], particularly with a view towards establishing ionic condensation steps. Very little dimer formation was observed, and this is probably in part attributable to the low dose rate employed, which would lead to longer free radical and ionic condensation chains than in the liquid phase, where parent-ion recapture is important. In the liquid the dimer and trimer are again thought to result from ionic condensation. Another study of liquid butene radiolysis[98] concerned itself largely with the mechanism of hydrogen formation, and a correlation of the yields of alternate processes with theoretical predictions by the optical approximation[127].

Ionic reactions are also highly important in isobutylene[168,169]. This system, as all other ionic systems, is exceedingly sensitive to the presence of impurities of high proton affinity, such as water or ammonia, and only meticulously dried starting materials give appreciable ionic chain lengths[161,169]. An excellent demonstration that polymerization in isobutylene occurs by a cationic mechanism has been given by the ion-injection technique[170,171]. Ions are produced in the gas phase by photoionization, and are drawn into the liquid with a weak electrostatic field. Only when positive ions are in contact with the liquid does polymerization occur.

The radiolysis of branched higher-molecular weight olefins, particularly that of 3,3-dimethyl-1-butene[162,163] gives skeletally isomerized products of the same molecular weight and of a specific structure, most likely by an ionic mechanism which presumably proceeds through a cyclopropyl ion derivative. This is particularly indicated because cyclopropane derivatives are actually observed in the product distribution. An intramolecular hydrogen rearrangement accompanies dimerization. The new double bond is formed between originally double bonded carbons end-to-end, but sometimes between the end carbon of one molecule and the second carbon of another. Ionic condensation appears to be not simply head to tail, but considerably less ordered. Studies of the effects of additives which can trap electrons or accept charge provided further evidence for the suggested mechanism. It is doubtful, however, that such processes account for the major portion of the radiolysis mechanism, since the isomerized products are formed with yields of less than 0·2 molecules/100 eV.

e. Neutralization. The neutralization process can be between positive ions and electrons or negative ions. In the gas phase it normally occurs homogeneously as a result of ambi-polar diffusion. However, at very low dose rates and therefore very small steady-state concentrations of charged species in the radiation vessel, the dominant neutralization process is diffusion to the walls and neutralization thereon. Neutralization rates have been studied extensively for inorganic gases because of the importance in upper-atmosphere chemistry, but only a few measurements have been made for hydrocarbons. Back[172] has used the application of intermittent voltages to determine the steady-state concentration of charged species, and hence the rate constants for recombination by physical means. Mahan and coworkers[173] used pulsed photoionization of gaseous lead and thallium halides. Lawrence and Firestone[174] on the other hand, have used chemical means to assess the rate of neutralization. As a general rule, rate constants for ion-electron neutralization processes are of the order of 10^{17} mole^{-1} cc sec^{-1}, while those for recombination of positive and negative ions are about one order of magnitude smaller[173,175]. Little is known about the chemical consequences of neutralization and this remains one of the major frontiers in the radiation chemistry of olefins.

Neutralization in the condensed phases requires distinction between two types of processes; parent-ion-electron recapture (geminate neutralization), and bulk or homogeneous neutralization[50-58,137]. While the electron is typically ejected with some excess energy, it normally does not travel far enough to escape the influence sphere of the positive ion, and only some 2–5% of the electrons escape the field of the parent ion when no external electrical fields are present. Even the recaptured electron sweeps out a considerable range or volume before it returns to its parent, and electron attachment reactions can therefore be observed at relatively low concentrations. An example of the participation of negative ions in radiolysis mechanisms is the radiation induced *cis–trans* isomerization of stilbene in solutions of benzene or cyclohexane[131]. G values of up to ca 210 isomerization/100 eV have been reported[131]. This can only be explained in terms of a chain process. A negative ion formed by attachment of 'free electrons' to the stilbene molecule was proposed as a chain carrier for 0·02 to 0·6 M solutions.

f. Electron Attachment Reactions. When the energy of the free electron falls below that of the lowest excited state of the compound under irradiation, further energy loss is slow. All electrons pass through the sub-excitation energy range and they may either be unreactive towards the substrate molecule, attach themselves with the formation of stable negative molecule ions, or undergo a dissociative attachment process. A thermodynamic equilibrium will be reached if the electron affinity of the molecule is not too large, and if the cross-section for the attachment process is not too small. Based on the assumption of equilibrium conditions, measurements have been made by pulsed removal of free electrons to estimate electron affinities[126]. Results were in adequate agreement with those obtained by other techniques[123].

Olefins do not attach electrons in the gas phase[177]. However, there appears to be substantial evidence that such a process is possible in the liquid, and has been suggested for propylene[58].

$$e^- + C_3H_6 \longrightarrow C_3H_6^- \tag{33}$$

This could be expected if the double bond is weakly attractive and collisional stabilization is required to retain the electron on the molecule. Therefore it seems plausible that in the low-temperature radiolysis of ethylene in very dilute solutions of liquid argon, electron attachment would contribute to the radiolysis mechanism[178].

g. Radical reactions. In olefin radiolysis radicals are chiefly produced by simple bond scission, and by hydrogen atom addition to the double bond. Radicals can either react with other radical species, or with substrate molecules. Their reactions with charged species are normally disregarded, but these can occur when the rate of primary radical formation is about the same as that of charged species (i.e. When G values are comparable). Radical-ion interactions should occur when electron-capture is the dominant process for electron disappearance, which would lead to equal probabilities of radical-charged species and radical-radical interactions. No evidence for such reactions exists.

The reaction of radicals with each other can lead to combination or to disproportionation. The relative extent of these reactions depends largely on the availability of hydrogen atoms at β positions, and on C—H bond strength in the radical. Thus hydrogen atoms at a tertiary carbon lead to easier disproportionation than at a primary one. These reactions lead to the destruction of the radical character of the interacting species and occur approximately with collision efficiency ($k_{R+R} \simeq 10^{11}$ l mole^{-1} sec) for most simple hydrocarbon radicals[99,179].

Free-radical species may also react with their molecular substrates[99]. This may be by an abstraction reaction (metathesis) which may change the nature of the radical. Activation energies for metathetical reactions vary from 1 to 2 to as much as 11 kcal/mole. In olefins, addition is an important process. Activation energies for addition are very low for the addition of hydrogen atoms to olefins[99-101], while the addition of alkane radicals is normally competitive with abstraction reactions. Radical addition reactions may lead to polymerization, or to telomerization, if the chain length is very short or if chain transfer is important. Such reactions are of interest because they provide a possible means for utilizing radiation for the synthesis of polymers and telomers at relatively high pressures, or in the condensed phases and at low dose rates.

Radical reactions in ethylene radiolysis are dominated by the formation of hydrogen atoms produced in reactions (6a), (6c), and (16) etc., and that of methyl radical possibly produced in reactions (21) through (30). Hydrogen atoms immediately add to ethylene

$$H + C_2H_4 \longrightarrow C_2H_5 \quad (34)$$

and this reaction accounts quantitatively for hydrogen atoms under most conditions. At very high dose rates

$$H + C_2H_5 \longrightarrow C_2H_6^* \quad (35)$$

may set in and lead to additional ethane and some methyl radical formation if the dissociation

$$C_2H_6^* \longrightarrow 2\,CH_3 \quad (36)$$

is not quenched at elevated pressures. The collision efficiency of reaction (34) is about 0.1[100], and the ethyl radical produced in this step has excess energy. Under normal conditions this is rapidly removed by collisions, so that the species participating in reactions (35) and (36) is in thermal equilibrium with its surroundings. The vinyl radical produced in reaction (6c) is most likely to add to ethylene yielding butenyl

$$C_2H_3 + C_2H_4 \longrightarrow C_4H_7 \tag{37}$$

and since the yield of this radical is normally minor, it will chiefly disappear by reaction with ethyl radicals:

$$C_2H_5 + C_4H_7 \longrightarrow C_6H_{12} \tag{38}$$
$$C_2H_5 + C_4H_7 \longrightarrow C_2H_4 + C_4H_8 \tag{39}$$
$$C_2H_5 + C_4H_7 \longrightarrow C_2H_6 + C_4H_6 \tag{40}$$

The direct combination of vinyl radicals cannot be disregarded as a possibility of butadiene formation[180]. Ethyl radicals not participating in reactions (38)–(40) combine or disproportionate

$$2\,C_2H_5 \longrightarrow n\text{-}C_4H_{10} \tag{41}$$
$$2\,C_2H_5 \longrightarrow C_2H_4 + C_2H_6 \tag{42}$$

with $k_{41}/k_{42} \simeq 0.14$[99,179,181].

At low dose rates and high pressures free-radical induced polymerization will be important. Kinetic analysis of pressure and dose rate effects on polymerization has shown that the rate constant for termination must be on the order of 10^7 times greater than that for addition to ethylene[182], in agreement with the known values for the rate constants of these radical processes, but at variance with estimates for possible corresponding ionic processes. This agreement has been extended[15] to cover the radiation-induced polymerization of ethylene at high pressure and very low dose rates. By taking the rate constants for radical propagation and termination to be 2×10^3 l mole^{-1} sec and 1×10^{11} l mole^{-1} sec respectively, the yield of ethylene disappearance $G(-C_2H_4)$ can be expressed by the equation

$$G(-C_2H_4) = 20 + 0.9 \times 10^3 \sqrt{[C_2H_4]/I} \tag{VII}$$

where $[C_2H_4]$ is the ethylene concentration in moles cc^{-1} and I is the dose rate in Mrad/hr. The constant term 20 in equation (VII) represents the substantial yield of ethylene disappearance in the absence of any propagation by radical addition to ethylene. This equation agrees extremely well with the polymerization data of Hayward and Bretton[183] which satisfies the relation,

$$G(-C_2H_4) = 16 + 1 \times 10^3 \sqrt{[C_2H_4]/I} \tag{VIII}$$

The constant term here is smaller because the formation of hydrogen and acetylene was not deducted when the rate of polymerization (ethylene disappearance) was followed by observing the decrease in pressure.

Thus it appears that although processes involving ions and excited molecules play an important role in determining the distribution of primary radicals in the radiolysis of ethylene, particularly in the gas phase, the stepwise formation of higher alkanes at low pressure and of polymer at high pressure (up to at least 40 kg/cm^2) are both adequately described by kinetic parameters characteristic of free-radical reactions in this system. This interpretation does not dispose of the eventual fates of the higher-molecular weight ions seen in the mass spectrometric studies [63,67].

In the condensed phases, or at very high pressures where densities approach those of the liquid, radicals will be produced in spurs directly and by geminate neutralization. Since the majority of ions appear to undergo geminate recombination in hydrocarbons, the disappearance of the ions should give rise to a distribution of free radicals generally similar to that envisaged by Samuels and Magee [137].

Magee [184] has summarized the chronological order of processes in a spur. As the time increases, the diffusion of the neutral radicals leads to an increase in spur size until ultimately a homogeneous distribution is reached (steady state). Thus the reactions between neutral radicals must be treated by diffusion theory to take account of the time dependent growth of the spur [137,185]. In particular, the competition between radical-radical and radical-solute (scavenging) reactions depends on the local radical concentration during the expansion of the spur, and it is clear that for any given solute concentration, radical scavenging assumes increasing importance in the approach to the steady state, where homogeneous kinetics apply. In other words, very high values of $k_s[S]$, where k_s is the rate constant of the scavenging reaction and $[S]$ is the scavenger concentration, are needed to scavenge those radicals which otherwise would undergo combination in the early history of the spur, whereas the scavenging of radicals undergoing homogeneous recombination is effective at $k_s[S]$ values which are lower by several orders of magnitude [186].

While the applicability of the diffusion model to reactive intermediates in the condensed phases is now well established, experimental tests have been largely restricted to aqueous systems [185] and to simple hydrocarbons [141], and little is known about the importance of nonhomogeneous initial-species formation in the radiolysis of olefins.

For example, geminate neutralization could be effectively inhibited in liquid and solid olefins if the double bond can act as an efficient electron trap as suggested by Freeman[58]. On the other hand, the apparently short lifetime of ethylene ions in the solid-phase radiolysis would strongly support rapid neutralization[9,10]. It is not surprising, therefore, that most radiolytic investigations have so far not attacked this problem in olefins, but have restricted themselves to seek evidence on particular intermediates.

5. Mixtures

a. Sub-excitation electrons. The degradation of electron energy once it is below the lowest electronic state of the molecule is slow. The sub-excitation electrons constitute a constant fraction of the energy loss and the addition of even a small amount of gas having energy levels lower than the major component, can lead to selective excitation and to product formation essentially independent of the concentration. It follows that a detailed estimate of the energy division must be made. Although approximate stopping powers may apply to the gross energy distribution between higher excited and ionic states of a binary system, the contribution of the lowest excited state of one compound to product formation will differ substantially from that in pure components. Mixtures of olefins and alkanes should show a preferential excitation of the π-bond system of the olefin since sub-excitation electrons in the energy interval between lowest excited states of the olefin and the alkane can induce excitation even to forbidden levels.

b. Energy transfer. Energy transfer may take any of three forms; excitation transfer, charge transfer, and the formation of a new ion pair. The last requires the existence of an excited state whose level is above the ionization potential of the substrate molecule. Alternately, an associative ionization process may occur as discussed in Section III.3.4.c and may involve preferentially a highly excited state of one molecule and a different ground-state molecule.

Excitation transfer from a super-excited state may decrease the overall ionization if the ionization efficiency of the energy acceptor is smaller than that of the energy donor. This is one mechanism by which reduction of ionization may occur in gaseous mixtures. An increase may result if the efficiencies are reversed. Interference with associative ionization is another posssible process. In this case, energy transfer from the excited state to a molecule will again lead to a decrease of ionization, or a negative Jesse effect[129].

14+c.a. 2

The processes summarized above constitute the principal non-trivial differences from pure system. In mixtures there will also be a wider variety of ionic and free-radical reactive species, and the sequence of reactions may become more complicated. A simplification may occur if the added component is a scavenger of free radicals, for example, nitric oxide, iodine, and hydrogen iodide, or a charge acceptor—that is, a species whose ionization potential is lower than that of the reactant ion. Both types of reactions can be used to advantage in attempts to elucidate mechanisms of radiation chemistry. Since free-radical scavengers normally also have low ionization potential, care must be taken in the interpretation of results.

A good example of the complication of the use of scavengers is the effect of nitric oxide on the radiolysis of ethylene. While it may be expected that this substrate will inhibit the formation of all products arising from reactions of free radicals in the system, its addition leads to a sizeable enhancement in the yield of butenes, which can in fact be correlated with the yield of ethylene ions originally produced[17,156]. This can be ascribed to the interaction of nitric oxide with an intermediate collision complex $C_4H_8^+$, as follows

$$C_4H_8^+ + NO \longrightarrow C_4H_8 + NO^+ \qquad (43)$$

which is a particular example of reaction (23). Strong support for this mechanism was presented from kinetic examinations, deuterium labelling, and the ability to correlate the butene yields with the ionization potentials of some twenty additives[17]. Charge-transfer techniques can only give a minimum value for the yield of butene and therefore ethylene ion, since the additive may reduce the extent of butene ion formation by charge transfer from parent ethylene ion, and because competitive reactions such as H_2^- transfer[187] and other ion-molecule reactions with the additive $C_4H_8^+$ not leading to neutral butenes, can affect yields. The charge-transfer reaction was confirmed for nitric oxide as the charge acceptor when $C_4H_8^+$ ion was created in the vacuum ultraviolet photolysis of cyclobutane at 1047–1065 Å[188]. Butenes were observed with an ion-pair yield (M/N) slightly above 1 when NO was added, but with a yield of less than 0·1 molecules/ion pair when it was absent. Also, the distribution of the isomers was identical to that in radiolysis. The effect of other additives in cyclobutane radiolysis provided further confirmation[189].

Mass spectrometric investigations, however, have indicated little charge transfer of butene ion to nitric oxide[67] and have pointed instead chiefly to the adduct $C_4H_8NO^+$. It is conceivable that this ion yields stable butene on neutralization as well, and the two mechanisms (charge transfer and addition-neutralization) would therefore give the same yield of butene. It is difficult to see how this could apply to all other additives as well. Sieck and Futrell[190] have recently demonstrated that olefin ions do indeed exchange charge with nitric oxide with an efficiency which depends on structure. Perhaps the observation of $C_4H_8NO^+$ arises from reactions of 2-butene ions for which the charge transfer is endothermic, and is not observed even in the vacuum u.v. photochemistry[191].

IV. YIELDS OF PRODUCTS AND INTERMEDIATES

The mechanisms described in the preceding section lead to a large variety of products. Only in one or two instances has an attempt been made to obtain a complete product distribution, or to arrive at a material balance. This is not surprising since even in the radiolysis of the simplest olefin, ethylene, more than twenty-nine compounds have been identified and measured quantitatively[15], but even then with a material balance of only some 60% to 70% of the ethylene consumed by radiation (since conversions are only on the order of one or two-tenths of a percent, more than 99·8% of the starting material are always recovered). A comparison of the yield for some systems where such estimates can be made are given in Tables 7–9.

TABLE 7. Some products in the radiolysis of ethylene

Product	Yield, molecules/100 eV absorbed in ethylene		
	Gas phase, 100 torr[a] 25°c	Liquid phase[b] −165°c	Solid phase[c] −196°c
Ethylene	−20		
Hydrogen	1·3	1·7	0·6
Methane	0·22		0·01
Ethane	0·85	0·7	0·6
Acetylene	3·5	1·8	0·6
Propane	0·56		
n-Butane	2·32	0·45	0·1
Isobutane	0·11		
Cyclobutane	0·10	0·12	
1-Butene	0·09	0·60	0·2

(*continued*)

TABLE 7. (*continued*)

Product	Yield, molecules/100 eV absorbed in ethylene		
	Gas phase, 100 torr[a] 25°C	Liquid phase[b] −165°C	Solid phase[c] −196°C
trans-2-Butene	0·03		trace
cis-2-Butene	0·03		trace
cis-2-Butene	0·03		trace
Butadiene	0·003		0·03
n-Pentane	0·04		
Isopentane	0·013		
1-Pentene	0·05		
n-Hexane	0·31		0·001
2-Methylpentane	0·02		
3-Methylpentane	0·05		0·018
1-Hexene	0·046	0·15	0·014
trans-2-Hexene	0·03	0·08	0·023
cis-2-Hexene		0·08	0·006
3-Hexene		0·30	0·001
3-Methyl-1-pentene		0·30	0·017
3-Methyl-2-pentene	0·01	0·30	0·019
2-Ethyl-1-butene		0·30	0·013
3-Ethylhexene structure olefin			0·03
3-Methylheptene structure olefin			0·03
n-Octene structure olefin			0·01
Other octenes			0·01

[a] From ref. 15; observed rate 2×10^{18} eV/g sec.
[b] From ref. 14; ca. 10^{17} eV/g sec.
[c] From ref. 9; ca. 6×10^{17} eV/g sec.

TABLE 8. Yields of primary processes ($C_2H_4 \longrightarrow$ primary products) in ethylene radiolysis at 100 torr (gas phase)[16,17,49]

Primary products	Energy required (eV)	Yield (Moles/100 eV)
$C_2H_2 + H_2$	1·8	0·5
$C_2H_3 + H$	4·6	≤0·1
$C_2H_2 + 2H$	6·3	2·8
$C_2H_4^+ + e^-$	10·5	1·5
$C_2H_2^+ + H_2$ (or 2 H) $+ e^-$	13·5	0·8
$C_2H_3^+ + H + e^-$	14·0	1·0

TABLE 9. Products of radiolysis of liquid propylene at $-78°C$[a]

Product	Yield (molecules/100 eV)
Hydrogen	0.6
Methane	0.04
Propane	0.2
Butane	0.02
1-Butene	0.02
2-Butene	0.02
Pentene	0.02
Pentadiene	0.02
2,3-Dimethylbutane	0.054
2-Methylpentane	0.080
n-Hexane	0.022
4-Methyl-1-pentene	0.375
4-Methyl-2-pentene	0.055
2-Methyl-1-pentene	0.033
1-Hexene	0.14
trans-2-Hexene	0.030
cis-2-Hexene	0.022
1,5-Hexadiene	0.17
Others[b]	0.08

[a] Ref. 11.
[b] Ref. 6.

V. ACKNOWLEDGEMENT

The preparation of this manuscript was made possible through the support of the United States Atomic Energy Commission under Grant Number AT(40–1)-3606; this is document number ORO-3606–10. We are greatly indebted to this agency for its support of our work.

VI. REFERENCES

1. S. C. Lind, *The Chemical Effects of Alpha Particles and Electrons*, The Chemical Catalogue Publishing Co., New York. 1st ed., 1921; *Radiation Chemistry of Gases*, Reinhold Publishing Co., New York, 1961.
2. W. Mund, *L'Action chimique des rayons alpha en phase gazeuze*, Hermann et Cie, Paris, 1935; W. Mund, *Bull. Soc. Chim. Belges*, **43,** 49 (1934).
3. H. Eyring, J. O. Hirschfelder and H. S. Taylor, *J. Chem. Phys.*, **4,** 479, 570 (1936).
4. B. M. Mikhailov, V. G. Kiselev and V. S. Bogdanov, *Izv. Akad. Nauk*, **1958,** No. 5, 545–549.

5. A. Chapiro, *J. Chim. Phys.*, **47**, 764 (1950).
6. V. L. Tal'rose, A. K. Lyubimova, *Dokl. Akad. Nauk SSSR.*, **86**, 909 (1952).
7. D. P. Stevenson and D. O. Schissler, *J. Chem. Phys.*, **24**, 926 (1956).
8. J. L. Franklin, F. H. Field and F. W. Lampe, *J. Am. Chem. Soc.*, **78**, 5697 (1956).
9. C. D. Wagner, *J. Phys. Chem.*, **65**, 2276–7 (1961).
10. C. D. Wagner, *J. Phys. Chem.*, **66**, 1158–62 (1962).
11. C. D. Wagner, *Tetrahedron*, **14**, 165–7 (1961).
12. M. C. Sauer, Jr. and L. M. Dorfman, *J. Phys. Chem.*, **66**, 322 (1962).
13. P. Ausloos and R. Gorden, Jr., *J. Chem. Phys.*, **36**, 5 (1962).
14. R. A. Holroyd and R. W. Fessenden, *J. Phys. Chem.*, **67**, 2743 (1963).
15. G. G. Meisels, *J. Am. Chem. Soc.*, **87**, 950 (1965).
16. G. G. Meisels and T. J. Sworski, *J. Phys. Chem.*, **69**, 2867 (1965).
17. G. G. Meisels, *Advan. in Chem. Ser.* **58**, 243 (1966).
18. K. Watanabe, T. Nakayama and J. Mottl, *J. Quant. Spectr. Radiative Transfer*, **2**, 369 (1962).
19. J. R. McNesby and H. Okabe, in *Advances in Photochemistry* (Ed., W. A. Noyes, G. S. Hammond and J. N. Pitts), Vol. 3, Interscience, N.Y., 1964., p. 157.
20. C. A. Jensen and W. F. Libby, *J. Chem. Phys.*, **49**, 2831 (1968); R. E. Rebbert and P. Ausloos, *J. Am. Chem. Soc.* **90**, 7370 (1968).
21. D. C. Walker and R. A. Back, *J. Chem. Phys.*, **38**, 1526 (1963).
22. R. Gorden, Jr., R. Doepker and P. Ausloos, *J. Chem. Phys.*, **44**, 3733 (1966).
23. P. Ausloos and S. G. Lias, *Radiation Rev.*, **1**, 75 (1968).
24. J. W. T. Spinks and R. J. Woods, *An Introduction to Radiation Chemistry*, John Wiley, New York, 1964.
25. L. M. Dorfman and M. S. Matheson, *Progress in Reaction Kinetics*, Vol 3, Pergamon Press, New York, N.Y. 1965, p. 237–301.
26. V. Auqilanti and G. G. Volpi, *J. Chem. Phys.*, **44**, 2307–2313 (1966).
27. R. Wolfgang, *J. Chem. Phys.*, **47**, 143–153 (1967).
28. W. P. Jesse and J. Sadauskis, *Phys. Rev.*, **97**, (1955).
29. G. G. Meisels, *J. Chem. Phys.*, **41**, 51 (1964).
30. R. A. Holroyd and R. W. Fessenden, *J. Phys. Chem.*, **67**, 2743 (1963).
31. Y. Toi, D. B. Peterson and M. Burton, *Radiation Res.*, **17**, 399 (1962).
32. G. G. Meisels, W. H. Hamill and R. R. Williams, Jr., *J. Phys. Chem.*, **61**, 1456 (1957).
33. T. W. Woodward and R. A. Back, *Can. J. Chem.*, **41**, 1463 (1963).
34. H. H. Carmichael, R. Gorden, Jr. and P. Ausloos, *J. Chem. Phys.*, **42**, 348 (1965).
35. J. Booz and H. G. Ebert, *Z. Angew. Phys.*, **13**, 376, 385 (1961).
36. G. N. Whyte, *Radiation Res.*, **18**, 265–271 (1963).
37. P. Adler and H. K. Bothe, *Z. Naturforschg.*, **20a**, 1700–1707 (1965).
38. R. M. LeBlanc and J. A. Herman, *J. Chim. Phys.*, **7–8**, 10353 (1966).
39. U. Buktas, *Z. Angew. Phys.*, **33**, 74 (1967).
40. A. O. Allen, *The Radiation Chemistry of Water and Aqueous Solutions*, Van Nostrand, Princeton, New Jersey, 1961.
41. R. H. Schuler and A. O. Allen, *J. Am. Chem. Soc.*, **77**, 507 (1955).
42. H. A. Dewhurst, *J. Phys. Chem.*, **61**, 1466 (1957).

43. F. T. Jones and T. J. Sworski, *J. Phys. Chem.*, **70**, 1546 (1966).
44. R. A. Back, T. W. Woodward and D. A. McLauchlan, *Can. J. Chem.*, **40**, 1380 (1962).
45. A. R. Anderson and T. V. F. Best, *Nature*, **216**, 576 (1967).
46. W. P. Jesse and J. Sadauskis, *Radiation Res.*, **7**, 167 (1957).
47. H. A. Bethe, in *Handbuch der Physik.*, Vol. 24, 1st ed., Julius Springer, Berlin, 1933, p. 273.
48. C. E. Klots, in *Fundamental Processes in Radiation Chemistry* (Ed., P. Ausloos), John Wiley, 1969, Chap. 1.
49. G. G. Meisels, Ref. 48, Chap. 6.
50. A. Hummel and A. O. Allen, *Disc. Faraday Soc.*, **36**, 95 (1961).
51. G. R. Freeman, *J. Chem. Phys.*, **39**, 988 (1963).
52. G. R. Freeman and J. M. Fayadh, *J. Chem. Phys.*, **43**, 86 (1965).
53. A. Hummel and A. O. Allen, *J. Chem. Phys.*, **44**, 3426 (1966).
54. A. Hummel, A. O. Allen and H. Watson, Jr., *J. Chem. Phys.*, **44**, 3431 (1966).
55. A. Hummel and A. O. Allen, *J. Chem. Phys.*, **46**, 1602 (1967).
56. W. H. Hamill, J. P. Guarino, M. R. Ronayne and J. A. Ward, *Disc. Faraday Soc.*, **36**, 169 (1963).
57. W. H. Hamill, *J. Am. Chem. Soc.*, **88**, 5376 (1966).
58. G. R. Freeman and M. G. Robinson, *J. Chem. Phys.*, **48**, 983 (1968).
59. R. H. Schuler, J. M. Warmant and K. D. Asmus, *Adv. in Chem.*, in press.
60. V. R. Fuchs, *Z. Naturforschg.*, **16a**, 1026 (1961).
61. V. L. Tal'rose, *Pure Appl. Chem.*, **5**, 455 (1962).
62a. P. S. Rudolph and C. E. Melton, *J. Chem. Phys.*, **32**, 586 (1960).
62b. C. E. Melton and P. S. Rudolph, *J. Chem. Phys.*, **32**, 1128 (1960).
63. F. H. Field, *J. Am. Chem. Soc.*, **83**, 1523 (1961).
64. S. Wexler, *J. Am. Chem. Soc.*, **86**, 781 (1964).
65. P. Kebarle and E. W. Godbole, *J. Chem. Phys.*, **39**, 1131 (1963).
66. P. Kebarle and A. M. Hogg, *J. Chem. Phys.*, **42**, 668, 798 (1965).
67. P. Kebarle, R. M. Haynes and S. Searles, *Adv. in Chem.*, **58**, 210 (1966).
68. S. Wexler, A. Lifschitz and A. Quattrochi, *Adv. in Chem.*, **58**, 193 (1966).
69. R. L. Platzman, in *Radiation Biology and Medicine* (Ed., W. Claus), Chap. 2. Addison-Wesley, Reading, Mass., 1958.
70. R. L. Platzman, *J. Phys. Radium*, **21**, 538 (1960).
71. R. L. Platzman, *Radiation Res.*, **17**, 419 (1962).
72. A. A. Scala and P. Ausloos, *J. Chem. Phys.*, **45**, 847 (1966).
73. P. Ausloos and S. G. Lias, *J. Chem. Phys.*, **45**, 524 (1966).
74. R. Gordon, Jr. and P. Ausloos, *J. Chem. Phys.*, **47**, 1799 (1967).
75. G. G. Meisels, *Nature*, **206**, 287 (1965).
76. R. S. Berry, *J. Chem. Phys.*, **45**, 1228 (1966).
77. D. P. Stevenson, *Radiation Res.*, **10**, 610 (1959).
78. W. A. Chupka and M. Kaminsky, *J. Chem. Phys.*, **35**, No. 6, 1991 (1961).
79. G. H. Wannier, *Phys. Rev.*, **90**, 807 (1953).
80. J. W. McGowan, M. A. Fineman, E. M. Clarke, H. P. Hanson, *Phys. Rev.*, **167**, 52 (1968).
81. C. E. Brion and G. E. Thomas, *Internat. J. Mass. Spectry. Ion Phys.*, **1**, 25 (1968).
82. J. T. Tate, P. T. Smith and A. L. Vaughan, *Phys. Rev.*, **48**, 525 (1935).

83. A. Adamczyk, A. J. H. Boerboom, B. L. Schram and J. Kistemaker, *J. Chem. Phys.*, **44,** 4640 (1966).
84. Most textbooks on kinetics give excellent summaries of the applicable theories. For example, see K. J. Laidler, *Chemical Kinetics*, McGraw-Hill, New York; 2nd ed., 1965, p. 150 ff., S. Benson, *The Foundations of Chemical Kinetics*, McGraw-Hill, N.Y., 1960, p. 225 ff, or H. S. Johnston, *Gas Phase Reaction Rate Theory*, Reinhold Press, N.Y., 1966, p. 263 ff., etc.
85. M. Vestal, A. L. Wahrhaftig and W. H. Johnston, *J. Chem. Phys.*, **37,** 1276 (1962).
86. B. S. Rabinovitch and D. W. Setser, *Adv. Photochem.*, Vol. 3, Chap. 1, 1964.
87. M. Vestal, *J. Chem. Phys.*, **43,** 1356 (1965).
88. H. M. Rosenstock, M. B. Wallenstein, A. L. Wahrhaftig and H. Eyring, *Proc. Natl. Acad. Sci. (U.S.)*, **38,** 667 (1952).
89. H. M. Rosenstock and M. Krauss, in *Mass Spectrometry of Organic Ions*, (Ed., F. W. McLafferty), Academic Press, New York, 1963.
90. J. C. Y. Chen and J. L. Magee, *J. Chem. Phys.*, **36,** 1407 (1962).
91. D. F. Evans, *J. Chem. Soc. (London)*, 1735 (1960).
92. K. J. Laidler, *The Chemical Kinetics of Excited States*, Clarendon Press, Oxford, 1955, p. 95.
93. R. B. Cundall, in *Progress in Reaction Kinetics*, Vol. II (Ed., G. Porter), Pergamon Press, Oxford, 1964.
94. D. A. Becker, H. Okabe and J. R. McNesby, *J. Phys. Chem.*, **69,** 528 (1965). E. Tschuikow-Roux, *J. Phys. Chem.*, **71,** 2355 (1967).
95. J. G. Calvert and J. N. Pitts, Jr., *Photochemistry*, John Wiley, New York 1966.
96. C. E. Klots, *J. Chem. Phys.*, **41,** 117 (1964).
97. M. Trachtman, *J. Phys. Chem.*, **70,** 3382 (1966).
98. Y. Hatano and S. Shida, *J. Chem. Phys.*, **46,** 4784 (1967).
99. S. W. Benson and Wl. B. DeMore, *Ann. Rev. Phys. Chem.*, **16,** 397 (1965).
100. K. Yang, *J. Am. Chem. Soc.*, **84,** 3795 (1962).
101. R. J. Cvetanovic and R. S. Irwin, *J. Chem. Phys.*, **46,** 1694 (1967).
102. T. J. Hardwick, *J. Phys. Chem.*, **64,** 1623 (1960).
103. J. P. Manion and M. Burton, *J. Phys. Chem.*, **56,** 560 (1952).
104. R. L. Platzman, *Vortex*, **23,** 8 (1962).
105. R. L. Platzman, *Energy Spectrum of Primary Activators in the Action of Ionizing Radiation*, Paper presented at the Third International Congress of Rad. Res. Cortina d'Ampezzo, Italy, June 27, 1966.
106. W. P. Jesse, *J. Chem. Phys.*, **41,** 2060 (1964).
107. G. L. Weissler, in *Handbuch der Physik.*, 2nd ed., Vol. 21, Julius Springer, Berlin, 1956, p. 304.
108. R. I. Schoen, *J. Chem. Phys.*, **40,** 1830 (1964).
109. J. C. Person, *J. Chem. Phys.*, **43,** 2553 (1965).
110. G. S. Hurst and T. E. Bortner, *J. Chem. Phys.*, **42,** 713 (1965).
111. W. P. Jesse and R. L. Platzman, *Nature*, **195,** 790 (1962).
112. W. P. Jesse, *J. Chem. Phys.*, **38,** 2774 (1963).
113. W. P. Jesse, *J. Chem. Phys.*, **46,** 4981 (1967).
114. J. C. Person and P. P. Nicole, *J. Chem. Phys.*, **49,** 5421 (1968).
115. T. A. R. Samson, F. F. Marmo and K. Watanabe, *J. Chem. Phys.*, **36,** 783 (1962).

116. C. E. Klots, *J. Chem. Phys.*, **46,** 1197 (1967).
117. M. Halman and I. Laulicht, *J. Chem. Phys.*, **43,** 1503 (1965).
118. J. H. Futrell, *J. Chem. Phys.*, **35,** 353 (1961).
119. M. Vestal, A. L. Wahrhafting and W. H. Johnston, A.R.L. 62–426, *Theoretical Studies in Basic Radiation Chemistry*, Report to the Office of Aerospace Research, U.S.A.F., September 1962.
120. P. Ausloos, S. G. Lias and I. B. Sandoval, *Disc. Faraday Soc.*, **36,** 66 (1963).
121. P. Ausloos, R. Gorden, Jr. and S. G. Lias, *J. Chem. Phys.*, **40,** 1854 (1964).
122. H. S. W. Massey and E. H. S. Burhop, *Electronic and Ionic Impact Phenomena*, Oxford Univ. Press, 1952.
123. L. G. Christophorou, R. N. Compton, G. S. Hurst and P. W. Reinhardt, *J. Chem. Phys.*, **45,** 536 (1966).
124. R. L. Platzman, *Radiation Res.*, **2,** 1 (1955).
125. S. G. El Komoss and J. L. Magee, *J. Chem. Phys.*, **36,** 256 (1962).
126. W. E. Wentworth and E. Chen, *J. Gas Chromatog.*, **5,** 170 (1967).
127. Y. Hatano, S. Shida and M. Inokuti, *J. Chem. Phys.*, **48,** 940 (1968).
128. J. Bednar, *Collection Czech. Chem. Commun.*, **30,** 1328 (1965).
129. C. E. Klots, *J. Chem. Phys.*, **44,** 2715 (1966).
130. C. E. Klots, *J. Chem. Phys.*, **39,** 1571 (1963).
131. *Nat. Bur. of Std. (U.S.) Handbook*, **79** (1961), **85** (1964).
132. D. E. Lea, *Actions of Radiations on Living Cells*, Cambridge Univ. Press, Cambridge, 1946.
133. B. G. Dzantiev and A. P. Shvedchikov, *Russ. J. Phys. Chem.*, **38,** 1494 (1964).
134. P. M. Mulas and R. C. Axtmann, *Phys. Rev.* **146,** 296 (1966).
135. W. J. McConnell, R. D. Birkhoff, R. A. Hamm and R. H. Ritchie, *Radiation Res.*, **33,** 216 (1968).
136. L. V. Spencer and V. Fano, *Phys. Rev.*, **73,** 1172 (1954).
137. A. H. Samuel and J. L. Magee, *J. Chem. Phys.*, **21,** 1080 (1953); A. K. Ganguly and J. L. Magee, *J. Chem. Phys.*, **25,** 129 (1956).
138. A. Mozumder and J. L. Magee, *J. Chem. Phys.*, **45,** 3332 (1966); **47,** 939 (1967), **47,** 1859 (1967).
139. L. Onsager, *Phys. Rev.*, **54,** 554 (1938).
140. J. W. Buchanan and F. Williams, *J. Chem. Phys.*, **44,** 4377 (1966).
141. G. R. Freeman, *J. Chem. Phys.*, **46,** 2822 (1967).
142. F. Williams, *Disc. Faraday Soc.*, **36,** 254 (1963).
143. R. A. Back, *J. Phys. Chem.*, **64,** 124 (1960).
144. P. Ausloos, S. G. Lias and R. Gorden, Jr., *J. Chem. Phys.*, **39,** 3341 (1963).
145. K. R. Ryan and J. H. Futrell, *J. Chem. Phys.*, **43,** 3009 (1965).
146. K. R. Ryan and J. H. Futrell, *J. Chem. Phys.*, **42,** 824 (1965).
147. M. T. Bowers, D. D. Elleman and J. L. Beauchamp, *J. Phys. Chem.*, **72,** 3599 (1968).
148. I. Koyana, I. Tanaka and I. Omura, *J. Chem. Phys.*, **40,** 2734 (1964).
149. M. S. B. Munson, *J. Chem. Phys.*, **69,** 572 (1965).
150. J. H. Futrell, Ref. 48, Chap. 4.
151. S. Wexler and N. Jesse, *J. Am. Chem. Soc.*, **84,** 3425 (1962).
152. F. H. Field and M. S. B. Munson, *J. Am. Chem. Soc.*, **87,** 3829 (1965).
153. M. S. B. Munson, *J. Am. Chem. Soc.*, **87,** 2332, 5313 (1965).
154. M. S. B. Munson, *J. Phys. Chem.*, **70,** 2034 (1966).

155. G. G. Meisels, *J. Chem. Phys.*, **42,** 2328 (1965).
156. G. G. Meisels, *J. Chem. Phys.*, **42,** 3237 (1965).
157. G. G. Meisels and T. J. Sworski, *J. Phys. Chem.*, **69,** 815 (1965).
158. J. J. Myher and A. G. Harrison, *Can. J. Chem.*, **46,** 101 (1968).
159. G. G. Meisels and H. F. Tibbals, *J. Phys. Chem.*, **72,** 3746 (1968).
160. I. Szabo, *Arkiv. Fysik.*, **33,** 57 (1966).
161. T. F. Williams, ref. 46, Ch. 9.
162. C. D. Wagner, *J. Phys. Chem.*, **71,** 3445 (1967).
163. C. D. Wagner, *Trans Faraday Soc.*, **64,** 163 (1968).
164. R. Gorden, Jr. and P. Ausloos, *156th National Meeting, Am. Chem. Soc.*, Atlantic City, N.J., Sept. 1968.
165. A. G. Harrison, *Can. J. Chem.*, **41,** 236 (1963).
166. F. P. Abramson and J. H. Futrell, *J. Phys. Chem.*, **72,** 1826 (1968).
167. P. C. Kaufman, *J. Phys. Chem.*, **67,** 1671 (1963).
168. H. Okamoto, K. Fueki and Z. Kuri, *J. Phys. Chem.*, **71,** 3222 (1967).
169. R. Taylor and F. Williams, *J. Am. Chem. Soc.*, **89,** 6359 (1967).
170. E. W. Schlag and J. J. Sparapany, *J. Am. Chem. Soc.*, **86,** 1875 (1964).
171. N. S. Viswanathan, L. Kevan, *J. Am. Chem. Soc.*, **90,** 1375 (1968).
172. C. J. Wood, R. A. Back and D. H. Dawes, *Can. J. Chem.*, **45,** 3071 (1967); D. A. Armstrong and R. A. Back, *Can. J. Chem.*, **45,** 3079 (1967).
173. G. A. Fisk, B. H. Mahan and E. K. Packs, *J. Chem. Phys.*, **47,** 2649 (1967).
174. R. H. Lawrence and R. F. Firestone, *Advan. Chem.*, **58,** 278 (1966).
175. B. H. Mahan and J. C. Person, *J. Chem. Phys.*, **40,** 392 (1964).
176. R. R. Hentz, D. B. Peterson, S. B. Srivastava, H. F. Barzynski and M. Burton, *J. Phys. Chem.*, **70,** 2362 (1966); R. R. Hentz, K. Shimer and M. Burton, *J. Phys. Chem.*, **71,** 461 (1967).
177. J. M. Warman, *J. Phys. Chem.*, **71,** 4066 (1967).
178. D. N. Klassen, *J. Phys. Chem.*, **71,** 2409 (1967).
179. J. A. Kerr and A. F. Trotman-Dickenson, *The Reactions of Alkyl Radicals*, in *Progress in Reaction Kinetics*, Vol. I, p. 107 (Ed., G. Porter), Pergamon Press, New York (1961).
180. N. L. Ruland, *Dissertation Abstr.*, **27,** 4296 (1967).
181. H. Umezawa and F. S. Rowland, *J. Am. Chem. Soc.*, **84,** 3077 (1962).
182. A. Chapiro, *Radiation Chemistry of Polymeric Systems*, Interscience, New York, 1962.
183. J. C. Hayward and R. H. Bretton, *Chem. Eng. Progr.*, **50,** 73 (1954).
184. J. L. Magee, *Ann. Rev. Phys. Chem.*, **12,** 389 (1961); *Disc. Faraday Soc.*, **36,** 232 (1963).
185. A. Kupperman, *The Chemical and Biological Actions of Radiations*, Vol. 5, Academic Press, New York, 1961, Chap. 3, p. 85.
186. H. A. Schwarz, *J. Am. Chem. Soc.*, **77,** 4960 (1955).
187. P. Ausloos and S. G. Lias, *Gas Phase Radiolysis of Hydrocarbons*, in *Actions Chimiques et Biologiques des Radiations*. (Ed., M. Haissinsky), Masson and Cie., Paris, Vol. 11, 1967.
188. R. D. Doepker and P. Ausloos, *J. Chem. Phys.*, **43,** 3814 (1965).
189. R. D. Doepker and P. Ausloos, *J. Chem. Phys.*, **44,** 1641 (1966).
190. L. W. Sieck and J. H. Futrell, *J. Chem. Phys.*, **48,** 1409 (1968).
191. G. G. Meisels, P. S. Gill and C. T. Chen, *24th Southwest Regional Meeting of the Am. Chem. Soc.*, Austin, Texas, December, 1968.

CHAPTER **9**

Polymers containing C=C bonds

MORTON A. GOLUB*

Stanford Research Institute, Menlo Park, California, U.S.A.

I. GENERAL COMMENTS	412
II. CLASSES OF POLYMERS CONTAINING C=C BONDS . . .	413
A. Diene homopolymers	413
1. Isoprene polymers	414
2. Butadiene polymers	416
3. Chloroprene polymers	418
4. Piperylene polymers	419
5. Polymers from other butadiene derivatives . .	421
6. Other diene homopolymers	421
a. Polycyclopentadiene	421
b. Vinylcyclohexene polymers	422
c. Various tetraene polymers	422
B. Diene copolymers	423
1. From butadiene	423
2. From isoprene	424
3. Ethylene-propylene terpolymer (EPT) . . .	424
4. Graft copolymers	426
C. Other polymers containing C=C bonds	426
1. From alkynes	426
2. From cyclic olefins	428
3. From conjugated trienes	429
4. Unsaturated polyesters	430
5. Unsaturated polymers formed through polymer reactions	431
a. Dehydrochlorination of polyvinyl chloride . .	431
b. Dechlorination of 1,2 vicinal units . . .	432
c. From polyacrylonitrile	432
d. Radiation induced formation of C=C bonds .	433
III. CHEMICAL REACTIONS OF POLYMERS CONTAINING C=C BONDS .	433
A. Halogenation and hydrohalogenation	434
1. Chlorination	434

* Present address: Ames Research Center, National Aeronautics and Space Administration, Moffett Field, California 94035.

           2. Bromination 437
           3. Hydrohalogenation 438
        B. Cyclization 441
           1. 1,4-Polyisoprene 441
           2. 3,4-Polyisoprene 443
           3. 1,4-Polybutadiene 444
           4. 1,2-Polybutadiene or 1,2-polyisoprene . . . 446
           5. Cyclopolymers from diene monomers 447
        C. Cis–Trans isomerization 449
           1. 1,4-Polybutadiene 449
              a. Photo- and radiation-induced reactions . . . 449
              b. Other isomerization methods 452
           2. 1,4-Polyisoprene 454
              a. Photo- and radiation-induced reactions . . . 455
              b. Other isomerization methods. 458
        D. Hydrogenation 460
        E. Vulcanization 463
           1. Peroxide vulcanization 464
              a. 1,4-Polyisoprene 464
              b. 1,4-Polybutadiene 467
           2. Sulphur vulcanization 468
              a. Unaccelerated reaction 468
              b. Accelerated reaction 473
           3. Vulcanization of neoprene 474
           4. Radiation vulcanization 474
        F. Oxidation 476
           1. 1,4-Polyisoprene 477
           2. Butadiene-containing polymers 483
        G. Ozonization 484
        H. Degradation 486
           1. Thermal degradation 487
              a. 1,4-Polyisoprene 487
              b. 1,4-Polybutadiene 488
           2. Mechanochemical degradation 489
        I. Miscellaneous reactions 491
           1. Epoxidation 492
           2. Thiol addition to double bonds 493
           3. Carbene additions 494
           4. Addition of ethylenic compounds 496
           5. Addition of carbonyl compounds 498
IV. REFERENCES 500

I. GENERAL COMMENTS

Although numerous review articles and many chapters in monographs have been written about individual reactions of polymers, both of the saturated and unsaturated types, and several recent texts [1-7] present useful surveys of various reactions of unsaturated polymers, there is apparently no unified review of the many different types of polymers formed with C=C bonds and of their chemical reactions. The present chapter aims to provide an up-to-date account emphasizing the latest developments in this field. Little or no attention will be paid to polymers with saturated main chains even if they do contain occasional olefinic double bonds in long branches. The macromolecules to be discussed here will be largely restricted to compounds of sufficiently high molecular weight that they display the customary physical properties of elastomers, or plastics. The topic of polymerization of alkenes or diolefins is outside the scope of this chapter, and will not be discussed here except to make brief mention of some of the synthetic routes to the various unsaturated polymers.

II. CLASSES OF POLYMERS CONTAINING C=C BONDS

A. Diene Homopolymers

By far the most important class of polymers containing C=C bonds are the natural and synthetic elastomers, the latter being derived from the polymerization of diolefin monomers either alone (homopolymerization) or with other monomers (copolymerization), generally of the monoolefin type. The term 'elastomer' is a generic one applied to all polymeric materials exhibiting 'long-range' elasticity, i.e., the ability to undergo gross elongation or distortion under stress and to return rapidly to virtually its original dimensions upon removal of the stress. The main classes of unsaturated elastomers, besides natural rubber (*Hevea*), are the stereoregular rubbers derived from isoprene or butadiene, the non-stereoregular homopolymers of these same monomers, the butadiene-styrene and butadiene-acrylonitrile copolymers, neoprene (or polychloroprene), butyl rubber (isobutylene-isoprene copolymer), and ethylene-propylene terpolymers. These unsaturated elastomers and the various classes of saturated elastomers are the subject of a very recent two-part monograph[3,8], and an encyclopaedia article[9].

Depending upon the particular structure or microstructure of the unsaturated polymer or copolymer, the material may have plastic instead of elastomeric properties. Thus, while *Hevea* (an all-*cis* 1,4-polyisoprene) is a rubber, its naturally occurring geometric isomer, balata or gutta-percha (an all-*trans* 1,4-polyisoprene), is a leathery, crystalline thermoplastic; likewise, depending upon the ratio of butadiene to styrene in the copolymer, the properties range from elastomeric to plastic, with the optimum elastomeric composition corresponding to around 25% styrene in the random butadiene-styrene copolymer. In this section, our attention is addressed to the variety of unsaturated polymers, whether elastomers or plastics, which can be obtained by the homopolymerization of different diolefin monomers.

I. Isoprene polymers

Depending upon the catalyst and the reaction conditions employed, isoprene (or 2-methyl-1,3-butadiene) can be made to undergo 1,4- or 1,2- (or 3,4-) addition polymerization to yield macromolecules in which the monomer units have one or more of the configurations 1–4. To add to the possibilities for stereoisomerism, these units can be linked

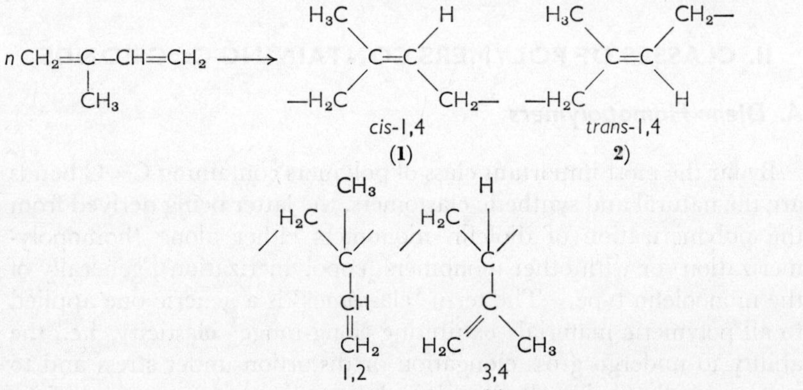

$$n\,CH_2=C(CH_3)-CH=CH_2 \longrightarrow$$

cis-1,4 (1)   trans-1,4 (2)

1,2 (3)   3,4 (4)

together in either head-to-tail or head-to-head configurations, while the head-to-tail 1,2- or 3,4-addition units, by virtue of the presence of an asymmetric carbon atom, can also exist in isotactic, syndiotactic or atactic (random) structures:

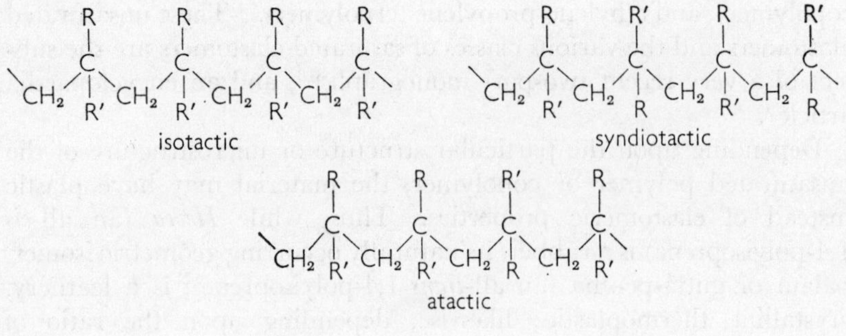

isotactic   syndiotactic

atactic

where R is —CH=CH$_2$ or —C(CH$_3$)=CH$_2$ and R' is CH$_3$ or H, respectively. The mechanisms and catalysts for the polymerization of isoprene and other monomers to various stereoregular as well as non-stereoregular polymer structures, have been reviewed recently[10-12] and need not be discussed here. It should be noted, however, that only three of the six possible stereoregular forms of polyisoprene have been synthesized: two have structures nearly identical to those of the natural polyisoprenes, i.e., > 98% *cis*-1,4 (corresponding to *Hevea*) and 99–100% *trans*-1,4 (corresponding to balata), and the third one has a very high ($\sim$95%) 3,4-configuration, of uncertain tacticity.

Polyisoprenes containing various proportions of all four kinds of addition units (**1–4**) have been prepared. The relative proportions of the different addition units in any soluble polyisoprene can be determined quite accurately by a combination of infrared[13-16] and n.m.r.[17,18] spectroscopy. Thus, units **1** and **2** are very well determined by means of characteristic n.m.r. peaks at 8·33 and 8·40 τ associated with the *cis* and *trans* methyl protons, respectively, and by means of infrared spectroscopy using the 12 μ absorption band (C—H out-of-plane deformation of the *cis*-C(CH$_3$)=CH— unit is about 1·6 times as intense as that of the corresponding *trans* unit) or the bands at around 7·5–7·7 μ[19] and in the 8–10 μ region[14,19]. On the other hand, units **3** and **4** are determined by their characteristic infrared bands at 11·0 and 11·3 μ, respectively, and are also indicated by n.m.r. peaks at $\sim$4·5–4·6 (—C$\underline{H}$=CH$_2$) and 5·05 τ (—CH=C$\underline{H}_2$), for the vinyl units, and at 5·34 τ ($>$C=CH$_2$), for the vinylidene units. Other features of the n.m.r. spectra of the different polyisoprenes worth noting here are peaks at $\sim$9·1 τ (CH$_3$—C$\diagdown$ in **3**); 8·75 τ (—CH$_2$—C$\diagdown$ in **3** and **4**); 8·45 τ (—C(CH$_3$)= in **4**); 8·05 τ (—CH$_2$C= in **1** and **2**; —CH—C= in **4**); and 4·95 τ (—CH= in **1** and **2**).

Although the essential *cis*- and *trans*-1,4 structures of *Hevea* and balata have been recognized for many years, there has been some controversy recently concerning their microstructural purity. Because the infrared spectra of these natural polyisoprenes show a weak absorption at 11·3 μ, it has been assumed by Binder[15,20,21] and others that the unsaturation in *Hevea* and balata consists of 2·2 and 1·3%, respectively, of the 3,4 units (**4**). However, this assumption has been contradicted by near infrared[22] and high resolution n.m.r. studies[18], which indicate that if such units are present they constitute less than 0·3% of the total unsaturation. Very recently, Chen[23], using even more sensitive n.m.r. analysis, showed that the 3,4 content in these polymers was below the minimum detectable level of 0·1%. At the same time he showed that there was less than 0·5% of the opposite geometric isomeric form present in their microstructures. Accordingly, *Hevea* and balata can really be considered to have the 100% *cis*- and *trans*-1,4 structures assumed originally, largely on the basis of the much less exact X-ray diffraction analysis[24].

Despite this compelling evidence for the microstructural purity of the natural polyisoprenes, Chakravarty and Sircar[25] have renewed the claim for the presence of structure **4** in the rubber molecule. Their claim, based on the observation that formic acid was a product of the ozonolysis of rubber followed by oxidative hydrolysis, was challenged by Bevilacqua[26] on the premise that small amounts of formic acid are an inevitable consequence of thermal and catalysed reactions of polyisoprene with molecular oxygen.

Polyisoprenes with more than 90% *cis*-1,4 content are rubbery, those with more than ∼90% *trans*-1,4 content are hard, crystalline materials, while those with intermediate *cis–trans* contents are resinous. The physical properties of vulcanized polyisoprenes as a function of the initial *cis–trans* content have been described[27], and the properties and applications of synthetic and natural *cis*-1,4[28] and *trans*-1,4[29] polyisoprenes have been reviewed recently.

2. Butadiene polymers

Four different stereoregular polymers are possible from butadiene, on the basis of 1,4- or 1,2-addition polymerization, and all have been achieved using Ziegler-Natta catalysts[10,30]. These are the (nearly) all-*cis* and all-*trans* 1,4-polybutadienes (**5,6**) and the isotactic and syndiotactic 1,2-polybutadienes based on the microstructural unit **7**. Variation in the Ziegler catalyst formulations, or the reaction conditions, or the use of different ionic and free radical catalysts[31], can lead to the preparation of polybutadienes containing various proportions of these units, as well as an atactic 1,2-polybutadiene. The *cis*-1,4-polybutadiene is a soft, easily soluble elastomer having properties comparable to those of *Hevea*, and enjoying increasing attention as a replacement for natural rubber in many applications. The *trans*-1,4-polybutadiene, on the other hand, is a hard, difficultly soluble crystalline polymer closely resembling balata. The isotactic and

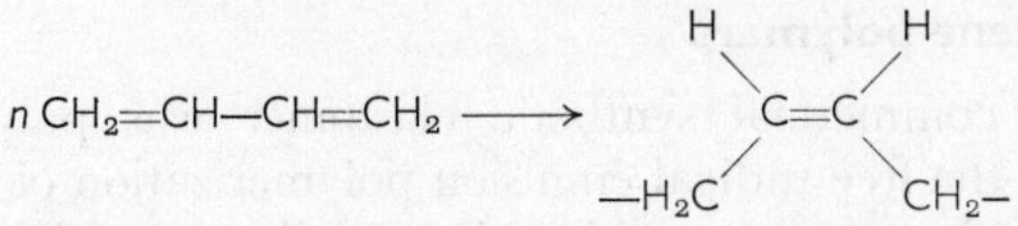

$$n\ CH_2{=}CH{-}CH{=}CH_2 \longrightarrow\ \underset{\substack{cis\text{-}1,4\\(\mathbf{5})}}{\overset{\displaystyle HH}{\underset{\displaystyle -H_2CCH_2-}{C{=}C}}}$$

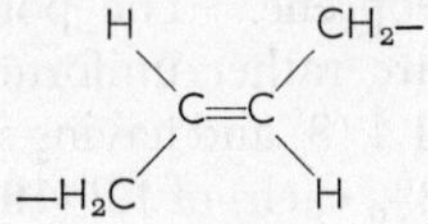

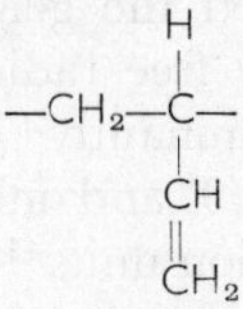

$$\underset{\substack{trans\text{-}1,4\\(\mathbf{6})}}{\overset{\displaystyle HCH_2-}{\underset{\displaystyle -H_2CH}{C{=}C}}} \qquad \underset{\substack{1,2\\(\mathbf{7})}}{-CH_2-\underset{\underset{CH_2}{\|}}{\overset{|}{\underset{|}{C}}}-}$$

syndiotactic 1,2-polybutadienes are rigid, crystalline and difficultly soluble, while the atactic ones are soft elastomers. Interestingly, it was also found that a nearly all-*trans* crystalline polybutadiene was formed in aqueous emulsion using rhodium chloride as a catalyst[32-34]. A survey of the various butadiene polymers and copolymers, their microstructures, properties and types of polymerization, has been presented[31]. The effect of *cis–trans* ratio on the physical properties of 1,4-polybutadiene vulcanizates has been described[35], as has also the use of *cis* polybutadiene-natural rubber blends[36].

The analysis for the microstructural units **5–7** in polybutadiene is based largely on infrared spectra[37,38] which show characteristic absorptions at 13·6 (*cis* —CH=CH—), 11·0 (vinyl unit) and 10·4 μ (*trans* —CH=CH—). Although high resolution n.m.r. spectroscopy has been used to analyse butadiene-isoprene copolymers[17], it is not possible to distinguish *cis* and *trans* butadiene units by this means. The n.m.r. spectra do show, however, olefinic proton peaks at ~4·7 τ (*cis* and *trans* —CH=CH— in **5** and **6**), and in the 4·5–5·3 τ region (for both terminal and non-terminal protons in —CH=CH$_2$ in **7**), besides the peaks at 8·05 τ (—CH$_2$—$\overset{|}{C}$= in **5** and **6**; —$\overset{|}{C}$H—$\overset{|}{C}$= in **7**) and 8·75 τ (—CH$_2$—C$\overset{\diagup}{\diagdown}$ in **7**). Although the olefinic proton peaks of the vinylene and vinyl units overlap considerably, thus complicating the determination of the relative amounts of 1,2- and 1,4-addition units in polybutadiene, Mochel[39] has recently provided an accurate n.m.r. method for the microstructural analysis of these and other units in butadiene-isoprene copolymers and butadiene-styrene block copolymers.

3. Chloroprene polymers

The first commercial synthetic elastomer was polychloroprene, obtained by the free radical emulsion polymerization of 2-chloro-1,3-butadiene (chloroprene). This polymer, discovered by Carothers[40], was introduced as DuPrene by the DuPont Company in 1931, and later assigned the generic name, neoprene. The polychloroprenes prepared by free radical initiators are rather uniform in structure, being predominantly (78–96%) *trans*-1,4 (**8**) and having some (4–18%) *cis*-1,4 units (**9**) and only traces (0.2–2% each) of 1,2 (**10**) and 3,4 (**11**) polymerization units[41,42].

$$\underset{\substack{trans\text{-}1,4\\(\mathbf{8})}}{\overset{\text{Cl}\quad\quad\text{CH}_2-}{\underset{-\text{H}_2\text{C}\quad\quad\text{H}}{\text{C}=\text{C}}}} \quad \underset{\substack{cis\text{-}1,4\\(\mathbf{9})}}{\overset{\text{Cl}\quad\quad\text{H}}{\underset{-\text{H}_2\text{C}\quad\quad\text{CH}_2-}{\text{C}=\text{C}}}} \quad \underset{\substack{1,2\\(\mathbf{10})}}{\begin{array}{c}-\text{CH}_2-\text{CCl}-\\|\\\text{CH}\\\|\\\text{CH}_2\end{array}} \quad \underset{\substack{3,4\\(\mathbf{11})}}{\begin{array}{c}-\text{CH}_2-\text{CH}-\\|\\\text{CCl}\\\|\\\text{CH}_2\end{array}}$$

Analysis by n.m.r. indicated[42] that there was sequence isomerism in the polychloroprenes as a result of 'head-to-tail' (**12**), 'head-to-head' (**13**) and 'tail-to-tail' (**14**) 1,4-addition of successive monomer units. The typical sequence isomer concentrations were found to be 70–80% **12**, and 10–15% each of **13** and **14**.

$$\underset{(\mathbf{12})}{\begin{array}{c}-\text{C}=\text{CHCH}_2\text{CH}_2\text{C}=\text{CH}-\\|\quad\quad\quad\quad\quad\quad|\\\text{Cl}\quad\quad\quad\quad\quad\quad\text{Cl}\end{array}} \quad \underset{(\mathbf{13})}{\begin{array}{c}-\text{CH}=\text{CCH}_2\text{CH}_2\text{C}=\text{CH}-\\|\quad\quad\quad\quad\quad\quad|\\\text{Cl}\quad\quad\quad\quad\quad\quad\text{Cl}\end{array}} \quad \underset{(\mathbf{14})}{\begin{array}{c}-\text{C}=\text{CHCH}_2\text{CH}_2\text{CH}=\text{C}-\\|\quad\quad\quad\quad\quad\quad|\\\text{Cl}\quad\quad\quad\quad\quad\quad\text{Cl}\end{array}}$$

The chlorinated olefin backbone of neoprene is less reactive than the backbone of natural rubber and polybutadiene, and its vulcanizates are thus more resistant to ozone and oxidative degradation than those of the other diene polymers. These properties combined with its good strength, non-flammability and good solvent resistance have kept polychloroprene in first place among the special purpose elastomers (as opposed to the general purpose elastomers like styrene-butadiene rubber).

In 1964 Aufdermarsh and Pariser[43] reported an elegant four-step synthesis of *cis*-1,4-polychloroprene starting with the monomer, proceeding via a Grignard reagent obtained from it to 2-(tri-n-butyltin)-1,3-butadiene which is made to undergo a *cis*-1,4 polymerization; the resulting polymer is then subjected to chlorinolysis to yield the desired polymer. By means of ozonolysis, and infrared and n.m.r. spectroscopy, the unsaturation of the *cis*-polychloroprene was shown to be at least 95%, and possibly as high as 99%, of type **9**, with negligible amounts of **8**, and only traces of **10** and **11**, with an overall unsaturation of 80–97% of the theoretical. Also, the sequence isomer concentrations were of the order of 50–60% **12** and 20–25% each of **13** and **14**.

The infrared analysis[41,42] for structures **8** and **9** has been based for a long time on the C=C stretching frequencies at 1660 and 1653 cm^{-1} (~6·02–6·05 μ) for the *trans*- and *cis*-1,4 units, respectively, but a more reliable analysis is based on the n.m.r. olefinic proton peaks at 4·65 and 4·49 τ respectively[42], which are more cleanly separated than are the corresponding infrared bands. The units **10** and **11** are determined by means of infrared bands at 10·8 and 11·3 μ, respectively.

4. Piperylene polymers

Considerable interest has been shown recently in the preparation of stereoregular polymers from 1,3-pentadiene (or piperylene), using anionic coordinate polymerization techniques[10,11]. The presence of the 1-methyl substituent on 1,3-butadiene admits the possibility of forming polymers with steric regularity of higher order than that of the polymers previously discussed, namely, those showing ditacticity due to the possibility of exhibiting both optical and geometric isomerism. Thus, on the basis of the 3,4-addition polymerization,

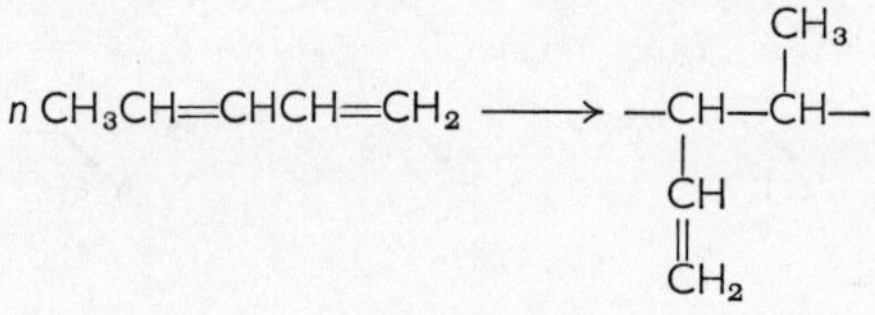

$$n\ CH_3CH=CHCH=CH_2 \longrightarrow -CH-\underset{\underset{\underset{CH_2}{\|}}{CH}}{\overset{CH_3}{\underset{|}{C}H}}-$$

three stereoisomeric polymers (**15**, **16**, **17**) are possible.

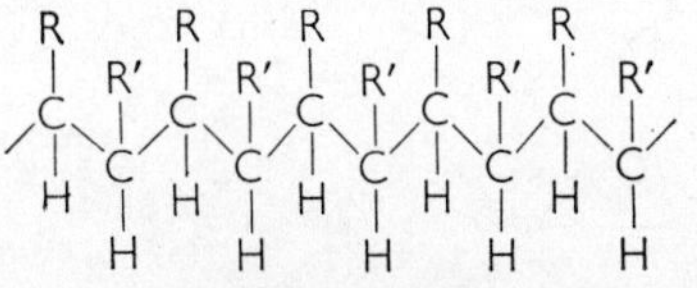

threo-diisotactic
(**15**)

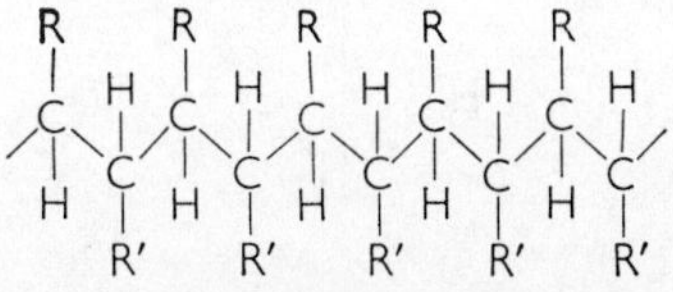

erythro-diisotactic
(**16**)

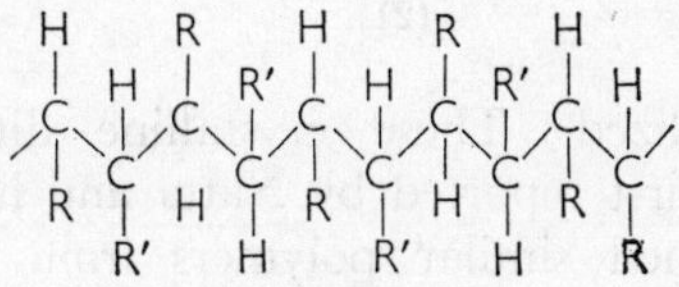

disyndiotactic
(**17**)

Also, there are four stereoisomers possible involving the 1,2-polymerization unit (**18**), namely, isotactic polymers with *cis* or *trans*

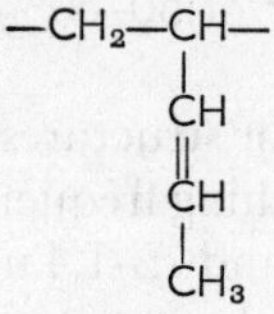

$$\begin{array}{c}-CH_2-CH-\\|\\CH\\||\\CH\\|\\CH_3\end{array}$$
(**18**)

double bonds in the side chain, and syndiotactic *cis* and *trans* polymers. Likewise, there are four stereoisomers possible from 1,4 polymerization: *cis*-1,4-isotactic or -syndiotactic; and *trans*-1,4-isotactic or -syndiotactic polymers. Of the eleven *stereoregular* possibilities, only the first three isomers from 1,4-addition (**19, 20** and **21**) have been

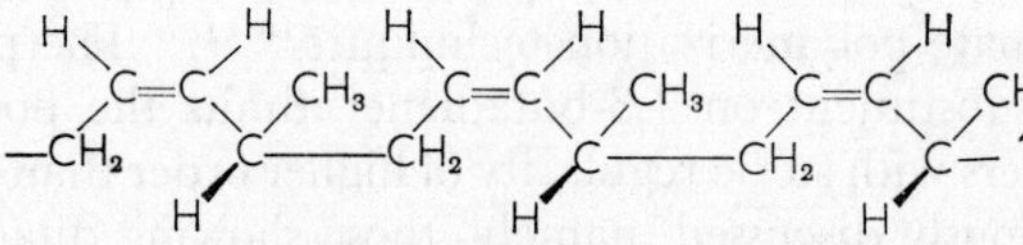

(**19**)

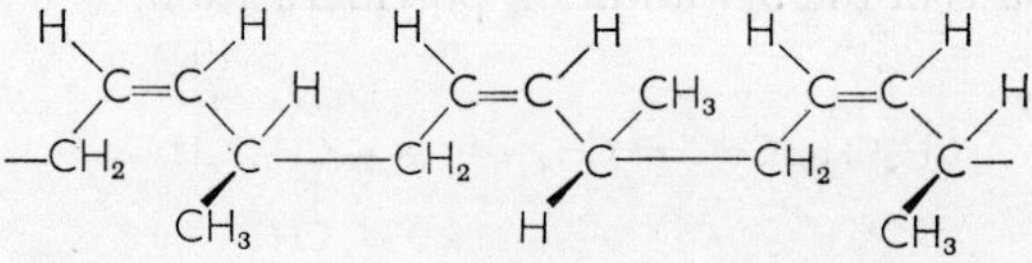

(**20**)

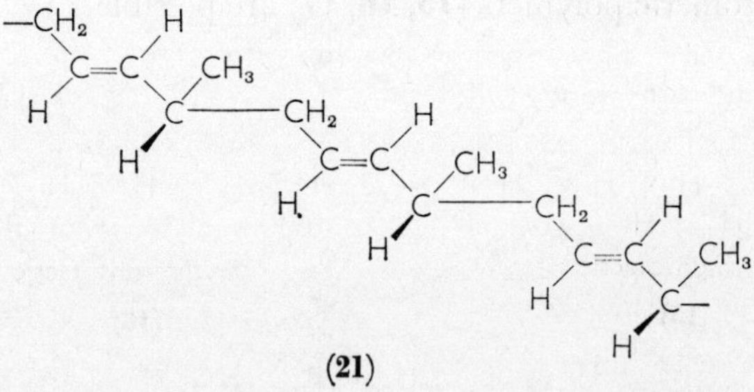

(**21**)

experimentally realized. These crystalline ditactic 1,4-poly(1,3-pentadiene)s were first reported by Natta and his school[44-46], who subsequently obtained similar polymers from 1,3-hexadiene (**22**, R=Et), 1,3-heptadiene (**22**, R = n-Pr) and other 1,3-alkadienes[47]. Polymers of 1,3-pentadiene which are predominantly *cis*- or *trans*-1,4 but which are amorphous because of insufficient order in the configuration of the asymmetric carbon atoms, have also been prepared[44].

$$CH_2=CHCH=CHR$$
(22)

A variety of plastic and rubbery butadiene-pentadiene copolymers have been prepared[48], and other copolymers of pentadiene with isoprene, ethylene, propylene and 1-butene reported[49].

5. Polymers from other butadiene derivatives

A general survey of the polymers obtained from various substituted butadienes besides those from isoprene or chloroprene has appeared[49]. Because of their close relationship to isoprene, the 2-alkylbutadienes have received much attention in the past as potential candidates for new synthetic rubbers. Thus, free radical emulsion polymerization (of the well-known butadiene-styrene or SBR type; see below) of 2-ethyl-, 2-isopropyl-, 2-n-amyl-, and 2-*tert*-butyl-1,3-butadienes, to indicate just a few monomers, have yielded polymers whose microstructures contained mixtures of all four kinds of addition units: 1,2; 3,4; and *cis*- and *trans*-1,4 units. With the advent of stereoregular polymerization techniques, it has become possible to obtain high *cis*-1,4 polymers of a variety of substituted butadienes, e.g., 2-ethyl-, 2-isopropyl- and 2-n-propylbutadienes[50]. These particular polymers are rubbery and display elastic properties comparable to those of natural rubber. Among the many other stereoregular polymers which have been prepared may be mentioned crystalline *cis*-1,4-poly-(2-*tert*-butylbutadiene)[51], crystalline *trans*-1,4 polymers of 1-ethylbutadiene[52], 1-cyanobutadiene[53] and 1-methoxybutadiene[54], *cis*-1,4 and *trans*-1,4 polymers of 2,3-dimethylbutadiene[55], and high *cis*-1,4-poly(2-phenylbutadiene)[56,57]. Non-stereoregular dimethylbutadiene polymers were the 'methyl rubbers' of World War I. Other diolefin monomers which have been polymerized include various halogen-substituted butadienes, such as 1-chloro-, 2-chloro-3-methyl-, 1,2-dichloro-, 2,3-dichloro- and 1-phenyl-2-chlorobutadienes[49].

6. Other diene homopolymers

a. Polycyclopentadiene. Another interesting diene monomer which can yield polymers having C=C bonds is cyclopentadiene (23), which has been shown to undergo both 1,4- and 1,2-opening of the double bonds. Thus, cationic polymerization[58] yields mixtures of 24 and 25, while a homogeneous Ziegler catalyst system[59] was reported to yield

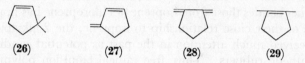

	1,4	1,2
(23)	(24)	(25)

predominantly structure **24**. A very recent n.m.r. study[60] has called attention to the possibility of additional microstructural units in polycyclopentadiene, namely **26-29**. Structures **24-27** are characterized by having two olefinic protons in each unit, while **28** and **29** have none.

(26)	(27)	(28)	(29)

b. Vinylcyclohexene polymers. 4-Vinylcyclohexene polymerizes to a structure having a C═C bond in the cyclohexyl ring of each repeating unit in the chain. Being a 1,5-diene, this monomer is capable of undergoing also a cyclopolymerization[61], and indeed with a cationic catalyst[62] it undergoes both vinyl polymerization and bicyclic ring formation, producing a polymer with structural units **30** and **31** in approximately equal amounts.

(30)	(31)

The use of a Ziegler catalyst[62], on the other hand, leads to a polymer which is made up almost entirely of the structural unit **31**.

c. Various tetraene polymers. Butler and Raymond[63] have recently shown that 1,3,6,8-nonatetraene, 1,3,9,11-dodecatetraene, and 3,6-dimethylene-1,7-octadiene undergo a cumulative 1,4–1,4 cyclic polymerization with a Ziegler catalyst, resulting in the formation of the novel, soluble, unsaturated polymers having predominantly the structures **32-34**, respectively. These findings on the tetraenes are

(32)

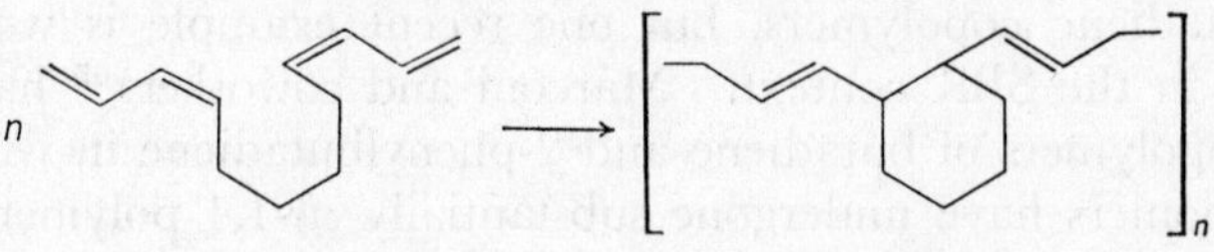

(33)

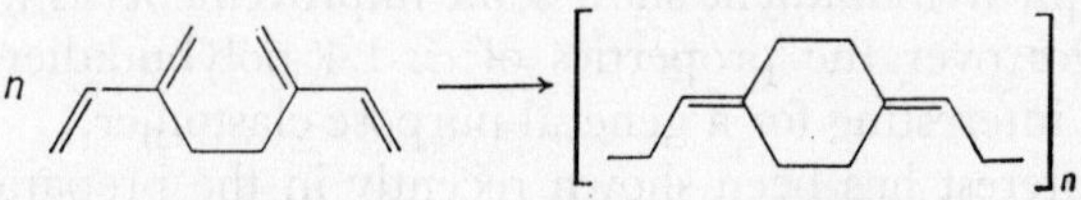

(34)

analogous to prior observations by Butler and Brooks[64] that 1,3,7-octatriene and 1,3,8-nonatriene undergo cumulative 1,2–1,4 cyclic polymerization to yield polymers possessing five- and six-membered rings, respectively.

B. Diene Copolymers

I. From butadiene

An extremely wide variety of copolymers containing C=C bonds in the backbone have been prepared from pairs of monomers one or both of which is a diolefin. One of the most important classes of such copolymers is that based on butadiene, and indeed the styrene-butadiene rubbers (known originally as Buna S, later as GR-S, and currently as SBR), a family of random copolymers consisting of about 25% styrene by weight, represent the most extensively employed synthetic rubber developed to date. The major use for this general purpose elastomer is in tyres for passenger automobiles. The linear backbone of the styrene-butadiene rubber contains —$CH_2CH(C_6H_5)$— units in combination with butadiene units 5–7, the relative amounts of which depend somewhat on the catalyst and conditions of polymerization.

Another important butadiene copolymer is the nitrile rubber, (known originally as Buna N, later as GR-A, and currently as NBR) consisting of about 25–35% acrylonitrile. Because of their excellent solvent and oil resistance, the nitrile rubbers find application where natural rubber and styrene-butadiene rubber are unsuitable.

Much attention has been given by polymer chemists in the past, to the random copolymerization of butadiene with many different conjugated dienes and vinyl and acrylic monomers, in order to obtain polymers having desirable and possibly even unique properties. There is little to be gained here in reciting a number of examples of such butadiene copolymers, but one recent example is worthy of mention in the SBR context. Marconi and coworkers[65] have prepared copolymers of butadiene and 2-phenylbutadiene in which the two monomers have undergone substantially *cis*-1,4 polymerization. The mechanical properties of a typical copolymer having 25% by weight of 2-phenylbutadiene show some improvements as well as some disadvantages over the properties of *cis*-1,4 polybutadiene, and on balance are interesting for a general purpose elastomer.

Special interest has been shown recently in the preparation of so-called A-B-A block copolymers containing an elastomeric segment (B = butadiene or isoprene, for example) flanked by two plastic segments (A = styrene, for example). The A-B-A block copolymers behave like vulcanized elastomers but flow like plastics above the glass transition temperature of the end blocks. These 'thermoplastic' elastomers, first reported by workers at the Shell Chemical Company[66], may be processed by conventional thermoplastics techniques, such as injection moulding, and without the usual chemical vulcanization step they display the high resilience, high tensile strength, highly reversible elongation and abrasion resistance characteristic of vulcanized rubber. The synthesis of A-B-A block copolymers is based on anionic polymerization procedures[67], which can yield polymers having predictable A and B block lengths.

2. From isoprene

The main and perhaps only important use of isoprene for copolymerization, is in the preparation of the predominantly isobutylene polymer, butyl rubber (GR-I or IIR), an elastomer having outstanding low permeability to gases. The function of the isoprene units, incorporated at a level of about 1·5–4·5% by weight and with an essentially 1,4 configuration[68], is to provide the necessary unsaturation sites for vulcanization. Because of the limited unsaturation, butyl rubbers have greater stability against ozone or oxygen attack than natural rubber or SBR.

3. Ethylene-propylene terpolymer (EPT)

A rather recent development in the synthesis of new elastomers is the preparation with Ziegler-type catalysts of random copolymers of ethylene and propylene, which are amorphous and elastomeric over a composition range of about 35–65% of either monomer[69]. Interest in the ethylene-propylene rubber stems from the appreciation that a 50/50 alternating copolymer of these two monomers is identical in structure with the polymers (**37**), obtained by hydrogenating 1,4-polyisoprene (**35**) or 1,4-polypiperylene (**36**). These elastomers (**37**)

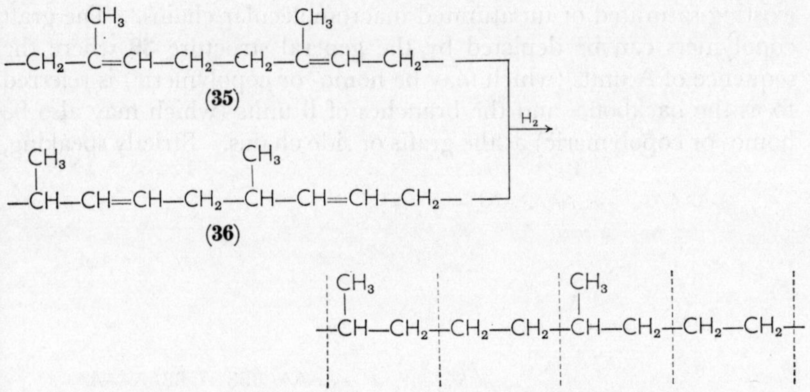

$$-CH_2-\underset{\underset{(35)}{|}}{\overset{CH_3}{C}}=CH-CH_2-CH_2-\overset{CH_3}{\underset{|}{C}}=CH-CH_2-$$

$$-\overset{CH_3}{\underset{|}{CH}}-CH=CH-CH_2-\overset{CH_3}{\underset{|}{CH}}-CH=CH-CH_2- \quad \xrightarrow{H_2}$$
(36)

$$-\!\!\!+\!\!\overset{CH_3}{\underset{|}{CH}}-CH_2\!\!+\!\!CH_2-CH_2\!\!+\!\!\overset{CH_3}{\underset{|}{CH}}-CH_2\!\!+\!\!CH_2-CH_2\!\!+\!\!-$$
(37)

are difficult to vulcanize, a fact which has led to the preparation of ethylene-propylene terpolymers (EPT rubbers) containing a small amount ($<5\%$) of a *non-conjugated* diene as a third monomer, to provide the required C=C bonds for conventional sulphur vulcanization[70]. Among the various non-conjugated dienes which have been employed for EPT rubbers are 1,4-hexadiene, dicyclopentadiene, 5-methylene-2-norbornene and 1,5-cyclooctadiene. The principal feature of these dienes is that one double bond enters into the polymerization while the other ends up in a pendant group; since the unsaturation is not in the main chain of the terpolymer, the excellent ozone resistance of the ethylene-propylene copolymers is retained. A typical EPT rubber containing dicyclopentadiene as the third monomer, has the structure **38**. Hank[71] has recently discussed the determination of double bonds in EPT rubbers, and Tyler[72] has reviewed the microstructural analysis of these and other rubbers.

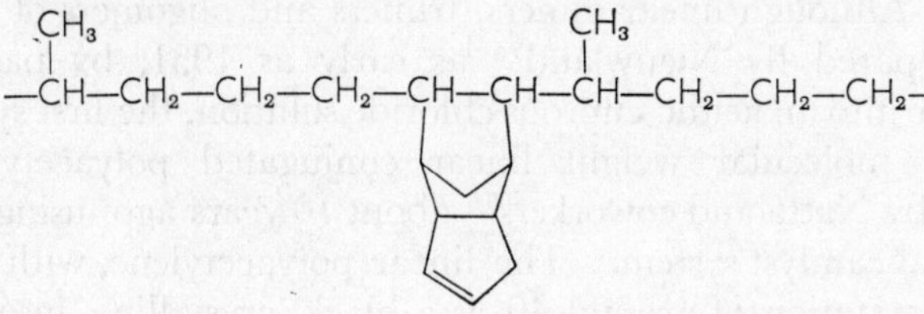

$$-\overset{CH_3}{\underset{|}{CH}}-CH_2-CH_2-CH_2-CH-CH-\overset{CH_3}{\underset{|}{CH}}-CH_2-CH_2-CH_2-$$
(38)

4. Graft copolymers

A wide assortment of polymers containing C=C bonds can be prepared by graft copolymerization of various diene monomers onto existing saturated or unsaturated macromolecular chains. The graft copolymers can be depicted by the general structure **39** where the sequence of A units (which may be homo- or copolymeric) is referred to as the backbone, and the branches of B units (which may also be homo- or copolymeric) as the grafts or side chains. Strictly speaking,

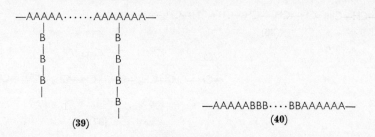

these polymers are 'grafted block copolymers' in contrast to the ordinary block copolymers (**40**) which can be regarded as 'linear block copolymers.' Using a variety of established grafting techniques, copolymer structures can be obtained having C=C bonds in the side chains or backbone or both. For a general survey of this topic and for leading references, reference 73 may be consulted.

C. Other Polymers Containing C=C Bonds

1. From alkynes

Acetylene and many substituted acetylenes undergo both addition and condensation homopolymerization, as well as copolymerization with different olefins. The products from the two kinds of homopolymerization are, respectively, linear unsaturated polymers, and cyclic trimers (benzene and various aromatic compounds) and tetramers. Although linear dimers, trimers and oligomers of acetylene were prepared by Nieuwland[74] as early as 1931, by passing the monomer into an acidic cuprous chloride solution, the first synthesis of a 'high' molecular weight linear conjugated polyacetylene was realized by Natta and coworkers[75] about 10 years ago, using a transition-metal catalyst system. The linear polyacetylene, with a degree of polymerization of around 40, is a black, crystalline, insoluble, infusible, oxygen-sensitive material having a single infrared absorption peak at 9·85 μ, indicative of a stereoregular *trans* configuration (**41**). Extraction of the black polyacetylene yields soluble, oily oligomers

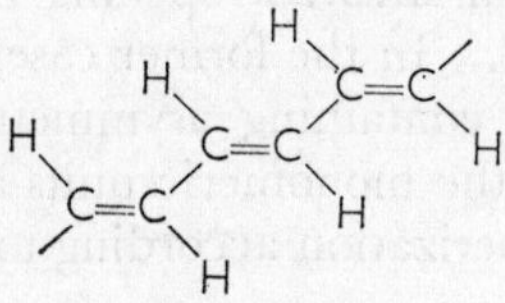

(**41**)

of structure **42**[76], which differ from the low molecular weight Nieuw-

$$CH_3CH_2-(CH=CH)_x-CH=CH_2$$
(**42**)

land polyacetylenes. The latter are generally of the ene-yne type, i.e., the dimer is vinylacetylene, $CH_2=CHC\equiv CH$, the trimer is divinylacetylene, $CH_2=CHC\equiv CCH=CH_2$, etc.

In the case of substituted acetylenes, such as but-1-yne and hex-1-yne, it is possible to obtain, using Ziegler-Natta catalysts, rubberlike polymers which are generally soluble in organic solvents[77].

Considerable attention has been given to the polyacetylenes and polyalkynes because of their interesting electronic properties arising from their long sequences of conjugated double bonds, and which suggest the possible use of these compounds as organic semiconductors[76]. It is generally accepted that electronic conductivity in organic systems with highly conjugated molecules is associated with delocalization of π-electrons. It is also noted that practically all semiconducting polymers exhibit an e.s.r. spectrum, but generalizations regarding the concentration of unpaired spins and the electrical conductivity are very difficult to establish. Thus, the Ziegler-type polyacetylenes mentioned above have conductivities much higher than those for the linear, soluble polyphenylacetylenes (**43**) of molecular weight ~1100–1500, even though the spin concentrations in these different polyalkynes are comparable (~10^{18}–10^{19} spins/g)[78].

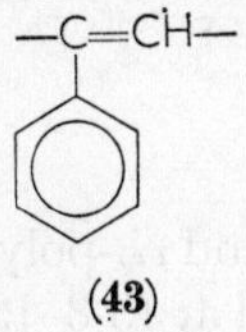

(**43**)

Another use for polyacetylenes, by virtue of their facility to add oxygen, is as oxygen scavengers.

2. From cyclic olefins

Cycloolefins can undergo two types of polymerization depending on whether the addition involves opening of the double bond or cleavage of the ring[11,79]. In the former case, this leads to saturated cyclic monomeric units containing asymmetric carbon atoms (**44**), while in the latter case the monomeric units are linear and unsaturated (**45**). Homopolymerization according to path I, can occur only

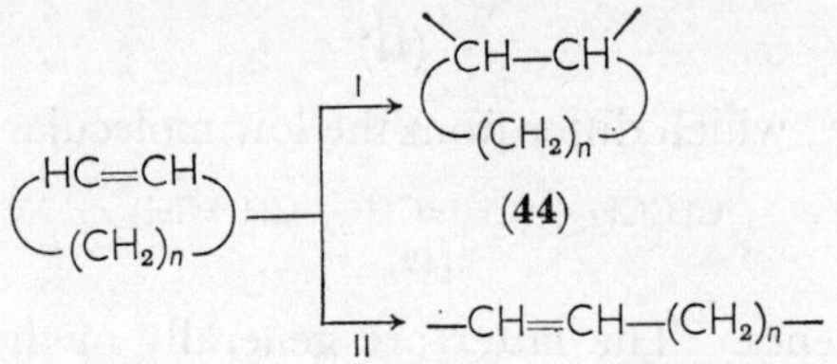

$$\underset{(CH_2)_n}{HC=CH} \longrightarrow \begin{cases} \overset{I}{\longrightarrow} \underset{(CH_2)_n}{CH-CH} \\ \quad (44) \\ \overset{II}{\longrightarrow} -CH=CH-(CH_2)_n- \\ \quad (45) \end{cases}$$

with cyclobutene, but where $n > 2$ copolymerization with ethylene is possible, such as **46** for $n = 3$, both cases in the presence of Ziegler-Natta catalysts.

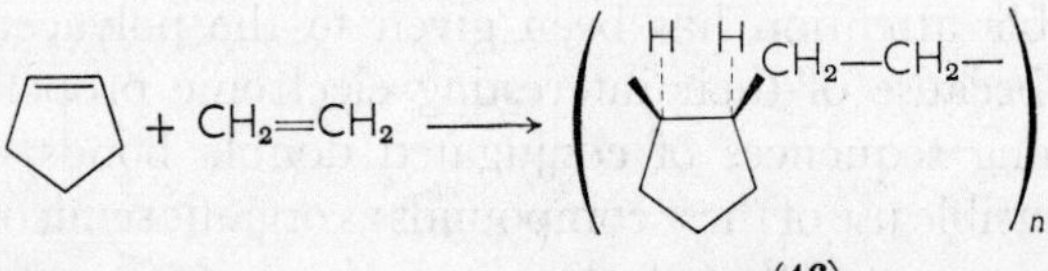

$$\bigcirc + CH_2=CH_2 \longrightarrow \left(\underset{}{\overset{H\ H}{\diagup}} \overset{CH_2-CH_2-}{\diagdown} \right)_n$$
(**46**)

Cyclopentene, again using Ziegler-Natta catalysts, yields by path II two different polypentenamers (**47, 48**) having *cis* and *trans* configuration, respectively, about the C=C bond. As with polybutadiene, the

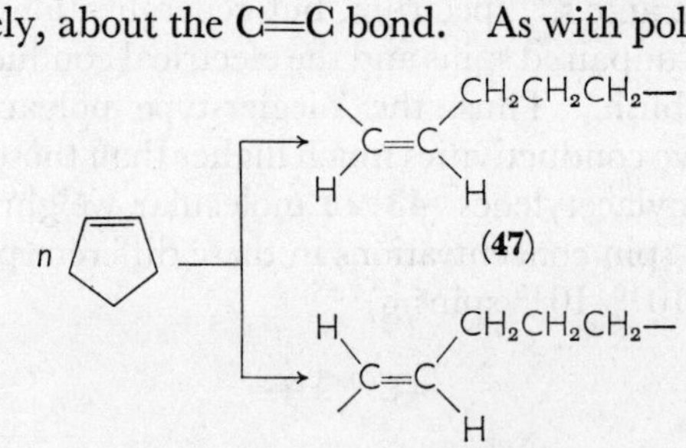

$$n \bigcirc \longrightarrow \begin{cases} \underset{H}{\overset{}{\diagup}} C=C \underset{H}{\overset{CH_2CH_2CH_2-}{\diagdown}} \\ \quad (47) \\ \underset{H}{\overset{H}{\diagdown}} C=C \underset{H}{\overset{CH_2CH_2CH_2-}{\diagup}} \\ \quad (48) \end{cases}$$

infrared spectra of the *trans*- and *cis*-polypentenamers show characteristic absorption at 10·35 μ and at 13·8–13·9 μ, respectively, due to the —CH=CH— unit in the corresponding geometric configurations. On the basis of ozonization and oxidation degradation analysis, the polypentenamers were considered to have a head-to-tail structure[11,79].

The presence of the additional methylene group in the repeat unit of the polypentenamers (**47,48**) over those in the repeat unit of the polybutadienes (**5,6**), is responsible for an appreciable depression in the melting points of the former polymers compared to the latter. Thus, the melting point of the *trans*-polypentenamer is 23°C. whereas that of *trans*-polybutadiene is 145°C., while the melting points of the corresponding *cis* polymers are < −50°C. and +2°C., respectively. Just as *cis*-polybutadiene is a very good elastomer, so also is the *trans*-polypentenamer which has properties approximating those of natural rubber (*cis*-polyisoprene).

Another type of cyclic olefin which can produce a polymer containing C=C bonds in the main chain is norbornene (**49**) which, on ring cleavage in the presence of an anionic coordinate polymerization catalyst, yields a flexible polymer containing vinyl cyclopentane groups (**50**)[80].

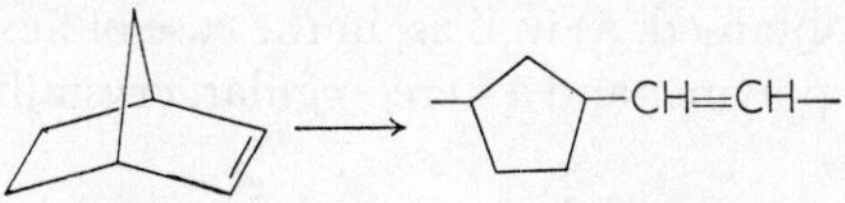

(49)   (50) —⟨cyclopentane⟩—CH=CH—

The fact that the infrared spectrum of such a polynorbornene shows the characteristic absorption band at 10·37 μ indicative of *trans* —CH=CH— unsaturation, but no band at 13·5 μ due to *cis* ethylenic units, demonstrates that the coordination polymerization of norbornene proceeds via a highly stereospecific reaction. On the other hand, with ruthenium halides as polymerization catalysts, polymers of structure **50** but having appreciable amounts of *cis* as well as *trans* —CH=CH— units can also be prepared[81].

3. From conjugated trienes

Bell[82] has recently described the polymerization of 1,3,5-hexatriene, 1,3,5-heptatriene and 2,4,6-octatriene with a variety of catalysts to give unique unsaturated polyolefins. Soluble, high molecular weight polymers with a predominantly 1,6-enchainment, and containing conjugated 1,3-dienyl groups within the backbone (**51**), were prepared with a soluble coordination catalyst system. Other triene polymers were obtained having mixtures of all three possible kinds of enchainment in the same macromolecule: 1,4 (**52**), 1,2 (**53**) and 1,6. Moreover, for each of the conjugated trienes, 1,6-(*trans,trans* and *cis,trans*)

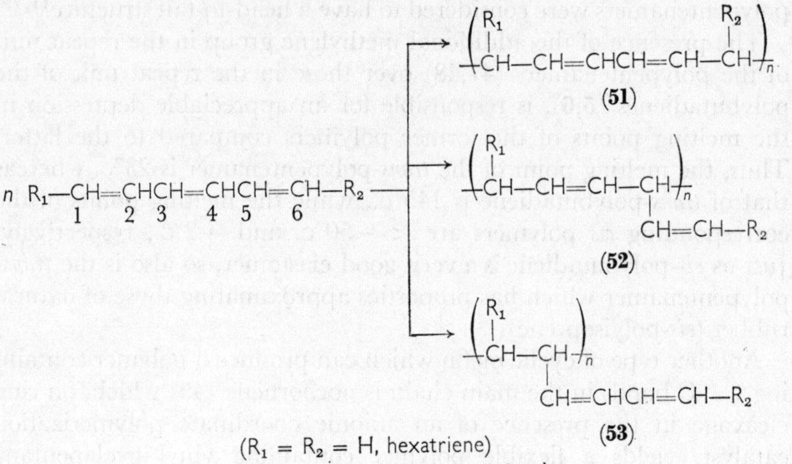

$$n\ R_1-CH=CHCH=CHCH=CH-R_2 \longrightarrow$$
$$\underset{1\quad 2\quad 3\quad 4\quad 5\quad 6}{}$$

$$\left(\begin{array}{c} R_1 \\ | \\ -CH-CH=CHCH=CH-CH- \\ \end{array} \begin{array}{c} R_2 \\ | \\ \\ \end{array} \right)_n$$
(51)

$$\left(\begin{array}{c} R_1 \\ | \\ -CH-CH=CH-CH- \\ | \\ CH=CH-R_2 \end{array} \right)_n$$
(52)

$$\left(\begin{array}{c} R_1 \\ | \\ -CH-CH- \\ | \\ CH=CHCH=CH-R_2 \end{array} \right)_n$$
(53)

($R_1 = R_2 = H$, hexatriene)
($R_1 = H; R_2 = CH_3$, heptatriene)
($R_1 = R_2 = CH_3$, octatriene)

structures were obtained, as well as, in the case of hexatriene, a 1,6-*cis, trans* amorphous polymer and a stereoregular, crystalline 1,6-*trans,trans*-polyhexatriene.

The infrared spectra of the various 1,6-polytrienes showed intense absorption bands at 10·15 and 10·31 μ which were identified with *trans,trans* and *cis,trans* diene structures, respectively, in **51**, partly on the basis of their response to Diels-Alder reactions of the polymers with reactive dienophiles[83]. Thus, the 10·15 μ band disappeared while the 10·31 μ band was unaffected by dienophile reactions. Likewise, a strong band at 11·0 μ was assigned to non-conjugated vinyl groups (as in **52**, $R_2=H$), since it was not affected by dienophiles, while a band at 11·1 μ was assigned to conjugated vinyls (as in **53**, $R_2=H$), since it was easily removed by Diels-Alder reactions.

4. Unsaturated polyesters

Condensation polymers of various types (e.g., polyesters, polyamides, polyurethanes) can be prepared having C=C bonds in the backbone from unsaturated condensation reagents. Thus, for example, unsaturated polyesters (**54**) can be formed if the R group in the dibasic acid and/or the R' group in the glycol, contains ethylenic units. Such compounds are of interest commercially since the C=C bonds in the R or R' groups can then be made to undergo radical chain polymerization with a vinyl monomer (e.g., styrene), to yield thermosetting polyester resins, crosslinked through the ethylenic unsaturation.

x HOOC—R—COOH + x HO—R'—OH ⟶

$$\text{H}\!-\!\!\left[\text{O—CO—R—CO—O—R}'\right]_x\!\!\text{OH} + (2x - 1)\,\text{H}_2\text{O}$$

(54)

5. Unsaturated polymers formed through polymer reactions

a. Dehydrochlorination of polyvinyl chloride. Polyvinyl chloride undergoes a progressive elimination of hydrogen chloride, with the formation of a conjugated polyene system:

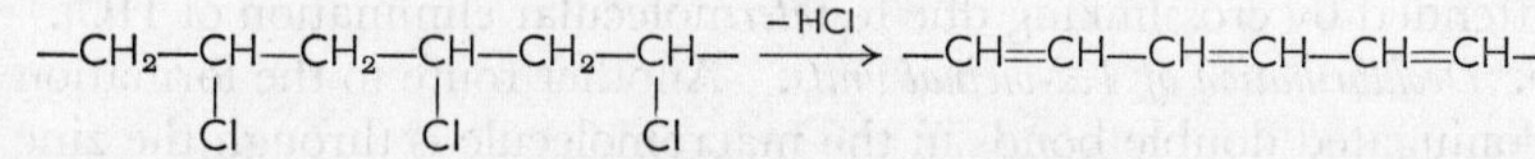

—CH₂—CH—CH₂—CH—CH₂—CH— $\xrightarrow{-\text{HCl}}$ —CH=CH—CH=CH—CH=CH—
         | | |
        Cl   Cl   Cl

under the influence of heat, ultraviolet or high energy radiation, or certain catalysts. Except in the catalysed case, this dehydrochlorination reaction, which has been extensively reviewed recently[84], is an autocatalytic, free radical chain reaction in which the unstable radical, —ĊHCHCl—, is considered to be the propagating species[85]:

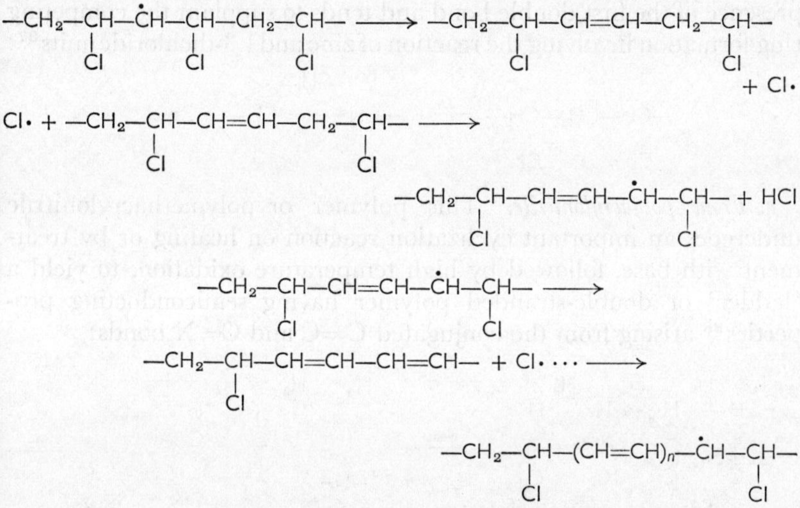

—CH₂—CH—ĊH—CH—CH₂—CH— ⟶ —CH₂—CH—CH=CH—CH₂—CH—
         | | | | |
        Cl   Cl   Cl   Cl   Cl
                                                              + Cl·

Cl· + —CH₂—CH—CH=CH—CH₂—CH— ⟶
         | |
        Cl   Cl

                           —CH₂—CH—CH=CH—ĊH—CH— + HCl
                                | |
                               Cl   Cl

                           —CH₂—CH—CH=CH—ĊH—CH— ⟶
                                | |
                               Cl   Cl

—CH₂—CH—CH=CH—CH=CH— + Cl· ··· ⟶
     |
    Cl

                           —CH₂—CH—(CH=CH)ₙ—ĊH—CH—
                                | |
                               Cl   Cl

Formation of a double bond in the backbone activates the adjacent methylenic hydrogen for abstraction by chlorine, with the result that the unsaturation takes the form of long sequences of conjugated double bonds, (—CH=CH—)$_n$. Since the latter structures absorb in the near ultraviolet through the visible spectrum, polyvinyl chloride becomes quite coloured as a result of the dehydrochlorination reaction. Similar non-chain scission reactions occur in various other vinyl polymers (e.g., polyvinyl acetate, polyvinyl alcohol, polyvinyl bromide) leading to C=C bonds in the backbone.

It is worth noting that polyvinylidene chloride undergoes an analogous thermal or catalysed dehydrochlorination to yield conjugated double bonds[86]. The resulting polyvinylene chloride (55) can be

$$-CH_2-CCl_2-CH_2-CCl_2- \longrightarrow -CH=CCl-CH=CCl-$$
$$(55)$$

further dehydrochlorinated to yield some triple bonds, but the reaction is attended by crosslinking due to *inter*molecular elimination of HCl.

b. Dechlorination of 1,2-vicinal units. Another route to the formation of conjugated double bonds in the macromolecule is through the zinc dechlorination of chlorine atoms on adjacent carbon atoms, if present in the polymer: The further dehydrochlorination is assisted by the

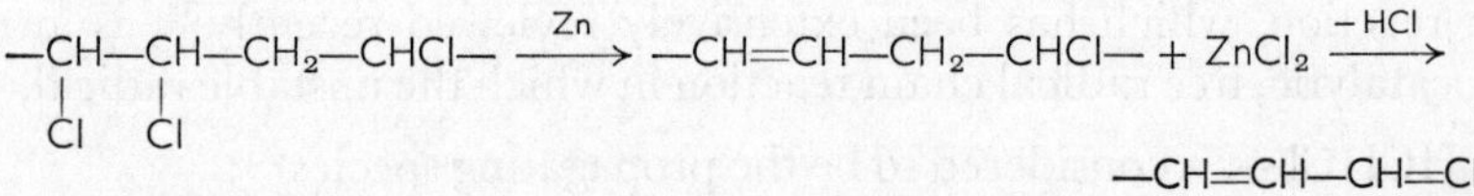

$$-\underset{\underset{Cl}{|}}{CH}-\underset{\underset{Cl}{|}}{CH}-CH_2-CHCl- \xrightarrow{Zn} -CH=CH-CH_2-CHCl- + ZnCl_2 \xrightarrow{-HCl}$$
$$-CH=CH-CH=CH-$$

presence of the first double bond and tends to supplant the competing ring formation involving the reaction of zinc and 1,3-dichloride units[87]:

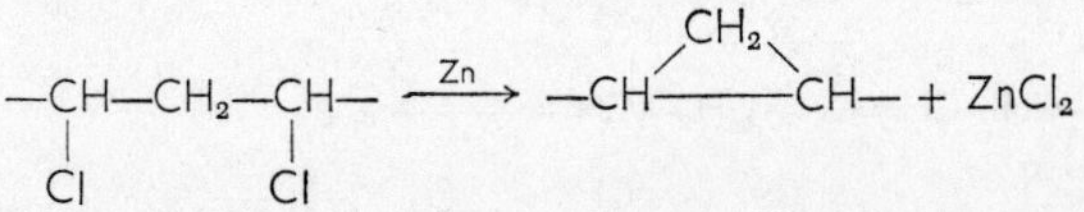

$$-\underset{\underset{Cl}{|}}{CH}-CH_2-\underset{\underset{Cl}{|}}{CH}- \xrightarrow{Zn} -CH\overset{CH_2}{\underset{}{\diagdown}}CH- + ZnCl_2$$

c. From polyacrylonitrile. This polymer or polymethacrylonitrile undergoes an important cyclization reaction on heating or by treatment with base, followed by high temperature oxidation, to yield a 'ladder' or double-stranded polymer having semiconducting properties[88] arising from the conjugated C=C and C=N bonds:

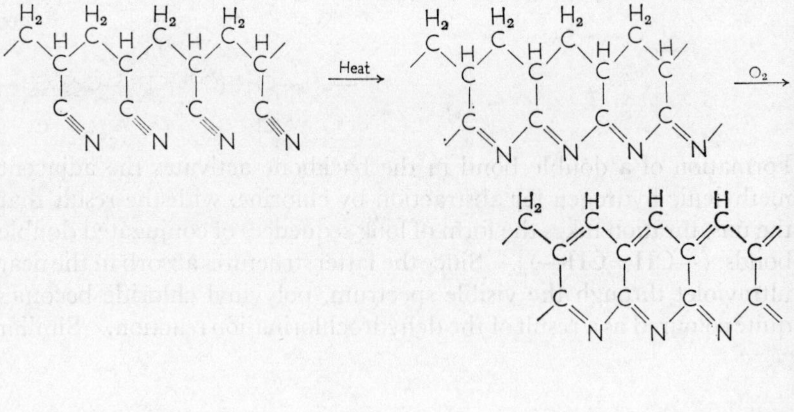

In this connection, it should be noted that Peebles and Brandrup[89] have discussed a chemical means for distinguishing between polymers containing conjugated $(C=C)_x$ bonds (polyenes) and polymers containing $(C=N)_x$ bonds (polyimines).

d. Radiation induced formation of $C=C$ bonds. When polyethylene is exposed to high energy radiation *in vacuo*, in addition to crosslinking, evolution of hydrogen and disappearance of vinylidene and vinyl unsaturation (formed during the free-radical polymerization of ethylene) there is an important formation of new *trans* (and presumably also some *cis*) vinylene unsaturation[90]. It is generally assumed that the formation of —CH=CH— units arises from the loss of hydrogen by molecular detachment:

$$-CH_2-CH_2- \longrightarrow -CH=CH- + H_2$$

by detachment and abstraction of atomic hydrogen:

$$-CH_2-CH_2- \longrightarrow -CH_2-\dot{C}H- + H\cdot$$
$$-CH_2-\dot{C}H- + H\cdot \longrightarrow -CH=CH- + H_2$$

or by disproportionation of chain radicals:

$$-CH_2-\dot{C}H- + -CH_2-\dot{C}H- \longrightarrow -CH=CH- + -CH_2CH_2-$$

Although *trans* —CH=CH— units are destroyed during irradiation (by hydrogen addition, crosslinking or other reactions), a steady state concentration of such $C=C$ bonds is approached at high doses. The radiation-induced production of —CH=CH— units can be adapted to the formation of a limited unsaturation in an essentially completely saturated polymer such as polymethylene.

Finally, unsaturation of various types, RCH=CHR', RR'C=CH$_2$ and RCH=CH$_2$, are also formed in the thermal degradation of polyethylene and other essentially saturated polymers, as well as, to a limited extent, in their polymerization. In fact, all three types of double bonds have been encountered in various high-pressure (low density) polyethylenes[91].

III. CHEMICAL REACTIONS OF POLYMERS CONTAINING C=C BONDS

On the assumption that the chemical reactivity of an organic functional group in a macromolecule is comparable to the reactivity of the same group in a small molecule, it may be expected that polymers containing olefinic double bonds will show many of the chemical reactions characteristic of simple alkenes and some special reactions by virtue of their large size. This expectation is essentially fulfilled, although some organic reactions involving high polymers are considerably slower (e.g., hydrogenation of an unsaturated polymer) or faster (e.g., enzymatic reactions) than analogous reactions on small molecules[92].

A. Halogenation and Hydrohalogenation

I. Chlorination

One of the earliest chemical reactions carried out on an unsaturated polymer was the chlorination of natural rubber (*cis*-1,4-polyisoprene). This reaction can be carried out by adding chlorine gas either to a solution of rubber in a chlorinated solvent, or to rubber latex, or passing the gas over thin sheets of rubber swollen with a solvent such as carbon tetrachloride. When the chlorination of rubber is carried to completion a product is obtained containing about 65% chlorine or about 3·5 chlorine atoms per isoprene residue, and this is indicative of the occurrence of several simultaneous reactions. On the basis of the work of many investigators[93], it is now realized that the overall chlorination process involves four distinct reactions: substitution, addition, cyclization and crosslinking. A reasonable scheme for the first three of these reactions is indicated in equations (1–4), although the specific mechanisms for equations (1–3) have not yet been determined. The substitution reaction is probably ionic, as in equation (5), although it could also be free radical, as in equation (6). In either case, the substitution results in the same evolution of HCl and the same overall

$$-CH_2-\underset{CH_3}{\overset{|}{C}}=CH-CH_2-CH_2-\underset{CH_3}{\overset{|}{C}}=CH-CH_2- + 2\,Cl_2 \longrightarrow$$

$$-CHCl-\underset{CH_3}{\overset{|}{C}}=CH-CH_2-CHCl-\underset{CH_3}{\overset{|}{C}}=CH-CH_2- + 2\,HCl \quad (1)$$

$$\begin{array}{c} \text{HC—CH}_2 \\ \text{CH}_3-\text{C} \quad\quad \text{CHCl} \\ \text{HCCl} \quad \text{C=CH—CH}_2 \\ \text{CH}_3 \end{array} \longrightarrow \begin{array}{c} \text{HC—CH}_2 \\ \text{CH}_3-\text{C} \quad\quad \text{CHCl} \\ \text{ClC—C} \\ \text{CH}_3 \quad \text{CH}_2-\text{CH}_2 \end{array} \quad (2)$$

$$\begin{array}{c} \text{CH—CH}_2 \\ \text{CH}_3-\text{C} \quad\quad \text{CHCl} \\ \text{CCl—C} \\ \text{CH}_3 \quad \text{CH}_2-\text{CH}_2 \end{array} + Cl_2 \longrightarrow \begin{array}{c} \text{HC—CHCl} \\ \text{CH}_3-\text{C} \quad\quad \text{CHCl} \\ \text{ClC—C} \\ \text{CH}_3 \quad \text{CH}_2-\text{CH}_2 \end{array} + HCl \quad (3)$$

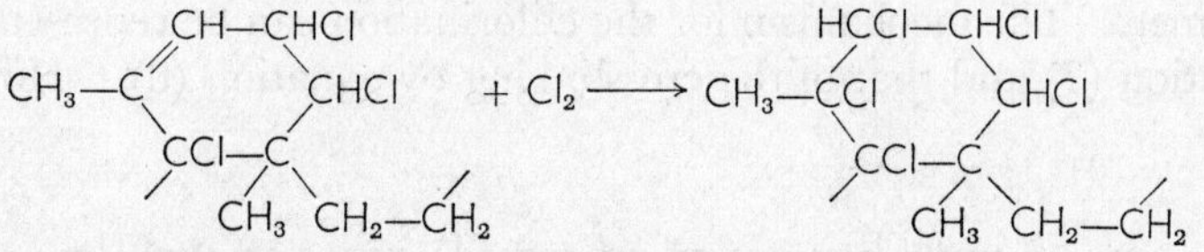

9. Polymers Containing C=C Bonds

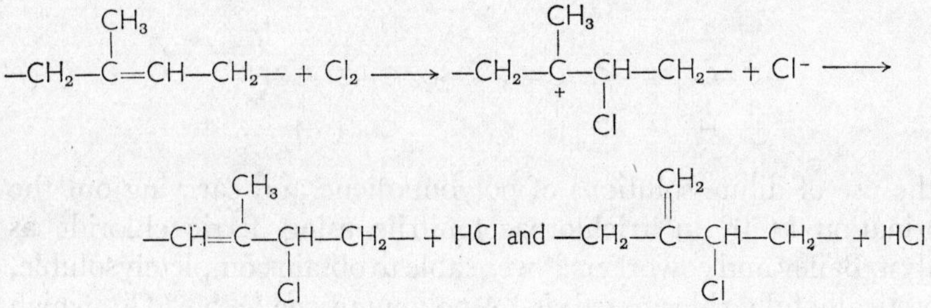

$$\underset{\substack{CH_3\ CH_2-CH_2 \\ (56)}}{\overset{CH-CHCl}{CH_3-C\diagdown CHCl}} + Cl_2 \longrightarrow \underset{CH_3\ CH_2-CH_2}{\overset{HCCl-CHCl}{CH_3-CCl\diagdown CHCl}} \quad (4)$$

$$-CH_2-\underset{\substack{|\\CH_3}}{C}=CH-CH_2- + Cl_2 \longrightarrow -CH_2-\underset{\substack{|\\Cl}}{\overset{CH_3}{\underset{+}{C}}}-CH-CH_2- + Cl^- \longrightarrow$$

$$-CH=\underset{\substack{|\\Cl}}{\overset{CH_3}{C}}-CH-CH_2- + HCl \text{ and } -CH_2-\underset{\substack{|\\Cl}}{\overset{CH_2}{\overset{\|}{C}}}-CH-CH_2- + HCl \quad (5)$$

$$-CH_2-\underset{\substack{|\\CH_3}}{C}=CH-CH_2- + Cl\cdot \longrightarrow -CH-\underset{\substack{|\\}}{\overset{CH_3}{C}}=CH-CH_2- + HCl \longrightarrow$$

$$-CH=\underset{\substack{\cdot\\}}{\overset{CH_3}{C}}-CH-CH_2- \xrightarrow{+Cl_2} -CH=\underset{\substack{|\\Cl}}{\overset{CH_3}{C}}-CH-CH_2- + Cl\cdot \quad (6)$$

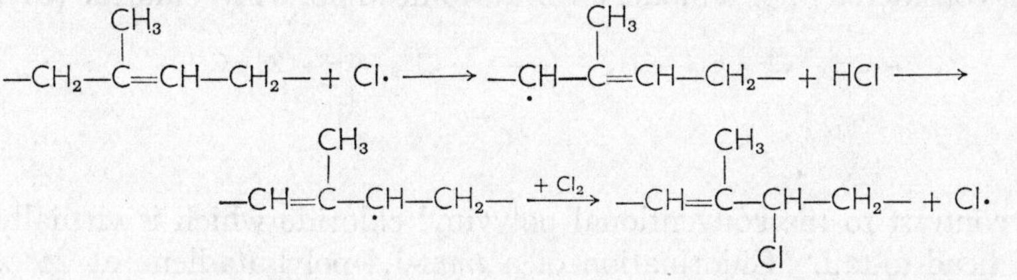

unsaturation. Likewise, the cyclization reaction could follow either path, although in view of the well-known cyclizing tendency of the carbonium ion on the isoprenic double bond (see Section III.B.1), it is probably mainly ionic in nature. The occurrence of crosslinking (and gelation) presumably is due either to the *inter*molecular counterpart of the (*intra*molecular) cyclization reaction, or to coupling of a macro-radical on one polymer chain with that on a nearby chain. To attain the high degree of chlorination noted above ($\sim 65\%$ Cl), structure **56** must undergo further substitutive chlorination. The relative participation of addition, substitution, cyclization and crosslinking in the chlorination process depends on the reaction conditions, such as solvent, temperature and type of halogenating agent[93].

Chlorine atom addition to the double bonds of polyisoprene has been observed with phenyl iododichloride[93a] or sulphuryl chloride[93b] in the presence of a peroxide.

In contrast to the situation with polyisoprene, the chlorination of polybutadiene proceeds mainly by addition, with little or no cyclization, although crosslinking tends to be more pronounced in the latter polymer. The mechanism for the chlorination can be represented by equation (7) and that of the crosslinking by equation (8). However,

$$-\overset{H}{\underset{|}{C}}=\overset{H}{\underset{|}{C}}- + Cl_2 \longrightarrow -\overset{Cl}{\underset{|}{C}}-\overset{+}{\underset{|}{C}}- + Cl^- \longrightarrow -\overset{Cl}{\underset{|}{C}}H-\overset{}{\underset{|}{C}}H- \quad (7)$$

$$-\overset{Cl}{\underset{|}{C}}-\overset{+}{\underset{|}{C}}- + -\overset{}{\underset{|}{C}}=\overset{}{\underset{|}{C}}- \longrightarrow -\overset{H}{\underset{|}{C}}-\overset{H}{\underset{|}{C}}-\overset{}{\underset{|}{C}}-\overset{}{\underset{|}{C}}- \quad (8)$$

by the use of dilute solutions of polybutadiene and carrying out the chlorination at 0° in trichloroacetonitrile using ferric chloride as catalyst, Bailey and coworkers[94] were able to obtain completely soluble, theoretically fully chlorinated *cis*-1,4-polybutadiene ($\sim 56\%$ Cl), which was considered to be a model for head-to-head polyvinyl chloride (**57**),

$$-CH_2-CH=CH-CH_2- \longrightarrow -CH_2-\underset{\underset{Cl}{|}}{CH}-\underset{\underset{Cl}{|}}{CH}-CH_2-$$
$$(57)$$

in contrast to the conventional polyvinyl chloride which is virtually all head-to-tail. Chlorination of a *trans*-1,4-polybutadiene at 25°c. yielded a similar polymer structure, ($\sim 58\%$ Cl) but here a small amount of substitution accompanied the simple addition of halogen. The preparation of fully chlorinated *cis*- and *trans*-polybutadienes was also reported by Murayama and Amagi[95] who likewise carried out the chlorination of polydichlorobutadiene to yield a head-to-head polyvinylidene chloride (**58**). These workers showed that the thermal degradation of the head-to-head polymers **57** and **58** differed re-

$$-CH_2-\underset{\underset{Cl}{|}}{C}=\underset{\underset{Cl}{|}}{C}-CH_2- \longrightarrow -CH_2-\underset{\underset{Cl}{|}}{\overset{\overset{Cl}{|}}{C}}-\underset{\underset{Cl}{|}}{\overset{\overset{Cl}{|}}{C}}-CH_2-$$
$$(58)$$

spectively from the degradation of the regular head-to-tail polyvinyl chloride and polyvinylidene chloride[96].

A large number of polymers and copolymers of diolefins have been chlorinated and a voluminous patent literature exists[93]. Chlorinated rubber has found wide application in chemically resistant paints, printing inks, coatings, adhesives and special finishes.

2. Bromination

Reaction of bromine with polyisoprene or polybutadiene could be almost entirely by radical addition if the reaction conditions are judiciously selected. On the other hand, substitution reactions become important with bromine radical sources such as N-bromosuccinimide.

Pinazzi and Gueniffey[97] have recently described the bromination of *cis*-1,4-polyisoprene and -polybutadiene followed by dehydrobromination, to yield polymers containing moderately long sequences of conjugated double bonds (equation 9, R = H or CH_3). In the case of

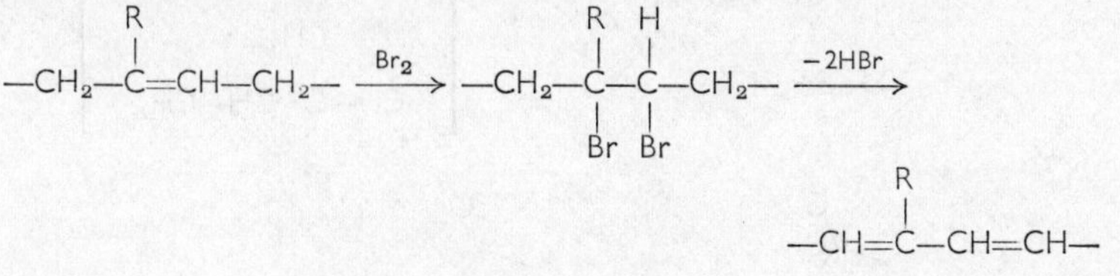

$$\begin{array}{c} R \\ | \\ -CH_2-C=CH-CH_2- \end{array} \xrightarrow{Br_2} \begin{array}{c} R\ H \\ |\ | \\ -CH_2-C-C-CH_2- \\ |\ | \\ Br\ Br \end{array} \xrightarrow{-2HBr}$$

$$\begin{array}{c} R \\ | \\ -CH=C-CH=CH- \end{array} \quad (9)$$

polyisoprene, dehydrobromination can also yield the isomeric structure **59**. The assessment of the length of the conjugated sequences in

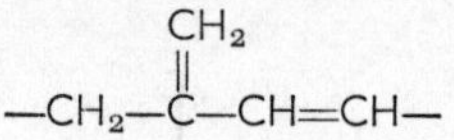

$$\begin{array}{c} CH_2 \\ \parallel \\ -CH_2-C-CH=CH- \end{array}$$
(**59**)

the products from polybutadiene and polyisoprene was based on their ultraviolet-visible spectra, which showed a number of characteristic absorption peaks in the 300–500 mμ range. The wavelengths for the individual peaks were comparable to those noted for the absorption peaks in polyacetylenes[75] and in dimethyl polyenes[98], with $(-CH=CH-)_n$ groups having n in the range 3–13.

In the case of N-bromosuccinimide and polyisoprene, various substitutions can occur in accordance with equation (10)[97] (see over).

Dehydrobromination of the resulting brominated polyisoprenes (**60**) can also yield various conjugated double bond structures, as for example,

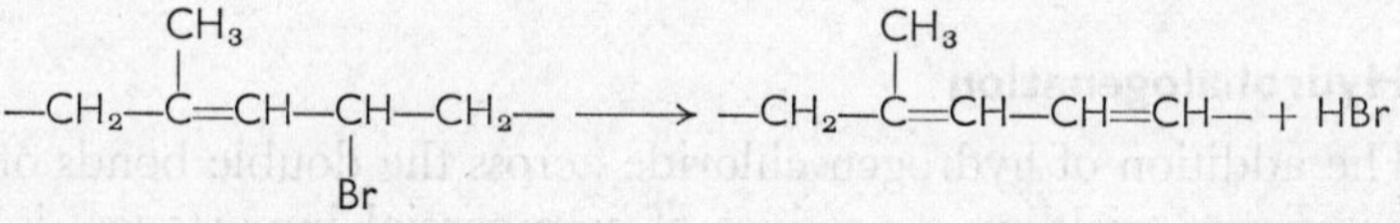

$$\begin{array}{c} CH_3 \\ | \\ -CH_2-C=CH-CH-CH_2- \\ | \\ Br \end{array} \longrightarrow \begin{array}{c} CH_3 \\ | \\ -CH_2-C=CH-CH=CH- \end{array} + HBr$$

Drefahl, Hörhold and Hesse[99] have also studied the formation of conjugated polymers by bromination of *cis*- and *trans*-polybutadienes

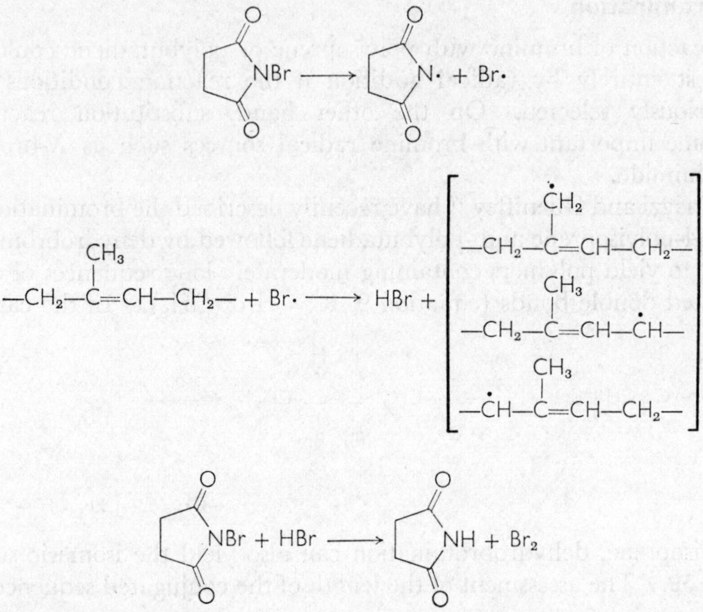

$$\text{(reaction schemes)} \tag{10}$$

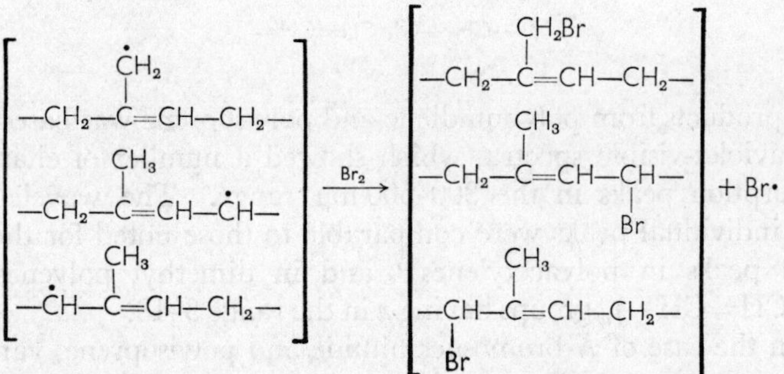

followed by dehydrobromination. The resulting brownish-black insoluble products showed a single narrow ESR signal and indications of semiconductivity.

3. Hydrohalogenation

The addition of hydrogen chloride across the double bonds of 1,4-polyisoprene, yielding a product of commercial importance, is considered to proceed through an ionic mechanism (equation 11) which entails a certain amount of cyclization (equation 12). The product,

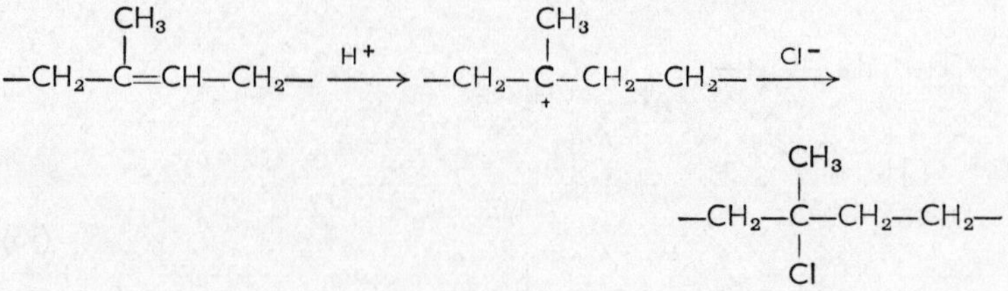

$$-CH_2-\underset{CH_3}{\overset{CH_3}{C}}=CH-CH_2- \xrightarrow{H^+} -CH_2-\underset{+}{\overset{CH_3}{C}}-CH_2-CH_2- \xrightarrow{Cl^-}$$

$$-CH_2-\underset{Cl}{\overset{CH_3}{C}}-CH_2-CH_2- \quad (11)$$

rubber hydrochloride (Pliofilm), is used as film for protective wrapping purposes.

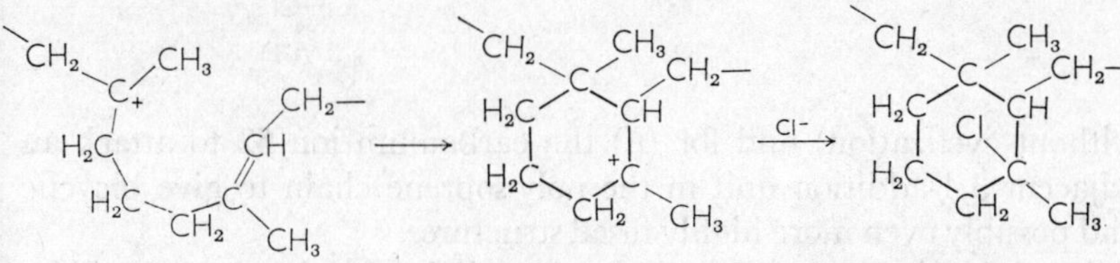

$$(12)$$

The assumption that some cyclization must attend the hydrochlorination reaction was advanced some time ago[100] to account for the fact that the maximum chlorine content obtainable in rubber hydrochloride always fell short of the theoretical limit. Recent work by Golub and Heller[101], involving a detailed analysis of the n.m.r. spectra of hydrochlorinated *Hevea* and balata, confirmed the occurrence of partial cyclization as well as the assumed addition of HCl in accordance with Markownikoff's Rule. More recently, Matsuzaki and Fujinami[102] reached similar conclusions for the reaction of *cis*- and *trans*-1,4-polyisoprenes with HBr and HI, as well as with HCl, while Tom[103] had reported earlier that the reaction of natural rubber with HF involves both addition and cyclization processes.

In contrast to the 1,4-polyisoprenes, when 3,4-polyisoprene is reacted with HCl there is virtually simultaneous cyclization and addition to yield nearly quantitatively a monocyclic structure containing one chlorine atom for each two double bonds consumed[104]. This novel cyclohydrochlorination reaction (equation 13) has been depicted as proceeding through a carbonium ion mechanism, similar to those indicated below for the cyclization of 1,4- and 3,4-polyisoprenes (Sections III.B.1 and III.B.2), except that no deprotonation is involved. In order to obtain the almost pure monocyclic structure **61**, the tendency for attachment of Cl⁻ to the monocyclic ion **62** must dominate over the competing tendencies for (*i*) Cl⁻ to be attached at the time of the initial protonation (to give simple hydrochlorination

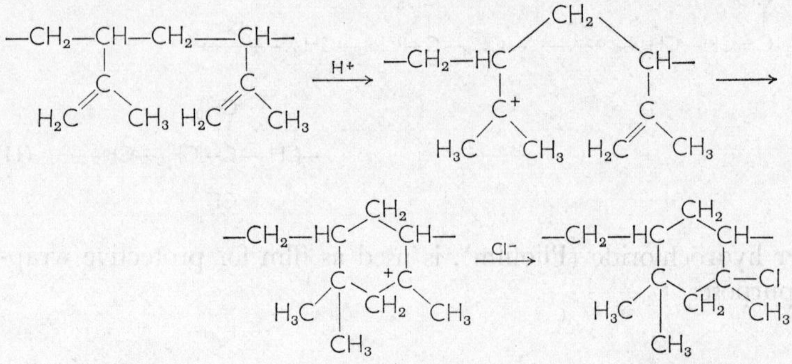

[Scheme showing equation (13) with structures and intermediates, leading to structures (62) and (61)]

(13)

(62)   (61)

without cyclization) and for (*ii*) the carbonium ion **62** to attack an adjacent 3,4-addition unit in the polyisoprene chain to give bicyclic and possibly even more highly fused structures.

Any 1,4 units present in a predominantly 3,4-polyisoprene would be expected to add HCl according to equation (11), while any 1,2 units which might be present, instead of adding HCl directly, would be attacked by adjacent 3,4 units to form another kind of monocyclic ion (**63**), followed by Cl⁻ addition (equation 14)[104]. In this connection it

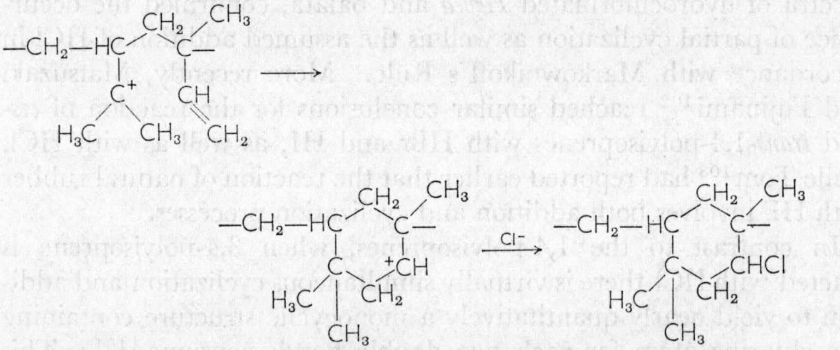

[Scheme showing equation (14) with structure (63)]

(14)

(63)

should be noted that neither 1,2-polybutadiene nor 1,2-polyisoprene adds HCl under the mild conditions employed for the cyclohydrochlorination of 3,4-polyisoprene or for the hydrochlorination of the 1,4-polyisoprenes. However, under rather rigorous conditions a 1,2-polybutadiene hydrochloride can be obtained having apparently a monocyclic structure[105].

B. Cyclization

I. 1,4-Polyisoprene

Naturally occurring *cis*- or *trans*-1,4-polyisoprene, or a synthetic polyisoprene having predominantly 1,4-addition units, when treated at somewhat elevated temperatures with strong acids (e.g., H_2SO_4), with organic acids and their derivatives (e.g., *p*-toluenesulphonic acid and its chloride), with Lewis acids (e.g., $SnCl_4$, $TiCl_4$, BF_3, $FeCl_3$), or with other catalysts of an acidic character, undergoes a progressive resinification with the formation of so-called cyclized rubber, of the same empirical composition as the initial polyisoprene, $(C_5H_8)_n$. Cyclization of natural rubber can be carried out on the solid rubber, in solution, or even on the latex; the resulting resin (produced commercially under the trade names Marbon, Plioform, Pliolite, Thermoprene) has found application in the formulation of adhesives, paints, printing inks and as a general compounding ingredient.

In this process a long linear macromolecule is converted into a greatly shortened polymer chain consisting of an assortment of mono-, bi-, tetra- and other polycyclic groups distributed at random throughout the polymer backbone, interrupted occasionally by unreacted isoprene units. The cyclization of the 1,4-polyisoprenes can be depicted by the sequence of reactions shown in equation (15)[106]. The carbonium ion formed on protonation of a double bond attacks the double bond in an adjacent isoprene unit, to produce a cyclic carbonium ion which either deprotonates (to give the monocyclic group **64**) or attacks the next adjacent double bond to form a bicyclic carbonium ion; the latter can likewise deprotonate (to give **65**) or go on to produce more highly fused ring structures (**66**). Evidently a competition exists between deprotonation and further attack on neighbouring double bonds, since the final product has neither a strictly monocyclic nor polycyclic structure; depending somewhat on the reaction conditions, the structure of the typical resin has, on the average, about three six-membered rings in each fused segment, with an unsaturation of the order of one residual double bond for every four C_5H_8 units, as compared to an initial unsaturation of one double bond per C_5H_8 unit[107,108].

The microstructure of cyclized rubber shows additional minor heterogeneity owing to the fact that the deprotonation step can give rise to three different double bonds: di- and trisubstituted double bonds (exemplified by **67** and **68**, respectively), in addition to the tetrasubstituted double bonds illustrated in structures **64–66**. Their relative amounts have been found to be in order tetra- > tri- >

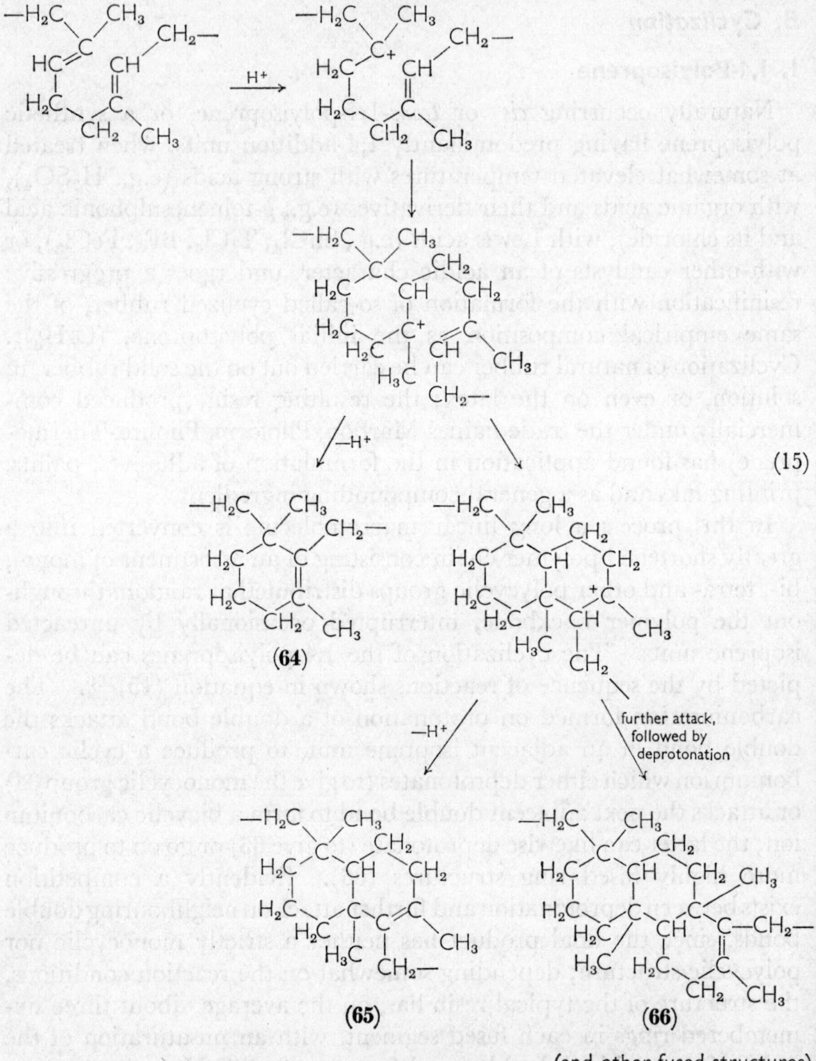

$$ \tag{15} $$

(64)

(65)

(66)

(and other fused structures)

disubstituted[109]. Moreover, a small percentage of the original double bonds end up as 'widows' or uncyclized, because their neighbours have already reacted, so that cyclized segments are sometimes separated by linear segments.

Cyclized rubber can also be obtained by subjecting rubber and synthetic polyisoprenes to a silent electric discharge[110]. The reaction

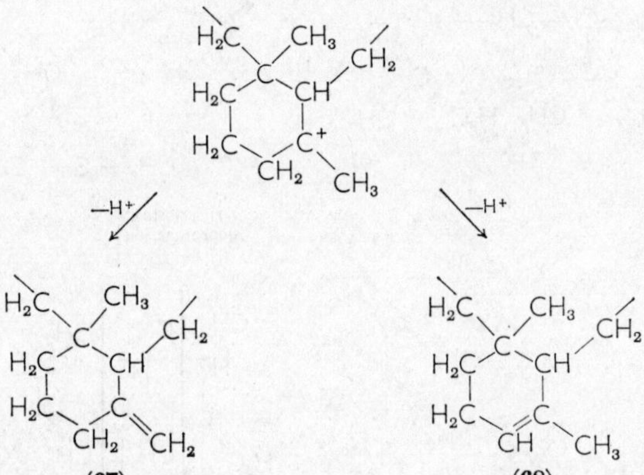

(67)   (68)

mechanism in this instance is probably similar to the cyclization mechanism in solid polyisoprene when exposed to high energy electrons or γ-rays[111].

The cyclization of the 1,4-polyisoprenes and the controversy concerning the microstructure of cyclized rubber have been reviewed recently[108,112].

2. 3,4,-Polyisoprene

The cyclization of a predominantly 3,4-polyisoprene, by a carbonium ion mechanism similar to that indicated above for the 1,4-polyisoprenes, leads to a product having the structural characteristics of an incomplete ladder or double-chain polymer (equation 16). Recent interest in the development of various ladder polymers stems from the expectation that their double backbone structure would make them much more stable at high temperatures than the analogous single-chain polymers.

The first indication of a reaction such as (16) was reported by Tocker[113] in connection with the cyclization, catalysed by BF_3 or $POCl_3$, of an ethylene-isoprene block copolymer (25:1 mole ratio) in which the isoprene was present initially almost entirely as 3,4-addition units. The resulting cyclized product was considered to contain blocks of linearly fused cyclohexane units, separated by blocks of methylene groups. Subsequently, Angelo reported the cyclization of a high 3,4-polyisoprene to a ladder-like polymer having the form **69** ($n = 2-6$[114], and lately $n = 1-4$[115]). A similar result was reported by two other groups[116,117], even though the 3,4 units in the

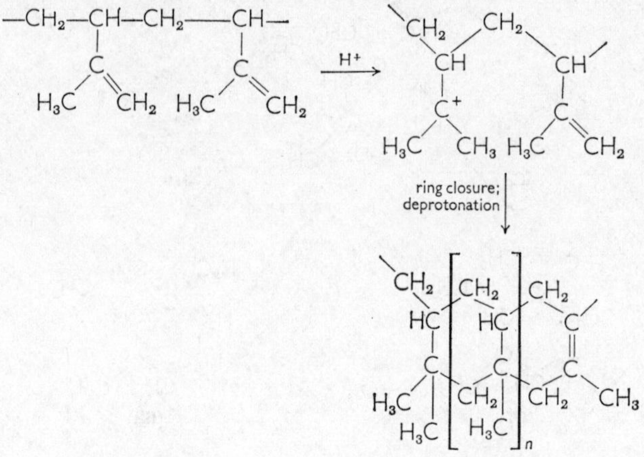

$$\begin{array}{c}\text{—CH}_2\text{—CH—CH}_2\text{—CH—}\\|\quad\quad\quad|\\\text{C}\quad\quad\quad\text{C}\\\text{H}_3\text{C}\quad\text{CH}_2\;\;\text{H}_3\text{C}\quad\text{CH}_2\end{array}\xrightarrow{\text{H}^+}$$

[structure with ring closure; deprotonation leading to structure (69)]

(16)

(69)

polymers employed in the several laboratories ranged from 79 to 97% of the total unsaturation. Indeed, it follows from this fact that the same relatively low cyclicity obtained for cyclized 3,4-polyisoprene by the different workers is not a direct result of the 'defects' (1,4 or 1,2 units) present in the starting polymers.

As with cyclized rubber, the structure of cyclized 3,4-polyisoprene was elucidated mainly with the aid of infrared and n.m.r. spectroscopy, and reinforced with aromatization experiments using Wallenberger's technique[118]. Thus, the ultraviolet spectra of the aromatization products disclosed the existence of derivatives of naphthalene, anthracene, naphthacene and possibly pentacene. These results correspond to a cyclicity of around 2–5 cycles per fused segment, in line with the conclusions from n.m.r. spectra, based on the relative areas of the peaks associated with methyl protons attached to saturated and unsaturated carbon atoms. In addition to the microstructural irregularities associated with the different lengths of the fused segments in cyclized 3,4-polyisoprene, there are also irregularities due to uncyclized 3,4 and other unsaturation units, as well as the occurrence of some trisubstituted along with tetrasubstituted double bonds formed in the final deprotonation step.

3. 1,4-Polybutadiene

Polymers and copolymers containing butadiene largely in the form of 1,4-addition units can also cyclize, but the required conditions are more severe. Thus Buizov and Kusov[119] heated a sodium polybutadiene (~40% 1,4- and 60% 1,2-addition units) to 230–270° and obtained 33–67% yields of cyclized rubber, whereas natural rubber experienced this same reaction but at a much lower temperature (160°). Also, addition of sulphuric acid, sulphonic acids, or other catalysts lowered the cyclization temperature to 100–140° in the case of natural rubber, but had little effect on polybutadiene. Again, Endres[120] found a styrene-butadiene rubber to be cyclized in solution in phenol, cresol or naphthalene at 160–180° with BF_3 or $SnCl_4$, while polyisoprene was cyclized with $SnCl_4$ at only 70–75°. The difference in ease of cyclization of the polymerized butadiene and isoprene units has been attributed to the methyl group, which confers on the —$C(CH_3)$=CH— unit a higher proton affinity (by ~13·5 kcal/mole)[121] than that in the —CH=CH— unit, and hence a lower activation energy for the formation of the requisite carbonium ion. Evidently, this effect is carried over to poly(2,3-dimethyl-1,3-butadiene), which has an even higher rate of cyclization than does polyisoprene[122].

Shelton and Lee[123] reported that the cyclization of a polybutadiene (80% 1,4 and 20% 1,2 units) in xylene, on treatment with concentrated H_2SO_4 at 140° for 18 hours, resulted in an average tricyclic structure **70**. Considerable crosslinking occurred too.

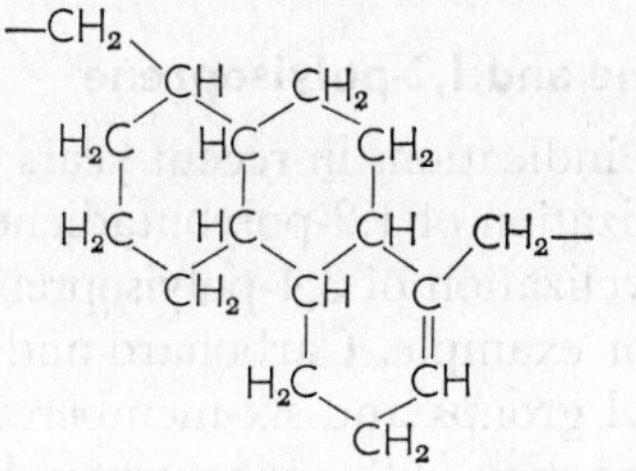

(**70**)

Kössler and coworkers[122] have very recently studied in detail the cyclization of *cis*- and *trans*-1,4-polybutadienes in a 3% solution in *p*-xylene at 145° using H_2SO_4 as the cyclization agent. By carrying out the reaction with intensive stirring they were able to obtain almost completely soluble products, but with low speed of stirring the cyclization products were insoluble. The infrared spectra of the highly cyclized polymers obtained from the isomeric polybutadienes were identical, and showed virtually total disappearance of the original unsaturation. In contrast not only to the relatively low cyclicity obtained in the cases discussed above, the cyclized 1,4-polybutadienes obtained by Kössler and coworkers were found to have the surprisingly high cyclicities of around 25–45 rings in each fused segment. These workers also reported that the cyclization of poly(2,3-dimethyl-1,3-butadiene) yielded only monocyclic structures. Thus, while the ease of cyclization for the different diene polymers decreases in the order, poly-(2,3-dimethyl-1,3-butadiene) > polyisoprene > polybutadiene, the average number of cyclohexane rings in each fused segment in the cyclized polymers increases in the inverse order.

The sulphuric acid-catalysed cyclization of the 1,3-diene polymers was assumed[122] to involve reversible addition of H_2SO_4 across the double bonds in forming the requisite carbonium ion (equation 17),

$$-CH_2-CH=CH-CH_2- + H_2SO_4 \rightleftharpoons -CH_2-\underset{OSO_3H}{CH}-CH_2-CH_2- \quad (17)$$

$$\text{polycyclic structures} \xleftarrow{\substack{\text{successive ring closure;} \\ \text{ultimate deprotonation} \\ \text{or attachment of } HSO_4^-}} -CH_2-\underset{+}{CH}-CH_2-CH_2- + HSO_4^-$$

which accounts for the observation of some *cis–trans* isomerization of polybutadiene, as well as the incorporation of some sulphur in the cyclized product.

4. 1,2-Polybutadiene and 1,2-polyisoprene

There have been indications in recent years of considerable interest in the possible cyclization of 1,2-polybutadiene and 1,2-polyisoprene, analogous to the cyclization of 3,4-polyisoprene, to produce a ladder polymer. Thus, for example, Carbonaro and Greco[124] reported the formation of methyl groups and six-membered cyclic systems in 1,2-polybutadiene on reaction with concentrated H_2SO_4, and suggested a carbonium ion mechanism (equation 18). Infrared and n.m.r.

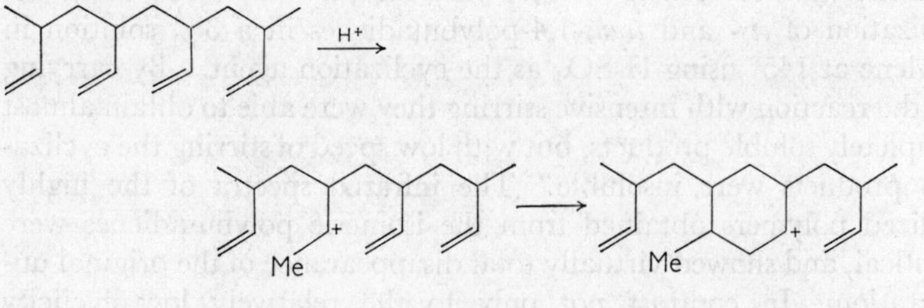

spectra of cyclized 1,2-polybutadiene[124,125] indicate that there can not be many rings in each fused segment.

5. Cyclopolymers from diene monomers[116,122,126-129]

Gaylord, Kössler and coworkers reported[116] that butadiene, isoprene and chloroprene polymerize with Ziegler-type catalysts (of low Mg/Ti or Al/Ti ratio) to insoluble, powdery polymers with high density and high heat resistance and having a polycyclic structure. Thus, these polymers had infrared spectra which were very similar to each other and to those of cyclized 1,4- or 3,4-polyisoprene. Their cyclic structure content ranged from 60–90% and was increased further by treating with dilute sulphuric acid in toluene. The reagents used in the preparation of the cyclopolymers were ineffective in cyclizing 1,4- or 3,4-polyisoprene under the same conditions.

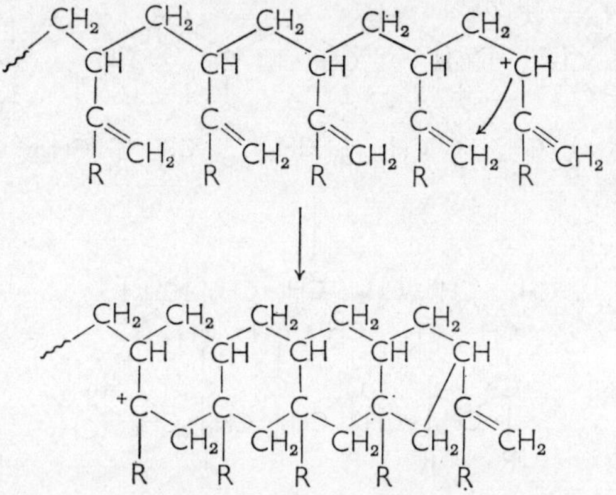

(19)

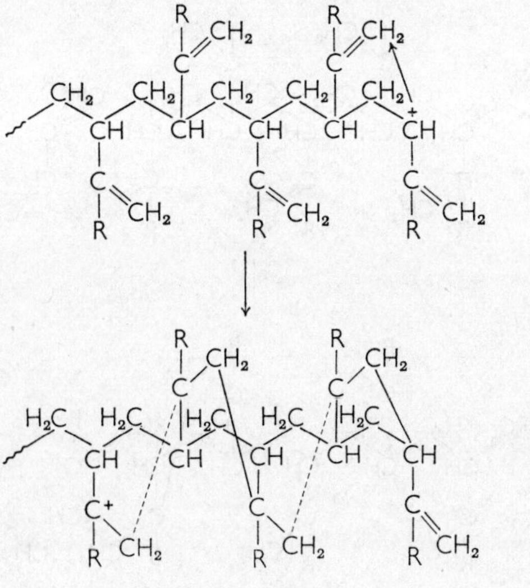

(20)

The cyclopolymers were considered to result from a cationic polymerization to isotactic or syndiotactic 3,4 units (in the case of isoprene or chloroprene) or 1,2 units (in the case of butadiene), followed by cyclization occurring in either two ways. Instead of the polymerization progressing further, an intramolecular cyclization takes place yielding, after the formation of the initial five-membered ring, a sequence of fused six-membered saturated rings (equations 19 and 20, $R = CH_3$, Cl or H). Alternatively, the cyclization could be initiated by the copolymerization of a monomer unit or a growing chain with the pendant 3,4- or 1,2-double bonds (equations 21 and

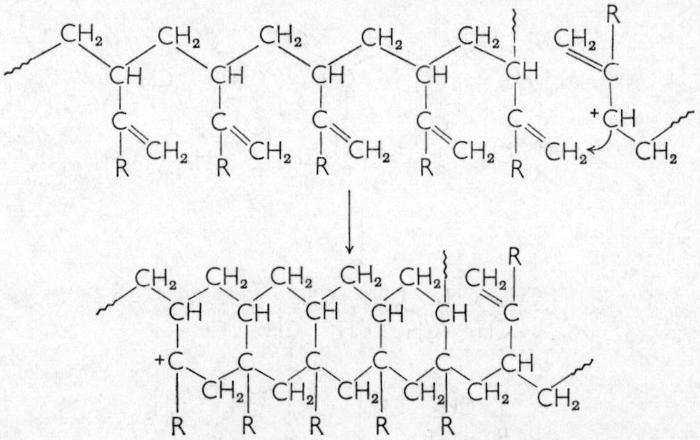

(21)

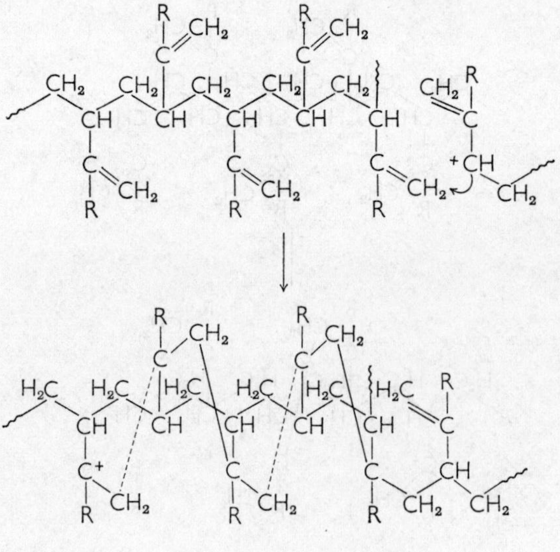

(22)

22). In the isotactic case, the structure of the resulting cyclopolymer is that of a linear ladder polymer (equations 19 and 21), whereas in the syndiotactic case the structure is that of a spiral ladder polymer in which one chain spirals around the backbone (equations 20 and 22). The length of the fused segment would be limited by the number of isotactic or syndiotactic units in a given sequence, the sequences being interrupted by the presence of 1,4 units or pendant groups of the alternative type (1,2 or 3,4 units). The cyclopolymers were all insoluble, probably due to crosslinking, and could not be examined by n.m.r. spectroscopy to determine the average length of their fused sequences. At any rate, the indications were that the 'cyclopolymers' were more highly fused than the corresponding 'cyclized polymers'.

C. Cis–Trans Isomerization

Although thermal, catalysed and photochemical *cis–trans* isomerizations of low molecular weight olefinic compounds had been known for many years[130,131] such a reaction in an unsaturated high polymer was unknown prior to 1957 when Golub[132] reported the photosensitized isomerization of polybutadiene. Before that time, several unsuccessful attempts were made to effect the interconversion of the naturally occurring isomeric 1,4-polyisoprenes, *Hevea* (*cis*) and balata (*trans*)[133,134], and as recently as 1962, the patent literature[135] made reference to isomerized rubbers or polymers which were actually cyclized modifications (Section III.B) of the starting polymers. The term 'isomerized rubber' should be reserved for the *cis–trans* isomerizates of polyisoprene or polybutadiene or other diene polymers[108,136].

I. 1,4-Polybutadiene

a. Photo- and radiation-induced reactions. Golub[132] discovered that a high *cis*-polybutadiene could be isomerized to a high *trans* structure by means of ultraviolet irradiation in dilute benzene solution in the presence of a suitable sensitizer, such as an organic bromide, sulphide, disulphide or mercaptan, or even elemental bromine. He later showed[137] that polybutadiene could also be isomerized in benzene when irradiated with γ-rays in the presence of the same sensitizers as in the photochemical case. The mechanism depicted with ultraviolet or ionizing radiation is essentially that advanced for the analogous *cis–trans* isomerization of low molecular weight olefins[138,139] and involves the reversible formation of a freely rotating radical adduct from the polybutadiene, either *cis* or *trans*, and the bromine atom or thiyl radical generated in the photolysis or radiolysis of the organic bromine or sulphur compound employed as sensitizer (equation 23). With release of the attached radical (X = Br· or RS·), the double bonds

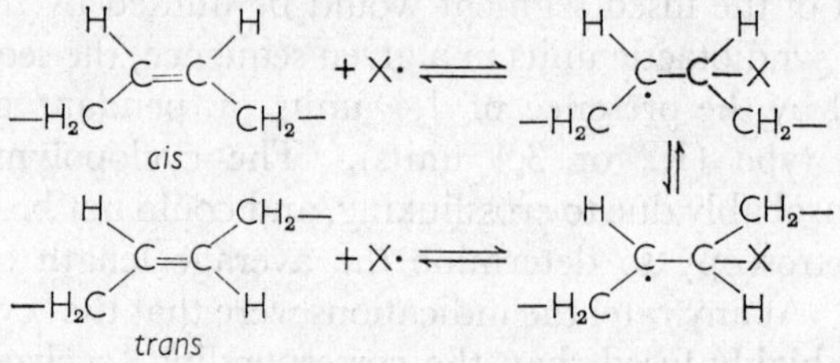

$$\begin{array}{c}\text{cis} \\ \text{trans}\end{array} \quad (23)$$

are reestablished with the thermodynamically more stable configuration, *trans*, being formed predominantly. The process is nearly free of side reactions.

Kinetic studies on the photosensitized[140] and radiation sensitized[137] isomerization of polybutadiene showed it to be a long chain reaction, and that the radiation-produced C_6H_5S· radical can isomerize about 1000–1300 *cis* double bonds[141], and the corresponding photoinduced radical about 750 double bonds[140,141], before they are terminated. The sensitized isomerization proceeds to an equilibrium[137,142,143] in which the *cis–trans* ratio of polybutadiene is approximately 1:4 to 1:3.

In addition to the radiation-sensitized *cis–trans* isomerization, polybutadiene was also found to isomerize when γ-irradiated, either in benzene (or toluene) solution in the absence of deliberately added sensitizer[137], or in the pure solid state[144]. The 'unsensitized' isomerization approached a radiostationary equilibrium in which the *cis–trans* ratio of the polymer is 1:2[142,144]. The fact that this ratio is higher than the thermodynamic ratio obtained in the photo- or radiation-sensitized isomerization is in line with a different excitation mechanism being operative in the 'unsensitized' reaction. The mechanism in the solid state γ-irradiation case was considered to involve direct as well as indirect excitation of the π-electrons of the double bonds to an antibonding state, where free rotation and hence geometric interconversion can readily occur. Although this process also occurs in the solution case, the dominant mechanism for excitation of the polymer double bonds involves energy transfer from lowest benzene triplet (3·6 e.v.) to the vinylene unit, which is thereby excited to its lowest triplet (3·2 e.v.), from which state it can return to the ground state either as a *cis* or *trans* double bond[145]. In benzene the polymer also becomes phenylated, while in the solid state it also undergoes a marked loss of unsaturation (apparently due to radiation-induced cyclization) as well as crosslinking and evolution of hydrogen[144].

Irradiation studies performed on $(-CH_2-CD=CD-CH_2-)_n$, cis-polybutadiene-2,3-d_2, showed unequivocally that the observed formation of trans —CH=CH— was due entirely to cis–trans isomerization and not to a double bond migration masquerading as an isomerization (as in equation 24). This result indicated also that radical processes were not implicated in the unsensitized radiation-induced isomerization in the solid state[144].

$$-CH_2-CH=CH-CH_2- \longrightarrow -CH_2-CH=CH-\dot{C}H- \rightleftarrows$$
$$-CH_2-\dot{C}H-CH=CH- \qquad (24)$$

Evidence was presented recently that pure polybutadiene in the solid state could also be isomerized when exposed *in vacuo* to krypton 1236 Å or mercury 2537 Å radiation[146]. In the former irradiation, the isomerization was accompanied by an even more pronounced reaction, viz., loss of unsaturation which was assumed to be due to a chain cyclization[111,144]. The rationale was that the photons from the krypton source have sufficient energy (~ 10 e.v.) to cause ionization in polybutadiene and thus bring about processes analogous to those attained by γ-radiation. In the mercury irradiation case, on the other hand, the isomerization was found to approach a photostationary cis–trans ratio of 3:2, starting from either cis- or trans-polybutadiene. Although it was originally assumed[146] that the photoisomerization was sensitized by adventitious impurities present in the polymer film, it is now believed[147] that the mechanism involves direct absorption of photons by the double bonds (whether as singlet-triplet or singlet-singlet transitions), which are then excited to a state in which free rotation and hence geometric alteration can take place. Irradiation of polybutadiene film with mercury light resulted also in a moderate consumption of double bonds (although much less severe than in the krypton irradiation), as well as the formation of vinyl groups, the latter arising through occasional chain scission (equation 25). By way of

$$-CH_2-CH=CH-CH_2-CH_2-CH=CH-CH_2- \xrightarrow{h\nu}$$
$$-CH_2-CH=CH-\dot{C}H_2 + \dot{C}H_2-CH=CH-CH_2- \qquad (25)$$
$$-CH_2-CH=CH-\dot{C}H_2 \rightleftarrows -CH_2-\dot{C}H-CH=CH_2 \longrightarrow$$
$$-CH_2-CH_2-CH=CH_2 \text{ (or crosslink or endlink)}$$

contrast, little or no chain scission or formation of vinyl double bonds occurs in the krypton 1236 Å irradiation or γ-irradiation of polybutadiene films.

The occurrence of a direct photoisomerization of *cis*-polybutadiene *in vacuo* at room temperature, has also been reported recently by Ho[148] in connection with a study of the photooxidation of that polymer.

Photosensitization with diphenyl disulphide affords a convenient means for preparing a series of polymer samples of different *cis–trans* content but of practically the same molecular weight which could be used in studying structure dependence of properties such as spectra, crystallinity, elongation and tensile strength[143,149–151].

b. Other isomerization methods. A wide variety of other methods exist for the *cis–trans* isomerization of polybutadiene. It can be expected that any bromine or thiyl radicals generated in the thermal dissociation of an appropriate organic compound (as opposed to its photolysis or radiolysis), can also serve to promote the reaction in accordance with equation (23). Thus, for example, Dolgoplosk and coworkers[152] found recently that a high *cis*-polybutadiene, on heating with diphenyl disulphide (to generate $C_6H_5S\cdot$ radicals), isomerized to an equilibrium *cis–trans* ratio of 2:7 at 60°, and 1:2 at 170°.

Cis–trans isomerization of polybutadiene brought about by heating with elemental sulphur at 140–160°, because of its implications in the vulcanization area, has received considerable attention[153–161]. No thermal isomerization of polybutadiene takes place in the absence of sulphur. Since the physical properties of polybutadiene vulcanizates have been shown to be critically dependent upon stereoregularity near 100% *cis* or *trans* content[36], any isomerization which accompanies the sulphur vulcanization is very undesirable. To illustrate, after 80 minutes of a conventional cure of gum stocks of high *cis*-(>97%) polybutadiene, the *cis* content was decreased by about 5% for a cure recipe having 4 phr (parts per hundred of rubber) sulphur; by about 2% for 1 phr sulphur, and by less than 1% for 0·5 phr sulphur[156], with substantial decreases in the tensile strength[36].

The sulphur-promoted *cis–trans* isomerization of polybutadiene was considered to proceed through a polar mechanism[155,159] similar to that generally postulated for sulphur-olefin reactions[162–165]: a persulphenium ion, TS_a^+, where T is an alkyl or alkenyl group, adds to a double bond to form the transitory structure **71**. The TS_a^+ ion is derived from the heterolysis of the polysulphide, $TS_x \cdot S_y T$, formed in the initial sulphuration of the olefin. In the subsequent steps leading to cyclic sulphide formation, the persulphenium ion remains attached to the double bond of the polymer at all times. However, if this ion is easily detached, the double bond could then be reestablished with isomerization to the more stable *trans* form.

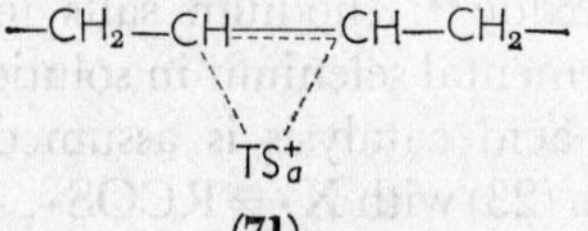

$$-CH_2-CH\cdots\cdots CH-CH_2-$$
$$TS_a^{\ddagger}$$
(71)

An alternative mechanism proposed by Bishop[154] (equation 26) involves the formation of a π-complex between sulphur (in the form of S_8 rings or polysulphides) and the polymer double bond, and is analogous to the π-complex formed between selenium and olefins[166]. After homolytic cleavage of the polysulphide at the complexed sulphur atom, the resulting polymeric radical is free to rotate about the σ-bond and, after splitting off a second sulphur-containing radical, yields mostly double bonds in either *cis* or *trans* configuration.

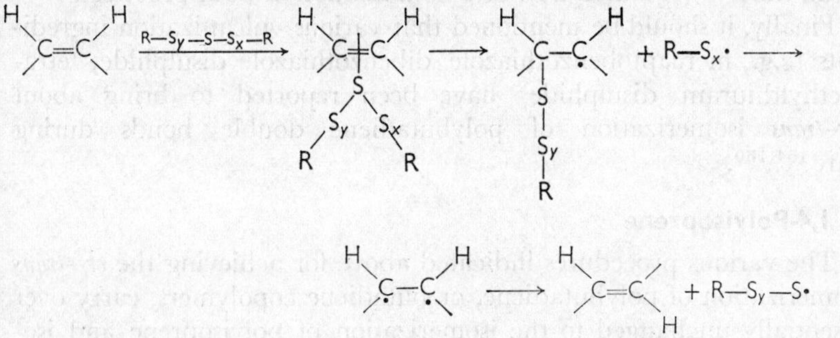

$$\underset{H}{\overset{H}{>}}C=C\underset{\diagdown}{\overset{H}{<}} \xrightarrow{R-S_y-S-S_x-R} \underset{H}{\overset{H}{>}}C\overset{+}{=}C\underset{\diagdown}{\overset{H}{<}} \longrightarrow \underset{H}{\overset{H}{>}}C-C\underset{\diagdown}{\overset{H}{<}} + R-S_x\cdot \longrightarrow$$
$$\underset{R}{\overset{S_y}{|}}\underset{R}{\overset{S_x}{|}}\underset{R}{\overset{S_y}{\overset{|}{\overset{S}{|}}}}$$

$$\underset{H}{\overset{H}{>}}C-C\underset{\diagdown}{\overset{H}{<}} \longrightarrow \underset{}{\overset{H}{>}}C=C\underset{H}{\overset{}{<}} + R-S_y-S\cdot \quad (26)$$

The minimum *cis* content which could be obtained in a high *cis*-polybutadiene on reaction with sulphur under vulcanizing conditions was reported to be 35–36%[154,155]. On the other hand, a high *trans*-polybutadiene treated with sulphur under the same conditions was isomerized to a *cis* content of about 22%[154]. Activation energies of 30.1[154] and 34.5[159] kcal/mole have been obtained for the sulphur-promoted *cis*→*trans* isomerization[154,159].

The use of polybutadiene-2,3-d_2 again made it possible to show that double bond migration (as in equation 24) was not involved in the formation of *trans* —CH=CH— units in polybutadiene isomerized (and vulcanized) with sulphur[155]; in other words, equation (27) was not implicated in the sulphur–polybutadiene-2,3-d_2 system.

$$-CH_2-CD=CD-CH_2- \longrightarrow -CH_2-CD=CD-\overset{\cdot}{CH}- \rightleftharpoons$$
$$-CH_2-\overset{\cdot}{CD}-CD=CH- \quad (27)$$

Other methods for the isomerization of polybutadiene include heating in benzene with nitrogen dioxide[167], treatment with various thiol acids[14], sulphur dioxide[14], rhodium salts in emulsion[32], iron carbonyls[168] or with elemental selenium in solution at elevated temperatures[19]. The thiol acid catalysis is assumed to involve an 'on-off' reaction like equation (23) with $X\cdot = RCOS\cdot$. In the case of selenium, a π-complex mechanism was invoked analogous to that indicated for the selenium-catalysed isomerization of oleic acid and stilbene[166]. The other four catalyst systems presumably function by forming unstable complexes with the polymer double bonds, and result in equilibrium *cis–trans* ratios in the neighbourhood of 1:3.

It has also been reported that *cis*-polybutadiene isomerizes on reaction with peroxides[154,161,169,170] with a marked depletion of double bonds. It remains to be determined, however, whether the formation of *trans* —CH=CH— units is a true *cis–trans* isomerization or a double bond shift[171] (equation 24), or a combination of both processes.

Finally, it should be mentioned that various vulcanization ingredients (e.g., mercaptobenzothiazole, dibenzothiazole disulphide, tetramethylthiuram disulphide) have been reported to bring about *cis–trans* isomerization of polybutadiene double bonds during cure[154,160].

2. 1,4-Polyisoprene

The various procedures indicated above for achieving the *cis–trans* isomerization of polybutadiene, or butadiene copolymers, carry over essentially unchanged to the isomerization of polyisoprene and isoprene copolymers. However, because the infrared spectroscopic features associated with *cis–trans* isomerism in polyisoprene are rather subtle, in contrast to the situation with polybutadiene[108] (see Sections II.A.1 and II.A.2), the early attempts to extend the bromine or thiyl radical photosensitized isomerization of polybutadiene to polyisoprene, were reported as unsuccessful[132]. Later it was discovered that *Hevea* and balata, on appropriate treatment with elemental selenium[19], or with sulphur dioxide or thiol acids[14], could be transformed into similar products having infrared spectra which were not only nearly identical to each other, but were effectively intermediate between those of the original *cis*- and *trans*-polyisoprenes. With the subsequent development of high resolution n.m.r. spectroscopy of polymers, it was possible to confirm the occurrence of *cis–trans* isomerization of polyisoprene and also to assess, more accurately than by infrared spectroscopy, small changes in its *cis* or *trans* content[108].

a. *Photo- and radiation-induced reactions*. Cunneen and coworkers[14] found that ultraviolet irradiation of a benzene solution of gutta-percha (or balata) in the presence of diphenyl disulphide, dibenzoyl disulphide or thiolbenzoic acid led to the conversion of some *trans* double bonds to the *cis* configuration. Comparable irradiations of milled natural rubber led to insoluble products which could not be analysed for *cis* content by infrared spectroscopy, but were considered to have been isomerized by virtue of the observed retardation in rate of their crystallization at moderately low temperature. The isomerization view was substantiated by the finding that *cis*- and *trans*-3-methyl-pent-2-ene, as model compounds for the two natural isomers of polyisoprene, were readily isomerized to an equilibrium *cis–trans* ratio of 9:16 by diphenyl disulphide photosensitization, presumably by the same kind of mechanism as in equation (23).

Recently, Tsurugi and coworkers[172] described the radiation-induced *cis–trans* isomerization of *Hevea* and gutta-percha in benzene solution with sensitizers similar to those used for polybutadiene[137], viz., ethyl bromide, ethylene bromide and n-butyl mercaptan. These workers reported that the polyisoprene isomerization approached an equilibrium *cis–trans* ratio at room temperature of about 1:3, which is similar to that indicated for polybutadiene. Also, they obtained an activation energy for the thiyl radical-sensitized isomerization of polyisoprene (2·7 kcal/mole) which was close to the corresponding value (2·3 kcal/mole) found for polybutadiene. However, in contrast to the polybutadiene case in which the equilibrium *cis–trans* ratio increased somewhat with temperature, the polyisoprene work showed the opposite effect, the ratio dropping from 1:3 at 22° to ~1:7 at 100°.

Just as with polybutadiene, purified films of *cis*- and *trans*-1,4-polyisoprene were found to undergo direct photochemical *cis–trans* isomerization when irradiated with mercury 2537 Å radiation *in vacuo* at room temperature[147]. Besides the isomerization, other important photochemical processes take place. The photoinduced *cis–trans* isomerization of this polymer, as well as that of polybutadiene noted earlier[146], was assumed to proceed through triplet excitation (and hence interconversion) of the isoprenic double bonds, ensuant on direct absorption of the Hg 2537 Å photons. Some of the absorbed energy undoubtedly is used to rupture the C—C bonds connecting successive isoprene units (equation 28). The fate of the radicals **72–75** is to recombine (in any of four different ways) or to add to double bonds in the same or other macromolecular chain. The overall effect would be to produce some endlinks (or crosslinks) and some new

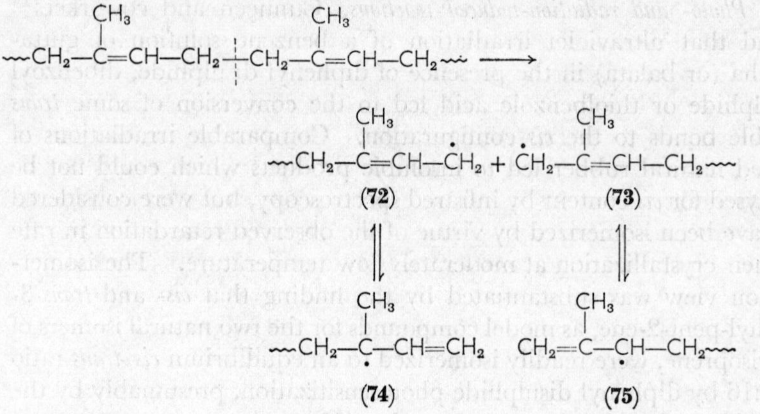

$$\sim\!CH_2-\underset{CH_3}{C}=CH-CH_2\!-\!\!\mid\!-CH_2-\underset{CH_3}{C}=CH-CH_2\!\sim \longrightarrow$$

$$\sim\!CH_2-\underset{CH_3}{C}=CH-\overset{\cdot}{CH_2} + \overset{\cdot}{CH_2}-\underset{CH_3}{C}=CH-CH_2\!\sim \quad (28)$$
$$(72) \qquad\qquad (73)$$

$$\Updownarrow \qquad\qquad \Updownarrow$$

$$\sim\!CH_2-\underset{\overset{\cdot}{}}{\underset{CH_3}{C}}-CH=CH_2 \qquad CH_2=\underset{\overset{\cdot}{}}{\underset{CH_3}{C}}-CH-CH_2\!\sim$$
$$(74) \qquad\qquad (75)$$

vinylidene and vinyl double bonds in the ultraviolet-irradiated polyisoprene (such as equations 29 and 30). Another reaction is the

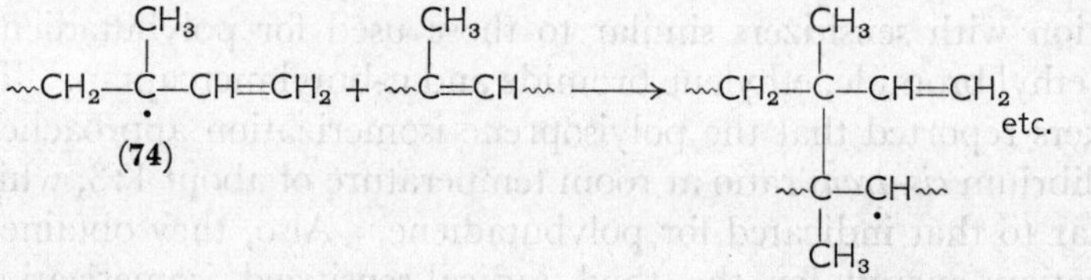

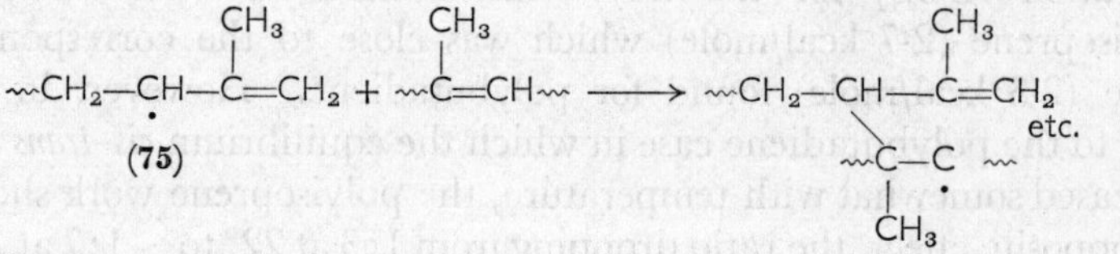

$$\sim\!CH_2-\underset{\overset{\cdot}{}}{\underset{CH_3}{C}}-CH=CH_2 + \sim\!\underset{CH_3}{\overset{\cdot}{C}}=CH\!\sim \longrightarrow \sim\!CH_2-\underset{\underset{\sim\!\underset{CH_3}{\underset{|}{C}}-CH\!\sim}{|}}{\underset{CH_3}{C}}-CH=CH_2 \quad (29)$$
$$(74) \qquad\qquad\qquad\qquad\qquad \text{etc.}$$

$$\sim\!CH_2-\underset{CH_3}{CH}-\underset{\overset{\cdot}{}}{C}=CH_2 + \sim\!\underset{CH_3}{\overset{\cdot}{C}}=CH\!\sim \longrightarrow \sim\!CH_2-\underset{CH_3}{CH}-\underset{\underset{\sim\!\underset{CH_3}{\underset{|}{C}-C}\!\sim}{|}}{C}=CH_2 \quad (30)$$
$$(75) \qquad\qquad\qquad\qquad\qquad\qquad \text{etc.}$$

formation of cyclopropyl groups, possibly involving photoinduced formation of a biradical, followed by a 1,2-hydrogen migration and then ring closure (equation 31). Although this novel *photocyclization*

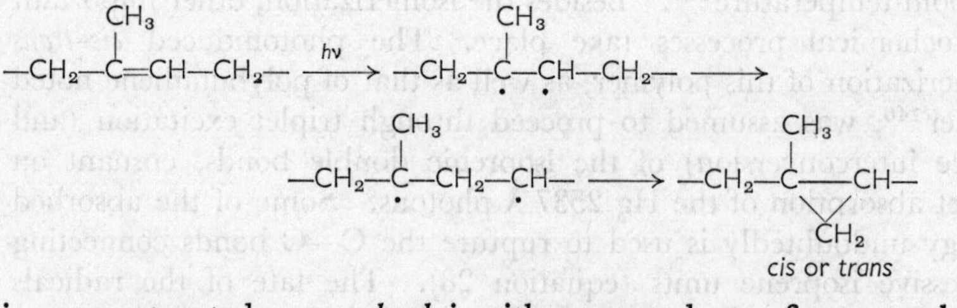

$$-CH_2-\underset{CH_3}{C}=CH-CH_2- \xrightarrow{h\nu} -CH_2-\underset{CH_3}{\overset{\cdot}{C}}-\overset{\cdot}{CH}-CH_2- \longrightarrow$$

$$-CH_2-\underset{CH_3}{C}-CH_2-CH- \longrightarrow -CH_2-\underset{\underset{CH_2}{|}}{\underset{CH_3}{C}}\!\!-\!\!\underset{}{CH}- \quad (31)$$
$$\qquad\qquad\qquad\qquad\qquad\qquad\qquad \textit{cis or trans}$$

in an unsaturated *macromolecule* is without precedent, a few examples of such a biradical reaction in small olefinic molecules have been reported recently. Thus, *trans*-1,3-diphenylpropene was found to undergo in solution, an unsensitized photocyclization to *cis*- and *trans*-1,2-diphenylcyclopropanes, along with isomerization to the *cis* form of the starting olefin[173], while 1-butene was found to undergo in the vapour phase a mercury photosensitized cyclization to methylcyclopropane[174].

The photochemical production of hydrogen from purified *Hevea* film (according to equation 32), observed over twenty years ago by Bateman[175], is relatively quite unimportant compared to the photo-induced *cis–trans* isomerization or loss of 1,4 double bonds in polyiso-

$$-CH_2-\underset{CH_3}{C}=CH-CH_2- \xrightarrow{h\nu} -CH_2-\underset{CH_3}{C}=CH-\overset{\bullet}{CH}- + H\cdot \quad (32)$$

$$H\cdot + -CH_2-\underset{CH_3}{C}=CH-CH_2- \longrightarrow H_2 + -CH_2-\underset{CH_3}{C}=CH-\overset{\bullet}{CH}-$$

prene, the quantum yields for the latter two processes being about 40–80 times as large as that of hydrogen production[147]. As a consequence, double bond migration (equation 33) is not observed in the

$$-CH_2-\underset{CH_3}{C}=CH-CH_2- \longrightarrow -CH_2-\underset{CH_3}{C}=CH-\overset{\bullet}{CH}- \rightleftharpoons$$

$$-\overset{\bullet}{CH_2}-\underset{CH_3}{C}-CH=CH- \quad (33)$$

photolysis of polyisoprene. Likewise, double bond migration in the opposite direction (equation 34) was ruled out on the basis of work with polyisoprene-3-*d*.

$$-CH_2-\underset{CH_3}{C}=CD-CH_2- \longrightarrow -\overset{\bullet}{CH}-\underset{CH_3}{C}=CD-CH_2- \rightleftharpoons$$

$$-CH=\underset{CH_3}{\overset{|}{C}}-\overset{\bullet}{CD}-CH_2- \quad (34)$$

Again, in common with polybutadiene, polyisoprene can be isomerized by γ-irradiation in benzene solution and in the solid state. This reaction had been suggested originally by Evans and coworkers[176] but unambiguous evidence was lacking. Subsequently, Golub and Danon[111] confirmed that suggestion. Starting with *cis*- or *trans*-polyisoprene, the isomerization either in solution or in the solid film approached an equilibrium in which the *cis–trans* ratio is 1:1. The energy transfer mechanism discussed previously for the radiation chemical isomerization of polybutadiene in benzene without added

458   Morton A. Golub

sensitizer, was assumed to hold for polyisoprene as well, and moreover it too was accompanied by a sharp decrease in unsaturation, which was attributed to cyclization. Electron irradiation of polyisoprene-3-d showed that various double bond shifts (equation 35) occur to only a very minor extent, if at all, thus ruling out free radical processes in the

$$\begin{array}{c} \\ -CH_2-C(CH_3)=CH-CH_2- \end{array} \Biggl\{ \begin{array}{l} \rightarrow -CH_2-C(CH_3)=CH-\overset{\bullet}{C}H- \rightleftharpoons -CH_2-\overset{\bullet}{C}(CH_3)-CH=CH- \\ \rightarrow -\overset{\bullet}{C}H-C(CH_3)=CH-CH_2- \rightleftharpoons -CH=C(CH_3)-\overset{\bullet}{C}H-CH_2- \\ \rightarrow -CH_2-C(\overset{\bullet}{C}H_2)-CH-CH_2- \rightleftharpoons -CH_2-C(=CH_2)-\overset{\bullet}{C}H-CH_2- \end{array} \tag{35}$$

radiation-induced 'unsensitized' isomerization. Nevertheless, disproportionation reactions involving the allylic radicals in equation (35) were assumed to be responsible for the formation of a very small amount of conjugated diene and triene groups, as revealed in the ultraviolet spectra of γ-irradiated *Hevea* films[176].

Squalene, a hexaisoprene often employed as a model compound for rubber, was also isomerized with loss of unsaturation on irradiation in the pure liquid state and in benzene solution[177,178].

b. Other isomerization methods. Cunneen and coworkers[14,27,179-182] succeeded in isomerizing *Hevea* and gutta-percha, squalene and *cis* and *trans* forms of 3-methyl-2-pentene by heating them with thiol acids, sulphur dioxide, butadiene sulphone and related materials. After slight isomerization of *Hevea*, its rate of crystallization at moderately low temperatures is greatly retarded (thus improving its service ability at sub-zero temperatures), while gutta-percha, which is partly crystalline at temperatures up to about 65°, is transformed, after somewhat more extensive isomerization, into a polymer which is rubbery at room temperatures[14,27,181].

The isomerization by SO_2, which can add reversibly to C=C bonds above the 'ceiling temperature'[183]* for the given olefin complex, is depicted in equation (36). At 140° this isomerization can be carried to an equilibrium *cis–trans* ratio in polyisoprene of 43:57, a value close

---

* Ceiling temperature is the temperature above which the olefin–SO_2 complex or sulphone is unstable and decomposes as soon as it is formed.

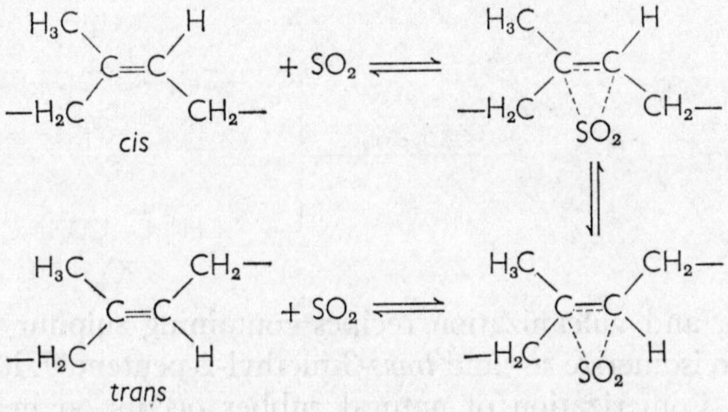

$$\begin{array}{c}
\underset{\text{cis}}{\overset{H_3C}{\underset{-H_2C}{\diagdown}}\mathrm{C}=\mathrm{C}\overset{H}{\underset{CH_2-}{\diagup}}} + SO_2 \rightleftharpoons \underset{-H_2C}{\overset{H_3C}{\diagdown}}\mathrm{C}\overset{H}{\underset{SO_2}{-}}\mathrm{C}\overset{H}{\underset{CH_2-}{\diagup}} \\
\Updownarrow \\
\underset{\text{trans}}{\overset{H_3C}{\underset{-H_2C}{\diagdown}}\mathrm{C}=\mathrm{C}\overset{CH_2-}{\underset{H}{\diagup}}} + SO_2 \rightleftharpoons \underset{-H_2C}{\overset{H_3C}{\diagdown}}\mathrm{C}\overset{CH_2-}{\underset{SO_2}{-}}\mathrm{C}\overset{}{\underset{H}{}}
\end{array}$$

(36)

to that obtained in the selenium-catalysed isomerization (47:53)[18,108]. The sulphur dioxide method is, however, much superior to the selenium method, since the latter involves some polymer chain scission while the former is virtually free of side-reactions. From the standpoint of practical applications butadiene sulphone is a very convenient reagent for the isomerization, since it is a solid and can be milled into rubber or mixed with the isoprenic polymer where it evolves SO_2 when the temperature is raised.

An interesting application of the SO_2 method concerns the isomerization of the decaisoprene side chain of coenzyme Q (**76**) from a nearly

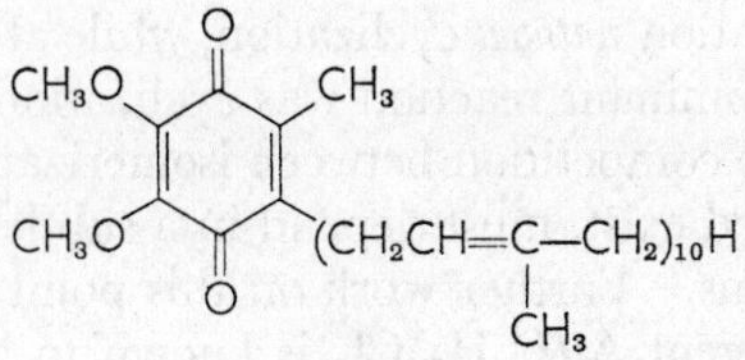

(**76**) CH_3O, CH_3O, CH_3, $(CH_2CH=C-CH_2)_{10}H$, CH_3

all-*trans* to a random *cis,trans* structure, without disturbing the quinone nucleus[184]. This is noteworthy because other isomerization techniques (e.g., photosensitization) generally produce other structural modifications in the coenzyme, such as intramolecular cycloaddition of the first isoprene unit onto the quinone nucleus.

Recently, Cunneen and coworkers[171] concluded, using model compounds, that polyisoprene isomerization according to equation (36a) is not accompanied by double bond migration (36b). This conclusion was substantiated by Golub[185] using polyisoprene-3-*d* and infrared and n.m.r. spectroscopy to distinguish between geometric and positional isomerization (equations 36a and 36b).

Since elemental sulphur isomerizes polybutadiene under vulcanizing

$$-CH_2-\underset{cis}{\overset{\overset{\displaystyle CH_3}{|}}{C}=CD}-CH_2- \xrightarrow{SO_2 \text{ catalysis}} \begin{cases} -CH_2-\underset{trans}{\overset{\overset{\displaystyle CH_3}{|}}{C}=CD}-CH_2- & (36a) \\ -CH=\underset{trans\ or\ cis}{\overset{\overset{\displaystyle CH_3}{|}}{C}}-CDH-CH_2- & (36b) \end{cases}$$

conditions, and vulcanization recipes containing sulphur have been reported to isomerize *cis*- and *trans*-3-methyl-2-pentene[27], it is possible that some isomerization of natural rubber occurs on heating with sulphur, but the evidence to date on this point is inconclusive[108,164].

Dolgoplosk and coworkers[186-188] have claimed that *cis–trans* isomerization of polyisoprene also takes place when the polymer is treated with $TiCl_4$, anhydrous HCl, or various organometallic compounds in benzene at 80–90°. These claims were disproved by Golub and Heller who showed that the effect of $TiCl_4$ on *Hevea* or balata is to induce cyclization almost exclusively[109], and that the reaction of HCl involves only addition across the double bonds along with a small amount of cyclization[101]. More recently, Dolgoplosk[189] reported that treating natural rubber in dilute benzene solution with ethyl-aluminum dichloride at quite low concentrations at 60°, produced *cis–trans* isomerization *without* cyclization, while at higher catalyst concentrations the dominant reaction was cyclization. It would appear from this that the competition between isomerization and cyclization, if such does indeed exist, must depend in a subtle way on the specific reaction conditions. Further work on this point is needed. At any rate, this same agent $Al(C_2H_5)Cl_2$ is known to be a potent catalyst for the cyclopolymerization of isoprene[128], so that its strong cyclizing action on polyisoprene itself can be anticipated.

D. Hydrogenation

The subject of hydrogenation of natural rubber, balata and gutta-percha dates back almost a century, although the first significant results were accomplished around 1920 by Harries[190], Staudinger[191,192], Pummerer[193] and others. Since an excellent review on the hydrogenation of various unsaturated polymers has been written by Wicklatz[194], the present treatment will discuss only the main features of some of the latest studies in this field.

Much attention had been given by early workers, especially in the patent literature, to the use of the destructive (degradative) hydrogenation process for preparing oils and other low molecular weight products[194]. Recently, however, emphasis has been placed on nondestructive hydrogenation of synthetic unsaturated polymers, with the aim either of preparing new polymers with interesting physical properties or of solving problems in polymer microstructure. Only a few diene homopolymers or copolymers have been actually hydrogenated. These include various polybutadienes (emulsion-polymerized[195,196], sodium-polymerized[197,198], cis-1,4 polymer[196,198,199]), polyisoprenes (cis-1,4 polymer[196,200,201], trans-1,4 polymer[201]), butadiene-styrene rubber[195,196], 1,4-polypiperylene[196], and various poly-2-alkylbutadienes[201].

The rates and ultimate degree of hydrogenation obtainable under given reaction conditions depend on the polymer composition[194]. In general, natural rubber and the synthetic hydrocarbon polymers are readily hydrogenated while gutta-percha is more difficult to hydrogenate. For reasons not yet clear, hydrocarbon copolymers appear to be less easily hydrogenated than homopolymers derived from the same diene monomers. Also, in contrast to the all-hydrocarbon copolymers, those unsaturated copolymers having a comonomer containing nitrogen (e.g., vinylpyridine, acrylonitrile) or halogen (e.g., vinylidene chloride) or having acrylates as comonomers, show great resistance to hydrogenation of their C=C bonds. While hydrogenation of a butadiene-acrylonitrile copolymer to a polymer amine has been reported[202], the double bonds in butadiene units probably did not react. Likewise, the halogen-containing diene polymer, polychloroprene, apparently cannot be hydrogenated[194,196].

The catalysts employed in the past have been based largely on platinum, palladium and nickel. Staudinger[191] reported the complete hydrogenation of natural rubber in dilute solution in a hydrocarbon solvent, using a nickel catalyst at high temperature and pressure. This method was very recently extended by Gregg[201] to cis-1,4- and trans-1,4-polyisoprenes and various 1,4-poly-2-alkylbutadienes. Since the molecular weight of the product derived from cis-1,4-polyisoprene was about 60% of that of the original polymer, a small amount of chain cleavage accompanied the hydrogenation, of the order of one cleavage per 1300 isoprene units reduced. Fractionation of a 90% hydrogenated polyisoprene showed that the unsaturation was concentrated in the low molecular weight fraction, the high molecular weight fraction being completely saturated. Moreover, infrared and n.m.r. spectroscopic examination of the unsaturated low molecular weight fraction indicated that it was a mixture of hydrogenated polyisoprene and essentially unreacted *cis*-1,4-polyisoprene. This interesting finding signified that the hydrogenation process involves hydrogenating a macromolecule entirely or not at all. Another interesting result was that no *cis–trans* isomerization accompanied the hydrogenation.

Gregg[201] also reported that a hydrogenated balata (with 9% residual unsaturation) had an infrared spectrum identical to that of a hydrogenated *cis*-1,4-polyisoprene having the same unsaturation. Since there was no infrared evidence for a difference in tacticity between the two hydrogenated polyisoprenes, it was presumed that the addition of hydrogen to the C=C bonds was non-stereoregular: *cis*-addition of hydrogen to *cis*- and *trans*-polyisoprenes should yield, respectively, syndiotactic and isotactic polymers. The virtually complete hydrogenation ($\geqslant 98.9\%$) of several 1,4-poly-2-alkylbutadienes (**77**), where R = Et, n-Pr, n-pentyl, or Me_2C=$CHCH_2CH_2$—,

$$(-CH_2-\underset{\underset{R}{|}}{C}=CH-CH_2-)_n$$
(**77**)

under the same conditions used for R = Me, was also reported. Although some polymer degradation occurs in all of these hydrogenations, high molecular weight polymers can be obtained by fractionating the hydrogenated materials.

Ramp, DeWitt and Trapasso[196] have recently described the homogeneous hydrogenation of various diene polymers (and small olefinic compounds) catalysed by triisobutyl borane. The mechanism depicted for the borane-catalysed hydrogenation, which is carried out in solution at high temperature and high pressure, is shown in equation (37). By means of this reaction, the olefinic C=C bonds in *cis*-1,4-

$$R_3B + 3H_2 \longrightarrow BH_3 + 3RH$$

$$BH_3 + -\overset{|}{C}=\overset{|}{C}- \longrightarrow \overset{H}{\underset{H}{\overset{|}{B}}}-\overset{|}{\underset{H}{C}}-\overset{|}{\underset{H}{C}}- \overset{H_2}{\longrightarrow} \overset{H}{\underset{H}{\overset{|}{B}}}-H + -\overset{|}{\underset{H}{C}}-\overset{|}{\underset{H}{C}}- \quad (37)$$

polyisoprene, *cis*-1,4-polybutadiene, emulsion 1,4-polybutadiene, 1,2-polybutadiene, butadiene-styrene copolymer and 1,4-polypiperylene were completely saturated. Some drop in molecular weight was encountered in the reduction, which was more severe for free radical-polymerized polymers than for polymers prepared with Ziegler catalysts. Neoprene resisted any hydrogen uptake but evolved HCl on treatment with the borane catalyst.

Cis-1,4-polybutadiene on quantitative hydrogenation yielded a crystalline polymer with an infrared spectrum and melt characteristics of a high density, moderate molecular weight polyethylene (equation 38). Analogously, synthetic *cis*-1,4-polyisoprene yielded a tough, rubbery polymer having an infrared spectrum which was very similar

$$-CH_2-CH=CH-CH_2- \xrightarrow{H_2} -CH_2-CH_2-CH_2-CH_2- \quad (38)$$

to that of a Ziegler-catalysed ethylene-propylene random copolymer (**37**). Likewise, the 1,4-polypiperylene was hydrogenated to a product having an infrared spectrum similar to that of hydrogenated polyisoprene (**37**).

Evidence for a rapid addition-elimination process involving B—H and the olefinic linkage was provided by the infrared spectrum of a *partially* reduced *cis*-1,4-polybutadiene which showed complete elimination of the 13·9 μ band (*cis* —CH=CH—) and the development of an intense 10·3 μ band (*trans* —CH=CH—)[196]. Further evidence for the addition-elimination reaction in the borane catalysis was seen in the fact that hydrogenation of *cis*-1,4-polyisoprene-3-*d* involved extensive removal of deuterium from the polymer chain.

Finally, it is interesting to note that microstructural studies on hydrogenated sodium polybutadiene[198] showed that external double bonds (1,2 units; **7**) are reduced much more rapidly than internal double bonds (1,4 units; **5,6**).

E. Vulcanization

The term 'vulcanization' originally referred to the process, discovered independently by Goodyear in 1839 and by Hancock in 1843, of heating natural rubber with sulphur to transform it from a soft, gummy, irreversibly deformable substance into a hard, elastic material capable of reversible deformation. Since other methods exist today for achieving a crosslinked polymer network which is the essential feature of a vulcanizate, the term has been broadened to include all types of controlled crosslinking reactions carried out on elastomers—saturated as well as unsaturated. Crosslinking of linear macromolecules can also be accomplished by high energy, and sometimes ultraviolet, irradiation or by various chemical means, such as mutual combination of nascent polymer radicals, graft copolymerization, reaction of backbone carbon atoms with various reagents, and reactions involving functional groups in, or attached to, the polymer backbone. A comprehensive survey of the various non-sulphur chemical and radiation chemical methods for crosslinking has been presented [203].

There are four main types of vulcanizing (or curing) systems for the unsaturated hydrocarbon elastomers, involving (1) organic peroxides, (2) elemental sulphur alone, (3) elemental sulphur in combination with accelerators and (4) organosulphur systems in the absence of elemental sulphur. Neoprene, however, requires a vulcanization system based on the reaction of the active chlorine atoms instead of the C=C bonds.

Most of our current understanding of the sulphur-rubber reactions is due to the outstanding contributions, especially in the last ten years, by workers at the Natural (formerly British) Rubber Producers' Research Association. Excellent reviews have recently appeared on sulphur vulcanization [164,205] and the chemistry of peroxide vulcanization [164,204]. The present discussion will be confined to a survey of the principal reactions involved in vulcanization to the extent that they concern the chemistry of the C=C bond in macromolecules.

I. Peroxide vulcanization

a. 1,4-Polyisoprene. The use of an organic peroxide as a crosslinking agent for natural rubber was first reported by Ostromislensky [206] in 1915 using benzoyl peroxide. This and other diaroyl peroxides are not very efficient in promoting vulcanization, and modern practice is based on the use of dialkyl (or diaralkyl) peroxides (e.g., di-t-butyl peroxide and dicumyl peroxide).

In the case of di-t-butyl peroxide reacting with polyisoprene at 140°, the first step is the unimolecular decomposition of the peroxide to peroxy radicals. These radicals for the most part abstract α-methylenic hydrogen atoms yielding t-butyl alcohol and polyisoprenyl radicals which then undergo mutual combination (Scheme I). The particular abstraction shown is that of the hydrogen on carbon atom a but abstraction of the hydrogen on carbon atom b or c is also possible, resulting in different isomeric structures for the crosslinked isoprene units. Evidence for double bond shifts to form some vinylene and even fewer vinylidene double bonds, the latter arising from abstraction at c, has been found in infrared spectroscopy [207]. However, the order of reactivity of the three different α-methylenic hydrogens is $a > b > c$ [208] so that the predominant structure for a crosslinked pair of isoprene units is **78**.

The polyisoprenyl radical also undergoes cyclization to a small extent. The network will thus consist of many C—C crosslinks with *occasional* cyclic structures vicinal to the crosslink as in **79**. Evidence

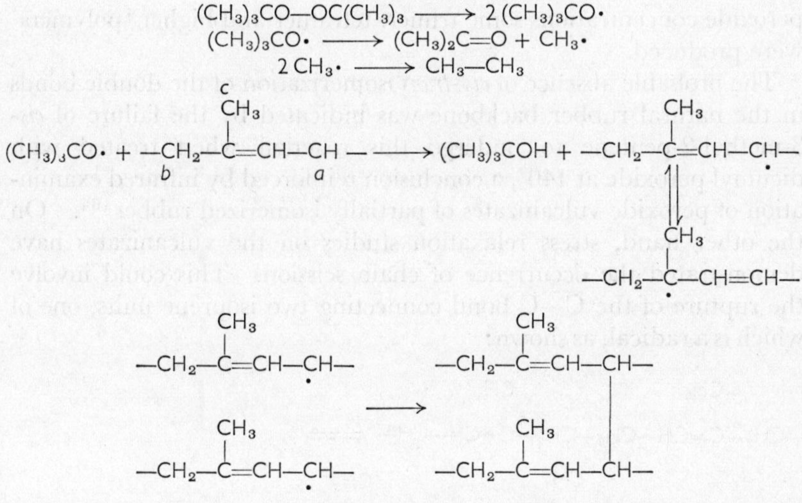

9. Polymers Containing C=C Bonds

$$(CH_3)_3CO-OC(CH_3)_3 \longrightarrow 2\,(CH_3)_3CO\cdot$$
$$(CH_3)_3CO\cdot \longrightarrow (CH_3)_2C=O + CH_3\cdot$$
$$2\,CH_3\cdot \longrightarrow CH_3-CH_3$$

```
                  c
                  CH₃                                         CH₃
                  |                                           |
(CH₃)₃CO· + —CH₂—C=CH—CH—  ———→  (CH₃)₃COH + —CH₂—C=CH—CH—
                b       a                                          ·
                                                            |
                                                            CH₃
                                                            |
                                                       —CH₂—C—CH=CH—

       CH₃                        CH₃
       |                          |
  —CH₂—C=CH—CH—              —CH₂—C=CH—CH—
               ·
                         ———→              |
       CH₃                        CH₃
       |                          |
  —CH₂—C=CH—CH—              —CH₂—C=CH—CH—
               ·
```
(78)

SCHEME I.

for the cyclization is the observation that the unsaturated polymers obtained from 2,6-dimethyl-2,6-octadiene (dihydromyrcene) and digeranyl retained only 82 and 72%, respectively, of the unsaturation of the original olefins, whereas the unsaturated polymers derived from the corresponding monoisoprenes[207,209] showed 100% retention of unsaturation. Furthermore, gas-liquid chromatographic examina-

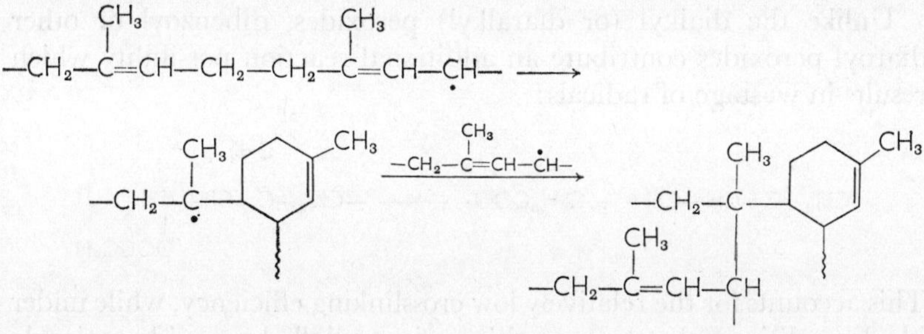

```
       CH₃                  CH₃
       |                    |
  —CH₂—C=CH—CH₂—CH₂—C=CH—CH—  ———→
                                   ·
```

(79)

tion of the products of the reaction of dihydromyrcene with dicumyl peroxide at 120–150° indicated that the diisoprenyl radicals (and presumably also the corresponding radicals in polyisoprene) undergo both inter- and intramolecular combination[210]. At large diene–peroxide ratios the product was almost entirely dimer, while at higher

16+C.A. 2

peroxide concentrations some trimer, tetramer and higher 'polymers' were produced.

The probable absence of *cis–trans* isomerization of the double bonds in the natural rubber backbone was indicated by the failure of *cis*-3-methyl-2-pentene to undergo this reaction when treated with dicumyl peroxide at 140°, a conclusion reinforced by infrared examination of peroxide vulcanizates of partially isomerized rubber[164]. On the other hand, stress relaxation studies on the vulcanizates have demonstrated the occurrence of chain scission. This could involve the rupture of the C—C bond connecting two isoprene units, one of which is a radical, as shown:

$$\begin{array}{c}CH_3CH_3\\-CH_2-\overset{|}{C}=CH-CH_2-CH_2-\overset{|}{C}=CH-\overset{\bullet}{C}H- \rightleftharpoons\end{array}$$

$$\begin{array}{c}CH_3CH_3\\-CH_2-\overset{|}{C}=CH-CH_2\!\mid\!CH_2-\overset{|}{C}-CH=CH- \rightleftharpoons\end{array}$$

$$\begin{array}{c}CH_3CH_3\\-CH_2-\overset{|}{C}=CH-CH_2\bullet\ +\ CH_2=\overset{|}{C}-CH=CH-\end{array}$$

However, under normal vulcanization conditions, scission is very minor compared to the crosslinking reaction. Sol-gel measurements on the dicumyl peroxide vulcanizates of natural rubber showed, in fact, that the ratio of chain scission to crosslinking is of the order of 0–0·2[204].

Unlike the dialkyl (or diaralkyl) peroxides, dibenzoyl or other diaroyl peroxides contribute an additional reaction possibility which results in wastage of radicals:

$$\begin{array}{c}CH_3CH_3\\-CH_2-\overset{|}{C}=CH-\underset{\bullet}{C}H-\ +\ C_6H_5COO\bullet\ \longrightarrow\ -CH_2-\overset{|}{C}=CH-CH-\\|\\OOCC_6H_5\end{array}$$

This accounts for the relatively low crosslinking efficiency, while under ideal conditions, the decomposition of one dialkyl peroxide molecule leads to the formation of one crosslink in natural rubber. Crosslinking efficiency of near unity is also obtained with synthetic *cis*-1,4-polyisoprene and natural *trans*-1,4-polyisoprene (balata)[204]. Polyisoprenes with rather high 3,4 contents show higher crosslinking efficiency[204].

Mention may be made of 1,4-polypiperylene which, on hydrogen abstraction, yields predominantly radical **80**; since this radical is virtually identical to the macroradical formed in 1,4-polyisoprene, it is not too surprising that polypiperylene shows a crosslinking efficiency close to unity[170].

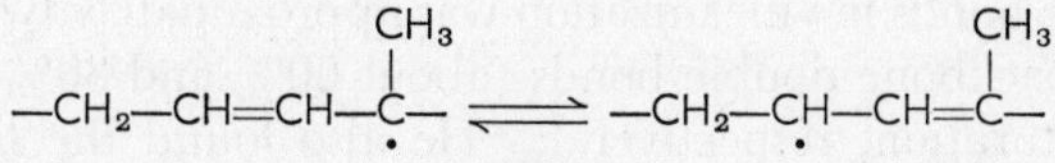

$$-CH_2-CH=CH-\underset{\bullet}{\overset{\underset{|}{CH_3}}{C}}- \rightleftharpoons -CH_2-\underset{\bullet}{CH}-CH=\overset{\underset{|}{CH_3}}{C}-$$

(80)

b. 1,4-Polybutadiene. In contrast to the 1,4-polyisoprenes, 1,4-polybutadiene and other butadiene-containing polymers, such as styrene-butadiene rubber, show crosslinking efficiencies considerably greater than one, with values of around 12 in the case of *cis-* and *trans-*1,4-polybutadienes[170], and even higher values for other configurations. Thus, the efficiency of a 79% 1,2-polybutadiene increases from 18 to 45 over the temperature range 115–160°, while that for a 10% 1,2-polybutadiene decreases from over 50 to 22 over the same temperature range[170]. These high crosslinking efficiencies are undoubtedly the result of chain reactions, as confirmed by the chain suppressing effect of antioxidants. Since there is a loss of double bonds approximately equal to the number of crosslinks formed, the chain reaction is presumed to consist of a series of addition (or crosspolymerization) steps followed by hydrogen transfer reactions (equation 39, in the case of 1,4-polybutadiene), before the macroradicals are terminated by combination. Thus, in contrast to the polyisoprenyl radicals, the polybutadienyl radicals, assumed to have higher reactivity, can attack

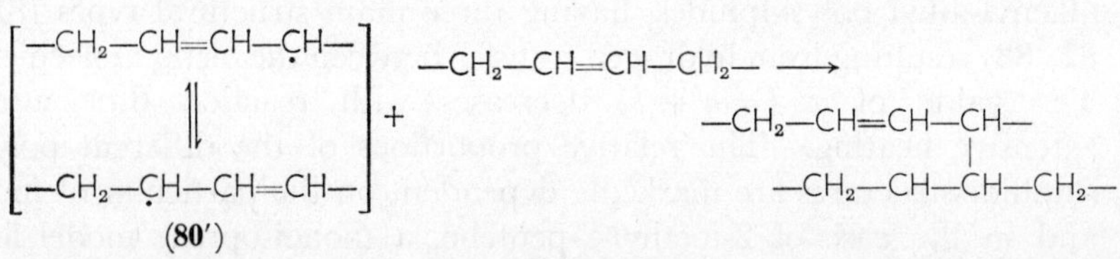

$$\begin{bmatrix} -CH_2-CH=CH-\overset{\bullet}{C}H- \\ \Updownarrow \\ -CH_2-\overset{\bullet}{C}H-CH=CH- \end{bmatrix} + \quad -CH_2-CH=CH-CH_2- \longrightarrow \quad \begin{matrix} -CH_2-CH=CH-CH- \\ | \\ -CH_2-CH-CH-CH_2- \end{matrix}$$

(80')

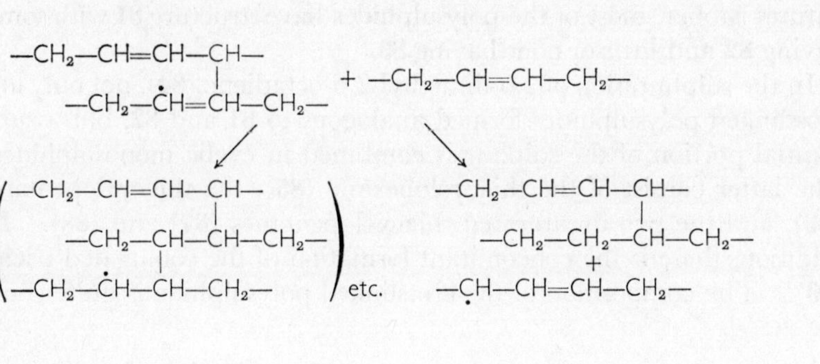

$$\begin{matrix} -CH_2-CH=CH-CH- \\ | \\ -CH_2-\overset{\bullet}{C}H-CH-CH_2- \end{matrix} + -CH_2-CH=CH-CH_2- \tag{39}$$

$$\left(\begin{matrix} -CH_2-CH=CH-CH- \\ | \\ -CH_2-CH-CH-CH_2- \\ | \\ -CH_2-\overset{\bullet}{C}H-CH-CH_2- \end{matrix} \right) \text{etc.} \quad \begin{matrix} -CH_2-CH=CH-CH- \\ | \\ -CH_2-CH_2-CH-CH_2- \\ + \\ -\overset{\bullet}{C}H-CH=CH-CH_2- \end{matrix}$$

double bonds and 'polymerize' the polymer. By means of infrared examination of a peroxide-cured polybutadiene (originally 93·5% cis-1,4; 3·0% trans; 3·5% 1,2), van der Hoff[170] showed that the loss of vinyl double bonds in vulcanization was approximately twice as fast as the loss of backbone double bonds (about 60% and 30% decrease in initial unsaturation, respectively). He also found the trans content increased by about 7%, which probably resulted from a double bond shift in the polybutadienyl radical **80′** rather than from a true cis–trans isomerization of a given C=C bond.

The crosslinking behaviour of two important butadiene-containing elastomers may be noted here[204]. The more important of the two, styrene-butadiene rubber, behaves in much the same way as cis-polybutadiene. However, the purified nitrile rubber (butadiene-acrylonitrile copolymer) has a crosslinking efficiency close to unity, so that it would appear that the chain reaction observed with the butadiene homopolymer is suppressed in this particular copolymer (~60% butadiene) by the acrylonitrile residues.

2. Sulphur vulcanization

a. Unaccelerated reaction. Modern views[164,205] on the structure of the natural rubber-sulphur vulcanizate network have been developed only after extensive research on the mechanisms and products of the reactions of sulphur with various monoolefins and with the diolefin, 2,6-dimethyl-2,6-octadiene. In the case of the sulphur-monoolefin reaction at ~140° it is now known that the principal products are alkenyl-alkyl polysulphides, having three main structural types (**81, 82, 83**) resulting from hydrogen transfer between the olefin molecules. The value of x $(= a + b)$ decreases with reaction time after extensive heating. The relative proportions of the different polysulphide structures are markedly dependent on the particular olefins, and in the case of 2-methyl-2-pentene, a monoisoprene model for natural rubber, most of the polysulphides have structure **81** with some having **82** and little or none having **83**.

In the sulphuration of 2,6-dimethyl-2,6-octadiene (**84**), not only are crosslinked polysulphides formed analogous to **81** and **82**, but a substantial portion of the sulphur is combined in cyclic monosulphides. The latter consist of the thiacyclohexane (**85**), the thiacyclopentane (**86**), and the two unsaturated thiacyclopentanes (**87**) and (**88**). In addition, there is the concomitant formation of the conjugated triene (**89**). The composition of the crosslinked polysulphide is quite com-

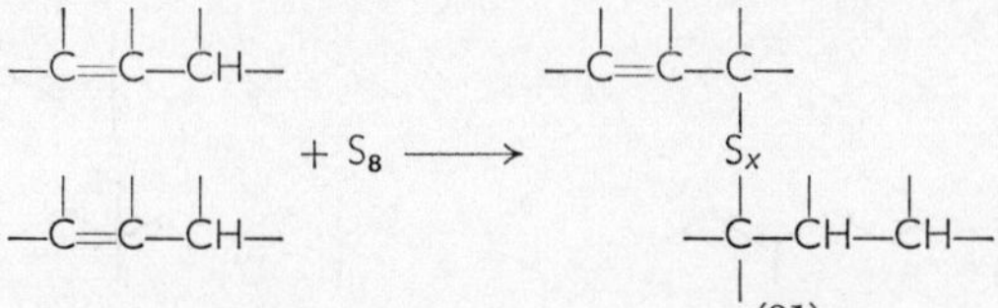

9. Polymers Containing C=C Bonds

(81)

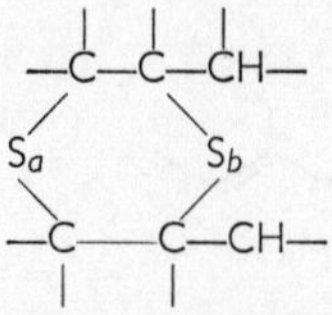

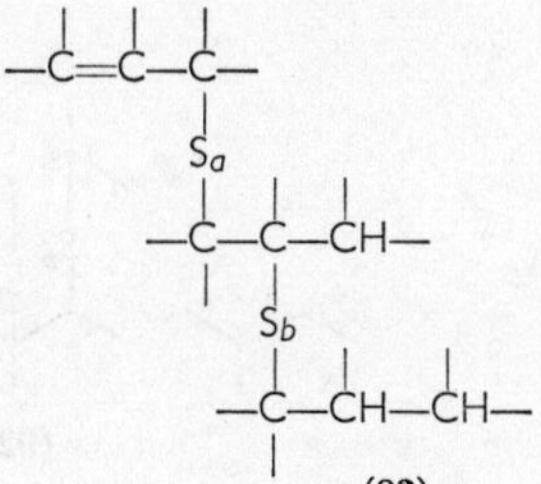

(82) (83)

plex and changes considerably with time of vulcanization; in the early stages it consists mainly of structures like **90**, **91** and **92** which later are transformed into structures containing cyclic monosulphide groups,

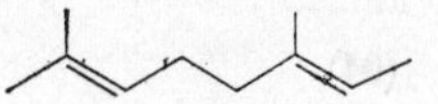

(84) 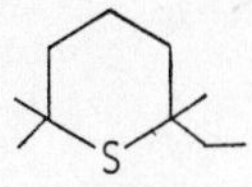 (85) 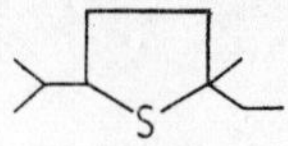 (86)

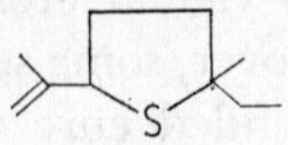

(87) 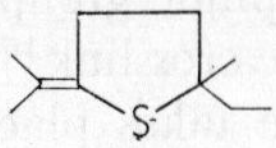 (88) 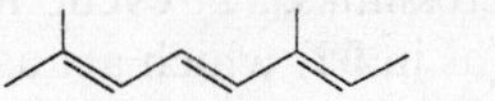 (89)

and eventually into structures such as **93** and **94** in which the crosslinks contain only one or two sulphur atoms. The dotted lines represent alternative positions of the C=C and C—S bonds. In the course of these transformations, the cyclic monosulphides are being formed continuously so that the proportion of total combined sulphur located in such structures increases steadily. Thus, for example, after reaction of the diisoprene **84** with sulphur (in a 10:1 weight ratio) at 140° for 5 hours, 9·4% of the initial sulphur is present in cyclic monosulphides and 48% in polysulphides, while after 40 hours the amounts are 38% and 61%, respectively.

The very poor crosslinking efficiency of the vulcanization of natural rubber with sulphur alone (i.e., without accelerator), in which some 40–55 sulphur atoms are combined for every chemical crosslink formed,

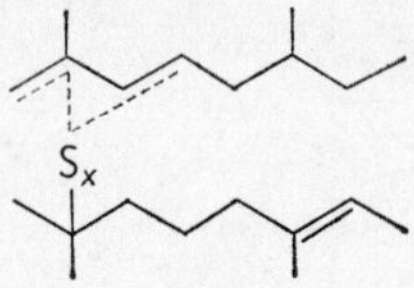

(90) 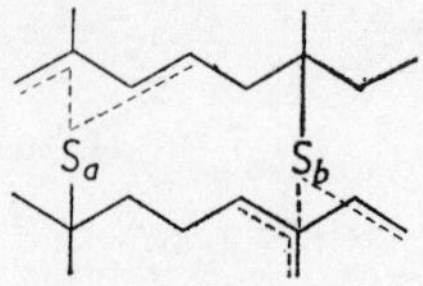   (91)

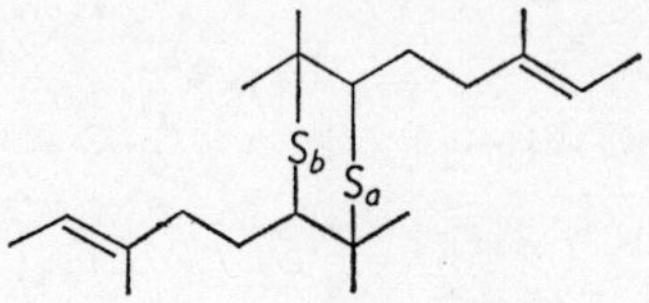

(92)

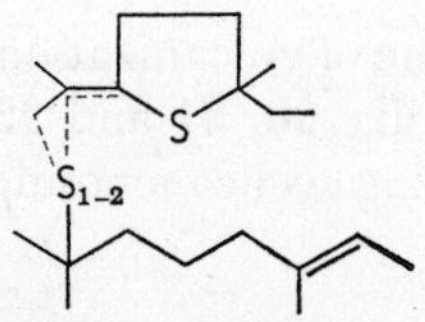

(93) 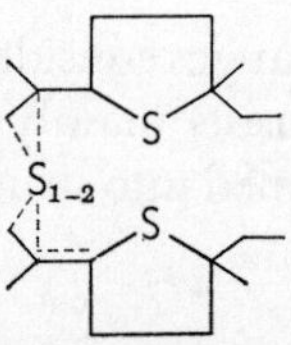   (94)

is evidently due to the wastage of sulphur in (*1*) long polysulphide crosslinks, (*2*) cyclic monosulphide groups and (*3*) vicinal crosslinks (as in **92**) which act as a single crosslink[164]. Moreover, some scission of the polyisoprene backbone takes place on extended cure which partially offsets the crosslinking. The function of an accelerator in sulphur vulcanization, therefore, is to minimize the formation of cyclic monosulphides and vicinal crosslinks while producing simple crosslinks with few sulphur atoms, and at the same time to inhibit chain scission. The effectiveness of some accelerated sulphur vulcanizations[205] can be gauged by the fact that as little as 1·6 sulphur atoms are required per crosslink formed in such systems.

By converting polysulphidic links into mono- or disulphide links on treatment with triphenyl phosphine, Moore and Trego[211] were able to estimate the fraction of combined sulphur present as cyclic monosulphide groups in natural rubber vulcanizates. Their structural data were in very good agreement with those indicated from the dihydromyrcene model compound approach. Thus they found, for example, that after 24 hours of an unaccelerated vulcanization (10 parts sulphur/100 parts rubber) at 140°, about 39% of the isoprene

9. Polymers Containing C=C Bonds

units in the original rubber were converted into cyclic monosulphide groups.

The likelihood that the polyisoprene vulcanizate has sustained some *cis–trans* isomerization is indicated by the easy *cis-trans* isomerization of 3-methyl-2-pentene under vulcanizing conditions[27], but definite data for this reaction in polyisoprene are lacking[164].

Another possible isomerization is that of a double bond shift suggested by early workers[212–214] on the grounds that the infrared spectrum of a natural rubber vulcanizate showed a prominent 10·4 μ band implicating *trans* —CH=CH— units. Such an assignment was in line with the then-held free radical mechanism of vulcanization[215], in which the initiation of sulphuration involves abstraction of the α-methylenic hydrogen (equation 40). However, infrared evidence obtained from sulphur vulcanizates of *cis*-polyisoprene and partially and fully deuterated *cis*-polyisoprenes[155] showed that the proposed double bond migration can occur to only a minor extent.

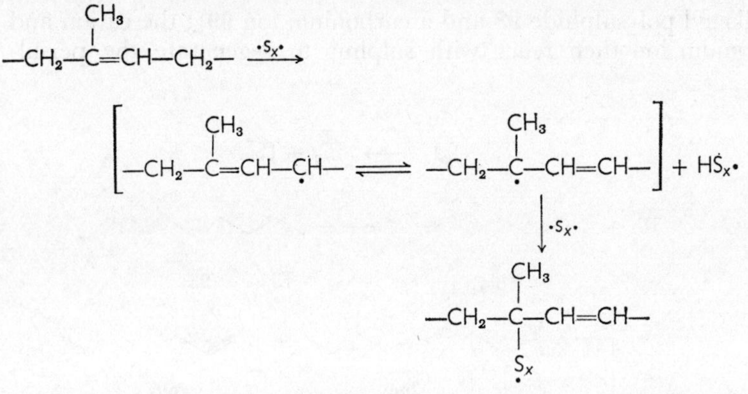

$$-CH_2-\underset{\underset{CH_3}{|}}{C}=CH-CH_2- \xrightarrow{\cdot S_x\cdot}$$

$$\left[-CH_2-\underset{\underset{CH_3}{|}}{C}=CH-\overset{\cdot}{C}H- \rightleftharpoons -CH_2-\underset{\underset{CH_3}{|}}{\overset{\cdot}{C}}-CH=CH-\right] + H\dot{S}_x \quad (40)$$

$$\downarrow \cdot S_x\cdot$$

$$-CH_2-\underset{\underset{\dot{S}_x}{|}}{\overset{\overset{CH_3}{|}}{C}}-CH=CH-$$

The 10·4 μ band was considered, instead, to be due largely to various saturated ring structures of the type indicated above (**85–88**) and to conjugated double bonds, presumably in triene structures like **89**, with a possible small contribution from isolated *trans* —CH=CH— units.

The detailed structures of the alkenyl-alkyl polysulphides and cyclic monosulphides formed in the sulphur-dihydromyrcene reaction, along with complementary kinetic data, prompted the conclusion that the sulphuration reaction is polar in nature[164]. Furthermore, sulphuration is insensitive to free radical initiators or ultraviolet light, while not being affected by radical inhibitors or retarders in the expected manner, and the reaction is promoted by polar compounds or by increasing the polarity of the reaction medium.

The polar mechanism for sulphur vulcanization[162-165] involves the heterolysis of the polarized S—S bond in either elemental sulphur or an already-formed polysulphide to give persulphenium ion, TS_a^+ (which initiates the sulphuration), and persulphenyl ion, TS_b^- (which acts as chain terminator). The TS_a^+ ion (where T is an alkenyl or alkyl group) then adds to a double bond to form a cyclic persulphonium ion **95**,

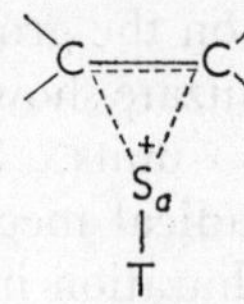

$$\underset{(95)}{\overset{\overset{\displaystyle C = C}{\underset{\underset{\displaystyle T}{\displaystyle |}}{\underset{\displaystyle S_a}{\displaystyle +}}}}{}}$$

which in the case of dihydromyrcene, and presumably also polyisoprene, undergoes mainly hydride ion transfer (to form an alkyl polysulphide **96** and an alkenyl cation **97**) and proton transfer (to form an alkenyl polysulphide **98** and a carbonium ion **99**); the cation and carbonium ion then react with sulphur to regenerate the persul-

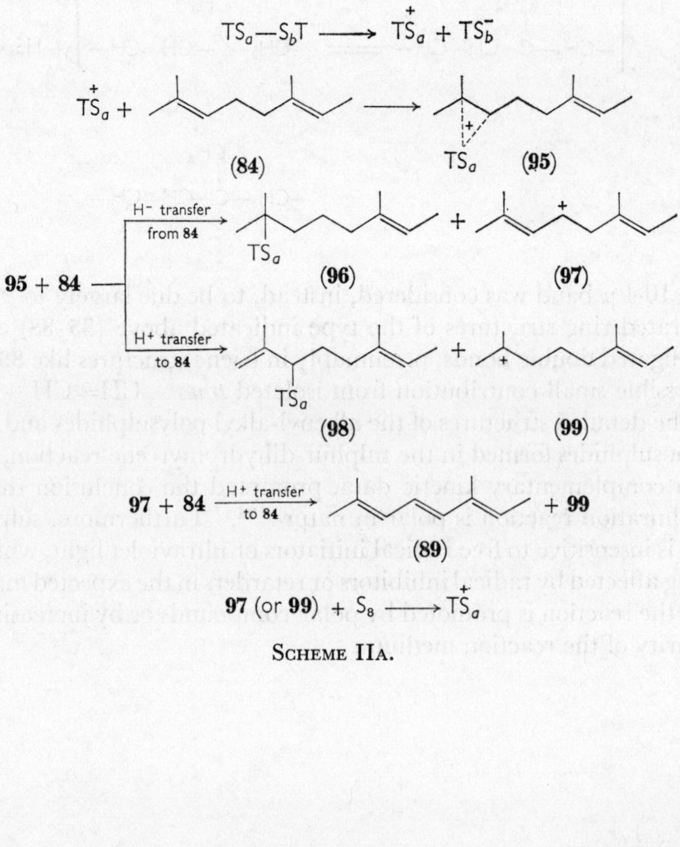

SCHEME IIA.

phenium ions, as shown in Scheme IIA, where the formation of cross-linked polysulphides and the conjugated triene are depicted. Scheme IIB depicts the formation of the cyclic monosulphides. In these schemes, the cyclic monosulphides are secondary products derived

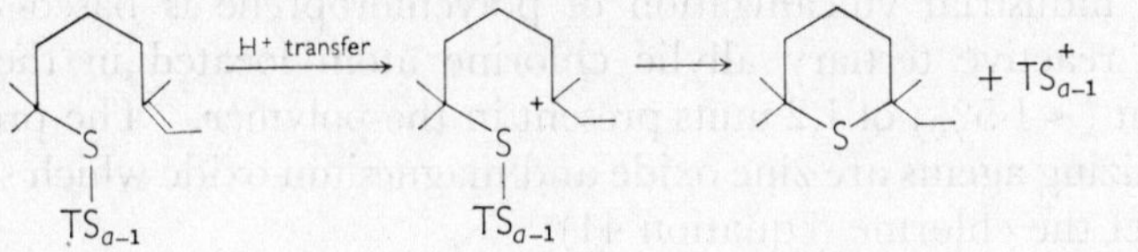

$$\underset{(96)}{\overset{TS_{a-1}}{\big|}} \xrightarrow{H^+ \text{ transfer}} \underset{(85)}{\overset{TS_{a-1}}{\big|}} \longrightarrow \quad + TS_{a-1}^+$$

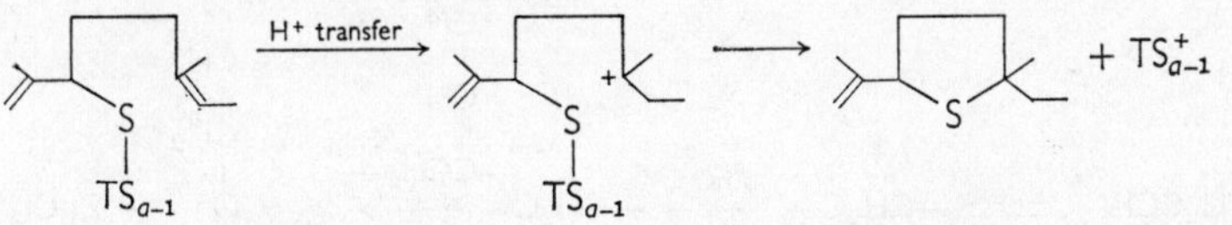

$$\underset{(98)}{\overset{TS_{a-1}}{\big|}} \xrightarrow{H^+ \text{ transfer}} \underset{(87)}{\overset{TS_{a-1}}{\big|}} \longrightarrow \quad + TS_{a-1}^+$$

SCHEME IIB.

from the polysulphides which are the initial sulphuration products.

The mechanism(s) of the unaccelerated sulphur vulcanization of polybutadiene and butadiene-styrene rubber have not been examined in any detail, but they may be expected to be similar to that indicated above for polyisoprene.

b. Accelerated reaction. The vulcanization systems employed industrially for the unsaturated elastomers include, in addition to sulphur, an organic accelerator (e.g., 2-mercaptobenzothiazole, or its disulphide, zinc salt or sulphenamide derivative), an activator (a metallic oxide such as zinc oxide), and a fatty acid (e.g., stearic or lauric acid) or its zinc salt. Elemental sulphur can also be replaced in these systems by certain organosulphur compounds, such as tetramethylthiuram disulphide and dithiobisamines.

The chemistry of the accelerated sulphur vulcanization is very complex and little understood, and is really beyond the scope of this chapter. Accordingly, we will note only that the high crosslinking efficiency at long cure time of ~1·6 sulphur atoms consumed per chemical crosslink in natural rubber, noted earlier (for an efficient mercaptobenzothiazole-accelerated system[164]), effectively rules out cyclic monosulphide structures while indicating a preponderance of mono- and disulphide crosslinks.

3. Vulcanization of neoprene[215a]

Polychloroprene can be crosslinked by dicumyl peroxide, but the efficiency of the reaction is low, of the order of 0·5 crosslink/peroxide molecule. Sulphur can also vulcanize this polymer but very slowly, and the usual rubber accelerators are generally not effective.

The industrial vulcanization of polychloroprene is based on the highly reactive tertiary allylic chlorine atom located in the small amount ($\sim 1\cdot5\%$) of 1,2 units present in the polymer. The preferred vulcanizing agents are zinc oxide and magnesium oxide which serve to abstract the chlorine (equation 41).

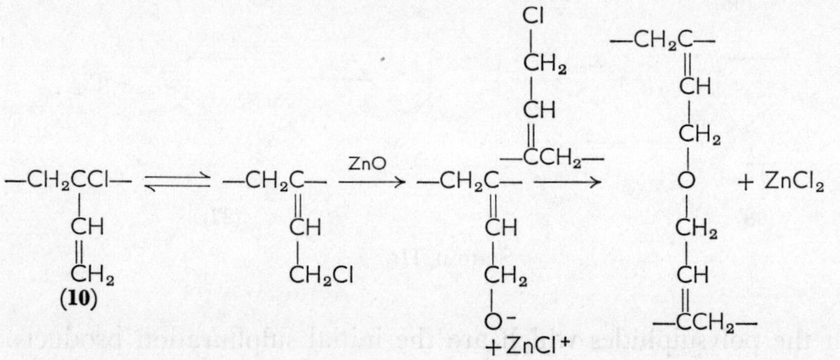

$$-CH_2CCl- \rightleftharpoons -CH_2C- \xrightarrow{ZnO} -CH_2C- \longrightarrow \cdots + ZnCl_2 \quad (41)$$

Alternatively, diamines may be used to crosslink neoprene through a bisalkylation reaction involving chlorine removal in the presence of magnesium oxide (equation 42).

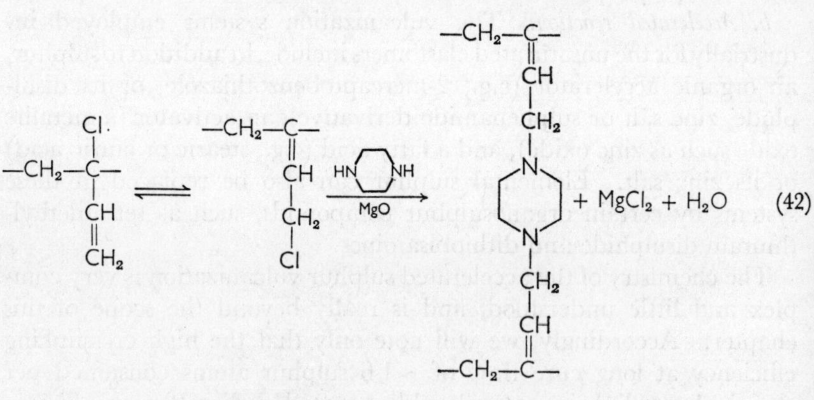

$$\cdots \rightleftharpoons \cdots \xrightarrow{MgO} \cdots + MgCl_2 + H_2O \quad (42)$$

4. Radiation vulcanization

Isoprene and butadiene homopolymers, and many of their copolymers, as well as various other polymers, can be crosslinked (and hence vulcanized) by exposure to high energy radiation. Still other polymers, on the other hand, undergo predominantly chain scission on irradiation. An enormous literature exists on the radiation chemistry of polymers, and much of this has been surveyed in several texts [4,5,216] and review articles. Gehman and Gregson [217], in a general review of the effects of ionizing radiation on elastomers, have given special attention to the chemical as well as technological aspects of radiation vulcanization; and Turner [218] has recently provided a comprehensive discussion of the radiation chemistry of natural rubber.

The mechanism(s) of radiation-induced crosslinking in polyisoprene and polybutadiene, and the exact nature of the resulting crosslinks, are not known with certainty. The general presumption [218] has been, however, that the crosslinks are formed principally through the mutual combination of allylic polymer radicals generated in the system (equation 43, where

$$R = -CH_2-\underset{|}{\overset{\overset{\displaystyle CH_3}{|}}{C}}=CH-CH- \quad \text{or} \quad -CH_2-CH=CH-\overset{|}{CH}-).$$

An alternative or competing, free radical mechanism, which may be even more important than equation (43), is indicated in equation

$$\begin{aligned} RH &\longrightarrow R\cdot + H\cdot \\ H\cdot + RH &\longrightarrow R\cdot + H_2 \\ 2\,R\cdot &\longrightarrow R{-}R \end{aligned} \quad (43)$$

(44). Since the crosslinked products obtained in the radiolysis of

$$\begin{aligned} RH &\longrightarrow R\cdot + H\cdot \\ H\cdot + RH &\longrightarrow RH_2\cdot \\ RH_2\cdot + R\cdot &\longrightarrow RH_2{-}R \end{aligned} \quad (44)$$

the model compound, squalene, even at very low doses, consisted not only of dimers but of higher molecular weight material as well, it can be expected that the alkyl radicals, $RH_2\cdot$, can undergo also a polymerization crosslinking reaction (equation 45). Turner [219] has found that free radical scavengers can suppress the radiation-induced crosslinking in polyisoprene by no more than $\sim 40\%$, so that in addition to reactions (43)–(45), some non-radical or

$$RH_2\cdot + RH \longrightarrow RH_2{-}RH\cdot \text{ etc.} \quad (45)$$

'molecular' process (e.g., ion-molecule or Stern–Volmer reaction) (equation 46) must also contribute to the crosslinking. In accord

$$2\,RH \longrightarrow R{-}R + H_2 \quad (46)$$

with this view was the complementary finding[219] that the hydrogen yield was reduced only 10% by the use of free radical scavengers.

The radiation chemical production of crosslinks in emulsion and cis-1,4-polybutadienes occurs with a higher yield ($G(X) \sim 3\cdot6\text{--}4\cdot0$ crosslinks/100 e.v. absorbed)[220-222] than in cis- or trans-polyisoprene ($G(X) \sim 0\cdot9\text{--}1\cdot0$ and $1\cdot3$, respectively)[218,223]. This difference in crosslinking efficiency may be related to the difference noted earlier for the crosslinking efficiency in peroxide vulcanization of the two different diene polymers.

Concomitant with the radiation-induced crosslinking and hydrogen evolution in polyisoprene and polybutadiene ($G(H_2) = 0\cdot48\text{--}0\cdot64$, and $0\cdot45$, respectively)[144,177,218,219], cis–trans isomerization and loss of double bonds also take place and with even higher yields[111,144] (G-values of ~ 10 and $7\text{--}10$, respectively, for polyisoprene; and $4\text{--}7$, and $8\text{--}14$ for polybutadiene). While some loss of unsaturation may be due to the crosslinking reaction (44), this reaction can account for no more than 5–6% of all the double bonds consumed, the overwhelming majority of them presumably being removed by cyclization (equation 47, using polybutadiene as example). The asterisks denote either

$$\begin{array}{c}\sim\!\!\!-\text{H}_2\text{C}\diagdown\\\text{CH*}\quad\text{CH}_2\\\text{H}_2\text{C}\quad\text{CH}\\|\quad\quad\|\\\text{H}_2\text{C}\diagdown\text{CH}\\\text{CH}_2\end{array}\quad\begin{array}{c}\sim\!\!\!-\text{H}_2\text{C}\diagdown\\\text{CH}\quad\text{CH}_2\\\text{H}_2\text{C}\quad\text{CH}\\|\quad\quad|\\\text{H}_2\text{C}\diagdown\text{*CH}\\\text{CH}_2\end{array}\longrightarrow\text{various cyclic structures}\quad(47)$$

radical or carbonium ion sites. In the latter case, the ion is presumed to arise and react further as in equation (48). In addition, a small amount of chain scission ($G \lesssim 0\cdot1$) also accompanies the crosslinking.

$$-\text{CH}_2-\text{CH}=\text{CH}-\text{CH}_2-\longrightarrow -\text{CH}_2-\overset{+}{\text{CH}}-\overset{\cdot}{\text{CH}}-\text{CH}_2- + e^-$$

$$-\text{CH}_2-\overset{+}{\text{CH}}-\overset{\cdot}{\text{CH}}-\text{CH}_2- + -\text{CH}_2-\text{CH}=\text{CH}-\text{CH}_2- \longrightarrow$$

$$-\text{CH}_2-\overset{+}{\text{CH}}-\text{CH}_2-\text{CH}_2- + -\text{CH}_2-\text{CH}=\text{CH}-\overset{\cdot}{\text{CH}}- \quad(48)$$

F. Oxidation

Oxidation of polymers—both saturated and unsaturated—is another category of polymer reactions which has been extensively investigated and also well reviewed. Particular attention is called to several recent review articles which are concerned mainly with the oxidation of polyisoprene but also, to some extent, with that of the butadiene-containing polymers[224-229].

I. 1,4-Polyisoprene

As in the case of vulcanization, much of our current understanding of the mechanism of oxidation of natural rubber, as well as of the accompanying microstructural changes in the polyisoprene macromolecule, is the result of extensive work on simple olefins and on appropriate model compounds of polyisoprene. However, unlike the situation with vulcanization involving sulphur (Section III.E.2), the originally proposed free radical mechanism of oxidation of hydrocarbons[230,231] is still accepted. The essential features of that mechanism, insofar as the diene polymers and copolymers are concerned, are that hydroperoxides are the primary products, the decomposition of which in turn is responsible for the reaction being autocatalytic, and that the rate of oxidation is largely insensitive to oxygen pressure above a certain moderate pressure but directly dependent on the lability of the α-methylenic C—H bond.

The oxidation mechanism comprises the following reaction steps, where RH is an olefin with an α-methylenic hydrogen atom:

Initiation	Production of R· or RO$_2$· radicals	(49a)
Propagation	R· + O$_2$ $\longrightarrow$ RO$_2$·	(49b)
	RO$_2$· + RH $\longrightarrow$ RO$_2$H + R·	(49c)
Termination	2 R· $\longrightarrow$ ⎫	(49d)
	R· + RO$_2$· $\longrightarrow$ ⎬ non-chain carriers	(49e)
	2 RO$_2$· $\longrightarrow$ ⎭	(49f)

At a sufficiently high oxygen pressure, step (49b) is much faster than (49c), the latter becoming rate-controlling, and the overall rate of oxidation is then independent of oxygen pressure. Under these conditions, the chain termination occurs principally, if not solely, by reaction (49f). The source of the radicals in the initiation step, assuming thermal oxidation, is the decomposition of the hydroperoxide which may proceed by a unimolecular or bimolecular path to yield alkoxy, peroxy or hydroxyl radicals (reaction 49g or 49h). The uni-

$$\text{ROOH} \longrightarrow \text{RO·} + \text{·OH} \quad (49g)$$
$$2 \text{ROOH} \longrightarrow \text{RO·} + \text{RO}_2\text{·} + \text{H}_2\text{O} \quad (49h)$$

molecular decomposition predominates in the initial stages of oxidation, while the bimolecular process becomes important as the hydroperoxides accumulate. Eventually the hydroperoxides reach a steady state concentration at which the autocatalytic stage of the oxidation is completed and the reaction attains a maximum rate. The kinetics of the autoxidation reaction have been reviewed by Tobolsky and coworkers[228].

In addition to hydrogen abstraction the peroxy radical can add to a double bond. This is especially manifested in 1,5 diene compounds, such as dihydromyrcene, squalene, gutta-percha and natural rubber. Diperoxide and hydroperoxide formation can thus proceed in the same molecule as shown in Scheme III[232]. Structure **100**, which has been isolated and characterized, conforms to the experimental finding that somewhat less than half of the combined oxygen in oxidized dihydromyrcene, squalene and digeranyl is in the form of hydroperoxide groups[233].

Due to the fact that chain scission attends the oxidation of polyisoprene, the absorption of a small amount of oxygen by the macromolecule has a large effect on its physical properties. Thus, natural rubber loses virtually all its useful elastic properties when only 1% oxygen by weight has been absorbed[234]. Absorption of as little as 0.05% oxygen is sufficient to reduce the molecular weight of natural rubber from around one million down to half a million. Oxidative chain scission is unimportant in a small molecule, but is magnified into importance in a macromolecule by virtue of its disproportionate effect on the molecular weight. While crosslinking can and does accompany chain scission in the oxidation of certain polymers, this is negligible in the case of polyisoprene.

The number of molecules of oxygen absorbed for each cut in the molecular chain, which is an inverse measure of the scission efficiency[226], was cited as ranging from 40 at 25° to 16 at 110° in the oxidation of *Hevea* latex[235,236], from 5 to 20 at 140° in the oxidation of gum rubber[237], and from around 28 at 60° to 18 at 100° in the oxidation of gutta-percha as a film or in solution[238]. The values, in the case of a particular sulphur vulcanizate of rubber[239], were 30 at 75°, 5 at 100° and 1 at 130°, while in the case of a peroxide vulcanizate[240,241], were around 20 at 100–120°. On the other hand, Bevilacqua[242] has recently claimed that both scission efficiency and mechanism are not materially affected by the extent or nature of vulcanization, or by the presence or absence of vulcanization residues, antioxidants or fillers, and he suggested that, to a close approximation, the scission efficiency is determined only by the temperature. However, Parks and Lorenz[243] showed that incorporation of sulphur in a peroxide vulcanizate of natural rubber significantly lowered the scission efficiency in oxidation at 100°. In this last case it would appear[242] that some

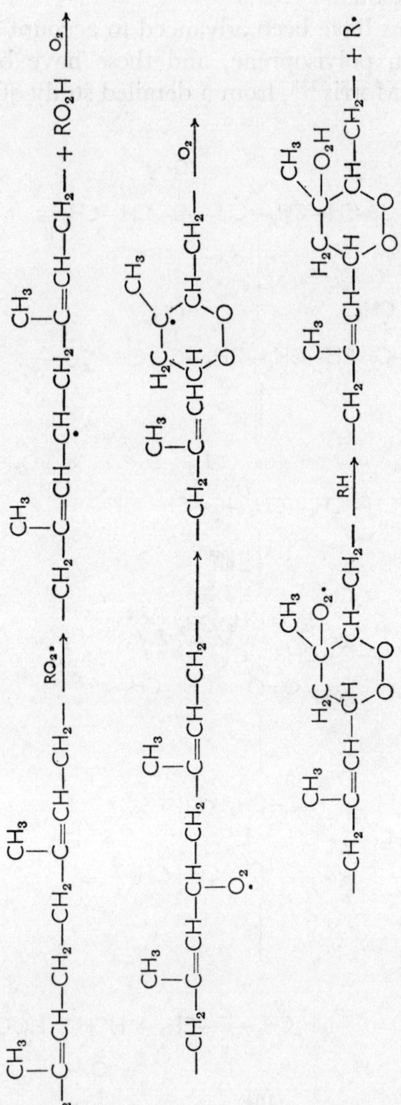

SCHEME III.

sulphur-promoted crosslinking occurred during oxidation to offset some of the chain scission.

Several mechanisms have been advanced to account for the oxidative chain scission in polyisoprene, and these have been discussed recently[226,228,244]. Morris[238], from a detailed study of the oxidation

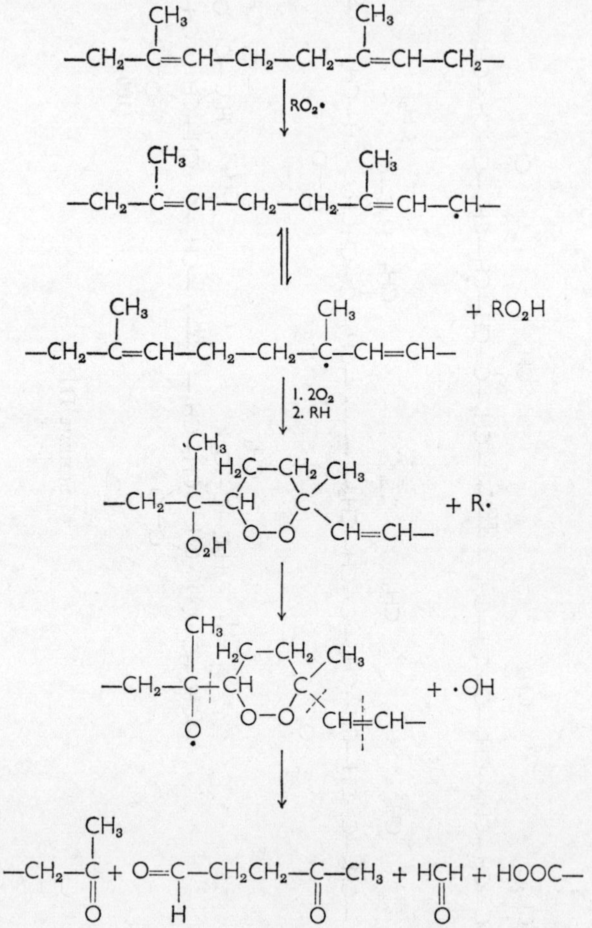

$$-CH_2-\underset{\underset{CH_3}{|}}{C}=CH-CH_2-CH_2-\underset{\underset{CH_3}{|}}{C}=CH-CH_2-$$

$$\downarrow RO_2\cdot$$

$$-CH_2-\underset{\underset{CH_3}{|}}{C}=CH-CH_2-CH_2-\underset{\underset{CH_3}{|}}{C}=CH-\overset{\cdot}{C}H-$$

$$\Updownarrow$$

$$-CH_2-\underset{\underset{CH_3}{|}}{\overset{\cdot}{C}}-CH=CH-CH_2-CH_2-\underset{\underset{CH_3}{|}}{\overset{\cdot}{C}}-CH=CH- \quad + RO_2H$$

$$\downarrow \begin{array}{l}1.\ 2O_2\\ 2.\ RH\end{array}$$

(intermediate peroxide structure) + R·

↓

(intermediate structure) + ·OH

↓

$$-CH_2-\underset{\underset{O}{\parallel}}{\overset{\underset{CH_3}{|}}{C}} + O=\underset{\underset{H}{|}}{C}-CH_2CH_2-\underset{\underset{O}{\parallel}}{C}-CH_3 + H\underset{\underset{O}{\parallel}}{C}H + HOOC-$$

(101)

↓ ↓

CH₃COOH + 3CO₂   HCOOH

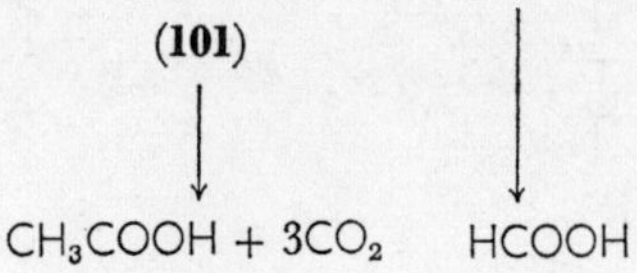

SCHEME IV.

9. Polymers Containing C=C Bonds

of gutta-percha, concluded that the scission process must be a minor side reaction, which is kinetically identical to one of the normal propagation steps in the autoxidation chain cycle, but no mechanism was given. Bevilacqua [236,237,241,245], on the basis of a quantitative study of the principal volatile products in rubber oxidation, proposed a chain scission mechanism which is a variation on the mechanism suggested by Bolland and Hughes [232] for the formation of structure **100**, and readily accounts (Scheme IV) for the actual formation of levulinaldehyde (**101**), formic acid, acetic acid and carbon dioxide, as well as the carbonyl endgroups on polymer residues revealed by infrared spectroscopy. This mechanism has been supported by Mayo [246] who pointed out the ease with which β-peroxyalkoxy radicals (such as the precursor to **101**) can disproportionate. According to Bevilacqua [247], the scission reaction is an alternative to a propagation step.

Tobolsky and Mercurio [248], on the other hand, concluded from a study of the free radical catalysed oxidation of natural rubber in dilute benzene solution that scission occurs during the chain termination reaction. In their work using various initiators with known decomposition rates, they found that each initiating radical produced approximately one scission. They proposed that peroxy radicals interact to give alkoxy radicals and oxygen (equation 49f′), followed

$$2\,RO_2\cdot \longrightarrow 2\,RO\cdot + O_2 \qquad (49f')$$

by cleavage of the resulting β-peroxyalkoxy radical (**102**) to produce chain scission (equation 50); the acetonyl radicals then dimerize to form 2,5-hexanedione and thus terminate the oxidation chain. It

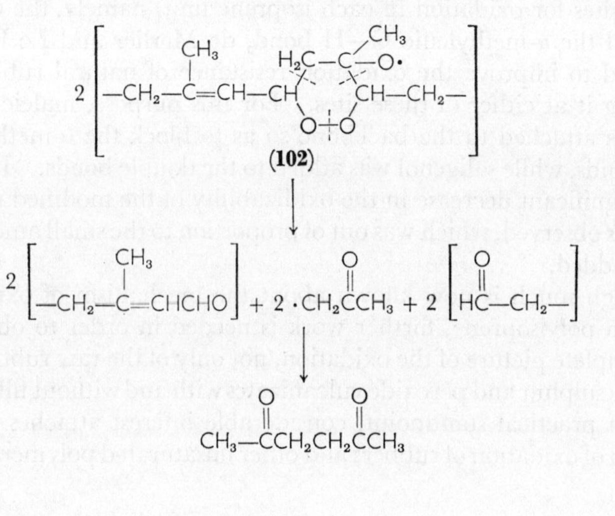

$$2\left[\begin{array}{c}CH_3\\|\\-CH_2-C=CH-CH\\\diagdown\,O\!-\!O\end{array}\begin{array}{c}CH_3\\H_2C-C-O\cdot\\\diagup\\CH-CH_2-\end{array}\right]$$

(**102**)

$$2\left[\begin{array}{c}CH_3\\|\\-CH_2-C=CHCHO\end{array}\right] + 2\cdot CH_2\overset{O}{\overset{\|}{C}}CH_3 + 2\left[\begin{array}{c}O\\\|\\HC-CH_2-\end{array}\right] \qquad (50)$$

$$\downarrow$$

$$CH_3-\overset{O}{\overset{\|}{C}}CH_2CH_2\overset{O}{\overset{\|}{C}}CH_3$$

should be noted that structure **102** is just the alkoxy radical counterpart of the peroxy radical, precursor to structure **100** proposed by Bolland and Hughes[232].

Early attempts to confirm the formation of 2,5-hexanedione were unsuccessful[249], but subsequently it was shown[250] that this compound was indeed formed in the oxidation of *cis*-polyisoprene, but at higher temperatures (150–180°) than that employed (120°) in the previous experiments. This observation demonstrated that there are two pathways to chain scission, one involving the β-peroxyalkoxy radical shown in Scheme IV and one involving the isomeric radical **102**. Moreover, since the yield of the hexanedione was about 10% of that of levulinaldehyde, the relative involvement of Schemes III and IV in forming, respectively, the two alternative β-peroxy hydroperoxide structures is probably in the ratio of 1:10.

Bell[251] studied the oxidative scission of natural rubber and gutta-percha in solution. He found that for a 100-fold increase in initiator concentration, the value of $1/\epsilon$ (ϵ = number O_2 molecules absorbed per scission) changed by only 50% whereas the rate of oxidation increased by a factor of 25. It was assumed that ϵ would be independent of chain length λ if scission is kinetically equivalent to the propagation step, and inversely proportional to λ if scission is associated with initiation or termination. The above results indicate that most scission occurs during propagation of the oxidation chain reaction, but some must occur during termination; in fact, it was concluded that scission occurs nearly every time an oxidation chain is terminated.

On the premise that the polyisoprene hydrocarbon presents two reactive sites for oxidation in each isoprene unit, namely, the double bond and the α-methylenic C—H bond, de Merlier and Le Bras[252] attempted to improve the oxidation resistance of natural rubber by modifying it at either of these sites. For this purpose, maleic anhydride was attached to the backbone so as to block the α-methylenic C—H bonds, while saligenol was added to the double bonds. In both cases a significant decrease in the oxidizability of the modified rubber at 80° was observed, which was out of proportion to the small amount of reagent added.

Although much is now known about the mechanism of oxidative scission in polyisoprene, further work is needed in order to obtain a more complete picture of the oxidation, not only of the raw rubber but also of the sulphur and peroxide vulcanizates with and without fillers[226].

From a practical standpoint, considerable interest attaches to the inhibition of oxidation of rubbers and other unsaturated polymers, and, on the basis of the mechanism given above (equation 49), this can be brought about in two ways: by interfering either with the initiation step or the propagation step. In the latter case, the most potent antioxidants (HA) are various mono- and polyhydric phenols and secondary aromatic amines, e.g., hydroquinone and N,N'-diphenyl-p-phenylene diamine, which operate by transferring hydrogen to the peroxy radicals, thereby converting them to hydroperoxides while being transformed themselves into stable antioxidant residues (A·) (equation 51). In the case of hydroquinone, for example, each anti-

$$HA + RO_2\cdot \longrightarrow RO_2H + A\cdot \tag{51}$$

oxidant molecule can interact with and thus terminate two chain-carrying peroxy radicals (equation 52). The other type of antioxidant serves to repress the initiation step by essentially interfering

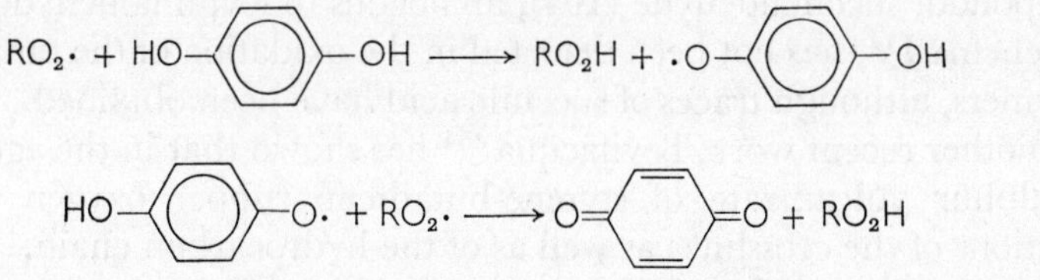

$$RO_2\cdot + HO{-}\langle\bigcirc\rangle{-}OH \longrightarrow RO_2H + \cdot O{-}\langle\bigcirc\rangle{-}OH$$

$$HO{-}\langle\bigcirc\rangle{-}O\cdot + RO_2\cdot \longrightarrow O{=}\langle\bigcirc\rangle{=}O + RO_2H \tag{52}$$

with the decomposition of hydroperoxides. Thus, for example, the catalytic action of certain metallic impurities on the hydroperoxide decomposition can be suppressed by the incorporation of specific chelating agents. Alternatively, various sulphur compounds may be employed to deactivate the hydroperoxides. A more complete account of antioxidants and antioxidant action can be found elsewhere [225].

2. Butadiene-containing polymers

Relatively little work has been carried out on the oxidation of butadiene polymers and copolymers, but the mechanisms considered for polyisoprene are presumed to carry over largely unchanged to these other unsaturated polymers. The butadiene polymers tend to be less readily oxidized than the polyisoprenes because of the relatively lower lability of their α-methylenic C—H bonds. This in turn enhances the chances for crosslinking, by making peroxy radical addition to a double bond (equation 53) in polybutadiene more competitive with hydrogen abstraction, (equation 49c) compared with the case of polyisoprene. This effect appears to be even more important in the case of pendant vinyl groups. Significantly, Bevilacqua [253] has recently noted that in

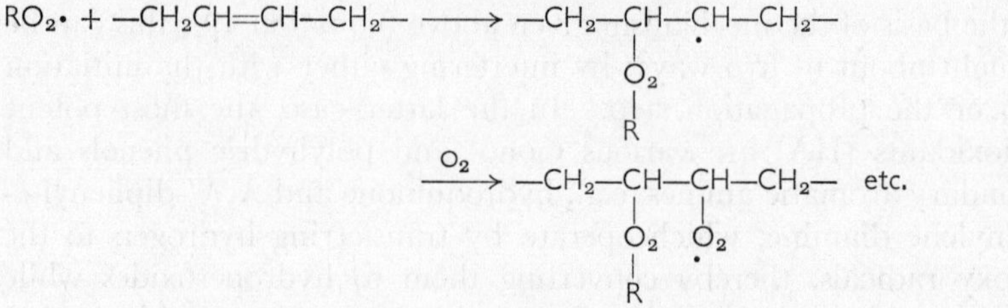

$$RO_2\cdot\ +\ -CH_2CH{=}CH-CH_2-\ \longrightarrow\ -CH_2-\underset{\underset{R}{\overset{\overset{O_2}{|}}{|}}}{\overset{\cdot}{C}H}-CH-CH_2-$$

$$\xrightarrow{O_2}\ -CH_2-\underset{\underset{R}{\overset{\overset{O_2}{|}}{|}}}{CH}-\underset{\overset{\cdot}{O_2}}{CH}-CH_2-\ \text{etc.} \qquad (53)$$

the oxidation of butadiene polymers and copolymers chain scission and crosslinking occur at comparable rates.

Formic acid is a major product from the oxidation of *cis*-1,4-polybutadiene as well as of polybutadienes containing various proportions of 1,2 units[253]; this product is assumed to arise through scission of either the main chain or the side chain. However, the four-carbon compound, succinaldehyde (**104**), analogous to levulinaldehyde (**101**) in Scheme IV, has not been detected in the oxidation of the butadiene polymers, although traces of succinic acid have been obtained.

In other recent work, Bevilacqua[254] has shown that in the ageing of a sulphur vulcanizate of styrene-butadiene rubber oxygen affects reactions of the crosslinks as well as of the hydrocarbon chain.

Lee, Stacy and Engel[255] have studied the effect of a number of metallic stearates and stearic acid on the autoxidation of *cis*-1,4-polybutadiene and several butadiene copolymers. Their results support the generally accepted theories concerning the metal-catalysed oxidation through a hydroperoxide intermediate.

Finally, it may be noted that infrared spectroscopy has been used recently to study the oxidation, under various conditions, of *cis*-polybutadiene[148,256,257], sodium polybutadiene, which had a high ($\sim 55\%$) 1,2 content[258], butadiene-acrylonitrile copolymer[259] and of polyalkynes[260].

G. Ozonization

Polymers containing C=C bonds are subject to attack and degradation by ozone, a reagent which has been used for many years to cleave olefinic double bonds. The early recognition by Pickles[261] that rubber has a regular 1,4-polyisoprene structure was based on the finding that on ozone cleavage it yielded levulinaldehyde (**101**) and levulinic acid (**103**) as the only important products. An irregular structure would have yielded also some succinaldehyde (**104**) and acetonyl acetone (**105**), as is the case in the ozonolysis of non-stereo-

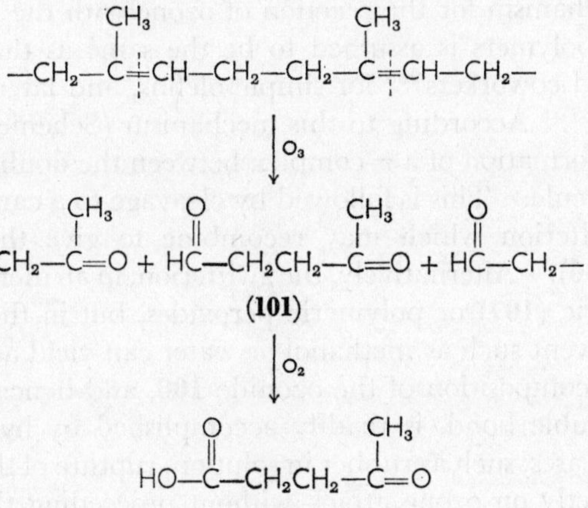

9. Polymers Containing C=C Bonds

$$-CH_2-\underset{CH_3}{\overset{|}{C}}\!\!=\!\!CH-CH_2-CH_2-\underset{CH_3}{\overset{|}{C}}\!\!=\!\!CH-CH_2-$$

$$\downarrow O_3$$

$$-CH_2-\underset{CH_3}{\overset{|}{C}}\!\!=\!\!O + H\overset{O}{\overset{\|}{C}}-CH_2CH_2-\underset{CH_3}{\overset{|}{C}}\!\!=\!\!O + H\overset{O}{\overset{\|}{C}}-CH_2-$$
(101)

$$\downarrow O_2$$

$$HO-\overset{O}{\overset{\|}{C}}-CH_2CH_2-\underset{CH_3}{\overset{|}{C}}\!\!=\!\!O$$
(103)

regular synthetic polyisoprenes. The use of ozone in microstructural analysis of unsaturated synthetic elastomers (as in the determination, for example, of 1,2- (or 3,4-) and 1,4-addition units in various butadiene and isoprene polymers and copolymers) has been reviewed

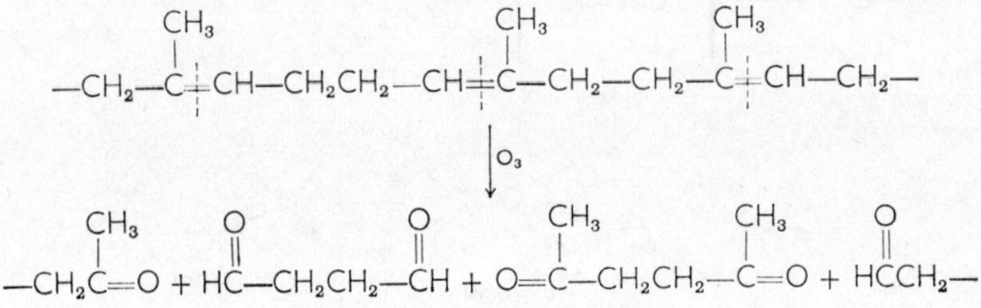

$$-CH_2-\underset{CH_3}{\overset{|}{C}}\!\!=\!\!CH-CH_2CH_2-CH\!\!=\!\!\underset{CH_3}{\overset{|}{C}}-CH_2-CH_2-\underset{CH_3}{\overset{|}{C}}\!\!=\!\!CH-CH_2-$$

$$\downarrow O_3$$

$$-CH_2\underset{CH_3}{\overset{|}{C}}\!\!=\!\!O + H\overset{O}{\overset{\|}{C}}-CH_2CH_2-\overset{O}{\overset{\|}{C}}H + O\!\!=\!\!\underset{CH_3}{\overset{|}{C}}-CH_2CH_2-\underset{CH_3}{\overset{|}{C}}\!\!=\!\!O + H\overset{O}{\overset{\|}{C}}CH_2-$$
$$\qquad\qquad\qquad\quad\text{(104)}\qquad\qquad\qquad\text{(105)}$$

recently[262]. Attention has been called to possible difficulties in interpretation when dealing with minor ozonolysis products[26].

Of further interest is the severe cracking encountered in (unprotected) rubber under stress when exposed to the small amounts of ozone generally present in the atmosphere. Normally, ozone has no visible effect on unstretched rubber, but definite physical effects can be observed in very thin films ($\sim$ 200 Å thick). Kendall and Mann[263] have, in fact, studied the formation of rubber ozonide by means of infrared spectral measurements on such films. Much research has been carried out in an effort to elucidate the mechanism of this phenomenon and to develop antiozonants for the protection of rubber articles. These aspects have been very well reviewed recently[262,264].

The mechanism for the reaction of ozone with the C=C bonds in the diene polymers is assumed to be the same as that proposed by Criegee and coworkers[265] for simple olefins, and later elaborated by Bailey[266,267]. According to this mechanism (Scheme V), the initial step is the formation of a π-complex between the double bond and an ozone molecule. This is followed by cleavage to a carbonyl structure and a zwitterion which may recombine to give the conventional ozonide (**106**). Alternatively, the zwitterion, in an inert medium, may form dimeric (**107**) or polymeric peroxides, but in the presence of a reactive solvent such as methanol or water can yield a hydroperoxide (**108**). Decomposition of the ozonide **106**, and hence scission of the original double bond, is readily accomplished by hydrolysis. In a number of cases, such as rubber in solution, rupture of the double bond occurs directly on ozone attack without proceeding through the decomposition of the ozonide. As evidence for this, carbonyl groups have been detected spectroscopically immediately on contact of rubber with ozone[268].

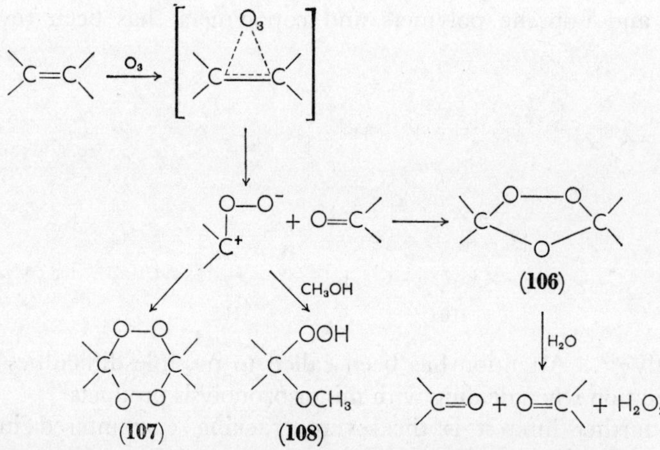

SCHEME V.

H. Degradation

In the broadest sense, degradation of a high polymer is any process which results in a serious deterioration of its useful physical properties, whether the chemical reactions involved lead to a net decrease in molecular weight (as in chain scission) or a net increase (as in crosslinking). In this sense, polymers containing C=C bonds, in common with polymers generally, degrade under the influence of heat, oxygen, ozone, ultraviolet or ionizing radiation, mechanical action, ultrasonic energy, and various chemical reagents, acting alone or in combination[269]. As discussed previously, the deteriorative effects of oxygen and ozone on the unsaturated polymers are principally those involving chain cleavage or 'degradation' in the more limited sense of a drop in molecular weight or a decomposition of the polymer. On the other hand, exposure of the unsaturated polymers to the 'degradative' effects of high energy (ionizing) radiation or ultraviolet light in vacuum produces, among other things, crosslinked material, although, to be sure, significant chain rupture is also observed in the ultraviolet irradiation case. Since photo- and radiation chemical effects in diene polymers have been touched on earlier, in connection with isomerization and vulcanization (and reviewed extensively[1,2,4,5,216-218]), and since oxidation and ozonization of these polymers have been treated in the preceding two sections, this section will be restricted to a brief discussion of two other important types of degradation, viz., thermal and mechanochemical degradation.

I. Thermal degradation

A formidable literature exists on the thermal degradation of all kinds of polymers in book[7,106,270] and review articles[269,271,272]. Here only the main features of the pyrolysis of the butadiene and isoprene polymers will be treated.

a. 1,4-Polyisoprene. The destructive distillation of natural rubber and of synthetic polyisoprenes under various temperature and pressure conditions has been studied for many years. Midgley and Henne[273] in 1929 found that isoprene and its dimer, dipentene (**109**), constituted the only major products from the pyrolysis of *Hevea*. These products were formed in approximately 1:2 ratio by weight, and were accompanied by some 15 minor products none of which was more than 1% of the isoprene yield. More recently, Straus and Madorsky[274] made a detailed comparative study of the pyrolysis of *Hevea*, gutta-percha and a synthetic polyisoprene, and likewise found dipentene and isoprene as the major volatile products. The yield of dipentene was about 4–5 times as large as that of isoprene in the case of the stereoregular polyisoprenes, and only about 2·5 times as large in the case of the structurally irregular synthetic polyisoprene.

Formation of monomer in the pyrolysis of isoprene polymers was assumed to involve mainly (though not exclusively) successive scission of C—C bonds connecting isoprene units in the chain (equation 54a). Formation of dipentene (**109**), on the other hand, was assumed to in-

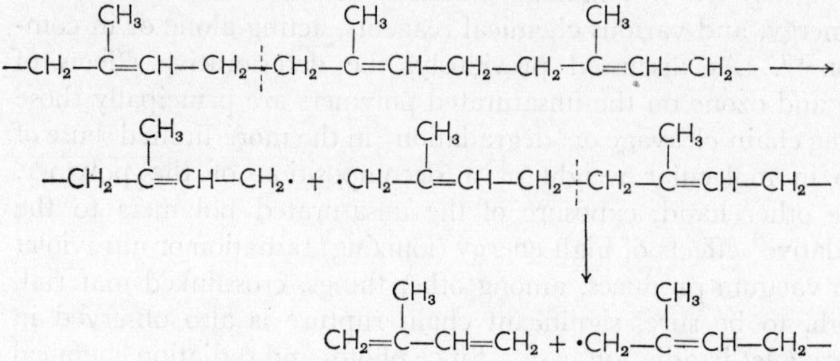

$$-CH_2-\underset{\underset{CH_3}{|}}{C}=CH-CH_2\!\!\mid\!\!CH_2-\underset{\underset{CH_3}{|}}{C}=CH-CH_2-CH_2-\underset{\underset{CH_3}{|}}{C}=CH-CH_2- \longrightarrow$$

$$-CH_2-\underset{\underset{CH_3}{|}}{C}=CH-CH_2\cdot \ + \ \cdot CH_2-\underset{\underset{CH_3}{|}}{C}=CH-CH_2\!\!\mid\!\!CH_2-\underset{\underset{CH_3}{|}}{C}=CH-CH_2- \quad (54a)$$

$$\downarrow$$

$$CH_2=\underset{\underset{CH_3}{|}}{C}-CH=CH_2 \ + \ \cdot CH_2-\underset{\underset{CH_3}{|}}{C}=CH-CH_2-$$

volve either unzipping or dimerization of monomer (equation 54b). Evidently, the higher yield of dipentene obtained from the *cis*- and *trans*-1,4-polyisoprenes compared to that from the synthetic polyisoprene was a consequence of the greater structural regularity of the

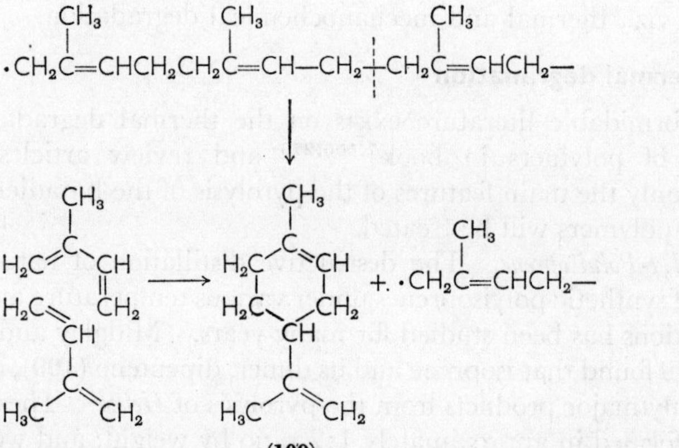

$$\cdot CH_2\underset{\underset{CH_3}{|}}{C}=CHCH_2CH_2\underset{\underset{CH_3}{|}}{C}=CH-CH_2\!\!\mid\!\!CH_2\underset{\underset{CH_3}{|}}{C}=CHCH_2-$$

$$\downarrow \qquad (54b)$$

(**109**)

former polymers which favours the appropriate unzipping of pairs of isoprene units.

b. 1,4-Polybutadiene. Straus and Madorsky[275] have also pyrolysed polybutadiene, obtaining mainly monomer and dimer (presumably vinyl cyclohexene, **110**) along with an assortment of minor products indicative of some accompanying *random* chain scission. As with polyisoprene, the formation of monomer in the pyrolysis of polybutadiene can be considered to involve scission of the C—C bonds connecting successive butadiene units, while the formation of dimer involves unzipping of the chain and/or dimerization of monomer (equation 55). Since the yield of butadiene monomer was very small

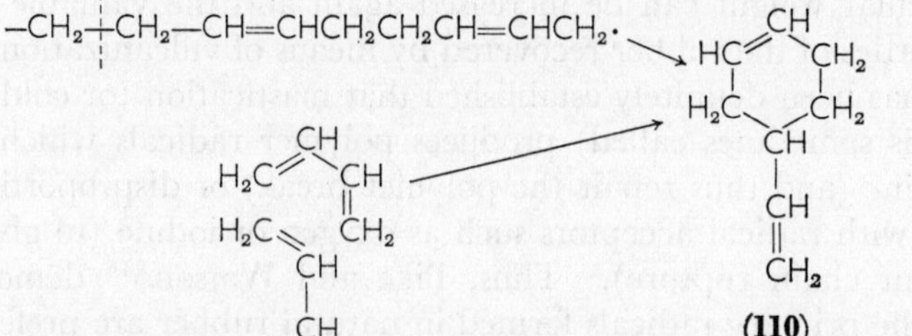

$$-CH_2 \!\!+\!\! CH_2-CH=CHCH_2CH_2CH=CHCH_2\cdot \longrightarrow \text{(110)} \tag{55}$$

relative to the formation of low molecular weight polymer fragments (average degree of polymerization ~ 14, and comprising $\sim 80\text{-}90\%$ of the pyrolysis products), it was assumed that many of the scissions of the C—C bonds in the chains involve hydrogen transfer (equation

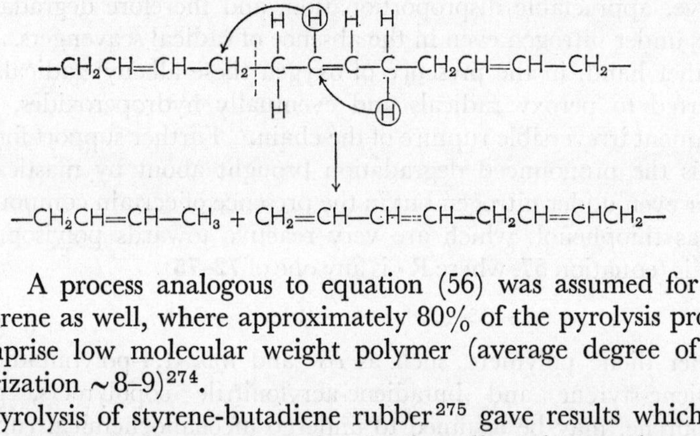

$$-CH_2CH=CH-CH_2 \!\!+\!\! \overset{H}{\underset{H}{C}} \!-\! \overset{H}{C} \!=\! \overset{H}{C} \!-\! \overset{H}{\underset{H}{C}} \!-\! CH_2CH=CH-CH_2-$$

$$\downarrow$$

$$-CH_2CH=CH-CH_3 + CH_2=CH-CH=CH-CH_2CH=CHCH_2- \tag{56}$$

56). A process analogous to equation (56) was assumed for polyisoprene as well, where approximately 80% of the pyrolysis products comprise low molecular weight polymer (average degree of polymerization $\sim 8\text{-}9$)[274].

Pyrolysis of styrene-butadiene rubber[275] gave results which were remarkably similar to those for the butadiene homopolymer with regard to the yield of monomer, dimer and low molecular weight polymer.

2. Mechanochemical degradation

A very useful kind of degradation is that of mechanical working or mastication of rubber, which is at the foundation of rubber technology. Mechanical energy is utilized to rupture the macromolecular chains, thereby reducing the molecular weight of the raw rubber from a level at which it is difficult to process, to a level at which the rubber can be readily mixed with appropriate compounding ingredients in conventional rubber manufacturing equipment. After processing, the molecular weight can be increased again and the valuable physical properties of the rubber recovered by means of vulcanization.

It has been definitely established that mastication (or cold milling, as it is sometimes called) produces polymer radicals which can recombine (and thus repair the polymer break) or disproportionate or react with radical acceptors such as oxygen or iodine (to give a permanent chain rupture). Thus, Pike and Watson[276] demonstrated that the primary radicals formed in natural rubber are preferentially those arising from rupture of the C—C bonds connecting successive isoprene units, viz., **72–75** (Section III.C.2.a). In the absence of oxygen or a radical scavenger, these radicals evidently recombine with little alteration in overall polymer size; they are apparently not reactive enough, by virtue of their allylic resonance energy, to undergo significant disproportionation. In the mastication of saturated polymers, where the resulting primary radicals are considerably more reactive, appreciable disproportionation, and therefore degradation, occurs under nitrogen even in the absence of radical scavengers. On the other hand, in the presence of oxygen these alkenyl radicals are converted to peroxy radicals and eventually hydroperoxides, with consequent irreversible rupture of the chain. Further support for this view is the pronounced degradation brought about by masticating rubber even under nitrogen but in the presence of certain compounds, such as thiophenol, which are very reactive towards polyisoprenyl radicals (equation 57, where R· is any one of **72–75**).

$$R\cdot + HS—C_6H_5 \longrightarrow RH + \cdot S—C_6H_5 \tag{57}$$

Other diene polymers, such as *cis*- and *trans*-1,4-polybutadienes, butadiene-styrene and butadiene-acrylonitrile copolymers, polychloroprene, may be assumed to undergo mechanochemical rupture in the manner depicted for polyisoprene.

Mechanochemical degradation can obviously be employed for the preparation of block copolymers (**40**) and, in certain cases, graft copolymers (**39**) (equations 58a and 58b). A branching reaction such as equation (58b) is observed in the mastication of natural rubber in an inert atmosphere in the absence of radical scavengers. This reaction could involve, for example, addition of a polyisoprenyl radical to a double bond in another polyisoprene chain and, depending on conditions, lead to gelled polymers. Alternatively, the reaction could proceed through a disproportionation, followed by radical recombination (equation 58c, where R· is one of **72–75**)[277,278].

$$A—A \longrightarrow 2A·$$
$$B—B \longrightarrow 2B· \qquad (58a)$$
$$A· + B· \longrightarrow A—B$$

$$A· + B—B \longrightarrow B\overset{A}{\underset{·}{-}}B \qquad (58b)$$

$$B· + A—A \longrightarrow A\overset{B}{\underset{·}{-}}A$$

$$R· + —CH_2\overset{CH_3}{\underset{|}{C}}\!\!=\!\!CHCH_2— \longrightarrow RH + —CH_2\overset{CH_3}{\underset{|}{C}}\!\!=\!\!CHCH—\overset{R·}{\longrightarrow}$$
$$—CH_2\overset{CH_3}{\underset{|}{C}}\!\!=\!\!CHCH—\underset{R}{|} \qquad (58c)$$

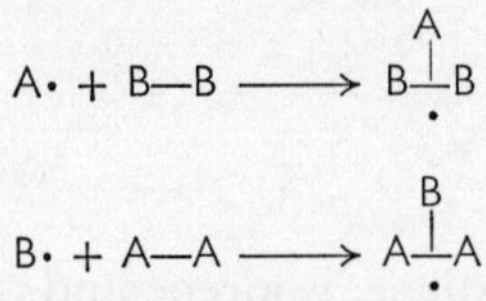

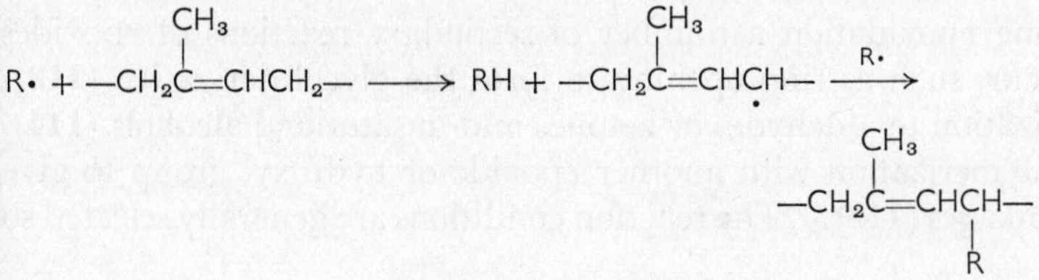

Another route to the preparation of block (and/or graft) copolymers involves the mastication of a polymer in the presence of an appropriate monomer which is an active radical scavenger. Typical structures for block and graft copolymers from polyisoprene and a vinyl monomer are indicated as **111** and **112**, respectively.

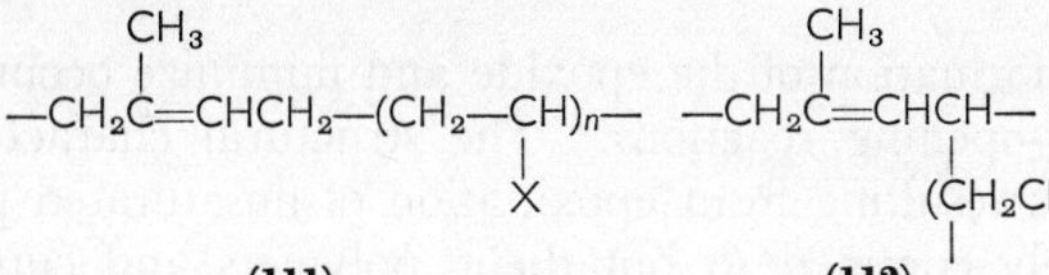

$$—CH_2\overset{CH_3}{\underset{|}{C}}\!\!=\!\!CHCH_2—(CH_2—\underset{\underset{X}{|}}{CH})_n— \qquad —CH_2\overset{CH_3}{\underset{|}{C}}\!\!=\!\!CHCH—\underset{\underset{|}{(CH_2CHX)_n}}{|}$$

(111) \qquad\qquad (112)

Excellent surveys of the mastication of polymers and of polymer modifications based on mechanochemical reactions have recently appeared[277,278].

I. Miscellaneous Reactions

The rather large topic of block and graft copolymerization will be deleted from this survey, partly because the various techniques and approaches applied to the unsaturated polymers are substantially the same as those for saturated polymers, and partly because this subject has been well treated in recent books[279,280] and review articles[73,281–284].

I. Epoxidation

Epoxidation leads to structures containing oxirane (or 1,2-epoxy) rings. In the case of polymers, this reaction is preferably carried out with organic peracids, usually peracetic acid (equation 59). The epoxidation of a wide variety of polymers, including homopolymers

$$RCOOOH + {-}\overset{|}{C}{=}\overset{|}{C}{-} \longrightarrow {-}\overset{|}{C}\underset{\diagdown O \diagup}{}\overset{|}{C}{-} + RCOOH \qquad (59)$$

and copolymers of butadiene, isoprene and cyclopentadiene, has been described in the literature, and recently reviewed by Greenspan[285].

During epoxidation a number of secondary reactions of epoxides can occur, such as ring opening to form the glycol derivative (**113**), isomerization to aldehydes or ketones and unsaturated alcohols (**114**), and polymerization with another epoxide or hydroxyl group to give ether linkages (**115**). The reaction conditions are generally selected so

$$\underset{(113)}{\overset{\displaystyle -\overset{|}{\underset{|}{C}}-\overset{|}{\underset{|}{C}}-}{\underset{\displaystyle \overset{|}{\underset{R}{C}{=}O}}{\overset{\displaystyle OO}{\overset{|}{H}}}}} \qquad \underset{(114)}{-\overset{|}{C}{=}\overset{|}{C}{-}CH_2OH} \qquad \underset{(115)}{-\overset{|}{\underset{|}{C}}-\overset{|}{\underset{|}{C}}-O-\overset{|}{\underset{|}{C}}-\overset{|}{\underset{|}{C}}-}$$

as to optimize formation of the epoxide and minimize occurrence of secondary ring-opening reactions. The structural characterization of the polymers resulting from epoxidation of unsaturated polymers has been largely confined to butadiene polymers and copolymers. Infrared analysis of an epoxidized emulsion polybutadiene[285] shows that the reaction occurs selectively at the internal double bonds ($-CH{=}CH-$), leaving the external double bonds ($-CH{=}CH_2$) largely unchanged.

The epoxidized polymers, because of the high reactivity of their oxirane rings, can be easily crosslinked to a three dimensional thermoset resin by a variety of reagents, such as polyamines, dibasic acids, anhydrides and boron trifluoride. The cured resins find use in a wide assortment of plastics applications.

The epoxidation of *cis*- and *trans*-1,4-polyisoprenes, *cis*-1,4-polybutadiene, butadiene-styrene copolymer and polychloroprene in dilute solution, using monoperphthalic acid, has also been described[286].

2. Thiol addition to double bonds

The reaction of thiols with various diene homopolymers and copolymers has been reviewed recently[287], and in this section we will note only the major features of the reaction. The anti-Markownikoff addition of an alkanethiol (RSH) to the polymer double bonds is shown in Scheme VI as a free radical chain reaction involving the use of an initiator (In·). Although some termination may occur through the formation of crosslinked polymer (**116**), the thiol addition reactions are normally carried out under conditions where that is not important. Thus, the effective termination involves either or both of the other two reaction steps shown. To the extent that oxygen may be present, oxidative chain scission may accompany the thiol addition reaction. With α,α'-azobisisobutyronitrile as initiator and methane-thiol as the reagent, Pierson and coworkers[288] were able to obtain nearly quantitative conversion of polybutadiene to the thiol-adduct (~ 97–98% saturation of the double bonds). Significantly, they were never able to obtain products having more than the theoretical quantity of sulphur, a result which indicated that the possible abstraction of any allylic hydrogen from the diene polymer (equation 60) presumably

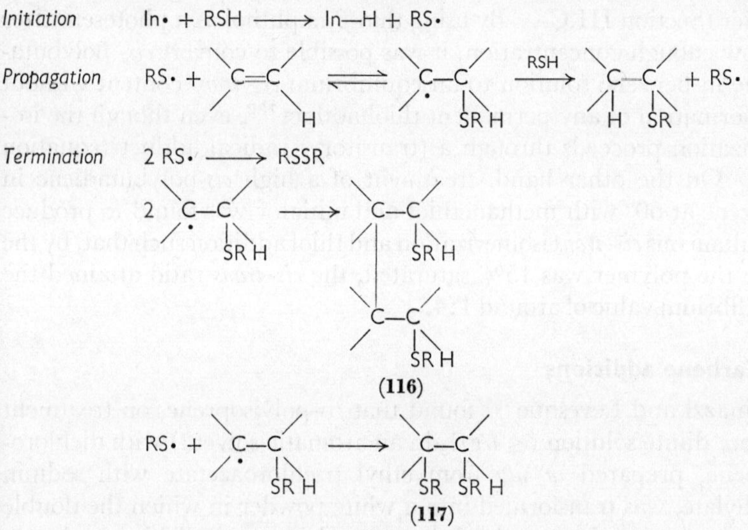

Initiation   In· + RSH ⟶ In—H + RS·

Propagation   RS· + \C=C/ ⇌ \C—C/ —(RSH)→ \C—C/ + RS·
                                  · \SR H   H SR

Termination   2 RS· ⟶ RSSR

2 \C—C/ ⟶ \C—C/
 · \SR H   \SR H
             \C—C/
             \SR H
             (**116**)

RS· + \C—C/ ⟶ \C—C/
      · \SR H   \SR SR H
                 (**117**)

Scheme VI.

does not occur, as attachment of more than one thiol molecule per monomer unit would then be expected. It was shown likewise, that in the butadiene-methanethiol addition reaction, at least, virtually all the

$$\text{In· (or RS·)} + -CH_2CH=CHCH_2- \longrightarrow \text{In-H (or RSH)} + -CH_2CH=CHCH-$$

$$\begin{array}{c} \overset{RS\cdot}{}\\ -CH_2CH=CH-CH- \\ | \\ SR \end{array} \quad (60)$$

termination proceeds through formation of disulphide rather than **117**. These workers also showed that addition of alkanethiol was not sufficiently selective between internal and external double bonds to give an accurate measure of the vinyl content in the diene polymers.

All primary alkanethiols up to dodecyl react sufficiently well with butadiene polymers and copolymers to yield highly saturated adducts. These materials can be expected to show greater stability towards oxygen, ozone and heat than the initial unsaturated polymer. However, the only important evaluation work on butadiene adducts was carried out on those prepared with methanethiol[288-290].

The reversibility of the thiyl radical addition to the olefinic double bond (shown in Scheme VI), especially in the case of a resonance-stabilized radical like $C_6H_5S\cdot$, is the basis for the thiyl-catalysed *cis–trans* isomerization of polybutadiene or polyisoprene discussed earlier (Section III.C). By using thio-β-naphthol as a photosensitizer in low enough concentration, it was possible to convert *cis*- polybutadiene in benzene solution to an equilibrium *cis–trans* content without the formation of any permanent thiol adduct[132], even though the isomerization proceeds through a (transitory) radical adduct (equation 23). On the other hand, treatment of a high *cis*-polybutadiene in benzene at 50° with methanethiol and initiator was found to produce simultaneous *cis–trans* isomerization and thiol addition such that, by the time the polymer was 15% saturated, the *cis–trans* ratio attained the equilibrium value of around 1:4.

3. Carbene additions

Pinazzi and Levesque[291] found that *cis*-polyisoprene, on treatment in very dilute solution (≤ 1 g/l, in an aromatic solvent) with dichlorocarbene, prepared *in situ* from ethyl trichloroacetate with sodium methylate, was transformed into a white powder in which the double bonds were nearly completely converted into *gem*-dichlorocyclopropane rings (equation 61). This was indicated by the disappearance of the characteristic infrared band at 6·0 μ (C=C) and replacement of the broad 12 μ band (—C(CH$_3$)=CH— unit) by a narrow intense band ascribed to geminal chlorine atoms. Polybutadiene treated

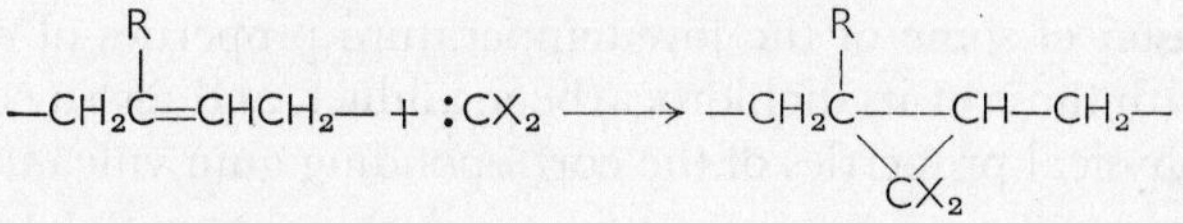

$$-CH_2\overset{\overset{R}{|}}{C}=CHCH_2- + :CX_2 \longrightarrow -CH_2\overset{\overset{R}{|}}{\underset{\underset{CX_2}{\diagdown\diagup}}{C}}\!\!-\!\!-\!\!-\!CH-CH_2- \quad (61)$$

R = CH$_3$, H, or Cl
X = Cl or Br

under the same conditions yielded about 60–70% dichlorocyclopropane formation, but polychloroprene was not affected. Also, treatment of polyisoprene with dibromocarbene, formed *in situ* by basic elimination from bromoform, resulted in 70–75% saturation of the double bonds.

In later work, Pinazzi and Levesque[292] were able to transform completely polyisoprene, polybutadiene and polychloroprene into their *gem*-dihalocyclopropane derivatives by treatment with carbenes produced by the thermolysis of phenyltrihalomethyl mercury. In these reactions :CCl$_2$ was much more reactive than :CBr$_2$. Those derivatives showing quantitative conversion were all highly crystalline at ambient temperatures, but those incompletely converted were amorphous.

They also treated the dibromocarbene adduct of 1,4-polyisoprene with methyllithium at -60 to $-80°$ and obtained the allenic structure **118**, as evidenced by an infrared peak at 5·13 μ due to cumulative

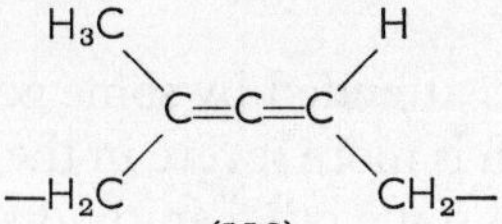

$$\overset{H_3C}{\underset{-H_2C}{\diagdown\diagup}}C=C=C\overset{H}{\underset{CH_2-}{\diagdown\diagup}}$$
(118)

double bonds, and by the diminution of the bands due to C—Br bonds. This conversion of a polycyclopropane to a polycumulene, to an extent of about 85%, constitutes a regular insertion of a carbon atom within nearly every isoprene unit in the chain and is apparently without precedent in the field of polymer reactions.

Pinazzi and coworkers[293] also carried out studies on the dihalocarbene addition to the model compounds, 4-methyl-3-heptene, dihydromyrcene and squalene, and obtained results completely analogous to those for polyisoprene.

Lishanskii, Tsitokhtsev and Vinogradova[294] reported on the addition of dichlorocarbene, carbethoxycarbene and carbomethoxycarbene to *cis*- and *trans*-1,4-polyisoprenes and to 1,4- and 1,2-polybutadienes. Very recently, Lal and Saltman[295] presented a comparison of some of the low temperature properties of *cis*-polyisoprene with those of its dichlorocarbene adduct, and also a comparison of the physical properties of the corresponding gum vulcanizates.

4. Addition of ethylenic compounds

Le Bras and coworkers[296] have summarized work, carried out at the French Institute of Rubber and elsewhere, on the reactions of maleic anhydride and other ethylenic compounds with diene polymers. Much of that work was directed towards obtaining chemically modified rubbers which, in addition to preserving the high elasticity of the starting material, display new and unusual physical (or chemical) properties. The addition of maleic anhydride to polyisoprene, initiated by various free radical catalysts, leads to structure **119** having an anhydrosuccinic residue attached to the α-methylene group[297].

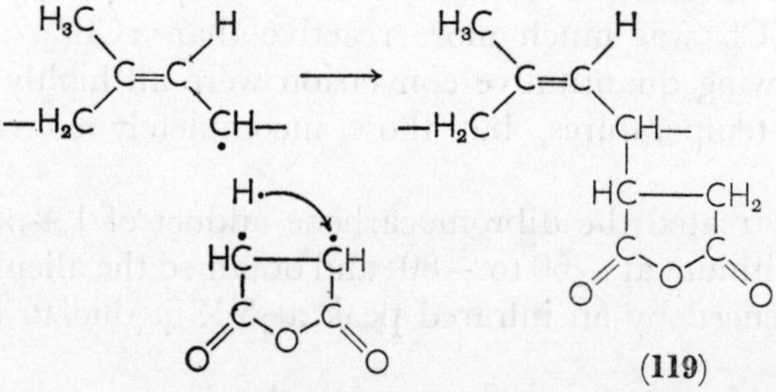

(119)

The addition reaction is attended by some gel formation (exemplified by structure **120**), which is more severe in the case of benzoyl peroxide initiation than it is with some other initiators. By way of illustration *p*-menthane hydroperoxide catalysis of the maleic anhydride addition

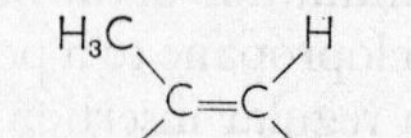

(120)

reaction at 130° resulted in about 20% of the isoprene units being converted to structure **119** along with about 5% gel structure, whereas benzoyl peroxide yielded 19% addition of maleic anhydride and 23% gel formation. Infrared spectroscopy confirmed the attachment of anhydride groups and showed also that there was negligible change in the unsaturation and no isomerization of the original double bonds. In the absence of initiator, maleic anhydride can add to polyisoprene at ~180–240°, mainly by a concerted electron-transfer mechanism (equation 62) involving direct addition to the double bond, but also to some extent by thermolysis of the α-methylenic C—H bond followed by attachment of the anhydride, yielding **119**. The formation of vinylidene units in equation (62) was confirmed by infrared and n.m.r.

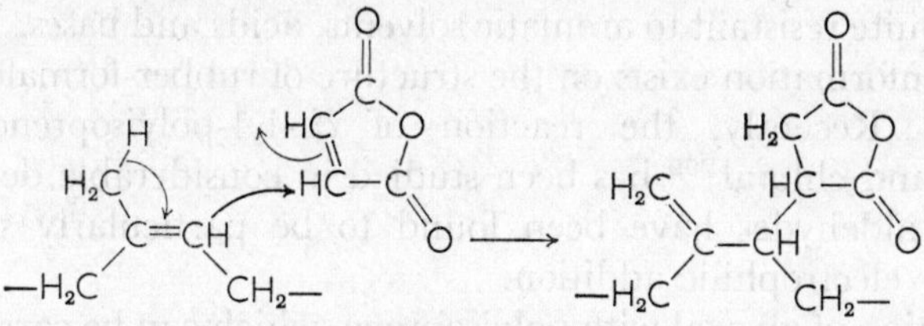

$$\text{(62)}$$

spectroscopy. The radical nature of the initiated reaction and the non-radical nature of the thermal reaction were substantiated by the observations that thiophenol had no effect on the latter process but a pronounced effect on the former.

The maleic anhydride addition to *Hevea* has been extended to other diene polymers, namely, gutta-percha, synthetic *cis*-1,4-polyisoprene and *cis*-1,4-polybutadiene, the products from the synthetic polymers having higher gel contents than those from natural rubber. The carboxyl and anhydride groups in these adduct polymers make it possible to prepare their amides, esters and urethanes. The maleic rubbers can be crosslinked by bifunctional reagents, alcohols and amines, as well as by certain metal oxides. While they can also be vulcanized with sulphur, noteworthy vulcanizates, at levels of ~5–10% combined anhydride, are obtained with oxides of calcium, zinc and especially magnesium.

Polyisoprene can be reacted with various other ethylenic compounds, but in order to avoid or minimize grafting, compounds showing a tendency towards homopolymerization must be excluded. A necessary (although not sufficient) condition for the ethylenic compound to show significant reactivity towards 1,4-polyisoprene is that its double bond have a low electron density, in contrast to the relatively high electron density at the extremity of the polyisoprene double bond. Thus, for example, replacement of a hydrogen on the C=C bond in maleic anhydride by a chlorine (chloromaleic anhydride) or by a methyl group (citraconic anhydride) gives a reactive substance in the former case and a non-reactive one in the latter case. A further requirement of the ethylenic compound is that it offer no steric hindrance opposing its reaction with polyisoprene. Other reactive ethylenic compounds which have been noted [296] are N-methylmaleimide, fumaric and maleic acids, γ-crotonolactone, p-benzoquinone and acrylonitrile.

5. Addition of carbonyl compounds

Various workers in the past have studied the reaction of rubber with aldehydes, especially formaldehyde, to yield thermoplastic compounds which are quite resistant to aromatic solvents, acids and bases. However, little information exists on the structure of rubber-formaldehyde derivatives. Recently, the reaction of *cis*-1,4-polyisoprene with glyoxal [298] and chloral [299] has been studied in considerable detail; in fact, these aldehydes have been found to be particularly suitable reagents for electrophilic addition.

The reaction of glyoxal with polyisoprene, which can be carried out either in solution or in the solid phase [296,298], is assumed to proceed mainly as shown in equation (63a), although some attachment may also occur at the α-methylenic site according to equation (63b). The

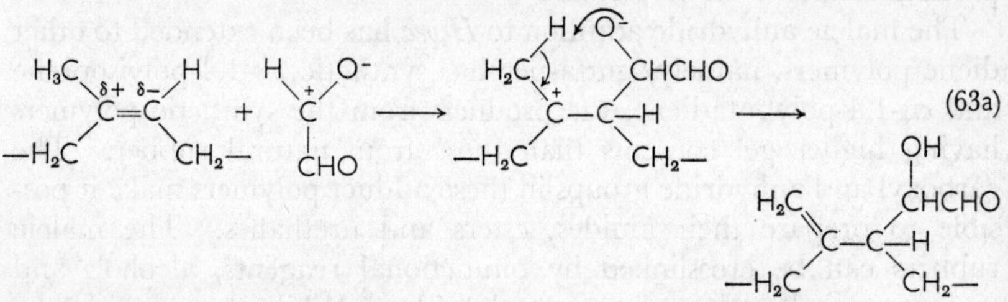

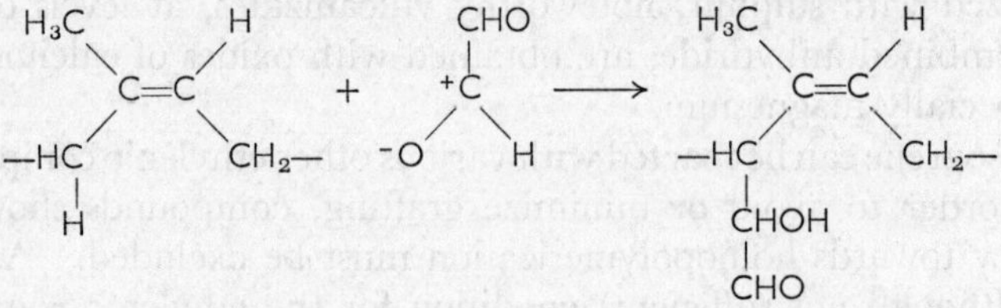

solution reaction is carried out with anhydrous glyoxal at 180° in the absence of a catalyst and at 120° in the presence of a catalyst (e.g.,

AlCl$_3$—NaCl). When more than ~11% of the isoprene units have been reacted the reaction products show significant crosslinking. In the solid phase, use is made of the hydrated trimer of glyoxal which is incorporated in the rubber on a mill, the reagent then being generated *in situ* (equation 64). Infrared spectra of the glyoxal-rubber reaction

$$\underset{\substack{\text{HOHC} \\ | \\ \text{HOHC}}}{\overset{O}{\diagup}} \underset{\substack{\text{CH} \\ | \\ \text{CH}}}{\overset{O}{\diagdown}} \underset{\substack{\text{CHOH} \\ | \\ \text{CHOH}}}{\overset{}{\diagdown}} \longrightarrow 3\,\text{CHO} + 2\,\text{H}_2\text{O} \quad (64)$$

products indicated the presence of ether (9·26 μ), carbonyl (5·8 μ) and hydroxyl (2·98 μ) groups, and these functional groups have been further confirmed by means of various characteristic reactions.

The addition of chloral (CCl$_3$—CHO) to polyisoprene[299], catalysed by AlCl$_3$ or BF$_3$, is analogous to the glyoxal reaction shown in equation (63a) and results in about 11% of the isoprene units being converted to structure **121**, as revealed by infrared spectroscopy (vinylidene at

$$-\text{CH}_2-\underset{\substack{| \\ \text{CHOH}-\text{CCl}_3}}{\overset{\substack{\text{CH}_2 \\ \|}}{\text{C}}}-\text{CH}-\text{CH}_2-$$
(121)

11·3 μ, hydroxyl at 2·85–3·05 μ and chlorine atom at 9·25, 13·7, 16·1 and 17·5 μ) and by characteristic chemical reactions involving the —CHOH—CCl$_3$ side chain. When BF$_3$ was used as catalyst, the reaction products also showed infrared bands at 9·6 and 13·0 μ, suggestive of concomitant cyclization.

Chloral reacts also with the isoprene units in butyl rubber and, only very slightly, with *cis*-1,4-polybutadiene. In the latter case, the double bonds are not polarized and so do not attack chloral in the manner depicted for polyisoprene (analogue of equation 63a). Instead, there is attachment of —CCl$_2$—CHO units to polybutadiene, presumably at α-methylenic sites, indicated by the characteristic infrared band at ~5·6 μ due to carbonyls.

Very recently Pautrat and Marteau[300] have studied the reaction of fluoral with *cis*-polyisoprene. Products having up to 10% of the isoprene units converted to structure **122**, were completely soluble, viscoelastic and thermally stable to 200°, whereas those having greater than 50% attachment were fibrous and relatively insoluble.

To conclude, mention may be made of the hydroformylation of

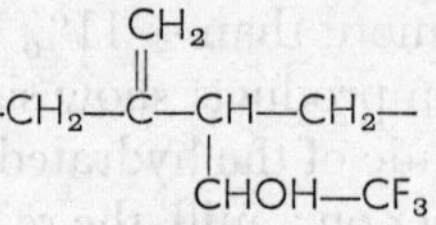

$$\begin{array}{c} \text{CH}_2 \\ \parallel \\ -\text{CH}_2-\text{C}-\text{CH}-\text{CH}_2- \\ | \\ \text{CHOH}-\text{CF}_3 \end{array}$$
(122)

diene polymers according to equation (65), which can be carried as far as 100% conversion of the C=C units. The chemical and physical properties of the resulting polymers have been discussed in a recent paper[301].

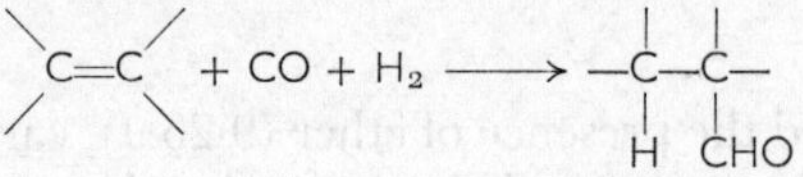

$$\diagup\!\!\!\diagdown\text{C}=\text{C}\diagdown\!\!\!\diagup + \text{CO} + \text{H}_2 \longrightarrow -\underset{\text{H}}{\overset{|}{\text{C}}}-\underset{\text{CHO}}{\overset{|}{\text{C}}}- \tag{65}$$

IV. REFERENCES

1. E. M. Fettes (Ed.), *Chemical Reactions of Polymers*, Interscience Publishers, New York, 1964.
2. L. Bateman (Ed.), *The Chemistry and Physics of Rubber-Like Substances*, Maclaren, London; John Wiley and Sons, New York, 1963.
3. J. P. Kennedy and E. G. M. Tornqvist (Ed.), *Polymer Chemistry of Synthetic Elastomers*, Part II, Interscience Publishers, New York, 1969.
4. A. Charlesby, *Atomic Radiation and Polymers*, Pergamon Press, New York, 1960.
5. A. Chapiro, *Radiation Chemistry of Polymeric Systems*, Interscience Publishers, New York, 1962.
6. M. B. Neiman (Ed.), *Aging and Stabilization of Polymers*, Consultants Bureau, New York, 1965.
7. S. L. Madorsky, *Thermal Degradation of Organic Polymers*, Interscience Publishers, New York, 1964.
8. Reference 3, Part I, 1968.
9. W. Cooper, in *Encycl. Polymer Sci. Technol.*, **5**, 406 (1966).
10. M. H. Lehr, in *Survey of Progress in Chemistry* (Ed. A. F. Scott), Vol. 3, Academic Press Inc., New York, 1966, p. 183.
11. G. Natta, *Pure Appl. Chem.*, **12**, 165 (1966).
12. G. Holden and R. H. Mann, in *Encycl. Chem. Technol.*, **12**, 70–83 (1967).
13. W. S. Richardson and A. J. Sacher, *J. Polymer Sci.*, **10**, 353 (1953).
14. J. I. Cunneen, G. M. C. Higgins and W. F. Watson, *J. Polymer Sci.*, **40**, 1 (1959).
15. J. L. Binder and H. C. Ransaw, *Anal. Chem.*, **29**, 503 (1957).
16. J. L. Binder, *J. Polymer Sci.*, **A1**, 37 (1963).
17. H. Y. Chen, *Anal. Chem.*, **34**, 1134, 1793 (1962).
18. M. A. Golub, S. A. Fuqua and N. S. Bhacca, *J. Am. Chem. Soc.*, **84**, 4981 (1962).
19. M. A. Golub, *J. Polymer Sci.*, **36**, 523 (1959).
20. J. L. Binder, *Rubber Chem. Technol.*, **35**, 57 (1962).

21. F. W. Staveley, P. H. Biddison, M. J. Forster, H. G. Dawson and J. L. Binder, *Rubber Chem. Technol.*, **34,** 423 (1961).
22. D. W. Fraga, *J. Polymer Sci.*, **41,** 522 (1959).
23. H. Y. Chen, *J. Polymer Sci.*, **B4,** 891 (1966).
24. C. W. Bunn, *Proc. Roy. Soc. (London) Ser. A:* **180,** 40 (1942).
25. S. N. Chakravarty and A. K. Sircar, *J. Appl. Polymer Sci.*, **11,** 37 (1967).
26. E. M. Bevilacqua, *J. Polymer Sci.*, **B5,** 601 (1967).
27. J. I. Cunneen and G. M. C. Higgins, in reference 2, Chap. 2.
28. J. D. D'Ianni, *Kautschuk Gummi Kunststoffe*, **19,** 138 (1966).
29. E. G. Kent and F. B. Swinney, *Ind. Eng. Chem., Prod. Res. Develop.* **5,** 134 (1966).
30. G. Natta, *Rubber Plastics Age*, **38,** 495 (1957).
31. W. M. Saltman, in *Encycl. Polymer Sci. Technol.*, **2,** 678 (1964).
32. R. E. Rinehart, H. P. Smith, H. S. Witt and H. Romeyn, Jr., *J. Am. Chem. Soc.*, **83,** 4864 (1961); **84,** 4145 (1962).
33. A. J. Canale, W. A. Hewett, T. M. Shryne and E. A. Youngman, *Chem. Ind. (London)*, 1054 (1962).
34. A. J. Canale and W. A. Hewett, *J. Polymer Sci.*, **B2,** 1041 (1964).
35. G. Kraus, J. N. Short and V. Thornton, *Rubber Plastics Age*, **38,** 880 (1957); *Rubber Chem. Technol.*, **30,** 1118 (1957).
36. J. N. Short, G. Kraus and R. P. Zelinski, *Rev. Gen. Caoutchouc*, **40,** 253 (1963).
37. R. S. Silas, J. Yates and V. Thornton, *Anal. Chem.*, **31,** 529 (1959).
38. J. L. Binder, *J. Polymer Sci.*, **A1,** 47 (1963).
39. V. D. Mochel, *Rubber Chem. Technol.*, **40,** 1200 (1967).
40. W. H. Carothers, *Ind. Eng. Chem.*, **26,** 30 (1934).
41. J. T. Maynard and W. E. Mochel, *J. Polymer Sci.*, **13,** 251 (1954).
42. R. G. Ferguson, *J. Polymer Sci.*, **A2,** 4735 (1964).
43. C. A. Aufdermarsh, Jr., and R. Pariser, *J. Polymer Sci.* **A2,** 4727 (1964).
44. G. Natta, L. Porri, P. Corradini, G. Zanini and F. Ciampelli, *J. Polymer Sci.*, **51,** 463 (1961).
45. G. Natta, L. Porri, A. Carbonaro, F. Ciampelli and G. Allegra, *Makromol. Chem.*, **51,** 229 (1962).
46. G. Natta, L. Porri, G. Stoppa, G. Allegra and F. Ciampelli, *J. Polymer Sci.*, **B1,** 67 (1963).
47. G. Natta, L. Porri and M. C. Gallazzi, *Chim. Ind. (Milan)*, **46,** 1158 (1964).
48. G. Natta, *Rev. Gen. Caoutchouc*, **40,** 784 (1963).
49. W. M. Saltman and J. N. Henderson, in *Encycl. Polymer Sci. Technol.*, **2,** 667 (1964).
50. I. A. Livshits and L. M. Korobova, *Dokl. Akad. Nauk SSSR*, **121,** 474 (1958); *Vysokomolekul. Soedin.*, **3,** 891 (1961).
51. W. Marconi, A. Mazzei, S. Cucinella and M. Cesari, *J. Polymer Sci.*, **A2,** 4261 (1964).
52. G. Perego and I. W. Bassi, *Makromol. Chem.*, **61,** 198 (1963).
53. U. Giannini, M. Cambrini and A. Cassata, *Makromol. Chem.*, **61,** 246 (1963).
54. R. F. Heck and D. S. Breslow, *J. Polymer Sci.*, **41,** 521 (1959).
55. T. F. Yen, *J. Polymer Sci.*, **35,** 533 (1959); **38,** 272 (1959).
56. J. K. Stille and E. D. Vessel, *J. Polymer Sci.*, **49,** 419 (1961).

57. R. Asami and A. Shoji, Paper presented at the International Symposium on Macromolecular Chemistry, Tokyo-Kyoto, Japan, 1966.
58. C. Aso, T. Kunitake, K. Ito and Y. Ishimoto, *J. Polymer Sci.*, **B4,** 701 (1966).
59. S. P. S. Yen, *A.C.S. Polymer Div.*, *Polymer Preprints*, **4,** No. 2, 82 (1963).
60. A. G. Davies and A. Wassermann, *J. Polymer Sci.*, **A-1, 4,** 1887 (1966).
61. G. B. Butler, *J. Polymer Sci.*, **48,** 279 (1960).
62. G. B. Butler and M. L. Miles, *J. Polymer Sci.*, **A3,** 1609 (1965).
63. G. B. Butler and M. A. Raymond, *J. Macromol. Chem.* **1,** 201 (1966).
64. G. B. Butler and T. W. Brooks, *J. Org. Chem.*, **28,** 2699 (1963).
65. W. Marconi, A. Mazzei, G. Lugli and M. Bruzzone, *J. Polymer Sci.*, **C16,** 805 (1967).
66. G. Holden and R. Milkovich, *Belg. Pat.* 627,652, July 29, 1963; *Chem. Abstr.*, **60,** 14714f (1964).
67. M. Szwarc, M. Levy and R. Milkovich, *J. Am. Chem. Soc.*, **78,** 2656 (1956).
68. H. Y. Chen and J. E. Field, *J. Polymer Sci.*, **B5,** 501 (1967).
69. G. Natta, G. Crespi, A. Valvassori and G. Sartori, *Rubber Chem. Technol.*, **36,** 1583 (1963).
70. S. Adamek, E. A. Dudley and R. T. Woodhams, *Brit. Pat.* 880,904 (to Dunlop Rubber Co., Ltd.), Oct. 25, 1961; *Chem. Abstr.*, **56,** 8903c (1962).
71. R. Hank, *Kautschuk Gummi*, **18,** 295 (1965); *Rubber Chem. Technol.*, **40,** 936 (1967).
72. W. P. Tyler, *Rubber Chem. Technol.*, **40,** 238 (1967).
73. R. J. Ceresa, in *Encycl. Polymer Sci. Technol.*, **2,** 485 (1964).
74. J. A. Nieuwland, W. S. Calcott, F. B. Downing and A. S. Carter, *J. Am. Chem. Soc.*, **53,** 4197 (1931).
75. G. Natta, G. Mazzanti and P. Corradini, *Atti Accad. Naz. Lincei, Rend., Classe Sci. Fis., Mat. Nat.* [8] **25,** 3 (1958).
76. G. Lombardi and L. Giuffre, *Atti Accad. Naz. Lincei, Rend., Classe Sci. Fis., Mat. Nat.* [8] **25,** 70 (1958).
77. H. A. Pohl and R. P. Chartoff, *J. Polymer Sci.*, **A2,** 2787 (1964).
78. A. Rembaum, J. Moacanin and H. A. Pohl, *Progress Dielectrics*, **6,** 41–102 (1965).
79. G. Natta, G. Dall'Asta, G. Mazzanti, *Angew. Chem., Intern. Ed. Engl.*, **3,** 723 (1964).
80. W. L. Truett, D. R. Johnson, I. M. Robinson and B. A. Montague, *J. Am. Chem. Soc.*, **82,** 2337 (1960).
81. F. W. Michelotti and W. P. Keaveney, *A.C.S. Polymer Div.*, *Polymer Preprints*, **4(2),** 293 (1963).
82. V. L. Bell, *J. Polymer Sci.*, **A2,** 5291 (1964).
83. V. L. Bell, *J. Polymer Sci.*, **A2,** 5305 (1964).
84. W. C. Geddes, *Rubber Chem. Technol.*, **40,** 178 (1967).
85. D. E. Winkler, *J. Polymer Sci.*, **35,** 3 (1959).
86. R. M. Aseeva, A. A. Berlin, V. I. Kasatochkin and Z. S. Smutkina, *Vysokomolekul. Soedin.*, **8,** 2171 (1966); *Chem. Abstr.*, **66,** 66084 (1967).
87. Ref. 1, p. 80.
88. M. A. Geĭderikh, B. A. Davydov, B. A. Krentsel, I. M. Kustanovich, L. S. Polak, A. V. Topchiev and R. M. Voĭtenko, *J. Polymer Sci.*, **54,** 621 (1961).
89. L. H. Peebles, Jr. and J. Brandrup, *Makromol. Chem.*, **98,** 189 (1966).

90. A. R. Schultz, in reference 1, Chap. IX.A.
91. F. M. Rugg, J. J. Smith and L. H. Wartman, *J. Polymer Sci.*, **11**, 1 (1953).
92. T. Alfrey, Jr., in reference 1, Chap. I.A.
93. P. J. Canterino, in reference 1, Chap. II.D.
93a. C. S. Ramakrishnan, D. Raghunath and J. B. Pande, *Trans. Inst. Rubber Ind.*, **30**, 129 (1954); *Rubber Chem. Technol.*, **28**, 598 (1955).
93b. G. J. van Amerongen and C. Koningsberger, *J. Polymer Sci.*, **6**, 653 (1950).
94. F. E. Bailey, Jr., J. P. Henry, R. D. Lundberg and J. M. Whelan, *J. Polymer Sci.*, **B2**, 447 (1964).
95. N. Murayama and Y. Amagi, *J. Polymer Sci.*, **B4**, 119 (1966).
96. N. Murayama and Y. Amagi, *J. Polymer Sci.*, **B4**, 115 (1966).
97. C. P. Pinazzi and H. Gueniffey, *Makromol. Chem.*, **93**, 109 (1966); *Rev. Gen. Caout. Plast., Ed. Plast.*, **44**, 777 (1967).
98. F. Bohlmann and H. J. Mannhardt, *Chem. Ber.*, **89**, 1307 (1956).
99. G. Drefahl, H. H. Hörhold and E. Hesse, *J. Polymer Sci.*, **C16**, 965 (1967).
100. S. D. Gehman, J. E. Field and R. P. Dinsmore, *Proc. Rubber Technol. Conf., London*, **1938**, p. 961.
101. M. A. Golub and J. Heller, *J. Polymer Sci.*, **B2**, 723 (1964).
102. K. Matsuzaki and K. Fujinami, *Kogyo Kagaku Zasshi*, **68**, 1456 (1965); *Chem. Abstr.*, **64**, 828c (1966).
103. D. H. E. Tom, *J. Polymer Sci.*, **20**, 381 (1956).
104. M. A. Golub and J. Heller, *J. Polymer Sci.*, **B2**, 523 (1964).
105. M. A. Golub and J. Heller, unpublished results.
106. N. Grassie, *Chemistry of High Polymer Degradation Processes*, Butterworths, London, and Interscience, New York, 1956, pp. 315–319.
107. M. A. Golub and J. Heller, *Tetrahedron Letters*, 2147 (1963).
108. M. A. Golub, in reference 3.
109. M. A. Golub and J. Heller, *Can. J. Chem.*, **41**, 937 (1963).
110. G. Fromandi, *Kautschuk*, **4**, 185 (1928).
111. M. A. Golub and J. Danon, *Can. J. Chem.*, **43**, 2772 (1965).
112. J. Scanlan, in reference 1, Chap. 2.B.
113. S. Tocker, *J. Am. Chem. Soc.*, **85**, 640 (1963).
114. R. J. Angelo, *A.C.S. Polymer Div. Polymer Preprints*, **4**, 32 (1963); *Chem. Abstr.*, **62**, 645g (1965).
115. R. J. Angelo, M. L. Wallach and R. M. Ikeda, *A.C.S. Polymer Div., Polymer Preprints*, **8**, 221 (1967).
116. N. G. Gaylord, I. Kössler, M. Štolka and J. Vodehnal, *J. Polymer Sci.*, **A2**, 3969 (1964).
117. M. A. Golub and J. Heller, *J. Polymer Sci.*, **B4**, 469 (1966).
118. F. T. Wallenberger, *Monatsh. Chem.*, **93**, 74 (1962); *Rubber Chem. Technol.*, **36**, 558 (1963).
119. B. V. Buizov and A. B. Kusov, *J. Rubber Ind. (USSR)*, **12**, 46 (1935).
120. H. A. Endres, *Rubber Age (NY)*, **55**, 361 (1944); *Rubber Chem. Technol.*, **17**, 903 (1944).
121. M. Gordon, *Ind. Eng. Chem.*, **43**, 386 (1951); *Rubber Chem. Technol.*, **24**, 940 (1951).
122. I. Kössler, J. Vodehnal, M. Štolka, J. Kálal and E. Hartlová, *J. Polymer Sci.*, **C16**, 1311 (1967); and references cited therein.

123. J. R. Shelton and L. H. Lee, *Rubber Chem. Technol.*, **31,** 415 (1958).
124. A. Carbonaro and A. Greco, *Chim. Ind. (Milan)*, **48,** 363 (1966); *Chem. Abstr.*, **65,** 5536h (1966).
125. J. L. Binder, *J. Polymer Sci.*, **B4,** 19 (1966).
126. M. Štolka, J. Vodehnal and I. Kössler, *J. Polymer Sci.*, **A2,** 3897 (1964).
127. I. Kössler, J. Vodehnal and M. Štolka, *J. Polymer Sci.*, **A3,** 2081 (1965).
128. I. Kössler, M. Štolka and K. Mach, *J. Polymer Sci.*, **C4,** 977 (1963).
129. N. G. Gaylord, I. Kössler, B. Matyska and K. Mach, *A.C.S. Polymer Div., Polymer Preprints*, **8,** 174 (1967).
130. R. B. Cundall, in *Progress in Reaction Kinetics* (Ed. G. Porter), Vol. 2, Macmillan Company, New York, 1964, Chap. 4.
131. K. Mackenzie, in *The Chemistry of Alkenes* (Ed. S. Patai), Interscience Publishers, London–New York, 1964, Chap. 7.
132. M. A. Golub, *J. Polymer Sci.*, **25,** 373 (1957); *U.S. Pats.* 2,878,175–6 (to The B. F. Goodrich Co.), Mar. 17, 1959.
133. K. H. Meyer and C. Ferri, *Helv. Chim. Acta*, **19,** 694 (1936); *Rubber Chem. Technol.*, **9,** 570 (1936).
134. C. Ferri, *Helv. Chim. Acta*, **20,** 393 (1937); *Rubber Chem. Technol.*, **11,** 350 (1938).
135. H. Bartl, Ger. Pat. 1,124,683 (to Farbenfabriken Bayer A.G.); *Chem. Abstr.*, **57,** 6092h (1962).
136. M. A. Golub, in reference 1, Chap. 2.A.
137. M. A. Golub, *J. Am. Chem. Soc.*, **80,** 1794 (1958); **81,** 54 (1959).
138. H. Steinmetz and R. M. Noyes, *J. Am. Chem. Soc.*, **74,** 4141 (1952).
139. C. Sivertz, *J. Phys. Chem.*, **63,** 34 (1959).
140. G. R. Seely, *J. Am. Chem. Soc.*, **84,** 4404 (1962).
141. M. A. Golub, *J. Phys. Chem.*, **68,** 2360 (1964).
142. M. A. Golub, *J. Phys. Chem.*, **66,** 1202 (1962).
143. M. Berger and D. J. Buckley, *J. Polymer Sci.*, **A1,** 2945 (1963).
144. M. A. Golub, *J. Am. Chem. Soc.*, **82,** 5093 (1960); *J. Phys. Chem.*, **69,** 2639 (1965).
145. R. B. Cundall and P. A. Griffiths, *Discussions Faraday Soc.*, **36,** 111 (1963); *Trans. Faraday Soc.*, **61,** 1968 (1965); *J. Phys. Chem.*, **69,** 1866 (1965).
146. M. A. Golub and C. L. Stephens, *J. Polymer Sci.*, **C16,** 765 (1967).
147. M. A. Golub and C. L. Stephens, *J. Polymer Sci.*, **A-1,6,** 763 (1968).
148. T. C. Ho, *K'o Hsueh Ch'u Pan She*, 365 (1963); *Chem. Abstr.*, **64,** 2253h (1966).
149. G. S. Trick, *J. Polymer Sci.*, **31,** 529 (1958); **41,** 213 (1959).
150. M. E. A. Cudby, *J. Polymer Sci.*, **B3,** 73 (1965).
151. M. A. Golub and J. J. Shipman, *Spectrochim. Acta*, **16,** 1165 (1960); **20,** 701 (1964).
152. I. A. Kop'eva, E. I. Tinyakova, B. A. Dolgoplosk, L. I. Red'kina and E. N. Zavadovskaya, *Vysokomolekul. Soedin. Ser. A.* **9(3),** 645 (1967); *Chem. Abstr.*, **67,** 33124 (1967).
153. G. Kraus, paper presented before the German Chemical Society, Plastics & Rubber Division, Bad Nauheim, May 1960.
154. W. A. Bishop, *J. Polymer Sci.*, **55,** 827 (1961).
155. J. J. Shipman and M. A. Golub, *J. Polymer Sci.*, **58,** 1063 (1962).
156. E. W. Madge, *Chem. Ind. (London)*, 1806 (1962).

157. C. A. Ceselli, T. Garlanda, M. Camia and G. Manza, *Chim. Ind.* (Milan), **44,** 1203 (1962); **45,** 347 (1963); *Chem. Abstr.*, **58,** 9307e (1963).
158. H. Blümel, *Kautschuk Gummi,* **16,** 571 (1963); *Rubber Chem. Technol.,* **37,** 408 (1964).
159. D. Reichenbach, *Kautschuk Gummi Kunststoffe,* **18,** 213 (1965); *Chem. Abstr.,* **63,** 3145c (1965).
160. K. V. Nelson and N. N. Novikova, *Kolebatel'nye Spektry i Molekul. Protsessy v Kauchukakh, Vyes. Nauchn.-Issled. Inst. Sintetich. Kauchuka,* 136 (1965), through *Chem. Abstr.,* **64,** 5273b (1966).
161. J. Hayashi, J. Furukawa and S. Yamashita, *Kobunshi Kagaku,* **23,** 531 (1966); *Chem. Abstr.,* **66,** 116461 (1967).
162. L. Bateman, R. W. Glazebrook and C. G. Moore, *J. Appl. Polymer Sci.,* **1,** 257 (1959).
163. L. Bateman and C. G. Moore, in *Organic Sulfur Compounds* (Ed. N. Kharasch), Vol. 1, Pergamon Press, New York, 1961.
164. L. Bateman, C. G. Moore, M. Porter and B. Saville, in reference 2, Chap. 15.
165. C. G. Moore and M. Porter, *J. Chem. Soc.,* 6390 (1965).
166. J. D. Fitzpatrick and M. Orchin, *J. Am. Chem. Soc.,* **79,** 4765 (1957); *J. Org. Chem.,* **22,** 1177 (1957).
167. I. I. Ermakova, B. A. Dolgoplosk and E. N. Kropacheva, *Dokl. Akad. Nauk SSSR,* **141,** 1363 (1961); *Rubber Chem. Technol.,* **35,** 618 (1962).
168. M. Berger and T. A. Manuel, *J. Polymer Sci.,* **A-1,4,** 1509 (1966).
169. D. Reichenbach, *Kautschuk Gummi Kunststoffe,* **18,** 9 (1965); *Chem. Abstr.,* **62,** 13354d (1965).
170. B. M. E. van der Hoff, *Ind. Eng. Chem., Prod. Res. Develop.,* **2,** 273 (1963); *A.C.S. Polymer Div., Polymer Preprints,* **8,** 1461 (1967).
171. J. I. Cunneen, G. M. C. Higgins and R. A. Wilkes, *J. Polymer Sci.,* **A3,** 3503 (1965).
172. J. Tsurugi, T. Fukumoto, M. Yamagami and H. Itatani, *J. Polymer Sci.,* **A-1,4,** 563 (1966).
173. G. W. Griffin, J. Covell, R. C. Petterson, R. M. Dodson and G. Klose, *J. Am. Chem. Soc.,* **87,** 1410 (1965).
174. R. J. Cvetanović and L. C. Doyle, *J. Chem. Phys.,* **37,** 543 (1962).
175. L. Bateman, *Trans. Inst. Rubber Ind.,* **21,** 118 (1945); *J. Polymer Sci.,* **2,** 1 (1947).
176. M. B. Evans, G. M. C. Higgins and D. T. Turner, *J. Appl. Polymer Sci.,* **2,** 340 (1959).
177. J. Danon and M. A. Golub, *Can. J. Chem.,* **42,** 1577 (1964).
178. D. T. Turner, *Polymer,* **1,** 27 (1960).
179. J. I. Cunneen and F. W. Shipley, *J. Polymer Sci.,* **36,** 77 (1959).
180. J. I. Cunneen, W. P. Fletcher, F. W. Shipley and R. I. Wood, *Trans. Inst. Rubber Ind.,* **34,** 260 (1958); *Chem. Abstr.,* **53,** 9711d (1959).
181. J. I. Cunneen and W. F. Watson, *J. Polymer Sci.,* **38,** 521, 533 (1959).
182. J. I. Cunneen, P. McL. Swift and W. F. Watson, *Trans. Inst. Rubber Ind.,* **36,** 17 (1960); *Chem. Abstr.,* **54,** 12631e (1960).
183. R. E. Cook, F. S. Dainton and K. J. Ivin, *J. Polymer Sci.,* **26,** 351 (1957).
184. M. A. Golub, unpublished results; cited in K. Folkers, H. W. Moore, G. Lenaz and L. Szarkowska, *Biochem. Biophys. Res. Commun.,* **23,** 386 (1966).

185. M. A. Golub, *J. Polymer Sci.*, **B4,** 227 (1966).
186. B. A. Dolgoplosk, E. N. Kropacheva and K. V. Nelson, *Dokl. Akad. Nauk. SSSR*, **123,** 685 (1958); *Rubber Chem. Technol.*, **32,** 1036 (1959).
187. I. I. Boldyreva, B. A. Dolgoplosk, E. N. Kropacheva and K. V. Nelson, *Dokl. Akad. Nauk SSSR*, **131,** 830 (1960); *Rubber Chem. Technol.*, **33,** 985 (1960).
188. B. A. Dolgoplosk, G. P. Belonovskaja, I. I. Boldyreva, E. N. Kropacheva, K. V. Nelson, Ja. M. Rosinoer and J. D. Chernova, *J. Polymer Sci.*, **53,** 209 (1961).
189. E. N. Kropacheva, I. I. Ermakova, B. A. Dolgoplosk, A. I. Koltsov and K. V. Nelson, Preprint P408, International Symposium on Macromolecular Chemistry, Prague, 1965.
190. C. D. Harries, *Chem. Ber.*, **56,** 1048 (1923); *Chem. Abstr.*, **17,** 2806 (1923).
191. H. Staudinger and E. O. Leupold, *Chem. Ber.*, **67,** 304 (1934).
192. H. Staudinger and G. V. Schulz, *Chem. Ber.*, **68,** 2320 (1935).
193. R. Pummerer and P. A. Burkard, *Chem. Ber.*, **55,** 3458 (1922).
194. J. Wicklatz, in reference 1, Chap. 2.f.
195. R. V. Jones, C. W. Moberly and W. B. Reynolds, *Ind. Eng. Chem.*, **45,** 1117 (1953).
196. F. L. Ramp, E. J. DeWitt and L. E. Trapasso, *J. Org. Chem.*, **27,** 4368 (1962).
197. A. I. Yakubchik, B. I. Tikhomirov and L. N. Mikhailova, *Zh. Prikl. Khim.*, **34,** 652 (1961); *Chem. Abstr.*, **55,** 18252a (1961).
198. A. I. Yakubchik, V. N. Reikh, B. I. Tikhomirov and A. V. Pavlikova, *Zh. Prikl. Khim.* **34,** 2501 (1961); *Rubber Chem. Technol.*, **35,** 1052 (1962).
199. B. I. Tikhomirov, A. I. Yakubchik and I. A. Klopotova, *Vysokomolekul. Soed.*, **3,** 486 (1961); **4,** 25 (1962); *Chem. Abstr.*, **56,** 570i; 15649a (1962).
200. A. I. Yakubchik, B. I. Tikhomirov and V. S. Sumilov, *Rubber Chem. Technol.*, **35,** 1063 (1962).
201. E. G. Gregg, Jr., *A.C.S. Polymer Div.*, *Polymer Preprints*, **8,** 851 (1967).
202. M. E. Cupery, *U.S. Pat.* 2,526,639 (to E. I. duPont de Nemours & Co.), Oct. 24, 1950; *Chem. Abstr.*, **45,** 1384a (1951).
203. S. C. Temin and A. R. Shultz, in *Encycl. Polymer Sci. Technol.*, **4,** 331 (1966).
204. L. D. Loan, *Rubber Chem. Technol.*, **40,** 149 (1967).
205. B. Saville and A. A. Watson, *Rubber Chem. Technol.*, **40,** 100 (1967).
206. I. I. Ostromislensky, *J. Russ. Phys. Chem. Soc.*, **47,** 1467 (1915); Engl. trans., *India Rubber J.*, **52,** 470 (1916).
207. E. H. Farmer and C. G. Moore, *J. Chem. Soc.*, 142 (1951).
208. J. L. Bolland, *Trans. Faraday Soc.*, **46,** 358 (1950).
209. E. H. Farmer and C. G. Moore, *J. Chem. Soc.*, 131 (1951).
210. C. R. Parks and O. Lorenz, *J. Polymer Sci.*, **50,** 287 (1961).
211. C. G. Moore and B. R. Trego, *J. Appl. Polymer Sci.*, **5,** 299 (1961).
212. N. Sheppard and G. B. B. M. Sutherland, *J. Chem. Soc.*, 1699 (1947).
213. G. Salomon and A. Chr. van der Schee, *J. Polymer Sci.*, **14,** 181 (1954).
214. F. J. Linnig and J. E. Stewart, *J. Res. Nat. Bur. Stds.*, **60,** 9 (1958); *Rubber Chem. Technol.*, **31,** 719 (1958).
215. E. H. Farmer and F. W. Shipley, *J. Polymer Sci.*, **1,** 293 (1946); *J. Chem. Soc.*, 1519 (1947).
215a. P. Kovacic, *Ind. Eng. Chem.*, **47,** 1090 (1955).

9. Polymers Containing C=C Bonds

216. F. A. Bovey, *The Effects of Ionizing Radiation on Natural and Synthetic High Polymers*, Interscience Publishers, New York, 1958.
217. S. D. Gehman and T. C. Gregson, *Rubber Chem. Technol.*, **33,** 1375 (1960).
218. D. T. Turner, in reference 2, Chap. 16.
219. D. T. Turner, *J. Polymer Sci.*, **B1,** 93, 97 (1963).
220. E. Witt, *J. Polymer Sci.*, **41,** 507 (1959).
221. G. Vaughan, D. E. Eaves and W. Cooper, *Polymer*, **2,** 235 (1961).
222. B. Jankowski and J. Kroh, *J. Appl. Polymer Sci.*, **9,** 1363 (1965).
223. D. T. Turner, *J. Polymer Sci.*, **B4,** 717 (1966).
224. E. M. Bevilacqua, in *Autoxidation and Antioxidants* (Ed. W. O. Lundberg), Vol 2, Interscience Publishers, New York, 1962, Chap. 18.
225. J. C. Ambelang, R. H. Kline, O. M. Lorenz, C. R. Parks, C. Wadelin and J. R. Shelton, *Rubber Chem. Technol.*, **36,** 1497 (1963).
226. D. Barnard, L. Bateman, J. I. Cunneen and J. F. Smith, in reference 2, Chap. 17.
227. W. L. Hawkins and F. H. Winslow, in reference 1, Chap. 13.
228. P. M. Norling, T. C. P. Lee and A. V. Tobolsky, *Rubber Chem. Technol.*, **38,** 1198 (1965).
229. A. S. Kuzminskii, in reference 6, Chap. 10; *Rubber Chem. Technol.*, **39,** 88 (1966).
230. J. L. Bolland, *Quart. Rev.*, **3,** 1 (1949).
231. L. Bateman, *Quart. Rev.*, **8,** 147 (1954).
232. J. L. Bolland and H. Hughes, *J. Chem. Soc.*, 492 (1949).
233. K. R. Hargrave and A. L. Morris, *Trans. Faraday Soc.*, **52,** 89 (1956).
234. E. H. Farmer and A. Sundralingham, *J. Chem. Soc.*, 125 (1943).
235. J. McGavack and E. M. Bevilacqua, *Ind. Eng. Chem.*, **43,** 475 (1951).
236. E. M. Bevilacqua, *J. Am. Chem. Soc.*, **77,** 5394 (1955); *J. Org. Chem.*, **21,** 369 (1956).
237. E. M. Bevilacqua, *J. Am. Chem. Soc.*, **79,** 2915 (1957).
238. A. L. Morris, *Ph.D. Thesis*, London University, 1952.
239. A. V. Tobolsky, D. J. Metz and R. B. Mesrobian, *J. Am. Chem. Soc.*, **72** 1942 (1950).
240. J. R. Dunn, unpublished work, cited in Ref. 226.
241. E. M. Bevilacqua, *J. Am. Chem. Soc.*, **80,** 5364 (1958).
242. E. M. Bevilacqua, *J. Polymer Sci.*, **B4,** 27 (1966).
243. C. R. Parks and O. Lorenz, *Ind. Eng. Chem., Prod. Res. Dev.* **2,** 279 (1964).
244. E. M. Bevilacqua and E. S. English, *J. Polymer Sci.*, **49,** 495 (1961).
245. E. M. Bevilacqua, *J. Am. Chem. Soc.*, **77,** 5396 (1955); *Sci.*, **126,** 396 (1957); *Rubber Age*, **80,** 271 (1956); *J. Am. Chem. Soc.*, **81,** 5071 (1959).
246. F. R. Mayo, *Ind. Eng. Chem.*, **52,** 614 (1960).
247. E. M. Bevilacqua, E. S. English, J. S. Gall and P. M. Norling, *J. Appl. Polymer Sci.*, **8,** 1029 (1964).
248. A. V. Tobolsky and A. Mercurio, *J. Am. Chem. Soc.*, **81,** 5355 (1959).
249. E. M. Bevilacqua, E. S. English and E. E. Philipp, *J. Org. Chem.*, **25,** 1276 (1960).
250. E. M. Bevilacqua and P. M. Norling, *Sci.*, **147,** 289 (1965).
251. C. L. M. Bell, *Trans. Inst. Rubb. Ind.*, **41,** 212 (1965); *Rubber Chem. Tech.*, **39,** 530 (1966).

252. J. de Merlier and J. Le Bras, *Ind. Eng. Chem., Prod. Res. Dev.*, **2**, 22 (1963); *Rubber Chem. Technol.*, **36**, 1043 (1963).
253. E. M. Bevilacqua, *A.C.S. Polymer Div., Polymer Preprints* **8**, 834 (1967); *Chem. Abstr.*, **66**, 116433 (1967).
254. E. M. Bevilacqua, *J. Appl. Polymer Sci.*, **10**, 1295 (1966).
255. L. H. Lee, C. L. Stacy and R. G. Engel, *J. Appl. Polymer Sci.*, **10**, 1699 (1966).
256. F. Kurihara, T. Kobayashi and K. Ishikawa, *Nippon Gomu Kyokaishi*, **38**, 569 (1965); through *Chem. Abstr.*, **64**, 901c (1966).
257. Z. Jedlinski and A. Janik, *Roczniki Chem.*, **40**, 1639 (1966); through *Chem. Abstr.*, **66**, 76431 (1967).
258. M. A. Salimov, *Azerbaidzhan. Khim. Zhur.*, **1**, 31 (1961); *Rubber Chem. Technol.*, **36**, 747 (1963).
259. H. C. Beachell and J. C. Spitsbergen, *J. Polymer Sci.*, **62**, 73 (1962).
260. P. M. Duncan and W. G. Forbes, *A.C.S. Polymer Div., Polymer Preprints*, **7**, 1035 (1966); *Chem. Abstr.*, **66**, 29316 (1967).
261. S. S. Pickles, *J. Chem. Soc.*, **97**, 1085 (1910).
262. R. W. Murray and P. R. Story, in reference 1, Chap. 8-D.
263. F. H. Kendall and J. Mann, *J. Polymer Sci.*, **19**, 503 (1956).
264. E. H. Andrews, D. Barnard, M. Braden and A. N. Gent, in reference 2, Chap. 12.
265. R. Criegee, *Record Chem. Progr.* (Kresge-Hooker Sci., Lib.), **18**, 111 (1957).
266. P. S. Bailey, *Chem. & Ind.*, 1148 (1957); *Chem. Rev.*, **58**, 925 (1958).
267. A. H. Riebel, R. E. Erickson, C. J. Abshire and P. S. Bailey, *J. Am. Chem. Soc.*, **82**, 1801 (1960).
268. G. M. C. Higgins, unpublished results, cited in ref. 264.
269. N. Grassie, in *Encycl. Polymer Sci. Technol.*, **4**, 647 (1966).
270. H. H. G. Jellinek, *Degradation of Vinyl Polymers*, Academic Press, New York, 1965.
271. N. Grassie, in reference 1, Chap. 8.B.
272. L. A. Wall and J. H. Flynn, *Rubber Chem. Technol.*, **35**, 1157 (1962).
273. T. Midgley, Jr. and A. L. Henne, *J. Am. Chem. Soc.*, **51**, 1215 (1929).
274. S. Strauss and S. L. Madorsky, *J. Research Nat. Bur. Stds.*, **50**, 165 (1953); *Ind. Eng. Chem.*, **48**, 1212 (1956).
275. S. Strauss and S. L. Madorsky, *J. Research Nat. Bur. Stds.*, **61**, 77 (1958).
276. M. Pike and W. F. Watson, *J. Polymer Sci.*, **9**, 229 (1952).
277. W. F. Watson, in reference 1, Chap. 14.
278. G. M. Bristow and W. F. Watson, in reference 2, Chap. 14.
279. W. J. Burlant and A. S. Hoffman, *Block and Graft Polymers*, Reinhold, New York, 1960.
280. R. J. Ceresa, *Block and Graft Copolymers*, Butterworths, Washington, D.C., 1962.
281. G. Smets and R. Hart, *Fortschr. Hochpolymer. Forsch.* (*Advances in Polymer Science*), **2**, 173 (1960).
282. N. G. Gaylord and F. S. Ang, in reference 1, Chap. 10.B.
283. D. J. Angier, in reference 1, Chap. 12.
284. P. W. Allen, in reference 2, Chap. 5.
285. F. P. Greenspan, in reference 1, Chap. 2.E.

286. C. Roux, R. Pautrat, R. Cheritat and C. Pinazzi, *Compt. Rend.*, **258,** 5442 (1964).
287. G. E. Meyer, L. B. Tewksbury and R. M. Pierson, in reference 1, Chap. 2.C.
288. R. M. Pierson, W. E. Gibbs, G. E. Meyer, F. J. Naples, W. M. Saltman, R. W. Schrock, L. B. Tewksbury and G. S. Trick, *Rubber Plastics Age*, **38,** 592, 708 (1957); *Rubber Chem. Technol.*, **31,** 213 (1958).
289. R. Harrington, *Rubber Age* (*N.Y.*), **85,** 963 (1959).
290. G. E. Meyer, F. J. Naples and H. M. Rice, *Rubber World*, **140,** 435 (1959).
291. C. Pinazzi and G. Levesque, *Compt. Rend.*, **260,** 3393 (1965); *Rev. Gen. Caout. Plast., Ed. Plast.*, **42,** 1012 (1965).
292. C. Pinazzi and G. Levesque, *Compt. Rend.*, **C264,** 288 (1967).
293. C. Pinazzi, G. Levesque and D. Reyx, *Compt. Rend.*, **C263,** 859 (1966).
294. M. S. Lishanskii, V. A. Tsitokhtsev and N. D. Vinogradova, *Vysokomolekul. Soed.*, **8,** 186 (1966); *Rubber Chem. Technol.*, **40,** 934 (1967).
295. J. Lal and W. M. Saltman, *J. Polymer Sci.*, **A-1,4,** 1637 (1966).
296. J. Le Bras, R. Pautrat and C. P. Pinazzi, reference 1, Chap. 2.G.
297. C. Pinazzi, J. C. Danjard and R. Pautrat, *Bull. Soc. Chim. France*, 2433 (1961); *Rev. Gen. Caoutchouc*, **39,** 600 (1962); *Rubber Chem. Technol.*, **36,** 282 (1963).
298. C. Pinazzi and R. Pautrat, *Compt. Rend.*, **254,** 1997 (1962); *Rev. Gen. Caoutchouc*, **39,** 799 (1962).
299. C. Pinazzi, R. Pautrat and R. Cheritat, *Compt. Rend.*, **256,** 2390, 2607 (1963); *Rubber Chem. Technol.*, **36,** 1054, 1056 (1963).
300. R. Pautrat and J. Marteau, *Compt. Rend.*, **C262,** 1561 (1966); *Chem. Abstr.*, **65,** 5632e (1966).
301. F. L. Ramp, E. J. DeWitt and L. E. Trapasso, *A.C.S. Polymer Div., Polymer Preprints*, **6,** 172 (1965).

CHAPTER 10

Olefinic Properties of Cyclopropanes

MARVIN CHARTON

Pratt Institute, Brooklyn, N.Y., U.S.A.

I. INTRODUCTION	512
II. BONDING IN CYCLOPROPANES	513
A. The Trigonally-Hybridized Model	513
B. The Bent-Bond Model	515
C. The Molecular-Orbital Model	518
D. Transformation of Cyclopropane Models	519
E. Other Models of Cyclopropane Bonding	520
F. Calculation of Ring Strain in Cyclopropane	520
III. MOLECULAR GEOMETRY OF CYCLOPROPANE DERIVATIVES	521
IV. CONFORMATION IN SUBSTITUTED CYCLOPROPANES	525
V. IONIZATION POTENTIALS AND SPECTRAL PROPERTIES OF CYCLOPROPANE DERIVATIVES	530
A. Ionization Potentials	530
B. Ultraviolet Spectra	532
C. Near Infrared, Infrared and Raman Spectra	538
D. Nuclear Magnetic Resonance, Nuclear Quadruple Resonance, and Electron Paramagnetic Resonance Spectra	544
VI. MOLAR REFRACTIVITIES AND DIPOLE MOMENTS	548
A. Molar Refractivities	548
B. Dipole Moments	549
VII. OTHER PHYSICAL PROPERTIES	552
VIII. THE CYCLOPROPYL GROUP AS A SUBSTITUENT	553
A. Electrical Effects	553
B. Substituent Effects on Ionization Constants	555
C. Effects on Reaction Rates	557
D. Orientation in Electrophilic Aromatic Substitution	559
E. Effects on Carbonium Ions	560
F. Stabilization of Carbanions	563
G. Stabilization of Radicals	564
IX. TRANSMISSION OF RESONANCE EFFECTS BY THE CYCLOPROPANE RING	566

X.	ADDITION REACTIONS OF CYCLOPROPANE	569
	A. Donor-Acceptor Complexes of Cyclopropanes	570
	B. Electrophilic Addition to Cyclopropane	571
	C. Electrophilic Addition of Halogen to Substituted Cyclopropanes	575
	D. Addition of Hydrogen Halide to Substituted Cyclopropanes	577
	E. Addition of Water and of Carboxylic Acids to Substituted Cyclopropanes	578
	F. Reaction of Substituted Cyclopropanes with Metal Salts	580
	G. Alkylation of Benzene by Substituted Cyclopropanes	581
	H. Electrophilic Addition to Bicyclo[n.1.0]alkanes	582
	I. Terminal Addition to Vinyl Cyclopropanes	585
	J. Relative Reactivity of the Cyclopropane Ring and the Double Bond Toward Electrophilic addition	585
	K. Radical Additions to Cyclopropanes	585
	L. Nucleophilic Additions to Cyclopropanes	587
	M. Hydrogenation of Cyclopropanes	588
XI.	REACTIONS OF THE C—H BOND IN CYCLOPROPANE	592
	A. Acidity of the C—H Bond in Cyclopropanes	592
	B. Configurational Stability of the Cyclopropyl Carbanion	595
	C. Hydrogen Abstraction by Radicals	595
XII.	RECENT DEVELOPMENTS	595
XIII.	REFERENCES	598

I. INTRODUCTION

It has long been known that the cyclopropane ring is, in its physical and chemical properties, somewhat analogous to a carbon–carbon double bond. This striking behaviour, atypical of cycloalkanes, has attracted the attention of many chemists. The purpose of this review is to describe those properties of cyclopropane which are similar to properties of olefins. This problem has been reviewed before[1,2], and most recently by Lukina[3].

There are three major areas in which we may compare the double bond and the cyclopropane ring:

1. Resonance interaction between a substituent and the cyclopropyl or vinyl group;
2. Transmission of the resonance effect of a substituent through the cyclopropane ring or double bond to a 'reaction site';
3. Reactions of cyclopropane derivatives which are analogous to reactions of ethylene derivatives.

The first two categories involve both physical properties and chemical reactions, whereas the third is concerned only with reactions. Papers have been included in this review only in so far as they provide evidence directly bearing on these topics. Reactions of cyclopropanes which are not typical of other cycloalkanes and are also not directly comparable with reactions of the double bond will not be examined here.

A number of abbreviations have been used in this review, that are not of widespread use. They are set forth in Table 1.

TABLE 1. Abbreviations used in this chapter

cy	cyclo (e.g. cyC_3H_6 is cyclopropane)
Hl	halogen atom
Ts	*p*-toluenesulphonyl
T	*trans*
C	*cis*
Vi	vinyl
1-Vn	vinylidene (e.g. 1-MeVnCl is $CH_2{=}CClMe$)
2-Vn	vinylene (e.g. *T*-2-MeVnCl is Me⟋Cl)
2-Pn	*ortho*-phenylene
3-Pn	*meta*-phenylene
4-Pn	*para*-phenylene

II. BONDING IN CYCLOPROPANES

In order to describe the alkenoic properties of cyclopropanes it is first necessary to consider the bonding in these molecules. Bonding in cyclopropanes has recently been reviewed by Bernett[4]. In 1946 and 1947 a number of notes appeared in which bonding in cyclopropane was discussed. This early work is reviewed by Lukina[3]. As Bernett has pointed out there are now three models available for bonding in cyclopropane: the trigonally-hybridized model, the bent-bond model and the molecular-orbital model.

A. The Trigonally-Hybridized Model

This concept is due to Sugden[12], and later, Walsh[14] who constructed a model of cyclopropane in which the carbon atoms are hybridized sp^2. The Sugden-Walsh model is shown in Figure 1. The C—H bond orbitals have not been included in this Figure. In Figure 2 the electron distribution of ring-bonding orbitals for the Sugden-Walsh model is shown. As can be seen in Figure 1, each carbon has an sp^2 hybrid orbital directed to the centre of the ring.

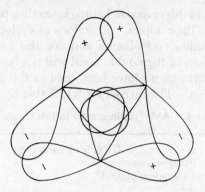

Figure 1. Ring orbitals of the trigonally-hybridized model.

The p orbital on each carbon atom overlaps to form molecular orbitals about the perimeter of the ring. The C—H bonds are formed from the remaining two sp^2 hybrid orbitals on each carbon atom. Thus the H—C—H angle will be 120°. The C—H bonds will lie in a plane normal to the plane of the ring. Assuming that the C—H orbitals are localized, and that the molecular orbitals formed from the overlap of atomic orbitals on carbon are not localized, the molecular orbitals are then given by

$$\frac{1}{\sqrt{3}} (\psi_{1H} + \psi_{2H} + \psi_{3H}) \quad (H1)$$

$$\frac{1}{\sqrt{2}} (\psi_{1H} - \psi_{2H}) \quad (H2)$$

$$\frac{1}{\sqrt{6}} (\psi_{1H} + \psi_{2H} - 2\psi_{3H}) \quad (H3)$$

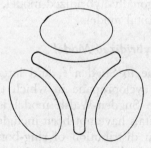

Figure 2. Electron distribution in the occupied orbitals of the trigonally-hybridized model.

where H refers to the sp^2 hybridized atomic orbitals on the carbon atom. Orbitals $H2$ and $H3$ are degenerate. Overlap of the p orbitals on the carbon gives rise to a set of molecular orbitals given by

$$\frac{1}{\sqrt{3}} (\psi_{1P} + \psi_{2P} + \psi_{3P}) \quad (P1)$$

$$\frac{1}{\sqrt{2}} (\psi_{1P} - \psi_{2P}) \quad (P2)$$

$$\frac{1}{\sqrt{6}} (\psi_{1P} + \psi_{2P} - 2\psi_{3P}) \quad (P3)$$

Again $P2$ and $P3$ are degenerate orbitals. $H1$ is a bonding orbital whereas $H2$ and $H3$ are antibonding; $P2$ and $P3$ are bonding orbitals whereas $P1$ is antibonding. Thus the orbitals which are filled are $H1$, $P2$ and $P3$. The resulting probability pattern is shown in Figure 2.

Dyatkina and Syrkin[15] have considered a trigonally-hybridized cyclopropane model. In this model it is assumed that each carbon atom forms C—H bonds using two of the sp^2 hybrid orbitals. The ring bonds are constructed from the remaining sp^2 hybrid orbital and the pure p orbital. The sp^2 hybrid orbitals are combined to form two three-centre orbitals, bonding and an antibonding, occupied by four electrons. The three p orbitals are combined to give two three-centre orbitals lying in the plane of the ring, antibonding and a bonding, of which the lower one is occupied. Bespalov[16] has carried out Hückel molecular-orbital calculations for vinylcyclopropane, divinyl cyclopropane and phenylcyclopropane based on the trigonally-hybridized model.

B. The Bent-Bond Model

Kilpatrick and Spitzer[6], assuming that the bond strength is proportional to the orbital strength in the desired direction, calculated bond strengths and H—C—H bond angles for various C—C—C bond angles. This paper is significant in that it marks the first recognition that the electron density in the bond between carbon atoms in the cyclopropane ring need not be symmetric with respect to the line joining the nuclei. Coulson and Moffit[17,18] extended these ideas, basing their approach on the following arguments: (1) compounds which do not have tetrahedral valence angles must have hybridization which differs from that of compounds with tetrahedral valence angles,

(2) the hybridization of the orbitals used to form the H—C bonds in cyclopropane must be different from those used to form the C—C bonds in the ring. Thus Coulson and Moffit consider the hybridization states of the atomic orbitals in cyclopropane used to form the C—C and C—H bonds as variable parameters. Calculations were then made by the 'valence-bond perfect-pairing approximation'; the hybridization being varied so as to maximize the bond strength (minimize the energy of the system). The resulting model can be described in terms of bent bonds, that is bonds in which the electron density is not cylindrically symmetric with respect to the line joining the bonded nuclei. Ingraham[19] has pointed out that the hybridization states obtained for carbon orbitals forming the C—H and C—C bonds in cyclopropane, are $sp^{2.28}$ and $sp^{4.12}$ respectively. The results of bent-bond model calculations by several authors are given in Table 2. The quantities reported in the Table include the hybridization

TABLE 2. Bent bond model calculations for cyclopropane

Method	Hybridization		% s character		Angles		
	C—H	C—C	C—H	C—C	θ_H	θ_C	ω
Valence-bond perfect-pairing approximation[18,19]	$sp^{2.28}$	$sp^{4.12}$	30.5	19.5	116°	104°	22°
Maximum overlap[20]	$sp^{2.14}$	$sp^{4.49}$	31.8	18.2	117°48′	102°52′	21°26′
MO with localized orthogonal bent MO, s[21]			12–14	0–4		102–110°	21–25°
MO del Re method for determining hybridization[23]	sp^2		33				21°7′
MO of maximum overlap[24]	sp^2	sp^5	33	17	119°38′		

states of the orbitals, the angle between the orbitals used to form the C—H bonds (θ_H), the angle between the orbitals used to form the bent bonds in the ring (θ_C) and the angle formed by the line joining the carbon nuclei and the direction of the bent-bond orbital ω, as shown in Figure 3.

Coulson and Goodwin[20] introduced the use of the method of maximum overlap for the study of cyclopropane bonding. Their

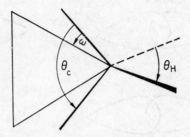

Figure 3. Angles involved in the bent bond model.

calculations based on this method agreed with the results they obtained previously by the valence-bond approach. Peters[21] obtained the cyclopropane structure from localized orthogonal MO's which were constructed from low principal quantum AO's of the atoms. The results are in agreement with Coulson and Moffit in that the C—C bonds are extensively bent and their carbon hybrid AO's contain very little 2s character. The C—H bonds resemble those of ethylene. Flygare[22] has discussed the general question of bent bonds. Veilard and del Re[23] have applied the del Re procedure for determining hybrid orbitals from the overlap matrix to the structure of cyclopropanes. These results are presented in Table 2.

Randic and Maksic[24] have used the method of maximum overlap in a study of cyclopropane, spiropentane, nortricyclene and 1,1-dimethylcyclopropane. The results for cyclopropane are in accord with a C—C bond hybridization in the carbon orbitals of about sp^5, and carbon orbitals in the C—H bond hybridized about sp^2. Trinajstic and Randic[25] have examined the hybridization in various methyl-substituted cyclopropanes by the method of maximum overlap. Their results are: (1) hybridization of the methyl group is unaffected by further introduction of methyl groups; (2) hybridization of one ring carbon atom does not affect the hybridization of other ring carbon atoms; (3) a methyl substituent causes a decrease in the s content of the hybrid involved in the substitution, while the remaining hybrids at the substitution site, which must therefore increase in s content, do so by approximately equal amounts.

Gassman[26] has recently suggested that reactivity in certain cyclopropane derivatives can be accounted for by the concept of twist bent bonds (Figure 4), which should be very reactive.

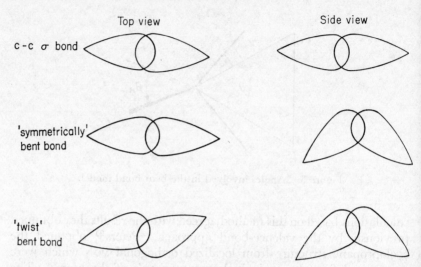

Figure 4. Appearance of atomic orbitals forming various types of C—C bonds.

C. The Molecular-Orbital Model

Hoffmann[27] has carried out an extended Hückel theory calculation on cyclopropane. The lowest occupied MO is almost entirely composed of carbon $2s$ orbitals (91% s character). There is very little contribution from the carbon $2p_x$ and $2p_y$ and the hydrogen $1s$ orbitals and no contribution from the carbon $2p_z$ orbitals. The next six occupied MO's are almost exclusively associated with the C—H bonds. The two highest energy occupied MO's which constitute a degenerate pair are composed almost entirely of carbon $2p_x$ and $2p_y$ orbitals in the plane of the ring (93% p character). Yonezawa, Shimizu and Kato[28] have carried out an extended Hückel calculation on cyclopropane and methylcyclopropane. They found that the C—C and C—H bonds in cyclopropane have 19·53 and 30·49% s character respectively. Also calculated were energies of the highest occupied and lowest vacant energy levels in cyclopropane and methylcyclopropane. These results are shown in Table 3. It is interesting to note that the gap between the highest occupied and lowest vacant levels is the smallest for cyclopropane.

Baird and Dewar[29b] have reported SCF MO calculations for cyclopropane and its *cis-* and *trans-*1,2-dimethyl derivatives. The charge distributions they have found are shown in Figure 5.

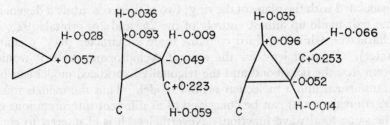

Figure 5. Net charges on atoms in cyclopropane, *cis*-1,2-dimethylcyclopropane and *trans*-1,2-dimethylcyclopropane.

Hoffmann[27] has carried out an extended Hückel MO calculation for protonated cyclopropane. The proton is in the plane of the ring and equidistant from the two nearest carbon atoms. An extended Hückel MO calculation for protonated cyclopropane has also been reported by Yonezawa, Shimizu and Kato[28a], who find that the most stable configuration has the proton in the plane of the ring. They report that the changes in total electronic energy and in charge distribution caused by protonation of cyclopropane are comparable to those caused by protonation of ethylene.

TABLE 3. Energies of the highest occupied (H.O.) and lowest vacant (L.V.) orbitals in cycloalkanes[28]

	H.O. (eV)		L.V. (eV)		Gap (eV)	
n	cyC_nH_{2n}	Mecy-C_nH_{2n-1}	cyC_nH_{2n}	Mecy-C_nH_{2n-1}	cyC_nH_{2n}	Mecy-C_nH_{2n-1}
3	−13·052	−12·612	−4·344	−4·166	8·708	8·446
4	−12·658	−12·400	−0·715	−0·659	11·943	11·741
5	−12·741	−12·626	0·338	−0·101	13·079	12·525
6	−12·534	−12·493	2·085	1·054	14·619	13·547

D. Transformation of Cyclopropane Models

Bernett[4] has shown that it is possible to transform the trigonally-hybridized model into the bent-bond model. He has further noted that it is possible to transform the bent-bond model into a MO model. In this transformation the three localized orbitals associated with the three bent C—C bonds which were formed from a linear combination of two sp^5 hybridized orbitals, have been transformed into delocalized symmetry orbitals. The three symmetry orbitals which result are associated with the plane of the ring; two of them constitute a degenerate pair made up almost entirely of the $2p_x$ and $2p_y$ orbitals (92% p character), while the third contains more s character (67% p character). Then at least for the case of cyclopropane itself, it would seem that the bent-bond and the trigonally-hybridized model can be transformed into a molecular-orbital model. Thus the models must be equivalent and can be conceived of as different interpretations of the same total wave function. Nevertheless, it is of interest to consider which of the models is more useful in describing the behaviour of cyclopropanes.

It must be noted however, that Kemp and Flygare[26a] have stated that the transformation of the trigonally-hybridized model into the bent-bond model is valid only on the basis of the assumption that:

1. AO's do not overlap with their neighbours;
2. There are an equal number of electrons in each combining hybrid orbital in the trigonally-hybridized model. On the basis of these assumptions, a linear transformation on the trigonally-hybridized orbitals then yields the bent-bond model hybrid orbitals. If the two in-ring trigonally hybridized orbitals do not have equal ionic parameters a linear combination will not yield the bent-bond orbitals.

E. Other Models of Cyclopropane Bonding

A fourth approach to the bonding in cyclopropanes had been proposed by Handler and Anderson[29]. In this approach it was proposed that the molecular orbitals used to form the ring bonds need not be orthogonal to each other. The orbitals used to form the C—H bonds were considered to be orthogonal to each other and to the C—C orbitals. This proposal has been criticized by Peters[21], on the basis that certain of the quantities required for a calculation would be difficult to estimate, at least at the present time. Jaffé[29c] has considered the significance of 'many centre' forces in the bonding in cyclopropane.

F. Calculation of Ring Strain in Cyclopropane

Kilpatrick and Spitzer[6] calculated the strain in cyclopropane to be 10 kcal per methylene group. It will be recalled that their work is based upon the use of hybrids of s and p orbitals and the assumption that bond strength is proportional to orbital strength in the desired direction. Bernett[4], in an extension of this approach, has calculated the C—C bond energy in cyclopropane to be 70·14 kcal/mole. Thus from the value of 78·82 kcal/mole for the C—C bond in ethane, each bond is strained by 8·68 kcal/mole corresponding to a total strain energy in cyclopropane of 26·04 kcal/mole. The experimental value reported is 27·15 kcal/mole[30]. Bernett reports a value of 55·7 kcal/mole for the strain energy in spiropentane, the experimental value being 63·13 kcal/mole[30,31]. Baird and Dewar[29b] have calculated a value of 31·0 kcal/mole for the strain energy in cyclopropane; they also report calculations of the strain energies of *cis*- and *trans*-1,2-dimethylcyclopropane.

III. MOLECULAR GEOMETRY OF CYCLOPROPANE DERIVATIVES

Considerable interest has been evinced over the years in the structure of cyclopropane derivatives. The earlier literature has been reviewed by Goldish[32]. It is now possible to describe the electron-density distribution in compounds by means of X-ray diffraction studies. Fritchie[54] in his X-ray diffraction examination of 2,5-dimethyl-7,7-dicyanonorcaradiene has reported that the electron density in the cyclopropane ring is outside of the line joining the nuclei, and therefore supports the bent-bond model. The observed electron density corresponds to a bending angle of 20°. This is in very good agreement with the various results for the bent-bond models reported in section II.C.

Further support for this evidence comes from the study of Hartman and Hirshfeld[55] on *cis*-1,2,3-tricyanocylopropane. A detailed analysis of the electron distribution shows that the residual bonding electron distribution is in good agreement with the bent-bond model.

The available data on structural determinations have been summarized in Table 4. The results reported therein are those which are believed to represent the best results now available. The values of the H—C—H angle reported in Table 4 support a hybridization state for the carbon orbitals used in the C—H bonds of approximately sp^2 character. The values observed for the X—$C_{(1)}$—X angles are also in agreement with sp^2 character for the orbitals used in forming these bonds.

Two interpretations have been proposed for the variation of C—C bond length. In one the length is considered to be solely a factor of hybridization changes, whereas in the other hybridization and delocalization effects are believed to be of comparable magnitude[58]. A comparison (as made in Table 5) of the ring to substituent bond length

TABLE 4. Molecular geometry

No.	Compound	Method[a]	$C_{(1)}$—$C_{(2)}$ length (Å)	C—H length (Å)
1	Cyclopropane	E	1·510	1·089
		I	1·524 ± ·014	
2	1,1,2,2-Tetramethylcyclopropane	X	1·52 ± ·03[b]	
3	Bicyclopropyl	E	1·517 ± ·005	1·094
4	Spirocyclopentane	E	1·48 ± ·03	
			1·51 ± ·04 ($C_{(2)}$—$C_{(3)}$)	
5	Chlorocyclopropane	E	1·52 ± ·02	
		M	1·513 ± ·001	1·105
			1·513	1·08
			1·515($C_{(2)}$—$C_{(3)}$)	
6	1,1-Dichlorocyclopropane	E	1·52 ± ·02	
		M	1·532	1·085
			1·534($C_{(2)}$—$C_{(3)}$)	
7	3-Methylene-*trans*-1,2-cyclopropanedicarboxylic acid	X	1·55	
			1·49($C_{(1)}$—$C_{(3)}$)	
8	Cyclopropanecarbohydrazide	X	1·52, 1·49, 1·48	
9	2,3-Dihydro-2,3-methylene-1,4-naphthoquinone	X	1·496	
			1·522($C_{(2)}$—$C_{(3)}$)	
10	Cyanocyclopropane	M	1·513 ± ·001	1·107 ± ·002
11	*cis*-1,2,3-Tricyanocyclopropane	X	1·518	
12	2,5-Dimethyl-7,7-dicyanonorcaradiene	X	1·554, 1·559	
			1·501	
13	Cyclopropane	M	1·515	1·087 ± ·004
14	Cyclopropanecarboxaldehyde	E	1·507	1·115
15	Methyl cyclopropyl ketone	E	1·510	1·126
16	Cyclopropanecarboxylic acid chloride	E	1·506	1·105
17	6,6-Diphenyl-3,3-diethyl-3-azabicyclo[3,1,0]hexane bromide monohydrate	X	1·521, 1·525, 1·514	
18	Nortricyclene	E	1·50	

in cyclopropane derivatives with the bond length in corresponding alkyl compounds and ethylene or benzene derivatives, should be indicative of the degree of hybridization in the orbitals used to form the carbon to substituent bond. When several bond lengths were available, as was the case for the C—H bond, those believed to be most reliable were chosen. The results show clearly that the hybridization state for the carbon orbital involved in the formation of the bond to the substituent, must be between sp^3 and sp^2 and probably is closer to of cyclopropanes

C—X length Å	∠CCX(°)	∠HCH(°)	∠CX¹X¹(°)	Ref.
		115·1		33, cf. 34–38
		120		37
1·52 ± ·03			114 ± 6	43
1·517 ± ·01	115·2 ± 1·5	117·3 ± 1·5	115·2 ± 1·5	41
		120 ± 8		42
1·76 ± ·02				44
1·778		114·6 ± 3	120·89 ± ·02	45, 46
1·740				47
1·76 ± ·02			112 ± 4	44
1·734		117·6	114·6	48
1·48	115, 118			49
1·50(C$_{(2)}$—C$_{(3)}$)	115, 117			
1·48	118			50
1·483	116·7, 118·3			51
1·472		114·6 ± 1	119·6 ± ·5	45
1·449				55
1·436[c]	117·5, 118[c]		115·2	54
1·475[d]				
		114·7		56
				52
				53
	111 ± 2·5			53
1·499, 1·50	121·2, 114·4		113·0	57
1·507				
1·59				57a

[a] E, electron diffraction; I, infrared spectroscopy; X, X-ray diffraction; M, microwave spectroscopy.
[b] Average for all C—C.
[c] C—CN.
[d] C—C—C.

sp^2 hybridization. Then overall, the literature on the molecular structure of cyclopropane derivatives is in good agreement with the models for the bonding in cyclopropane reported in section II. The

TABLE 5. Comparison of C—X bond lengths (Å) in similarly substituted alkanes, cyclopropanes and alkenes

X	Compound	C—X[a]	Cpd. No.[b]	C—X	Compound	C—X[a]
H	propane[c]	1·096	1	1·089	ethylene	1·084
			3	1·094		1·086
			5	1·08		
			6	1·085		
			13	1·087		
Me	methylpropane	1·540	2	1·52	propene	1·488
Cl	2-chloropropane	1·775 ± ·02	5	1·76, 1·74	chloroethylene	1·72, 1·74
Cl	2,2-dichloropropane	1·776 ± ·02	6	1·76 1·74	1,1-dichloroethylene	1·727, 1·710, 1·707
CHO	methylpropanal[d]	1·528 (avg)	14	1·507	acraldehyde	1·45
COR	cyclohexanone	1·54	15	1·510		
CO₂H	pentanoic acid[e]	1·53	7	1·48	acrylic acid	1·44 ± ·03
CONHNH₂	dodecanoic acid hydrazide	1·49	8	1·48	benzamide	1·48
COCl	chloroacetyl chloride	1·52 ± ·04	16	1·506	acryloyl chloride	1·44 ± ·03
COC^{sp2}	sodium pyruvate[f]	1·518	9	1·483	1,4-benzoquinone	1·49 ± ·04
CN	t-butylcyanide	1·46	10	1·472	acrylonitrile	1·426
			11	1·449		
			12	1·436		
Vi	propene	1·488	12	1·475	1,3-butadiene	1·483 ± ·01
Ph	toluene	1·51 ± ·02	17	1·499	2-bromo-1,1-di-(4'-tolyl)-ethylene	1·499

[a] *Interatomic Distances, Interatomic Distances Supplement*, Special Publications No. 11 and No. 18, The Chemical Society, London, 1958 and 1965.
[b] Numbers refer to Table 4.
[c] D. R. Lide Jr., *J. Chem. Phys.*, **33**, 1514 (1960).
[d] L. S. Bartell and J. P. Guillory, *J. Chem. Phys.*, **43**, 654 (1965).
[e] R. F. Scheuerman and R. L. Sass, *Acta Crystallog.*, **15**, 1244 (1962).
[f] S. S. Tarale, L. M. Pant and H. B. Biswas, *Acta Crystallog.*, **14**, 1281 (1961).

X-ray diffraction studies on *cis*-1,2,3-tricyanocyclopropane[55] and 2,5-dimethyl-7,7-dicyanononorcaradiene[54] provide strong support for the bent-bond model. As all three models are in agreement with the remaining structural evidence, it may be concluded that the bent-bond model provides the best interpretation of the molecular structure of cyclopropanes.

IV. CONFORMATION IN SUBSTITUTED CYCLOPROPANES

The existence of preferred conformations in π-electron systems bearing π-electron or lone-pair substituents, has been known for some time[59]. Unfortunately no comprehensive review of this type of conformational equilibrium exists. That analogous conformational preference is to be expected in substituted cyclopropanes is an obvious consequence of the trigonally-hydridized model. As Bernett[4] has pointed out the sp^5 orbitals used to form the ring bonds in the most recent bent-bond model of cyclopropane (Section II.B) are capable of overlap with appropriately oriented adjacent p orbitals. The extent of overlap between the sp^5 ring orbitals and an adjacent p orbital will not be as great, however, as that between two adjacent p orbitals which are parallel. Thus, while both the trigonally-hydridized and bent-bond models predict the existence of preferred conformations, the interaction should be greater from the trigonally-hydridized model than from the bent-bond model. A considerable number of reports concerning the conformation of cyclopropyl derivatives in which the ring is conjugated with a π-bonded substituent (such as Ph, Vi or —C=O), have appeared in the literature. All of the available conformational studies on cyclopropane derivatives are summed in Table 6. A number of authors have reported values of ΔH for the conformational equilibrium, other authors have reported values of ΔE, the energy difference between lower-energy and higher-energy conformations. These values are also reported in Table 6.

Vinylcyclopropane may exist in three possible conformations; the *s-trans* (**1**), the *s-cis* (**2**) and the gauche (**3**). The preferred conforma-

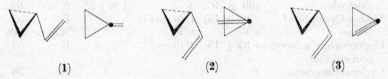

(**1**)   (**2**)   (**3**)

tion generally is the *s-trans*. Steric effects may cause the *s-cis* or gauche conformations to be preferred. Thus Ketley, Berlin and Fisher[64] have proposed that 1-aryl-1-cyclopropylpropenes, in which the cyclopropyl and methyl groups are *trans* (**4**) exist preferentially in a bisected *s-cis* conformation, whereas the corresponding compounds in

Table 6. Conformation in substituted cyclopropanes

Compound	Conformation[a] Low energy	High energy	H or E^b (kcal/mole)	Method[c]	Ref.
Vinylcyclopropane	t	g (and c?)	·66 − ·93	N.m.r.	60
	t	c	1·1 ± ·2	N.m.r.	61
	t	g		E	62
trans-1,2-Dicyclopropyl-ethylene (solid)	c,c or t,t			I.r	61
(liquid)	c,c or t,t	c,t		I.r.	61
trans-1-Isocyanato-2-vinylcyclopropane	t	g	·6	N.m.r.	63
trans-1-Carbomethoxy-2-vinylcyclopropane	t	g	·7	N.m.r.	63
trans-1-Cyclopropyl-1-(substituted phenyl)-propene	c			N.m.r.	64
cis-1-Cyclopropyl-1-(substituted phenyl)-propene	g			N.m.r.	64
Phenylcyclopropane	b			E	65
	b	s	1·4		67
4-Chlorophenylcyclo-propane	b	s		N.m.r.	67
4-Methylphenyl-cyclopropane	b	s		N.m.r.	67
4-Methoxyphenyl-cyclopropane	b	s		N.m.r.	67
1-Phenyl-2,2-dimethyl-cyclopropane	g			N.m.r.	68
Bicyclopropyl	t			I.r,R	69
	g(60)	t(40)	≤ 0·5	E	40,41
1,1'-Dimethylbicyclo-propyl	t			I.r,R	69
Cyclopropanecarbox-aldehyde	c(55 ± 10)	t(45)		E	63a,52
	c80 ± 70°)	g	1 to 1·5	N.m.r.	69a
Methyl Cyclopropyl ketone	c(80 ± 15)	t(20 ± 15)		E	53
	c(70)	t(30)		N.m.r.	70a
Cyclopropanecarboxylic acid chloride	c(85 ± 15)	t(15 ± 15)		E	53
Cyclopropanecarbo-hydrazide	c			X	50
2,3-Methylenecyclopro-pane dicarboxylic acid (Feist's acid)	c,c			X	49
trans-1-Acetyl-2-methyl cyclopropane	c(70)	t(30)		N.m.r.	70a

TABLE 6. (continued)

Compound	Conformation[a] Low energy	High energy	H or E[b] (kal/mole)	Method[c]	Ref.
1-Acetyl-2,2-dimethyl-cyclopropane	c(70)	t(30)		N.m.r.	70a
1-Acetyl-1-methylcyclo-propane	c(50)	t(50)		N.m.r.	70a
trans-1-Acetyl-2-iso-propylcyclopropane	c(70)	t(30)		I.r.	70b
trans-1-Acetyl-2-n-butylcyclopropane	c(70)	t(30)		N.m.r.	70a
7-Acetyl-7-methyl-bicyclo[4·1·0]heptane	c(65)	t(35)		N.m.r.	70a
6-Acetyl-6-methyl-bicyclo[3·1·0]hexane	c(62)	t(38)		N.m.r.	70a
1-Acetylbicyclo[4·1·0]-heptane	c(55)	t(45)		N.m.r.	70a
trans-1-Acetyl-2-phenylcyclopropane	c(100)	—		I.r.	70b
1-Acetyl-1-phenyl-cyclopropane	t(100)	—		I.r.	70b
Methyl-2,2,3,3-tetra-methylcyclopropane carboxylate	g	c and/or t		N.m.r.	70
9-Cyclopropylanthra-cene radical anion	s		·9	E.s.r.	71
9-Cyclopropylanthra-cene radical cation	b		1·0	E.s.r.	71
1,4-Dicyclopropylnaph-thalene radical anion	s		·9	E.s.r.	71
Dimethylcyclopropyl-methylcarbonium ion	b			N.m.r. Other	73,73a, 74 74a
Dicyclopropylmethyl carbonium ion	t,t			N.m.r.	74
Tricyclopropylmethyl carbonium ion	t,t,t			N.m.r.	74
Tris(4-cyclopropyl-phenyl)methyl carbonium ion	b			N.m.r.	76

[a] t = s-trans; c = s-cis; g = gauche; b = bisected; s = symmetrical; 't' = 's-trans like'. Percents are given in parentheses when available.

[b] H values are in italics.

[c] N.m.r. = Nuclear magnetic resonance; E.s.r. = electron spin resonance; I.r. = infrared spectroscopy; R = Raman spectroscopy; E = electron diffraction; X = X-ray diffraction.

which the cyclopropyl and methyl groups are *cis* (**5**) exist in a gauche conformation. These conformations are said to be in accord with

(**4**) (**5**)

long-range coupling constants in the n.m.r. spectrum. Results for a range of carbonyl compounds including formyl, acyl, carboxamido, carboxyl, alkoxycarbonyl and halocarbonyl groups show that in general the *s-cis* conformation is preferred, although in some cases steric effects may be responsible for a preference for the *s-trans* or gauche conformations. This is apparently the case in methyl 2,2,3,3-tetramethylcyclopropane carboxylate which prefers the gauche conformation and in 1-acetyl-1-phenylcyclopropane which prefers the *s-trans* conformation.

Aromatic rings bonded to the cyclopropane ring are considered to exist in three possible conformations; a bisected (**6a**) form, in which the plane of the aromatic ring bisects the cyclopropane ring, a symmetric conformation (**6b**) in which the aromatic ring has been rotated through 90° from its position in the bisected conformation, and a gauche

(**6a**) Top view   (**6b**) Top view

conformation, in which the aromatic ring has been rotated between 0° and 90° from its position in the bisected conformation. The bisected conformation is generally preferred. Again, steric effects may cause a preference for the symmetric or gauche conformations. Thus, 1-phenyl-2,2-dimethylcyclopropane prefers the gauche conformation. The preference of the 9-cyclopropylanthracene and 1,4-dicyclopropyl-naphthalene radical anions for the symmetric conformation may be due at least in part to steric effects. The preference of the 9-cyclopropyl-anthracene cation for a bisected conformation may be accounted for in terms of resonance stabilization by the cyclopropyl groups (an effective electron donor by resonance, see section VIII.A) outweighing the steric effect.

While it is not possible on the basis of the available information to come to a definitive conclusion, the results obtained for bicyclopropyl

10. Olefinic Properties of Cyclopropanes

suggest that the energy difference between the preferred and higher-energy conformations is considerably smaller than that between the *s-trans* and *s-cis* conformations of 1,3-butadiene. This result would suggest that the bent-bond model for the bonding in cyclopropane is to be preferred to the trigonally-hybridized model. This argument is supported by the magnitude of the energy difference between conformations in vinylcyclopropane and in substituted vinylcyclopropanes (see Table 6). These energy differences, while apparently larger than those observed between conformations in bicyclopropyl, are nevertheless smaller than that which is observed in butadiene.

The question of the conformation in cyclopropyl carbonium ions has received considerable attention. Dialkyl cyclopropyl carbonium ions may exist in a bisected (**7**), symmetric (**8**) and gauche conformations, analogous to those for phenylcyclopropane. The bisected con-

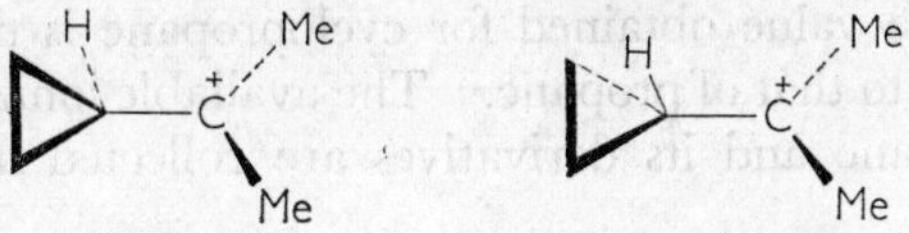

(**7**)   (**8**)

formation seems to be preferred. Dicyclopropyl and tricyclopropyl carbonium ions may exist in *s-trans*, *s-cis* and gauche conformations. The preferred *s-trans* conformations are shown (**9,10**).

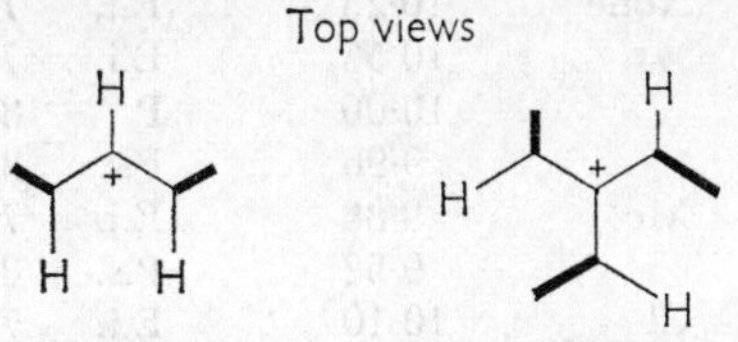

Top views

(**9**)   (**10**)

These results are in accord with the extended Hückel MO calculation made by Hoffmann[27], who found that a bisected conformation should be preferred for the cyclopropyl carbonium ion. In a later paper, Hoffmann used the extended Hückel MO method to predict an energy minimum for the bisected conformations of cyclopropanecarboxaldehyde[75].

The conformational preference of the cyclopropyl group in semidiones has been studied by means of e.s.r. spectra[72]. The results show the presence of a single conformation or of a time averaged pair of enantiomeric conformations.

Evidence has been presented for a solvent dependence of the conformational equilibrium in the case of cyclopropanecarboxaldehyde[69a].

V. IONIZATION POTENTIALS AND SPECTRAL PROPERTIES OF CYCLOPROPANE DERIVATIVES

A. Ionization Potentials

The first report of an ionization potential for cyclopropane is that of Field[77], using the electron-impact method. In a later work Field, again using this method, examined the ionization potentials of methyl cyclopropane, chlorocyclopropane, and 1,1-dichlorocyclopropane[78]. Pottie, Harrison and Lossing[79] carried out a determination of ionization potentials of cycloalkanes by the electron-impact method. They noted that the value obtained for cyclopropane is closer to that of ethylene than to that of propane. The available ionization potentials for cyclopropane and its derivatives are collected in Table 7. In

TABLE 7. Ionization potentials of substituted cyclopropanes

Substituent	I(eV)	Method[a]	Ref.
None	10·23	E.i.	77
	10·53	E.i.	79
	10·06	P	80
	9·96	P.s.	82
Me	9·88	E.i.	78
	9·52[b]	P.s.	82
Cl	10·10	E.i.	78
1,1-Cl$_2$	10·30	E.i.	78
CN	11·2	E.i.	81

[a] E.i. = electron impact; P = photoionization; P.s. = photoelectron spectroscopy.
[b] Higher ionization potentials were also determined.

Table 8 a comparison is made of ionization potentials for cyclopropane and its derivatives with those of corresponding isopropyl, propyl, vinyl and phenyl derivatives. Unfortunately, it is impossible to make a comparison with other cycloalkane derivatives as there seems to be a paucity of data on these compounds in the literature.

TABLE 8. Comparison of ionization potentials

X	Y	1,1-cyC$_3$H$_4$XY[a]	I-PrX[81]	PrX[81]	ViX[81]	PhX[81]
H	H	10·23	11·07	11·07	10·51	9·56
Cl	H	10·10	10·78	10·82	10·00	9·60
Me	H	9·88	10·56	10·63	9·73	9·18
CN	H	11·2		11·67	10·91	10·09
Cl	Cl	10·30				

[a] From table 7.

The electron-impact ionization potentials of cyclopropane and its methyl, chloro and cyano derivatives have been correlated with the extended Hammett equation[83]

$$Q_\text{X} = \alpha\sigma_{\text{I},\text{X}} + \beta\sigma_{\text{R},\text{X}} + h \qquad (1)$$

A statistically significant correlation was obtained. The values of α and β are set forth in Table 9, as are the values of α and β for correlations of the ionization potentials of ethylene and benzene derivatives. The results show that the ionization potentials of substituted

TABLE 9. Values of α, β and ε for the correlation of ionization potentials of GX with equation (1)

G	α	β	ε	Ref.
cyC$_3$H$_5$	1·19	2·95	2·47	84
Vi	1·32	2·51	1·90	85
Ph	1·04	1·50	1·44	85

cyclopropanes behave in a manner analogous to those of substituted ethylenes and substituted benzenes. They establish the importance of the resonance effect of the substituent on the ionization potential. This is further borne out by a comparison of the values of the parameter ε for the cyclopropyl, vinyl and phenyl derivatives, where ε is defined as

$$\varepsilon = \frac{\beta}{\alpha} \qquad (2)$$

Electrical substituent effects may be resolved into contributions from σ_I, the localized (field and inductive) and σ_R, the delocalized (resonance) effects. ε, the ratio of β to α (the coefficients of σ_R and σ_I in equation 1) is a convenient parameter for describing the composition of the substituent effect. The values of ε in Table 9 show that the resonance effect of a substituent on the ionization potential of a substituted cyclopropane is comparable to its effect on the ionization potential of a substituent in a substituted ethylene or benzene.

B. Ultraviolet Spectra

The ultraviolet spectra of cyclopropane derivatives have been reviewed by Pete[86]. As early as 1918, Carr and Burt[87] reported that the ultraviolet spectra of a series of derivatives of benzoyl and anisoyl cyclopropanes resemble those of the corresponding compounds with conjugated double bonds. Klotz[88] examined the ultraviolet spectra of steroids and terpenes containing cyclopropane rings adjacent to double bonds or carbonyl groups. He observed that these spectra were indicative of conjugation effects. Rogers has compared the spectra of ethylbenzene, phenylcyclopropane and styrene (Table 10). These

TABLE 10. U.v. spectra of substituted benzenes (PhX)[89]

X	λ_{max}	$10^{-3}\varepsilon$	λ_{max}	$10^{-3}\varepsilon$
cyC$_3$H$_5$	220	8·4	274	0·28
Et	206	32·0	259	0·17
Vi	245·5	16·0	290·4	0·55

results show that the cyclopropyl group exerts a bathochromic effect on the absorption between that of the ethyl and vinyl groups[89]. Robertson, Music and Matson[90] have examined the ultraviolet spectra of ethylbenzene, cycloalkyl benzenes and styrene. They have defined spectroscopic bond orders by means of the expression

$$\text{SBO}_X = \frac{1 + \nu_B - \nu_X}{\nu_B - \nu_S}$$

where ν_B, ν_S and ν_X are the frequencies of the 00 bands for benzene, styrene and PhX (Table 11). They indicate a behaviour for the cyclopropyl group which is again intermediate between that for alkyl

TABLE 11. Spectroscopic bond orders (SBO) of substituted benzenes (PhX)[90]

X	Spectroscopic Bond Order
Et	1·17
Me	1·18
cyC$_3$H$_5$	1·67
Vi	2·00

groups and that for the vinyl group. Music and Matson[91] were able to calculate the spectrum of phenylcyclopropane by means of MO theory on the basis of the trigonally-hybridized model of Sugden and Walsh. The effect of *para* substituents upon the 'primary' ($^1A_{1g} - {}^1B_{1u}$) forbidden transition in benzene has been studied by Strait and his coworkers[92]. Their results are presented in Table 12.

The existence of conformations such as *s-cis*, *s-trans* and bisected indicates that the degree of overlap between the cyclopropane ring and an adjacent double bond or lone pair will be dependent on the geometry of the molecule, and may therefore be expected to have an effect on the nature of the ultraviolet spectrum. This point was first noted by Music and Matson[91] who used a bisected conformation of phenylcyclopropane in their MO calculation of its u.v. spectrum. It was also suggested by Cromwell and Hudson[93] that a preferred geometry is required for maximal effect of the cyclopropyl group on the ultraviolet spectrum. The importance of molecular geometry on the ultraviolet spectrum of phenylcyclopropanes has been investigated by Goodman and Eastman[94] who studied phenylcyclopropane and compounds **11, 12** and **13**. They found that the spectra of all four compounds are very similar and report that the results are explicable on the postulate that the cyclopropane ring is a source of high electron

TABLE 12. 'Primary' ($^1A_{1g} - {}^1B_{1u}$) transition in 1,4-disubstituted benzenes (4-XPnY)[92]

Y	X= H		OMe		NO$_2$	
	λ_{max}	$10^{-3}\varepsilon$	λ_{max}	$10^{-3}\varepsilon$	λ_{max}	$10^{-3}\varepsilon$
cyC$_3$H$_5$	220	8·4	225	7·5	280	11·0
Me	206	3·2	223	7·7	264	10·4

density, whose polarizability is devoid of stereochemical bias. For this to be the case, the localized electrical effect ('inductive' or field effect) of the cyclopropyl group would have to be considerably more electron donor in character than that of an ordinary alkyl group. It will be pointed out later that in fact, as measured by the σ_I constants, the cyclopropyl group has about the same electron donor effect as alkyl groups. It is difficult to account for the observed behaviour of compounds **11, 12,** and **13**. Bernett[4] has suggested that, at least in part, the results may be explained in terms of the variation of overlap between a p orbital and the cyclopropane ring if the bent-bond model is used. While in the case of the bent-bond model, maximum overlap will result when the bisected conformation is attained, rotation of the plane of the cyclopropyl group will result in decreasing overlap of one of the sp^5 ring orbitals with the p orbital, with increasing overlap of the other sp^5 orbital with the p orbital. Then this suggests that the magnitude of overlap between an adjacent p orbital and the cyclopropane ring should not only be smaller than that observed between two p orbitals, but should also be less sensitive to the dihedral angle (Figure 6). Pete[86], in his review, has noted that there is a definite dependence in the spectra of vinylcyclopropanes upon the dihedral angle (the angle made by the plane of the cyclopropane ring with the plane of the double bond). The bisected conformation has a dihedral angle of 90°. The spectrum of tricyclo[2·2·2·0$^{2-6}$]oct-7-ene (**14**) shows a bathochromic shift of 23mμ with respect to cyclohexene[95]. In this compound, of course, the dihedral angle is 90°. Many other examples are cited by Pete, who gives the value 16 mμ to the effect of the cyclopropyl group on a double bond for a dihedral angle of 90°; 12mμ for

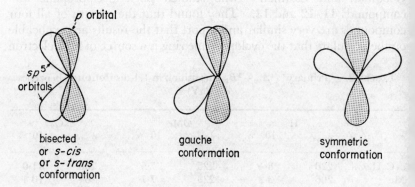

Figure 6. Overlap of p and sp^5 orbitals.

109° and 6mμ for 115°. Perhaps the results obtained by Goodman and Eastman[94] can be explained, at least in part, by steric effects. Stuart-Briegleb models suggest that the spiro compound **11**, which would otherwise be expected to exhibit a dihedral angle 90° and therefore the greatest bathochromic effect of the cyclopropyl group, may show some steric strain.

The effect of a cyclopropyl group on the ultraviolet spectrum of an alkene is shown in Table 13 where λ_{max} and molar extinction co-

TABLE 13. Effect of the cyclopropyl group on the u.v. spectrum of cyclohexene in ethanol[86]

	Cyclohexene	1,3-Cyclohexadiene	Bicyclo[4·1·0]hept-2-ene
λ_{max}	190	257	222
$10^{-3}\varepsilon$	7·25	8	5

efficients of cyclohexene, 1,3-cyclohexadiene and bicyclo[4·1·0]hept-2-ene are compared. The maximum for the compound containing the cyclopropane ring is about half way between that of cyclohexene and that of the diene.

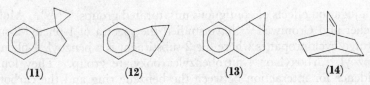

(11)   (12)   (13)   (14)

A sharp band in the ultraviolet spectrum of allenic cyclopropanes has been observed at 203–4mμ (ε 2400), ascribed to conjugation of π-electrons with the cyclopropane ring[96], as shown in Figure 7. It has been reported[97] that 2-cyclopropylpyridine absorbs a wavelength in between those of 2-propylpyridine and 2-vinylpyridine. It has since

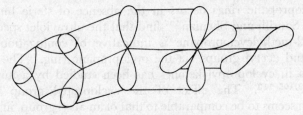

Figure 7. Overlap in cyclopropylallene.

been shown[98] that the sample of 2-cyclopropylpyridine studied might not have been pure. In contrast with 4-phenylpyridine, the ultraviolet spectra of 4-cyclopropylpyridine dissolved in 0·1N NaOH solution shows no significant difference from those of 4-picoline or 4-benzylpyridine. However, the 4-cyclopropylpyridinium cation shows a considerably enhanced intensity of absorption although there is little difference in the position of the peak[98]. The spectral results obtained for derivatives of cyclopropylpyridine substituted in the cyclopropane ring, are considered to be evidence for conjugation between the cyclopropane ring and the pyridine ring.

Considerable attention has been devoted to cyclopropane derivatives in which a carbonyl group is bonded to the cyclopropane ring[87–89]: phenyl and styryl cycloalkyl ketones have been studied[99]. Eastman has proposed on the basis of the study of compounds 15 and 16 and the semi-carbazone of 16, that the cyclopropane ring cannot transmit the

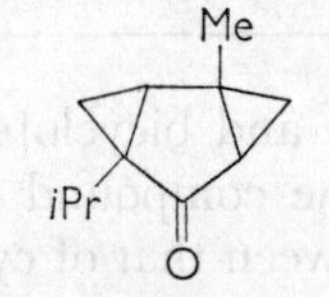

(15) 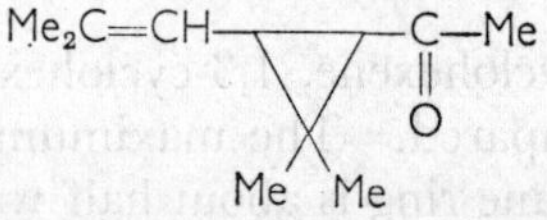 (16)

'conjugation effects of contiguous unsaturated groups'[100,101]. Mohrbacher and Cromwell[102] have studied the spectra of 1-phenyl-2-substituted cyclopropanes where the 2-substituent is a benzoyl, 4-phenylbenzoyl, carboxyl or 4-nitrobenzylcarboxylate group. They found evidence for interaction between the benzene ring and the carbonyl group. Thus they claim transmission of a resonance effect by the cyclopropane ring. The ultraviolet spectrum of compound 17 has been interpreted as being indicative of transmission of the resonance effect through the cyclopropane ring[103]. Perold[104–106] has studied the ultraviolet absorption spectra of 2-arylcyclopropanecarboxylic acids, amides and esters. He finds that conjugation of the benzene and cyclopropane rings occurs in the absence of steric hindrance. Cannon, Santilli and Shenian[107] find that the ultraviolet spectrum of 1-acetyl-2-methylcyclopropane is indicative of conjugation of the methyl and acetyl groups via the cyclopropane ring. The n → π* transition in cyclopropyl ketones has been studied by a number of authors[89,108–110]. The effect of the cyclopropyl group on this transition seems to be comparable to that of an alkyl group, in so far as the position of the absorption is concerned. The intensities of the absorption in the cyclopropyl ketones are generally closer to those observed for the vinyl ketones than to those observed for the alkyl ketones.

The importance of molecular geometry in determining the effect of the cyclopropyl group on the ultraviolet spectrum of a cyclopropyl ketone, was noted by Kosower and Ito[111] in their study of compounds **18** and **19**. Dauben and Berezin[112] have examined an extensive series of compounds of this type. Pete has reviewed the importance of the dihedral angle between the carbonyl and cyclopropyl groups on cyclopropyl ketone absorption spectra[86]. Of particular interest is the study by Lorenc and coworkers[113] of the absorption spectra of compounds **20–23**. In these rigid molecules, the cyclopropyl group possesses a favourable conformation for overlap with the carbonyl group and the double bond. The results suggest the existence of a conjugated system with the cyclopropane ring in the centre. The spectra of cyclopropyl ketone 2,4-dinitrophenylhydrazones is indicative of conjugation between the cyclopropane ring and the carbon-nitrogen double bond[114,115]. Reports have appeared of conjugation between the cyclopropane

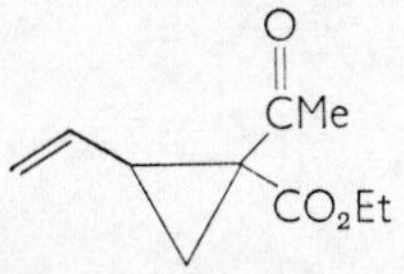

(17) 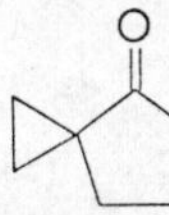 **(18)** 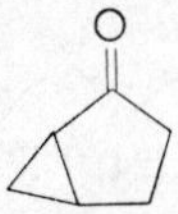 **(19)** 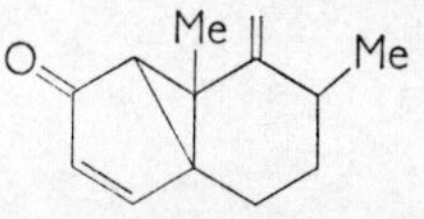 **(20)**

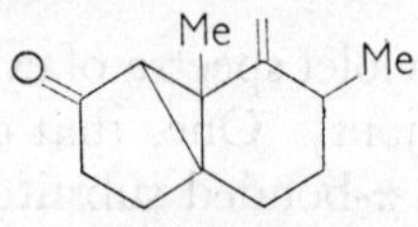

(21) 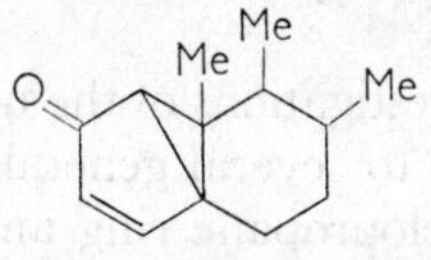 **(22)** 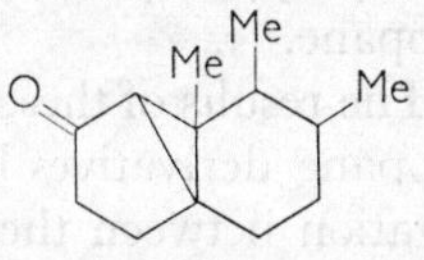 **(23)**

ring and a cyano group based on the u.v. spectrum[89,102]. Cannon and coworkers[107] were not able, however, to find any evidence for conjugation of methyl and cyano groups in appropriately substituted cyclopropanes. In a study of the ultraviolet spectra of the anions of nitroalkenes and nitrocycloalkanes, Williams and coworkers[116] observed a bathochromic shift in the case of cyclopropyl groups, attributed to conjugation. The far ultraviolet spectrum of chlorocyclopropane has been reported to show conjugation between the chlorine and the cyclopropane ring[117]. Skattebol[118] has interpreted the effect of methyl groups on the u.v. spectra of geminal dibromocyclopropanes as evidence for the ability of the cyclopropane ring to transmit resonance effects. Deno and coworkers[119] have studied the effect of the cyclopropyl group on the ultraviolet spectrum of carbonium ions of types **24** and **25**. Their results are shown in Table 14.

TABLE 14. Carbonium ion u.v. spectra[119]

Compound	X^1	X^2	λ_{max}	$10^{-3}\varepsilon$
24	Me	cyC$_3$H$_5$	309	78
	Me	Me	275	11
	Me	Ph	378	23·5
	Ph	Ph	446	17·4
25	cyC$_3$H$_5$	cyC$_3$H$_5$	364	33
	Me	cyC$_3$H$_5$		56
	Me	Me	314	9·1
	Me	Ph	411	44·8

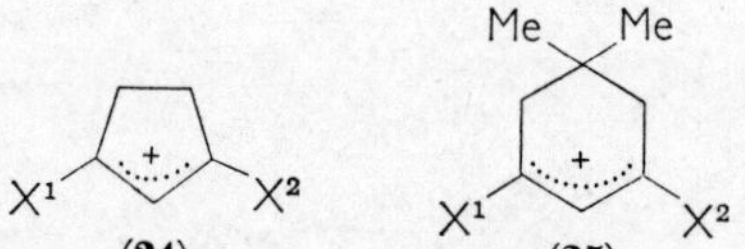

Smith and Rogier[120] find no evidence for conjugation in *trans*-1-cyclopropyl-2-phenylcyclopropane above that extant in phenylcyclopropane.

The results of these investigations of the ultraviolet spectra of cyclopropane derivatives lead to several generalizations. One, that conjugation between the cyclopropane ring and a π-bonded substituent or lone-pair substituent generally occurs. Two, the cyclopropane ring appears to be capable of transmitting the resonance effect of a substituent. Three, in accord with the discussion in Section IV, the dihedral angle between the cyclopropane ring and a double bond or benzene ring is of considerable importance in determining the effect of the cyclopropane ring on the ultraviolet spectrum of the compound.

C. Near Infrared, Infrared and Raman Spectra

The near infrared spectra of a number of cyclopropane derivatives were reported by Gassman[121] who observed a correlation between the position of the maximum and the Taft σ^* constants. The correlation of this data with the Hammett equation[122] using σ_I, σ_m and σ_p constants was carried out by Charton[123] who reported that the best results were obtained with σ_p. On the basis of additional data, Gassman and Zalar[124] found that a better correlation could be obtained with σ^*. They further concluded, on the basis of substituent effects upon the near infrared spectra, that the cyclopropyl group may function either as an electron donor or as an electron acceptor. The data reported by Gassman and Zalar have been correlated with the extended Hammett equation (equation 1, Section V.A). The results of the correlation of the data for the substituted cyclopropanes show no significant dependence upon σ_R when the amino substituent is included in the set; when, however, this substituent is excluded from the set, a significant dependence upon the σ_R constants is in fact observed, with $\varepsilon = 0.80$ (equation 2, section V.A). The value of ε observed for σ_p constants is 1.0. Analogous results are obtained when the data reported by Gassman and Zalar for *trans* 1-substituted 2-phenylcyclopropanes are correlated with equation (1). Inclusion of the amino compound in the set again results in no dependence on σ_R, whereas on exclusion of the amino compound a definite dependence on σ_R is patent. It is noteworthy that best correlation is obtained in those sets from which the amino substituent has been excluded. Thus the nature of substituent effects on the first overtone of the fundamental C—H stretching vibration in substituted cyclopropanes remains an open question.

Considerable effort has been expended on the infrared spectra of substituted cyclopropanes. Linnett[126] has noted that the C—H stretching frequency in cyclopropane is similar to that in ethylene and the stretching force constant of 5.0×10^5 dyne/cm is close to that found in ethylene (Table 15). Values of the C—H and C—C force constants have been reported by Wiberg and Nist[128]. These authors have also reported results on the C—H stretching and ring-breathing frequencies of the cycloalkanes (Table 16). The absorption intensities of the OH frequency in substituted carbinols have been studied by Brown, Sandri and Hart[129] (Table 17). The carbonyl stretching frequency of cyclopropyl ketones and the cyano stretching frequency in cyanocyclopropanes have been studied[107]. The cyclopropyl group decreases the carbonyl stretching frequency (see Table 18). The results suggest conjugation between the cyclopropane ring and the carbonyl group. The infrared spectra of a number of *trans* 1-substituted 2-phenylcyclopropanes have been reported by Mohrbacher and Cromwell[102]. The results are believed to indicate

TABLE 15. Variation of C—H force constant K with hybridization at the carbon atom

Species	Hybridization at C	$10^{-5} K_{C-H}$ (dyne/cm)	Ref.	$10^{-5} K_{C-C}$ (dyne/cm)	Ref.
CH	p	4·09	127		
MeH	sp^3	4·79	127		
ViH	sp^2	5·1	127,128		
cyC$_3$H$_5$H	$\sim sp^2$	5·0	126	4·27	128
HC$_2$H	sp	5·85	127		
cyC$_4$H$_7$H		4·5	128	4·21	128
cyC$_5$H$_9$H	sp^3	4·6	128		
cyC$_6$H$_{11}$H	sp^3	4·6	128	4·00	128

TABLE 16. I.r. spectra of cycloalkanes

Ring Size	A_1' or A_{1g} CH stretching (cm^{-1})	A_2'' or A_{2u} CH stretching (cm^{-1})	A_1' or A_{1g} ring-breathing (cm^{-1})
3	3009	3103	1188
4	2870	2896	1003
5	2868	2965	886
6	2853	2960	802

TABLE 17. Intensities of the OH frequency in X^{1}X^{2}X^{3}COH

X^1	X^2	X^3	A^a
Me	Me	Me	0·33
Et	Et	Et	0·25
i-Pr	i-Pr	i-Pr	0·22
i-Pr	i-Pr	cyC$_3$H$_5$	0·28
i-Pr	cyC$_3$H$_5$	cyC$_3$H$_5$	0·33
cyC$_3$H$_5$	cyC$_3$H$_5$	cyC$_3$H$_5$	0·37
Vi	H	H	0·48
H	H	H	0·45

^{a} Intensities in units 10^4 l/mole cm^2.

10. Olefinic Properties of Cyclopropanes

TABLE 18. Carbonyl and cyano stretching frequencies in compounds XY

X	Y = Ac	Bz	CO_2Et	GN
Me		1691		
Et	1720		1738	
Pr	1720		1738	
Vi	1684		1728	2230
1-MeVn				2230
T-2-MeVn	1676		1722	2223
cyC_3H_5	1704	1673	1730	2245
1-MecyC_3H_4		1680	1725	
2-MecyC_3H_4	1699	1675	1728	2245
1,2,2-Me$_3$cyC_3H_2		1696	1722	2234

conjugation of the benzene ring and substituent through the cyclopropane ring. Smith and Rogier[120] have examined the infrared spectrum of 2-phenylbicyclopropyl. They find that as was the case for the ultraviolet absorption spectrum, there is no evidence of conjugation in this molecule above that exhibited by phenylcyclopropane. The infrared spectra of silyl derivatives of 1,1-dichlorocyclopropane have been interpreted as indicating the absence of resonance effects involving the d orbitals on the silicon atom and the cyclopropane ring. It is suggested that the silyl group has a purely localized effect. Gray and Kraus[98] have reported the infrared carbonyl stretching frequencies of carbalkoxy-substituted cyclopropanes bearing pyridyl substituents. They suggest that the frequencies obtained were indicative of conjugation between the cyclopropane ring and the carbonyl group.

The available studies of infrared absorption spectra of substituted cyclopropanes provide support for conjugation involving the cyclopropane ring and adjacent π-electron substituents such as a double bond, a carbonyl group or an aromatic ring. There is some indication that the cyclopropane ring can transmit a resonance effect. Finally, the C—H force constant provides further support for the models described in Section II, for the bonding in cyclopropane.

A number of papers have appeared reporting on the Raman spectra of cyclopropane derivatives[131-138]. Most of this work has been on alkyl, vinyl and aryl derivatives of cyclopropane. The results are collected in Table 19. It has been suggested that they are indicative of conjugation between aryl or vinyl substituents and the cyclopropane ring. This conclusion is based upon the enhanced intensity of the absorption. The arylcyclobutanes which have been studied, however, show absorptions of about the same intensity as those of many of the aryl cyclopropanes (see Table 19). Thus the results cannot be regarded as conclusive evidence for the existence of conjugation. One

TABLE 19. Intensities of Raman spectra of alkyl, vinyl and aryl cyclopropanes and related compounds

Compound	Region			Ref
	~1200 cm^{-1}	~1600 cm^{-1}	1640–1650 cm^{-1}	
1-Pentene			340	131
Dicyclopropylmethane	700			138
Ethylcyclpropane	290			131
Vinylcyclopropane	430		660	131
Isopropenylcyclopropane	410			131
Isopropylcyclopropane	260			131
2-Methyl-1-pentene			290	131
Benzene		215		134
Toluene	220	290		131
Mesitylene		470		134
Cumene	190	330		131
Isopropenylbenzene		~2000		135
Phenylcyclopropane	800	1150		131
4-Methylphenylcyclopropane		1355		137
4-Ethylphenylcyclopropane	375	1340		136
4-vinylphenylcyclopropane	1530	2464		136
4-Isopropylphenylcyclopropane	300	1585		136
4-Isopropenylcyclopropane	920	4680		136
1,4-Dicyclopropylbenzene	660	1970		136
4-Isopropenylcumene	560	~2700		135
1,4-Diisopropylbenzene	158	668		135
2,4-Dimethylphenylcyclopropane		1135		137
Mesitylcyclopropane		532		134
2-Bromo-4-methylphenyl-cyclopropane		748		137
4-Bromophenylcyclopropane		2040		137
5-Bromo-2,4-Dimethylphenyl-cyclopropane		1080		137
Phenylcyclobutane	334	690		133
3-Methylphenylcyclobutane	245,400	560,670		137
4-Methylphenylcyclobutane	440	1100		133
2-Methoxyphenylcyclobutane	340,645	1340,1435		133
4-Methoxyphenylcyclobutane	360,490	858,900		133
1,1-Diphenylcyclopropane		1300		132a

paper has appeared[138] in which the carbonyl stretching frequencies and intensities in cyclopropyl ketones are indicative of conjugation between the cyclopropane ring and the carbonyl group (see Table 20).

TABLE 20. Raman spectra of ketones[138] X^1COX^2

X^1	X^2	ν_{CO}	Intensity
Me	Me	1710	135
Me	Et	1713	140
Me	Pr	1715	155
Pr	Pr	1712	140
i-Pr	i-Pr	1713	145
Me	cyC_3H_5	1692	310
cyC_3H_5	cyC_3H_5	1674,1684	315

Conjugation of a double bond with the carbonyl group results in a six to tenfold increase in the intensity of the carbonyl absorption, whereas the cyclopropyl group caused only about a twofold increase.

The introduction of a cyclopropyl group into the *para* position of an alkyl benzene seems to result, generally, in a five fold increase in the intensity of the Raman line at about 1600 cm^{-1} (Table 19). It is interesting to note that this is not the case for mesitylcyclopropane which has about the same intensity for this Raman line as does mesitylene itself. This observation suggests the existence of steric inhibition of resonance. The conformation in which maximal conjugation with the ring occurs is the bisected conformation. Such a conformation is not possible in the case of mesitylcyclopropane due to the methyl groups in the 2 and 6 positions. This molecule must exist in a symmetrical conformation. Although possibly fortuitous, if the difference in intensities between benzene and cumene is added to the intensity of mesitylene, the resulting sum is approximately equal to the observed intensity of mesitylcyclopropane (585 calculated vs. 582 observed).

It has been suggested [132] that the higher boiling forms of 1,2-diphenylcyclopropane and 1-cyclopropyl-2-phenylcyclopropane must have the *trans* configuration, as their Raman lines at about 1200 and 1600 cm^{-1} show very much higher intensities than do those of the lower boiling forms. The higher boiling forms are capable of existing in configurations which would result in maximal conjugation.

Overall, while the results obtained for the intensities of Raman lines are not conclusive, they are suggestive of the existence of conjugation between the cyclopropane rings and double bonds, carbonyl groups and aromatic rings.

D. Nuclear Magnetic Resonance, Nuclear Quadruple Resonance, and Electron Paramagnetic Resonance Spectra

On the basis of the ^{13}C—^{1}H coupling constant, Muller and Pritchard[139] report that the carbon orbitals used to form C—H bonds in cyclopropane have somewhat less than but close to 33% s character. Their results for the coupling constants are shown in Table 21; these results are supported by the work of Patel, Howden and Roberts[140]. The value of the coupling constant for 1,1-dicyanocyclopropane indicates 35% s character carbon orbitals[141]. The value however, must

TABLE 21. ^{13}C—^{1}H coupling constants

Compound	J_{13C-1H}	Ref.
cyC$_3$H$_6$	161	139
	162	142
	160·45	143
cyC$_4$H$_8$	134·6	144
	136	142
cyC$_5$H$_{10}$	128·5	144
	128	139
	131	142
cyC$_6$H$_{12}$	123	139
	127	142
cyC$_7$H$_{14}$	126	142
cyC$_8$H$_{16}$	127	142
cyC$_9$H$_{18}$	125	142
cyC$_{10}$H$_{20}$	126	142
1,1-cyC$_3$H$_4$(CN)$_2$	170	141
1,1-cyC$_3$H$_4$(CN)CO$_2$Me	173	141
PhH	159	139
Me$_4$C	124	139
MeC≡^{13}CH	248	139
PhC≡^{13}CH	251	139
ViH	157	139
Bicyclo[1·1·1]pentane	144 (C$_{(2)}$—H)	150
	160 (C$_{(1)}$—H)	150
Bicyclo[1·1·0]butane	152 (C$_{(2)}$—H)	150
	202 (C$_{(1)}$—H)	150
Hybridization	167 (sp^2)	127
	250 (sp)	127

be viewed with some degree of caution: while the relationship between the extent of s character and coupling constant seems reasonably reliable for hydrocarbons, the coupling constants are known to be affected by the electronegativity of the substituents attached to the carbon. Weigert and Roberts[145] have made use of the relationship between carbon–carbon coupling constants and % s character[146,147] (equation 3, where S_1 and S_2 are the % s character of the orbitals on $C_{(1)}$ and $C_{(2)}$)

$$550\ S_1 S_2 = J_{CC} \qquad (3)$$

to calculate % s character for a number of cyclopropane derivatives. The results (Table 22) show a hybridization state of about sp^5 in the ring bonds and of about sp^2 in the C—Me bond, in excellent agreement with the bent-bond model.

TABLE 22. C—C coupling constants in cyclopropane derivatives[145]

Compound	Bond	J_{C-C}
Bromocyclopropane	1,2	13·3
Iodocyclopropane	1,2	12·9
Methylcyclopropane	1,1'	44·0
1,1-Dichlorocyclopropane	1,2	15·5
Bond type sp^3-sp^3		34·6[a]
Bond type sp^5-sp^5		15[a]

[a] These coupling constants belong to C atoms in these hybridization states.

Proton chemical shifts are smaller for cyclopropane than they are for other cycloalkanes[144]. ^{13}C chemical shifts are largest for cyclopropane and then decrease to an approximately constant value of about 167 for other cycloalkanes[142]. A number of authors[128,140,148] have proposed a ring-current model to account for the chemical shifts in cyclopropane. Thus a ring current of 3 electrons precessing in a ring of radius of 1·46Å has been reported to account for the observed chemical shift of the protons in cyclopropane[128,148]. The long-range shielding effect of a cyclopropane ring has been estimated from the McConnell equation[149].

A study of the ^{13}C n.m.r. spectra of cyclopropyl ketones has been interpreted as indicative of conjugation between the cyclopropane ring and the carbonyl group[151]. The chemical shift of the hydroxyl proton in protonated cyclopropyl ketones has been examined. The results are shown in Table 23[152]. In a number of cases two chemical

TABLE 23. Proton chemical shifts of the OH group in protonated carbonyl compounds[152]

	a) Acyclic compounds XCOY		
X	Y = Me	cyC$_3$H$_5$	Ph
H	14·78 15·14	12·62	13·29
CH$_3$	13·33 14·24	12·52 13·18	13·03
Ph	13·03	11·78	12·23
cyC$_3$H$_5$	12·52 13·18	12·08	11·78

b) Cyclic ketones

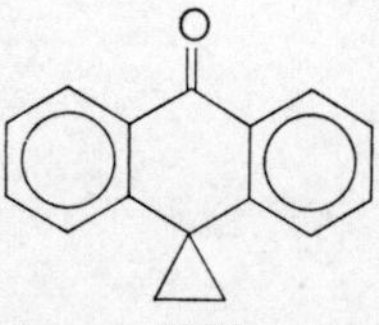

10·78 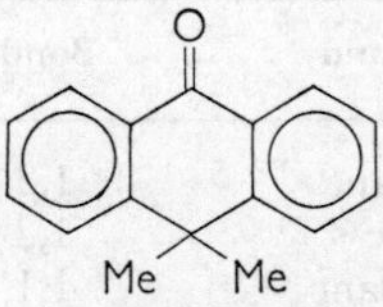   11·50

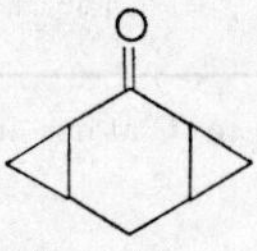

11·77 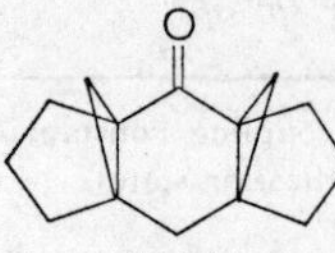   10·98 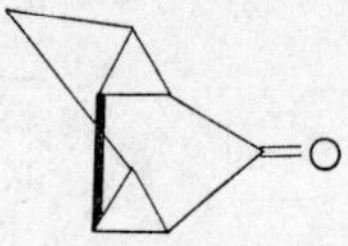   12·17

shifts were observed, this being ascribed to the presence of *syn* and *anti* protonated forms. The cyclopropyl derivatives, in general, show a chemical shift considerably lower than that of the corresponding methyl derivatives and quite close to that of the corresponding phenyl derivatives. Williamson, Lanford and Nicholson[153] have compared the chemical shifts and coupling constants of 2-substituted 1,1-dichlorocyclopropanes with those of saturated and unsaturated compounds and have concluded that the cyclopropane ring system has some π-electron character that places it in a position intermediate between saturated and unsaturated compounds. In an extension of this work, Williamson and Braman[154] have examined the effect of substituents on the coupling constants of 3-substituted 1,1-dichloro-2,2,3-trifluorocyclopropanes and again find that their behaviour is roughly intermediate between that in alkanes and that in alkenes.

Wittstruck and Trachtenberg[154a] on the basis of a comparison of the chemical shifts observed in *trans*-2-(substituted phenyl)-cyclopropane-1-carboxylic acids with chemical shifts in substituted phenyl propanoic acids, *trans*-substituted cinnamic acids, substituted ethylbenzenes and substituted styrenes report that the cyclopropane ring is intermediate between —CH_2CH_2— and —CH=CH— in its ability to transmit resonance effects.

A monotonic relationship between nuclear quadrupole resonance frequency and the effective electronegativity of the carbon orbital (calculated by the equation of Whitehead and Jaffé) has been reported[155]. The compounds studied were chloro derivatives of cyclopropanes. It is argued that the results are in accord with the bent-bond model but do not agree with the trigonally-hybridized model for cyclopropane.

An electron paramagnetic resonance study of the radical anion for phenylcyclopropane has shown that not less than 60% of the unpaired electron is delocalized in the cyclopropane ring[156]. The nitrogen splitting constants show that delocalization of the unpaired electron in nitrocyclopropane radical anion is comparable to that in the corresponding radical anions of nitroolefins[157]. These results are particularly interesting in view of the report of Bauld, Gordon and Zoeller[71] cited previously, to the effect that the 9-cyclopropylanthracene and 1,4-dicyclopropylnaphthalene anion radicals have the symmetrical conformation. In order to observe the extensive delocalization reported by Kostyanovskii and his coworkers[157] it should be necessary for the nitrocyclopropane and phenylcyclopropane radical anions to possess the bisected conformation.

The results of the nuclear magnetic resonance, nuclear quadruple resonance and electron paramagnetic resonance investigations have produced the following generalizations. The hybridization of the carbon orbitals used in forming bonds to hydrogen or substituents in the cyclopropane ring is approximately sp^2. This is in good agreement with both bent-bond and trigonally-hybridized models. The hybridization of the ring bonds in cyclopropane is approximately sp^5, in excellent agreement with the bent-bond model. Conjugation between the cyclopropane ring and substituents exists in substituted cyclopropanes, in cyclopropyl-substituted carbonium ion analogues (protonated cyclopropyl ketones) and in cyclopropane-substituted radical anions. Chemical shifts in cyclopropane derivatives can be accounted for by a ring-current model. Bernett[4] has suggested that in this respect, the trigonally-hybridized model is superior to the bent-bond model as it permits the existence of a ring current. Finally, evidence has been presented to show that the cyclopropane ring is capable of transmitting resonance effects, although to a lesser extent than the vinylene group.

VI. MOLAR REFRACTIVITIES AND DIPOLE MOMENTS

A. Molar Refractivities

It is well known that conjugated unsaturated systems usually, although not always, show an exaltation of refraction; that is, the molar refraction observed is larger than that which is calculated from additivity with due allowance for the unsaturation of the groups composing the conjugated system[158]. Molar refractivities of cyclopropane derivatives were first investigated by Tschugaeff[159] who observed an average exaltation of 0·7. Further work gave values for the average exaltation of 0·8[159a] and 0·7[160]. Exaltations of 0·3 to 0·4 were obtained for di- tri- and tetrachlorocyclopropanes[160a]. Jeffery and Vogel[161] determined the contribution of the cyclopropyl group by subtracting the observed molar refraction of a structurally similar acyclic compound from the sum of the observed molar refraction of the cyclopropane derivative and two hydrogen atomic refractivities, added to account for the additional two hydrogen atoms in the acyclic structure. From results obtained principally with alkyl cyclopropane mono- and dicarboxylates, they assigned a value of 0·614 to the ring contribution. The refractive indices of isopropenylcyclopropane[162] and of vinylcyclopropane[163] show exaltations.

The molar refraction of diethyl 2-vinyl-1,1-cyclopropanedicarboxylate has been interpreted as an indication of conjugation between the vinyl group and the cyclopropane ring[163a]. A small exaltation of 0·14 has been found for methylenecyclopropane[164] after accounting for the effect of the cyclopropane ring by a factor of 0·69[164]. This is ascribed to behaviour analogous to that of allene systems. Molar refractivities of cyclopropyl ketones have been studied[108]. The molar refraction of 1-phenyl-2-cyclopropylcyclopropane does not indicate any conjugative effect other than that shown by phenyl cyclopropane itself[120]. This is in accord with results obtained from the ultraviolet and infrared spectra. Slabey[165] has studied the molar refractivities of thirty cyclopropane derivatives. While twenty-four of the thirty derivatives give exaltations of 0·3 or greater, no constant value for the exaltation can be chosen which would be applicable to any cyclopropane derivative. The observed values for the molar refractivity of fused ring bromocyclopropanes are lower than the values calculated on the basis of atomic refractions. It is suggested that this is accounted for by the high s character in the carbon–bromine bonds. This would be in accord with the models for the bonding in cyclopropane discussed previously.

While by themselves these results would not constitute conclusive evidence for the unsaturated nature of the cyclopropane ring, taken in conjunction with other physical data they do support the existence of conjugation between the cyclopropane ring and various substituents.

B. Dipole Moments

Dipole moments of substituted cyclopropanes have received considerable attention in the literature. The available dipole moments of cyclopropane derivatives are listed in Table 24. A comparison of the dipole moments of monosubstituted cyclopropanes with those of the corresponding cyclobutanes, cyclopentanes, cyclohexanes, ethylenes and benzenes (Table 25) points to their similarity with the latter two groups.

The dipole moments reported for 4-methylphenylcyclopropane and 4-bromophenylcyclopropane suggest that the cyclopropane ring donates electrons to the benzene ring. Dipole moments of *para* disubstituted benzenes may be calculated from the equation

$$\mu_{4-X^1PnX^2} = |\mu_{X^1Ph} + \mu_{X^2Ph}| \tag{4}$$

where the sign assigned to the dipole moments is the same as that of the σ_p constant of the X substituent. Values of the calculated dipole moment for 4-methyl-, 4-nitro-, 4-chloro- and 4-bromophenylcyclopropane are given in Table 26. These calculations are based on the assumption that the cyclopropyl group is an electron donor with respect to the benzene ring and has therefore been given a negative value of μ. Thus μ_{X^2Ph} was assigned a value of -0.51. The agreement between calculated and observed values of dipole moment is excellent. As the value for 4-carboxyphenylcyclopropane in benzene seemed anomalously low, no attempt was made to calculate a dipole moment for this compound in this solvent; this is believed to be due to the formation of 'polymeric species.' In dioxane, however, the

TABLE 24. Dipole moments of substituted cyclopropanes

Substituents at $C_{(1)}$[a]	$C_{(2)}$[a]	μ	Solvent	T°C	Ref.
		0	pure gas	23–95	167
CN	Me	3·78	Benzene	25	89
Ac		2·84	Benzene	25	89
Ph		0·49	Benzene	25	89
		0·51	Benzene	25	168
		0·48	Benzene	25	168a
4-MePn		0·20	Benzene	25	168
4-ClPn		2·07	Benzene	25	168a
4-BrPn		2·02	Benzene	25	168
		2·03	Benzene	25	168a
4-HO$_2$CPn		1·16	Benzene	25	168
		1·36	Benzene	25	168a
		2·03	Dioxane	25	168a
		2·08	Dioxane	35	168a
		2·11	Dioxane	45	168a
4-O$_2$NPn		4·61	Benzene	25	168a
2-O$_2$NPn		3·61	Benzene	25	168a
CN		3·75	Benzene	25	169
Cl		1·76	Benzene	25	169
		1·76	Benzene	25	171
Cl,Cl		2·04	Benzene	25	169
		1·58	(g)		48
Br		1·69	Benzene	25	170
Cl	Cl(trans)	1·18	Benzene	25	171
Pr		0·75	Benzene	20	172
Et		0·18	(l)	20	173
Ph,Ph		0·54	Benzene	25	174a
		0·2–0·5	p-Xylene	23·7	174
Ph	Ph	0·52	Benzene	25	174a
4-ClPn,4-ClPn		2·05	Benzene	23·7	174
		2·09	Benzene	25	174a
		2·01	p-Xylene	30·6	174
		1·99	C$_2$Cl$_4$	23·7	174
4-ClPn	4-ClPh	1·46	Benzene	25	174a

[a] Only substituents other than H are given.

observed value is in good agreement with the calculated value (assuming that μ for phenylcyclopropane in dioxane is equal to its value in benzene)[168a].

The results of the calculations support the donor effect of the cyclopropyl group relative to benzene and suggest that it is a more effective donor than is the methyl group.

10. Olefinic Properties of Cyclopropanes

TABLE 25. Dipole moments of substituted cycloalkanes, ethylenes and benzenes

X	$cyC_3H_5X^a$	$cyC_4H_7X^{123}$	$cyC_5H_9X^{123}$	$cyC_6H_{11}X^{123}$	ViX^{176}	PhX^{276}
Ph	0·49				0·13	0
Ac	2·84				3·00	3·03
H	0	0	0	0	0	0
Et	0·18	0·05		0	0·37	0·58
Cl	1·76		2·08	2·3	1·44	1·72
CN	3·75	3·48	3·71		3·89	4·14
Pr	0·75				0·37	
Br	1·69	2·09	2·20	2·31	1·417	1·72

^{a} From Table 24.

TABLE 26. Dipole moment for 4-substituted phenylcyclopropanes ($4\text{-}X^1PncyC_3H_5$)

X^1	μ_{X^1Pn}	$\mu_{calc}{}^a$	$\mu_{obs}cyC_3H_5$
Me	0·43	0·08	0·20
Br	1·55	2·06	2·02
NO_2	3·93	4·44	4·61
Cl	1·58	2·09	2·07

^{a} According to equation (4) with $\mu_{X^2Ph} = -0.51$.

Charton[123] has shown that dipole moments of substituted cyclopropanes, as those of substituted ethylenes[176], are best correlated by σ_p constants, whereas the dipole moments of substituted cyclobutanes, cyclopentanes and cyclohexanes are best correlated by σ_I constants[123]. The data have been reexamined by applying the extended Hammett equation (equation 1 Section V.A) to the cyclopropane dipole moments in Table 25. Best results were obtained on the exclusion of the value for propylcyclopropane, but even with all eight values included in the set, a highly significant correlation was obtained. The values of α and β found in this correlation, are given in Table 27. The large, statistically significant value of β shows an important dependence on the σ_R constants. It is thus clearly shown that dipole moments of monosubstituted cyclopropanes involve a significant resonance effect, the magnitude of which is comparable to that observed for the dipole moments of substituted ethylenes and benzenes.

TABLE 27. Correlation of dipole moments of compounds G—X with extended Hammett equation[a]

G	α	β
cyC_3H_5	4·48	3·99
	4·98[b]	4·21[b]
Vi	4·96	3·19
Ph	5·65	3·55

[a] Equation (1), Section V.A, applied on data on Table 25.
[b] Excluding value for X = Pr.

VII. OTHER PHYSICAL PROPERTIES

A study of the heats of combustion of phenylcyclopropane and 1,1-, cis-1,2- and trans-1,2-diphenylcyclopropane has been reported[176a]. The results are interpreted as indicating conjugation between the cyclopropane ring and the benzene rings. The highest degree of conjugation is shown by trans-1,2-diphenylcyclopropane; the resonance energy due to benzene ring-cyclopropane conjugation in phenylcyclopropane is 1·8 ± 0·08 kcal/mole. Values of the stabilization energy in the reaction

$$MeX + e^- \rightarrow XCH_2^+ + 2e^- + H\cdot$$

are 0, 36, 55 and 58 kcal/mole for X = H, Me, Ph, and cyC_3H_5

TABLE 28. Conformation in substituted N,N-dimethylamides and N,N-Dimethylthioamides[176b]

	$\Delta G^{\ddagger}$ (kcal/mole)	
X	$XCONMe_2$	$XCSNMe_2$
H	21·0	24·0
Me	17·4	21·6
Et	16·7	
i-Pr	16·2	19·3
Vi	16·1	
Ph	15·3, 14·9	18·4
cyC_3H_5	16·4	18·4

respectively. The value for X = Ph is high due to rearrangement to the tropylium ion[119]. These results show clearly the effect of the cyclopropyl group in stabilizing a carbonium ion. A study of the conformation of the dimethylamino group in N,N-dimethylamides and thioamides shows conjugation between the thioamido group and the cyclopropane ring[176b]. Values of $\Delta G^{\ddagger}$ for rotation around the C—N bond are collected in Table 28. As conjugation between the substituent X and the amido or thiamido group increases, $\Delta G^{\ddagger}$ decreases. This is because increased conjugation between X and the carbonyl group results in a higher X—C bond order at the expense of the C—N bond order.

VIII. THE CYCLOPROPYL GROUP AS A SUBSTITUENT

A. Electrical Effects

There have been a number of papers reporting substituent constants for the cyclopropyl group. These results are set forth in Table 29. Values of σ_I can be calculated from the equation

$$\sigma_I = \sigma^*/6 \cdot 28 \tag{5}$$

TABLE 29. Substituent constants of the cyclopropyl group

		Source	Ref.
σ^*	−0·15	$E_{1/2}{}^a$	177
	0·11	OH intensity, I.r.	178
σ_I	−0·08	N.m.r., ^{19}F	179
	0·02	Calculated from equation (5) with $\sigma^* = 0·11$	179
	−0·02	Calculated from equation (5) with $\sigma^* = -0·15$	
σ_m	−0·102	pK_a, XPnCO$_2$H, 50% EtOH–H$_2$O	180
	−0·07	pK_a, XPnCO$_2$H, H$_2$O, 25°	181
σ_p	−0·21	pK_a, XPnCO$_2$H, H$_2$O, 25°	181
	−0·24	K_a, XPnCO$_2$H, H$_2$O, 20°	182
	−0·19	$K_r{}^b$, XPnCO$_2$Et, 85% EtOH–H$_2$O	186
σ_p^+	−0·56^c	$K_r{}^b$ XPnCMe$_2$Cl, 90% Dioxane-water	183
σ_R^0	−0·13	N.m.r., ^{19}F	179
σ_R	−0·19	Calculated from equation (6)^d	

^{a} Polarographic half-wave potential.
^{b} Solvolysis rate constant.
^{c} Calculated from data in the reference.
^{d} This work.

Using the cyclopropyl values of Zuman[177] and Brown[178], σ_I values of -0.02 and 0.02 are obtained. The value of -0.08 obtained for σ_I by Pew[179] from ^{19}F chemical shift measurements, seems to be much too low. This is true of the σ_I values of other groups determined by this method. Thus, the values of σ_I for NH_2 and Vi are 0.10 and 0.09 respectively[179], whereas the n.m.r. method gives 0.01 for both NH_2 and Vi. The best present value of σ_I for cyC_3H_5 seems to be -0.02. A value of -0.19 for σ_R can be calculated using the equation

$$\sigma_R = \sigma_p - \sigma_I \tag{6}$$

and the σ_p value obtained from the pK_a of 4-cyclopropylbenzoic acid in water at 25°. Then for σ_m we may calculate from the equation

$$\sigma_m = \sigma_I + \sigma_R/3 \tag{7}$$

a value of -0.08, in good agreement with the observed values in Table 29. From the relationship

$$\sigma_p^+ = \sigma_I + 1.6\sigma_R \tag{8}$$

a value of -0.32 for σ_p^+ is obtained in poor agreement with the value reported in Table 29. The quantity σ_R^+ may be calculated from the equation

$$\sigma_R^+ = \sigma_p^+ - \sigma_I \tag{9}$$

The values of the ratio σ_R^+/σ_R for Me, Et, i-Pr, t-Bu and Ph are 2.18, 2.45, 2.08, 1.00 and 2.54 respectively. Neglecting the value for t-Bu, an average value of σ_R^+/σ_R for these groups is 2.3. Then from this value and the values of σ_I and σ_R for the cyclopropyl group, we may calculate a value of -0.56 for σ_p^+ of the cyclopropyl group, in fair agreement with the value observed.

The equation

$$\sigma_R^0 = 0.67\,\sigma_R \tag{10}$$

gives $\sigma_R^0 = -0.13$, in good agreement with the value reported by Pew[179]. Thus the values of $\sigma_I = -0.02$, $\sigma_R = -0.19$ are recommended for the cyclopropyl substituent as they are in good accord with the available data.

For purposes of comparison, values of σ_I and σ_R for the cyclopropyl, isopropyl, cyclohexyl, vinyl, phenyl and ethynyl groups are collected in Table 30. The most striking feature of the substituent constants in this table is the σ_R value of the cyclopropyl group. The other hydrocarbon substituents have σ_R values of -0.11 ± 0.02. The value of σ_R for the cyclopropyl group is almost twice this. Obviously this

TABLE 30. Substituent constants for hydrocarbon groups

	i—Pr	cyC_6H_{11}	cyC_3H_5	Vi	Ph	C_2H
$\sigma_I{}^a$	−0·03	−0·02	−0·02^c	0·09	0·10	0·35
$\sigma_R{}^b$	−0·12^d		−0·19^c	−0·11^e	−0·11^d	−0·09^f

^{a} M. Charton, *J. Org. Chem.*, **29**, 1222 (1964).
^{b} Calculated from equation (6).
^{c} This work.
^{d} σ_p from D.H. McDaniel and H. C. Brown, *J. Org. Chem.*, **23**, 420 (1958).
^{e} M. Charton, *J. Org. Chem.*, **31**, 2991 (1966).
^{f} σ_p calculated from the correlation for *trans* 3-substituted acrylic acids reported by M. Charton, *J. Org. Chem.*, **30**, 979 (1965) using the pK_a for *trans*-$HC_2CH=CHCO_2H$ of G. H. Mansfield and M. C. Whiting, *J. Chem. Soc.*, 4701 (1956).

cannot be accounted for in terms of more effective overlap between the cyclopropyl group and the structure to which it is attached. The trigonally-hybridized model would predict the same degree of overlap between the cyclopropyl group and some other π-bonded group as between a vinyl, aryl or ethynyl group and the other π-bonded group. The bent-bond model would predict less overlap for cyclopropyl as compared with vinyl, aryl or ethynyl.

B. Substituent Effects on Ionization Constants

Perhaps the most frequently investigated examples of the cyclopropyl substituent effect are the ionization constants of various types of cyclopropyl substituted acids and bases. Ionization constants of carboxylic acids XCO_2H where X is alkyl, cycloalkyl, vinyl or phenyl have been studied in a number of solvents[170,184]. Table 31 shows that the cyclopropyl group has almost the same effect as alkyl and cycloalkyl groups. This is in agreement with the observation that pK_a values for XCO_2H are well correlated by the σ_m constants, which are about the same for all alkyl and cycloalkyl groups. Ionization constants of cyclopropyl and *N,N*-dimethylcyclopropyl amines have been reported by Roberts and Chambers[170]. There seems to be a significant decrease in basicity of the cyclopropylamines as compared with the other cycloalkylamines. The effect can best be accounted for in terms of delocalization of the electrons in the non-bonding orbital of the nitrogen atom over the cyclopropane ring. The pK_a of tricyclopropyl phosphine has been studied[185], it is far closer to that of the trialkylphosphines than it is to that of triphenylphosphine (Table 31).

TABLE 31. Substituent effects on ionization constants of compounds XGY

X	GY = CO_2H[a] $10^5 K_a$	CO_2H[b] $10^{10} K_a$	CO_2H[c] $10^{10} K_a$	CO_2H[d] $10^{10} K_c$	CO_2H[e] $10^7 K_a$	NH_2[e] $10^6 K_b$	NMe_2 $10^6 K_b$	P[f] pK_a	4-C_5H_4N[g] pK_a	4-C_5H_4N[h] pK_a	4-$PnNH_2$[i] pK_b	4-$PnCO_2H$[i] pK_a	(4-Pn)$_3$COH[j] pK_{R+}
Vi	5·66	5·36		530·0	2·96	4·6	0·50	7·60	5·62[k]				
cyC$_3$H$_5$	1·49	1·48	1·24	138·0	0·62	22·0	5·9		6·44	5·99	9·10	4·52	−4·9
cyC$_4$H$_7$	1·64	1·29	0·345	131·0	0·62	89·0	8·6			9·20	4·35		
cyC$_5$H$_9$	1·03	0·715	0·236	89·7	0·33	68·0	15·0	9·70					
cyC$_6$H$_{11}$	1·25	0·923	0·173	95·0	0·32								
i-Pr	1·447[b] 1·301		0·170						6·02[k]		9·15	4·31	−6·5
Pr	1·515[l]	2·0[m]						8·64	5·97[k]				
Ph	6·320[l]	5·2						2·73	5·55[k]				
Me									6·07	5·71			
i-Bu								7·97					

[a] At 25° in water[184].
[b] At 25° in methanol[184].
[c] At 25° in ethanol[184].
[d] At 25° in ethylene glycol[184]. k_c is the classical equilibrium constant.
[e] At 25° in 50% aqueous ethanol[170].
[f] In water, temperature not given, probably 25°[185].
[g] At 25° in water[98].
[h] At 25° in 25% aqueous methanol[98].
[i] At 20° in water[186].
[j] At 25° in sulphuric acid[187].
[k] From reference 190.
[l] From reference 188.
[m] From reference 189.

This behaviour is possibly due to a competition between delocalization of the lone pair on the phosphorus atom and electron donation to the phosphorus atom by virtue of the overlap of the empty $3d$ orbital on phosphorus with the ring orbitals. While this could also occur with the phenyl group, it should be more important for the cyclopropyl group, as the σ_R constants show this group to be the better electron donor by resonance.

Ionization constants of 4-substituted pyridinium ions[98] show the cyclopropyl group to be a more effective electron donor than alkyl groups when it is bonded to the pyridine ring. This is also observed to a slight extent (0·05 pK units) in the ionization of the 4-substituted anilines[186] and to a greater extent (0·2 pK units) in the ionization constants of 4-substituted benzoic acids[186]. The effect of the cyclopropyl group on the pK_{R^+} values of 4,4',4"-trisubstituted triphenyl carbinols is considerably greater (1·6 pK units, or 0·5 pK unit per cyclopropyl group). The results show that the cyclopropyl group is involved in resonance interactions with the pyridine ring in the substituted pyridinium ions, and the benzene ring in the substituted benzoic acids and triphenylcarbinols.

Eastman and Selover[190a] have concluded that as the pK_a of **26** in 50% aqueous alcohol is about 9·8, almost the same as that reported by other workers for **27**, the cyclopropane ring cannot function as an electron acceptor in the ground state.

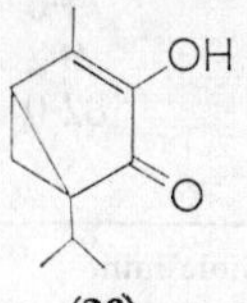

(26) 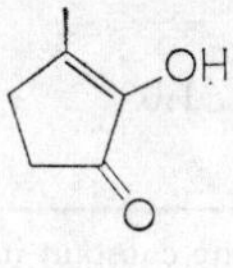   (27)

As the dihedral angle of the cyclopropane ring in **26** is 105°[86] maximum resonance interaction is not possible. Furthermore, the effect of the isopropyl group in **26** is ignored in this comparison. Thus the conclusion drawn by these authors does not seem warranted.

C. Effects on Reaction Rates

Rates of the reaction of carboxylic acids with diphenyldiazomethane in ethanol have been studied by Roberts and Chambers[170]. In this reaction the cyclopropyl group shows about the same substituent effect as other cycloalkyl groups. This is in accord with the observation that rates for this reaction generally parallel the acidities of the carboxylic acids involved. Rates of esterification of carboxylic acids in ethanol, catalysed by hydrogen chloride, have been reported; in this reaction the cyclopropyl group is from two to twenty times more reactive than other cycloalkyl groups. Its effect on the reactivity of the acids is about half that of the vinyl group. Some very early rate studies on the reaction of ketones with phenyl hydrazine [192] and hydroxylamine [193] suggest that the cyclopropyl group is much closer in its effect to a vinyl or phenyl group than it is to alkyl or cycloalkyl groups. This is reasonable in view of the large resonance effect of the cyclopropyl group which would tend to reduce the electrophilicity of the carbonyl carbon. The results are presented in Table 32. The alkaline hydrolysis of vinyl and cycloalkyl acetates has been investigated [194]. The

TABLE 32. Substituent effects on reaction rates of compounds XGY

GY= Reagent= X	CO_2H^a Ph_2CN_2 k_2	CO_2H^b k_{rel}	Ac^c $PhNHNH_2$ % reaction/ hr.	Ac^d H_2NOH % reaction/ hr.	OAc^e OH^- k_2	OAc^f OH^- k_{rel}
Vi	1·36	54·8			3·28	
cyC_3H_5	0·522	24·9	5·6	9·1	0·178	0·717
cyC_4H_7	0·502	1·46				
cyC_5H_9	0·762	4·06			0·026	0·089
cyC_6H_{11}	0·385	9·83				
i-Pr		5·11	15·0	33·0		
Pr			38·0	74·6		
Ph				9·2		
Me		1·0	66·0	82·0		
$Me_2C{=}CH$			3·6			

[a] At 30° in ethanol [170]. Rate constant in units l/mole min.
[b] At 25° in ethanol [191].
[c] At 25° in 50% aqueous ethanol [192].
[d] At room temperature in 50% aqueous ethanol [193].
[e] At 20° in water [194]. Rate constant in units l/mole sec.
[f] At 40° in water [194]. Rate constant in units l/mole sec.

results show that the cyclopropyl group is considerably more rate enhancing than is the cyclopentyl group; nevertheless it is only about one-tenth as effective as the vinyl group [194] (Table 32).

Rates of oxidation and bromination in 99% acetic acid in the presence of 0·1M perchloric acid, have been investigated for 3-methyl-2-butanone, acetylcyclopropane and 1-phenyl-2-methyl-1-propanone. The cyclopropyl group seems to be intermediate between the alkyl and phenyl groups in its effect on the rate [195].

D. Orientation in Electrophilic Aromatic Substitution

Although the results on the electrophilic aromatic substitution reaction of aryl cyclopropanes (Table 33) are limited, they show that

TABLE 33. Orientation in electrophilic aromatic substitution of arylcyclopropanes

Ar	Reagent	Product	Ref.
Ph	Br_2	4-BrPncyC_3H_5	177,197
4-MePn	Br_2	4-Me-2-BrC_6H_3cyC_3H_5	177
2,4-$Me_2C_6H_3$	Br_2	5-Br-2,4-$Me_2C_6H_3$cyC_3H_5	177
Ph	fuming HNO_3	4-NO_2PncyC_3H_5	196
Ph	AcCl + $AlCl_3$	4-AcPncyC_3H_5	197

the cyclopropyl substituent directs *ortho-para* and is more effective in determination of orientation than is a single methyl group, but less effective than two methyl groups. This is in accord with the order of the resonance effect. Thus, the σ_R values for methyl, cyclopropyl, and dimethyl are -0.12, -0.19 and -0.24 respectively.

Ortho-para ratios for the nitration of phenylcyclopropane have been examined (Table 34). The results are in fairly good agreement with predictions based on the σ^+ value for the cyclopropyl group[198].

It is interesting to note that whereas phenylcyclopropane, 1,1-diphenylcyclopropane[199], 1-methyl-2-phenylcyclopropane and 1-cyclohexyl-2-phenylcyclopropane[200] are nitrated in the benzene ring, the nitration of *cis*- or *trans*-1,2-diphenylcyclopropane results in the formation of 1-nitro-*trans*-1,2-diphenylcyclopropane.[200]

TABLE 34. *Ortho-para* product ratios for the nitration of alkylbenzene (PhX)[198].

X	H_2SO_4–HNO_3	$AcONO_2{}^a$	$AcONO_2{}^b$
Me	1·57	1·78	1·96
Et	0·93	0·86	0·90
i-Pr	0·48	0·41	0·27
t-Bu	0·217	0·17	0·066
cyC_3H_5	2·10	1·99	4·0–4·7

^{a} At $-4°$.
^{b} At room temperature.

E. Effects on Carbonium Ions

Hanack and Schneider[201] discussed the effect of the cyclopropyl group on carbonium ions, in a review of neighbouring group effects and rearrangements in the reactions of cyclopropylmethyl, cyclobutyl and homoallyl systems. Cyclopropylcarbinyl derivatives have long been known to be highly reactive in solvolyses[202,203]. Typical data are presented in Table 35. Roberts and Mazur[204] have pro-

TABLE 35. Solvolysis of cyclopropylcarbinyl derivatives

Compound	Solvent	T(°C)	$k_1{}^a$	Ref.
cyC$_3$H$_5$CH$_2$Cl	50% v/vEtOH–H$_2$O	50	0·45	202
cyC$_4$H$_7$Cl	50% v/vEtOH–H$_2$O	50	0·017	202
1-MeVnCH$_2$Cl	50% v/vEtOH–H$_2$O	50	0·011	202
cyC$_3$H$_5$Br	50% v/vEtOH–H$_2$O	25	0·34	202
cyC$_4$H$_7$Br	50% v/vEtOH–H$_2$O	25	0·015	202
ViCH$_2$Br	50% v/vEtOH–H$_2$O	25	0·013	202
cyC$_3$H$_5$OSO$_2$Ph	absolute EtOH	20·13	2·3 × 10^5	203
ViCH$_2$	absolute EtOH	20·1	0·15 × 10^5	203

^{a} In hr^{-1}.

posed the bicyclobutonium ion, **28**, as an intermediate in the solvolysis of cyclopropylcarbinyl derivatives. Hückel calculations have been carried out on this ion by Howden and Roberts[205]. The studies of Olah[73,74], Deno and coworkers[73a] and Ritchie and Ritchie[74a] seem to have established the existence of the cyclopropylmethyl-carbonium ion. The solvolysis of trisubstituted carbinyl *p*-nitro-benzoates (**29**) has been studied by Hart and Sandri[206], whose results

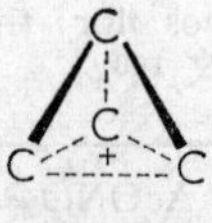

(28) 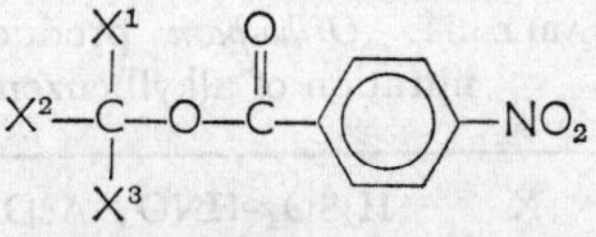   (29)

are shown in Table 36. The results show not only a rate enhancement when isopropyl is replaced by cyclopropyl, but that a second cyclopropyl group causes a rate enhancement nearly equal to that of the first one. The rate enhancement for a third cyclopropyl group is comparable or greater in magnitude than that for the first and second cyclopropyl groups, as found in the solvolysis of tricyclopropylcar-

TABLE 36. Solvolysis of trisubstituted carbinyl 4-nitrobenzoates in 80% aqueous dioxane at 60°[206]

X^1	X^2	X^3	k_{rel}
i-Pr	i-Pr	i-Pr	1
H	cyC_3H_5	cyC_3H_5	60·7
cyC_3H_5	i-Pr	i-Pr	246
i-Pr	cyC_3H_5	cyC_3H_5	23500
i-Pr	2-Me-cyC_3H_4	2-Me cyC_3H_4	124000

binyl benzoate in dioxane–water, implying that each cyclopropyl group is involved in stabilizing the charge on the tricyclopropylcarbonium ion. The exclusive hydrolysis product observed is tricyclopropylcarbinol. These results are more readily rationalized in terms of a reaction proceeding through the cyclopropylmethylcarbonium ion than through a bicyclobutonium ion (**28**). Solvolysis of the dicyclopropylcarbinyl compounds **30a** and **30b**, in 80% aqueous acetone containing $NaHCO_3$, results in the same product mixture containing 99·5% **30b** and 0·5% **31b**. The reaction is regarded as proceeding

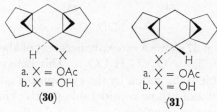

a. X = OAc
b. X = OH
(**30**)

a. X = OAc
b. X = OH
(**31**)

through a bisected dicyclopropylmethyl carbonium ion[208]. A study of the effect of methyl substituents upon solvolysis of cyclopropylcarbinyl 3,5-dinitrobenzoates in 60% aqueous acetone at 100°, shows that the 2- and 3-methyl substituents have a remarkably constant effect, each additional group enhancing the rate independently of the number and location of the other groups. Results are considered to be consistent with the bisected structure for the cyclopropylmethyl carbonium ion transition state, and inconsistent with the bicyclobutonium ion structure[209].

The exceptional reactivity of these compounds is in accord with the high values of σ_R and σ_p^+ reported in Table 29 and demonstrate the remarkable ability of the cyclopropyl group to function as an electron donor by resonance.

Further evidence for the existence of cyclopropylmethyl carbonium ions comes from a measurement of the i factor for tricyclopropylcarbinol dissolved in 96% sulphuric acid. The value of 4·1 obtained is in accord with the reaction [210]

$$\text{ROH} + 2\,\text{HSO}_4^- \rightleftharpoons \text{R}^+ + \text{H}_3\text{O}^+ + 2\,\text{HSO}_4^-$$

Tricyclopropylcarbinol has been reacted with tetrafluoroboric acid to form an isolatable **32**[119].

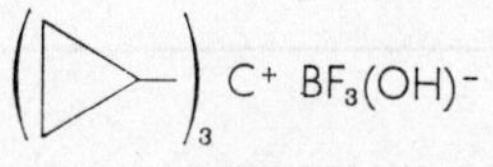

$$\left(\triangleright\!-\right)_3 \text{C}^+ \; \text{BF}_3(\text{OH})^-$$

(**32**)

The solvolysis of **33** and **34** in aqueous dioxane has been found to be 2×10^4 faster than that of **35** and **36** when X = 4-ClPnCO$_2$. The products have the same structure as the reactants [208a].

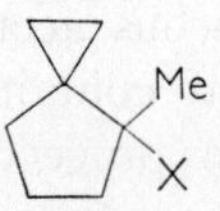

(**33**) 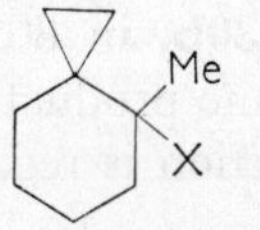   (**34**) 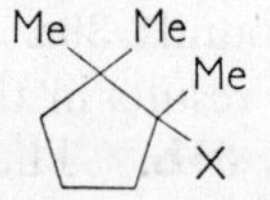   (**35**) 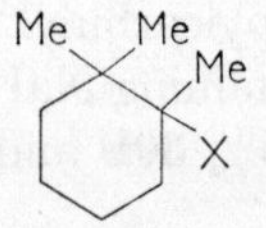   (**36**)

X = 4-ClPhCO$_2$

The solvolysis of **37** shows a rate enhancement of about 10^3 over that of **38**, with X = 3,5-(NO$_2$)$_2$C$_6$H$_3$CO$_2$. This behaviour is ascribed to the stabilization of the cation by the cyclopropyl group. Extended Hückel MO calculations show considerable positive charge on C$_{(2)}$ and C$_{(3)}$ of the cyclopropane ring [208b].

A number of interesting aryl bridged cations have been investigated [211-214]. Examples are **39-42**.

The importance of the geometry of the cyclopropyl group in determining its effectiveness at stabilizing a positive charge is shown by the work of Brown and Cleveland [183] who studied the solvolysis of substi-

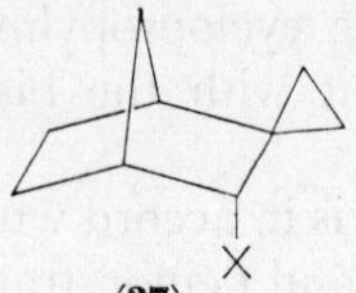

(**37**) 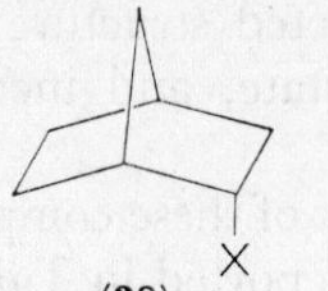   (**38**)   (**39**)

X = 3,5-(O$_2$N)$_2$C$_6$H$_3$CO$_2$

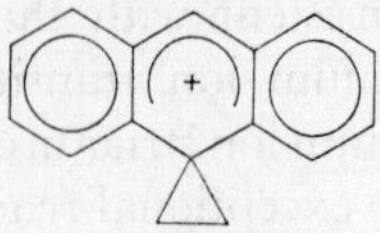

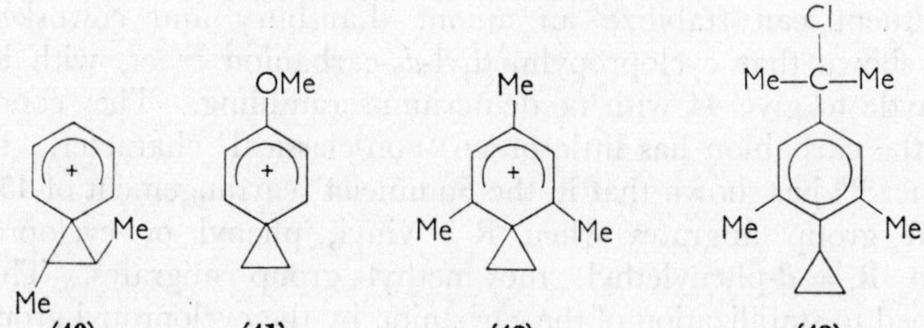

         OMe   Me   Cl
                                       |
                                   Me—C—Me

    +   +   +

      Me   Me   Me   Me   Me
 Me
  (40)   (41)   (42)   (43)

tuted cumyl chlorides in aqueous dioxane. (Table 37). The effect of introducing a single methyl group in the 3-position of 4-cyclopropylcumyl chloride on the rate of solvolysis, is to decrease the relative

TABLE 37. Solvolysis of substituted cumyl chlorides in 90% aqueous dioxane at 25°[183]

	H	4-i-Pr	3-Me	3,5-Me$_2$	4-cyC$_3$H$_5$	3-Me-4-cyC$_3$H$_5$	3,5-Me$_2$-4-cyC$_3$H$_5$
k$_{rel}$	1	17·8	2	3·9	157	172	37·1

rate to a value of about 60% of that which would otherwise be expected on the basis of additivity of substituent effects. Methyl groups in both 3 and 5 positions, **43**, decrease the rate to a value considerably less than that of 4-cyclopropylcumyl chloride. This is believed to be due to the inability of **43** to exist in a bisected conformation. A second example of this steric inhibition of resonance is shown by the pK_{R^+} values of substituted triarylcarbinols[76], set forth in Table 38.

TABLE 38. pK_{R^+} of substituted triphenylcarbinols

	H	i-Pr	3,5-Me$_2$	4-cyC$_3$H$_5$	3,5-Me$_2$-4-cyC$_3$H$_5$
pK_{R^+}	−6·6	−6·5	−6·5	−4·9	−6·5
Ref	187	187	76	187	76

F. Stabilization of Carbanions

One of the characteristic properties of phenyl, vinyl and ethynyl groups is their ability to stabilize cations, anions and radicals. It is of great interest therefore to determine whether or not the cyclopropyl substituent can stabilize an anion. Lansbury and coworkers[215] have shown that cyclopropylmethyl-d_2-carbanion reacts with benzaldehyde to give **44** with no deuterium scrambling. They conclude that the carbanion has little or no 'non-classical' character. Bumgardner[216] has shown that in the Sommelet rearrangement of **45**, the CH_2R group migrates when R = vinyl, phenyl or cyclopropyl. When R = β-phenylethyl, the methyl group migrates. This is ascribed to stabilization of the carbanion by the cyclopropyl group.

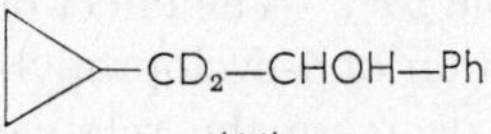

▷—CD₂—CHOH—Ph   NC—C(Me)(R)—N=N—C(Me)(R)—CN   ▷—C(Me)(CN)—C(Me)(CN)—◁
(**44**) 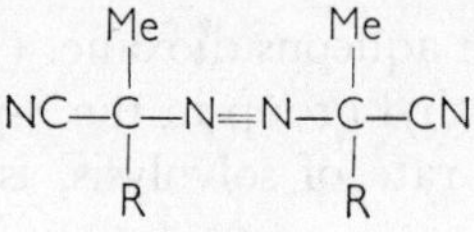   (**46**) 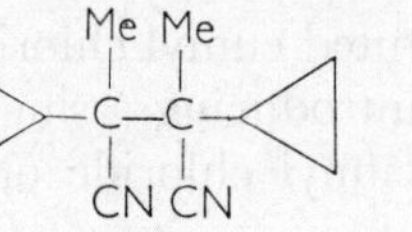   (**47**)

PhCH₂N⁺Me₂CH₂R   ▷—CH₂CO₂CH₂—◁
(**45**)   (**48**)

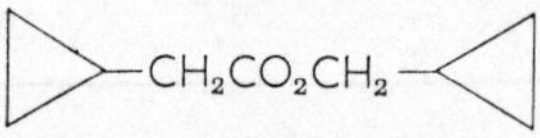

Relative rates for the addition of isopropyllithium to α-substituted styrenes in ether have been studied; the results are given in Table 39. They show clearly that the cyclopropyl group can stabilize a carbanion[180].

TABLE 39. Addition of isopropyllithium to α-substituted styrenes (1-XVnPh) in ether at −45°[180]

X	Et	i-Pr	Et₂CH	cyC₃H₅	T-2,3-Me₂cyC₃H₃
k_{rel}	28·0	1·0	0·6	310	115

G. Stabilization of Radicals

The cyclopropyl group accelerates the rate of decomposition of compounds of the type **46** (Table 40). The major product of the decomposition of the compound in solution (R = cyclopropyl) is a low molecular weight polymer; decomposition of the solid gives predominantly **47**[219].

Hart and Cipriani[220] have shown that the decomposition rate of cyclopropylacetyl peroxide is 55 times greater than that of cyclohexylacetyl peroxide, the principal product being **48**. Relative re-

TABLE 40. Rates of decomposition of azo compounds (46) in toluene at 80·2–80·5°

R	m.p. of 46	$10^4 k$	Ref.
cyC$_3$H$_5$	64–5	33	217
	76–7	25	217
cyC$_4$H$_7$	81·5–82·5	1·51	218
	38–42	1·51	218
cyC$_5$H$_9$	96·3–97·6	1·30	218
	72·2–74·5	1·31	218
cyC$_6$H$_{22}$		2·27	218
i-Bu	56–57	10	217
	74–6	7·1	217
t-Bu	114–6	0·77	218
	116–8	1·09	218

activities of phenylalkyl and phenylcycloalkyl ketones toward the addition of H· have been studied by investigating the reaction of the ketones with t-butyl peroxide and 2-butanol at 125°. Results are shown in Table 41. The effect of a cyclopropyl group is comparable to that of the phenyl group. Methyl and phenyl substituents in the 2 position of the cyclopropyl group show a considerable enhancement of the rate[221,222]. On the other hand phenylcyclopropyl ketone, dicyclopropyl ketone and methylcyclopropyl ketone do not show any significant rate enhancement over the rate observed for cyclohexanone (Table 41).

It has been observed that chlorination of methylcyclopropane in the presence of ultraviolet light produces as the major product, cyclopropyl methyl chloride[224]. A later report found that cyclopropylmethyl chloride and 4-chloro-1-butene were formed in the ratio 65:35, together with considerable quantities of other products[225]. When the reactant was labelled with ^{13}C in the methyl group, cyclopropyl methyl chloride contained ^{13}C in the chloromethyl group, and 4-chloro-1-butene contained all of its ^{13}C at carbon 1. The higher reactivity of the methyl group in methylcyclopropane as compared to the tertiary hydrogen atom, suggests that there is some stabilization of a radical carbon atom by the cyclopropyl group. Obviously this is not the only factor in determining the reactivity of the methyl group as compared to the ring hydrogen atoms, since the carbons bonded to the ring hydrogen atoms are using orbitals hybridized approximately sp^2, whereas the carbon in the methyl group is hybridized sp^3.

TABLE 41. Relative rates for the reaction of ketones with H[222]

Ketone	T°C	$k_{BzR^2}/k_{Bz\text{-}i\text{-}Pr}$	$k_{R^1COR^2}/k_{cyclohexanone}$
BzMe	125	7·60	
BzEt	125	3·33	
Bz cyC$_3$H$_5$	125	36·00	
Bz cyC$_3$H$_5$	130		~1·1·10^4
Bz cyC$_4$H$_7$	130	11·30	
Bz cyC$_5$H$_7$	130	1·84	
Bz cyC$_6$H$_{11}$	130	1·23	
Bz-i-Pr	130	1·00	
Bz-s-Bu	130	3·16	
BzPh	130	28·4	
1-Bz-2-Me cyC$_3$H$_4$	130	70·3	
Cis-1-Bz-2-Ph cyC$_3$H$_4$	130	515	
T-1-Bz-2-Ph cyC$_3$H$_4$	130	450	
Bz-t-Bu	130		33·4
(cyC$_3$H$_5$)$_2$CO	130		1·29
Ac cyC$_3$H$_5$	130		1·22
Cyclohexanone	130		1·00

While the results presented above can hardly be considered conclusive, they do suggest that the cyclopropyl group can stabilize a radical carbon atom.

IX. TRANSMISSION OF RESONANCE EFFECTS BY THE CYCLOPROPANE RING

Evidence for transmission of resonance effects by the cyclopropane ring obtained from the i.r., u.v. and n.m.r. spectra of substituted cyclopropanes has been discussed in Section V. Here evidence related to the chemical activity of substituted cyclopropanes will be considered.

Jaffé[122], in his now classic review of the Hammett equation, pointed out that for sets of compounds of the structural type XPnZY the ratio of ρ values ρ_Z/ρ_0 (where ρ_Z is the ρ value obtained from a correlation of data for XPnZY and ρ_0 is the ρ value obtained for a correlation of data for XPnY, Y being the same in both cases) is a measure of the ability of the Z group to transmit the substituent effect to the reaction site. Trachtenberg and Odian[226] made use of this method in an attempt to determine the ability of the cyclopropyl group to transmit substituent effects. They came to the conclusion that transmission through the cyclopropane ring as determined by the ρ value for the ionization of the *trans*-substituted phenylcyclopropane carboxylic acids in water, was no more effective than transmission through a —CH_2CH_2— group. Fuchs and Bloomfield[227-230], in a series of papers, examined ionization constants, in aqueous alcohol, of the phenylcyclopropanecarboxylic acids, and the hydrolysis in aqueous alcohol of their ethyl esters. They conclude on the basis of the ρ values obtained, that the cyclopropane ring is intermediate in its transmission of substituent effects between a vinylene group and a —CH_2CH_2— group. These results are collected in Table 42. The results of these

TABLE 42. Transmission of substituent effects through Z in XPnXY

Z	$Y = CO_2H^a$			$Y = CO_2H^b$			$Y = CO_2H^c$		
	ρ	β^d	Ref.	ρ	β^d	Ref.	ρ	β^d	Ref.
CH_2CH_2	−0·212	−0·189	226	−0·344	−0·295	229	0·591	0·427	230
C-cyC$_3$H$_4$				−0·436	−0·341	229	1·02	1·06	230
T-cyC$_3$H$_4$	−0·182	−0·161	226	−0·473	−0·435	229	0·812	0·701	230
C-2-Vn	−0·643		125				1·122		230
T-2-Vn	−0·466		122	−0·807^{d,e}	−0·873	231	1·314	1·33	228
							1·30^{d,f}	1·53	234
$C\equiv C$				−0·41		233	1·10	1·03	235
				−0·69		232			

[a] Measured parameter: pK_a in water at 25°.
[b] Measured parameter: pK_a in 50% ethanol–water at 25°.
[c] Measured parameter: rate constant of alkaline hydrolysis 87·8% aqueous ethanol at 30°.
[d] Calculated from the data in the reference.
[e] In 78·1% EtOH–H$_2$O at 25°.
[f] In 85·4% EtOH–H$_2$O at 24·8°.

investigations cannot be regarded as conclusive. One objection is that correlation of substituted 3-phenylpropanic acids should require the use of σ° constants, whereas correlation of cinnamic acids should require σ_p constants. It is not possible to predict which of the two σ constants would be preferable for the correlation of the cyclopropane derivatives. A more serious objection is that the ρ values do not answer the question of greatest interest, whether and to what extent can the cyclopropane ring transmit resonance effects. The magnitude of ρ will be determined largely by the localized effect. The localized effect, in turn, appears to be a function of molecular geometry (field effect). Charton[123] has shown that ionization constants of *trans*-2-substituted cyclopropanecarboxylic acids are best correlated by the σ_m constants and 1-substituted cyclopropane carboxylic acids are best correlated by the σ_p constants. As the σ_m and σ_p constants are given by the equations[83]

$$\sigma_m = \sigma_I + 0 \cdot 33\, \sigma_R \tag{11}$$

$$\sigma_p = \sigma_I + \sigma_R \tag{12}$$

successful correlation with these constants is an obvious indication of transmission of substituent resonance effect by the cyclopropane ring. These results answer the question of transmission of resonance effects through a cyclopropane qualitatively but not quantitatively. Correlation with the extended Hammett equation (equation 1, Section V.A.), is potentially capable of providing a quantitative measure of the transmission of the resonance effect. Such correlations have been carried out with sets of data for which ρ values are listed in Table 42[125]. While there is a definite trend, the results cannot be considered conclusive as the sum of β and its standard deviation for the cyclopropane compounds, is not statistically different from the value of β for the sets in which the benzene ring and reaction site are separated by two methylene groups in three of the five cases studied. One of the other two cases, that of the *trans*-phenylcyclopropanecarboxylic acid in water, has a value of β smaller than that for phenylpropanoic acid in water.

The quantity γ_R, defined as

$$\gamma_R \equiv \beta_G/\beta_{G^0} \tag{13}$$

may be taken as a convenient measure of the degree of transmission of the resonance effect in a set of compounds of the type XGY. For the reference group, G^0, the phenylene group was chosen. Correlation of data for the ionization constants of 2-substituted cyclopropanecarboxylic acids in water and in 50% 'methyl cellosolve' (MCS) and of 1-substituted cyclopropanecarboxylic acids in water, permits the calculation of values of γ_R. In Table 43 values of γ_R for various groups G are set forth. The results show that the *trans*-cyclopropylene group is about as effective in transmitting resonance effects as is the phenylene group.

The correlation of data for the solvolysis of *trans*-2,3-disubstituted cyclopropylmethyl 3,5-dinitrobenzoates[209] (**49**) with the equation

$$Q_X = \alpha \Sigma \sigma_I + \beta \Sigma \sigma_R + h \tag{14}$$

10. Olefinic Properties of Cyclopropanes

TABLE 43. Transmission of substituent effects through G in XG[125]

G	$pK_a{}^a$	$pK_a{}^b$	γ_R
trans-Cyclopropylene	−0·751	−1·55[c]	0·828
	(−0·828)[d]		0·923
Cyclopropylidene	−5·47		5·47
trans-Vinylene	−2·08	−3·34	2·08
cis-Vinylene	−1·77		1·77
Vinylidene	−0·873		0·873
Ethynylene	1·9		
p-Phenylene	1·00	−1·68	1·000

[a] In water at 25°. β values
[b] In 80% MCS–H$_2$O at 25°. β values
[c] From correlation with equation (14).
[d] 3,3-Dimethyl-trans-1,2-cyclopropylene.

gives a value of β of −7·45. Unfortunately no comparable set of data for the p-phenylene reference group is available. A value of β of −10·0 can be estimated for the solvolysis of substituted benzyl fluorides in a 2M solution of potassium formate in 20% acetone–formic acid at 50°C[209]. A value of β of −7·2 can be estimated for the solvolysis of substituted cumyl chlorides in 90% aqueous acetone at 25°C. This suggests a value of γ_R for the $trans$-cyclopropylene group between 0·75 and 1·0, in good agreement with the values reported in Table 43.

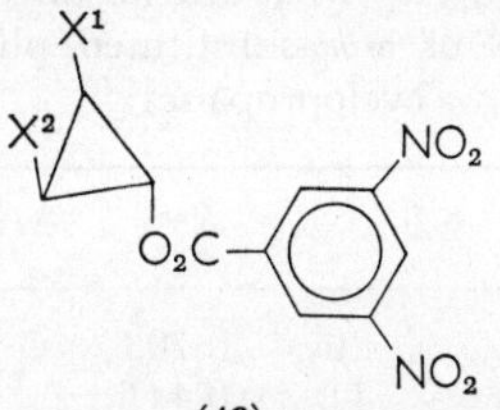

(49)

X. ADDITION REACTIONS OF CYCLOPROPANE

Addition reactions are perhaps the most characteristic reactions of the cyclopropane ring. Furthermore, it is in these reactions that cyclopropane shows the strongest analogy to the properties of olefins. Cyclopropane has been found to undergo addition with electrophilic, nucleophilic and radical reagents.

A. Donor-acceptor Complexes of Cyclopropanes

Charge-transfer complexes of olefins have long been proposed as possible intermediates in electrophilic addition. It is of some interest then to consider the possibility of complex formation with cyclopropanes. The first instance of a possible charge-transfer complex of a cyclopropane is the report by Filipov[243] that ethylcyclopropane gives a yellow colour with tetranitromethane. The ultraviolet spectrum of a solution of iodine in cyclopropane is indicative of charge-transfer complex formation. The value of ΔH_f obtained for the formation of the complex is 0·5 kcal/mole. This value is comparable to the values obtained for propene, cis- and trans-2-butene, 0·5, 0·5 and 0·2 kcal/mole respectively. A platinum complex of cyclopropane (50) has been

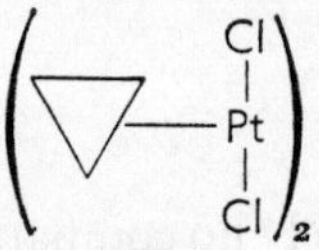

(50a) 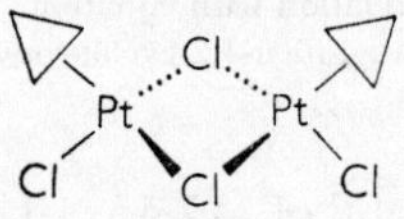   (50b)

prepared. It reacts with potassium cyanide to liberate cyclopropane[245]. Mercuric acetate has been shown to form complexes with substituted phenylcyclopropanes. Values of the equilibrium constants for complex formation are reported in Table 44.

Obviously the results available on charge-transfer complex forma-

TABLE 44. Mercuric acetate complexes of ortho-substituted phenylcyclopropanes[246]

X	T(°C)	K^a	ΔH^b
H	20	0·703	5·70
	11	0·441	
	0	0·302	
	−14	0·166	
Cl	35	0·249	9·77
	27	0·171	
	20	0·109	
NO$_2$	35	0·200	
	27	0·122	
	15	0·059	

^{a} Formation constant in l/mole.
^{b} In kcal/mole.

tion of substituted cyclopropanes are rather sparse. Much remains to be done in this area. Nevertheless it seems apparent that cyclopropanes can and do form charge-transfer complexes.

It is convenient to consider here, hydrogen-bond formation in which the cyclopropane ring acts as the electron donor. Infrared spectral evidence [246a] has been presented for the existence of hydrogen bonds between the hydroxyl group in 4-fluorophenol and the cyclopropane ring in 1,1-dimethylcyclopropane or dicyclopropylmethane. Intramolecular hydrogen bonding was observed in *endo-syn*-tricyclo-[3·2·1·0^{2,4}]octan-8-ol. These results confirm previous reports of the possible existence of hydrogen bonds between the cyclopropane ring and the hydroxyl group [246b,c,d,e].

B. Electrophilic Addition to Cyclopropane

The first known example of an electrophilic addition to cyclopropane is the report by Freund[247] of the formation of propyl iodide on treatment of cyclopropane by hydrogen iodide. Freund also reported the reaction of cyclopropane with bromine to give 1,3-dibromopropane. The reaction of cyclopropane with bromine was studied under various conditions by Gustavson[248], who found the products to contain both 1,2- and 1,3-dibromopropane. The reaction has recently been reinvestigated by Deno and Lincoln[249]. They find that $FeBr_3$, $AlCl_3$ or $AlBr_3$ is required to obtain a reasonable reaction rate; the products include 1,1-, 1,2- and 1,3-dibromopropane and 1,1,2-tribromopropane. The reaction path and product distribution obtained are summarized in Scheme I and Table 45. The addition of hydrogen fluoride to cyclopropane in anhydrous HF has been reported to give propyl fluoride as the major product, with small amounts of isopropyl fluoride and polymers[250]. A number of papers have reported on the alkylation of benzene with cyclopropane[252-256]. The results are summarized in Table 45. With aluminium chloride–hydrogen chloride[258] or aluminium chloride in nitromethane or with anhydrous hydrogen fluoride as catalysts, n-propylation occurs. With sulphuric acid as a catalyst, at 0°C, the product obtained is n-propylbenzene whereas at 65°C, isopropylbenzene is obtained. Cyclopropane has been polymerized in the presence of $AlBr_2$ and HBr to a low molecular weight polymer(molecular weight < 700). The polymer consists of long chain alkanes with a terminal double bond. The rate law observed is

$$v = k[\text{HBr} \cdot \text{AlBr}_3][\text{cyC}_3\text{H}_6]$$

TABLE 45. Products of electrophilic addition to cyclopropane

Reagent	Solvent	T(°C)	Time (min)	Catalyst	1,1	1,2	1,3	other	Ref.
HF		−15	12			Small amount	pre-dom.		250
Br_2		60	1	$Fe(1g/100gBr_2)$		46	39	2.5^a	249
Br_2		65	0.25	$Fe(1g/100gBr_2)$		60	10	12^a	249
Br_2		25	240	$AlCl_3 (6.7g/100gBr_2)$	25	25	30	12^a	249
Br_2		20–50		$AlCl_3(0.5g/100gBr_2)$	3.8	2.8	15	3^a	249
PhH	$MeNO_2$			$AlCl_3–MeNO_2$		b	c		251
HI							c		247
PhH		0		H_2SO_4			c		252,253
PhH		65		H_2SO_4		c	c		252,253
PhH		any		$AlCl_3 + HCl$			c		252,254,258
PhH		0		HF			c		255
$MecyC_6H_{11}$		−10		$AlCl_3 + HCl$			c		256
AcCl	CCl_4			$AlCl_3$	4	41	35	20^d	261
AcCl	CH_2Cl_2			$AlCl_3$	4	56	24	16^d	261
$ClCH_2COCl$	$CHCl_3$			$AlCl_3$			c		261
TsCl	$CHCl_3$			$AlCl_3$			32^e		262
AcOH				BF_3			c		264
Cl_2CHCO_2H				BF_3			c		264
BzOH				BF_3			c		264
H_2SO_4							95		263,264
B_2H_6							c		266,267
									268
$LiAlH_4$	Et_2O						c		269

^{a} $CHBR_2CHBrMe$.
^{b} No 1,2 product could be detected.
^{c} Only product observed.
^{d} $AcMeC=CH_2$.

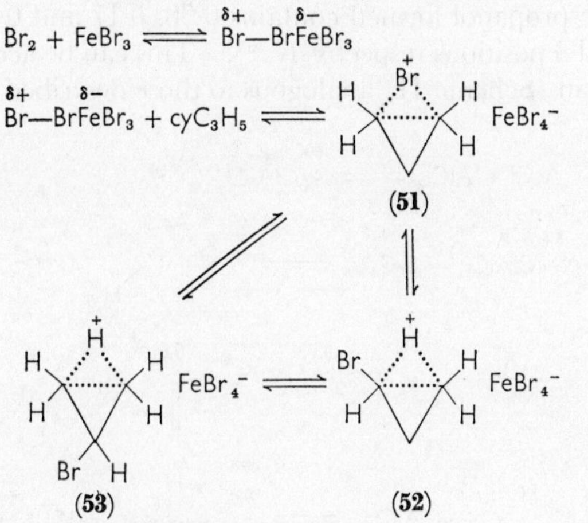

$$Br_2 + FeBr_3 \rightleftharpoons \overset{\delta+}{Br}-\overset{\delta-}{BrFeBr_3}$$

$$\overset{\delta+}{Br}-\overset{\delta-}{BrFeBr_3} + cyC_3H_5 \rightleftharpoons \text{(51)} \quad FeBr_4^-$$

(51)

(53) ⇌ (52) FeBr$_4^-$

(51) ⟶ BrCH$_2$CH$_2$CH$_2$Br + FeBr$_3$
(52) ⟶ Br$_2$CHCH$_2$CH$_3$ + BrCH$_2$CH$_2$CH$_2$Br + FeBr$_3$
(53) ⟶ BrCH$_2$CHBrCH$_3$ + FeBr$_3$

SCHEME I. Electrophilic addition of bromine to cyclopropane

A value of 6 kcal/mole has been determined for E_A. The molecular weight obtained is inversely proportional to the temperature (°c). In the presence of branch-chain alkanes the molecular weight is decreased, probably due to hydride transfer. The comparatively low molecular weight of the polymer is believed to be due to elimination[257]. Hart and coworkers[259-261] have studied the reaction of cyclopropane with acyl halides in the presence of aluminium chloride. Their results are presented in Table 45. 1,1-, 1,2- and 1,3-addition products are all obtained[261], with 1,2-addition probably predominating. The proposed mechanism is similar to that suggested by Deno and Lincoln for the addition of bromine.

Cyclopropane has been reacted with *para*-toluenesulphonyl chloride in the presence of AlCl$_3$. Vapour phase chromatography showed the major components of the reaction mixture to be 1-tosyl-3-chloropropane and unreacted *p*-toluenesulphonyl chloride[262]. Cyclopropane reacts with sulphuric acid to give propyl sulphate, dipropyl sulphate and propyl alcohol[263-267]. Results are given in Table 45. The rate of the reaction is given by

$$v = k[cyC_3H_6][HA]^n$$

For perchloric or sulphuric acid in water, n equals 6 to 10; for sulphuric acid in acetic acid, n equals 2·5 to 3·5[265]. When reacted with

D_2SO_4, the propanol formed contains 0·38, 0·17 and 0·46 D atom in the 1, 2 and 3 positions respectively [267]. This can be accounted for by a mechanism (Scheme II) analogous to those described previously for

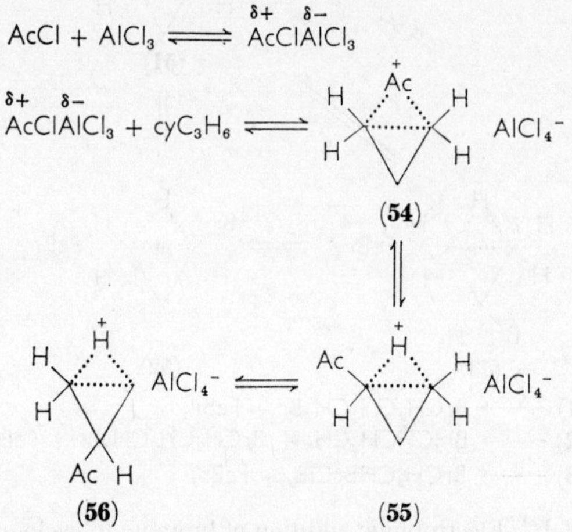

$$AcCl + AlCl_3 \rightleftharpoons \overset{\delta+}{Ac}\overset{\delta-}{Cl}AlCl_3$$

$$\overset{\delta+}{Ac}\overset{\delta-}{Cl}AlCl_3 + cyC_3H_6 \rightleftharpoons (54) \; AlCl_4^-$$

(54) ⇌

(56) ⇌ (55)

(54) ⟶ $AcCH_2CH_2CH_2Cl + AlCl_3$
(55) ⟶ $AcCH_2CH_2CH_2Cl + AcCHClEt + AlCl_3$
(56) ⟶ $AcCHMeCH_2Cl + AlCl_3$

SCHEME II. Electrophilic addition of acetyl chloride to cyclopropane

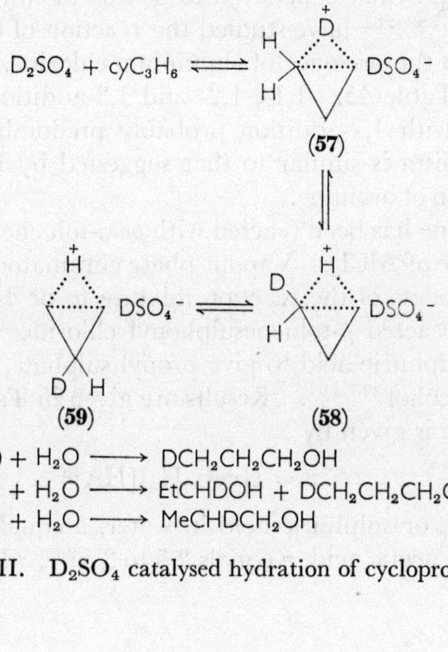

$$D_2SO_4 + cyC_3H_6 \rightleftharpoons (57) \; DSO_4^-$$

(59) ⇌ (58)

(57) + H_2O ⟶ $DCH_2CH_2CH_2OH$
(58) + H_2O ⟶ $EtCHDOH + DCH_2CH_2CH_2OH$
(59) + H_2O ⟶ $MeCHDCH_2OH$

SCHEME III. D_2SO_4 catalysed hydration of cyclopropane

the reaction of cyclopropane with bromine (Scheme I) and with acetyl chloride (Scheme III).

With carboxylic acids in the presence of boron trifluoride, cyclopropane gives the corresponding esters[264]. Cyclopropane has been reported to react with diborane to form tri-n-propyl boron and with lithium aluminium hydride to form tri-n-propyl aluminium[268,269]. The data collected in Table 45 show that cyclopropane undergoes a number of ring opening addition reactions with electrophilic reagents. Contrary to the impression conveyed in most elementary organic texts. the product is not always the result of 1,3-addition. In particular, the recent work of Deno and Lincoln[249], Hart and Schlosberg[261] and Baird and Aboderin[267] shows that a range of products is often obtained. It would undoubtedly be very useful if some of the reactions reported in Table 45 were reinvestigated with the assistance of modern instrumentation.

C. Electrophilic Addition of Halogen to Substituted Cyclopropanes

An unsymmetrically substituted cyclopropane undergoing the addition of a symmetrical reagent may give rise to more than one product, depending upon which of the ring bonds is broken in the course of the addition. All of the available evidence on this point is summarized in Table 46. Only one example of the addition of chlorine to a

TABLE 46. Electrophilic addition of halogen to substituted cyclopropanes

Substituents $C_{(1)}$	$C_{(2)}$	Reagent	Products	Ref.
Me		Br_2	$BrCH_2CH_2CHBrMe$	270
Me,Me		Br_2	$MeCHBrCBrMe_2$ (40)% $EtMeCBrCH_2Br$ (60%)	271
Me,Me	Me	Br_2	$Me_2CBrCH_2CHBrMe$	272a
CO_2H		Cl_2	$ClCH_2CH_2CHClCO_2H$	273
CO_2H		Br_2	$BrCH_2CH_2CHBrCO_2H$	273

substituted cyclopropane is known. It is especially important to note that while the reaction conditions for these reactions suggest that an electrophilic addition is involved, in no case has it been positively established that the reaction does not proceed by a radical path. Only one example[271] in Table 46 represents a recent study in which modern instrumentation has been used to help effect a product analysis. It would seem that this reaction needs considerable investigation before any meaningful conclusions can be drawn.

The products obtained from the addition of bromine to 1,1-dimethylcyclopropane can be rationalized in terms of Scheme IV.

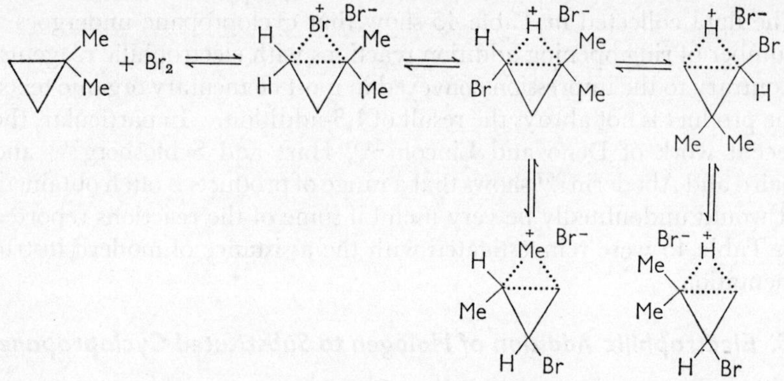

(60) ⟶ MeCHBrCBrMe$_2$ (61) ⟶ EtCHBrCHBrMe

SCHEME IV. Addition of bromine to 1,1-dimethylcyclopropane

Rates of addition of bromine to substituted cyclopropanes in 37·5% aqueous acetic acid at 0, 24 and 40° have been reported[272] (Table 47). Cyclopropanecarboxylic acid was found not to react under these conditions. The kinetics of the reaction were similar to those for the addition of bromine to olefins in aqueous acetic acid. As all of the substituents in Table 47 are of the type CH$_2$Z the rate constants have

TABLE 47. Rates of addition^a of bromine to substituted cyclopropanes (cyC$_3$H$_5$X) in 37·5% aqueous acetic acid[272]

X	0°	24°	40°
CH$_2$CO$_2$H	3·12	24·4	106
CH$_2$CH$_2$CO$_2$H	62·7	645	1840
CH$_2$CH$_2$CH$_2$CO$_2$H	278	3280	9210
(CH$_2$)$_4$CO$_2$H	597	5480	171000
Bu	1290	12100	36300

^{a} Rate constants in sec^{-1}.

been correlated with the σ_I values for the Z substituent, by means of equation (15).

$$Q_X = \alpha\sigma_{IX} + h \qquad (15)$$

Excellent correlations were obtained with $\alpha \sim -7$. The value of α obtained graphically for the addition of bromine to olefins bearing CH_2Z substituents in acetic acid at 24° is approximately -6.5. Thus both addition reactions seem to show about the same sensitivity to the localized effect of a substituent.

It would be interesting to study the stereochemistry of the addition of bromine to substituted cyclopropanes. An examination of the configurations of the products obtained from the addition of bromine to *cis*- and *trans*-1,2-dimethylcyclopropane for example, could be useful in providing further mechanistic evidence. Unfortunately, no such results are available at the present time.

D. Addition of Hydrogen halide to Substituted Cyclopropanes

A number of studies of the addition of hydrogen halide to substituted cyclopropanes have been reported; the results are gathered in Table 48. The trend in orientation is toward 1,3-addition following

TABLE 48. Hydrogen halide addition to substituted cyclopropanes

Substituents				
$C_{(1)}$	$C_{(2)}$	HHl	Product	Ref.
Me		HI	s-BuI	270
Et		HBr	Et_2CHBr	274
		HI	Et_2CHI	274
Me,Me		HBr	Me_2CBrEt	275
		HI	Me_2EtCI	276
Me,Me	Me	HBr	Me_2CBr-i-Pr	277
CO_2H		HBr	$Br(CH_2)_3CO_2H$	278
Cl,Cl		HF	$[CCl_2FEt]^a$	279
2,4,6-$Me_3C_6H_2CO$		HBr	2,4,6-$Me_3C_6H_2CO(CH_2)_3Br$	280
Bz		HBr	$Bz(CH_2)_3Br$	281
Ac		HBr	$Ac(CH_2)_3Br$	282
CO_2H,CO_2H		HBr	$BrCH_2CH_2CH(CO_2H)_2$	282
$PhSO_2$		HBr	no reaction	283
		HI	no reaction	283
4-$MePnSO_2$		HBr	no reaction	283
$EtSO_2,EtSO_2$		HI	$(EtSO_2)_2CHCH_2CHI$	284

^{a} This is probably the product of addition. The product obtained is CF_3Et, resulting from Cl displacement by F under the reaction conditions.

the Markownikoff rule. This reaction is seriously in need of reinvestigation with the aid of modern instrumentation: most of the work reported was done before 1950. It is uncertain whether the products reported are the predominant products, the major products or the sole products. The available data can be interpreted in terms of Scheme V.

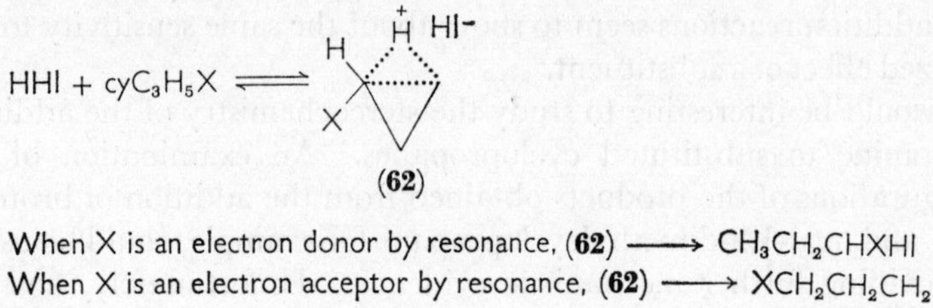

$$HHl + cyC_3H_5X \rightleftharpoons \begin{array}{c} H \\ \diagup \\ X \end{array} \begin{array}{c} \overset{+}{H} \quad Hl^- \\ \end{array}$$

(62)

When X is an electron donor by resonance, (62) $\longrightarrow$ CH_3CH_2CHXHI
When X is an electron acceptor by resonance, (62) $\longrightarrow$ $XCH_2CH_2CH_2HI$

SCHEME V. Substituent effects on hydrogen halide addition to cyclopropane

It seems that cyclopropanes bearing electron-donor substituents react more readily than do those which bear electron-acceptor substituents, in accord with what would be expected for an electrophilic attack. Unfortunately no quantitative data are available on this point.

E. Addition of Water and of Carboxylic Acids to Substituted Cyclopropanes

The results of studies of hydration of substituted cyclopropanes are gathered in Table 49. The dominant trend in orientation appears to be toward 1,3-addition in accord with the Markownikoff rule. The hydration of cyclopropanecarboxaldehyde leads to some rather sur-

TABLE 49. Hydration of substituted cyclopropanes

Substituents				
$C_{(1)}$	$C_{(2)}$	Reagent	Product	Ref.
Me		50% H_2SO_4–H_2O	s-BuOH	270
Et		80% H_2SO_4–H_2O	Et_2CHOH	274
Me,Me	Me	HNO_3(Sp.Gr. 1·4) in AcOH	Me_2COH-i-Pr	277
Ac		5% HCl–H_2O	$HO(CH_2)_3Ac$	282
CHO		92% H_2SO_4, 0°	EtCHOHCHO	285
		60% H_2SO_4, 120°	AcCHOHMe	285

prising results however[285]. Possibly these may be accounted for by the formation of the intermediate **63**, in which the carbonyl group has been converted to a substituent which should have a small donor resonance effect and therefore is capable of undergoing 1,3-addition to

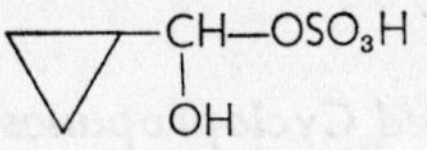

▽—CH—OSO₃H
    |
    OH
   (63)

give 2-hydroxybutanal at lower temperatures. Formation of acetoin at a higher temperature and more dilute acid is perhaps accountable in terms of an initial formation of 2-hydroxybutanal, followed by a carbonium ion rearrangement.

It is of interest to consider at this point the results obtained by De Puy and his coworkers[286] for the acid-catalysed ring opening of *cis*-2-phenyl-1-methylcyclopropanol in 50% aqueous dioxane. The products obtained are 4-phenyl-2-butanone (60%) and 3-phenyl-2-butanone (40%). The reaction carried out in dioxane–D₂O is bimolecular. The products obtained were 4-deutero-4-phenyl-2-butanone and 4-deutero-3-phenyl-2-butanone. The formation of the 4-deutero-2-butanone results in the generation of a new asymmetric centre. This occurred with retention of configuration; the results obtained may be accounted for by Scheme VI.

Benzoylcyclopropane has been reported to react with acetic acid in the presence of sulphuric acid to form 4-acetoxy-1-phenyl-1-butanone[286a]. With hydrogen bromide, *trans*-caronic acid (**64**) produces the lactone **65**[286b].

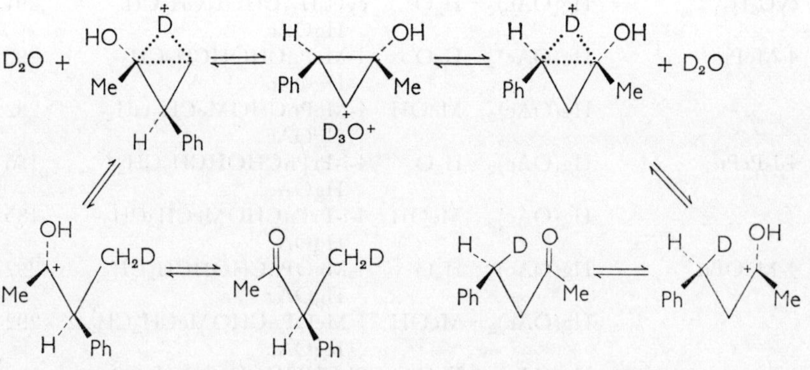

SCHEME VI. Addition of D₂O to 2-phenyl-1-methyl-cyclopropanol

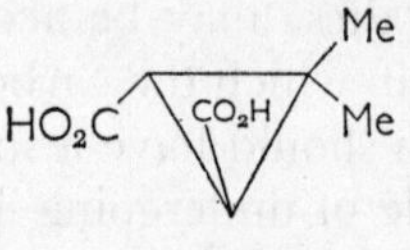

(64) 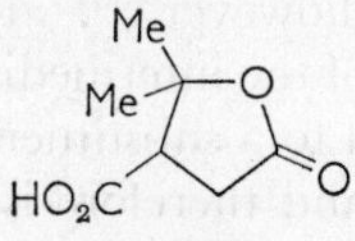   (65)

F. Reaction of Substituted Cyclopropanes with Metal Salts

The reaction of olefins with mercuric salts has been known for some time[286c]. Levina and Gladshtein[286d] first observed the reaction of cyclopropanes with mercuric salts. Levina and coworkers' extensive investigations on this reaction are summarized in Table 50. The

TABLE 50. Mercuration of substituted cyclopropanes

Substituents					
$C_{(1)}$	$C_{(2)}$	Reagent	Solvent	Product	Ref.
Me		$Hg(OAc)_2$	H_2O	$MeCHOHCH_2CH_2HgOAc$	287
		$Hg(OAc)_2$	MeOH	$MeCHOMeCH_2CH_2HgOAc$	287
Me,Me		$Hg(OAc)_2$	H_2O	$Me_2COHCH_2CH_2HgOAc$	287
		$Hg(OAc)_2$	MeOH	$Me_2COMeCH_2CH_2HgOAc$	287
Me,Me	Me	$Hg(OAc)_2$	H_2O	$Me_2COHCHMeCH_2HgOAc$	288
Me,Me	Me,Me	$Hg(OAc)_2$	H_2O	$Me_2COHCMe_2CH_2HgOAc$	289
		$Hg(OAc)_2$	MeOH	$Me_2COMeCMe_2CH_2HgOAc$	289
		$Hg(OAc)_2$	EtOH	$Me_2COEtCMe_2CH_2HgOAc$	289
		$HgCl_2$	H_2O	$Me_2COHCMe_2CH_2HgCl$	289
Me,Et	Me,Me	$Hg(OAc)_2$	H_2O	$Me_2COHCMeEtCH_2HgOAc$ (58%) + $MeEtCOHCMe_2CH_2HgOAc$ (42%)	290
cyC_6H_{11}		$Hg(OAc)_2$	H_2O	$cyC_6H_{11}CHOHCH_2CH_2$-$HgOAc$	291
4-MePn		$Hg(OAc)_2$	H_2O	4-$MePnCHOHCH_2CH_2$-$HgOAc$	292
		$Hg(OAc)_2$	MeOH	4-$MePnCHOMeCH_2CH_2$-$HgOAc$	292
4-i-PrPn		$Hg(OAc)_2$	H_2O	4-i-$PrPnCHOHCH_2CH_2$-$HgOAc$	135
		$Hg(OAc)_2$	MeOH	4-i-$PrPnCHOMeCH_2CH_2$-$HgOAc$	135
4-MeOPn		$Hg(OAc)_2$	H_2O	4-$MeOPnCHOHCH_2CH_2$-$HgOAc$	292
		$Hg(OAc)_2$	MeOH	4-$McOPnCHOMeCH_2CH_2$-$HgOAc$	292
Ph		$Hg(OAc)_2$	H_2O	$PhCHOHCH_2CH_2HgOAc$	196

observed orientation is in accord with the Markownikoff rule. The products obtained are the result of 1,3-addition. Thus far the reaction has been studied only in the case of donor substituents on the cyclopropane ring. The reaction of cyclopropanes with lead (IV) acetate[293] and with thallium (III) acetate[234] have been studied[293] and results are given in Table 51. 4-Bromophenylcyclopropane reacts

TABLE 51. Reaction of lead(IV) acetate and thallium(III) acetate with substituted cyclopropanes

Substituent	Reagent[a]	Products	Ref.
Ph	Pb(OAc)$_4$	PhCHOAcCH$_2$CH$_2$OAc (65%) + PhCH(CH$_2$OAc)$_2$ (5%) + 2-PhVnCH$_2$OAc	293
	Tl(OAc)$_3$	PhCHOAcCH$_2$CH$_2$OAc (92%) + 2-PhVnCH$_2$OAc (8%)	294
Et	Pb(OAc)$_4$	EtCHOAcCH$_2$CH$_2$OAc predom. + EtCH(CH$_2$OAc)$_2$	293
	Tl(OAc)$_3$	EtCHOAcCH$_2$CH$_2$OAc predom. + EtCH(CH$_2$OAc)$_2$ + acetoxyolefin	294

[a] In acetic acid.

with Pb(OAc)$_4$ more slowly than phenylcyclopropane by a factor of 2. Ethylcyclopropane also reacts more slowly than phenylcyclopropane. It is suggested that the reaction with Tl(OAc)$_3$ proceeds in two steps via the formation of an organothallium compound.

G. Alkylation of Benzene by Substituted Cyclopropanes

Several examples of this reaction are known; they are collected in Table 52. The results indicate 1,3-addition with orientation in

TABLE 52. Alkylation of Benzene by substituted cyclopropanes[a]

Substituent	Product	Ref.
Ph	Ph$_2$CHEt (52%) + polymer	295
4-MePn	4-MePnPhCHEt (34%) + polymer	295
4-MeOPn	4-MeOPnPhCHEt (5·5%) + polymer	295
2,4,6-Me$_3$C$_6$H$_2$CO	2,4,6-Me$_3$C$_6$H$_2$CO(CH$_2$)$_3$Ph	280

[a] Catalysed by AlCl$_3$.

accord with the Markownikoff rule. The polymer obtained is presumably the result of cationic polymerization of the aryl cyclopropane. A number of other examples of cationic polymerization of substituted cyclopropanes are summarized in Table 53. No studies of the polymer

TABLE 53. Cationic polymerization of substituted cyclopropanes[a]

Substituent $C_{(1)}$	$C_{(2)}$	T(°c)	Catalyst	Products	Ref.
Me,Me		24	$AlBr_3$ + HBr	100% polymer	296
		−50	$AlBr_3$	100% polymer	296
Et		25	$AlCl_3$ + HCl	100% polymer	296
Pr		24	$AlBr_3$ + HBr	90% polymer + 3·5% PrH +5·4% i-PrPr	
		24	$AlCl_3$	100% polymer	
Me	Me		$AlCl_2(OH)$	100% polymer	296
Ph	Me		90% H_2SO_4	polymer	297
Ph	Et		90% H_2SO_4	polymer	297
Ph	i-Pr		90% H_2SO_4	1,1,2-trimethylindane	297

[a] See also Table 52.

structure seem to have been made. Again, it would seem that a reinvestigation of this reaction is called for.

H. Electrophilic Addition to Bicyclo[n·1·0]alkanes

Bicycloalkanes containing a cyclopropane ring have long been of interest, as a number of terpene derivatives fall into this category. Bicyclobutanes will not be included in this review as their properties are not typical of the cyclopropane ring. They have been reviewed by Wiberg[297a]. Pertinent results are collected in Table 54. In general, the reactions are 1,3-additions. The bicyclo[2·1·0]pentane products are invariably the result of the cleavage of the bond common to both rings, the products being substituted cyclopentanes. Addition to bicyclo[n·1·0]alkanes with $n \geq 3$ results in the opening of one of the cyclopropane ring bonds which is not common to both rings. LaLonde and coworkers[298,302,305-307] have attempted to account for the products obtained in terms of polarization and non-bonded interactions in the transition state and strain energy in the ground state.

In the tricyclenes the addition results in the cleavage of the 0 bridge. Thus when $n \geq 3$, the cleavage obeys the Markownikoff rule, whereas

10. Olefinic Properties of Cyclopropanes

TABLE 54. Addition reactions of bicyclo[$n \cdot 1 \cdot 0$]alkanes

Bicyclane	Reagent	Products	Ref.
Bicyclo[$2 \cdot 1 \cdot 0$]pentane	TsOH in AcOH	Cyclopentyl acetate	298
	HBr	Cyclopentyl bromide	298
	Hg(OAc)$_2$ in H$_2$O	3-Acetoxymercuricyclopentanol	300
	Pb(OAc)$_4$	1,3-Diacetoxycyclopentane	299
Bicyclo[$3 \cdot 1 \cdot 0$]hexane	TsOH in AcOH	trans-2-Methylcyclopentylacetate[a]	298
	HgOAc	1-Acetoxymercurimethyl-3-cyclopentanol	300
2,6,6-Trimethylbicyclo[$3 \cdot 1 \cdot 0$]hexane	HBr	1-Methyl-2-(1'-bromo-1'-methylethyl)cyclopentane	301
exo-6-Methylbicyclo[$3 \cdot 1 \cdot 0$]hexane	H$_2$SO$_4$ in AcOH	Cyclopentenes[a], cyclohexenes and cyclohexyl acetates	302
1-Isopropyl-4-methylbicyclo[$3 \cdot 1 \cdot 0$]hexane	HBr	1-Bromo-1-isopropyl-2,3-dimethylcyclopentane	303
Bicyclo[$4 \cdot 1 \cdot 0$]heptane	H$_2$SO$_4$ in AcOH	trans-2-Methylcyclohexyl acetate[a]	298
	TsOH in AcOH	trans-Methylcyclohexyl acetate[a]	298
	Hg(OAc)$_2$ in H$_2$O	trans-2-Acetoxymercurimethylcyclohexanol	300
	Pb(OAc)$_4$	trans-2-(acetoxymethyl)-cyclohexyl acetate[a] + 1,3-diacetoxycycloheptane	293
	Tl(OAc)$_3$	trans-2-Acetoxymethylcyclomethyl acetate[a] + 1,3-diacetoxycycloheptane	294
1-Methylbicyclo[$4 \cdot 1 \cdot 0$]heptane	diborane	Tris[(2-methylcyclohexyl)methyl]borane	304
exo-7-Methylbicyclo[$4 \cdot 1 \cdot 0$]heptane	H$_2$SO$_4$ in AcOH	Cyclohexenes[a] cyclohexyl acetates[a], cycloheptanes and cycloheptyl acetates	302
Bicyclo[$5 \cdot 1 \cdot 0$]octane	H$_2$SO$_4$ in AcOH	Cycloheptanes, cycloheptyl acetates[a], some cyclohexyl and cyclooctyl derivatives	305
Tricyclo[$5 \cdot 4 \cdot 0 \cdot 0^{3,5}$]undecane(2,3-methanodecalin)	H$_2$SO$_4$ in AcOH	Acetoxydecalins and octalins	306
exo-Tricyclo[$3 \cdot 2 \cdot 1 \cdot 0^{2,4}$]octane	D$_2$SO$_4$ in AcOD	exo(axial)-Bicyclo[$3 \cdot 2 \cdot 1$]-octan-2-ol-4-d_1 acetate[a]	307
Tricyclo[$2 \cdot 2 \cdot 1 \cdot 0^{2,6}$]heptane	AcCl-AlCl$_3$	2-Acetyl-7-chlorobicyclo[$2 \cdot 2 \cdot 1$]heptane	308

TABLE 54. (continued)

Bicyclane	Reagent	Products	Ref.
1-Acetoxytricyclo-[2.2.1.0^{2,6}]heptane (1-acetoxynortricyclene)	D$_2$SO$_4$ in AcOD–D$_2$O	endo-6-Deuterobicyclo-[2.2.1]heptan-2-one	310
2,3,3-Trimethyltricyclo-[2.2.1.0^{2,6}]heptane	HF	2-Fluoro-2,3,3-trimethyl-bicyclo[2.2.1]heptane (camphene hydrofluoride) + 2-fluoro-1,7,7-trimethyl-bicyclo[2.2.1]heptane (isobornyl fluoride) + camphene	309

^{a} Predominant product.

when $n = 2$ or in the case of the tricyclene system, other factors become sufficiently important to be product determining. The reaction of exo-tricyclo[3.2.1.0^{2,4}] octane (**66**) with D$_2$SO$_4$ in AcOD results in inversion of configuration leading to the conclusion that deuteron attack on the C$_{(2)}$—C$_{(4)}$ bond must be 'end-on'[307]. This is believed due to the geometry of the molecule which prevents the more usual edge-on attack.

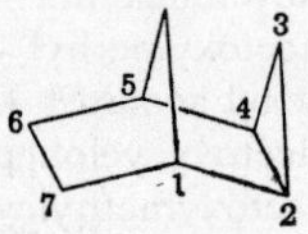

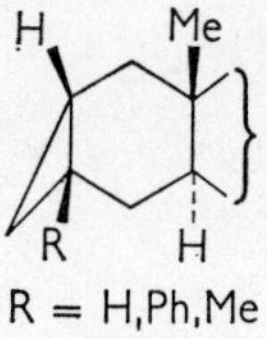

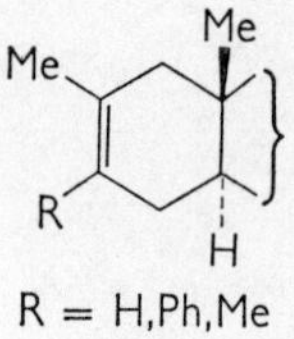

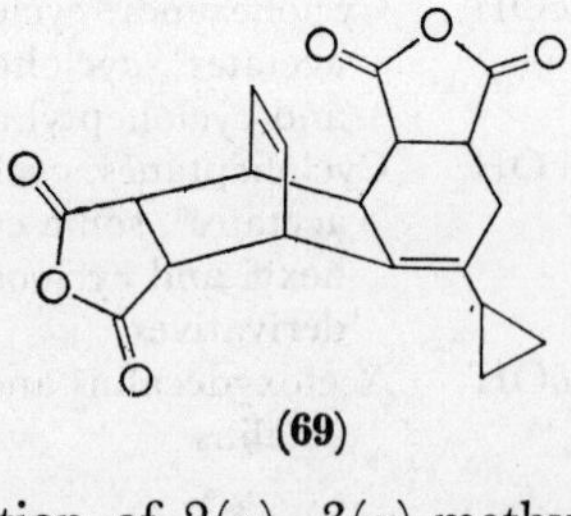

The reaction of 2(α), 3(α)-methylenecholestane (**67**) and its 3β-methyl and phenyl derivatives with perchloric acid in aqueous acetic acid, opens the cyclopropane ring in accord with the Markownikoff rule, to form the corresponding 3-substituted 2-cholestenes[311] (**68**).

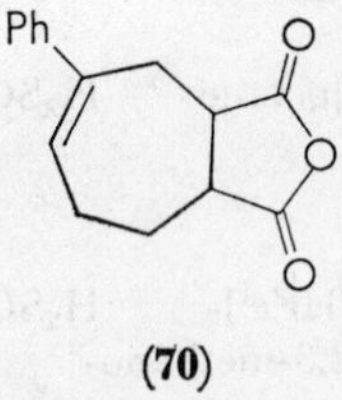

I. Terminal Addition to Vinyl Cyclopropanes

The reaction of α-gurjenene with diborane followed by oxidation and its reaction with peracids are believed to be examples of addition to a conjugated system of the vinyl cyclopropane type [312,313]. 4-substituted α-cyclopropylstyrenes have been shown to react with acetic and trifluoroacetic acids to give 2-substituted phenyl-2-pentene-5-ol acetates and trifluoroacetates. The yield is greatest in the case of the 4-methoxy substituent and least in the case of the 4-chloro [314]. The reaction of α-cyclopropylstyrene with maleic anhydride has been studied. The products obtained are **69** and **70**. The formation of the latter is suggestive of a Diels-Alder type of reaction involving the vinylcyclopropane system. As the mechanism of this reaction is not known, however, such a comparison is merely formal at the present time [315].

J. Relative Reactivity of the Cyclopropane Ring and the Double Bond Toward Electrophilic Addition

Cyclopropane has been reported to be more reactive than propene toward sulphuric acid. However, propene is said to be much more readily absorbed by bromine or bromine and iodine solutions in water, than is cyclopropane [316]. Methylenecyclopropane has been reported to undergo normal olefinic reactions leaving the cyclopropane ring unaffected [316a]. The reaction of bicyclo[3·1·0]hex-2-ene with methanol catalysed by acid, produces as the major products, cis- and trans-2-methoxy-bicyclo[3·1·0]hexane and only small amounts of 4-methoxycyclohexene. With HCl, 31% of cis- and 66% of trans-2-chlorobicyclo[3·1·0]hexane are formed with 3% 4-chlorocyclohexene [317,318]. Hydroboration of spiro[2·5]oct-4-ene occurs only at the double bond [319]. Apparently electrophilic reagents which are known to attack the cyclopropane ring and the double bond generally attack the double bond more readily.

K. Radical Additions to Cyclopropanes

Radical additions of bromine and iodine to cyclopropane have been observed. With chlorine, substitution occurs in preference to addition [160a,331-334]. The reaction of spiro pentane with chlorine produces chlorospiropentane, 1,1-bis(chloromethyl)cyclopropane, 2-chloromethyl-4-chloro-1-butene and 2-chloromethyl-2,4-dichlorobutane as major products. The reaction is believed to proceed through a radical path[335]. Cyclopropane has undergone free radical polymerization to give a low molecular weight polymer. Attempts at polymerizing substituted cyclopropanes by radical catalysts were unsuccessful[328]. Studies of radical addition to cyclopropanes are collected in Table 55.

TABLE 55. Radical additions to substituted cyclopropanes

Substituent $C_{(1)}$	$C_{(2)}$	Reagent	Initiation	Product	Ref.
		$Br_2(g)$	$h\nu$	$(CH_2)_3Br_2$	320
		$Br_2(g)$	$h\nu + O_2$	$(CH_2)_3Br_2$	321
		$I_2(g)$	heat or $h\nu$	$(CH_2)_3I_2$	322
			$h\nu$	low mol. wt. polymer	323
			α-radiation	low mol. wt polymer	324
		Hg	$Hg(^3P_1)$	low mol. wt polymer	325
			$h\nu$	low mol. wt polymer	326
CN		Br_2	$h\nu$	no reaction	327
			$(AcO)_2$, $(BzO)_2$, $[Et_2C(CN)]_2N_2$	no reaction	328
Ac			$(AcO)_2$, $(BzO)_2$, $[Et_2C(CN)]_2N_2$	no reaction	328
CO_2H, CO_2H		Br_2	$h\nu$	$BrCH_2CH_2CBr(CO_2H)_2$	330
Ph			$(AcO)_2$, $(BzO)_2$, $[Et_2C(CN)]_2N_2$	no reaction	328
i-Pr	Ph	N-Bromosuccinimide	$(BzO)_2$	probably $PhCHBrCH_2CHBr$-i-Pr	329

The reactions of the methylthiyl and trichloromethyl radicals with isopropenylcyclopropane suggest that they may involve addition to the vinylcyclopropane conjugated system. Thus the methylthiyl radical reacts with isopropenylcyclopropane to give 1-methylthio-2-cyclopropylpropane and 1-methylthio-2-methyl-2-pentene. The trichloromethyl radical reacts with isopropenylcyclopropane to form 1,1,1,6-tetrachloro-3-methyl-3-hexene[335a].

Radical additions to cyclopropanes do not seem to occur as readily as electrophilic additions do. Very much more work remains to be done in this area before any firm conclusions can be reached. Our present state of knowledge of the radical addition to cyclopropanes is perhaps summed up in the statement that such radical additions do in fact exist.

L. Nucleophilic Additions to Cyclopropanes

A number of nucleophilic additions to substituted cyclopropanes have been observed (Table 56). Generally, the reaction requires two

TABLE 56. Nucleophilic addition to substituted cyclopropanes

Substituents $C_{(1)}$	$C_{(2)}$	Reagent	Products	Ref
CO_2Et, CO_2Et		$NaCHCO_2Et$	$EtO_2CCHCH_2CH_2CH-(CO_2Et)_2$	336
CO_2Et, CO_2Et	Vi	$NaCHCO_2Et$	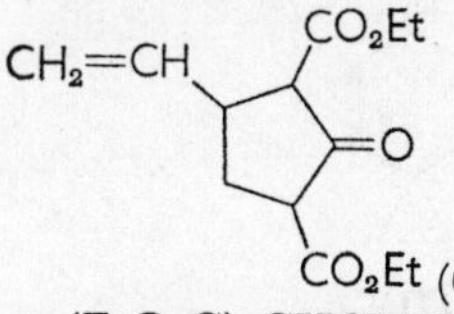 [cyclopentanone with $CH_2=CH$, CO_2Et, =O, CO_2Et substituents] (60%) + $(EtO_2C)_2CHCHViCH_2$-$CH(CO_2Et)_2$ + $(EtO_2C)_2$-$CHCH_2CH=CHCH_2$-$CH(CO_2Et)_2$	337
CN, CO_2Et		$NaCHCN(CO_2Et)$	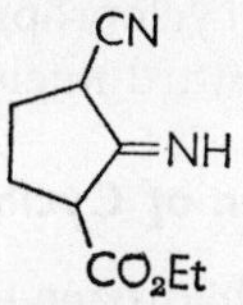 [cyclopentane with CN, =NH, CO_2Et substituents]	338
Ac		NaSPh	$PhS(CH_2)_3Ac$ (58%)	283
$PhSO_2$		NaSPh	no reaction	283
$PhSO_2$		t-BuO$^-$	no reaction	283
4-MePnSO$_2$		KOH	no reaction	283
t-BuSO$_2$		NaOMe	elimination occurs	283
CO_2Et, CO_2Et		Et_2NH	$Et_2N(CH_2)_2CH(CO_2Et)_2$	339
CO_2Et, CO_2Et		$C_5H_{11}N^a$	$C_5H_{10}NCH_2CH_2CH-(CO_2Et)_2$	339
CO_2Et, CO_2Et		BuSH	$BuSCH_2CH_2CH(CO_2Et)_2$	339
CO_2Et, CO_2Et		PhSH	$PhSCH_2CH_2CH(CO_2Et)_2$	339
CO_2Et, CO_2Et		PhOH	$PhOCH_2CH_2CH(CO_2Et)_2$	339
CO_2Et, CN		$C_5H_{11}N^a$	$C_5H_{10}NCH_2CH_2CH-(CO_2Et)CN$	339
CN, CN		Me_2NH	$Me_2NCH_2CH_2CH(CN)_2$	339
CN, CN		$C_5H_{11}N^a$	$C_5H_{10}NCH_2CH_2CH(CN)_2$	339
CN, $CONH_2$		$C_5H_{11}N^a$	$C_5H_{10}NCH_2CH_2CH-(CONH_2)CN$	339
$CONH_2$, $CONH_2$		$C_5H_{11}N^a$	$C_5H_{10}NCH_2CH_2CH-(CONH_2)_2$	339

[a] Piperidine.

electron-acceptor substituents on a carbon of the cyclopropane ring. The addition appears to be 1,3 and the orientation observed is analogous to that observed in the Michael addition to olefins. The reaction of spiro[2·5]octa-3,6-diene-5-one (**71**) with methoxide ion has been shown to give 4-(2'-methoxyethyl)phenoxide ion. The reaction has been shown to be bimolecular, first order in **71** and first order in methoxide ion [340]. Reaction of 4,6-di-t-butylspiro[2·5]-octa-3,-diene-5-one (**72**) with methyl or ethyl Grignard reagent gives **73** after hydrolysis.

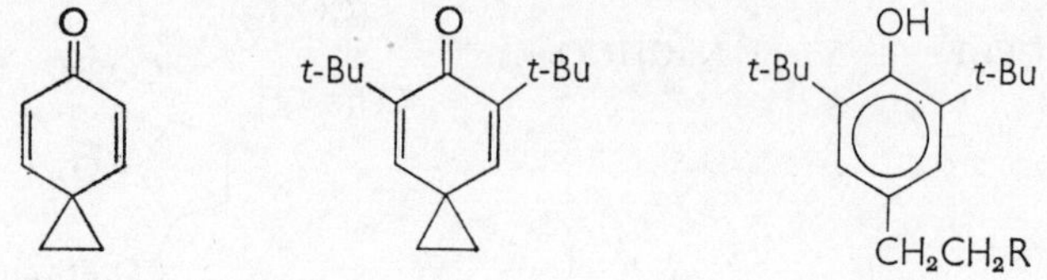

(**71**)   (**72**)   (**73**)

The sparse data available do not permit any conclusions to be reached concerning nucleophilic addition to cyclopropanes. It would seem however, that the cyclopropane ring is more reactive toward electrophiles than it is toward nucleophiles.

M. Hydrogenation of Cyclopropane

The addition of hydrogen to cyclopropane appears to have been first observed by Willstatter and Bruce [342]. Studies have appeared since then on the addition of hydrogen to various cyclopropane derivatives. The reaction is complicated by the observation that cyclopropane derivatives are often isomerized to olefins under the reaction conditions, thus ethylcyclopropane has been reported to isomerize to *cis*- and *trans*-2-pentene on being passed through a column of silica gel. Ethylcyclopropane is isomerized almost completely over alumino silicate at 50°, over Kieselguhr at 120° it is 75% isomerized; over pumice it is isomerized 0% at 120°, 20% at 170° and 45% at 220°. At temperatures up to 150°, platinum on charcoal does not cause isomerization [344]. Platinum and palladium on charcoal at 220° do not cause isomerization in alkylcyclopropanes [345]. Vinylcyclopropane has been found to isomerize over Kieselguhr to 2-methyl-1,3-butadiene [346]. 1,1,2-Trimethylcyclopropane over palladium on charcoal 220° isomerizes to a mixture containing 55% olefin [357,358]. Isomerization also takes place over platinum at this temperature [358].

A comparison of the deuterium distribution in the propane obtained by deuteration of cyclopropane and of propylene over palladium, shows that hydrogenation of cyclopropane over palladium does not involve a prior isomerization to propene[346a].

The results of rate studies on the hydrogenation of cyclopropane over various catalysts are summarized in Table 57. In general, over

TABLE 57. Rate laws for the hydrogenation and deuteration of cyclopropane

Catalyst	Reagent	Reaction Order[a]		T(°c)	E_A(kcal/mole)	Ref.
		m	n			
Ni on alumino-silicate	H_2	0.3[b]	−0.1[b]	75	15.2	347
Pd on pumice	H_2	0.3	−0.8	0		348
Pd on pumice	H_2	1.0	0	200		348
Pd on pumice	D_2	1.0	0	0	8.0	349
Pd on pumice	D_2	1.0	0.35	200		349
Ni	H_2	1.0	0		10.6	351
Pd	H_2	1.0	0		8.1	351
Pt	H_2	1.0	0		8.9	351

[a] Exponents in the equation $v = k[\text{cyC}_3\text{H}_5]^m[\text{H}_2]^n$.
[b] Exponents in the equation $v = kP^m[\text{cyC}_3\text{H}_5] \cdot P^n[\text{H}_2]$, P = partial pressure.

nickel, palladium and platinum, the reaction is first order in cyclopropane and zero order in hydrogen.

Results for the catalytic hydrogenation of substituted cyclopropanes are collected in Table 58. The results of catalytic hydrogenation of spiranes, bicycloalkanes and methylenecyclopropanes are collected in Table 59. As noted above, the results are complicated by the possibility of isomerization to olefins in the course of the reaction. The following generalizations can be made, however. Cyclopropanes bearing alkyl groups at $C_{(1)}$ and $C_{(2)}$ are usually opened at the 2–3 bond. Cyclopropanes bearing π-bond substituents such as vinyl or phenyl, at $C_{(1)}$ or $C_{(1)}$ and $C_{(2)}$ are usually opened at the 1–2 bond. Spiranes containing a cyclopropane ring generally behave like dialkylcyclopropanes opening at the 2–3 bond. The bicyclo[3·1·0]hexane derivatives studied open at the $C_{(2)}$—$C_{(3)}$ bond.

Ullmann has proposed that the hydrogen of vinylcyclopropanes and methylenecyclopropanes proceeds through a common carbanion intermediate formed by an initial hydride-ion transfer from the catalyst[373].

TABLE 58. Catalytic hydrogenation of substituted cyclopropanes

Substituents $C_{(1)}$	$C_{(2)}$		Catalyst	T(°C)	Products	Ref.
			Ni		PrH	342,351
			Ni on aluminosilicate		PrH	347
			Pd on pumice		PrH	348
			Pt on pumice		PrH	349,350
Et			Pt on charcoal	50		302
i-Pr			Pd		i-Pr$_2$	353
i-Pr			Pt		i-Pr$_2$	353
i-Pr			Ni	150	i-Pr$_2$	162
i-Pr			Ni on Kieselguhr	150–200	i-Pr$_2$ + small amount of i-PrPr	354
Me, Me			Ni	150	Et-i-Pr	271
Et, Et			Ni on Kieselguhr	180	Et$_2$CMe$_2$	355
Et, Et			Ni		i-Pr$_2$	272
Me, Me	Me		Pd on charcoal	120	t-BuEt(~80%) + (i-Pr$_2$ + i-PrPr)(20%)	350
Me, Me	Me		Pt on charcoal	150	t-BuEt (100%)	344
Me, Me	Me		Pt	100	t-BuEt	357
Me, Me	Me		Pt	60	t-BuEt	357
Me, Me	Me		Ni on Kieselguhr	250–320	i-Pr$_2$ (73%) + Me$_2$PrCH (18%) + t-BuEt (9%)	360
Vi			Pd	2–20	PrEt	361
1-MeVn			Pd	20	i-PrPr	361
1-MeVn			Raney Ni	35–40	i-PrPr(predominant) + cyC$_3$H$_5$-i-Pr	162
1-MeVn			Pt		cyC$_3$H$_5$-i-Pr (70%) + i-PrPr (30%)	362
1-MeVn			copper chromite		cyC$_3$H$_5$-i-Pr (98%) + i-PrPr (2%)	364
1-EtVn			copper chromite		cyC$_3$H$_5$-s-Bu (99%) + s-BuPr (1%)	364
1-PrVn			copper chromite		cyC$_3$H$_5$MeCHPr (99%) + Pr$_2$CHMe	364
1-BuVn			copper chromite		cyC$_3$H$_5$CHMeBu (99%) + PrCHMeBu	364
MeCH=CMe			copper chromite		cyC$_3$H$_5$-s-Bu (72%) + s-BuPr (15%)	364
EtCH=CMe			copper chromite		cyC$_3$H$_5$MeCHPr (71%) + Pr$_2$CHMe (17%)	364
PrCH=CMe			copper chromite		cyC$_3$H$_5$MeCHBu (79%) + PrCHMeBu (16%)	364

TABLE 58. (continued)

Substituents $C_{(1)}$	$C_{(2)}$	Catalyst	T(°c)	Products	Ref.
Ph		Pd	20	PhPr	365
Ph, Ph		Pd	20	no reaction	366
Ph	C-Ph	Pd	20	$Ph_2(CH_2)_3$	366
Ph	T-Ph	Pd	20	$Ph_2(CH_2)_3$	366
Ac		Zn		$cyC_3H_5CHOHMe$	367
		Zn-Cu		cyC_3H_5CHOMe	367
		Cu		ring opens	367
		Pt		$cyC_3H_5CHOHMe$ + PrMeCHOH	368
		Pd		PrAc	368
		Raney Ni		$cyC_3H_5CHOHMe$ + PrAc	369
		copper chromite		$cyC_3H_5CHOHMe$	369
CO_2Et, CO_2Et	Vi	Adams catalyst		$BuCH(CO_2Et)_2$	163a

TABLE 59. Catalytic hydrogenation of spiranes, bicycloalkanes and methylene cyclopropanes

Compound	Catalyst	Product	Ref.
Spiro[2·2]pentane	Pd	t-BuMe + Me_2PrCH + 1,1-$Me_2cyC_3H_4$	370
Spiro[5·2]octane	Ni on Kieselguhr	1,1-Dimethylcyclohexane[a]	355
2-Methylspiro[5·2]octane	Ni on Kieselguhr	1,1,2-Trimethylcyclohexane[a]	355
1-Isopropyl-4-methylenebicyclo[3·1·0]hexane	Pd	1-Isopropyl-2,3-dimethylcyclopentane	371
	Pt	5-Isopropyl-1,2-dimethylcyclopentene	371
1-Isopropyl-4-methylenebicyclo[3·1·0]hexan-3-ol	Pd	5-Isopropyl-1,2-dimethylcyclopenten-3-ol	372
	Pt	1-Isopropyl-4-methylbicyclo[3·1·0]hexan-3-ol	372
3-Acetoxy-1-isopropyl-4-methylenebicyclo[3·1·0]hexane	Pd	3-acetoxy-5-isopropyl-1,2-dimethylcyclopentene	372
Methylenecyclopropane	copper chromite	BuH	164

[a] At 155°c.

The mechanism of the deuteralysis of 1,1-dimethylcyclopropane on platinum, palladium, nickel and rhodium films and of 1,1,2-trimethylcyclopropane on nickel, platinum, and palladium films, has been studied [274,275].

With respect to the ease of hydrogenation of cyclopropane as compared with propene, it should be noted that an analytical method for determining the composition of mixtures of these compounds consists of hydrogenation over nickel on Kieselguhr partially poisoned with mercury but still active enough to react with the propene, and then over unpoisoned catalyst in order to react with the cyclopropane [376].

While catalytic hydrogenation of cyclopropanes has received most of the attention in the literature, several attempts at reduction by other reagents have been reported; they are collected in Table 60. The results seem to show that the cyclopropane ring is comparatively resistant to reduction.

TABLE 60. Reduction of substituted cyclopropanes

Substituents			Reagent	Products	Ref.
C(1)	C(2)	C(3)			
CO_2Me, CO_2Me	Bz	Ph	Zn, AcOH	$BzCH_2CHPhCH$-$(CO_2Me)_2$	377
Ac			$LiAlH_4$, Et_2O	$cyC_3H_5CHOHMe$	369
Ac			Na, EtOH	$cyC_3H_5CHOHMe$	369
1-PrVn			Na, MeOH, NH_3 (liquid)	$cyC_3H_5CHMePr$	378
1-PrVn			Na, NH_3 (liquid)	no reaction	378
1-PrVn			Na, BH_4Br, NH_3 (liquid)	no reaction	378

XI. REACTIONS OF THE C—H BOND IN CYCLOPROPANE

In Section II it was pointed out that the carbon orbitals used to bond to hydrogen in cyclopropane are approximately sp^2 hybridized. It is of interest to examine the reactivity of these C—H bonds in comparison with the reactivity of olefinic C—H on the one hand and of parafinic C—H on the other.

A. Acidity of the C—H Bond in Cyclopropanes

The acidity of hydrocarbons has recently been reviewed by Cram [379]. The determination of the acidity of hydrocarbons is a generally difficult problem as these compounds are as a rule too feebly acidic to be examined by the methods used with more familiar acid types. Rate constants and equilibrium constants for reactions which constitute a measure of acid strength have been utilized in studying the acidity of hydrocarbons. The results obtained by a number of authors are reported in Table 61. Many of these results have been correlated by

TABLE 61. Acidity of hydrocarbons

R	a	b	c	d	e	f	g	h
cyC$_6$H$_{11}$								1·1 × 10^{-8}
cyC$_5$H$_9$	6·9							
cyC$_4$H$_7$								10^{-6}
i-Pr		6·0	3·27					
i-Bu	4·6	4·3						
Pr	3·9							
Et	3·5	4·0	3·25					
Me		1·8	3·10					
PhCH$_2$CH$_2$		1·0	3·04					
cyC$_3$H$_5$	1·0	0·7	3·01					10^{-3}
Vi	−2·43	0·3	2·94					
Ph	0	0	2·92	1				
PhMe$_2$C				37		0·84	1·34	
ViCH$_2$		−0·4	2·32					
PhMeEtC						0·31	0·39	0·43
PhMeCH		−0·4						
PhCH$_2$		−0·7	2·08		1·1 × 10^2			
PhC≡C		−3·0		21				
Ph$_2$CH				35	2·9 × 10^4			
Ph$_3$C				32·5	1·2 × 10^5			

^{a} log K for RLi + PhI^{380}.
^{b} log K for R$_2$Hg + Ph$_2$Mg^{381}.
^{c} $E_{1/2}$ (half-wave polarographic potential) for RHgX^{381}.
^{d} pK_a of RH^{382}.
^{e,f} Relative rate of tritium exchange, at room temperature, catalysed by LiNHcyC$_6$H$_{11}$, at 25° (e) and 50° (f)^{383}.
^{g,h} Relative rate of tritium exchange, at room temperature, catalysed by LiNHcyC$_6$H$_{11}$, at 25° (g)^{383} and 50° (h)^{383,384}.

Cram to give the McEwen-Streitwieser-Applequist-Dessy scale (MSAD scale) shown in Table 62. On the MSAD scale cyclopropane has pK_a 39, as compared with ethylene at 36·5 and the C$_{(2)}$-hydrogen atoms in propane at 44. Thus the C—H bond in cyclopropane is much closer in acidity to that of ethylene or benzene than it is to that of propane, pentane or cyclohexane.

TABLE 62. The McEwen-Streitweiser-Applequist-Dessy Scale
(MSAD scale)[379]

X in XH	pK_a	X in XH	pK_a	X in XH	pK_a
PhC≡C	18·5	Vi	36·5	Et	42
HC≡C	25	Ph	37	cyC_4H_7	43
Ph_3C	32·5	$PhMe_2C$	37	i-Pr	44
$PhCH_2$	35	cyC_3H_5	39	cyC_5H_9	44
$ViCH_2$	35·5	Me	40	cyC_6H_{11}	45

Confirmation of the acidity of protons bonded to the cyclopropane ring comes from a study of hydrogen–deuterium exchange[385]. The results are presented in Table 63; the results clearly show the greater

TABLE 63. Hydrogen–deuterium exchange in KND_2–ND_3[385]

Conditions	Numbers of Hydrogens exchanged					
	Et-i-Pr	EtcyC_4H_7	EtcyC_3H_5	Ph-i-Pr	PhcyC_3H_5	VicyC_3H_5
0·05M KND_2 25°, 6 hr					5	2
1·0M KND_2 25°, 8 hr			0·3		10	8
1·0M KND_2 25°, 240 hr	0		4	6	10	8
1·0M KND_2 120°, 6 hr		0·1	5			
1·0M KND_2 120°, 200 hr	2	2	6	12		

acidity of the cyclopropane protons as compared with the methylbutane protons and the ethylcyclobutane protons.

Cram[379] has observed an interesting linear relationship between pK_a on the MSAD scale and percent *s* character of the carbon orbital used in forming the C—H bond. The results are in accord with approximately sp^2 character for the cyclopropane protons.

Isopropyl phenyl sulphone and cyclopropyl phenyl sulphone have been found to be about equally acidic[386,387]. This can be accounted for by a difference in the effect of the phenylsulphonyl substituent upon the acidity of the cyclopropyl proton as compared with its effect upon the acidity of the isopropyl proton.

B. Configurational Stability of the Cyclopropyl Carbanion

The cyclopropane carbanion has been reviewed by Walborsky[388]. Whereas it has been established that carbanions derived from sp^3 hybridized carbons are not configurationally stable, carbanions obtained from sp^2 hybridized carbons are in fact configurationally stable, as are also a number of sp^2 hybridized nitrogen derivatives, for example, oximes, axomethines and azo compounds. It has now been shown that 1-methyl-2,2-diphenylcyclopropyllithium[389] and the analogous magnesium bromide[390], are configurationally stable. The results obtained are in accord with calculated inversion barriers[391] (Table 64). It would seem therefore, that cyclopropane carbanions

TABLE 64. Inversion barriers for carbanions[391]

Species	$V_{calculated}$	$V_{observed}$
Me^-	0·0465	
cyC_3H_5	19	
cyC_4H_7	19	
$=CH^-$	38	> 30 for HlCH=CHl$^-$
CR_3^-	0·0484	
1-R-$cyC_3H_4^-$	20	
1-R-$cyC_4H_6^-$	20	
$=CR^-$	40	10–20 for PhCH=CPh$^-$

are configurationally stable, although probably less so than carbanions derived from sp^2 hybridized carbon in olefins.

C. Hydrogen Abstraction by Radicals

The reaction of cyclopropane with methyl and trideuteromethyl radicals has an activation energy of 10·3 and 13·1 kcal/mole respectively[392,393], as compared with activation energies of 8·5 and 9·3 kcal/mole for cyclopentane. These results support a difference in hybridization of the carbon atoms in cyclopropane and cyclopentane.

XII. RECENT DEVELOPMENTS

An *ab initio* calculation for cyclopropane has been carried out by the SCF–MO–LC Gaussian Orbitals method[394]. The calculated ionization potential is 10·33 e.V. Calculations by the CNDO method have been carried out for methylcyclopropane and for the bisected and symmetric conformations of the cyclopropylmethyl carbonium ion[395]. The results obtained are shown in Fig. 8.

The u.v. spectra of nitrocyclopropanes show an appreciable conjugative effect; values of λ_{max} are for nitroalkanes ~ 210 mμ, nitrocyclopropanes 213–219 mμ and nitroalkenes 220–250 mμ[396].

A value of σ_R^0 for the cyclopropyl group of -0.175 has been determined from the integrated intensity of the ν_{16} mode in the i.r. spectrum of cyclopropylbenzene[397].

The cyclopropyl group has been found to have a moderate activating effect (when compared with the isopropyl and phenyl groups) upon the vapour phase pyrolysis of 2-substituted isopropyl acetates (XCMe$_2$-OAc)[398].

In FSO$_3$H–SbF$_5$–SO$_2$ClF or HF–SbF$_5$–SO$_2$ClF above $-80°$c cyclopropane forms the t-butyl cation through various di- and trimerization processes and fragmentations, and i-PrMe$_2$C$^+$. At $-100°$c the n.m.r. spectrum suggests that cyclopropane is protonated, either on the ring 'face' or as an equilibrating 'edge' protonated mixture. Ethylcyclopropane forms exclusively t-amyl cation (Me$_2$EtC) in these media, whereas isopropylcyclopropane undergoes hydride loss to form the dimethylcyclopropylmethylene cation[399].

Rates of reaction of 3-substituted 1-cyclopropylbutanes with tri-

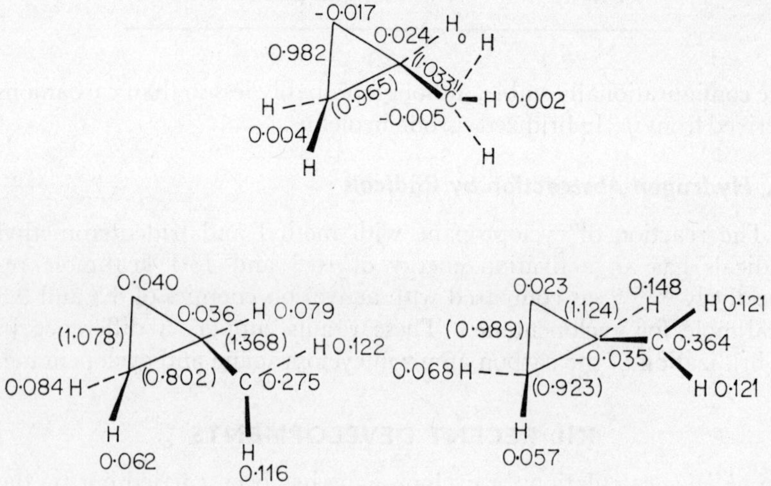

Figure 8. Bond indices (in brackets) and charge densities of methylcyclopropane and cyclopropylmethyl carbonium ion.

fluoroacetic acid have been measured[400] (Table 65). Both addition and elimination occur. Rate constants show a value of α (equation 15) of −3·5. The cyclopropane reacted about 300 times faster than comparable alkenes. The cyclopropanes studied were unreactive to IBr.

TABLE 65. Rates of reaction of 3-substituted 1-cyclopropylbutanes ($cyC_3H_5CH_2$-CH_2CHXCH_3) with trifluoroacetic acid at 0°C

X	$10^6 k (sec^{-1})$	% Elimination
H	3260·0	39·2
Cl	98·1	32·2
O_2CCF_3	10·2	27·7

Bicyclo[2·1·0]pentane reacts with dimethyl acetylenedicarboxylate[401] to give **74** and **75**. Rate studies show a negligible solvent effect. Results are most consistent with the formation of a diradical intermediate[402].

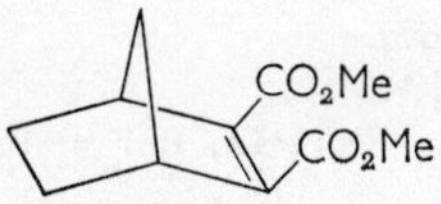

(74) 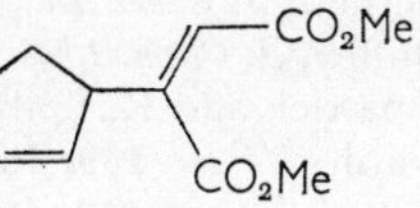   (75)

The kinetics of the addition of mercuric acetate in acetic acid to ring substituted phenylcyclopropanes has been studied[403]. Results are correlated by the Hammett equation with a ρ value of −3·2 at 50°. The results are given in Table 66.

The reaction of *cis,trans*- and *trans,trans*-2,3-dimethyl-1-phenylcyclopropanol and their acetates with a number of brominating agents has been studied[404]. The products are bromoketones with complete inversion of configuration at the site to which the bromine is attached. The mechanism is thought to be 'ionic or concerted'. The *cis,trans* isomer gives only *erythro*-α-methyl-β-bromobutyrophenone; the *trans,trans* isomer gives exclusively the *threo* configuration. With chlorinating agents the same 50:50 mixture of *threo* and *erythro* chloroketones is obtained from either isomer. The chlorination is believed to proceed

TABLE 66. Rate of addition of $Hg(OAc)_2$ to $XPncyC_3H_5$ in acetic acid

X	$10^4 \, k(1/\text{mole sec})$		
	25·0°	50·1°	75·6°
4-MeO	570	5000	
4-Me	22	220	
3-Me	3·3	40	370
H	1·6	22	210
4-Cl	0·47	7·0	75
3-Cl	0·067	1·2	16

by a free radical path. Halogenating agents give with 1,2,2-trimethylcyclopropanol solely 1,3-bond breaking, with *trans*-2-phenyl-1-methylcyclopropanol only 1,2-bond breaking is involved in the ring opening. The chemistry of the cyclopropanols has been reviewed [405].

XIII. REFERENCES

1. R. A. Raphael, in *Chemistry of the Carbon Compounds* (Ed. E. H. Rodd), [Vol. **IIA**.] Elsevier, Amsterdam, 1953, p. 23.
2. E. Vogel, *Fortschr. Chem. Forsh.*, **3**, 430 (1955); *Angew. Chem.*, **72**, 4 (1960).
3. M. Y. Lukina, *Usp. Khim.*, **31**, 419 (1962).
4. W. A. Bernett, *J. Chem. Ed.*, **44**, 17 (1967).
5. G. H. Duffey, *J. Chem. Phys.*, **14**, 342 (1946).
6. J. E. Kilpatrick and R. Spitzer, *J. Chem. Phys.*, **14**, 463 (1946).
7. A. D. Walsh, *Nature*, **159**, 165 (1947).
8. R. Robinson, *Nature*, **159**, 400 (1947).
9. C. A. McDowell, *Nature*, **159**, 508 (1947).
10. A. D. Walsh, *Nature*, **159**, 712 (1947).
11. R. Robinson, *Nature*, **160**, 162 (1947).
12. T. M. Sugden, *Nature*, **160**, 367 (1947).
13. A. D. Walsh, *Nature*, **160**, 902 (1947).
14. A. D. Walsh, *Trans. Faraday Soc.*, **45**, 179 (1949).
14a. J. F. Music and R. A. Matson, *J. Am. Chem. Soc.*, **72**, 5256 (1950).
15. M. E. Dyatkina and Y. K. Syrkin, *Dokl. Akad. Nauk. SSSR.*, **122**, 837 (1958)
16. V. Y. Bespalov, *Vestn. Leningr. Univ.*, **22**, Ser. Fiz. Khim., **1**, 117 (1967).
17. C. A. Coulson and W. E. Moffit, *J. Chem. Phys.*, **15**, 151 (1947).
18. C. A. Coulson and W. E. Moffit, *Phil. Mag.*, **40**, 1 (1949).
19. L. L. Ingraham, *Steric Effects in Organic Chemistry*, John Wiley and Sons, New York, 1956, p. 519.
20. C. A. Coulson and T. H. Goodwin, *J. Chem. Soc.*, 2851 (1962); errata 3161 (1963).
21. D. Peters, *Tetrahedron*, **19**, 1539 (1963).
22. W. H. Flygare, *Science*, **140**, 1179 (1963).

23. A. Veilard and G. del Re, *Theoret. Chim. Acta*, **2,** 55 (1964).
24. M. Randić and Z. Maksić, *Theoret. Chim. Acta*, **3,** 59 (1965).
25. N. Trinajstić and M. Randić, *J. Chem. Soc.*, 5621 (1965).
26. P. G. Gassman, *Chem. Commun.*, 793 (1967).
26a. M. K. Kemp and W. H. Flygare, *J. Am. Chem. Soc.*, **89,** 3925 (1967).
27. R. Hoffmann, *J. Chem. Phys.*, **49,** 2480 (1964).
28. T. Yonezawa, K. Shimizu and H. Kato, *Bull. Chem. Soc. Japan*, **40,** 456 (1967).
28a. T. Yonezawa, K. Shimizu and H. Kato, *Bull. Chem. Soc. Japan*, **40,** 1302 (1967).
29. G. S. Handler and J. H. Anderson, *Tetrahedron*, **2,** 345 (1958).
29a. H. A. Skinner and G. Pilcher, *Quart. Rev.*, **17,** 264 (1963).
29b. N. C. Baird and M. J. S. Dewar, *J. Am. Chem. Soc.*, **89,** 3960 (1967).
29c. H. H. Jaffé, *Ber. Bunsenges.*, **59,** 823 (1955).
29d. F. M. Fraser and E. M. Prosen, *J. Res. Natl. Bur. Stds.*, **54,** 143 (1955).
30. J. W. Knowlton and F. D. Rossini, *J. Res. Natl. Bur. Stds.*, **43,** 113 (1949).
31. R. B. Turner, P. Gocbel, W. v. E. Doering and J. F. Coburn, Jr., *Tetrahedron Letters*, 997 (1965).
32. E. Goldish, *J. Chem. Ed.*, **36,** 408 (1959).
33. O. Bastiansen, F. N. Fritsch and R. Hedberg, *Acta Cryst.*, **17,** 538 (1964).
34. O. Bastiansen and O. Hassel, *Tidsskr. Kjemi. Bergvesen Met.*, **6,** 71 (1946); *Chem. Abstr.*, **40,** 6059 (1946).
35. O. Hassel and H. Vierroll, *Acta Chem. Scand.*, **1,** 149 (1947).
36. L. Pauling and L. O. Brockway, *J. Am. Chem. Soc.*, **59,** 1223 (1937).
37. H. H. Gunthard, R. C. Lord, and T. K. McCubbin, *J. Chem. Phys.*, **25,** 768 (1956).
38. S. P. Sinha, *J. Chem. Phys.*, **18,** 217 (1950).
39. S. Meiboom and L. C. Snyder, *J. Am. Chem. Soc.*, **89,** 1038 (1967).
40. O. Bastiansen and A. de Meijere, *Angew. Chem.*, **78,** 142 (1966).
41. O. Bastiansen and A. de Meijere, *Acta Chem. Scand.*, **20,** 516 (1966).
42. J. Donohue, G. L. Humphrey and V. Schomaker, *J. Am. Chem. Soc.*, **67,** 332 (1945).
43. H. P. Lemaire and R. L. Livingston, *Acta Cryst.*, **5,** 817 (1952).
44. J. M. O'Gorman and V. Schomaker, *J. Am. Chem. Soc.*, **68,** 1138 (1946).
45. J. P. Friend and B. P. Dailey, *J. Chem. Phys.*, **29,** 577 (1958).
46. J. P. Friend, R. F. Schneider and B. P. Dailey, *J. Chem. Phys.*, **23,** 1557 (1955).
47. R. H. Schwendeman, G. D. Jacobs and T. M. Krigas, *J. Chem. Phys.*, **40,** 1022 (1964).
48. W. H. Flygare, A. Narath and W. D. Gwinn, *J. Chem. Phys.*, **36,** 200 (1962).
49. D. R. Peterson, *Chem. and Ind.*, 904 (1956).
50. D. B. Chesnut and R. E. Marsh, *Acta Cryst.*, **11,** 413 (1958).
51. W. G. Grant and S. C. Speakman, *J. Chem. Soc.*, 3753 (1958).
52. L. S. Bartell and J. P. Guillory, *J. Chem. Phys.*, **43,** 647 (1965).
53. L. S. Bartell, J. P. Guillory and A. P. Parks, *J. Phys. Chem.*, **69,** 3043 (1965).
54. C. J. Fritchie, Jr., *Acta Cryst.*, **20,** 27 (1966).
55. A. Hartman and F. L. Hirshfeld, *Acta Cryst.*, **20,** 80 (1966).

56. D. H. Kasai, R. J. Myers, D. F. Eggers and K. B. Wiberg, *J. Chem. Phys.*, **30,** 512 (1959).
57. F. R. Ahmed and E. J. Gube, *Acta Cryst.*, **17,** 603 (1964).
57a. E. Heilbronner and V. Schomaker, *Helv. Chim. Acta*, **35,** 1385 (1952).
58. N. S. Ham, *Rev. Pure Appl. Chem.*, **11,** 159 (1961).
59. E. L. Eliel, N. L. Allinger, S. J. Angyal and G. A. Morrison, *Conformational Analysis*, John Wiley and Sons, New York, 1965, 19–22.
60. G. R. DeMare and J. S. Martin, *J. Am. Chem. Soc.*, **88,** 5033 (1966).
61. W. Lüttke and A. de Meijere, *Angew. Chem., Int. Ed. English*, **5,** 512 (1966).
62. W. Lüttke, private communication.
63. H. Gunther and D. Wendisch, *Angew. Chem.*, **78,** 266 (1966).
63a. L. S. Bartell, B. L. Carroll and J. P. Guillory, *Tetrahedron Letters*, 705 (1964).
64. A. D. Ketley, A. J. Berlin and L. P. Fisher, *J. Org. Chem.*, **31,** 2648 (1966).
65. L. V. Vilkov and N. I. Sadova, *Dokl. Akad. Nauk. SSSR.*, **162** [3], 565 (1965).
66. M. Stiles, V. Burckhardt and G. Freund, *J. Org. Chem.*, **32,** 3718 (1967).
67. G. L. Closs and H. B. Klinger, *J. Am. Chem. Soc.*, **87,** 3265 (1965).
68. G. L. Closs and R. A. Moss, *J. Am. Chem. Soc.*, **86,** 4044 (1964).
69. W. Lüttke, A. de Meijere, H. Wolff, H. Ludwig and H. W. Schrötter, *Angew. Chem.*, **78,** 141 (1966).
69a. G. J. Karabatsos and N. Hsi, *J. Am. Chem. Soc.*, **87,** 2864 (1965).
70. P. S. Wharton and T. I. Bair, *J. Org. Chem.*, **30,** 1681 (1965).
70a. J. L. Pierre and P. Arnaud, *Bull. Soc. Chim. France*, 1690, 1966.
70b. J. L. Pierre, R. Barlet and P. Arnaud, *Spectrochim. Acta*, **23A,** 2297 (1967).
71. N. L. Bauld, R. Gordon and J. Zoeller, *J. Am. Chem. Soc.*, **89,** 3448 (1967).
72. G. A. Russell and H. Malkus, *J. Am. Chem. Soc.*, **89,** 160 (1967).
73. C. U. Pittman, Jr. and G. A. Olah, *J. Am. Chem. Soc.*, **87,** 2998 (1965).
73a. N. C. Deno, J. S. Liu, J. O. Turner, D. N. Lincoln and R. E. Fruit, *J. Am. Chem. Soc.*, **87,** 3000 (1965).
74. C. U. Pittman, Jr. and G. A. Olah, *J. Am. Chem. Soc.*, **87,** 5123 (1965).
74a. H. G. Richey, Jr. and J. M. Richey, *J. Am. Chem. Soc.*, **88,** 4971 (1966).
75. R. Hoffmann, *Tetrahedron Letters*, 3819 (1965).
76. T. Sharpe and J. C. Martin, *J. Am. Chem. Soc..* **88,** 1815 (1966).
77. F. H. Field, *J. Chem. Phys.*, **20,** 1734 (1952).
78. F. H. Field, private communication to J. L. Franklin, *J. Chem. Phys.*, **22,** 1304 (1954).
79. R. F. Pottie, A. G. Harrison and F. P. Lossing, *J. Am. Chem. Soc.*, **83,** 3204 (1961).
80. K. Watanabe, T. Nakayama and J. Mottl, *Final Report on Ionization Potentials of Molecules by a Photoionization Method*, Univ. of Hawaii, Dec. 1954.
81. R. W. Kiser, *Introduction to Mass Spectrometry and its Applications*, Prentice-Hall, Englewood Cliffs, 1965, p. 311.
82. D. W. Turner, *Advan. Phys. Org. Chem.*, **4,** 31 (1966).
83. R. W. Taft, Jr., and I. C. Lewis, *J. Am. Chem. Soc.*, **80,** 2436 (1958).
84. M. Charton, unpublished results.
85. M. Charton, *Abstr. 2nd Middle Atlantic Reg. Mtg. Am. Chem. Soc.*, 88 (1967).
86. J. P. Pete, *Bull. Soc. Chim. France*, 357 (1967).

87. E. P. Carr and C. P. Burt, *J. Am. Chem. Soc.*, **40,** 1590 (1918).
88. J. M. Klotz, *J. Am. Chem. Soc.*, **66,** 88 (1944).
89. M. T. Rogers, *J. Am. Chem. Soc.*, **69,** 2544 (1947).
90. W. W. Robertson, J. F. Music and F. A. Matson, *J. Am. Chem. Soc.*, **72,** 5260 (1950).
91. J. F. Music and F. A. Matson, *J. Am. Chem. Soc.*, **72,** 5256 (1950).
92. L. A. Strait, R. Ketchum, D. Jambotkar and V. P. Shah, *J. Am. Chem. Soc.*, **86,** 4682 (1964).
93. N. A. Cromwell and G. V. Hudson, *J. Am. Chem. Soc.*, **75,** 872 (1953).
94. A. L. Goodman and R. H. Eastman, *J. Am. Chem. Soc.*, **86,** 908 (1964).
94a. R. W. Kierstead, R. P. Linstead and B. C. L. Weedon, *J. Chem. Soc.*, 1803 (1953).
95. C. A. Grob and J. Hostynek, *Helv. Chim. Acta*, **46,** 1676 (1963).
96. S. R. Landor and D. F. Whiler, *J. Chem. Soc.*, 5625 (1965).
97. R. P. Mariella, L. F. A. Peterson and R. C. Ferris, *J. Am. Chem. Soc.*, **70,** 1494 (1948).
98. A. P. Gray and H. Kraus, *J. Org. Chem.*, **31,** 399 (1966).
99. R. P. Mariella and R. R. Raube, *J. Am. Chem. Soc.*, **74,** 521 (1952).
100. R. H. Eastman, *J. Am. Chem. Soc.*, **76,** 4115 (1954).
101. R. H. Eastman and S. K. Freeman, *J. Am. Chem. Soc.*, **77,** 6642 (1955).
102. R. J. Mohrbacher and N. H. Cromwell, *J. Am. Chem. Soc.*, **79,** 401 (1957).
103. R. W. Kierstead, R. P. Linstead and B. C. L. Weedon, *J. Chem. Soc.*, 1799 (1953).
104. G. W. Perold, *J. S. African Chem. Inst.*, **6,** 22 (1953).
105. G. W. Perold, *J. S. African Chem. Inst.*, **8,** No. 1, 1 (1955).
106. G. W. Perold, *J. S. African Chem. Inst.*, **10,** 11 (1957).
107. G. W. Cannon, A. A. Santilli and P. Shenian, *J. Am. Chem. Soc.*, **81,** 1660 (1959).
108. R. P. Mariella and R. R. Raube, *J. Am. Chem. Soc.*, **74,** 518 (1952).
109. J. L. Pierre and P. Arnaud, *Compt. Rend. Acad. Sci.* (*Paris*), **263,** 557 (1966).
110. A. Padwa, L. Hamilton and L. Norling, *J. Org. Chem.*, **31,** 1244 (1966).
111. E. M. Kosower and M. Ito, *Proc. Chem. Soc.*, 25 (1962).
112. W. G. Dauben and G. H. Berezin, *J. Am. Chem. Soc.*, **89,** 3449 (1967).
113. L. J. Lorenc, M. Miliković, K. Schaffner and O. Jeger, *Helv. Chim. Acta*, **49,** 1183 (1966).
114. J. D. Roberts and C. Green, *J. Am. Chem. Soc.*, **68,** 214 (1966).
115. M. F. Hawthorne, *J. Org. Chem.*, **21,** 1523 (1956).
116. F. T. Williams, Jr., P. W. K. Flanagan, W. J. Taylor and H. Shechter, *J. Org. Chem.*, **30,** 2674 (1965).
117. E. C. Eberlin and L. W. Pickett, *J. Chem. Phys.*, **27,** 1439 (1957).
118. L. Skattebol, *J. Org. Chem.*, **29,** 2951 (1964).
119. N. C. Deno, H. G. Richey, Jr., J. S. Liu, D. N. Lincoln and J. O. Turner, *J. Am. Chem. Soc.*, **87,** 4533 (1965).
119a. G. A. Olah, C. V. Pittman, Jr., R. Waack and M. Doran, *J. Am. Chem. Soc.*, **88,** 1488 (1966).
120. L. I. Smith and E. R. Rogier, *J. Am. Chem. Soc.*, **73,** 3840 (1951).
121. P. G. Gassman, *Chem. Ind.*, 740 (1962).
122. H. H. Jaffé, *Chem. Rev.*, **53,** 191 (1953).
123. M. Charton, *J. Chem. Soc.*, 1205 (1964).

124. P. G. Gassman and F. V. Zular, *J. Org. Chem.*, **31,** 166 (1966).
125. M. Charton, unpublished results.
126. J. W. Linnett, *Nature.*, **160,** 162 (1947).
127. A. D. Walsh, *Disc. Faraday Soc.*, **2,** 18 (1947).
128. K. B. Wiberg and B. J. Nist, *J. Am. Chem. Soc.*, **83,** 1226 (1961).
129. T. L. Brown, J. M. Sandri and H. Hart, *J. Phys. Chem.*, **61,** 698 (1957).
130. M. Jakoubková, M. Horák, and V. Chavalovsky, *Coll. Czech. Chem. Communs.*, **31,** 979 (1966).
131. V. T. Aleksanyan, K. E. Sterin, M. Y. Lukina, L. G. Galnikova and I. L. Safonova, *Repts. 10th All-Union Conf. on Spectroscopy*, **1,** 1957, p. 64.
132. V. T. Aleksanyan, K. E. Sterin, M. Y. Lukina, I. L. Safonova and B. A. Kazanskii, *Optika i Spektr.*, **7,** 178 (1959).
132a. V. T. Aleksanyan and K. E. Sterin, *Dokl. Akad. Nauk.*, *SSSR,* **131,** 1373 (1960).
133. E. G. Treshchova, Y. N. Panchenko, N. I. Vasilev, M. G. Kuzmin, Y. S. Shabarov and R. Y. Levina, *Opt. i Spektroskopiya.*, **8,** 371 (1960).
134. R. Y. Levina, V. N. Kostin, P. A. Genbitski, S. M. Shostakovskii and E. G. Treschkova, *Zh. Obshch. Khim.*, **30,** 2435 (1960).
135. R. Y. Levina, V. N. Kostin, P. A. Gembitski, S. M. Shostakovskii and E. G. Treschkova, *Zh. Obshch. Khim.*, **31,** 1185 (1961).
136. R. Y. Levina, P. A. Gambitskii, V. N. Kostin, S. M. Shostakovskii and E. G. Treshchkova, *Zh. Obshch. Khim.*, **33,** 365 (1963).
137. R. Y. Levina, P. A. Gembitskii, and E. G. Treshchkova, *Zh. Obshch. Khim.*, **33,** 371 (1963).
138. E. V. Sobalev, V. T. Aleksanyan, K. E. Sterin, M. Y. Lukina and L. G. Cherkashina, *Zh. Strukt. Khim.*, **2,** 145 (1961).
139. N. Muller and D. E. Pritchard, *J. Chem. Phys.*, **31,** 768 (1959).
140. D. J. Patel, M. E. H. Howden and J. D. Roberts, *J. Am. Chem. Soc.*, **85,** 3218 (1963).
141. E. Ciganek, *J. Am. Chem. Soc.*, **88,** 1979 (1966).
142. J. J. Burke and P. C. Lauterbur, *J. Am. Chem. Soc.*, **86,** 1870 (1964).
143. V. S. Watts and J. H. Goldstein, *J. Chem. Phys.*, **46,** 4615 (1967).
144. E. Lippert and H. Prigge, *Ber. Bursenges.*, **67,** 415 (1963).
145. F. J. Weigert and J. D. Roberts, *J. Am. Chem. Soc.*, **89,** 5962 (1967).
146. K. Frei and H. J. Bernstein, *J. Chem. Phys.*, **38,** 1216 (1963).
147. J. A. Pople and D. P. Santry, *Mol. Phys.*, **8,** 1 (1964).
148. J. R. Lacher, J. W. Pollack, and J. P. Park, *J. Chem. Phys.*, **20,** 1047 (1952).
149. K. Tori and K. Kitahonki, *J. Am. Chem. Soc.*, **87,** 386 (1965).
150. K. B. Wiberg, *Record. Chem. Progr.*, **26,** 143 (1965).
151. D. H. Marr and J. B. Stothers, *Can. J. Chem.*, **45,** 225 (1967).
152. M. Brookhart, G. C. Levy and S. Winstein, *J. Am. Chem. Soc.*, **89,** 1735 (1967).
153. K. L. Williamson, C. A. Lanford and C. R. Nicholson, *J. Am. Chem. Soc.*, **86,** 762 (1964).
154. K. L. Williamson and B. A. Braman, *J. Am. Chem. Soc.*, **89,** 6183 (1967).
154a. T. A. Wittstruck and E. N. Trachtenberg, *J. Am. Chem. Soc.*, **89,** 3810 (1967).
155. J. E. Todd, M. A. Whitehead and K. W. Weber, *J. Chem. Phys.*, **39,** 404 (1963).

156. R. G. Kostyanovskii, S. P. Solodovnikov and O. A. Yushakova, *Izv. Akad. Nauk. SSSR, Ser. Khim.*, 735, 1966.
157. A. I. Prokofev, V. M. Chibrikin, O. A. Yuzhakova and R. G. Kostyanovskii, *Izv. Akad. Nauk. SSSR, Ser. Khim.*, 1105 (1966).
158. C. K. Ingold, *Structure and Mechanism in Organic Chemistry*, Cornell Univ., Ithaca, 1953, p. 126.
159. L. Tschugaeff, *Chem. Ber.*, **33**, 3118 (1900).
159a. N. Zelinsky and J. Zelikow, *Chem. Ber.*, **34**, 2856 (1901).
160. G. Ostling, *J. Chem. Soc.*, **101**, 457 (1912).
160a. P. G. Stevens, *J. Am. Chem. Soc.*, **68**, 620 (1946).
161. J. H. Jeffery and A. J. Vogel, *J. Chem. Soc.*, 1804 (1948).
162. R. van Volkenburgh, K. W. Greenlee, J. M. Derfer and C. E. Boord, *J. Am. Chem. Soc.*, **71**, 172 (1949).
163. R. van Volkenburgh, K. W. Greenlee, J. M. Derfer and C. E. Boord, *J. Am. Chem. Soc.*, **71**, 3595 (1949).
163a. R. W. Kierstead, R. P. Linstead, and B. C. L. Weedon, *J. Chem. Soc.*, 3610 (1952).
164. J. T. Gragson, K. W. Greenlee, J. M. Derfer and C. E. Boord, *J. Am. Chem. Soc.*, **75**, 3344 (1953).
165. V. Slabey, *J. Am. Chem. Soc.*, **76**, 3603 (1954).
166. W. D. Kumler, R. Boikess, P. Bruck and S. Winstein, *J. Am. Chem. Soc.*, **86**, 3126 (1964).
167. K. L. Ramaswamy, *Proc. Indian Acad. Sci.*, **A4**, 108 (1936).
168. I. B. Mazheiko, S. Hillers, P. A. Gembitskii and R. Y. Levina, *Zh. Obshch. Khim.*, **33**, 1698 (1963).
168a. S. Nishida, I. Moritani and T. Sato, *J. Am. Chem. Soc.*, **89**, 6885 (1967).
169. M. T. Rogers and J. D. Roberts, *J. Am. Chem. Soc.*, **68**, 843 (1946).
170. J. D. Roberts and V. C. Chambers, *J. Am. Chem. Soc.*, **73**, 5030 (1951).
171. B. I. Spinrad, *J. Am. Chem. Soc.*, **68**, 617 (1946).
172. J. Böeseken and H. V. Takes, *Rec. trav. chim.*, **56**, 858 (1937).
173. A. E. van Arkel, P. Meerburg and C. R. van den Handel, *Rec. trav. chim.*, **61**, 767 (1942).
174. M. Chodsmith and G. W. Wheland, *J. Am. Chem. Soc.*, **70**, 2632 (1948).
174a. T. Fujita and M. Hamada, *Betyu. Kagaku.*, **19**, 80 (1954).
175. A. L. McClellan, "Tables of Experimental Dipole Moments", Freeman, San Francisco, 1963.
176. M. Charton, *J. Org. Chem.*, **30**, 552 (1965).
176a. M. P. Kozina, M. Y. Lukina, N. D. Zubarova, J. L. Safonova, S. M. Skuratov and B. A. Kazanskii, *Dokl. Akad. Nauk. SSSR.*, **138**, 843 (1961).
176b. G. Isakson and J. Sandstrom, *Acta Chem. Scand.*, **21**, 1605 (1967).
177. P. Zuman, "Substituent Effects in Organic Polarography", Plenum Press, New York, 1966, p. 325.
178. T. L. Brown, *J. Am. Chem. Soc.*, **80**, 6489 (1958).
179. R. G. Pews, *J. Am. Chem. Soc.*, **89**, 5605 (1967).
180. J. A. Landgrebe and J. D. Shoemaker, *J. Am. Chem. Soc.*, **89**, 4465 (1967).
181. J. Smejkal, J. Jonás and J. Furkas, *Coll. Czech. Chem. Communs.*, **29**, 2950 (1964).
182. R. Y. Levina, P. A. Gembitskii, L. P. Guseva and P. K. Agasyan, *Zh. Obshch. Khim.*, **34**, 146 (1964).

183. H. C. Brown and J. D. Cleveland, *J. Am. Chem. Soc.*, **88**, 2051 (1966).
184. M. Kilpatrick and J. G. Morse, *J. Am. Chem. Soc.*, **75**, 1854 (1953).
185. D. B. Denney and F. J. Gross, *J. Org. Chem.*, **32**, 2445 (1967).
186. Y. S. Shabarov, V. K. Potapov and R. Y. Levina, *Zh. Obshch. Khim.*, **34**, 2832 (1964).
187. N. Deno and A. Schriesheim, *J. Am. Chem. Soc.*, **77**, 3051 (1955).
188. G. Kortüm, W. Vogel and K. Andrussov, *Dissociation Constants of Organic Acids in Aqueous Solution*, Butterworths, 1961, London.
189. J. Juillard, *Theses*, Faculte des Sciences de l'Universite de Clermont, 1965.
190. D. D. Perrin, *Dissociation Constants of Organic Bases in Aqueous Solution*, Butterworths, London, 1965.
190a. R. H. Eastman and J. C. Selover, *J. Am. Chem. Soc.*, **76**, 4118 (1954).
191. B. V. Bhide and J. J. Sudborough, *J. Indian Inst. Sci.*, **8A**, 89 (1925).
192. P. Petrenko-Kritschenko and E. Eltschaninoff, *Ann. Chem.*, **341**, 155 (1905).
193. P. Petrenko-Kritschenko and W. Kantscheff, *Chem. Ber.*, **39**, 1454 (1906).
194. C. H. DePuy and L. R. Mahoney, *J. Am. Chem. Soc.*, **86**, 2653 (1964).
195. J. Rocek and A. Riehl, *J. Am. Chem. Soc.*, **89**, 6691 (1967).
196. R. Y. Levina, Y. S. Shabarov and V. K. Potapov, *Zh. Obshch. Khim.*, **29**, 3233 (1959).
197. R. Y. Levina and P. A. Gembitskii, *Zh. Obshch. Khim.*, **31**, 3480 (1961).
198. R. Ketcham, R. Cavestri and D. Jambotkar, *J. Org. Chem.*, **28**, 2139 (1963).
199. Y. S. Shabarov, V. K. Potapov, N. M. Koloskova, A. A. Podterebkova, V. S. Svirina and R. Y. Levina, *Zh. Obshch. Khim.*, **34**, 2829 (1964).
200. Y. S. Shabarov, V. K. Potapov and R. Y. Levina, *Zh. Obshch. Khim.*, **33**, 3893 (1963).
201. M. Hanack and H. J. Schneider, *Angew. Chem. Intern. Ed. English*, **6**, 666 (1967).
202. J. D. Roberts and R. H. Mazur, *J. Am. Chem. Soc.*, **73**, 2509 (1951).
203. G. G. Bergstrom and S. Siegel, *J. Am. Chem. Soc.*, **74**, 145 (1952).
204. J. D. Roberts and R. H. Mazur, *J. Am. Chem. Soc.*, **73**, 3542 (1951).
205. M. E. H. Howden and J. D. Roberts, *Tetrahedron Suppl.*, **2**, 403 (1963).
206. H. Hart and J. M. Sandri, *J. Am. Chem. Soc.*, **81**, 320 (1959).
207. H. Hart and P. A. Law, *J. Am. Chem. Soc.*, **84**, 2462 (1962).
208. L. Birladeamu, T. Hanafusa, B. Johnson and S. Winstein, *J. Am. Chem. Soc.*, **88**, 2316 (1966).
208a. T. Tsuji, I. Moritani, S. Nishida and G. Tadokoro, *Tetrahedron Letters*, 1207 (1967).
208b. C. F. Wilcox, Jr. and R. G. Jesaitis, *Tetrahedron Letters*, 2567 (1967).
209. P. v. R. Schleyer and G. W. Van Dine, *J. Am. Chem. Soc.*, **88**, 2321 (1966).
210. N. C. Deno, H. G. Richey, Jr., J. S. Liu, J. D. Hodge, J. J. Houser and M. J. Wisotsky, *J. Am. Chem. Soc.*, **84**, 2016 (1962).
211. L. Eberson and S. Winstein, *J. Am. Chem. Soc.*, **87**, 3506 (1965).
212. G. A. Olah and C. V. Pittman, Jr., *J. Am. Chem. Soc.*, **87**, 3509 (1965).
213. G. A. Olah, M. B. Comisarow, E. Namanworth and B. Ramsey, *J. Am. Chem. Soc.*, **89**, 5259 (1967).
214. G. A. Olah, E. Namanworth, M. B. Comisarow and B. Ramsey, *J. Am. Chem. Soc.*, **89**, 711 (1967).

10. Olefinic Properties of Cyclopropanes

215. P. T. Lansbury, V. A. Pattison, W. A. Clement and J. D. Sidler, *J. Am. Chem. Soc.*, **86,** 2247 (1964).
216. C. L. Bumgardner, *J. Am. Chem. Soc.*, **85,** 73 (1963).
217. C. G. Overberger and M. B. Berenbaum, *J. Am. Chem. Soc.*, **73,** 2618 (1951).
218. C. G. Overberger and A. Lebovits, *J. Am. Chem. Soc.*, **76,** 2722 (1954).
219. C. G. Overberger, M. Tobkes and A. Zweig, *J. Org. Chem.*, **28,** 620 (1963).
220. H. Hart and R. A. Cipriani, *J. Am. Chem. Soc.*, **84,** 3697 (1962).
221. D. C. Neckers, J. Hardy and A. P. Schaap, *J. Org. Chem.*, **31,** 622 (1966).
222. D. C. Neckers, A. P. Schaap and J. Hardy, *J. Am. Chem. Soc.*, **88,** 1265 (1966).
223. D. C. Neckers and A. P. Schaap, *J. Org. Chem.*, **32,** 22 (1967).
224. H. C. Brown and M. Borkowski, *J. Am. Chem. Soc.*, **74,** 1894 (1952).
225. E. Renk, P. R. Shafer, W. H. Graham, R. H. Mazur and J. D. Roberts *J. Am. Chem. Soc.*, **83,** 1987 (1961).
226. E. N. Trachtenberg and G. Odian, *J. Am. Chem. Soc.*, **80,** 4018 (1958).
227. R. Fuchs and J. J. Bloomfield, *J. Am. Chem. Soc.*, **81,** 3158 (1959).
228. J. J. Bloomfield and R. Fuchs, *J. Org. Chem.*, **26,** 2991 (1961).
229. R. Fuchs, C. A. Kaplan, J. J. Bloomfield and L. F. Hatch, *J. Org. Chem.* **27,** 733 (1962).
230. R. Fuchs and J. J. Bloomfield, *J. Org. Chem.*, **28,** 910 (1963).
231. C. C. Price and E. A. Dudley, *J. Am. Chem. Soc.*, **78,** 68 (1956).
232. I. Benghiat and E. Becker, *J. Org. Chem.*, **23,** 1885 (1958).
233. J. D. Roberts and R. A. Carboni, *J. Am. Chem. Soc.*, **77,** 5554 (1955).
234. B. Jones and J. G. Watkinson, *J. Chem. Soc.*, 4064 (1958).
235. R. Fuchs, *J. Org. Chem.*, **28,** 3209 (1963).
236. M. Charton, *Abstr., 140th Mtg. Am. Chem. Soc.*, 1961, p57T.
237. M. Charton, *Abstr., 137th Mtg. Am. Chem. Soc.*, 1960, p. 920.
238. M. J. S. Dewar and Y. Takeuchi, *J. Am. Chem. Soc.*, **89,** 390 (1967).
239. W. Adcock and M. J. S. Dewar, *J. Am. Chem. Soc.*, **89,** 379 (1967).
240. C. Beguin and A. Meary-Tetrian, *Bull. Soc. Chim. France*, 795 (1967).
241. H. C. Brown and Y. Okamoto, *J. Am. Chem. Soc.*, **80,** 4979 (1958).
242. L. J. Andrews and R. M. Keefer, "Molecular Complexes in Organic Chemistry", Holden-Day, San Francisco, 1964, p. 157.
243. O. Filipov, *J. Russ. Phys. Chem. Soc.*, **46,** 1179 (1914); *J. Prakt. Chem.* [2], **93,** 176 (1916).
244. S. Freed and K. M. Sancier, *J. Am. Chem. Soc.*, **74,** 1273 (1952).
245. C. F. H. Tipper, *J. Chem. Soc.*, 2045 (1955).
246. V. K. Potapov, Y. S. Shabarov and R. Y. Levina, *Zh. Obshch. Khim.*, **34,** 2512 (1964).
246a. L. Joris, P. v. R. Schleyer and R. Gleiter, *J. Am. Chem. Soc.*, **90,** 327 (1968).
246b. P. v. R. Schleyer, D. S. Trifan and R. Bacskai, *J. Am. Chem. Soc.*, **80,** 6691 (1958).
246c. M. Davis, S. Julia and G. H. R. Summers, *Bull. Soc. Chim. France*, 742 (1960).
246d. P. v. R. Schleyer, *J. Am. Chem. Soc.*, **83,** 1368 (1961).
246e. M. Hanack and H. Allmendinger, *Chem. Ber.*, **97,** 1669 (1964).
247. A. Freund, *Monatsh. Chem.*, **2,** 642 (1881).
248. G. G. Gustavson, *J. Prakt. Chem.*, **62** [2], 273 (1900).

249. N. C. Deno and D. L. Lincoln, *J. Am. Chem. Soc.*, **88,** 5357 (1966).
250. A. V. Grosse and C. B. Linn, *J. Org. Chem.*, **3,** 26 (1938).
251. L. Schmerling, *Ind. Eng. Chem.*, **40,** 2072 (1948).
252. V. N. Ipatieff, H. Pines and L. Schmerling, *J. Org. Chem.*, **5,** 253 (1940).
253. V. N. Ipatieff, H. Pines and B. B. Corson, *J. Am. Chem. Soc.*, **60,** 577 (1938).
254. A. V. Grosse and V. N. Ipatieff, *J. Org. Chem.*, **2,** 447 (1937).
255. J. H. Simons, S. Archer and E. Adams, *J. Am. Chem. Soc.*, **60,** 2955 (1938).
256. A. V. Grosse, *U.S. Pat.*, 2182557 (1937); *Chem. Abstr.*, **34,** 1995 (1940).
257. C. F. H. Tipper and D. A. Walker, *J. Chem. Soc.*, 1352 (1959).
258. C. F. H. Tipper and D. A. Walker, *J. Chem. Soc.*, 1199 (1957).
259. H. Hart and O. E. Curtis, *J. Am. Chem. Soc.*, **79,** 931 (1957).
260. H. Hart and G. Levitt, *J. Org. Chem.*, **24,** 1261 (1959).
261. H. Hart and R. H. Schlosberg, *J. Am. Chem. Soc.*, **88,** 5030 (1966).
262. D. J. Abraham and W. E. Truce, *J. Org. Chem.*, **28,** 2901 (1963).
263. A. J. Kranzfelder and F. J. Sowa, *J. Am. Chem. Soc.*, **59,** 1490 (1937).
264. T. B. Dorris and F. J. Sowa, *J. Am. Chem. Soc.*, **60,** 358 (1938).
265. C. D. Lawrence and C. F. H. Tipper, *J. Chem. Soc.*, 713 (1955).
266. A. A. Aboderin and R. L. Baird, *Tetrahedron Letters*, 235 (1965).
267. R. L. Baird and A. A. Aboderin, *J. Am. Chem. Soc.*, **86,** 252 (1964).
268. W. A. G. Graham and F. G. A. Stone, *Chem. Ind.* 1096 (1957).
269. C. F. H. Tipper and D. A. Walker, *Chem. Ind.*, 730 (1957).
270. N. Y. Demyanov, *Chem. Ber.*, **28,** 22 (1895); *J. Russ. Phys. Chem. Soc.*, **34,** 217 (1902).
271. Y. A. Slobodin, V. I. Grigoreva and Y. E. Shmulyalovskii, *Zh. Obshch. Khim.*, **23,** 1480 (1953).
272. J. H. Turnbull and E. S. Wallis, *J. Org. Chem.*, **21,** 663 (1956).
272a. N. M. Kizhner, *J. Russ. Phys. Chem. Soc.*, **44,** 165 (1912).
273. N. Kizhner, *J. Russ. Phys. Chem. Soc.*, **41,** 653 (1909).
273a. P. B. D. De la Mare, *Quart. Rev.*, **3,** 126 (1949).
274. N. A. Rozanov, *J. Russ. Phys. Chem. Soc.*, **48,** 168 (1916).
275. G. G. Gustavson, *J. Prakt. Chem.* [2], **62,** 270 (1900).
276. G. G. Gustavson and O. Popper, *J. Prakt. Chem.*, **58** [2], 458 (1898).
277. N. M. Kizhner and G. V. Khonin, *J. Russ. Phys. Chem. Soc.*, **45,** 1772 (1913).
278. W. H. Perkin and W. A. Bone, *J. Chem. Soc.*, **67,** 118 (1895).
279. E. T. McBee, A. B. Hass, R. M. Thomas, W. G. Toland, Jr. and A. Truchan, *J. Am. Chem. Soc.*, **69,** 944 (1947).
280. R. C. Fuson and F. N. Baumgartner, *J. Am. Chem. Soc.*, **70,** 3255 (1948).
281. W. H. Perkin, *J. Chem. Soc.*, **47,** 801 (1885).
282. T. R. Marshall and W. H. Perkin, *J. Chem. Soc.*, **59,** 853 (1891).
283. W. E. Truce and L. B. Lindy, *J. Org. Chem.*, **26,** 1463 (1961).
284. E. Rothstein, *J. Chem. Soc.*, 1560 (1940).
285. E. D. Venus-Danilova and V. F. Kazimirova, *Zh. Obshch. Khim.*, **8,** 1438 (1938).
286. C. H. De Puy, F. W. Breitbeil and K. R. de Bruin, *J. Am. Chem. Soc.*, **88,** 3347 (1966).
286a. C. Allen and R. Boyer, *Can. J. Res.*, **9,** 159 (1933).
286b. P. Barbier, *Compt. rend.*, **153,** 188 (1911).

286c. P. B. D. De la Mare and R. Bolton, "Electrophilic Additions to Unsaturated Systems", Elsevier, Amsterdam, 1966, p. 203.
286d. R. Y. Levina and B. M. Gladshtein, *Dokl. Akad. Nauk. SSSR.*, **71,** 65 (1950).
287. R. Y. Levina, V. N. Kostin and A. Tartakovskii, *Zh. Obshch. Khim.*, **26,** 2998 (1956).
288. R. Y. Levina and V. N. Kostin, *Dokl. Akad. Nauk. SSSR*, **97,** 1027 (1954)
289. R. Y. Levina and V. N. Kostin, *Zh. Obshch. Khim.*, **23,** 1054 (1953).
290. R. Y. Levina, V. N. Kostin and A. Tartakovskii, *Vestnik. Moskow Gos. Univ.*, No. 2, 77 (1956).
291. R. Y. Levina, V. N. Kostin, P. A. Gembitskii, and E. G. Treshchova, *Zh. Obshch. Khim.*, **31,** 829 (1961).
292. R. Y. Levina, V. N. Kostin and A. Tartakovskii, *Zh. Obshch. Khim.*, **29** 40 (1959).
293. R. J. Ouellette and D. L. Shaw, *J. Am. Chem. Soc.*, **86,** 1651 (1964).
294. R. J. Ouellette, D. L. Shaw and A. South, Jr., *J. Am. Chem. Soc.*, **86,** 2744 (1964).
295. R. Y. Levina, Y. S. Shabarov and I. M. Shanazarova, *Zh. Obshch. Khim.*, **29,** 44 (1959).
296. V. N. Ipatieff, H. Pines and W. Huntsman, *J. Am. Chem. Soc.*, **75,** 2315 (1953).
297. D. Davidson and J. Feldman, *J. Am. Chem. Soc.*, **66,** 488 (1944).
297a. K. B. Wiberg, *Record. Chem. Progr.*, **26,** 143 (1965).
298. R. T. LaLonde and L. S. Forney, *J. Am. Chem. Soc.*, **85,** 3767 (1965).
299. R. Criegee and A. Rimmelin, *Chem. Ber.*, **90,** 414 (1957); *Chem. Ber.*, **90,** 417 (1957).
300. R. Y. Levina, V. N. Kostin and T. K. Ustynyuk, *Zh. Obshch. Khim.*, **30,** 359 (1960).
301. N. Kizhner. *J. Russ. Phys. Chem. Soc.*, **44,** 849 (1912).
302. R. T. LaLonde and M. A. Tobias, *J. Am. Chem. Soc.*, **86,** 4086 (1964).
303. N. M. Kizhner, *J. Russ. Phys. Chem. Soc.* **43,** 1132 (1911).
304. B. Rickborn and S. E. Wood, *Chem. Ind. (London)*, 162 (1966).
305. R. T. LaLonde and L. S. Forney, *J. Org. Chem.*, **29,** 2911 (1964).
306. R. T. LaLonde and M. A. Tobias, *J. Am. Chem. Soc.*, **85,** 3771 (1963).
307. R. T. LaLonde, J. Ding and M. A. Tobias, *J. Am. Chem. Soc.*, **89,** 6651 (1967).
308. H. Hart and R. A. Martin, *J. Org. Chem.*, **24,** 1267 (1959).
309. M. Hanack and H. Eggensperger, *Ann. Chem.*, **648,** 3 (1961).
310. A. Nickon, J. L. Lambert, R. O. Williams and N. H. Werstiuk, *J. Am. Chem. Soc.*, **88,** 3354 (1966).
311. R. C. Cookson, D. P. G. Hamon and J. Hudec, *J. Chem. Soc.*, 5782 (1963).
312. M. Palmade, P. Pesnelle, J. Streith and G. Ourisson, *Bull. Soc. Chim. France*, 1950 (1963).
313. P. Pesnelle, *Thesis*, Faculte des Sciences, Strasbourg, April (1965).
314. S. Sarel and R. E. Shoshan, *Tetrahedron Letters*, 1053 (1965).
315. S. Sarel and E. Breuer, *J. Am. Chem. Soc.*, **81,** 6522 (1959).
316. F. F. Rathman and S. Z. Roginsky, *J. Am. Chem. Soc.*, **55,** 2800 (1933); F. F. Rathman, *Zh. Obschch. Khim.*, **1,** 14 (1937).
316a. B. C. Anderson, *J. Org. Chem.*, **27,** 2720 (1962).

317. P. K. Freeman, M. F. Grostic and F. A. Raymond, *J. Org. Chem.*, **30,** 1047 (1965).
318. P. K. Freeman, F. A. Raymond and M. F. Grostic, *J. Org. Chem.*, **32,** 24 (1967).
319. S. Nishida, I. Moritani, K. Ito and K. Sakai, *J. Org. Chem.*, **32,** 939 (1967).
320. R. A. Ogg, Jr. and W. J. Priest, *J. Am. Chem. Soc.*, **60,** 217 (1938).
321. M. S. Kharasch, M. Z. Fineman and F. R. Mayo, *J. Am. Chem. Soc.*, **61,** 2139 (1939).
322. R. A. Ogg, Jr. and W. J. Priest, *J. Chem. Phys.*, **7,** 736 (1939).
323. L. Harris, A. A. Ashdown and R. T. Armstrong, *J. Am. Chem. Soc.*, **58,** 850 (1936); **58,** 852 (1936).
324. G. B. Heisig, *J. Am. Chem. Soc.*, **54,** 2331 (1932).
325. H. E. Gunning and E. W. R. Steacie, *J. Chem. Phys.*, **17,** 351 (1949).
326. K. J. Ivin, *J. Chem. Soc.*, 2241 (1956).
327. B. H. Nicolet and L. Sattler, *J. Am. Chem. Soc.*, **49,** 2070 (1927).
328. G. S. Hammond and R. W. Todd, *J. Am. Chem. Soc.*, **76,** 4081 (1954).
329. H. G. Kuivila, S. C. Gaywood, W. F. Boyce and F. L. Langevin, Jr., *J. Am. Chem. Soc.*, **77,** 5175 (1955).
330. R. Fittig and R. Marburg, *Chem. Ber.*, **18,** 3413 (1895).
331. G. G. Gustavson, *J. Prakt. Chem.*, **43** [2], 396 (1891).
332. G. G. Gustavson, *J. Prakt. Chem.*, **42** [2], 496 (1890).
333. G. G. Gustavson, *J. Prakt. Chem.*, **50** [2], 380 (1894).
334. J. D. Roberts and P. H. Dirstine, *J. Am. Chem. Soc.* **67,** 1281 (1945).
335. D. E. Applequist, G. F. Fanta and B. W. Henrikson, *J. Am. Chem. Soc.* **82,** 2368 (1960).
335a. E. S. Huyser and J. D. Taliferro, *J. Org. Chem.*, **28,** 3442 (1963).
336. W. A. Bone and W. H. Perkin, *J. Chem. Soc.*, **67,** 108 (1895).
337. R. W. Kierstead, R. P. Linstead and B. C. L. Weedon, *J. Chem. Soc.*, 3616 (1952).
338. J. R. Best and J. F. Thorpe, *J. Chem. Soc.*, 685 (1909).
339. J. M. Stewart and H. H. Westburg, *J. Org. Chem.*, **30,** 1951 (1965).
340. S. Winstein and R. Baird, *J. Am. Chem. Soc.*, **79,** 756 (1957).
341. V. V. Ershov, I. S. Belostotskaya and A. A. Volodkin, *Izv. Akad. Nauk., SSSR, Ser. Khim.*, 1496 (1966).
342. R. Willstatter and J. Bruce, *Chem. Ber.*, **40,** 4456 (1907).
343. B. A. Kazanskii, V. T. Aleksanyan, M. Y. Lukina, A. I. Malyshev and K. E. Sterin, *Izv. Akad. Nauk. SSSR. Otd. Khim. Nauk.*, 1118 (1955).
344. M. Y. Lukina, S. V. Zotova and B. A. Kazanskii, *Dokl. Akad. Nauk. SSSR.*, **123,** 105 (1958).
345. M. Y. Lukina, S. V. Zotova and B. A. Kazanskii, *Dokl. Akad Nauk. SSSR.*, **127,** 341 (1959).
346. B. A. Kazanskii, M. Y. Lukina and L. G. Cherkashina, *Iav. Akad. Nauk. SSSR., Otd. Khim. Nauk.*, 553 (1959).
346a. J. Addy and G. C. Bond, *Trans. Faraday Soc.*, **53,** 377 (1957).
347. J. E. Benson and T. Kwan, *J. Phys. Chem.*, **60,** 1601 (1957).
348. J. Addy and G. C. Bond, *Trans. Faraday Soc.*, **53,** 368 (1957).
349. G. C. Bond and J. Turkevich, *Trans. Faraday Soc.*, **49,** 281 (1953); **50,** 1335 (1954).

350. J. Boeseken, O. B. vander Weide and C. P. Mome, *Rec. trav. chim.*, **35,** 282 (1916).
351. G. C. Bond and J. Sheridan, *Trans. Faraday Soc.*, **48,** 713 (1952).
352. M. Y. Lukina, V. A. Ovodova and B. A. Kazanskii, *Dokl. Akad. Nauk. SSSR.*, **47,** 683 (1954).
353. B. A. Kazanskii, M. Y. Lukina and A. I. Malyshev, *Izv. Akad. Nauk. SSSR, Otd. Khim. Nauk.*, 1399 (1956).
354. R. V. Volkenburgh, K. W. Greenlee, J. M. Derfer and C. E. Boord, *J. Am. Chem. Soc.*, **71,** 172 (1949).
355. R. W. Shortridge, R. A. Craig, K. W. Greenlee, J. M. Derfer and C. E. Boord, *J. Am. Chem. Soc.*, **70,** 946 (1948).
356. M. Y. Lukina, S. V. Zotova and B. A. Kazanskii, *Dokl. Akad. Nauk. SSSR.*, **116,** 793 (1957).
357. M. Y. Lukina, S. V. Zotova and B. A. Kazanskii, *Dokl. Akad. Nauk. SSSR.*, **114,** 792 (1957).
358. M. Y. Lukina, S. V. Zotova and B. A. Kazanskii, *Izv. Akad. Nauk. SSSR, Otd. Khim. Nauk.*, 300 (1958).
359. B. A. Kazanskii, M. Y. Lukina and V. A. Ovodova, *Izv. Akad. Nauk. SSSR, Otd. Khim. Nauk.*, 878 (1954).
360. R. G. Kelso, K. W. Greenlee, J. M. Derfer and C. E. Boord, *J. Am. Chem. Soc.*, **74,** 287 (1952).
361. B. A. Kazanskii, M. Y. Lukina and L. G. Salnikova, *Dokl. Akad. Nauk. SSSR.*, **115,** 301 (1957).
362. B. A. Kazanskii, M. Y. Lukina, A. I. Malyshev, V. T. Aleksanyan and K. E. Sterin, *Izv. Akad. Nauk. SSSR, Otd. Khim. Nauk.*, 36 (1956).
363. B. A. Kazanskii, M. Y. Lukina, A. I. Malyshev, V. T. Aleksanyan and K. E. Sterin, *Izv. Akad. Nauk. SSSR, Otd. Khim. Nauk.*, 1102 (1956).
364. V. A. Slabey and P. H. Wise, *J. Am. Chem. Soc.*, **74,** 3887 (1952).
365. B. A. Kazanskii, M. Y. Lukina and I. L. Safonova, *Izv. Akad. Nauk. SSSR., Otd. Khim. Nauk.*, 102 (1958).
366. B. A. Kazanskii, M. Y. Lukina and I. L. Safonova, *Dokl. Akad. Nauk. SSSR.*, **130,** 322 (1960).
367. L. K. Freidlin, A. P. Meshcheryakov, V. I. Gorshkov, and V. G. Glukhovtsev, *Izv. Akad. Nauk SSSR., Otd. Khim. Nauk.*, 2237 (1959).
368. B. A. Kazanskii, M. Y. Lukina and L. G. Salnikova, *Izv. Akad. Nauk. SSSR., Otd. Khim. Nauk.*, 1401 (1957).
369. V. A. Slabey and P. H. Wise, *J. Am. Chem. Soc.*, **71,** 3252 (1949).
370. V. A. Slabey, *J. Am. Chem. Soc.*, **69,** 475 (1947).
371. F. Richter, W. Wolff and W. Presting, *Chem. Ber.*, **64,** 876 (1931).
372. A. G. Short and J. Read, *J. Chem. Soc.*, 1040 (1939).
373. E. F. Ullmann, *J. Am. Chem. Soc.*, **81,** 5386 (1959).
374. J. C. Prudhomme and F. G. Gault, *Bull. Soc. Chim. France*, 827 (1966).
375. J. C. Prudhomme and F. G. Gault, *Bull. Soc. Chim. France*, 832 (1966).
376. E. S. Corner and R. N. Pease, *Ind. Eng. Chem. Anal. Ed.*, **17,** 564 (1945).
377. E. P. Kohler and J. B. Conant, *J. Am. Chem. Soc.*, **39,** 1404 (1917).
378. H. Greenfield, R. Friedel and M. Orchin, *J. Am. Chem. Soc.*, **76,** 1257 (1954).
379. D. J. Cram, *Fundamentals of Carbanion Chemistry*, Academic Press, New York, 1965, p. 19.

380. D. E. Applequist and D. F. O'Brien, *J. Am. Chem. Soc.*, **85,** 743 (1963).
381. R. E. Dessy, W. Kitching, T. Psarras, R. Salinger, A. Chen and T. Chivers, *J. Am. Chem. Soc.*, **88,** 460 (1966).
382. W. K. McEwen, *J. Am. Chem. Soc.*, **58,** 1124 (1936).
383. A. Streitweiser Jr., R. A. Caldwell and M. R. Granger, *J. Am. Chem. Soc.*, **86,** 3578 (1964).
384. A. Streitweiser Jr., personal communication cited by Ref. 379.
385. A. I. Shatenshtein, *Adv. Phys. Org. Chem.*, **1,** 176 (1963).
386. H. E. Zimmerman and B. S. Thygagorajan, *J. Am. Chem. Soc.*, **82,** 2505 (1960).
387. A. Ratajczak, F. A. L. Anet and D. J. Cram, *J. Am. Chem. Soc.*, **89,** 2072 (1967).
388. H. M. Walborsky, *Record. Chem. Progr.*, **23,** 75 (1962).
389. H. M. Walborsky, F. J. Impastato and A. E. Young, *J. Am. Chem. Soc.*, **86,** 3283 (1964).
390. H. M. Walborsky and A. E. Young, *J. Am. Chem. Soc.*, **86,** 3288 (1964).
391. G. W. Koeppl, D. S. Sagatys, G. S. Krishnamurthy and S. I. Miller, *J. Am. Chem. Soc.*, **89,** 3396 (1967).
392. A. F. Trotman-Dickinson and E. W. R. Steacie, *J. Chem. Phys.*, **19,** 329 (1951).
393. J. R. McNesby and A. S. Gordon, *J. Am. Chem. Soc.*, **79,** 825 (1957).
394. H. Preuss and G. Diercksen, *Intern. J. Quantum Chem.*, **1,** 361 (1967).
395. K. B. Wiberg, *Tetrahedron*, **24,** 1083 (1968).
396. J. Asunskis and H. Shechter, *J. Org. Chem.*, **33,** 1164 (1968).
397. R. T. C. Brownlee, R. E. J. Hutchinson, A. R. Katritzky, T. T. Tidwell and R. D. Topsom, *J. Am. Chem. Soc.*, **90,** 1757 (1968).
398. K. K. Lum and G. G. Smith, *Abstr., 155th Mtg. Am. Chem. Soc.*, San Francisco, 1968, p. 35.
399. G. A. Olah and J. Lukas, *J. Am. Chem. Soc.*, **90,** 933 (1968).
400. P. E. Peterson and G. Thompson, *J. Org. Chem.* **33,** 968 (1968).
401. P. G. Gassman and K. T. Mansfield, *J. Am. Chem. Soc.*, **90,** 1517 (1968).
402. P. G. Gassman and K. T. Mansfield, *J. Am. Chem. Soc.*, **90,** 1524 (1968).
403. R. J. Ouellette, R. D. Robins and A. South Jr., *J. Am. Chem. Soc.*, **90,** 1619 (1968).
404. C. H. DePuy, W. C. Arney, Jr. and D. H. Gibson, *J. Am. Chem. Soc.*, **90,** 1830 (1968).
405. C. H. DePuy, *Acc. Chem. Res.*, **1,** 33 (1968).

Author Index

This author index is designed to enable the reader to locate an author's name and work with the aid of the reference numbers appearing in the text. The page numbers are printed in normal type in ascending numerical order, followed by the reference numbers in parentheses. The numbers in *italics* refer to the pages on which the references are actually listed. (x) indicates a text reference which does not have a number.

Aboderin, A. A. 572, 573 (266, 267), 574, 575 (267), *606*
Abraham, D. J. 572, 573 (262), *606*
Abrahamson, E. W. 281 (41b), *318*
Abraitys, V. Y. 312 (254), *324*
Abramson, F. P. 395 (166), *410*
Abshire, C. J. 486 (267), *508*
Acton, N. 255 (233c), *264*
Adamczyk, A. 373 (83), *408*
Adamek, S. 425 (70), *502*
Adams, D. G. 54, 74 (53), *108*
Adams, E. 571, 572 (255), *606*
Adams, R. 193 (53), 196 (61), *212*
Adams, R. W. 236 (131c), *260*
Adams, W. R. (49), *318*
Adcock, W. (239), *605*
Addy, J. 589 (346a, 348), 590 (348), *608*
Adkins, H. 246 (185), *262*
Adler, P. 366, 367 (37), *406*
Adrian, F. J. 48 (26), *107*
Agami, C. 163 (160), *173*
Agasyan, P. K. 553 (182), *603*
Agneux, A. G. 149 (115), *171*
Aguiló, A. 216, 237 (10a), *256*
Ahmed, F. R. 523 (57), *600*
Akasaki, Y. 316 (273, 274), *325*
Akentijevich, R. I. 299 (156), *321*
Akhmedov, V. M. 250 (219), *263*
Aladzhyeva, I. M. 159 (146), *172*
Alderson, T. 247–249, 253, 254 (194), *262*
Aleksanyan, V. T. 541 (131, 132, 132a, 138), 542 (131, 132a, 138), 543 (132, 138), 590 (362), (343), (363), *602*, *608*, *609*
Alexander, E. R. 154 (132), *172*
Alexander, S. 21 (77), *37*
Alfrey, T., Jr. 434 (92), *503*

Allegra, G. 249 (206b), 250 (206b, 213), *262*, *263*, 420 (45, 46), *501*
Allen, A. O. 362 (20), 366 (40), 367 (40, 41), 368 (50, 53–55), 375 (53), 387 (53–55), 396 (50, 53–55), *406*, *407*
Allen, C. 579 (286a), *606*
Allen, G. R., Jr. 203 (91), *213*
Allen, P. W. 491 (284), *508*
Allen, R. G. 118 (13), *168*
Allinger, N. L. 54 (48), 105 (222), *108*, *114*, 525 (59), *600*
Allmendinger, H. 571 (246e), *605*
Altman, L. J. 8 (46), *36*
Amagi, Y. 436 (95, 96), *503*
Amano, A. 160 (151), *172*
Ambelang, J. C. 477, 483 (225), *507*
American Institute of Physics 23(x)
American Petroleum Institute 116 (1a), *167*
Amerongen, G. J. van 435 (95b), *503*
Ames, D. E. 131 (59), *169*
Anderson, A. R. 367, 375 (45), *407*
Anderson, B. C. 585 (316a), *607*
Anderson, C. B. 241 (156), *261*
Anderson, D. H. 3 (10), *35*
Anderson, E. W. 100 (205), *114*
Anderson, H. L. 2 (1), *35*
Anderson, J. H. 520 (29), *599*
Anderson, N. H. 94 (185), *113*
Andes, B. A., Jr. 97 (195), *113*
Ando, T. 67 (107), *110*
Ando, W. 119 (15, 16), *168*
Andrac, M. 150 (118), *171*
Andrac-Taussig, M. 163 (160), *173*
Andreetta, A. 250 (220), *263*
Andrews, E. H. 485 (264), *508*
Andrews, L. J. (242), *605*
Andrussov, K. 556 (188), *604*

Author Index

Anet, F. A. L. 32 (111), *38*, 594 (387), *610*
Ang, F. S. 491 (282), *508*
Angelo, R. J. 443 (114, 115), *503*
Angier, D. J. 491 (283), *508*
Angus, H. J. F. 289 (87), *319*
Angyal, S. J. 105 (222), *114*, 525 (59), *600*
Anhalt, J. P. 313 (262), *324*
Anner, G. 233 (250), *264*
Ansell, M. F. 145 (97), *171*
Aoki, D. 253, 254 (225b), *263*
Appelbaum, A. 151 (120), *171*
Applequist, D. E. 50, 105 (31), *108*, 586 (335), 593 (380), *608*, *610*
Arai, H. 196 (62), *212*
Archer, S. 571, 572 (255), *606*
Arcus, C. L. 178 (20), *211*
Arguelles, M. 284 (66), 291 (104), *318*, *320*
Arkel, A. E. van 550 (173), *603*
Armstrong, D. A. 396 (172), *410*
Armstrong, R. T. 586 (323), *608*
Arnaud, P. 526 (70a), 527 (70a, 70b), 536 (109), *600*, *601*
Arney, W. C., Jr. 597 (404), *610*
Arnold, D. R. 255 (235a), *264*, 309 (222), 310 (235), 312 (254), *323*, *324*
Arrington, J. P. 277 (24), *317*
Asami, R. 421 (57), *502*
Aseeva, R. M. 432 (86), *502*
Ashby, E. C. 42 (11), *107*
Ashdown, A. A. 586 (323), *608*
Ashe, A. J. 97 (195), *113*
Asinger, F. 165 (166), 167 (174), *173*, 225 (54, 58), 228 (66–69), 229 (69), 230 (58, 69), 244 (54, 58), 246 (68), *257*, 278 (33), *317*
Askani, R. 291 (108), 307 (218), *320*, *323*
Asmus, K. D. 368 (59), *407*
Aso, C. 421 (58), *502*
Astaf'ev, I. V. 73 (130), *111*
Asunskis, J. 596 (396), *610*
Atkinson, J. G. 296, 311 (135), 312 (250), *321*, *324*
Audier, H. 343, 344 (33), *358*
Aufdermarsh, C. A., Jr. 418 (43), *501*
Auilanti, V. 363 (26), *406*
Ausloos, P. 360 (13), 362 (22), 366 (34), 367 (13, 34), 371 (72–74), 374 (13, 22), 377 (23, 73), 380 (120, 121), 388 (144), 390 (72, 74), 394 (164), 395 (22), 402 (187–189), *406*, *407*, *409*, *410*
Avram, M. 295 (129), *321*
Axtmann, R. C. 385 (134), *409*
Ayer, D. E. 296, 311 (135), 312 (250), *321*, *324*
Azuma, K. 236 (105), *259*
Azumi, T. 270 (6), *316*

Babos, B. 236 (112b), *259*
Back, R. A. 362 (21), 366 (33), 367 (33, 44), 388 (143), 396 (172), *406*, *407*, *409*, *410*
Bacskai, R. 571 (246b), *605*
Baenziger, N. C. 240 (150a, 150b), *261*
Bagby, M. O. 145 (99, 100), *171*
Baggiolini, E. 300, 301 (167), *322*
Bailar, J. C., Jr. 236 (124, 128, 130, 131a–131d), *260*
Bailer, J. C., Jr. 201 (82–84), *213*
Bailey, D. M. 86, 91, 93, 94 (170), *112*
Bailey, F. E., Jr. 436 (94), *503*
Bailey, G. F. 154 (130), *172*
Bailey, P. S. 486 (266, 267), *508*
Bair, T. I. 527 (70), *600*
Baird, M. C. 224 (44), *257*
Baird, M. D. 132, 133 (67), *170*, 282 (52), *318*
Baird, M. S. 67 (107), *110*
Baird, N. C. 518, 521 (29b), *599*
Baird, R. 588 (340), *608*
Baird, R. L. 572, 573 (266, 267), 574, 575 (267), *606*
Baird, W. C. 240 (149), *261*
Baird, W. C., Jr. 210 (117, 118), *214*
Bais, F. 219 (26), *256*
Baker, R. H. 179 (30, 31), 183–185 (30), *211*
Baldeschwieler, J. D. 15 (63), *36*
Baldwin, J. E. 298 (147), *321*
Baldwin, M. 327 (5, 6), 338 (5), *357*
Balueva, G. A. 41 (9), *107*
Bank, S. 68 (115, 118), 69 (115), 71 (118, 125), 83 (163, 164), 84 (164), 85 (163, 164), *111*, *112*, 141 (86), *170*
Banks, G. R. 157 (137), *172*
Banks, R. G. S. 231 (83), *258*
Bannister, W. D. 243 (166), *261*
Banwell, C. N. 6, 7 (37), 18 (69), *36*, *37*
Baranova, T. I. 248, 249 (199), *262*
Barber, M. S. 126 (45), *169*
Barbier, P. 579 (286b), *606*
Barborak, J. 300 (164), *322*
Barlet, R. 527 (70b), *600*

Barltrop, J. A. 307 (214), *323*
Barnard, D. 477, 478, 480, 482 (226), 485 (264), *507, 508*
Barneis, Z. J. 277 (24), *317*
Barney, A. L. 253, 254 (228b), *263*
Barrett, J. H. 285 (78), *319*
Bartell, L. S. 523 (52, 53), 524(x), 526 (52, 53, 63a), *599, 600*
Bartl, H. 449 (135), *504*
Bartlett, P. D. 80 (151), 83 (163, 164), 84 (164–166), 85 (163–165), 86 (168, 169), 87 (168, 172), 93 (165, 166), 94 (166, 168), 101 (207), *112, 114*
Bartley, W. J. 196 (63), *212*
Bartz, K. W. 18 (67), *36*
Barzynski, H. F. (176), *410*
Basolo, F. 243 (165, 170), *261*
Bass, J. D. 304 (202), *323*
Bassi, I. W. 421 (52), *501*
Bastian, B. N. 255 (235b), *264*
Bastiansen, O. 118 (12), *168*, 523 (33, 34, 41), 526 (40, 41), *599*
Bateman, L. 412 (2), 452 (162–164), 457 (175), 460, 464, 466, 468, 470, 471 (164), 472 (162–164), 473 (164), 477 (226, 331), 478, 480, 482 (226), 487 (2), *500, 505, 507*
Bates, P. 22 (84), *37*
Bates, R. B. 76 (142), *111*, 143 (91a), 145 (100), *170, 171*
Batley, G. E. 236 (131c), *260*
Bauld, N. L. 206 (97), *213*, 527, 547 (71), *600*
Baumgartner, F. N. 577, 581 (280), *606*
Bauslaugh, G. 303 (197), *322*
Baylar, J. C. 225, 230, 236, 244 (64), *257*
Beachell, H. C. 484 (259), *508*
Beauchamp, J. L. 389–391, 393 (147), *409*
Beaudet, R. A. 15 (63), *36*
Becconsall, J. K. 222 (36), *257*
Bechter, M. 219, 237 (28), *257*
Becker, D. A. 374, 395 (94), *408*
Becker, E. 567 (232), *605*
Becker, R. S. 290 (99c), *320*
Beckham, M. E. 306 (212), *323*
Bednar, J. 383 (128), *409*
Beeck, O. 177 (13), *211*
Beereboom, J. J. 308 (227), *323*
Beguin, C. (240), *605*
Bekkum, H. N. 234 (246), *264*
Bell, C. L. M. 482 (251), *507*

Bell, H. M. 100 (202), *114*
Bell, I. 292 (113), *320*
Bell, R. A. 86, 91, 93, 94 (170), *112*
Bell, R. P. 52 (40), *108*
Bell, V. L. 429 (82), 430 (83), *502*
Belonovskaja, G. P. 460 (188), *506*
Belostotskaya, I. S. (341), *608*
Benghait, I. 567 (232), *605*
Bennett, M. A. 216 (11), 236 (122), *256, 260*
Benson, J. E. 589, 590 (347), *608*
Benson, R. E. 4 (22), *35*, 77 (143a), *112*
Benson, S. 373, 376 (84), *408*
Benson, S. W. 116 (1b, 3), 117 (3, 5, 6), 135 (78), 136 (80), *167, 170*, 375, 398, 399 (99), *408*
Berenbaum, M. B. 565 (217), *605*
Berezin, G. H. 537 (112), *601*
Bergelson, L. D. 128, 129 (49), *169*
Berger, M. 450, 452 (143), 454 (168), *504, 505*
Bergmann, E. D. 233 (97), *258*
Bergstrom, G. G. 560 (203), *604*
Berlin, A. A. 432 (86), *502*
Berlin, A. J. 525, 526 (64), *600*
Bernett, W. A. 513, 519, 520, 525, 534, 548 (4), *598*
Bernstein, H. J. 545 (146), *602*
Berry, R. S. 161 (154b), *172*, 372, 376 (76), *407*
Bersohn, R. 2 (3), *35*
Berson, A. 159 (148), *172*
Berson, J. A. 79, 98 (150), *112*
Bespalov, V. Y. 515 (16), *598*
Best, J. R. 587 (338), *608*
Best, T. V. F. 367, 375 (45), *407*
Bethe, H. A. 368, 371 (47), *407*
Bethell, D. 41, 87, 89 (2a), *107*
Beugelmans, R. 303, 304 (198), *322*
Berilacqua, E. M. 416 (26), 477 (224), 478 (235–237, 241, 242), 480 (244), 481 (236, 237, 241, 242, 245, 247), 482 (249, 250), 483 (253), 484 (253, 254), 485 (26), *501, 507, 508*
Bhacca, N. S. 415, 416, 459 (18), *500*
Bhide, B. V. 558 (191), *604*
Bianchi, M. 243, 246 (172), *261*
Biddison, P. H. 415 (21), *501*
Bieber, J. B. 163 (159), *173*
Biellmann, J. F. 232, 233 (89, 91), 234 (100), 236 (89), *258, 259*
Biemann, K. 348 (37), *358*
Bigam, G. 296 (130a), *321*

Billig, F.   132 (65), *170*
Binder, J. L.   415 (15, 16, 20, 21), 417 (38), 446 (125), *500, 501, 504*
Binkley, R. W.   296 (132), *321*
Birch, A. J.   199 (74–76), 200 (75), 203 (90), 205 (96), *213*, 232 (90, 92), 233 (90, 92, 94, 95), 234 (247), 236 (95), *258, 264*
Bird, C. W.   247 (189), 255 (234b), *262, 264*, 302 (173), *322*
Birkhoff, R. D.   385, 391 (135), *409*
Birladeamu, L.   561 (208), *604*
Bishop, W. A.   452–454 (154), *504*
Biswas, H. B.   524(x)
Black, D. K.   157 (141), 158, 160 (144), *172*
Bladon, P.   304 (200), *322*
Blair, J. M.   289 (87), *319*
Blake, J.   311 (242), *324*
Blomstrom, D. C.   162 (155), *173*
Bloomfield, J. J.   567 (227–230), *605*
Blum, J.   233 (97), *258*
Blume, H.   277 (25a), *317*
Blümel, H.   452 (158), *505*
Bly, R. S.   96 (192), *113*
Bock, E.   24 (90), *37*
Boecke, R.   255 (233b), *264*
Boer, F. P.   348 (43), *358*
Boerboom, A. J. H.   373 (83), *408*
Boeseken, J.   590 (350), *609*
Böeseken, J.   550 (172), *603*
Bogdanov, V. S.   360 (4), *405*
Bogdanović, B.   216 (17), 219 (29), 221 (34), 222 (34, 37), 248 (17, 196), 249 (196), 250 (17), *256, 257, 262*
Bohlmann, F.   126 (43), *169*, 437 (98), *503*
Boikers, R. S.   282 (51c), *318*
Boikess, R.   (166), *603*
Boikess, R. S.   132, 133 (66), *170*
Boldyreva, I. I.   460 (187, 188), *506*
Bolland, J. L.   464 (208), 477 (230), 478, 481, 482 (232), *506, 507*
Bollinger, J.   58, 59, 64 (76), *109*
Bolon, D. A.   84, 93, 94 (166), *112*
Bolton, R.   580 (286c), *607*
Bond, F. T.   299 (155), *321*
Bond, G. C.   190, 192 (49), 193 (54a), *212*, 216 (12), 225 (63), 230 (63, 74), 234 (248), 244 (63), *256–258, 264*, 589 (346a, 348, 349, 351), 590 (348, 349, 351), *608, 609*
Bone, W. A.   577 (278), 587 (336), *606, 608*
Bönnemann, H.   222 (37), *257*

Boord, C. E.   548 (162, 163, 164), 590 (162, 354, 355, 360), 591 (164, 355), *603, 609*
Booth, B. I.   256 (239b), *264*
Booth, G. E.   35 (127), *38*
Booz, J.   366 (35), *406*
Bopp, R. J.   46 (21), *107*
Borcic, S.   85 (167), *112*
Borden, G. W.   284 (69), 285 (75a), 296 (138), *318, 319, 321*
Borg, A. P. ter   281 (46), 282 (51a), 284 (51a, 67c, 70), *318*
Borisov, A. E.   54 (48, 56), *108*, 163 (161), *173*
Borkman, R. F.   119 (17), *168*
Borkowski, M.   565 (224), *605*
Bortner, T. E.   376 (110), *408*
Bory, S.   343, 344 (33), *358*
Bos, H. J. T.   22 (86), *37*
Bothe, H. K.   366, 367 (37), *406*
Bothner-By, A. A.   7 (40), 13, 14 (56), 26 (96–98, 100), 27 (96), 28 (96–98), *36, 37*
Bott, R. W.   45 (15), *107*
Bouchaudon, J.   130 (51), *169*
Bovey, F. A.   475, 487 (216), *507*
Bowers, M. T.   389–391, 393 (147), *409*
Bowers, V. A.   48 (26), *107*
Boyce, C. B. C.   204 (92), *213*
Boyce, W. F.   586 (329), *608*
Boyd, D. B.   64 (94), *110*
Boyer, R.   579 (286a), *606*
Braden, M.   485 (264), *508*
Bradshaw, J. S.   120 (19), *168*, 274, 276, 277 (17), 311 (241, 243), *317, 324*
Bragole, R. A.   162 (156), *173*
Braman, B. A.   546 (154), *602*
Brandrup, J.   433 (89), *502*
Brash, J. L.   120 (23), *168*
Brauman, J. I.   68 (111), *110*, 332, 348, 350, 353 (11), *357*
Bream, J. B.   178 (23), *211*
Breckoff, W. E.   163, 164 (163), *173*
Breitbeil, F. W.   579 (286), *606*
Breitner, E.   196 (60), *212*
Bremner, J. B.   268, 290 (1d), *316*
Brenner, W.   252 (222a, 222b), 255 (232b), *263, 264*
Breslow, D.   225, 230, 244 (56), *257*
Breslow, D. S.   202 (88), *213*, 236 (135), 245, 246 (183), *260, 261*, 421 (54), *501*
Breslow, R.   58 (74a), 59, 60 (77), *109*

Bretton, R. H.   399 (183), *410*
Breuer, E.   585 (315), *607*
Brewer, J. P. N.   296 (133), *321*
Brewis, S.   247 (191a, 191b), *262*
Brey, W. S., Jr.   28 (101), *37*
Brion, C. E.   (81), *407*
Bristow, G. M.   491 (278), *508*
Broadbent, H. S.   196 (63), *212*
Broaddus, C. D.   67 (109, 110), *110*, 141 (87), 144 (93), *170*, *171*
Brockway, L. O.   523 (36), *599*
Brodee, H. J.   178 (22), *211*
Brookhart, M.   100 (204, 206), 101 (206), *114*, 546 (152), *602*
Brooks, T. W.   423 (64), *502*
Brown, H. C.   100 (202), *114*, 224 (50), *257*, 553 (183), 555(x), 562, 563 (183), 565 (224), (241), *604*, *605*
Brown, J.   58 (74a), *109*
Brown, J. M.   103 (213), *114*
Brown, M.   300 (160), *321*
Brown, R. F. C.   282 (48), *318*
Brown, T. L.   42 (10), *107*, 539 (129), 553, 554 (178), *602*, *603*
Brown, W. G.   121 (30), *168*, 206 (98), *213*
Browne, M. W.   48 (27), *108*
Brownlee, R. T. C.   596 (397), *610*
Bruce, J.   588, 590 (342), *608*
Bruce, J. M.   268, 290, 291 (2h), *316*
Bruck, P.   (166), *603*
Bruin, K. R. de   579 (286), *606*
Bruzzone, M.   424 (65), *502*
Bryce, W. A.   335, 341 (18), *357*
Bryce-Smith, D.   289 (87, 88), 296 (134), 308 (224), 309 (231), 311 (236, 240, 244, 247, 249), 312 (252), *319*, *321*, *323*, *324*
Buchanan, J. W.   387 (140), *409*
Buchi, G.   296 (135), 300 (158), 308 (221), 309 (230), 311 (135), 312 (250), *321*, *323*, *324*
Buckingham, A. D.   2 (8), 22 (87), *35*, *37*
Buckley, D. J.   450, 452 (143), *504*
Budzikiewicz, H.   327 (4), 332 (11), 337 (21), 348, 350 (11), 353 (11, 40), (34), (38), *357*, *358*
Buizov, B. V.   444 (119), *503*
Buktas, U.   366 (39), *406*
BuLoc, Le.   178 (20), *211*
Bumgardner, C. L.   564 (216), *605*
Bunn, C. W.   416 (24), *501*
Bunnett, J. F.   223 (41), *257*
Burckhardt, V.   (66), *600*

Burdon, J.   104 (216), *114*
Burger, G.   244 (177), *261*
Burhop, E. H. S.   380 (122), *409*
Burkard, P. A.   460 (193), *506*
Burke, J. J.   544, 545 (142), *602*
Burkoth, T. L.   295 (125), *321*
Burlant, W. J.   491 (279), *508*
Burness, D. M.   145 (102), *171*
Burnett, M. G.   231 (82), 236 (125), *258*, *260*
Burreson, B. J.   241 (156), *261*
Burt, C. P.   532, 536 (87), *601*
Burton, D. H. R.   204, 206 (93), *213*
Burton, M.   121 (25), *168*, 366 (31), 375 (103), (176), *406*, *408*, *410*
Burwell, R. L.   191 (50), 195 (56), *212*
Büthe, H.   234 (253), *264*
Butler, G. B.   422 (61–63), 423 (64), *502*
Butterfield, R. O.   202 (85), *213*, 236 (108b, 115b), *259*
Buyle, R.   208 (112), *214*
Byers, G. W.   277 (23), *317*
Bylina, A.   121 (29), *168*
Bystrov, V. F.   35 (125), *38*

Cais, M.   64 (93), *110*, 236 (255), *264*
Calcott, W. S.   426 (74), *502*
Calderazzo, F.   243 (163, 169, 173), *261*
Calderon, N.   256 (241), *264*
Caldwell, R. A.   52, 53 (42, 43), *108*, 593 (383), *610*
Calvert, J.   268, 290 (3a), *316*
Calvert, J. G.   375, 389 (95), *408*
Camaggi, G.   278 (30), *317*
Cambie, R. C.   151 (123), 157 (139), *172*
Cambrini, M.   421 (53), *501*
Camia, M.   452 (157), *505*
Campanelli, M.   125 (39), *169*
Campbell, B. K.   177 (17), *211*
Campbell, G. C.   196 (63), *212*
Campbell, K. N.   177 (17), *211*
Canale, A. J.   417 (33, 34), *501*
Candlin, J. P.   234 (249), *264*
Cannon, G. W.   536, 537, 539 (107), *601*
Canterino, P. J.   434–436 (93), *503*
Capon, B.   77, 85 (145), *112*
Carbonaro, A.   249 (205), *262*, 420 (45), 446 (124), *501*, *504*
Carboni, R. A.   81, 99 (155), *112*, 567 (233), *605*
Cardnell, P. C.   355 (41), *358*

Cargill, R. 306 (212), *323*
Cargill, R. L. 287, 288 (82), 299 (151), 304 (203a, 203b), *319*, *321*, *323*
Carlough, K. H. 297, 299 (141), *321*
Carlson, R. G. 186 (42), *211*
Carmichael, H. H. 366, 367 (34), *406*
Carnahan, J. C. 255 (233b), *264*
Carnighan, R. H. 143 (91a), *170*
Carothers, W. H. 418 (40), *501*
Carpenter, C. 240 (150a), *261*
Carr, E. P. 532, 536 (87), *601*
Carr, M. D. 71 (123), *111*, 136 (80), *170*, 278 (34), *317*
Carroll, B. L. 526 (63a), *600*
Carroll, R. D. 278, 304 (29), *317*
Carter, A. S. 426 (74), *502*
Caserio, M. C. 219 (25), *256*
Casey, C. 46 (22), *107*
Casey, C. P. 164 (164), *173*
Cassar, L. 216, 243–245 (14), *256*
Cassata, A. 249 (206a), *262*, 421 (53), *501*
Cassuto, A. 60 (79), *109*
Castellano, S. 26, 28 (98), *37*
Castelli, R. 250 (220), *263*
Castro, C. E. 208 (105), *213*
Caswell, L. R. 183 (37), *211*
Catchpole, A. G. 137, 138 (82), *170*
Caubere, P. 21 (75), *37*
Cavestri, R. 559 (198), *604*
Cawley, J. D. 145 (101), *171*
Cawley, S. 12 (52), 22 (84), *36*, *37*
Ceresa, R. J. 426 (73), 491 (73, 280), *502*, *508*
Cesari, M. 421 (51), *501*
Ceselli, C. A. 452 (157), *505*
Chakravarty, S. N. 416 (25), *501*
Chalk, A. J. 166 (170), 167 (172), *173*, 225, 226 (52, 53), 228 (52), 230 (52, 53), 243 (162), 244 (52, 53), *257*, *261*
Chambers, V. C. 550, 555–558 (170), *603*
Chang, H. W. 59, 60 (77), *109*
Chapiro, A. 360 (5), 399 (182), *406*, *410*, 412, 475, 487 (5), *500*
Chapman, G. V. 31 (104), *37*
Chapman, O. L. 268 (1b, 2i), 284 (69), 285 (75a, 75b), 290 (1b, 2i), 291 (2i), 296 (138), (49), *316*, *318*, *319*, *321*
Charlesby, A. 412, 475, 487 (4), *500*
Charlton, J. L. 290 (98), 299 (153), *320*, *321*
Chartoff, R. B. 427 (77), *502*

Charton, M. 531 (84, 85), 539 (123), 551 (123, 176), 555(x), 567 (125), 568 (123, 125), 569 (125), (236, 237), *600–603*, *605*
Chatt, J. 176 (6), *210*, 217 (20), 240 (154), *256*, *261*
Chauvin, Y. 148 (112), *171*, 230 (76a, 76b), 248, 249 (197, 202a), *258*, *262*
Chavalovsky, V. (130), *602*
Chemical Society 524(x)
Chen, A. 51, 105 (35), *108*, 593 (381), *610*
Chen, C. T. 403 (191), *410*
Chen, E. 381, 397 (126), *409*
Chen, H. Y. 256 (241), *264*, 415 (17), 416 (23), 417 (17), 424 (68), *500–502*
Chen, J. C. Y. 373, 380 (90), *408*
Chen, S. C. 303 (195), *322*
Cheritat, R. 492 (286), 498, 499 (299), *509*
Cherkashina, L. G. 541–543 (138), 588 (346), *602*, *608*
Chernova, J. D. 460 (188), *506*
Chesick, J. P. 312 (253), *324*
Chesnut, D. B. 523, 526 (50), *599*
Chibrikin, V. M. 547 (157), *603*
Chieffi, G. 308 (220), *323*
Chien, J. C. W. (38), *257*
Chin, C. G. 295 (130), *321*
Chiusoli, G. P. 216 (13, 14), 243 (14), 244 (13, 14), 245 (14, 182), *256*, *261*
Chivers, T. 51, 105 (35), *108*, 593 (381), *610*
Chodsmith, M. 550 (174), *603*
Chollar, B. 282 (52), *318*
Chollar, D. 132, 133 (67), *170*
Chopard-dit-Jean, L. 130 (52), *169*
Chow, Y. L. 303 (194, 195), *322*
Christ, H. 239 (143), *260*
Christophorou, L. G. 380, 381, 397 (123), *409*
Chumaevskii, N. A. 54 (48), *108*
Chupka, W. A. 372 (78), *407*
Ciamician, G. 299, 300 (157), *321*
Ciampelli, F. 420 (44–46), 421 (44), *501*
Ciganek, E. 544 (141), *602*
Cignarella, G. 144 (95), *171*
Cipriani, R. A. 564 (220), *605*
Clafferton, E. T. 151 (122), *171*
Clark, B. C., Jr. 94 (186), *113*
Clarke, E. M. 372 (80), *407*

Clarke, J. R. P. 71 (123), *111*
Clayton, R. B. 93 (184), *113*
Clement, W. A. 564 (215), *605*
Clement, W. H. 237 (139), *260*
Cleveland, J. D. 553, 562, 563 (183), *604*
Closs, G. L. 526 (67, 68), *600*
Closson, W. D. 47 (24), 86, 87 (168), 88 (174), 94 (168), *107, 112, 113*
Coates, G. E. 41 (5, 8), *107*
Coates, R. M. 287, 288 (82), 290 (92), 308, 310 (225), *319, 323*
Coburn, J. F., Jr. 521 (31), *599*
Cochran, E. L. 48 (26), *107*
Cocker, W. 186 (41), 187(x), *211*
Coffey, R. S. 225, 230 (61), 236 (123), 244 (61), *257, 260*
Cogdell, T. J. 86, 87, 94 (168), *112*
Cohen, D. 157 (137, 140), *172*
Cole, T. W. 300 (163), *322*
Colinese, D. L. 255 (234b), *264*
Collin, G. 228 (66), *257*
Collin, J. 361(x)
Collins, J. H. 361(x)
Colombo, A. 250 (213), *263*
Colon, C. 303 (195), *322*
Colpa, J. P. 57 (68), *109*
Cometti, G. 245 (182), *261*
Comisarow, M. B. 62(x), 562 (213, 214), *604*
Compton, R. N. 380, 381, 397 (123), *409*
Conant, J. B. 592 (377), *609*
Connolly, P. J. 231 (82), *258*
Conrow, R. B. 203 (91), *213*
Conti, F. 230 (75), 232 (243), *258, 264*
Cook, R. E. 458 (183), *505*
Cooks, R. G. 335, 342 (23), *357*
Cookson, R. C. 134 (76a), 135 (76b), *170*, 247 (189), 255 (234b), *262, 264*, 282 (47, 48), 300 (162, 166), 302 (173), *318, 322*, 584 (311), *607*
Cooley, J. H. 105 (224), *114*
Cooper, M. M. 304 (203a), *323*
Cooper, W. 413 (9), 476 (221), *500, 507*
Cope, A. C. 94 (189), *113*, 159 (150), *172*
Corey, E. J. 129 (50), *169*, 208 (108, 113, 114), 209, 210 (114), *214*, 278 (35), 279 (97), 285 (77), 290 (97), 300 (161), 304 (201, 202, 203c), 305 (205), *317, 319, 322, 323*
Corn, J. 306 (212), *323*
Corner, E. S. 592 (376), *609*

Corradini, P. 420, 421 (44), 426, 437 (75), *501, 502*
Corson, B. B. 571, 572 (253), *606*
Cort, L. A. 178 (20), *211*
Cortes, L. 291 (104), *320*
Cossee, P. 223 (42, 43), 239, 245 (148a, 148b), *260*
Cotton, F. A. 243 (169, 173), *261*
Cottrell, T. L. 338 (25), *357*
Coulson, C. A. 50 (33), *108*, 275 (21), *317*, 515 (17, 18), 516 (18, 20), *598*
Coulson, D. R. 302 (179), *322*
Counsell, R. C. 120 (19), *168*, 274, 276, 277 (17), *317*
Courtot, P. 290 (94), *319*
Covell, A. N. 131 (59), *169*
Covell, J. 457 (173), *505*
Cowan, D. O. 120 (19), *168*, 274, 276, 277 (17), *317*
Cowan, J. C. 231 (79d), *258*
Cowherd, F. G. 229 (71b), *258*
Cox, A. 307 (215), *323*
Cozort, J. 184 (39), 183–186(x), *211*
Craft, L. 309 (232), *324*
Craig, J. C. 145 (99), *171*
Craig, P. J. 243 (164), *261*
Craig, R. A. 590, 591 (355), *609*
Cram, D. J. 41 (3), 52 (40), 53 (45), 55 (59), 56 (45, 60), 68 (113), 70 (119–122), 71 (120–122), 72 (45, 113, 126), 77 (113, 120, 121), *107–109, 111*, 122, 123 (35), 136 (81), 138 (83–85), 139 (84, 85), 141 (85), *168, 170*, 592 (379), 594 (379, 387), *609, 610*
Cramer, R. 166 (170), *173*, 220 (30), 225 (51), 247–249 (195), 253, 254 (229), *257, 262, 263*
Cramer, R. D. 236 (129), *260*
Crandall, J. K. 86 (170), 91 (170, 181), 93, 94 (170), *112, 113*, 275 (18b), 277 (24), *317*
Crawford, J. W. 304 (203b), *323*
Crawford, R. J. 83–85 (164), *112*
Crespi, G. 424 (69), *502*
Cresson, P. 157 (142), *172*
Criegee, R. 291 (108), 306 (211), 307 (216), *320, 323*, 486 (265), *508*, 583 (299), *607*
Cristol, S. J. 67 (107), *110*, 291 (102), *320*
Crombie, L. 132 (61), *169*, 189, 190 (x, 48), *212*
Cromwell, N. A. 533 (93), *601*

Cromwell, N. H.   536, 537, 539 (102), *601*
Crowley, K. J.   133 (72), 146 (105), *170*, *171*, 279 (117), 283 (53, 56, 57, 60), 286, 287, 289 (53), 292 (116, 117), 294 (116, 117, 119), 299 (150), *318*, *320*, *321*
Crump, J. W.   54, 55 (46), *108*
Crundwell, E.   300 (166), 302 (173), *322*
Cseh, G.   48 (25), *107*
Cucinella, S.   421 (51), *501*
Cudby, M. E. A.   452 (150), *504*
Cundall, R. B.   119 (16, 17), 120 (23), *168*, 374 (93), *408*, 449 (130), 450 (145), *504*
Cunneen, J. I.   415 (14), 416 (27), 454 (14, 171), 455 (14), 458 (14, 27, 179–182), 459 (171), 460, 471 (27), 477, 478, 480, 482 (226), *500*, *501*, *505*, *507*
Cupery, M. E.   461 (202), *506*
Curtin, D. Y.   54 (46, 57), 55 (46, 58), *108*
Curtis, O. E.   573 (259), *606*
Cuscia, C. J.   203 (91), *213*
Cuts, H.   301 (170), *322*
Cvetanovic, R. J.   121 (27), *168*, 220 (32), *257*, 375, 398 (101), *408*, 457 (174), *505*

Dailey, B. P.   7 (41), *36*, 523 (45, 46), *599*
Dainton, F. S.   458 (183), *505*
Dalby, L. J.   277 (23), *317*
Dale, J.   118 (10), 121 (32), *168*, 278 (35), *317*
Dall'Agata, G.   249 (205), *262*
Dall'Asta, G.   428, 429 (79), *502*
Dalton, C.   120 (19), *168*, 274, 276, 277 (17), *317*
Dalton, J. C.   310 (235), *324*
Damewood, J. R.   304 (203a), *323*
Damico, R.   229 (71a), *258*
Dancanson, L. A.   176 (6), *210*
Daniels, F.   159 (150), *172*
Danjard, J. C.   496 (297), *509*
Danon, J.   443, 451, 457 (111), 458 (177), 476 (111, 177), *503*, *505*
Danti, A.   21 (74), *37*
Danyluk, S. S.   12 (52), 22 (84), *36*, *37*
Darling, S. D.   205 (94), *213*
Das, T. P.   2 (3), *35*
Dauben, W. G.   268 (2b), 274, 281 (15), 285 (72, 75b), 287 (82), 288 (15, 82, 83), 290 (2b, 92), 291 (2b), 292 (113), 293 (118), 299 (15, 151), 301 (171), 303, 304 (15), 308, 310 (225), *316*, *317*, *319–323*, 537 (112), *601*
Davidson, D.   582 (297), *607*
Davidson, J. A.   (225), *114*
Davies, A. G.   422 (60), *502*
Davies, N. R.   164, 165 (165), 166 (171), *173*, 230 (73), *258*
Davis, J. C., Jr.   33 (117), *38*
Davis, M.   571 (246c), *605*
Davis, S. B.   177 (19), *211*
Davison, V. L.   202 (85, 86), *213*, 236 (108b, 108c, 115a), *259*
Davydov, B. A.   432 (88), *502*
Dawes, D. H.   396 (172), *410*
Dawson, H. G.   415 (21), *501*
De Boer, C. D.   291 (101b), *320*
Degani, C.   122 (33a), *168*
Dehm, H. C.   (38), *257*
De la Mare, P. B. D.   64 (92), *110*, 580 (286c), (273a), *606*, *607*
De Mare, G. R.   526 (60), *600*
De More, Wl. B.   375, 398, 399 (99), *408*
Demyanov, N. Y.   575, 577, 578 (270), *606*
Denney, D. B.   555, 556 (185), *604*
Denny, D. B.   151 (120), *171*
Denny, D. Z.   151 (120), *171*
Deno, N.   556, 563 (187), *604*
Deno, N. C.   58 (75, 76), 59 (76), 60 (75, 78), 62 (84), 63 (78, 88), 64 (76, 84, 94), 65 (95, 97, 103), 66 (103, 105), 67 (105, 108), 89, 100 (78), *109*, *110*, 527 (73a), 538, 553 (119), 560 (73a), 562 (119, 210), 571, 572, 575 (249), *600*, *601*, *604*, *606*
De Puy, C. H.   67 (107), 98 (198), *110*, *113*, 558 (194), 579 (286), 597 (404), 598 (405), *604*, *606*, *610*
Derfer, J. M.   548 (162, 163, 164), 590 (162, 354, 355, 360), 591 (164, 355), *603*, *609*
Dessy, R. E.   51, 105 (35), *108*, 593 (381), *610*
Dewar, M. J. S.   217 (19), *256*, 281 (40), *317*, 518, 521 (29b), (238, 239), *599*, *605*
Dewey, R. S.   208 (110a, 110b, 111), 210 (110b), *214*
Dewhurst, H. A.   367 (42), *406*
Dewhurst, K. C.   249 (209), *263*

DeWitt, E. J. 461–463 (196), 500 (301), *506*, *509*
De Wolfe, R. H. 63 (89, 90), 73, 74 (131), *110*, *111*
Diana, G. 253 (231), *263*
D'Ianni, J. D. 416 (28), *501*
Diaz, A. 100 (204), *114*
Diehl, H. 256 (239a), *264*
Diehl, P. 22 (88), *37*
Diercksen, G. 595 (394), *610*
Dietrich, H. 221 (35), *257*
Dietrich, M. W. 18 (66), *36*
Dilling, W. L. 268, 290, 297 (2j), *316*
Dine, G. W. van 67 (107), *110*
Ding, J. 582–584 (307), *607*
Dinh-Nguyen, N. 342 (30), 343 (30, 32), *358*
Dinne, E. 279, 290, 296 (93), *319*
Dinsmore, R. P. 439 (100), *503*
Di Paquo, V. J. 300 (165), *322*
Dirstine, P. H. 585 (334), *608*
Ditmer, D. C. 149 (114), *171*
Djerassi, C. 199 (77), *213*, 233 (93, 251), *258*, *264*, 327 (4), 332 (11), 335 (19), 337 (21), 348, 350 (11), 351 (19), 353 (40), 353 (11), (34), (38), *357*, *358*
Dmuchovsky, B. 179, 180 (25), 184(x), *211*
Dobson, N. A. 195 (55), *212*
Dodson, R. M. 457 (173), *505*
Doepker, R. 362, 374 (22), *406*
Doepker, R. D. 402 (188, 189), *410*
Doering, E. von E. 162 (156), *173*
Doering, W. E. 177 (19), *211*
Doering, W. von E. 160 (152, 154a), *172*, 284 (67b), 295 (123a, 123b), *318*, *320*, 521 (31), *599*
Dolgoplosk, B. A. 452 (152), 454 (167), 460 (186–189), *505*, *506*
Donati, M. 230 (75), *258*
Donohue, J. 523 (42), *599*
Doran, M. (119a), *601*
Doran, M. A. 52 (38), 73 (127, 128), *108*, *111*, 143 (91b), *170*
Doree, C. 203 (89), *213*
Dorfman, L. M. 360 (12), 363, 365 (25), 367 (12), *406*
Dorp, D. A. van 131 (57, 58), *169*
Dorris, T. B. 572, 573, 575 (264), *606*
Dougherty, R. C. 348 (42), *358*
Douglas, A. W. 6, 8 (34), *36*
Douglass, D. C. 100 (205), *114*
Downing, F. B. 426 (74), *502*

Doyle, J. R. 240 (150a, 150b, 151a), *261*
Doyle, L. C. 457 (174), *505*
Doyle, M. P. 86 (169), *112*
Drefahl, G. 437 (99), *503*
Dreiding, A. S. 282 (51b), *318*
Drenth, W. 45 (17–19), *107*
Druckrey, E. 283 (55), 284 (66), 291 (104), *318*, *320*
Duck, E. W. 236 (256), *265*
Duddey, J. E. 46 (20), *107*
Dudley, E. A. 425 (70), *502*, 567 (231), *605*
Duffey, G. H. (5), *598*
Duncan, F. J. 220 (32), *257*
Duncan, P. M. 484 (260), *508*
Duncanson, L. A. 217 (20), *256*
Dunkel, M. 183(x), 184(x, 39), 185, 186(x), *211*
Dunn, G. L. 300 (165), *322*
Dunn, J. R. 478 (240), *507*
DuPont, Y. 192 (52), *212*
Durham, L. 306 (210), *323*
Durham, L. J. 33 (116), *38*
Dürr, H. G. 283 (64), *318*
Dutton, H. J. 231 (79d), 236 (108a, 115a), *258*, *259*
Dyatkina, M. E. 515 (15), *598*
Dzantiev, B. G. 384 (133), *409*

Eaborn, C. 45 (15), *107*
Eakin, M. A. 88 (177), *113*
Eastman, R. H. 533, 535 (94), 536 (100, 101), 557 (190a), *601*, *604*
Eaton, D. C. 178 (23), *211*
Eaton, D. R. 234 (245), *264*
Eaton, P. E. 300 (163), 304 (203e), 305 (204, 206), 313 (263), *322–324*
Eaves, D. E. 476 (221), *507*
Eberbach, W. 292 (109), *320*
Eberhardt, G. G. 225, 230, 236, 244 (62), *257*
Eberlin, E. C. 537 (117), *601*
Eberson, L. 562 (211), *604*
Ebert, H. G. 366 (35), *406*
Eckell, A. 279, 292 (117), 294 (117, 120), *320*
Edman, J. R. 291 (103), *320*
Edwards, A. G. 156 (135), *172*
Edwards, J. A. 306 (209, 210), *323*
Eggensperger, H. 584 (309), *607*
Egger, K. W. 116 (1b, 3), 117 (3, 5, 6), 135 (78), 136 (80), *167*, *170*, 282 (51d), *318*
Eggers, D. F. 523 (56), *600*

Eglinton, G. 195 (55), *212*
Eigen, M. 52 (41), *108*
Eisch, J. J. 10, 11 (51), *36*, 151 (119), *171*
Eisen, O. 32 (108), *38*
Eiter, K. 153 (127), *172*
Ela, S. W. 70 (119–121), 71, 77 (120, 121), *111*, 138, 139 (84, 85), 141 (85), *170*
Elad, D. 303 (192), *322*
Eley, D. E. 176 (11), *211*
Eliel, E. L. 105 (222), *114*, 525 (59), *600*
Elix, J. A. 285 (76), *319*
El Komoss, S. G. 381 (125), *409*
Elleman, D. D. 389–391, 393 (147), *409*
Ellis, R. J. 132 (68), 136 (79), *170*
Elphimoff-Felkin, I. 144 (94), *171*
Eltschaninoff, E. 558 (192), *604*
Emerson, G. F. 216 (5), *256*
Emken, E. A. 201 (82), 202 (85, 86), *213*, 236 (108b, 108c, 115a, 115b), *259*
Endres, H. A. 445 (120), *503*
Engel, R. G. 484 (255), *508*
English, E. S. 480 (244), 481 (247), 482 (249), *507*
Ergun, S. 246 (188), *262*
Erickson, K. 279, 292 (117), 294 (117, 120), *320*
Erickson, R. E. 486 (267), *508*
Ermakova, I. I. 454 (167), 460 (189), *505, 506*
Erman, W. F. 281 (42, 43), *318*
Ershov, V. V. 130 (56), *169*, (341), *608*
Eschenmoser, A. 92 (182), *113*, 158 (145), *172*
Espy, H. H. 88 (175), *113*
Evans, D. 201 (80, 81), *213*, 236 (102, 104, 258), *259*, *265*
Evans, D. F. 121 (28), *168*, 270, 299 (5), *316*, 373 (91), *408*
Evans, M. B. 457, 458 (176), *505*
Evans, T. R. (27), *317*
Evnin, A. B. 291 (103), *320*
Eyring, H. 360, 369 (3), 373, 379 (88), *405, 408*

Fagherazzi, G. 249, 250 (206b), *262*
Falbe, S. 247 (192), *262*
Falconer, W. E. 220 (32), *257*
Fano, V. 385 (136), *409*
Fanta, G. F. 586 (335), *608*

Farkas, A. 175 (1), 176 (1, 9), 177 (16), *210, 211*
Farkas, L. 175 (1), 176 (1, 9), 177 (16), *210, 211*
Farmer, E. H. 464 (207), 465 (207, 209), 471 (215), 478 (234), *506, 507*
Favini, G. 161 (154c), *172*
Fayadh, J. M. 368, 396 (52), *407*
Fedorova, A. V. 49 (28), *108*
Fedulova, V. V. 130 (55), *169*
Feldblyum, N. S. H. 248, 249 (199), *262*
Feldman, J. 246 (184), *262*, 582 (297), *607*
Felix, D. 92 (182), *113*, 158 (145), *172*
Fell, B. 165 (166), 167 (174), *173*, 225 (54, 58), 228 (66–69), 229 (69), 230 (58, 69), 244 (54, 58), 246 (68), *257*, 278 (33), *317*
Fellenberger, K. 67 (107), *110*, 163 (159), *173*
Feneslau, C. 337 (21), *357*
Fenical, W. 295 (126), *321*
Ferguson, R. G. 418, 419 (42), *501*
Ferri, C. 449 (133, 134), *504*
Ferris, R. C. 535 (97), *601*
Fessenden, R. W. 48 (26), *107*, 360 (14), 365 (14, 30), 394, 404 (14), *406*
Fetizon, M. 343, 344 (33), *358*
Fettes, E. M. 412 (1), 432 (87), 487 (1), *500, 502*
Field, F. H. 44(x) 337, 340 (22), *357*, 360 (8), 369 (8, 63), 390 (63, 152), 391, 393, 394, 400 (63), *406, 407, 409*, 530 (77, 78), *600*
Field, J. E. 424 (68), 439 (100), *502, 503*
Fieser, L. F. 98 (197), *113*
Fieser, M. 98 (197), *113*
Filipov, O. 570 (243), *605*
Fineman, M. A. 372 (80), *407*
Fineman, M. Z. 586 (321), *608*
Firestone, R. F. 396 (174), *410*
Fisch, M. 303 (193), *322*
Fischer, A. 313, 314 (264), *324*
Fischer, E. 121 (26), *168*, 291 (100), *320*
Fischer, E. O. 244 (177), *261*
Fischer, H. 68 (112), *110*
Fischer, L. P. 248, 249 (201), *262*
Fisher, H. 49 (28), *108*
Fisher, L. P. 525, 526 (64), *600*
Fishman, M. 304 (203f), *323*
Fisk, G. A. 396 (173), *410*

Author Index

Fittig, R. 586 (330), *608*
Fitzpatrick, J. D. 453, 454 (166), *505*
Flanagan, P. W. K. 537 (116), *601*
Flautt, T. J. 67 (109), *110*
Fletcher, F. J. 119 (16), *168*
Fletcher, W. P. 458 (180), *505*
Floy, F. 208 (112), *214*
Flygare, W. H. 517 (22), 520 (26a), 523, 550 (48), *598, 599*
Flynn, J. H. 487 (272), *508*
Foffani, A. 125 (39), *169*
Folkers, K. 459 (184), *505*
Fonken, G. J. 268 (2a), 281 (45), 285 (72), 290, 291 (2a), 296 (45), 303 (2a), *316, 318, 319*
Fontanelli, R. 118 (11), *168*
Foote, R. S. (258), *324*
Forbes, W. G. 484 (260), *508*
Forman, A. 4 (21), *35*
Forney, L. S. 582, 583 (298, 305), *607*
Forsén, S. 66 (105a), *110*
Forster, M. J. 415 (21), *501*
Fosselius, G. A. 255 (235b), *264*
Foster, E. G. 159 (150), *172*
Fotis, P., Jr. 236 (111), *259*
Fraenkel, G. 54, 74 (53), *108*
Fraga, D. W. 416 (22), *501*
Franchimont, E. 118 (9, 10), *168*
Frank, F. 162 (158), *173*
Franke, W. 250 (218), *263*
Frankel, E. N. 201 (82), 202 (85, 86), *213*, 236 (108a–c, 115a, 115b, 225), *259, 264*
Frankevich, Ye. L. 338 (25), *357*
Frankiss, S. G. 21 (81), *37*
Franklin, J. L. 44(x), 337, 340 (22), *357*, 360, 369 (8), *406*
Franklin, J. S. 62 (85), *109*
Franzus, B. 34 (119), *38*, 207 (99), 210 (117, 118), *213, 214*
Fraser, F. M. (29d), *599*
Frater, G. 302 (182), *322*
Fray, G. I. 309 (231), *323*
Freed, S. (244), *605*
Freedman, H. H. 123 (36), *168*
Freeman, G. R. 368 (51, 52, 58), 387 (58, 141), 396 (51, 52, 58), 397 (58), 400 (141), 401 (58), *407, 409*
Freeman, P. K. 292 (109), *320*, 585 (317, 318), *608*
Freeman, R. R. 22 (88), *37*
Freeman, S. K. 536 (101), *601*
Frei, K. 545 (146), *602*
Freidlin, L. K. 591 (367), *609*
Freund, A. 571, 572 (247), *605*

Freund, G. 307 (216), *323*, (66), *600*
Frey, A. J. 92 (182), *113*
Frey, H. M. 132 (68), 133 (71), 136 (79), 154 (129), 160 (153), *170*, *172*, 219 (24), *256*
Freyschlag, H. 128 (48), 153 (125), *169, 172*
Friebolin, H. 32 (109), *38*
Fried, J. H. 306 (208–210), *323*
Friedel, R. 592 (378), *609*
Friedman, N. 58, 59 (76), 62 (84), 64 (76, 84), 67 (108), *109, 110*
Friedrich, E. C. 132 (69), *170*
Friend, J. P. 523 (45, 46), *599*
Fritchie, C. J., Jr. 521, 523, 524 (54), *599*
Fritsch, F. N. 523 (33), *599*
Fromandi, G. 442 (110), *503*
Frost, A. A. 52 (40), *108*
Fruit, R. E. 527, 560 (73a), *600*
Fruit, R. E., Jr. 65, 66 (103), *110*
Frye, H. 252 (223), *263*
Fuchs, R. 567 (227–230, 235), *605*
Fuchs, V. R. 369, 395 (60), *407*
Fucks, B. 188 (46), *212*
Fueki, K. 395 (168), *410*
Fujii, T. 236 (109a, 109b), *259*
Fujimaki, T. 236 (109c), *259*
Fujimoto, H. 155 (134), *172*
Fujinami, K. 439 (102), *503*
Fujita, T. 550 (174a), *603*
Fujomoto, M. 281 (41a), *318*
Fukui, K. 155 (134), *172*, 281 (41a), *318*
Fukumoto, T. 455 (172), *505*
Fukushima, D. K. 176, 191 (4), *210*
Funasaka, W. 67 (107), *110*
Fuqua, S. A. 415, 416, 459 (18), *500*
Furkas, J. 553 (181), *603*
Furrer, H. 306 (211), 307 (216), *323*
Furstoss, R. 301 (172), *322*
Furukawa, J. 452, 454 (161), *505*
Fuson, R. C. 577, 581 (280), *606*
Fuss, P. G. 307 (219), *323*
Futrell, J. H. 380 (118), 388 (145, 146), 390 (145, 146, 150), 395 (166), 403 (190), *409, 410*

Gagarina, M. I. 236 (132), *260*
Gajewski, J. J. 58 (74a), *109*
Gale, L. H. 303 (184), *322*
Gall, J. S. 481 (247), *507*
Gallagher, T. F. 176, 191 (4), *210*
Gallazzi, M. C. 250 (213), *263*, 420 (47), *501*

Gallo, G. G. 144 (95), *171*
Galnikova, L. G. 541, 542 (131), *602*
Gambitskii, P. A. 541, 542 (136), *602*
Gandini, A. 124 (38), *169*
Ganguly, A. K. 386, 396, 400 (137), *409*
Garbers, C. F. 127 (46), 128 (47), *169*
Gardner, P. D. 304 (199), *322*
Garlanda, T. 452 (157), *505*
Garratt, P. J. 77 (143a), *111*
Garvick, L. V. 338 (25), *357*
Gassman, P. G. 102 (211), *114*, 288 (84), *319*, 517 (26), 538 (121), 539 (124), 597 (401, 402), *599, 601, 602, 610*
Gault, F. G. 176, 181 (7), *211*, 592 (374, 375), *609*
Gaylord, N. G. 443 (116), 447 (116, 129), 491 (282), *503, 504, 508*
Gaywood, S. C. 586 (329), *608*
Geddes, W. C. 431 (84), *502*
Gehman, S. D. 439 (100), 475, 487 (217), *503, 507*
Geïderikh, M. A. 432 (88), *502*
Gembitski, P. A. 541, 542, 580 (135), *602*
Gembitskii, P. A. 541, 542 (137), 550 (168), 553 (182), 559 (197), 580 (291), *602–604, 607*
Genbitski, P. A. 541, 542 (134), *602*
Gent, A. N. 485 (264), *508*
Gerberich, H. R. 148 (108), *171*
Gerbers, C. F. 128, 129 (49), *169*
Gertler, S. 122 (33b), *168*
Gianni, M. H. 15 (61), *36*
Giannini, U. 249 (206a, 206b), 250 (206b), *262*, 421 (53), *501*
Gibbs, W. E. 493, 494 (288), *509*
Gibson, D. H. 597 (404), *610*
Gibson, F. 157 (136a), *172*
Giessner, B. G. 335 (44), *358*
Gilbert, A. 289 (88), 308 (224), 309 (231), 311 (240, 249), *319, 323, 324*
Gilbert, J. C. 160 (152), *172*
Gill, P. S. 403 (191), *410*
Gillard, R. D. 236 (118), *259*
Ginsberg, D. 303 (191), 312 (257), *322, 324*
Giuffre, L. 427 (76), *502*
Givens, R. S. 283 (64), 296 (132), *318, 321*
Gladshtein, B. M. 580 (286d), *607*
Glass, C. A. 236 (255), *264*
Glass, D. S. 132 (66, 67, 69), 133 (66, 67), *170*, 282 (51c), *318*

Glazebrook, R. W. 452, 472 (162), *505*
Gleiter, R. 571 (246a), *605*
Glick, A. H. 309 (222), 310 (235), *323, 324*
Glukhovtsev, V. G. 591 (367), *609*
Gocbel, P. 521 (31), *599*
Godbole, E. W. 369, 389–391, 394 (65), *407*
Goering, H. L. 60 (81), 88 (174, 175), *109, 113*, 154 (131), *172*
Goetz, R. W. 165 (169), *173*, 207 (100), *213*, 236 (114a, 114b), *259*
Gogte, V. N. 134 (76a), *170*, 282 (47), *318*
Gold, E. H. 58, 59, 62, 63 (74), *109*, 303 (191), 312 (257), *322, 324*
Gold, V. 41, 87, 89 (2a), *107*
Golden, D. M. 116 (1b, 3), 117 (3, 5), 135 (78), *167, 170*
Goldish, E. 521 (32), *599*
Goldman, I. M. 300 (158), *321*
Goldman, N. L. 203 (91), *213*
Goldsmith, D. J. 94 (186), *113*
Goldstein, J. H. 4, 5 (25), 5 (31), 6 (25, 31–34), 7 (31, 32, 42–44), 8 (32, 34), 9 (48), 10, 11 (48, 49), 12 (48, 53), 13, 14 (53), 19 (70, 71), 20 (71), 21 (72, 73), 22 (82, 83, 85), 23, 24 (83), 25 (85, 92), 26 (92, 94), 35–37, 54 (55), *108*, 544 (143), *602*
Gollnick, K. 303 (183), *322*
Golub, M. A. 120 (23), *168*, 415 (18, 19), 416 (18), 439 (101, 104), 440 (104, 105), 441 (107, 108), 442 (109), 443 (108, 111, 117), 449 (108, 132, 136, 137), 450 (137, 141, 142, 144), 451 (111, 144, 146, 147), 452 (151, 155), 453 (155), 454 (19, 108, 132), 455 (137, 146, 147), 457 (111, 147), 458 (177), 459 (18, 108, 184, 185), 460 (101, 108, 109), 471 (155), 476 (111, 144, 177), 494 (132), *500, 503, 505, 506*
Goodburn, T. G. 131 (59), *169*
Goodman, A. L. 533, 535 (94), *601*
Goodwin, T. H. 516 (20), *598*
Gorden, R., Jr. 360 (13), 362 (22), 366 (34), 367 (13, 34), 371 (74), 374 (13, 22), 380 (121), 388 (144), 390 (74), 394 (164), 395 (22), *406, 407, 409, 410*
Gordon, A. S. 595 (393), *610*
Gordon, M. 445 (121), *503*
Gordon, R. 527, 547 (71), *600*
Gorman, E. H. 248, 249 (201), *262*

Gorshkov, V. I.   591 (367), *609*
Gosselink, D. W.   76 (142), *111*, 143 (91a), *170*
Gostunskaya, I. V.   195 (59), *212*
Gozzo, F.   278 (30), *317*
Grabowski, Z. R.   121 (29), *168*
Gragson, J. T.   548, 591 (164), *603*
Graham, D. M.   4 (27), *35*, 119 (14), *168*
Graham, W. A. G.   572, 575 (268), *606*
Graham, W. H.   565 (225), *605*
Granat, H. M.   195 (59), *212*
Granger, M. R.   52, 53 (42), *108*, 593 (383), *610*
Grant, D. M.   6 (36), 8 (36, 45), *36*
Grant, W. G.   523 (51), *599*
Grassie, N.   441 (106), 487 (106, 269, 271), *503, 508*
Grassner, H.   128 (48), 153 (125), *169, 172*
Gray, A. P.   536, 541, 556, 557 (98), *601*
Gray, H. B.   223 (40), *257*
Gream, G. E.   79 (149), *112*
Greco, A.   249 (205), *262*, 446 (124), *504*
Greeley, R. H.   298 (147), *321*
Green, C.   537 (114), *601*
Green, M.   240 (153b), 243 (164, 166, 167), *261*
Green, M. L. H.   216, 221 (7), *256*
Greene, R. N.   162 (157), *173*
Greenfield, H.   246 (188), *262*, 592 (378), *609*
Greenhalgh, R. K.   177 (18), *211*
Greenlee, K. W.   548 (162, 163, 164), 590 (162, 354, 355, 360), 591 (164, 355), *603, 609*
Greenspan, F. P.   492 (285), *508*
Gregg, E. G., Jr.   461, 462 (201), *506*
Gregorio, G.   232 (243), *264*
Gregson, T. C.   475, 487 (217), *507*
Griesbaum, K.   49 (28), *108*
Griffen, G. W.   121 (31), *168*
Griffin, G. W.   278 (36), 281 (44), *317, 318*, 457 (173), *505*
Griffith, W. P.   231 (80), *258*
Griffiths, P. A.   119 (17), 120 (23), *168*, 450 (145), *504*
Grigoreva, V. I.   575, 590 (271), *606*
Grimme, W.   279, 290, 296 (93), *319*
Grimwood, B. E.   145 (97), *171*
Grisar, J. M.   94 (189), *113*
Griswold, A. A.   285 (75a), *319*

Grob, C. A.   48 (25), *107*, 534 (95), *601*
Gronowitz, J. S.   233 (244), *264*
Gronowitz, S.   3 (20), 4 (24), 15 (57), *35, 36*, 233 (244), *264*
Gross, B.   74 (136), *111*
Gross, F. J.   555, 556 (185), *604*
Grosse, A. V.   571, 572 (250, 254, 256), *606*
Grostic, M. F.   585 (317, 318), *608*
Grovenstein, E.   296 (134), 311 (237), *321, 324*
Grovenstein, E., Jr.   312 (251), *324*
Gruenbaum, Z.   122 (34), *168*
Grüner, H.   291 (108), *320*
Grunewald, G. L.   296 (131), *321*
Grunwald, E.   60 (83), *109*
Grushko, I. E.   193 (54), *212*
Guarino, J. P.   368, 387, 396 (56), *407*
Gube, E. J.   523 (57), *600*
Gueniffey, H.   437 (97), *503*
Guenther, H.   32 (110), *38*
Guichard-Loudet, N.   230 (76a), *258*
Guillory, J. P.   523 (52, 53), 524(x), 526 (52, 53, 63a), *599, 600*
Gunning, H. E.   586 (325), *608*
Gunstone, F. D.   131 (60), *169*
Gunthard, H. H.   523 (37), *599*
Gunther, H.   26 (96–98), 27 (96), 28 (96–98), *37*, 526 (63), *600*
Gurudata.   34 (123), *38*
Guseva, L. P.   553 (182), *603*
Gustavson, G. G.   551 (276), 571 (248), 577 (275, 276), 585 (331–333), *605, 606, 608*
Güsten, H.   277 (25a), *317*
Gut, M.   92 (182), *113*, 178 (21, 22), *211*
Gutowsky, H. S.   8 (45), *36*
Gutzwiller, J.   199 (77), *213*
Gwinn, W. D.   523, 550 (48), *599*
Gwynn, B. H.   246 (184), *262*

Haag, W. O.   71 (123), *111*
Haber, R. G.   188 (46), *212*
Haddad, Y. M. Y.   236 (121a), *260*
Hafer, K.   58, 59, 64 (76), *109*
Hafner, W.   236 (136a), 237 (136a, 140), 238 (141), 239, 240 (136a), *260*
Hagemann, H.   292 (111), *320*
Hageveen, H.   15 (62), *36*
Hagihara, N.   236 (105), 249 (210), *259, 263*

Hagmann, D. L. 73, 74 (131), *111*
Hall, W. K. 148 (108), *171*
Hallam, H. E. 25 (91), *37*
Haller, I. 272 (12), 298 (148), *317, 321*
Hallman, P. S. 201 (81), *213*, 236 (102, 254), *259, 264*
Halman, M. 379 (117), *409*
Halpern, J. 196 (65), 197 (66), *212*, 218 (23), 230 (77), 232 (242), 235 (101), *256, 258, 259, 264*
Halpern, W. 183(x), 184 (x, 39), 185, 186(x), *211*
Halsall, T. G. 353 (39), *358*
Ham, N. S. 521 (58), *600*
Hamada, M. 550 (174a), *603*
Hamada, S. 196 (62), *212*
Hamanaka, E. 278 (35), *317*
Hameka, H. F. 2 (4), *35*
Hamill, W. H. 366, 367 (32), 368, 387, 396 (56, 57), *406, 407*
Hamilton, L. 536 (110), *601*
Hamm, R. A. 385, 391 (135), *409*
Hammond, G. S. 41 (1), *107*, 120 (19, 20, 22), *168*, 268 (1a, 4a), 272 (13), 274 (4a, 17), 276 (17, 22), 277 (17, 23), 285 (73), 286 (81), 290 (1a, 4a), 291 (101b), 299 (149), 311 (239), 312 (256), 313 (264, 265), 314 (264–266), 315 (268), *316, 317, 319–321, 324, 325*, 586 (328), *608*
Hammons, J. H. 52, 53 (39), 68 (111), *108, 110*
Hamon, D. P. G. 584 (311), *607*
Hanack, M. 560 (201), 571 (246e), 584 (309), *604, 605, 607*
Hanafusa, T. 561 (208), *604*
Hancock, R. I. 240 (153b), *261*
Handel, C. R. van den 550 (173), *603*
Handler, G. S. 520 (29), *599*
Hank, R. 425 (71), *502*
Hanna, M. W. 32 (107), *37*
Hansen, R. L. 88 (178), *113*
Hanson, H. P. 372 (80), *407*
Happe, J. A. 10, 11 (50), *36*
Hardman, W. M. 311 (239), *324*
Hardwick, T. J. 375 (102), *408*
Hardy, J. 565 (221, 222), 566 (222), *605*
Hargrave, K. R. 478 (233), *507*
Harries, C. D. 460 (190), *506*
Harrington, J. K. 32 (107), *37*
Harrington, R. 494 (289), *509*

Harris, L. 586 (323), *608*
Harrison, A. G. 44(x), 391 (158), 395 (165), *410*, 530 (79), *600*
Harrod, J. F. 166 (170), 167 (172), *173*, 197 (66), *212*, 225, 226 (52, 53), 228 (52), 230 (52, 53), 235 (101), 244 (52, 53), *257, 259*
Harrter, D. R. 124 (37), *168*
Hart, H. 154 (132), *172*, 303 (187), *322*, 539 (129), 560, 561 (206), 564 (220), 572 (261), 573 (259–261), 575 (261), 583 (308), (207), *602, 604–607*
Hart, R. 491 (281), *508*
Hartgenstein, J. H. 292 (111), *320*
Hartlová, E. 445–447 (122), *503*
Hartman, A. 521, 523, 524 (55), *599*
Hartmann, W. 308 (226), *323*
Hass, A. B. 577 (279), *606*
Hassel, O. 523 (34, 35), *599*
Haszeldine, R. H. 243 (166), *261*
Haszeldine, R. N. 256 (239b), *264*
Hata, G. 248, 249 (198), 253, 254 (225a, 225b), *262, 263*
Hatanaka, A. 333 (16), *357*
Hatano, Y. 375 (98), 382 (127), 395 (98, 127), *408, 409*
Hatch, L. F. 567 (229), *605*
Hausser, J. W. 67 (107), *110*
Havinga, E. 290 (90), 299 (152), *319, 321*
Hawkins, W. L. 477 (227), *507*
Hawthorne, M. F. 537 (115), *601*
Hawton, I. D. 117, 118 (7), *167*
Hayano, M. 178 (21, 22), *211*
Hayashi, J. 452, 454 (161), *505*
Haynes, R. 282 (50), *318*
Haynes, R. M. 369, 390, 391, 393, 400, 402 (67), *407*
Hayward, J. C. 399 (183), *410*
Heaney, H. 56 (63), *109*, 296 (133), *321*
Heasley, V. L. 150 (116), *171*
Heathcock, C. H. 283 (63), *318*
Hebiguchi, K. 188 (44), *212*
Heck, R. F. 197 (67), *212*, 216 (9), 225, 230 (56), 236 (117), 244 (56, 178), 245 (178, 183), 246 (183), *256, 257, 259, 261*, 421 (54), *501*
Hedberg, R. 523 (33), *599*
Heilbronner, E. 57, 62 (69), *109*, 523 (57a), *600*

Heimbach, P. 252 (222a–c), 255 (222c, 232a, 232b), 263, 264, 278 (33), 317
Heisig, G. B. 586 (324), 608
Hekkert, G. L. 45 (17, 18), 107
Heller, J. 439 (101, 104), 440 (104, 105), 441 (107), 442 (109), 443 (117), 460 (101, 109), 503
Hellier, M. 225 (63), 230 (63, 74), 244 (63), 257, 258
Hembest, H. B. 178 (23), 211
Hems, M. A. 311 (244), 324
Henbest, H. B. 236 (121a, 121b), 260
Henderson, J. N. 421 (49), 501
Henderson, W. A. 271 (9), 290, 292, 294 (96), 316, 319
Hendrickson, J. B. 93 (184), 113
Hendrix, W. T. 229 (71b), 258
Henmo, E. 290 (99a), 320
Henne, A. L. 487 (273), 508
Henrici-Olivé, G. 216 (15), 256
Henrikson, B. W. 586 (335), 608
Henry, J. P. (258), 324, 436 (94), 503
Henry, P. M. 220 (31), 237 (31, 137), 257, 260
Hentz, R. R. 121 (25), 168, (176), 410
Herkstroeter, G. 276 (22), 317
Herkstroeter, W. G. 120 (20, 22), 168
Herman, J. A. 366, 367 (38), 406
Hermann, R. B. 54 (48), 108, 291 (106), 320
Herschbach, D. R. 26 (99), 37
Herzog, E. G. 300, 301 (167), 322
Hesse, E. 437 (99), 503
Hewett, W. A. 417 (33, 34), 501
Heyer, E. W. 162 (157), 173
Hidai, M. 249 (204a, 204b), 262
Hidri, M. 250 (221b), 263
Higgins, G. M. C. 415 (14), 416 (27), 454 (14, 171), 455 (14), 457 (176), 458 (14, 27, 176), 459 (171), 460, 471 (27), 486 (268), 500, 501, 505, 508
Hikino, H. 305 (207), 323
Hilbers, C. W. 239, 245 (148a, 148b), 260
Hill, E. A. 104, 105 (217), (225), 114
Hill, K. A. 298 (145), 303 (190), 311 (246), 321, 322, 324
Hill, M. 256 (239b), 264
Hill, R. K. 156 (135), 172
Hill, R. R. 300 (166), 322

Hillers, S. 550 (168), 603
Hillman, M. E. D. 224 (49), 257
Hillyard, R. A. 234 (248), 264
Himelstein, N. 196 (61), 212
Hine, J. 218 (21), 256
Hinman, R. L. 309 (222), 323
Hinrichs, H. H. 32 (110), 38
Hiroike, H. 3 (12, 13), 35
Hirokami, S. 121 (27), 168
Hirschfelder, J. O. 360, 369 (3), 405
Hirshfeld, F. L. 521, 523, 524 (55), 599
Hirst, D. M. 57 (70), 109
Hirst, R. 6, 8 (36), 36
Ho, T. C. 452, 484 (148), 504
Hobgood, R. T. 6, 7, 8 (32), 9 (48), 10, 11 (48, 49), 12 (48), 36
Hobgood, R. T., Jr. 19, 20 (71), 37, 54 (55), 108
Hobson, M. C. 148 (111), 171
Hochmuth, U. 250 (218), 263
Hochstein, F. A. 206 (98), 213
Hochstetler, A. R. 278 (30), 317
Hodge, J. D. 58, 59 (76), 62 (84), 63 (88), 64 (76, 84, 94), 67 (108), 109, 110, 562 (210), 604
Hoff, B. M. E. van der 454, 467, 468 (170), 505
Hoffman, A. S. 491 (279), 508
Hoffman, R. J. 3, 4 (16), 35
Hoffmann, R. 43, 57 (12), 57 (71), 66 (106), 72 (71), 82 (159), 107, 109, 110, 112, 132, 133 (66), 155 (134), 170, 172, 255 (237), 264, 275 (20), 279 (37, 38), 280 (38, 39), 281 (39), 317, 518, 519 (27), 529 (27, 75), 599, 600
Hoffmann, R. A. 3 (20), 4 (24), 15 (57), 35, 36
Hofmann, J. E. 68 (114), 111
Hogben, M. G. 301 (170), 322
Hogen-Esch, T. E. 41 (7), 107
Hogeveen, H. 45 (19), 107, 256 (239c, 240), 264
Hogg, A. M. 369, 389–391, 394 (66), 407
Hojo, K. 295 (130), 321
Holden, G. 415 (12), 424 (66), 500, 502
Holloway, C. E. 4, 5 (27), 35
Holroyd, R. A. 360 (14), 365 (14, 30), 394, 404 (14), 406
Holt, J. T. 180, 181 (33), 182(x), 211
Honig, R. E. 361(x)
Honkanen, E. 333 (16), 357

Hoover, J. R. E.   300 (165), 302 (175), *322*
Horák, M.   (130), *602*
Hörhold, H. H.   437 (99), *503*
Horike, A.   67 (107), *110*
Horiuti, J.   175, 176 (2), *210*
Horner, L.   234 (253), *264*
Hörnfelt, A. S.   233 (244), *264*
Horspool, W. M.   302 (181), *322*
Hortmann, A. G.   279, 290 (97), *319*
Hosaka, S.   247 (190a, 190b), *262*
Hostynek, J.   534 (95), *601*
House, H. O.   186 (42), *211*
Houser, J. J.   58, 59 (76), 62 (84), 64 (76, 84), 66 (105), 67 (105, 108), *109*, *110*, 562 (210), *604*
Höver, H.   59, 60 (77), *109*
Howard, T. J.   178 (20), 188 (43), *211*, *212*
Howard, V. K.   234 (248), *264*
Howden, M. E. H.   81 (153), 104 (216), *112*, *114*, 544, 545 (140), 560 (205), *602*, *604*
Howe, R.   103 (212), *114*
Hruska, F.   24 (90), *37*
Hsi, N.   526, 530 (69a), *600*
Hudec, J.   134 (76a), *170*, 247 (189), 255 (234b), *262*, *264*, 282 (47, 48), 300 (162, 166), *318*, *322*, 584 (311), *607*
Hudson, G. V.   533 (93), *601*
Huet, J.   144 (94), *171*
Huffman, K. R.   290, 292, 294 (96), *319*
Hughes, E. D.   60 (82), *109*, 137, 138 (82), *170*
Hughes, H.   478, 481, 482 (232), *507*
Hughes, P. R.   247 (191a, 191b), *262*
Huitric, A. C.   31 (105), *37*
Hummel, A.   362 (20), 368 (50, 53–55), 375 (53), 387, (53–55), 396 (50, 53–55), *406*, *407*
Humphlett, W. J.   145 (101), *171*
Humphrey, G. L.   523 (42), *599*
Humski, K.   85 (167), *112*
Hünig, S.   208 (107), 209 (107, 115), 210 (107), *214*
Hunter, D. H.   55 (59), 70, 71 (122), *109*, *111*, 122, 123 (35), 138 (83), *168*, *170*
Huntsman, W.   582 (296), *607*
Hurst, G. S.   376 (110), 380, 381, 397 (123), *408*, *409*
Hurst, J. J.   281 (42), *318*
Hurst, W. S.   313 (263), *324*

Husbands, J.   236 (121a), *260*
Husk, G. R.   151 (119), *171*
Hussey, A. S.   179 (30, 31), 183–185 (30), *211*, 234 (248), *264*
Hutchinson, R. E. J.   596 (397), *610*
Huttel, R.   219, 237 (28), 239 (143), *257*, *260*
Hutton, T. W.   292 (113), *320*
Huyser, E. S.   586 (335a), *608*
Hymens, W. E.   288 (84), *319*
Hyne, J. B.   35 (126), *38*

Ichikawa, K.   241 (158), *261*
Igaki, H.   253, 254 (227a), *263*
Iguchi, M.   231 (78), *258*
Iijima, K.   188 (47), *212*
Ikeda, R. M.   443 (115), *503*
Ikeda, S.   250 (221a), *263*
Imamura, S.   (179a, 179b), *261*
Impastato, F. J.   595 (389), *610*
Imura, K.   147 (107), *171*
Ingold, C. K.   137, 138 (82), *170*, 328 (7), *357*, 548 (158), *603*
Ingraham, L. I.   78 (147), *112*
Ingraham, L. L.   516 (19), *598*
Inman, C. G.   308 (221), *323*
Inokuti, M.   382, 395 (127), *409*
Ioffe, S. T.   41 (9), *107*
Ipatieff, V. N.   571, 572 (252–254), 582 (296), *606*, *607*
Irwin, R. S.   220 (32), *257*, 375, 398 (101), *408*
Isakson, G.   552, 553 (176b), *603*
Ishii, Y.   241 (157), *261*
Ishikawa, K.   484 (256), *508*
Ishimoto, Y.   421 (58), *502*
Isler, O.   130 (51, 52), 152 (124), 154 (128), *169*, *172*
Itatani, H.   201 (82, 83), *213*, 236 (124, 128, 130, 131b, 131d), *260*, 455 (172), *505*
Ito, K.   421 (58), *502*, 585 (319), *608*
Ito, M.   537 (111), *601*
Ivin, K. J.   458 (183), *505*, 586 (326), *608*
Iwamoto, M.   253 (224a, 224b, 227a, 227b), 254 (227a, 227b), *263*
Iwamoto, N.   245 (181), *261*
Iwamura, H.   296 (136), *321*
Iwasaki, S.   300, 301 (167), *322*

Jackman, L. M.   15, 16 (58), *36*, 126 (45), 157 (136a), *169*, *172*
Jacobs, G. D.   523 (47), *599*
Jacobs, T. L.   45 (17), *107*

Jacques, B. 86, 91, 93, 94 (170), 112
Jaffé, H. H. 520 (29c), 539, 566, 567 (122), 599, 601
Jakoubková, M. (130), 602
Jambotkar, D. 533 (92), 559 (198), 601, 604
James, B. R. 197 (66), 212, 235 (101), 236 (119), 259
James, F. C. 274 (16), 317
Janik, A. 484 (257), 508
Jankowski, B. 476 (222), 507
Jankowski, S. 121 (30), 168
Jardine, F. H. 196 (64), 199 (64, 73), 200 (64), 201 (80), 212, 213, 232 (88), 233 (88, 98, 99), 234 (88b, 98), 236 (104), 258, 259
Jardine, I. 176, 193 (5), 194(x), 210, 236 (103), 259
Jaspar, P. P. 284 (67b), 318
Jedlinski, Z. 484 (257), 508
Jefford, C. W. 31 (106), 37, 67 (107), 110
Jeffrey, J. H. 548 (161), 603
Jeger, O. 537 (113), 601
Jellinek, H. H. G. 487 (270), 508
Jenkin, G. I. 177 (14), 211
Jenkins, P. A. 189, 190 (x, 48), 212
Jenner, E. L. 236 (129), 260
Jensen, F. R. 32 (112), 38
Jensen, S. L. 130 (53a), 169
Jesaitis, R. G. 562 (208b), 604
Jesse, N. 390 (151), 409
Jesse, W. P. 363 (28), 368 (46), 376 (106), 377 (106, 111), 381 (106), (112, 113), 406-408
Jira, M. 236, 237, 239, 240 (136a), 260
Jira, R. 237 (140), 238 (141), 240 (144), 260
Jirkovshy, I. 305 (203f), 323
Job, B. E. 222 (36), 257
Johnson, B. 561 (208), 604
Johnson, C. R. 224 (48), 257
Johnson, C. S., Jr. 54 (49), 74 (137), 108, 111
Johnson, D. R. 429 (80), 502
Johnson, H. W., Jr. 54 (57), 108
Johnson, J. H. 196 (63), 212
Johnson, M. 225, 230, 244, 246 (57), 257
Johnson, W. S. 86 (170, 171), 91 (170, 181), 92 (183), 93, 94 (170), 112, 113
Johnston, H. S. 373, 376 (84), 408

Johnston, W. H. 373, 379 (85), 380 (119), 408, 409
Joice, B. J. 148 (111), 171
Jolly, P. W. 165 (167), 173, 229 (70), 255 (234c), 257, 264
Jonás, J. 553 (181), 603
Jonassen, H. B. 225, 244 (55), 257
Jones, B. 567, 581 (234), 605
Jones, E. J. 236 (115a), 259
Jones, E. P. 236 (108a), 259
Jones, E. R. H. 151 (121), 158 (143), 171, 172
Jones, F. T. 367 (43), 406
Jones, G. T. 21 (79), 37
Jones, H. L. 299 (155), 321
Jones, L. B. 120 (22), 135 (77), 168, 170, 284 (69), 318
Jones, M. 159 (148), 172, 295 (124), 321
Jones, R. V. 461 (195), 506
Jones, V. K. 135 (77), 170, 284 (69), 318
Jones, W. M. 49 (29), 108
Jorgenson, M. J. 134 (75), 170, 283 (58), 309 (233), 318, 324
Jorgenson, M. T. 283 (63), 284 (65), 318
Joris, L. 571 (246a), 605
Juillard, J. 556 (189), 604
Jula, T. F. 75 (140), 111
Julia, M. 130 (51), 169
Julia, S. 571 (246c), 605
Jung, M. J. 234 (100), 259
Just, G. 303 (197), 322

Kabuss, S. 32 (109), 38
Kaczynski, J. A. 76 (142), 111, 143 (91a), 170
Kafka, T. M. 145 (97), 171
Kaiser, E. W. 118 (8), 167
Kálal, J. 445-447 (122), 503
Kalechits, J. V. 236 (134), 260
Kamat, R. J. 47, 48 (23), 107
Kaminsky, M. 372 (78), 407
Kan, R. O. 268, 290 (3c), 315 (270), 316, 325
Kane, V. V. 136 (80), 170, 278 (34), 317
Kang, J. W. 197 (70), 212, 231 (79b), 258
Kantscheff, W. 558 (193), 604
Kaplan, C. A. 567 (229), 605
Kaplan, L. 289 (85, 86, 87a, 89), 311 (89, 248), 312 (89), 319, 324
Kaplan, M. L. 284 (68), 318

Karabatsos, G. J.   34 (121), *38*, 526, 530 (69a), *600*
Karapinka, G. L.   165 (168), *173*, 243, 246 (174), *261*
Karmas, G.   153 (126), *172*
Karplus, M.   3 (10, 11, 19), 8 (45), *35*, *36*
Karrer, P.   207 (103), *213*
Karz, T. J.   255 (233c), *264*
Kasai, D. H.   523 (56), *600*
Kasai, P. H.   48 (26), *107*
Kasatochkin, V. I.   432 (86), *502*
Kasha, K.   270 (6), *316*
Kato, H.   518 (28), 519 (28, 28a), *599*
Katritzky, A. R.   596 (397), *610*
Katz, T. J.   58, 59, 62, 63 (74), 77 (143a), *109*, *111*, 255 (233a, 233b), *264*
Kaufman, P. C.   395 (167), *410*
Kawajiri, K.   236 (109d, 109e), *259*
Kawanisi, M.   275 (18a), *317*
Kazanskii, B. A.   193 (54), 195 (59), *212*, 541, 543 (132), 552 (176a), 588 (344–346, 357, 358), 590 (344, 353, 357, 361, 362), 591 (365, 366, 368), (343), (352, 356, 359, 363), *602*, *603*, *608*, *609*
Kazimirova, V. F.   578, 579 (285), *606*
Kealy, T. J.   253, 254 (228b), *263*
Kearns, D. R.   119 (17), *168*
Keating, V. T.   87 (173), *112*
Keaveney, W. P.   429 (81), *502*
Kebarle, P.   335, 341 (18), *357*, 369 (65–67), 389 (65, 66), 390, 391 (65–67), 393 (67), 394 (65, 66), 400, 402 (67), *407*
Keefer, R. M.   (242), *605*
Keller, R. A.   277 (23), *317*
Keller, R. E.   18 (66), *36*
Kelliher, J. M.   121 (31), *168*
Kelso, R. G.   590 (360), *609*
Kemball, C.   176, 181 (7), 185 (40), *211*, 231 (82), *258*
Kemp, M. K.   520 (26a), *599*
Kendall, F. H.   485 (263), *508*
Kende, A. S.   316 (272, 275), *325*
Kennedy, J. P.   412 (3), 413 (3, 8), *500*
Kent, E. G.   416 (29), *501*
Kern, L. W.   193 (53), *212*
Kerr, J. A.   116, 117 (3), *167*, 398, 399 (179), *410*
Ketcham, R.   559 (198), *604*

Ketchum, R.   533 (92), *601*
Ketley, A. D.   248, 249 (201), *262*, 525, 526 (64), *600*
Kevan, L.   395 (171), *410*
Kharasch, M. S.   219 (27), *257*, 586 (321), *608*
Kharasch, N.   303 (196), *322*
Khonin, G. V.   577, 578 (277), *606*
Kiefer, E. F.   58 (73), *109*
Kierstead, R. W.   536 (103), 548 (163a), 587 (337), 591 (163a), (94a), *601*, *603*, *608*
Kiji, J.   247 (190a, 190b), (179a), *261*, *262*
Kikuchi, T.   249 (207a), *262*
Kikuchi, Y.   54 (52), *108*
Kilpatrick, J. E.   515, 520 (6), *598*
Kilpatrick, M.   555, 556 (184), *604*
Kilroy, M.   196 (61), *212*
Kimata, T.   249 (203a), *262*
Kimoto, W. I.   154 (131), *172*
Kindler, H.   247 (193), *262*
King, N. K.   198 (71), *213*
King, N. N.   231 (84), *258*
King, R. W.   296 (138), *321*
Kingsbury, C. A.   56 (60), *109*
Kinnel, R. B.   92 (183), *113*
Kinstle, T. H.   277 (24), *317*, 337, 350, 354 (20), *357*
Kirkien-Konasiewicz, A.   327 (5, 6), 338 (5), *357*
Kirmse, W.   218 (22), *256*
Kiselev, V. G.   360 (4), *405*
Kiser, R. W.   338 (24), *357*, 530, 531 (81), *600*
Kistemaker, J.   373 (83), *408*
Kitahonki, K.   545 (149), *602*
Kitching, W.   51, 105 (35), *108*, 240 (146), *260*, 593 (381), *610*
Kitzing, R.   292 (111), *320*
Kizhner, N.   575 (273), 583 (301), *606*, *607*
Kizhner, N. M.   575 (272a), 577, 578 (277), 583 (303), *606*, *607*
Klager, K.   216 (2), *256*
Klassen, D. N.   397 (178), *410*
Kline, R. H.   477, 483 (225), *507*
Klinger, H. B.   526 (67), *600*
Kloosterziel, H.   281 (46), 282 (51a), 284 (51a, 67c, 70), *318*
Klopotova, I. A.   461 (199), *506*
Klose, G.   457 (173), *505*
Klots, C. E.   368 (48), 375 (96), 379 (116), 382 (48), 383 (129, 130), 394 (116), 401 (129), *407*, *409*

Klotz, J. M.   532, 536 (88), *601*
Kluiber, R. W.   154 (132), *172*
Kniepp, K. G.   120 (21), *168*
Knight, S. A.   300 (162), *322*
Knowles, W. S.   234 (253), *264*
Knowlton, J. W.   521 (30), *599*
Kobayashi, T.   484 (256), *508*
Kobrich, G.   163, 164 (163), *173*
Köbrich, G.   56 (63), *109*
Koch, K.   285 (75b), *319*
Kochi, J. K.   151 (120), *171*
Kock, R. J. de   290 (90), *319*
Koehl, W. J., Jr.   55 (58), *108*
Koeppl, G. W.   595 (391), *610*
Kofler, M.   130 (52), *169*
Kofron, J. T.   309 (230), *323*
Kohler, E. P.   592 (377), *609*
Kohlhaupt, R.   162 (158), *173*
Kojer, H.   236, 237, 239, 240 (136a), *260*
Koller, E.   309 (230), *323*
Koloskova, N. M.   559 (199), *604*
Koltsov, A. I.   460 (189), *506*
Koltzenburg, G.   307 (219), *323*
Komora, L.   230 (72), *258*
Kondakow, J. L.   219 (26), *256*
Kondratiev, V. N.   338 (25), *357*
Kong, N. P.   48 (27), *108*
König, J.   132, 133 (67), 133 (70), *170*
Koningsberger, C.   435 (93b), *503*
Konno, K.   188 (47), *212*
Koper, D. G.   292 (109), *320*
Kop'eva, I. A.   452 (152), *504*
Kopple, K. D.   208 (104), *213*
Kordo, H.   236 (255), *264*
Kornegay, R. L.   100 (205), *114*
Korobova, L. M.   421 (50), *501*
Korte, F.   303 (188), 312 (255), *322, 324*
Kortüm, G.   556 (188), *604*
Kosower, E. M.   537 (111), *601*
Kossa, W. C.   104 (219), *114*
Kössler, I.   443 (116), 445, 446 (122), 447 (116, 122, 126–129), 460 (128), *503, 504*
Koster, D. F.   21 (74), *37*
Kostin, V. N.   541, 542 (134–136), 580 (135, 287–292), 583 (300), *602, 607*
Kostyanovskii, R. G.   547 (156, 157), *603*
Kovacic, P.   474 (215a), *506*
Koyana, I.   389 (148), *409*
Kozina, M. P.   552 (176a), *603*
Kraihanzel, C. S.   (168), *261*

Kranzfelder, A. J.   572, 573 (263), *606*
Kratzer, J.   219, 237 (28), *257*
Krauch, C. H.   301 (168), *322*
Kraus, H.   536, 541, 556, 557 (98), *601*
Krauss, G.   417 (35, 36), 452 (36, 153), *501, 504*
Krauss, H. J.   278, 304 (28), *317*
Krauss, M.   373, 379 (89), *408*
Krebs, A. W.   58, 81 (72), *109*
Krentsel, B. A.   432 (88), *502*
Kretschmar, H. C.   281 (43), *318*
Krigas, T. M.   523 (47), *599*
Kriloff, H.   31 (102), *37*
Krings, P.   167 (174), *173*, 225 (54, 58), 228, 229 (69), 230 (58, 69), 244 (54, 58), *257*
Krishnamurthy, G. S.   595 (391), *610*
Krishner, L. C.   26 (99), *37*
Kristinsson, H.   278 (36), 281 (44), *317, 318*
Kroh, J.   476 (222), *507*
Kroll, W. R.   (133), *260*
Kroner, M.   125 (41), *169*
Kröner, M.   219 (29), *257*
Kropacheva, E. N.   454 (167), 460 (186–189), *505, 506*
Kröper, H.   216, 245 (4), *256*
Kropp, P. J.   268 (2f), 278 (28), 290, 291 (2f), 292 (110), 304 (28), *316, 317, 320*
Krsek, G.   246 (185), *262*
Krumholz, P.   244 (175), *261*
Kudo, S.   196 (62), *212*
Kuhls, J.   301 (168), *322*
Kuhn, R.   68 (112), *110*
Kuivila, H. G.   586 (329), *608*
Kuljian, E.   252 (223), *263*
Kumler, W. D.   (166), *603*
Kunitake, T.   421 (58), *502*
Kupperman, A.   400 (185), *410*
Kuri, K.   395 (168), *410*
Kurihara, F.   484 (256), *508*
Kurland, R. J.   33 (114), *38*
Kurtz, P.   21 (80), *37*
Kurzer, F.   203 (89), *213*
Kushner, A. S.   65 (102), *110*
Kusov, A. B.   444 (119), *503*
Kustanovich, I. M.   432 (88), *502*
Kutepow, N. von   247 (193), *262*
Kuwata, K.   73 (129), *111*
Kuzmin, M. G.   541, 542 (133), *602*
Kuzminskii, A. S.   477 (229), *507*
Kwaitkowski, G. T.   129 (50), *169*
Kwan, T.   589, 590 (347), *608*

Kwiatch, J. 197 (68, 69), 198 (72), 212, 213
Kwiatek, J. 231 (79a, 79c), 232 (87), 258

Lacher, J. R. 545 (148), 602
Laidler, K. J. 373 (84), 374 (92), 376 (84), 408
Lal, J. 495 (295), 509
LaLancette, E. A. 77 (143a), 112
LaLonde, R. T. 299 (156), 321, 582, 583 (298, 302, 305–307), 584 (307), 590 (302), 607
LaMahieu, R. 304 (202), 305 (205), 323
Lambert, J. L. 102 (208–210), 114, 584 (310), 607
Lamola, A. A. 120 (19), 168, 272 (13), 274, 276 (17), 277 (17, 23, 25b), 317
Lampe, F. W. 360, 369 (8), 406
Lancaster, J. E. 316 (275), 325
Landgrebe, J. A. 553, 564 (180), 603
Landolt, R. G. 162 (157), 173
Landor, S. R. 157 (141), 158, 160 (144), 172, 535 (96), 601
Lanford, C. A. 546 (153), 602
Langevin, F. L., Jr. 586 (329), 608
Langford, C. H. 223 (40), 257
Langguth, H. 132 (65), 170
Lanpher, E. J. 74 (132), 77 (144), 111, 112
Lansbury, P. T. 105 (221), 114, 163 (159), 173, 564 (215), 605
Lapporte, S. J. 236 (126), 260
Larson, J. G. 148 (108), 171
Lassila, J. D. 285 (75b), 319
Laszlo, P. 18 (68), 22 (86), 31 (103), 34 (118), 37, 38
Laulicht, I. 379 (117), 409
Lauterbur, P. C. 544, 545 (142), 602
Lavina, R. Ya. 130 (56), 169
Law, P. A. (207), 604
Lawrence, C. D. 573 (265), 606
Lawrence, R. H. 396 (174), 410
Laws, G. F. 292 (113), 320
Lawton, R. G. 83–85 (162), 112
Lazzaroni, R. 243, 246 (172), 261
Lea, D. E. 384 (132), 409
Leal, G. 89 (179), 113
Leass, M. F. 208 (110b), 214
LeBlanc, R. M. 366, 367 (38), 406
Lebovits, A. 565 (218), 605

Le Bras, J. 482 (252), 496, 498 (296), 508, 509
Lee, C. C. 85 (167), 97 (196), 112, 113
Lee, L. H. 445 (123), 484 (255), 504, 508
Lee, T. C. P. 477, 478, 480 (228), 507
Lee, W. G. 56 (61), 109
Leermakers, P. A. 274 (16), 277 (23), 313 (263), (27), 317, 324
Lee-Ruff, E. 303 (197), 322
Lefebvre, G. 148 (112), 171, 230 (76a, 76b), 248, 249 (202a, 202b), 258, 262
Lefevre, G. 248, 249 (197), 262
Leftin, H. P. 148 (111), 171
Legendre, P. 146 (103, 104), 171
Lehmann, H. P. 121 (26), 168
Lehn, M. M. 301 (172), 322
Lehr, M. H. 415, 416, 419 (10), 500
Leitich, J. 307 (219), 308 (229), 323
Lemaire, H. P. 523 (43), 599
Lemal, D. 255 (234a), 264
Lemal, D. M. 291 (108), 320
Lemmon, R. M. 33 (115), 38
Lemper, A. L. 159 (147), 172
Lenaz, G. 459 (184), 505
Lenz, G. 269, 290, 291 (2i), 316
Lenz, G. R. 290 (99b), 320
Leshcheva, A. I. 248, 249 (199), 262
Leupold, E. O. 460, 461 (191), 506
Levasseur, L. A. 159 (149), 172
Levesque, G. 494 (291), 495 (292, 293), 509
Levina, R. Y. 541, 542 (133–137), 550 (168), 553 (182, 186), 556, 557 (186), 559 (196, 197, 199, 200), 570 (246), 580 (135, 196, 286d, 287–292), 581 (295), 583 (300), 602–605, 607
Levine, P. 177 (19), 211
Levitt, G. 573 (259), 606
Levy, G. C. 546 (152), 602
Levy, M. 424 (67), 502
Lewicki, E. 105 (221), 114
Lewin, A. H. 99 (201), 114
Lewis, G. N. (26), 317
Lewis, I. C. 531, 568 (83), 600
Lewis, P. 303 (196), 322
Lewis, R. G. 283 (64), 318
Lias, S. G. 371 (73), 377 (23, 73), 380 (120, 121), 388 (144), 402 (187), 406, 407, 409, 410
Libit, L. 305 (205), 323
Lide, D. R., Jr. 524(x)

Liesenfelt, H. 232, 233 (89, 91), 236 (89), *258*
Lifschitz, A. 369, 390, 391 (68), *407*
Lin, K. 305 (204, 206), *323*
Lincoln, D. L. 571, 572, 575 (249), *606*
Lincoln, D. N. 65, 66 (103), *110*, 527 (73a), 538, 553 (119), 560 (73a), 562 (119), *600*, *601*
Lind, S. C. 360 (1), *405*
Lindegren, C. R. 78 (147), *112*
Lindsey, R. V. 166 (170), *173*, 236 (129), 247–249, 253, 254 (194), *260*, *262*
Lindy, L. B. 577, 587 (283), *606*
Lini, D. C. 239, 245 (147c), *260*
Linn, C. B. 571, 572 (250), *606*
Linnett, J. W. 57 (70), *109*, 539, 540 (126), *602*
Linnig, F. J. 471 (214), *506*
Linstead, R. P. 177 (19), *211*, 536 (103), 548 (163a), 587 (337), 591 (163a), (94a), *601*, *603*, *608*
Lipinsky, E. S. 308 (221), *323*
Lipkin, D. (26), *317*
Lipovich, V. G. 236 (134), *260*
Lippert, E. 544, 545 (144), *602*
Lippmaa, E. 32 (108), *38*
Liptak, K. 35 (127), *38*
Lishanskii, M. S. 495 (294), *509*
Lister, D. H. 160 (153), *172*
Liu, J. S. 65, 66 (103), *110*, 527 (73a), 538, 553 (119), 560 (73a), 562 (119, 210), *600*, *601*, *604*
Liu, R. S. H. 278 (31), 286 (81), 287 (31), 299 (149, 154), 312 (256), 313 (265), 314 (265, 266), *317*, *319*, *321*, *324*
Livingston, R. L. 523 (43), *599*
Livshits, I. A. 421 (50), *501*
Lloyd, D. 58, 81 (72), *109*
Loan, L. D. 464, 466, 468 (204), *506*
Locke, M. M. 236 (256), *265*
Loder, J. D. 158 (143), *172*
Lodge, J. E. 296 (134), 311 (236), 312 (252), *321*, *324*
Loemker, J. 22, 25 (85), *37*
Loeschen, R. 310 (234), *324*
Logan, T. J. 67 (109), *110*, 229 (71a), *258*
Logiudice, F. 249, 250 (206b), *262*
Lokengaard, J. P. 291 (108), *320*
Lombardi, E. 33 (115), *38*
Lombardi, G. 427 (76), *502*
Longevaille, P. 343, 344 (33), *358*
Longiave, G. 250 (220), *263*

Longuet-Higgins, H. C. 281 (41b), 289 (88), *318*, *319*
Lord, R. C. 523 (37), *599*
Lorenc, L. J. 537 (113), *601*
Lorenz, O. 465 (210), 478 (243), *506*, *507*
Lorenz, O. M. 477, 483 (225), *507*
Lossing, F. P. 44(x), 60 (79), *109*, 361 (x), 530 (79), *600*
Loudon, A. G. 327 (6), 349 (28), 355 (41), *357*, *358*
Loy, M. 290, 292, 294 (96), *319*
Ludwig, H. 526 (69), *600*
Luettringhaus, A. 32 (109), *38*
Lugli, G. 424 (65), *502*
Lukas, J. 596 (399), *610*
Lukina, M. Y. 512, 513 (3), 541 (131, 132, 138), 542 (131, 138), 543 (132, 138), 552 (176a), 588 (344–346, 357, 358), 590 (344, 353, 357, 361, 362), 591 (365, 366, 368), (343) (352, 356, 359, 363), *598*, *602*, *603*, *608*, *609*
Lukina, M. Yu. 50 (34), *108*
Lum, K. K. 596 (398), *610*
Lundberg, R. D. 436 (94), *503*
Lundin, R. E. 154 (130), *172*
Lustgarten, R. K. 100 (203, 206), 101 (206), *114*
Lüttke, W. 526 (61, 62, 69), *600*
Lydon, J. E. 250, 252 (212a, 212c), *263*
Lynden-Bell, R. M. 4, 5 (26), *35*
Lyubimova, A. K. 360, 369, 405 (6), *406*

Mabrouk, A. F. 231 (79d), *258*
Mabry, T. J. 234 (248), *264*
Mabuchi, K. 249 (208), *263*
Maccagnani, G. 15 (62), *36*
Maccoll, A. 44(x), 144 (96), *171*, 327 (5, 6), 338 (5), 349 (28), *357*
MacGregor, W. S. 151 (122), *171*
Mach, K. 447 (128, 129), 460 (128), *504*
Macho, V. 230 (72), *258*
Maciel, G. E. 9 (47), *36*
Mackenzie, K. 64, 68 (91), *110*, 165 (167), *173*, 229 (70), 255 (234c), *257*, *264*, 278 (32), *317*, 449 (131), *504*
Mackor, E. L. 57 (68), *109*
MacLean, C. 57 (68), *109*, 239, 245 (148a), *260*
MacMillan, W. G. 291 (106), *320*

Macoll, A.   116 (1b), *167*
Madge, E. W.   452 (156), *504*
Mador, I. L.   197 (68, 69), *212*, 231 (79a, 79c), *258*
Madorsky, S. L.   412 (7), 487 (7, 274), 488 (275), 489 (274, 275), *500*, *508*
Maercker, A.   104 (216), 105 (220), *114*
Magee, J. L.   373, 380 (90), 381 (125), 386 (137, 138), 396 (137), 400 (137, 184), *408–410*
Magel, T. T.   (26), *317*
Mahan, B. H.   396 (173, 175), *410*
Mahoney, L. R.   558 (194), *604*
Maitlis, P. M.   216 (8), 256 (239a), *256*, *264*
Maitte, P.   54 (50), *108*
Majer, J. R.   134 (73), *170*
Maksić, Z.   516, 517 (24), *599*
Malkus, H.   529 (72), *600*
Mallinson, C. J.   236 (256), *265*
Malrieu, J. P.   273, 288 (14), *317*
Malyshev, A. I.   590 (353, 362), (343), (363), *608*, *609*
Manabe, Y.   54 (52), *108*
Manchand, P. S.   126 (45), *169*
Mandell, L.   7 (42), 21 (73), *36*, *37*
Mango, F. D.   255 (236), *264*
Manion, J. P.   375 (103), *408*
Mann, J.   485 (263), *508*
Mann, R. H.   415 (12), *500*
Mannhardt, H. J.   126 (43), *169*, 437 (98), *503*
Manning, T. D. R.   151 (123), 157 (139), *172*
Mansfield, G. H.   555(x)
Mansfield, K. T.   597 (401, 402), *610*
Mantecon, J.   291 (104), *320*
Manuel, T. A.   165 (166), *173*, 228, 246 (65), *257*, 454 (168), *505*
Manza, G.   452 (157), *505*
Maples, P. K.   (168), *261*
Marburg, R.   586 (330), *608*
Marcantonio, A. F.   149 (114), *171*
Marconi, W.   421 (51), 424 (65), *501*, *502*
Mariella, R. P.   535 (97), 536 (99, 108), 548 (108), *601*
Markby, R.   244 (176), *261*
Marko, L.   236 (112a, 112b), *259*
Marks, J.   162 (158), *173*
Marmo, F. F.   378 (115), *408*
Marr, D. H.   545 (151), *602*
Marsh, R. E.   523, 526 (50), *599*

Marshall, H.   78 (147), *112*
Marshall, J. A.   94 (185), *113*, 278 (29, 30), 304 (29), *317*
Marshall, T. R.   577, 578 (282), *606*
Marteau, J.   499 (300), *509*
Martin, G. J.   21 (75), *37*, 54 (51, 54), *108*, 163 (162), *173*
Martin, J.   88 (177), *113*
Martin, J. C.   527, 563 (76), *600*
Martin, J. S.   526 (60), *600*
Martin, M. L.   21 (75), *37*, 54 (51, 54), *108*, 163 (162), *173*
Martin, R. A.   583 (308), *607*
Martin, W.   290 (95), *319*
Marvell, E. N.   154 (133), *172*
Masamune, S.   295 (130), 296 (130a), 301 (170), *321*, *322*
Massey, H. S. W.   380 (122), *409*
Massingill, J. L.   192 (51), *212*
Masuko, T.   188 (47), *212*
Matesich, M. A.   43 (13), 45 (13, 16), *107*
Matheson, M. S.   363, 365 (25), *406*
Matlack, A. S.   202 (88), *213*, 236 (135), *260*
Matson, F. A.   532 (90), 533 (90, 91), *601*
Matson, R. A.   (14a), *598*
Matsubara, I.   21 (81), *37*
Matsuda, A.   246 (187), *262*
Matsuoka, S.   34 (124), *38*
Matsuzaki, K.   439 (102), *503*
Matyska, B.   447 (129), *504*
Mawby, R. J.   243 (165, 170), *261*
Mayer, C. F.   275 (18b), *317*
Maynard, J. T.   418, 419 (41), *501*
Mayo, F. R.   219 (27), *257*, 481 (246), *507*, 586 (321), *608*
Mayo, P. de.   146 (104), *171*, 290 (98, 99a), 299 (153), 305 (207), 307 (213, 215), 316 (271), *320*, *321*, *323*, *325*
Mayo, R. E.   5, 6, 7 (31), *36*
Mazheiko, I. B.   550 (168), *603*
Mazur, R. H.   97 (196), *113*, 560 (202, 204), 565 (225), *604*, *605*
Mazzanti, G.   426 (75), 428, 429 (79), 437 (75), *502*
Mazzei, A.   421 (51), 424 (65), *501*, *502*
Mazzocchi, P. H.   292 (114, 115b), 293, 294 (114), 295 (121, 122), *320*
Mazzucato, U.   125 (39), *169*
McBee, E. T.   577 (279), *606*

McCalel, G. V.   179, 183 (28), *211*
McClellan, A. L.   (175), *603*
McClelland, M. J.   344, 345 (35), *358*
McClure, D. S.   270 (7), *316*
McClure, G. R.   26 (94), *37*
McCluskey, J. A.   344, 345 (35, 36), 347 (36), *358*
McCollum, J. D.   236 (111), *259*
McConnell, H. M.   3 (14, 17, 18), 4 (17), 15 (64), *35, 36*
McConnell, W. J.   385, 391 (135), *409*
McCubbin, T. K.   523 (37), *599*
McDaniel, D. H.   555(x)
McDaniel, J. C.   98 (198), *113*
McDowell, C. A.   (9), *598*
McEwen, W. K.   593 (382), *610*
McFadden, W. H.   333–335 (12), *357*
McGavack, J.   478 (235), *507*
McGhie, J. F.   203 (89), *213*
McGlynn, S. P.   270 (6), *316*
McGowan, J. W.   372 (80), *407*
McGravey, B. R.   236 (254), *264*
McKeon, J. E.   (258), *324*
McLachlen, A. D.   4 (23), *35*
McLafferty, F. W.   327 (1, 2), 328 (2, 8), 331 (10), 333, (15), 348 (43), *357, 358*
McLauchlan, D. A.   367 (44), *407*
McLean, S.   282 (50), *318*
McNeil, M. W.   94 (188), *113*
McNesby, J. R.   362 (19), 374, 395 (19, 94), *406, 408*, 595 (393), *610*
McOmbie, J. T.   151 (121), *171*
McQuillin, F. J.   176, 193 (5), 194(x), *210*, 236 (103), *259*
Meary-Tetrain, A.   (240), *605*
Mebane, A. D.   153 (126), *172*
Mecke, R.   32 (109), *38*
Medary, R.   67 (107), *110*
Medvedev, V. A.   338 (25), *357*
Meerburg, P.   550 (173), *603*
Megarity, E. D.   120 (21), *168*, 277 (25b), *317*
Meiboom, S.   (39), *599*
Meier, J.   16, 17 (65), *36*, 92 (182), *113*
Meijere, A. de   523 (41), 526 (40, 41, 61, 69), *599, 600*
Meinwald, J.   279 (117), 283 (60), 292 (114, 115b, 117), 293 (114), 294 (114, 117, 120), 295 (121, 122), 297 (142), 300 (159), *318, 320, 321*
Meisels, G. G.   335 (44), *358*, 360 (15–17), 363, 364 (29), 366, 367 (29, 32), 368 (49), 371 (75), 372 (49), 374 (16), 380 (17), 383 (29), 390 (15, 17, 155–157), 391 (17, 155, 159), 392 (17), 393 (16, 155), 394 (15), 395 (49), 399 (15), 402 (17, 156), 403 (15, 191), 404 (15–17, 49), *406, 407, 410*
Meister, H.   216 (2), *256*
Melton, C. E.   369, 390, 393, 394 (62a, 62b), *407*
Merck, W.   256 (238), *264*
Mercurio, A.   481 (248), *507*
Merlier, J. de   482 (252), *508*
Merwe, J. P. van der   128 (47, 49), 129 (49), *169*
Meshcheryakov, A. P.   591 (367), *609*
Mesrobian, R. B.   478 (239), *507*
Metlin, S.   246 (184, 188), *262*
Metz, D. J.   478 (239), *507*
Meurs, N. van   282, 284 (51a), *318*
Meyer, E. F.   191 (50), *212*
Meyer, G. E.   493 (287, 288), 494 (288, 290), *509*
Meyer, K. H.   449 (133), *504*
Michelotti, F. W.   429 (81), *502*
Michl, J.   290 (99c), *320*
Midgley, T., Jr.   487 (273), *508*
Mieville, R. L.   119 (14), *168*
Miginiac, P.   150 (117), *171*
Miginiac-Groizeleau, L.   150 (117), *171*
Mikaye, A.   248, 249 (198), *262*
Mikhailov, B. M.   360 (4), *405*
Mikhailova, L. N.   461 (197), *506*
Mikolajczak, K. L.   145 (99, 100), *171*
Miles, F. B.   124 (37), *168*
Miles, M. L.   422 (62), *502*
Miliković, M.   537 (113), *601*
Milkovich, R.   424 (66, 67), *502*
Millard, B. J.   333–335, 343 (14), *357*
Miller, C. E.   208 (106), *213*
Miller, F. W.   49 (29), *108*
Miller, L. S.   302 (175, 176), *322*
Miller, R. G.   253, 254 (228a, 228b), *263*
Miller, S. I.   56 (61), *109*, 595 (391), *610*
Milne, C. B.   41 (4), *107*
Milne, D. G.   119 (16), *168*
Milner, D. L.   236 (122), *260*
Minaard, N. G.   290 (90), *319*
Mironov, V. A.   35 (125), *38*
Mirza, N. A.   134 (76a), *170*, 282 (47), *318*

Misoni, A.   249 (207b), *263*
Misono, A.   202 (87), *213*, 236 (116a, 116b), 249 (204a, 204b), 250 (221a, 221b), 253 (226), *259*, *262*, *263*
Mitchard, D. A.   189, 190 (x, 48), *212*
Mitchell, D.   310 (234), *324*
Mitchell, T. R. B.   236 (121a), *260*
Mitra, R. B.   304 (201, 202), *322*, *323*
Mitsui, S.   179, 183 (29), 188 (44, 45, 47), *211*, *212*
Miyake, A.   236 (255), *264*
Miyashi, T.   285 (75b), *319*
Moacanin, J.   427 (78), *502*
Moberly, C. W.   461 (195), *506*
Mochel, V. D.   417 (39), *501*
Mochel, W. E.   418, 419 (41), *501*
Mock, W. L.   208 (108, 113, 114), 209, 210 (114), *214*
Moffit, W. E.   50 (33), *108*, 515 (17, 18), 516 (18), *598*
Mohrbacher, R. J.   536, 537, 539 (102), *601*
Moiseev, I. I.   237 (138), 240 (145a, 145b), *260*
Moiso, T.   333 (16), *357*
Moje, S.   208 (105), *213*
Mome, C. P.   590 (350), *609*
Montague, B. A.   429 (80), *502*
Montavon, M.   130 (51), *169*
Moore, C. G.   452 (162–165), 460 (164), 464 (164, 207), 465 (207, 209), 466, 468 (164), 470 (164, 211), 471 (164), 472 (162–165), 473 (164), *505*, *506*
Moore, D. W.   10, 11 (50), *36*, 225, 244 (55), *257*
Moore, H. W.   459 (184), *505*
Moore, J. A.   285 (79), *319*
Morelli, D.   232 (243), *264*
Morgan, C. R.   248, 249 (201), *262*
Morgan, R. A.   240 (151a, 151b), *261*
Mori, T.   283 (62), *318*
Morifuji, K.   250 (221a), *263*
Morikawa, M.   241 (159), *261*
Moritani, I.   313 (260), *324*, 550 (168a), 562 (208a), 585 (319), *603*, *604*, *608*
Morris, A. L.   478 (233, 238), 480 (238), *507*
Morrison, G. A.   105 (222), *114*, 525 (59), *600*
Morrison, H.   120 (18), *168*, 277 (24), 302 (178), *317*, *322*
Morse, J. G.   555, 556 (184), *604*

Mortelatici, S.   234 (252), *264*
Mortimer, F. S.   32 (113), *38*
Morton, A. A.   77 (144), *112*
Moss, R. A.   526 (68), *600*
Mottl, J.   360, 361 (18), *406*, 530 (80), *600*
Moussebois, C.   121 (32), *168*, 278 (35), *317*
Mousseron, M.   146 (103, 104), 147 (106), *171*, 268, 290, 291 (2c), *316*
Mousseron-Canet, M.   146 (103, 104), *171*
Mozumder, A.   386 (138), *409*
Mrowca, J. J.   255 (233a), *264*
Muhs, M. A.   220 (33), *257*
Muizebelt, W. J.   116 (4), *167*
Mukai, T.   285 (75b), 316 (273, 274), *319*, *325*
Mulas, P. M.   385 (134), *409*
Muller, H.   186 (42), *211*, 216, 250, 252 (18), *256*
Muller, J. C.   6 (38), *36*
Muller, N.   5 (29, 30), *36*, 544 (139), *602*
Müller, R. H.   208 (107), 209 (107, 115), 210 (107), *214*
Muller, R. J.   68, 69, (115), *111*
Mulliken, R. S.   275 (19), *317*
Mullineaux, R. D.   236 (113), *259*
Mund, W.   360 (2), *405*
Munro, H. D.   283 (61), *318*
Munson, M. S. B.   389 (149), 390 (152–154), *409*
Murakami, M.   197 (70), *212*, 231 (79b), *258*
Murayama, N.   436 (95, 96), *503*
Murov, S.   309 (233), *324*
Murphy, J. W.   306 (208), *323*
Murray, R. W.   284 (68), *318*, 485 (262), *508*
Murrell, J. N.   4 (21), *35*
Musco, A.   253 (231), *263*
Musgrave, O. D.   283 (61), *318*
Music, J. F.   532 (90), 533 (90, 91), (14a), *598*, *601*
Muszkat, K. A.   291 (100), *320*
Myers, R. J.   523 (56), *600*
Myher, J. J.   391 (158), *410*

Naar-Colin, C.   7 (40), 13, 14 (56), 26 (96, 100), 27, 28 (96), *36*, 37
Nagahisa, Y.   179, 183 (29), *211*
Nagy, F.   231 (85), *258*
Nagy, P. L. I.   216, 221 (7), *256*

Nakanishi, K.   304 (203d), *323*
Nakatsuka, N.   296 (130a), *321*
Nakayama, T.   360, 361 (18), *406*, 530 (80), *600*
Namanworth, E.   562 (213, 214), *604*
Namikawa, K.   15, 16 (59), *36*
Naples, F. J.   493 (288), 494 (288, 290), *509*
Narasimhan, P. T.   13 (55), *36*
Narath, A.   523, 550 (48), *599*
Nashund, L.   142 (90), *170*
Naslund, L.   68–70 (116), *111*
National Bureau of Standards (U.S.) 23(x), 328, 332, 333, 348, 354 (9), *357*, 383, 384, 397 (131), *409*
Natta, G.   249 (206a, 206b), 250 (206b) *262*, 415 (11), 416 (30), 419 (11), 420 (44–47), 421 (44, 48), 424 (69), 426 (75), 428, 429 (11, 79), 437 (75), *500–502*
Nealy, D. L.   96 (192), *113*
Nechvatal, A.   92 (182), *113*
Neckers, D. C.   565 (221, 222), 566 (222), (223), *605*
Neiman, M. B.   412 (6), *500*
Nelson, K. V.   452, 454 (160), 460 (186–189), *505, 506*
Nelson, P. H.   306 (209, 210), *323*
Nenitzescu, C. D.   295 (129), *321*
Nesmeyanov, A. N.   54 (48, 56), *108*, 163 (161), *173*
Neugebauer, F. A.   68 (112), *110*
Newhall, W. F.   195 (58), *212*
Nicholas, H. J.   93 (184), *113*
Nicholson, C. R.   546 (153), *602*
Nicholson, E. M.   87 (172), *112*
Nicholson, J. K.   225, 230, 244 (60), 245 (180), 248, 249 (200), 250, 252 (212a, 212b), *257, 261–263*
Nicholson, J. M.   103 (214), *114*
Nickon, A.   102 (208–210), *114*, 584 (310), *607*
Nicole, P. P.   377, 378 (114), *408*
Nicolet, B. H.   586 (327), *608*
Nieuwland, J. A.   426 (74), *502*
Nigam, S. S.   130 (56), *169*
Nishida, S.   550 (168a), 562 (208a), 585 (319), *603, 604, 608*
Nishino, K.   236 (105), *259*
Nisikawa, Y.   275 (18a), *317*
Nist, B. J.   31 (105), *37*, 539, 540, 545 (128), *602*
Nivard, R. J. F.   116 (4), *167*
Noack, K.   243 (163), *261*
Nogi, T.   247 (190b), *262*

Noltes, A. W.   186 (42), *211*
Nordlander, J. E.   74 (138, 139), 75 (138, 139, 141), 76, 77 (143), 104 (215), *111, 114*
Norin, T.   66 (105a), *110*
Norling, L.   536 (110), *601*
Norling, P. M.   477, 478, 480 (228), 481 (247), 482 (250), *507*
Normant, H.   54 (50), 56 (62), *108, 109*
Norris, W. P.   333, 334 (13), *357*
Norton, C.   99 (199), *113*
Nosworthy, J.   120 (24), *168*
Novikova, N. N.   452, 454 (160), *505*
Novikova, N. V.   54 (48), *108*, 163 (161), *173*
Noyce, D. S.   45 (16), *107*, 124 (37), *168*
Noyes, R. M.   449 (138), *504*
Noyes, W. A.   268, 290 (1a), *316*
Noyori, R.   275 (18a), 283 (62), *317, 318*
Nozaki, H.   275 (18a), 283 (62), *317, 318*
Nozoe, S.   304 (203c), *323*
Nurrenbach, A.   128 (48), 153 (125), *169, 172*
Nussim, M.   302 (179), 309 (233), *322, 324*
Ny, G. Le   88 (176), *113*
Nyholm, R. S.   200 (78), *213*
Nyman, C. J.   224 (44), *257*

Oae, S.   119 (15, 16), *168*
Obata, N.   313 (260), *324*
O'Brien, D. F.   50, 105 (31), *108*, 593 (380), *610*
O'Brien, S.   222 (36), *257*
Occolowitz, J. L.   103 (213), *114*
O'Connell, E. J.   121 (31), *168*
O'Conner, D. E.   141 (87), *170*
O'Connor, C.   236 (258), *265*
Odaira, Y.   242 (161), *261*, 308 (223), *323*
Odian, G.   566, 567 (226), *605*
Oediger, H.   153 (127), *172*
Ogata, I.   202 (87), *213*, 236 (116b), *259*
Ogawa, I. A.   98 (198), *113*
Ogg, R. A., Jr.   586 (320, 322), *608*
Ogliaruso, M.   103 (214), *114*
O'Gorman, J. M.   523 (44), *599*
Ohloff, G.   303 (183), *322*
Ohno, K.   224 (45), 233 (96), 243 (171), *257, 258, 261*

Ohno, M.   209 (116), *214*, 333 (16), *357*
Ohsawa, Y.   250 (221b), *263*
Oishi, T.   242 (161), *261*
Okabe, H.   362 (19), 374, 395 (19, 94), *406, 408*
Okamoto, H.   395 (168), *410*
Okamoto, M.   209 (116), *214*
Okamoto, Y.   (241), *605*
Olah, E. A.   527, 560 (74), *600*
Olah, G. A.   41 (2), 62(x), 65 (99, 104), 66 (104), 67 (99), 87, 89 (2), *107, 110*, 527, 560 (73), 562 (212–214), 596 (399), (119a), *600, 601, 604, 610*
Oldham, A. R.   234 (249), *264*
Olechowski, J. R.   225, 244 (55), *257*
Olivé, S.   216 (15), *256*
Oliver, J. E.   102 (209), *114*
Olofson, R. A.   57, 72 (71), *109*
Olsen, A. R.   116 (2), *167*
Olson, A. C.   88 (175), *113*
Omura, I.   389 (148), *409*
Ong, J.   154 (133), *172*
Ono, T.   249 (207b), *263*
Onsager, L.   386 (139), *409*
Oppenheimer, E.   233 (97), *258*
Orchin, M.   164 (165), 165 (165, 168, 169), 167 (173), *173*, 207 (100), *213*, 225, 230 (59), 236 (114a, 114b), 243 (174), 246 (174, 184), *257, 259, 261, 262*, 453, 454 (166), *505*, 592 (378), *609*
Ordronneau, C.   99 (200), *113*
Orgel, L. E.   4 (21), *35*
Orger, B. H.   311 (249), *324*
Orlando, C. M.   15 (61), *36*
Oroshnik, W.   153 (126), *172*
Osbond, J. M.   130 (54), *169*
Osborn, J. A.   196, 199, 200 (64), 201 (80, 81), *212, 213*, 232 (88), 233 (88, 99), 234 (88b, 252), 236 (102, 104, 118), *258, 259, 264*
Ostling, G.   548 (160), *603*
Ostromislensky, I. I.   464 (206), *506*
Osvorn, J. A.   199 (73), *213*
Oth, J. F. M.   290 (95), *319*
Otsuka, S.   249 (207a), *262*
Ouellette, R. J.   35 (127), *38*, 581, 583 (293, 294), 597 (403), *607, 610*
Ourisson, G.   303 (193), *322*, 585 (312), *607*
Overberger, C. G.   564 (219), 565 (217, 218), *605*
Ovodova, V. A.   (352, 359), *609*
Owyang, R.   86 (170, 171), 87 (172), 91, 93, 94 (170), *112*
Oxford, A. W.   353 (39), *358*
Ozaki, A.   147 (107), 148 (109), *171*

Pabon, H. J. J.   131 (57, 58), *169*
Packs, E. K.   396 (173), *410*
Padwa, A.   536 (110), *601*
Pagni, R. M.   296 (132), *321*
Paiaro, G.   253 (231), *263*
Paiaro, J.   240 (155), *261*
Palmade, M.   585 (312), *607*
Palumbo, R.   240 (251), *261*
Panchenko, Y. N.   541, 542 (133), *602*
Pande, J. B.   435 (93a), *503*
Pande, K. C.   99 (201), *114*
Pant, L. M.   524(x)
Pappas, B.   295 (127), *321*
Pappas, S. P.   285, 290 (74), *319*
Paquette, L. A.   285 (78), 315 (269), *319, 325*
Pariser, R.   418 (43), *501*
Park, J. P.   545 (148), *602*
Park, J. Y.   335 (44), *358*
Parker, W.   88 (177), *113*
Parks, A. P.   523 (53), *599*
Parks, C. R.   465 (210), 477 (225), 478 (243), 483 (225), *506, 507*
Pascual, C.   16, 17 (65), *36*
Pasqualuuci, C. R.   144 (95), *171*
Passarge, W.   207 (102), *213*
Pasto, D. J.   208 (108, 114), 209, 210 (114), *214*, 285 (75a), *319*
Patai, S.   122 (33a, 34), *168*
Patel, D. J.   544, 545 (140), *602*
Paterno, G.   308 (220), *323*
Patsch, M.   67 (107), *110*
Pattenden, G.   128 (47, 49), 129 (49), *169*
Pattenden, G. E.   157 (140), *172*
Pattison, V. A.   163 (159), *173*, 564 (215), *605*
Pauling, L.   50 (32), *108*, 523 (36), *599*
Paulson, P. L.   302 (181), *322*
Pautrat, R.   492 (286), 496 (296, 297), 498 (296, 298, 299), 499 (299, 300), *509*
Pavlikova, A. V.   461, 463 (198), *506*
Pavlis, R. R.   303 (186), *322*
Payo, E.   291 (104), 308 (228), *320, 323*
Pearson, R. G.   52 (40), *108*, 243 (165, 170), *261*

Pease, R. N. 592 (376), *609*
Peebles, L. H., Jr. 433 (89), *502*
Peltzer, B. 296 (139), *321*
Perego, G. 421 (52), *501*
Pereyre, M. 207 (101), *213*
Perkin, W. H. 577 (278, 281, 282), 578 (282), 587 (336), *606, 608*
Perkins, M. J. 33 (116), *38*
Perold, G. W. 536 (104–106), *601*
Perrin, D. D. 556 (190), *604*
Perrins, N. C. 304, 311 (245), *324*
Person, J. C. 376 (109), 377, 378 (114), 396 (175), *408, 410*
Pesnelle, P. 585 (312, 313), *607*
Pete, J. P. 532, 534, 535, 537, 557 (86), *600*
Peters, D. 516, 517, 520 (21), *598*
Peters, H. M. 202 (85), *213*, 236 (108a, 108b), *259*
Peterson, D. B. 366 (31), (176), *406, 410*
Peterson, D. R. 523, 526 (49), *599*
Peterson, L. F. A. 535 (97), *601*
Peterson, P. E. 45 (16), 46 (20–22), 47, 48 (23), 94 (189), *107, 113*, 597 (400), *610*
Petrenko-Kritschenko, P. 558 (192, 193), *604*
Petrov, A. A. 49 (28), *108*
Petterson, R. C. 457 (173), *505*
Pettit, R. 89 (179), *113*, 216 (5), 256 (238), *256, 264*, 300 (164), *322*
Pews, R. G. 553, 554 (179), *603*
Philipp, E. E. 482 (249), *507*
Phillips, W. D. 4 (22), *35*
Phillips, W. G. 224 (48), *257*
Phung, N. H. 230 (76a, 76b), 248, 249 (202a, 202b), *258, 262*
Piacenti, F. 243, 246 (172), *261*
Piccolini, R. J. 81 (154, 156), *112*
Pickett, L. W. 537 (117), *601*
Pickles, S. S. 484 (261), *508*
Pidacks, C. 203 (91), *213*
Pierre, J. L. 526 (70a), 527 (70a, 70b), 536 (109), *600, 601*
Pierson, R. M. 493 (287, 288), 494 (288), *509*
Pignataro, S. 60 (79), *109*
Pike, M. 490 (276), *508*
Pilcher, G. (29a), *599*
Pinazzi, C. 492 (286), 494 (291), 495 (292, 293), 496 (297), 498 (298, 299), 499 (299), *509*
Pinazzi, C. P. 437 (97), 496, 498 (296), *503, 509*

Pines, H. 71 (123), 105 (221), *111, 114*, 571, 572 (252, 253), 582 (296), *606, 607*
Pinkard, J. F. T. 134 (73), *170*
Pino, P. 249 (206a, 206b), 250 (206b), *262*
Pino, R. 243, 246 (172), *261*
Pinto, G. de 291 (104), *320*
Pirkles, W. H. 208 (110b), *214*
Pitcher, R. 285 (78), *319*
Pittman, C. P., Jr. 527, 560 (74), *600*
Pittman, C. U., Jr. 62 (84), 64 (84, 94), 65 (95, 97, 99, 104), 66 (104), 67 (99), *109, 110*
Pittman, C. V., Jr. 527, 560 (73), 562 (212), (119a), *600, 601, 604*
Pitts, J. N. 268 (1a, 3a), 283 (59), 290 (1a, 3a), 302 (180), *316, 318, 322*
Pitts, J. N., Jr. 375, 389 (95), *408*
Placzec, D. W. 134 (74), *170*
Planta, C. von 130 (51, 52), 152 (124), *169, 172*
Platzman, R. L. 370 (69), 371 (69–71), 375 (69, 104, 105), 376 (104), 377 (111), 379 (104), 381 (104, 124), 382 (70, 104, 105), 383 (104, 105), *407–409*
Plesch, P. H. 124 (38), *169*
Podterebkova, A. A. 559 (199), *604*
Pohl, H. A. 427 (77, 78), *502*
Polak, L. S. 432 (88), *502*
Polanyi, M. 175, 176 (2), 177 (18), *210, 211*
Poletto, J. F. 203 (91), *213*
Polievka, M. 230 (72), *258*
Pollack, J. W. 545 (148), *602*
Pomerantz, M. 292 (115a), *320*
Pommer, H. 128 (48), 153 (125), *169, 172*
Ponaras, A. A. 105 (224), *114*
Pope, B. M. 133 (71), 154 (129), *170, 172*
Pople, J. A. 3 (15), *35*, 545 (147), *602*
Popper, O. 551, 577 (276), *606*
Porri, L. 250 (213), *263*, 420 (44–47), 421 (44), *501*
Porter, M. 452 (164, 165), 460, 464, 466, 468, 470, 471 (164), 472 (164, 165), 473 (164), *505*
Potapov, V. K. 553, 556, 557 (186), 559 (196, 199, 200), 570 (246), 580 (196), *604, 605*
Pottie, R. F. 530 (79), *600*
Poulter, C. D. 288 (83), *319*

Pourcelot, G.   21 (78), *37*
Powell, H. M.   240 (152), *261*
Powell, J.   245 (180), 253 (230), *261, 263*
Praat, A. P.   239, 245 (148b), *260*
Prahl, H.   141 (88), *170*
Pratt, J. M.   231 (83, 86), *258*
Pregaglia, G.   230 (75), *258*
Pregaglio, G.   232 (243), *264*
Preobrzhenskii, N. A.   130 (55), *169*
Presting, W.   591 (371), *609*
Preuss, H.   595 (394), *610*
Prévost, C.   74 (136), *111*, 150 (117), 163 (160), *171, 172*
Price, C. C.   567 (231), *605*
Priest, W. J.   586 (320, 322), *608*
Prigge, H.   544, 545 (144), *602*
Prinzbach, H.   283 (55), 284 (66), 291 (104), 292 (109, 111), *318, 320*
Pritchard, D. E.   5 (29, 30), *36*, 544 (139), *602*
Prokofev, A. I.   547 (157), *603*
Prosen, E. M.   (29d), *599*
Prudhomme, J. C.   592 (374, 375), *609*
Psarras, T.   51, 105 (35), *108*, 593 (381), *610*
Pudjaatmaka, A. H.   68 (111), *110*
Pudovik, A. N.   159 (146), *172*
Pummerer, R.   224 (47), *257*, 460 (193), *506*
Puskar, J.   32 (108), *38*
Pusset, J.   303, 304 (198), *322*
Putte, T. V. de   234 (246), *264*
Pyatnova, Yu. B.   130 (55), *169*

Quattrochi, A.   369, 390, 391 (68), *407*

Rabinovich, E. A.   73 (130), *111*
Rabinovitch, B. S.   134 (74), *170*, 272 (11), *317*, 373 (86), *408*
Racah, E. J.   105 (224), *114*
Radlick, P.   295 (126), *321*
Raghunath, D.   435 (93a), *503*
Ralls, J. W.   154 (130), *172*
Ramakrishnan, C. S.   435 (93a), *503*
Ramaswamy, K. L.   550 (167), *603*
Ramey, B. J.   304 (199), *322*
Ramey, K.   31 (106), *37*
Ramey, K. C.   28 (101), *37*, 239, 245 (147a–c), *260*
Ramp, F. L.   461–463 (196), 500 (301), *506, 509*

Ramsey, B.   562 (213, 214), *604*
Ramsey, N. F.   2 (5), 3, 5 (9), *35*
Randic, M.   21 (79), *37*, 516 (24), 517 (24, 25), *599*
Rang, S.   32 (108), *38*
Ransaw, H. C.   415 (15), *500*
Rantwijk, F. V.   234 (246), *264*
Rao, D. V.   296 (134), 311 (237), 312 (251), *321, 324*
Rao, V. N. M.   292 (109), *320*
Raphael, K. A.   195 (55), *212*
Raphael, R. A.   512 (1), *598*
Rapoport, Z.   240 (146), *260*
Rappoport, Z.   122 (33a, 33b, 34), *168*
Ratajczak, A.   594 (387), *610*
Rathke, M. W.   224 (50), *257*
Rathman, F. F.   585 (316), *607*
Raube, R. R.   536 (99, 108), 548 (108), *601*
Raustenstrauch, V.   77 (143a), *112*
Ray, T. C.   25 (91), *37*
Raymond, F. A.   585 (317, 318), *608*
Raymond, M. A.   422 (63), *502*
Re, G. del   516, 517 (23), *599*
Read, J.   591 (372), *609*
Reddy, G. S.   4, 5 (25), 6 (25, 32), 7 (32, 42–44), 8 (32), 9, 10 (48), 12 (48, 53), 13, 14 (53), 19 (70), 22 (82), *35–37*
Red'kina, L. I.   452 (152), *504*
Redman, L. M.   77 (144), *112*
Reed, H. W. B.   250 (214), *263*
Reed, S. F.   157 (136b), *172*
Rees, T. C.   74 (135), 104, 105 (135, 217, 218), 106 (218), *111, 114*
Reese, C. B.   67 (107), *110*
Reichenbach, D.   452, 453 (159), 454 (169), *505*
Reichstein, I.   207 (103), *213*
Reid, S. T.   307 (213), *323*
Reikh, V. N.   461, 463 (198), *506*
Reilly, C. A.   15 (60), *36*
Reinhardt, P. W.   380, 381, 397 (123), *409*
Rejoan, A.   236 (255), *264*
Rembaum, A.   427 (78), *502*
Rempel, G. L.   236 (119), *259*
Renk, E.   565 (225), *605*
Renzi, A. de   240 (155), *261*
Reppe, W.   216 (2–4), 245 (4), *256*
Reynolds, W. B.   461 (195), *506*
Reyx, D.   495 (293), *509*
Rheiner, A.   292 (113), *320*

Rhodes, R. E. 201 (79), *213*, 236 (120a), *259*
Riccobono, P. X. 311 (242), *324*
Rice, H. M. 494 (290), *509*
Rice, M. R. 101 (207), *114*
Richards, C. F. 240 (150b), *261*
Richards, J. H. 90 (180), 93 (184), *113*
Richardson, W. S. 415 (13), *500*
Richey, H. G., Jr. 43, 45, 47–49 (14), 62 (84), 63 (88), 64 (84), 65 (102), 74 (134, 135), 75 (134), 94 (188), 97, 99 (194), 100 (194, 203), 104 (135, 217–219), 105 (135, 217, 218, 223), 106 (218), *107*, *109–111*, *113*, *114*, 527 (74a), 538, 553 (119), 560 (74a), 562 (119, 210), *600*, *601*, *604*
Richey, J. M. 527, 560 (74a), *600*
Richter, F. 591 (371), *609*
Rickborn, B. 56 (60), *109*, 583 (304), *607*
Rideal, E. K. 175 (1), 176 (1, 3, 10), 177 (14), 185 (40), *210*, *211*
Riebel, A. H. 486 (267), *508*
Rief, W. 128 (48), 153 (125), *169*, *172*
Riehl, A. 558 (195), *604*
Rimmelin, A. 583 (299), *607*
Rinehart, R. E. 417, 454 (32), *501*
Ring, W. 250 (218), *263*
Ringold, H. J. 178 (21), *211*
Ritchie, R. H. 385, 391 (135), *409*
Ritscher, J. S. 289 (86, 89), 311, 312 (89), *319*
Rivas, C. 291 (104), 308 (228), *320*, *323*
Roa, G. S. R. Subba 234 (247), *264*
Robb, E. W. 296, 311 (135), 312 (250), *321*, *324*
Robb, J. C. 134 (73), *170*
Robert, J.-M. 290 (94), *319*
Roberts, J. D. 8 (46), 21 (76), 36, 37, 58 (73), 71 (124), 74 (138, 139), 75 (138, 139, 141), 76, 77 (143), 81 (153, 155), 95 (190), 96 (190, 191), 97 (191, 196), 99 (155), 104 (215, 216), 105 (220), *109*, *111–114*, 219 (25), *256*, 537 (114), 544 (140), 545 (140, 145), 550 (169, 170), 555–558 (170), 560 (202, 204, 205), 565 (225), 567 (233), 585 (334), *601–605*, *608*
Roberts, R. M. 162 (157), *173*

Robertson, W. W. 532, 533 (90), *601*
Robeson, C. D. 145 (101, 102), *171*
Robins, R. D. 597 (403), *610*
Robinson, C. H. 204, 206 (93), *213*
Robinson, I. M. 429 (80), *502*
Robinson, M. G. 368, 387, 396, 397, 401 (58), *407*
Robinson, R. (8, 11), *598*
Robson, R. 307 (214), *323*
Rocek, J. 558 (195), *604*
Roedig, A. 162 (158), *173*
Rogers, M. T. 10, 11 (51), 13 (55), 36, 532, 536, 537 (89), 550 (89, 169), *601*, *603*
Rogier, E. R. 538, 541, 548 (120), *601*
Roginskii, E. 196 (60), *212*
Roginsky, S. Z. 585 (316), *607*
Rokach, J. 303 (192), *322*
Rokstad, O. A. 144 (92), *171*
Roll, D. B. 31 (105), *37*
Roman, S. A. 47 (24), *107*
Romeyn, H., Jr. 417, 454 (32), *501*
Ronayne, M. R. 368, 387, 396 (56), *407*
Rooney, J. J. 148 (111), *171*, 176 (7, 8), 181 (7, 8, 34, 35), 182(x), *211*
Roos, L. 167 (173), *173*, 225, 230 (59), *257*
Roothaan, C. C. J. 275 (19), *317*
Roquitte, B. C. 291 (101a, 107), *320*
Rosen, J. D. 302 (174), *322*
Rosen, P. 203 (91), *213*
Rosenbaum, J. 59(x)
Rosenberg, J. L. von 229 (71b), *258*
Rosenstock, H. M. 373, 379 (88, 89), *408*
Rosenthal, D. 309 (230), *323*
Rosenthal, J. W. 295 (123a, 123b), *320*
Rosinoer, Ja. M. 460 (188), *506*
Ross, I. G. 23 (89), *37*
Ross, R. A. 116 (1b), 144 (96), *167*, *171*
Rossini, F. D. 521 (30), *599*
Roth, J. A. 183 (38), *211*
Roth, W. R. 132, 133 (67), 133 (70), 160 (154a), *170*, *172*, 282 (50), 284 (67a), 296 (139), *318*, *321*
Rothman, A. M. 105 (223), *114*
Rothman, E. S. 157 (138), *172*
Rothstein, E. 577 (284), *606*
Roux, C. 492 (286), *509*
Rowe, C. A. 141 (86), 142 (89, 90), *170*

Rowe, C. A., Jr. 68 (114–118), 69 (115, 116), 70 (116), 71 (117, 118), *111*
Rowland, F. S. 399 (181), *410*
Rozanov, N. A. 577, 578 (274), *606*
Rüchardt, C. 104 (215), *114*
Rudolph, P. S. 369, 390, 393, 394 (62a, 62b), *407*
Rüegg, R. 130 (51, 52), 152 (124), *169*, *172*
Rüesch, R. 234 (248), *264*
Rugg, F. M. 433 (91), *503*
Ruhlen, J. L. 313 (263), *324*
Ruland, N. L. 399 (180), *410*
Russell, G. A. 529 (72), *600*
Ruttinger, R. 236, 237, 239, 240 (136a), *260*
Ryan, K. R. 388, 390 (145, 146), *409*
Ryhage, R. 342 (30), 343 (30, 32), *358*
Rylander, P. N. 196 (60, 61), *212*

Sabacky, M. J. 234 (253), *264*
Sabel, A. 238 (141), *260*
Sacher, A. J. 415 (13), *500*
Sack, R. A. 23 (89), *37*
Sadauskis, J. 363 (28), 368 (46), *406*, *407*
Sadova, N. I. 526 (65), *600*
Safonova, I. L. 541 (131, 132), 542 (131), 543 (132), 591 (365, 366), *602*, *609*
Safonova, J. L. 552 (176a), *603*
Sagatys, D. S. 595 (391), *610*
Saida, A. 2 (6), *35*
Saito, H. 188 (44, 45), *212*
Saito, T. 236 (116a), 249 (207b), 250 (221a), 253 (226), *259*, *263*
Sajus, L. 236 (257), *265*
Sakai, K. 585 (319), *608*
Sakai, M. 103 (214), *114*
Sakai, S. 241 (157), *261*
Salimov, M. A. 484 (258), *508*
Salinger, R. 51, 105 (35), *108*, 593 (381), *610*
Salnikova, L. G. 590 (361), 591 (368), *609*
Salomon, G. 47 (213), *506*
Saltiel, J. 120 (19, 21), *168*, 268 (4c), 274, 276 (17), 277 (17, 25b), 287, 288 (82), 290 (4c), 308, 310 (225), *316*, *317*, *319*, *323*
Saltman, W. M. 416, 417 (31), 421 (49), 493, 494 (288), 495 (295), *501*, *509*

Samson, T. A. R. 378 (115), *408*
Samuel, A. H. 386, 396, 400 (137), *409*
Sancier, K. M. (244), *605*
Sandel, V. R. 123 (36), *168*
Sandoval, I. B. 380 (120), *409*
Sandri, J. M. 539 (129), 560, 561 (206), *601*, *604*
Sandstrom, J. 552, 553 (176b), *603*
Santilli, A. A. 536, 537, 539 (107), *601*
Santry, D. P. 3 (15), *35*, 545 (147), *602*
Sarel, S. 585 (314, 315), *607*
Sargent, G. D. 48 (27), 79 (148), 84, 85, 93 (165), *108*, *112*
Sargent, M. V. 285 (76), *319*
Sarnecki, N. 153 (125), *172*
Sarnecki, W. 128 (48), *169*
Sartori, G. 424 (69), *502*
Sarycheva, I. K. 130 (55), *169*
Sass, R. L. 524(x)
Sato, S. 119 (16), 121 (27), *168*
Sato, T. 550 (168a), *603*
Sattar, A. B. M. A. 290 (99a), *320*
Sattler, L. 586 (327), *608*
Saucy, G. 130 (51), *169*
Sauer, J. 141 (88), *170*
Sauer, M. C., Jr. 360, 367 (12), *406*
Sauvage, J. -F. 179 (30, 31), 183–185 (30), *211*
Saville, B. 327, 338 (5), *357*, 452, 460 (164), 464 (164, 205), 466 (164), 468, 470 (164, 205), 471–473 (164), *505*, *506*
Savitsky, G. B. 15, 16 (59), *36*
Scala, A. A. 371, 390 (72), *407*
Scanlan, J. 443 (112), *503*
Scerbo, L. 299 (155), *321*
Schaap, A. P. 565 (221, 222), 566 (222), (223), *605*
Schachtschneider, J. H. 255 (236), *264*
Schade, G. 303 (183), *322*
Schaefer, T. 6 (35, 39), 8 (35), 24 (90), 26 (93), *36*, *37*
Schäfer, W. 291 (108), *320*
Schaffner, K. 268, 290, 291 (2d, 2e), 300, 301 (167), *316*, *322*, 537 (113), *601*
Scharf, E. 216, 250, 252 (18), *256*
Scharf, H. D. 303 (189), 312 (255), *322*, *324*
Schaub, R. F. 203 (91), *213*
Scheckendieck, W. J. 216 (2), *256*
Schee, A. Chr. van der 471 (213), *506*

Schenck, G. O.   301 (168, 169), 303 (183), 308 (226), 311 (238), *322–324*
Schenck, O.   287 (81a), *319*
Scheuerman, R. F.   524(x)
Schiavelli, M. D.   45 (16), *107*
Schissler, D. O.   360, 369 (7), *406*
Schlag, E. W.   118 (8), *167*, 395 (170), *410*
Schlatmann, J. L. M. A.   299 (152), *321*
Schleyer, P. V. R.   18 (68), *37*
Schleyer, P. von R.   31 (103), 34 (118), *37*, *38*, 41 (2), 67 (107) 87, 89 (2), *107*, *110*, 296 (140), *321*, 561, 568, 569 (209), 571 (246a, 246b, 246d), *604*, *605*
Schlichting, O.   216 (2), *256*
Schlosberg, R. H.   572, 573, 575 (261), *606*
Schlosser, M.   41 (6), *107*, 165 (158), *173*
Schmerling, L.   571 (252), 572 (251, 252), *606*
Schmid, G. H.   83 (164), 84 (164, 166), 85 (164), 93, 94 (166), *112*
Schmid, H.   302 (182), *322*
Schnack, L. G.   67 (107), *110*
Schneider, D. F.   127 (46), 128 (47, 49), 129 (49), *169*
Schneider, H. J.   560 (201), *604*
Schneider, R. A.   283 (60), 300 (159), *318*, *321*
Schneider, R. F.   523 (46), *599*
Schneider, W. G.   26 (93), *37*
Schoen, R. I.   376, 378 (108), *408*
Schollkopf, U.   163 (159), *173*
Schöllkopf, U.   67 (107), *110*
Schomaker, V.   523 (42, 44, 57a), *599*, *600*
Schoolery, J. N.   5 (28), *36*
Schor, R.   333, 334 (13), *357*
Schorta, R.   300, 301 (167), *322*
Schrage, K.   165 (166), *173*, 228 (67, 68), 246 (68), *257*, 278 (33), *317*
Schram, B. L.   373 (83), *408*
Schrauzer, G. N.   216, 221 (6), 255 (235b), *256*, *264*
Schriesheim, A.   68 (114–118), 69 (115, 116), 70 (116), 71 (117, 118), *111*, 141 (86), 142 (89, 90), *170*, 556, 563 (187), *604*
Schrock, R. W.   493, 494 (288), *509*
Schroder, G.   290 (95), 295 (128), *319*, *321*
Schroeter, S.   303 (183), *322*

Schroth, W.   132 (64, 65), *169*, *170*
Schrötter, H. W.   526 (69), *600*
Schudel, P.   154 (128), *172*
Schuett, W. R.   236 (126), *260*
Schuetz, R. D.   183 (37), *211*
Schuler, R. H.   48 (26), *107*, 367 (41), 368 (59), *406*, *407*
Schulte-Frohlinde, D.   277 (25a), *317*
Schultz, A. R.   433 (90), 464 (203), *503*, *506*
Schultz, R. G.   240 (153a), (39), *257*, *261*
Schulz, G. V.   460 (192), *506*
Schunes, R. L.   193 (53), *212*
Schwarz, H. A.   400 (186), *410*
Schweiter, U.   130 (51, 52), 152 (124), *169*, *172*
Schwendeman, R. H.   523 (47), *599*
Scott, K. W.   256 (241), *264*
Scott, L. T.   295 (124), *321*
Searles, S.   369, 390, 391, 393, 400, 402 (67), *407*
Searles, S., Jr.   45 (17), *107*
Sedlmeier, J.   237 (140), 238 (141), 240 (144), *260*
Seebach, D.   307 (216, 217), *323*
Seely, G. R.   450 (140), *504*
Seibt, H.   216, 250, 252 (18), *256*
Seidner, R. T.   295 (130), 296 (130a), *321*
Selke, E.   236 (255), *264*
Selman, L. H.   234 (248), *264*
Selover, J. C.   557 (190a), *604*
Selwitz, C. M.   237 (139), *260*
Semeluk, G. P.   117, 118 (7), *167*
Semenow, D. A.   97 (196), *113*
Senda, Y.   188 (44, 47), *212*
Sequeira, R. M.   67 (107), *110*
Servis, K. L.   95 (190), 96 (190, 191), 97 (191), *113*
Sestanj, K.   300 (161), *322*
Setser, D. W.   272 (11), *317*, 373 (86), *408*
Seyferth, D.   49, 53 (30), 54 (47, 49), 56 (30), 74 (133, 137), 75 (140), *108*, *111*, 163, 164 (163), *173*
Seyforth, D.   291 (103), *320*
Seyler, J. K.   197 (68, 69), 198 (72), *212*, *213*, 231 (79a, 79c), 232 (87), *258*
Seyler, R. C.   219 (27), *257*
Shabarov, Y. S.   541, 542 (133), 553, 556, 557 (186), 559 (196, 199, 200), 570 (246), 580 (196), 581 (295), *602*, *604*, *605*, *607*

Shabarov, Yu. S.   130 (56), *169*
Shafer, P. R.   104 (215), *114*, 565 (225), *605*
Shah, V. P.   533 (92), *601*
Shanazarova, I. M.   581 (295), *607*
Shannon, J. S.   327, 328 (3), *357*
Shannon, P. V. R.   186 (41), 187(x), *211*
Shannon, T. W.   348 (43), *358*
Sharma, R. K.   303 (196), *322*
Sharman, S. H.   62 (85), *109*, 148, 149 (110), *171*
Sharp, J. C.   224 (48), *257*
Sharpe, T.   527, 563 (76), *600*
Shatavsky, M.   99 (199), *113*
Shatenshtein, A. I.   53, 68 (44), 73 (130), *108*, *111*, 594 (385), *610*
Shaw, B. L.   225, 230, 244 (60), 245 (180), 248, 249 (200), 250, 252 (212a, 212b), 253 (230), *257*, *261–263*
Shaw, D. F.   333–335, 343 (14), *357*
Shaw, D. L.   581, 583 (293, 294), *607*
Shechter, H.   537 (116), 596 (396), *601*, *610*
Shelton, J. R.   445 (123), 477, 483 (225), *504*, *507*
Shemyakin, M. M.   128, 129 (49), *169*
Shenian, P.   536, 537, 539 (107), *601*
Sheppard, N.   4, 5 (26), 6, 7 (37), 18 (69), *35–37*, 471 (212), *506*
Sheridan, J.   589, 590 (351), *609*
Sherwin, M. A.   296 (132), *321*
Shibano, T.   249 (210), *263*
Shida, S.   375 (98), 382 (127), 395 (98, 127), *408*, *409*
Shier, W.   208 (107), *214*
Shikata, K.   236 (105), *259*
Shim, K. S.   255 (234a), *264*
Shima, K.   121 (25), *168*
Shimazaki, N.   249 (208), *263*
Shimer, K.   (176), *410*
Shimizu, K.   518 (28), 519 (28, 28a), *599*
Shionoya, M.   179, 183 (29), *211*
Shipley, F. W.   458 (179, 180), 471 (215), *505*, *506*
Shipman, J. J.   452 (151, 155), 453, 471 (155), *504*
Shmidt, F. K.   236 (134), *260*
Shmulyalovskii, Y. E.   575, 590 (271), *606*
Shoemaker, J. D.   553, 564 (180), *603*
Shoji, A.   421 (57), *502*

Shoolery, J. N.   7 (41), *36*
Short, A. G.   591 (372), *609*
Short, J. N.   417 (35, 36), 452 (36), *501*
Shortridge, R. W.   590, 591 (355), *609*
Shoshan, R. E.   585 (314), *607*
Shostakovskii, S. M.   541, 542 (134–136), 580 (135), *602*
Shryne, T. M.   417 (33), *501*
Shugemitsu, Y.   308 (223), *323*
Shumate, K. M.   281, 296 (45), *318*
Shvedchikov, A. P.   384 (133), *409*
Sianesi, D.   118 (11), *168*
Sidler, J. D.   564 (215), *605*
Sieber, R.   236, 237 (136a), 238 (141), 239, 240 (136a), *260*
Siebert, A. E.   306 (212), *323*
Sieck, L. W.   403 (190), *410*
Siegel, H.   234 (253), *264*
Siegel, S.   179 (25–28, 32), 180 (x, 25, 26, 33), 181 (x, 33), 182(x), 183 (x, 26, 28), 184 (x, 26, 39), 185, 186(x), 188 (26), *211*, 560 (203), *604*
Sih, N. C.   105 (221), *114*
Silas, R. S.   417 (37), *501*
Silber, P.   299, 300 (157), *321*
Siler, J. D.   163 (159), *173*
Silver, M. S.   97 (196), 104 (215), *113*, *114*
Simandi, L.   231 (85), *258*
Simmons, H. E.   162 (155), *173*
Simmons, J. W.   272 (11), *317*
Simon, W.   16, 17 (65), *36*
Simonetta, M.   57, 62 (69), 81, 95 (152), *109*, *112*, 161 (154c), *172*
Simonnin, M. P.   21 (78), *37*
Simons, J. H.   571, 572 (255), *606*
Simons, J. P.   304, 311 (245), *324*
Sims, J. J.   234 (248), *264*
Singer, L. A.   48 (27), *108*
Sinha, S. P.   523 (38), *599*
Sircar, A. K.   416 (25), *501*
Sivertz, C.   449 (139), *504*
Sivetrz, C.   119 (14), *168*
Skattebol, L.   160 (153), *172*, 290 (98), 299 (153), *320*, *321*, 537 (118), *601*
Skell, P. S.   87 (173), *112*, 118 (13), *168*, 303 (186), *322*
Skinner, H. A.   (29a), *599*
Skuratov, S. M.   552 (176a), *603*
Slabey, V.   548 (165), *603*
Slabey, V. A.   590 (364), 591 (369, 370), 592 (369), *609*

Slates, C. D.   186 (42), *211*
Sliam, E.   295 (129), *321*
Slichter, C. P.   2 (6), *35*
Sloan, M. F.   202 (88), *213*, 236 (135), *260*
Slobodin, Y. A.   575, 590 (271), *606*
Slomp, G.   34 (120), *38*, 315 (269), *325*
Smadja, W.   145   (98, 99), *171*
Smedvik, L.   118 (12), *168*
Smejkal, J.   553 (181), *603*
Smets, G.   491 (291), *508*
Smid, J.   41 (7), *107*
Smidt, J.   236 (136a, 136b), 237 (136a, 136b, 140), 238 (141), 239 (136a), 240 (136a, 136b, 144), *260*
Smith, A. H.   74, 75 (134), *111*
Smith, D.   327 (6), *357*
Smith, G. G.   596 (398), *610*
Smith, G. V.   31 (102), *37*, 178 (24), 179 (26, 27), 180 (26), 182 (36), 183 (x, 26, 38), 184 (x, 26, 39), 185, 186(x) 188 (26), 195 (56), *211*, *212*
Smith, G. W.   297 (142), *321*
Smith, H.   203 (90), 205 (96), *213*
Smith, H. P.   417, 454 (32), *501*
Smith, J. F.   477, 478, 480, 482 (226), *507*
Smith, J. J.   433 (91), *503*
Smith, L. A.   32 (112), *38*
Smith, L. I.   538, 541, 548 (120), *601*
Smith, P. T.   373 (82), *407*
Smith, S. L.   285 (75b), *319*
Smith, S. R.   333, 334 (13), *357*
Smith, W. B.   192 (51), *212*
Smutkina, Z. S.   432 (86), *502*
Smutny, E. J.   249 (211), *263*
Smythe, J. A.   224 (46), *257*
Sneen, R. A.   97 (193), *113*
Snell, R. L.   291 (102), *320*
Snyder, E. I.   8 (46), 21 (76), 34 (119), *36–38*, 207 (99), 210 (117), *213*, *214*
Snyder, L. C.   100 (205), *114*, (39), *599*
Sobalev, E. V.   541–543 (138), *602*
Solodovnikov, S. P.   547 (156), *603*
Solomon, S.   160 (153), *172*
Sondheimer, F.   285 (76), *319*
Song, K. M.   236 (116a), *259*
Sonntag, F. I.   284 (71), 297 (144), 315 (267), *319*, *321*, *324*
Sonogashira, K.   236 (105), *259*

Sorenson, T. S.   62 (86), 65, 67 (96, 98–101), *109*, *110*
South, A., Jr.   581, 583 (294), 597 (403), *607*, *610*
Sowa, F. J.   572, 573 (263, 264), 575 (264), *606*
Spangler, S. W.   126 (42), *169*
Sparapany, J. J.   395 (170), *410*
Speakman, S. C.   523 (51), *599*
Spencer, L. V.   385 (136), *409*
Spencer, R.   116, 117 (3), *167*
Spinks, J. W. T.   362, 386, 387 (24), *406*
Spinrad, B. I.   550 (171), *603*
Spitsbergen, J. C.   484 (259), *508*
Spitz, R. P.   285 (78), *319*
Spitzer, R.   515, 520 (6), *598*
Springall, H. D.   157 (137), *172*
Srinivasan, R.   271   (8), 272 (12), 283 (54), 284 (71), 286, 287 (80), 290 (91), 291 (105), 297 (141, 143, 144), 298 (145, 146, 148), 299 (141), 302 (177), 303 (190), 311 (246), 315 (267), *316–322*, *324*
Srivastava, S. B.   (176), *410*
Stacey, C. L.   484 (255), *508*
Stadler, P.   92 (182), *113*
Stadler, P. A.   92 (182), *113*
Ställberg-Stenhagen, S.   342 (30), 343 (30, 32), *358*
Stamhuis, E. J.   45 (17), *107*
Staniland, P. A.   186 (41), 187(x), *211*
Staples, C. E.   143 (91a), *170*
Stark, R. E.   337, 350, 354 (20), *357*
Statton, G. L.   239, 245 (147a, 147b), *260*
Staudinger, H.   460, 461 (191, 192), *506*
Staveley, F. W.   415 (21), *501*
Steacie, E. W. R.   586 (325), 595 (392), *608*, *610*
Steadman, T. R.   248, 249 (201), *262*
Stechl, H.   313 (259), *324*
Stedman, R. J.   302 (175, 176), *322*
Steel, C.   117 (7), *167*
Steen, D. van der   131 (57, 58), *169*
Steen, K.   158 (145), *172*
Stehling, F. C.   18 (67), *36*
Stein, G.   121 (26), *168*
Steiner, E. G.   54 (57), *108*
Steinmetz, H.   449 (138), *504*
Steinmetz, R.   287 (81a), 301 (169), 308 (226), 311 (238), *319*, *322–324*

Stenhagen, D. E.   343 (32), *358*
Stepanyants, A. U.   35 (125), *38*
Stephen, M. J.   2 (2, 7), *35*
Stephens, C. L.   120 (23), *168*, 451, 455 (146, 147), 457 (147), *504*
Stephens, R. D.   208 (105), *213*
Stephenson, J. L.   154 (133), *172*
Sterin, K. E.   541 (131, 132, 132a, 138), 542 (131, 132a, 138), 543 (132, 138), 590 (362), (343), (363), *602, 608, 609*
Stermitz, F.   268, 290, 291 (2g), *316*
Stern, E. W.   216, 237 (10b), *256*
Stern, R.   236 (257), *265*
Sternberg, H. W.   244 (176), 246 (188), *261, 262*
Stettiner, H. M. A.   244 (175), *261*
Stevens, P. G.   548, 585 (160a), *603*
Stevens, R. D.   94 (187), *113*
Stevenson, D. P.   334 (17), 338 (26), 340 (27), 342 (29), *357*, 360, 369 (7), 372, 380 (77), *406, 407*
Stevenson, P. E.   52 (36), *108*
Stewart, E. T.   275 (21), *317*
Stewart, J. E.   471 (214), *506*
Stewart, J. M.   587 (339), *608*
Stewart, W. E.   7 (44), 21 (72), *36, 37*
Stiles, M.   (66), *600*
Stille, J. K.   240 (151a, 151b), *261*, 421 (56), *501*
Stockwell, P. B.   236 (118), *259*
Stoessl, A.   290 (99a), *320*
Stogryn, E. L.   15 (61), *36*
Stolberg, U. G.   236 (129), *260*
Štolka, M.   443 (116), 445, 446 (122), 447 (116, 122, 126–128), 460 (128), *503, 504*
Stone, F. G. A.   165 (167), *173*, 229 (70), 255 (234c), *257, 264*, 572, 575 (268), *606*
Stoppa, G.   420 (46), *501*
Stork, G.   203 (91), 205 (94), *213*
Story, P. R.   83 (161), 100 (205), *112, 114*, 485 (262), *508*
Stothers, J. B.   34 (123), *38*, 146 (104), *171*, 545 (151), *602*
Strait, L. A.   533 (92), *601*
Strauss, S.   487 (274), 488 (275), 489 (274, 275), *508*
Streith, J.   285 (77), *319*, 585 (312), *607*
Streitwieser, A., Jr.   52, 53 (39, 42), 57 (65–67), 58 (72), 63 (87), 68 (111), 76 (66), 77 (146), 81 (72), 82 (158), 87 (146), *108–110, 112*, 593 (383, 384), *610*
Strohmeier, W.   33 (115), *38*
Studebaker, J.   33 (116), *38*
Su, T.   67 (107), *110*
Suart, S. R.   234 (245), *264*
Sudborough, J. J.   558 (191), *604*
Sugden, T. M.   513 (12), *598*
Sugimoto, K.   119 (15, 16), *168*
Sumilov, V. S.   461 (200), *506*
Summers, G. H. R.   571 (246c), *605*
Summitt, R.   10, 11 (51), *36*
Sunder-Plassmann, P. 306 (208, 210), *323*
Sundralingham, A.   478 (234), *507*
Sunko, D. E.   85 (167), *112*
Surridge, J. H.   210 (117, 118), *214*
Susuki, T.   247 (190a), *262*
Sutherland, G. B. B. M.   471 (212), *506*
Suzuki, K.   231 (79b), *258*
Suzuki, R.   163, 164 (163), *173*
Svirina, V. S.   559 (199), *604*
Swift, P. McL.   458 (182), *505*
Swindell, R.   284 (69), *318*
Swindell, R. T.   96 (192), *113*
Swinney, F. B.   416 (29), *501*
Swoap, J. R.   182 (36), *211*
Sworski, T. J.   360 (16), 367 (43), 374 (16), 390 (157), 393, 404 (16), *406, 410*
Sykes, P. J.   131 (60), *169*
Symons, M. C. R.   59(x)
Syrkin, Y. K.   515 (15), *598*
Syrkin, Ya. K.   237 (138), 240 (145a), *260*
Szabo, I.   393 (160), *410*
Szarkowska, L.   459 (184), *505*
Szwarc, M.   424 (67), *502*

Tada, M.   305 (205), *323*
Taddei, F.   15 (62), *36*
Tadokoro, G.   562 (208a), *604*
Taft, R. W., Jr.   531, 568 (83), *600*
Tai, S.   249 (203b), *262*
Taiman, J. D.   34 (123), *38*
Takahashi, H.   241 (159, 160), 249 (203a, 203b), 250 (217), *261–263*
Takahashi, S.   249 (210), *263*
Takahashi, Y.   241 (157), *261*
Takahiro, T.   316 (274), *325*
Takegami, Y.   236 (105, 109a–e), 246 (186), *259, 262*
Takes, H. V.   550 (172), *603*

Taketomi, T.   249 (207a), *262*
Takeuchi, Y.   234 (248), *264*, (238), *605*
Taliferro, J. D.   586 (335a), *608*
Taller, R. A.   34 (121), *38*
Tal'rose, V. L.   360 (6), 369 (6, 61), 390 (61), 405 (6), *406*, *407*
Tamai, K.   249 (204b), *262*
Tamelan, E. E. van   285, 290 (74), 295 (125, 127), *319*, *321*
Tanaka, I.   389 (148), *409*
Tanaka, M.   119 (16), *168*
Tanaka, S.   249 (208), *263*
Tani, K.   253, 254 (227a), *263*
Tao, E. V. P.   46 (22), *107*
Tarale, S. S.   524(x)
Tartakovskii, A.   580 (287, 290, 292), *607*
Tate, J. T.   373 (82), *407*
Tayim, H. A.   201 (84), *213*, 225, 230 (64), 236 (64, 131a), 244 (64), *257*, *260*
Taylor, D. R.   49 (28), *108*
Taylor, E. C.   315 (270), *325*
Taylor, H. S.   360, 369 (3), *405*
Taylor, J. W.   311 (237), *324*
Taylor, R.   395 (169), *410*
Taylor, T. I.   177 (12), *211*, 342 (31), *358*
Taylor, W. J.   537 (116), *601*
Teegenand, J. P.   361(x)
Temin, S. C.   464 (203), *506*
Tenge, R. O.   342 (31), *358*
Terabe, S.   67 (107), *110*
Terao, T.   119 (16), 121 (27), *168*
Testa, E.   144 (95), *171*
Tewksbury, L. B.   493 (287, 288), 494 (288), *509*
Tezuka, T.   284 (69), *318*
Tezuki, T.   316 (273), *325*
Theuer, W. J.   285 (79), *319*
Thill, B. P.   123 (36), *168*
Thoai, N.   292 (112), *320*
Thomas, D. A.   132 (62), *169*
Thomas, G. E.   (81), *407*
Thomas, P. A.   180, 181 (33), 182(x), *211*
Thomas, R. M.   577 (279), *606*
Thompson, G.   597 (400), *610*
Thornton, R. E.   205 (96), *213*
Thornton, V.   417 (35, 37), *501*
Thorpe, J. F.   587 (338), *608*
Thygagorajan, B. S.   594 (386), *610*
Tibbals, H. F.   391 (159), *410*
Tidd, B. K.   34 (122), *38*
Tidwell, T. T.   596 (397), *610*
Tieckelmann, H.   159 (147), *172*
Tikhomirov, B. I.   461 (179–200), 463 (198), *506*
Tillieu, J.   13 (54), *36*
Timmons, R. J.   208 (109), *214*
Tinyakova, E. I.   452 (152), *504*
Tipper, C. F. H.   570 (245), 571 (258), 572 (258, 269), 573 (257, 265), 575 (269), *605*, *606*
Tobias, M. A.   582, 583 (302, 306, 307), 584 (307), 590 (302), *607*
Tobkes, M.   564 (219), *605*
Tobolsky, A. V.   477 (228), 478 (228, 239), 480 (228), 481 (248), *507*
Tocker, S.   443 (113), *503*
Todd, J. E.   547 (155), *602*
Todd, R. W.   586 (328), *608*
Toepel, T.   216 (2), *256*
Toi, Y.   366 (31), *406*
Toland, W. G., Jr.   577 (279), *606*
Tom, D. H. E.   439 (103), *503*
Tomomura, G.   179, 183 (29), *211*
Topchiev, A. V.   432 (88), *502*
Topsom, R. D.   596 (397), *610*
Tori, K.   545 (149), *602*
Torimitsu, S.   209 (116), *214*
Tornqvist, E. G. M.   412 (3), 413 (3, 8), *500*
Toscano, V.   160 (152), *172*
Toubiana, R.   343, 344 (33), *358*
Trachtenberg, E. N.   547 (154a), 566, 567 (226), *602*, *605*
Trachtman, M.   375, 395 (97), *408*
Trahanovsky, W. S.   83 (160), 84 (166), 86 (169), 93, 94 (166), *112*
Trainor, J. T.   10, 11 (51), *36*
Trapasso, L. E.   461–463 (196), 500 (301), *506*, *509*
Traube, W.   207 (102), *213*
Trebellas, J. C.   225, 244 (55), *257*
Trecker, D. J.   255 (235a), *264* (258), *324*
Trego, B. R.   470 (211), *506*
Treshchova, E. G.   541, 542 (133–137), 580 (135, 291), *602*, *607*
Trick, G. S.   452 (149), 493, 494 (288), *504*, *509*
Trifan, D. S.   571 (246b), *605*
Trinajstić, N.   517 (25), *599*
Trippett, S.   126 (43), *169*
Trocha-Grimshaw, J.   236 (121b), *360*
Trotman-Dickenson, A. F.   116, 117 (3), *167*, 398, 399 (179), *410*

Trotman-Dickinson A. F.   595   (392), 610
Truce, W. E.   572, 573 (262), 577, 587 (283), 606
Truchan, A.   577 (279), 606
Truett, W. L.   429 (80), 502
Truscheit, E.   153 (127), 172
Truter, M. R.   250, 252 (212a, 212c), 263
Tschugaeff, L.   548 (159), 603
Tschuikow-Roux, E.   374, 395 (94), 408
Tsitokhtsev, V. A.   495 (294), 509
Tsuchiya, S.   148 (109), 171
Tsuji, J.   224 (45), 233 (96), 241 (159, 160), 243 (171), 245 (181), 247 (190a, 190b), (179a, 179b), 257, 258, 261, 262, 562 (208a), 604
Tsurugi, J.   455 (172), 505
Tsutsumi, S.   308 (223), 323
Tuley, W. F.   196 (61), 212
Tulupov, V. A.   236 (106, 107, 110), 259
Tupulov, V. A.   236 (132), 260
Turkevich, J.   589, 590 (349), 608
Turnbull, J. H.   576, 590 (272), 606
Turner, A.   178 (21), 211
Turner, D. T.   457 (176), 458 (176, 178), 475 (218, 219), 476 (218, 219, 223), 487 (218), 505, 507
Turner, D. W.   530 (82), 600
Turner, J. J.   21 (79), 37
Turner, J. O.   64 (94), 65 (97, 103), 66 (103), 110, 527 (73a), 538, 553 (119), 560 (73a), 562 (119), 600, 601
Turner, R. B.   194(x) 521 (31), 599
Turro, N. J.   120 (19), 168, 268 (3b), 274, 276, 277 (17), 285 (73), 286 (81), 290 (3b), 291 (101b), 310 (235), 312 (256), 313 (264), 314 (264, 266), 315 (268), 316, 317, 319, 320, 324, 325
Tutsumi, S.   242 (161), 261
Twigg, G. H.   176 (3, 10), 210, 211
Twigg, G. T.   177 (15), 211
Tyler, W. P.   425 (72), 502

Uchida, K.   253 (226), 263
Uchida, T.   249 (207b), 263
Uchida, Y.   236 (116a), 249 (204a, 204b), 250 (221a, 221b), 253 (226), 259, 262, 263
Uchino, M.   248, 249 (197), 262
Uchiyama, M.   160 (151), 172

Uda, H.   304 (201, 203d), 322, 323
Uemura, S.   241 (158), 261
Ueno, T.   236 (109a, 109b, 109d, 109e), 259
Ugelstad, J.   144 (92), 171
Ugo, R.   232 (243), 264
Ulery, H. E.   90 (180), 113
Ullman, E. F.   271 (9), 290, 292, 294 (96), 316, 319, 589 (373), 609
Umezawa, H.   399 (181), 410
Ungvary, F.   236 (112b), 259
Untch, K. G.   33 (114), 38
Urschler, E.   292 (113), 320
Ustynyuk, T. K.   583 (300), 607
Uttech, R.   221 (35), 257
Uyeda, R. T.   72 (126), 111, 136 (81), 170

Valade, J.   207 (101), 213
Valentine, D.   315 (268), 325
Valkanas, G.   149 (113), 171
Vallarino, L. M.   240 (154), 261
Valvassori, A.   424 (69), 502
Van Auken, T. V.   33 (117), 38
Van der Ent, A.   234 (252), 264
Van Dine, G. W.   561, 568, 569 (209), 604
Vane, F. M.   34 (121), 38
Van Overstractem, A.   208 (112), 214
Van Tamelen, E. E.   208 (109, 110a, 110b, 111), 210 (110b), 214
Vargaftik, M. N.   237 (138), 240 (145a, 145b), 260
Vasilev, N. I.   541, 542 (133), 602
Vaska, L.   201 (79), 213, 225, 230 (62), 236 (62, 120a, 120b), 244 (62), 257, 259
Vaughan, A. L.   373 (82), 407
Vaughan, G.   476 (221), 507
Vaughan, L. G.   54 (47), 108, 163, 164 (163), 173
Vedeneyev, V. I.   338 (25), 357
Veh, G. von   292 (109), 320
Veilard, A.   516, 517 (23), 599
Venanzi, L. M.   240 (152, 154), 261
Venner, E. L.   247–249, 253, 254 (194), 262
Venus-Danilova, E. D.   578, 579 (285), 606
Vermont, G. B.   311 (242), 324
Vernon, C. A.   60 (80), 109
Vessel, E. D.   421 (56), 501
Vest, R. D.   162 (155), 173
Vestal, M.   373 (85, 87), 379 (85), 380 (119), 408, 409

Viebrock, J. 252 (223), *263*
Viehe, H. G. 118 (9, 10), *168*, 303 (185), *322*
Vierroll, H. 523 (35), *599*
Vilkov, L. V. 526 (65), *600*
Vinogradova, N. D. 495 (294), *509*
Viola, A. 159 (149), *172*
Viswanathan, N. S. 395 (171), *410*
Vit, L. 219 (26), *256*
Vobeshchalova, N. 248, 249 (199), *262*
Vodehnal, J. 443 (116), 445, 446 (122), 447 (116, 122, 126, 127), *503*, *504*
Voelter, W. 233 (93, 251), *258*, *264*
Voge, H. H. 334 (17), *357*
Vogel, A. J. 548 (161), *603*
Vogel, E. 279, 290, 296 (93), *319*, 512 (2), *598*
Vogel, W. 556 (188), *604*
Vogt, V. 120 (19), *168*, 274, 276, 277 (17), *317*
Voĭtenko, R. M. 432 (88), *502*
Volger, H. C. 256 (239c, 240), *264*
Vol'kenau, N. A. 54 (56), *108*
Volkenburgh, R. V. 590 (354), *609*
Volkenburgh, R. van 548 (162, 163), 590 (162), *603*
Volodkin, A. A. (341), *608*
Volpi, G. G. 363 (26), *406*
Vries, B. de 231 (81), *258*
Vrieze, K. 239, 245 (148a, 148b), *260*

Waack, R. 52 (36–38), 73 (127, 128), *108*, *111*, 143 (91b), *170*, (119a), *601*
Wade, K. 41 (5, 8), *107*
Wadelin, C. 477, 483 (225), *507*
Waegell, B. 31 (106), *37*
Wagner, C. D. 334 (17), 342 (29), *357*, 360 (9–11), 394 (9, 10), 395 (11, 162, 163), 401 (9, 10), 404 (9), 405 (11), *406*, *410*
Wahrhaftig, A. L. 373, 379 (85, 88), 380 (119), *408*, *409*
Waight, E. S. 149 (113), *171*
Wakefield, B. J. 42 (11), *107*
Walborsky, H. M. 595 (388–390), *610*
Walch, S. P. 289 (87a), *319*
Walker, D. A. 571 (258), 572 (258, 269), 573 (257), 575 (269), *606*
Walker, D. C. 362 (21), *406*
Walker, K. A. M. 199 (74–76), 200 (75), *213*, 232 (90, 92), 233 (90, 92, 94, 95), 236 (95), *258*
Wall, L. A. 487 (272), *508*
Wallach, M. L. 443 (115), *503*
Wallenberger, F. T. 444 (118), *503*
Wallenstein, M. B. 373, 379 (88), *408*
Walling, C. 105 (224), *114*
Wallis, E. S. 576, 590 (272), *606*
Walsh, A. D. 361(x), 513 (14), 540, 544 (127), (7, 10, 13), *598*, *602*
Walton, D. R. M. 45 (15), *107*
Walz, H. 21 (80), *37*
Wan, J. K. S. 283 (59), 302 (180), *318*, *322*
Wannier, G. H. 372 (79), *407*
Warburton, W. K. 132 (62), *169*
Ward, J. A. 368, 387, 396 (56), *407*
Warman, J. M. 397 (177), *410*
Warmant, J. M. 368 (59), *407*
Warrener, R. N. 268, 290 (1d) *316*
Wartman, L. H. 433 (91), *503*
Wassermann, A. 422 (60), *502*
Watanabe, K. 360, 361 (18), 378 (115), *406*, *408*, 530 (80), *600*
Watanabe, Y. 246 (186), *262*
Watkins, R. J. 277 (24), *317*
Watkinson, J. G. 567, 581 (234), *605*
Watson, A. A. 464, 468, 470 (205), *506*
Watson, H., Jr. 368, 387, 396 (54), *407*
Watson, W. F. 415, 454, 455 (14), 458 (14, 181, 182), 490 (276), 491 (277, 278), *500*, *505*, *508*
Watts, L. 300 (164), *322*
Watts, V. S. 6 (33), 22 (82, 83, 85), 23, 24 (83), 25 (85, 92), 26 (92), *36*, *37*, 544 (143), *602*
Waugh, J. S. 54 (49), 74 (137), *108*, *111*
Webb, G. 181 (34), 190, 192 (49), *211*, *212*
Weber, K. W. 547 (155), *602*
Weber, M. 250 (218), *263*
Wechter, W. J. 34 (120), *38*
Weedon, B. C. L. 126 (45), 128 (47, 49), 129 (49), 130 (53b, 56), *169*, 536 (103), 548 (163a), 587 (337), 591 (163a), (94a), *601*, *603*, *608*
Weide, O. B. van der 590 (350), *609*
Weigert, F. J. 545 (145), *602*

Weimann, J. 292 (112), *320*
Weinberg, D. S. 335, 351 (19), *357*
Weiner, M. A. 54 (49), 74 (133, 137), *108*, *111*
Weir, M. R. S. 35 (126), *38*
Weis, L. D. (27), *317*
Weisbuch, F. 292 (112), *320*
Weisgerber, G. 303 (189), *322*
Weiss, F. T. 220 (33), *257*
Weiss, M. J. 203 (91), *213*
Weissler, G. L. 376, 378 (107), *408*
Weitkemp, A. W. 195 (57), *212*
Wells, P. B. 190, 192 (49), *212*
Wender, I. 244 (176), 246 (184, 188), *261*, *262*
Wendisch, D. 526 (63), *600*
Wendler, N. L. 98 (197), *113*
Wentworth, W. E. 381, 397 (126), *409*
Werf, C. A. V. 150 (116), *171*
Werstiuk, N. H. 102 (210), *114*, 584 (310), *607*
West, P. 52 (37), *108*
Westburg, H. H. 587 (339), *608*
Westman, T. L. 94 (187), *113*
Westphal, Y. L. 284 (70), *318*
Weth, E. 282 (51b), *318*
Wexler, S. 369, (64, 68), 390 (64, 68, 151), 391 (64, 68), 393, 394 (64), *407*, *409*
Whalen, D. L. 301 (171), *322*
Wharton, P. S. 527 (70), *600*
Wheeler, D. M. S. 277 (24), *317*
Whelan, J. M. 436 (94), *503*
Wheland, G. W. 57 (64), *109*, 550 (174), *603*
Whetsone, R. R. 177 (19), *211*
Whiler, D. F. 535 (96), *601*
Whipple, E. B. 7 (44), 21 (72, 73), 26 (94, 95), 29–31 (95), *36*, *37*, 48 (26), *107*, 255 (235a), *264*
White, E. H. 313 (262), *324*
White, W. N. 97 (196), *113*
Whitear, B. R. D. 300 (162), *322*
Whitehead, M. A. 547 (155), *602*
Whitehurst, D. D. 240 (151a), *261*
Whitehurst, J. S. 204 (92), *213*
Whitesides, G. M. 74, 75 (139), 76, 77 (143), *111*, 164 (164), *173*
Whitham, G. H. 67 (107), *110*, 281 (42), *318*
Whiting, M. C. 71 (123), *111*, 132 (63), 136 (80), 158 (143), *169*, *170*, *172*, 278 (34), *317*, 555 (x)
Whitla, W. A. 240 (152), *261*

Whyte, G. N. 366, 367, 385 (36), *406*
Wiberg, K. B. 97 (195), *113*, 238 (142), *260*, 523 (56), 539, 540 (128), 544 (150), 545 (128), 582 (297a), 596 (395), *600 602*, *607*, *610*
Wick, A. E. 158 (145), *172*
Wicklatz, J. 460, 461 (194), *506*
Wiedemann, W. 67 (107), *110*
Wieland, P. 233 (250), *264*
Wiesel, M. 296 (130a), *321*
Wiesner, K. 304 (203f), *323*
Wilcox, C. F. 291 (106), *320*
Wilcox, C. F., Jr. 96 (192), *113*, 562 (208b), *604*
Wiley, R. N. 15, 16 (58), *36*
Wilke, G. 216 (16, 17), 219 (29), 221 (34), 222 (34, 37), 248 (16, 17, 196), 249 (196), 250 (16, 17, 216), 252 (17), 255 (16), *256*, *257 262*, *263*
Wilkes, R. A. 454, 459 (171), *505*
Wilkinson, G. 196 (64), 199 (64, 73), 200 (64), 201 (80, 81), *212*, *213*, 224 (44), 231 (80), 232 (88), 233 (88, 98, 99), 234 (88b, 98, 252), 236 (102, 104, 118, 254, 258), *257–259*, *264*, *265*
Wilkinson, G. W. 236 (258), *265*
Williams, C. A. J. 304 (203f), *323*
Williams, D. H. 327 (4), 335, 342 (23), 353 (40), (34, 38), *357*, *358*
Williams, F. 387 (140, 142), 395 (169), *409*, *410*
Williams, F. T., Jr. 537 (116), *601*
Williams, I. H. 304 (200), *322*
Williams, J. 54, 74 (53), *108*
Williams, J. C. 132 (61), *169*, 189, 190 (x, 48), *212*
Williams, K. L. 546 (154), *602*
Williams, R. D. 247 (189), *262*
Williams, R. J. P. 231 (86), *258*
Williams, R. O. 102 (210), *114*, 584 (310), *607*
Williams, R. R. 255 (234b), *264*
Williams, R. R., Jr. 366, 367 (32), *406*
Williams, T. F. 394, 395 (161), *410*
Williamson, K. L. 546 (153), *602*
Willstatter, R. 588, 590 (342), *608*
Wils, R. G. 195 (55), *212*
Wilson, J. W. 272 (10), *316*
Wilson, K. W. 71 (124), *111*
Wilzbach, K. E. 289 (85, 86, 87a, 89), 311 (89, 248), 312 (89), *319*, *324*

Winfield, M. E. 198 (71), *213*, 231 (84), *258*
Wingler, F. 77 (143a), *112*
Winkler, B. 296 (138), *321*
Winkler, D. E. 431 (85), *502*
Winslow, F. H. 477 (227), *507*
Winstein, S. 60 (81, 83), 62 (85), 78 (147), 81 (152, 154, 156), 88 (178), 95 (152), 99 (199–201), 100 (204, 206), 101 (206), 103 (212, 214), *109*, *112–114*, 132, 133 (66, 67), 148 (110), 149 (110, 115), *170*, *171*, 240 (146), *260*, 282 (51c), 291 (106), 296 (137), *318*, *320*, *321*, 546 (152), 561 (208), 562 (211), 588 (340), (166), *602–604*, *608*
Winterbottom, J. M. 190, 192 (49), *212*
Winters, R. E. 361(x)
Wipke, W. T. 274, 281, 288, 299, 303, 304 (15), *317*
Wise, P. H. 590 (364), 591, 592 (369), *609*
Wise, W. B. 239, 245 (147c), *260*
Wiskott, E. 296 (140), *321*
Wisotsky, M. J. 63 (88), *110*, 562 (210), *604*
Witt, E. 476 (220), *507*
Witt, H. S. 417, 454 (32), *501*
Wittenberg, D. 216, 250, 252 (18), *256*
Witteneau, M. S. von 308 (227), *323*
Wittig, G. 77 (143a), *112*, 126 (44), *169*
Wittstruck, T. A. 547 (154a), *602*
Wolff, G. 344, 345, 347 (36), *358*
Wolff, H. 526 (69), *600*
Wolff, I. A. 145 (99, 100), *171*
Wolff, R. E. 344, 345, 347 (36), *358*
Wolff, W. 591 (371), *609*
Wolfgang, R. 362, 383 (27), *406*
Wolinsky, J. 132, 133 (67), *170*, 282 (52), *318*
Wong, E. W. C. 85 (167), *112*
Wong, L. Y. 232 (242), *264*
Wong, S. K. 349 (28), *357*
Wood, C. J. 396 (172), *410*
Wood, D. C. 243 (167), *261*
Wood, R. I. 458 (180), *505*
Wood, S. E. 583 (304), *607*
Woodhams, R. T. 425 (70), *502*
Woods, G. F. 126 (42), *169*
Woods, R. J. 362, 386, 387 (24), *406*
Woods, W. G. 81, 99 (155), *112*

Woodward, R. B. 66 (106), 99 (199), *110*, *113*, 132, 133 (66), 155 (134), *170*, *172*, 255 (237), *264*, 279 (37), 280, 281 (39), *317*
Woodward, T. W. 366 (33), 367 (33, 44), *406*, *407*
Wriede, P. 310 (235), *324*
Wright, A. N. 41 (4), *107*
Wright, M. 67 (107), *110*
Wyatt, P. 291 (101b), *320*
Wylde, J. 146 (104), *171*

Yada, S. 196 (62), *212*
Yagupsky, G. 236 (258), *265*
Yakubchik, A. I. 461 (197–200), 463 (198), *506*
Yamada, Y. 304 (203d), *323*
Yamagami, M. 455 (172), *505*
Yamagishi, K. 196 (62), *212*
Yamaguchi, M. 249 (203a, 203b), 250 (217), *262*, *263*
Yamamoto, A. 250 (221a), *263*
Yamanaka, H. 67 (107), *110*
Yamashita, S. 452, 454 (161), *505*
Yamauchi, K. 196 (62), *212*
Yang, K. 375, 398 (100), *408*
Yang, N. C. 283 (58), 284 (65), 290 (99b), 302 (179), 309 (233), 310 (234), *318*, *320*, *322*, *324*
Yates, J. 417 (37), *501*
Yen, S. P. S. 421 (59), *502*
Yen, T. F. 421 (55), *501*
Yen, Y. 207 (103), *213*
Yip, R. W. 307 (213, 215), 316 (271), *323*, *325*
Yipp, R. W. 146 (104), *171*
Yokokawa, C. 246 (186), *262*
Yokoo, H. 196 (62), *212*
Yonezawa, T. 518 (28), 519 (28, 28a), *599*
Yoshino, T. 54 (52), *108*
Young, A. E. 595 (389, 390), *610*
Young, J. F. 196 (64), 199 (64, 73), 200 (64), *212*, *213*, 232, 233 (88), 234 (88b), *258*
Young, R. H. 303 (187), *322*
Young, R. J. 145 (99), *171*
Young, W. G. 60 (81), 62 (85), 63 (89, 90), 71 (124), 73, 74 (131), 75 (141), *109–111*, 148 (110), 149 (110, 115), *171*, 240 (146), *260*
Youngman, E. A. 417 (33), *501*
Yuguchi, S. 253 (224a, 224b, 227a, 227b), 254 (227a, 227b), *263*

Yukawa, T.   242 (161), *261*
Yuzhakova, O. A.   547 (157), *603*

Zafiriov, O. C.   277 (25b), *317*
Zahler, R. E.   223 (41), *257*
Zakharkin, L. I.   250 (215, 219), *263*
Zalar, F. V.   102 (211), *114*, 539 (124), *602*
Zanini, G.   420, 421 (44), *501*
Zaradovskaya, E. N.   452 (152), *504*
Zderic, J.   306 (208), *323*
Zechmeister, L.   130 (53c), *169*
Zeise, W. E.   216, 219 (1), *256*
Zelikow, J.   548 (159a), *603*
Zelinski, R. P.   417, 452 (36), *501*
Zelinsky, N.   548 (159a), *603*
Zenda, H.   296 (130a), *321*
Zhigareva, G. G.   250 (215), *263*

Ziegenbein, W.   125 (40), *169*
Zimmerman, H. E.   82 (157), *112*, 205 (95), *213*, 268 (4b), 272, (10), 281 (4b), 283 (64), 290 (4b), 296 (131, 132, 136), 304 (4b), 309 (232), *316*, *318*, *321*, *324*, 594 (386), *610*
Zirner, J.   132, 133 (67), *170*, 296 (137), *321*
Zirngibl, U.   307 (216), *323*
Zoeller, J.   527, 547 (71), *600*
Zotova, S. V.   588 (344, 345, 357, 258), 590 (344, 357), (356), *608*, *609*
Zubarova, N. D.   552 (176a), *603*
Zuman, P.   553, 554, 559 (177), *603*
Zweig, A.   82 (157), *112*, 564 (219), *605*

Subject Index

Acetylcyclohexenes, mass spectra, retro-Diels-Alder reaction 352
Acrolein, nuclear magnetic resonance spectra 8
Acrylic acid, reduction 231, 235
Acrylonitrile-butadiene copolymer—see Butadiene-acrylonitrile copolymer
Addition, electrophilic, to cyclopropanes 571–577
  nucleophilic, to cyclopropanes 587, 588
Addition-elimination, metal-hydride, alkene-metal complexes 225–228
  in alkene rearrangements 166
  isomerization 225–228
Alka-1,5-dienes, Cope rearrangement 159
Alkenes, carbonylation—see Carbonylation
  conjugated, rearrangements 145
  cyclic, coupling parameters and ring size 31
    mass spectrometry 335, 336
      double bond location 344
      expulsion of alkyl radicals 353–357
      long range coupling 31, 32
      retro-Diels-Alder reaction 332
    see also individual compounds
  halogen substituted, palladium catalysed oxidation, reaction mechanism 239
    stereochemistry 117
  hindered rotation of single bonds 26
  hydrogen acidity 142
  hydrogenation—see Hydrogenation
  hydrogen transfer 133
  kinetic acidity 140
  mass spectrometry—see Mass spectrometry
  nuclear magnetic resonance spectra —see Nuclear magnetic resonance spectra
  oligomerization—see Oligomerization
  oxidation—see Oxidation
  photochemistry—see Photochemistry
  photoisomerization 120
  polymers—see Polyalkenes
  radiolysis—see Radiolysis
  reaction with sulphur 468
  rearrangements—see Rearrangements
  relative ground-state energies 140
  stereoselective synthesis 129
  substituted, incremental substituent shift 17
  see also Dialkenes; individual compounds
Alkene-transition metal complexes, in carbonylation reactions 242–247
  double bond migration—see Double bond migration
  dynamic stereochemistry—see Stereochemistry, dynamic
  in hydrogenation of alkenes 230–236
  nucleophilic reactions 236–242
  oxidation 236–240
  oligomerization 247–256
  stability 219–221
  structure 216–222
    π-allyl complexes 221, 222
    analogy to a singlet carbene 218
    Dewar's MO picture 217
    electronic 217
    similarity to cyclopropane 218
Alkylation of benzene by substituted cyclopropanes 581, 582
Alkylcyclohexenes, nuclear magnetic resonance spectra 32
Alkylpropenes, vicinal coupling constants 27
Alkyne polymers—see Polyalkynes; individual polymers
Allene, formation by 1,5-shift during photo-isomerization 283

Allene—*continued*
  proton magnetic resonance spectra 21
  proton–proton coupling, determination from ^{13}CH satellite pattern 21
  substituted, nuclear magnetic resonance parameters 21
Allyl anion 40, 56–58, 67–77
  collapse ratios for *cis* and *trans* 138
  description, by molecular orbital theory 57
    by resonance theory 56, 57
  effect of asymmetry of environment 72
  formation and reactions 77
  geometry 70
  infrared spectra 74
  proton magnetic resonance spectra 74
  relative stability of *cis* and *trans* 71
  rotational barrier in 70
  stability 57, 67–70
    effect of substituent 69, 70
    from equilibrium studies 67, 68
    from kinetic acidity measurements 68
    from rate of alkene isomerization 69
  ultraviolet spectra 72, 73
Allylbenzene, catalytic isomerization 230
Allyl cation 40, 56–67
  charge distribution in 62
  description, by molecular orbital theory 57
    by resonance theory 56, 57
  formation and reactions 63
  geometry 61–63
  infrared spectra 63
  rotational barriers in proton magnetic resonance spectra 61
  stability 57–60
    effect of substitution 60
    from equilibrium data 58
    from isomerization studies 67
    from kinetic studies 60
    from mass spectral studies 44, 60
  ultraviolet spectra 63
π-Allyl complex mechanism 228–230
π-Allyl complexes, σ-allyl structure 222
  carbonylation 244, 245
  hydrogenation, heterogeneous 176, 181, 192
  isomerization 228–230
  nickel hybridization 221
  oligomerization 248
  oxidation 239
  structure 221, 222
Allyl compounds, coupling parameters 28
Allylic alcohols, carbonylation with dicobaltoctacarbonyl 247
Allylic chlorides, carbonylation of palladium complexes 245
  catalytic oxidation 239
Allylic radical, stabilization energy 136
Allylidene fluoride, H—H, H—F couplings 28
Allyllithium, proton magnetic resonance spectra 74, 75
Appearance potential of ions 337–342, 348, 349, 353, 354

Balata—*see* Gutta-percha
Benzene, photoaddition to 296
  photochemical isomerization 289, 291, 292
Benzenediazonium ions, *p*-substituted, substituent effect on decomposition 223
Benzoylcyclopropane, addition of acetic acid 579
Benzyl cinnamate, selective reduction 233
Bicyclo[$n \cdot 1 \cdot 0$]alkanes, electrophilic addition to 582–584
Bicyclobutane, production during photoisomerization 281, 285, 288, 289
Bicyclobutonium ion, intermediate in solvolysis of cyclopropylcarbinyl derivatives 560
Bicycloheptadiene, nuclear magnetic resonance spectra 32
Bicyclooct-2-enes, substituted, decoupling 31
Bicyclopropyl group, conformation 528, 529
Bis(π-allyl)nickel complexes, symmetrical structure 221
Block copolymers, A—B—A 424
Bullvalene, photochemical isomerization 295, 296
Buna N—*see* Butadiene-acrylonitrile copolymer
Buna S—*see* Styrene-butadiene copolymer

Butadiene, condensation with alkenes 253, 254
  dimerization, photochemical 313–315
    to divinylcyclobutane 255
  heterogeneous catalytic hydrogenation, 1,2-addition 190–192
    1,4-addition 191, 192
    mechanism, π-allylic species 192
      σ-π-diadsorbed form 192
    stereochemistry, conformation of adsorbed C_4H_6 192
  homogeneous catalytic hydrogenation 198, 199
  mass spectra, ionization potential 349
  molecular orbitals 273, 274
  nuclear magnetic resonance spectra 19, 20
  oligomerization, cyclic 251
    in presence of palladium 252
  ultraviolet absorption spectra, 0–0 band 271
Butadiene-acrylonitrile copolymer 413
  hydrogenation 461
  mechanochemical degradation 490
  oxidation, studied by infrared spectroscopy 484
  peroxide vulcanization, cross linking 468
  uses 423
Butadiene copolymers 417, 423, 424
  epoxidation 492
  oxidation 483, 484
    chain scission and crosslinking 484
  radiation vulcanization 474, 475
  see also Butadiene-acrylonitrile copolymer; Butadiene-pentadiene copolymer; Isoprene-butadiene copolymer; Styrene-butadiene copolymer
Butadiene homopolymers—see Polybutadiene
Butadiene-pentadiene copolymer 421
1,3-Butadiene-silver complex, stability 220
Butadiene-styrene copolymer—see Styrene-butadiene copolymer
Butatriene, nuclear magnetic resonance spectra 21
1-Butene, isomerization, metal hydride addition-elimination mechanism 226
  vicinal coupling constants 26
2-Butene, rotational barriers 116
  stereomutation, iodine catalysed 136
  thermal 115
Butene-metal complexes 220
Butenes, deuterated, mass spectra 335, 341
  hydrogen scrambling 335
  isomerization, rhodium catalysed 225
  liquid-phase radiolysis, theoretical predictions 395
  mass spectra, ionization potential 338, 339
  oxidation, palladium catalysed, rate constant and complex formation constant 237
2-Butenyl cations, rotational barrier 148
Butyl rubber 413, 424
  reaction with chloral 499

Carbanion reaction, participation by non-conjugated double-bond 81–83, 104, 105
  molecular orbital calculations 81–83
Carbanions 39–106
  allyl, see Allyl anion
  ambident, halogenated, rearrangements 144
  cyclopropyl, configurational stability 595
  nuclear magnetic resonance spectra 142
  organometallic compounds of Groups IA and IIA 41
  rotational barrier 140
  stability from kinetic acidity studies 52
  stabilization by cyclopropyl group 563, 564
  steric and inductive effects 138, 139
  vinylic, geometry 123
Carbene, singlet, analogy to transition-metal derivative 218
  used in production of cyclopropane 218
Carbon radical reaction, molecular orbital calculations for participation by non-conjugated double bond 81–83

Carbonium ion reaction, participation by non-conjugated double bond 83–101
  effect of nucleophile 89
  effect of solvent 83–86
  effect of substituent 85, 88
  kinetic evidence 83
  molecular orbital calculations 81–83
  product evidence 83
  ring size formed in 87
  stereochemical consequences 90
Carbonium ions 39–106
  cyclopropylmethyl, intermediate in solvolysis of cyclopropylcarbinyl derivative 561
  stabilization by cyclopropyl group 552
  substituent effect of cyclopropyl group 560–563
Carbonylation 242–247
  reactions of alkene-metal complexes 245–247
    with cobalt hydrocarbonyl 245, 246
    with iron carbonyl 246
  reaction of π-allyl complexes 244, 245
    production of β,γ-unsaturated acids 244
  reaction of carbon monoxide within alkene-transition metal complex 243, 244
  stereochemistry 247
Carboxymethylation, photochemical 303
Catalysts, heterogeneous half-hydrogenation 195
  heterogeneous hydrogenation, irridium 190
    osmium 190
    palladium 178, 179, 181, 185–188, 190–192
      hindrance 186, 188
      rate limiting step 188
      study of deuterium exchange 181
    platinum 178, 181, 190
    platinum oxide 177–179
      in acetic acid 183
    Raney nickel 188
    rhodium 190
    ruthenium 190
    selective 192–196

homogeneous hydrogenation, irridium complex 200, 201
  metal carbonyls 202
  pentacyanocobaltate 197–199, 231, 232
    mechanism, complexes involved 198, 199
  platinum complexes 201, 202
    mechanism involving metal hydride 201
  rhodium complex 201
  ruthenium chloride 196, 197, 235, 236
    mechanism, complexes involved 197
  tris(triphenylphosphine)dichlororuthenium 201, 236
  tris(triphenylphosphine)halogenorhodium 199, 200, 232–234
    cis addition 200
  Ziegler-type 202, 203
isomerization, π-allyl complex mechanism, iron carbonyl 228
  palladium 230
  $CoH(CO)_4$ 230
  rhodium 225
oligomerization, cobalt 253, 255
  iron 253, 255
  nickel 248, 249, 253
  palladium 248
  rhodium 248, 253, 255
  ruthenium 248
rearrangements, silica-alumina 147
  transition metals 166
  Ziegler-type 148
Chemical shifts 2
  correlation charts, configuration of alkenes 17
  cyclic compounds 33
  differential proton and ^{13}C shift, symmetrical disubstituted ethylenes 15
  explained by d_π–p_π bonding 10
  incremental substituent, substituted alkenes 17
  medium effect 22–24
  $\nu_{C(\beta)}$ (vinyl) vs. ν_{Cortho} (phenyl) 9
α-Chloroacrylonitrile, nuclear magnetic resonance parameters, effect of medium 23, 24
2-Chloro-1,3-butadiene—*see* Chloroprene
1-Chloro-2-butene, silver ion-catalysed hydrolysis 149

Subject Index

Chlorocyclopropane, ionization potential 530
Chloroprene, condensation with ethylene 254
Chloroprene polymers—see Polychloroprene
Chodkiewiecz coupling 131
Cholestene, mass spectra, double bond location 344
5α-2-Cholestene, selective reduction 232
Claisen rearrangement 154–162
 carbon analogue 162
 cyclic transition states, conformational effects 155
 orbital repulsion 155
 oxy analogue 134, 135
Coenzyme Q, isomerization 459
Conjugated double bond, photocyclo addition to 304–308
Cope rearrangement 154–162
 cyclic transition states, conformational effects 155
 internal rotations 160
 orbital repulsion 155
 transition-state geometry 161
Copolymers—see Block copolymers; Graft copolymers; Ladder copolymers; individual compounds
Coupling constants 3
 additivity relations, substituted ethylenes 18
 $^{13}C-^{1}H$, cyclopropane 544
 determination using Fermi contact term 5
 effect of substituents 6
 long range 3, 8, 31, 32
  H—H in $H_2C=CY-CH_2X$ 28, 29
  methyl and vinyl protons in propene 13
 medium effect 22–24
  explained by 'reaction field' theory 22
  on $J_{13_{CH}}$ 25
 multiply substituted ethylenes 18
 ring size in cyclic alkenes 31
 sum of, in vinyl compounds bonded to Si, Hg, Al, Sn 11
 vicinal 26, 27
Crotonic acid, reduction 235
Cubane synthesis, photoaddition 300, 301
Cumulenes, nuclear magnetic resonance spectra 21

Cyanoethylene-rhodium complex, stability 220
Cyclic olefins—see Alkenes, cyclic; individual compounds
Cyclizations, polymers containing C=C bond 441–449, 451
 accompanied by vulcanization 464, 465
Cyclobutanation 255, 256
Cyclobutane, photodimers 312–315
 production, catalytic dimerization of alkenes 255
Cyclobutanone, production by photoisomerization 281
Cyclobutene, production by photochemical isomerization 284–288
 mechanisms 286–288
Cyclodecadiene, production by cyclic condensation 255
Cyclododecatriene, production by cyclic condensation 255
Cyclododecatriene-metal complex, production during oligomerization 251, 252
Cycloheptatrienes, nuclear magnetic resonance spectra 32
1,3-Cyclohexadiene, nuclear magnetic resonance spectra 32
1,4-Cyclohexadiene, selective reduction 233
1,4-Cyclohexadiene-silver complex, stability 220
Cyclohexene, mass spectra 348–355
 appearance potentials 348, 349, 353, 354
 ionization potential 348
 retro-Diels-Alder fragmentation 332, 348, 351, 353, 355
 thermal 349
 reduction 235
 selective 232
cis,cis,cis-1,4,7-Cyclononatriene, nuclear magnetic resonance spectra 33
1,5-Cyclooctadiene, production from butadiene 252
 stereochemistry, carbonylation 247
1,5-Cyclooctadiene-metal complex 219, 220
Cyclooctatetraene, production by photoaddition 296, 312
 sensitized photolysis 296
Cyclooctene, mass spectra 355, 356
Cyclopentadiene, chemical shift 33

Cyclopentadiene polymers, epoxidation 492
Cyclopentene, mass spectra 354, 355
  polymerization 428
Cyclopropanecarboxylic acid, electrophilic addition of halogen 576
Cyclopropane derivatives, addition of hydrogen halide 577, 578
  addition of water 578–580
  alkylation of benzene 581, 582
  conformation 525–530
  dipole moments 549–552
    resonance effect 551
  electron paramagnetic resonance spectra 544–548
  electrophilic addition of halogen 575–577
  heat of combustion 552
  infrared spectra 538–541
  ionization potentials 530–532
  molecular geometry 521–525
    electron density distribution from X-ray diffraction 521
    structural determinations 521–523
  nuclear magnetic resonance spectra 544–548
  nuclear quadruple resonance 544–548
  Raman spectra 541–544
    steric inhibition of resonance 543
  reaction with metal salts 580, 581
  ultraviolet spectra 532–538
    dependence on dihedral angle 534
    spectroscopic bond orders 532
Cyclopropane ring, addition reactions, reactivity relative to the double bond 585
  conjugation with the benzene ring 536, 552
  transmission of resonance effects 566–569
    localized effect 567
Cyclopropanes, *ab initio* calculation 595
  acidity of the C—H bond 592–594
    studied by hydrogen-deuterium exchange 594
  addition reactions 569–592
    donor-acceptor complexes 570, 571
    electrophilic 571–577
    nucleophilic 587, 588
  bonding in 513–521

  bent-bond model 515–518
    hybridization states 516
  'many centre' forces 520
  molecular-orbital model 518, 519
  ring strain 520, 521
  studied by maximum overlap method 516, 517
  transformation of models 519, 520
  trigonally-hybridized model 513–515
  twist bent bonds 517, 518
  bonded to aromatic ring, bisected conformation 528
  carbonyl derivatives, conformation 528
  coupling constant $^{13}C-^{1}H$ 544
  hydrogen abstraction by radicals 595
  hydrogenation 588–592
    accompanied by isomerization 588
    rate studies 589
  molar refractivities 548, 549
  production from a carbene 218
  proton chemical shifts 545
  radical additions 585, 586
  reaction with lead(IV) acetate 581
  reaction with mercuric salts 580
  reaction with thallium(III) acetate 581
  SCF MO calculations 518, 519
  structure related to transition-metal-alkene complex 218
  transmission of resonance effects 538
Cyclopropanols, review 598
Cyclopropenyl anion 58
Cyclopropenyl cation 57
Cyclopropylbutanes, 3-substituted, rate of reaction with trifluoroacetic acid 596
Cyclopropyl carbanion, configurational stability 595
Cyclopropylcarbinyl derivatives, solvolyses 560, 561
Cyclopropyl group, in azo compounds, polymerization on decomposition 564
  orientation in electrophilic aromatic substitution 559
  stabilization of carbanions 563, 564

stabilization of carbonium ions 552
stabilization of radicals 564–566
  effect of chlorination 565
substituent constants 553–555
substituent effect on, carbonium ions 560–563
  ionization constants 555–557
  reaction rates 557, 558
Cyclopropylmethyl carbonium ion, intermediate in solvolysis of cyclopropylcarbinyl derivatives 561

Decarbonylation during catalytic hydrogenation 233
1-Decene, selective reduction 232
Dialkenes, conjugated, catalytic hydrogenation 195
N,N-Dialkylallylamines, rearrangement 141
3,3-Dialkylpropenes, vicinal coupling constants 27
Dibenzoylethylene, stereomutation 124
Dibromoethylenes, nuclear magnetic resonance parameters, reaction field estimation 24
Dichloroethylenes, nuclear magnetic resonance parameters, reaction field estimation 24
  structure determination 352, 353
Diene homopolymers 413–423
  see also individual polymers
Dienes, conjugated, isomerization by palladium 252
  oligomerization 249–252
    cyclic 250–252
      transition-metal catalyst 250
      zero valent nickel catalyst 250, 251
    linear 249, 250
  reduction 231
1,5-Diene–metal complexes, addition trans to metal 240
1,4-Dihydrobenzenes, coupling constants 33
1,4-Dihydrobenzoic acid, nuclear magnetic resonance spectra 33
1,4-Dihydrobiphenyl, nuclear magnetic resonance spectra 33
Dihydromyrcene, dichlorocarbene addition 495
  oxidation 478
  reaction with peroxides 465
  reaction with sulphur 471

Diisobutylaluminium alkenes, rearrangements with 151
Dimercaptobutadiene, photoisomerization 132
Dimercaptoethylene, isomerization 132
Dimerization, catalysed by metals 255–256
  photochemical 312–316
    mechanism, interpretation from stereoisomeric triplet states 314
    triplet intermediate 313
    product dependence on triplet energy 314
1,3-Dimethylallyl phenyl ether, rearrangement 154
1,2-Dimethylcyclohexene, mass spectra, double bond migration 350
  retro-Diels-Alder rearrangement 332, 350
3,4-Dimethylhexa-1,5-diene, rearrangement 161
Dimethylmaleic anhydride, reduction 235
2,6-Dimethyl-2,6-octadiene, reaction with sulphur 468, 469
4,5-Dimethyl-1-pentene, mass spectra, McLafferty rearrangement 332
Dimethylsulphoxide, stereomutation 123
1,3-Dioxepenes, conformational mobility 32
Diphenylacetylene, selective reduction 236
Diphenylbutenes, equilibria 123
Dipole moments, substituted cyclopropanes 549–552
1,3-Dithiepenes, conformational mobility 32
cis-1,2-Divinylcyclobutadiene, production from butadiene 252
Divinylcyclobutane, production by dimerization of butadiene 255
Dodecatetraene, production during oligomerization of butadiene 252
Double bond, conjugated—see Conjugated double bond
  non-conjugated—see Non-conjugated double bond
Double bond migration, alkene-transition metal complexes 224–230

Double bond migration—*continued*
  π-allyl mechanism 228–230
  metal-hydride addition-elimination mechanism 225–228
 catalytic hydrogenation 177, 342
 mass spectroscopy 336, 342, 347–350
 photochemical isomerization 277, 278
 photolysis 457

Electron paramagnetic resonance spectra, cyclopropane derivatives 544–548
Epoxidation, polymers containing C=C bond 492
Ethylcyclopropane, donor-acceptor complexes 570
Ethylene, alkyl substituted, nuclear magnetic resonance spectra 12, 13
 carbonylation, production of acyl-cobalt carbonyl 246
  production of ethylcobalt carbonyl 246
 condensation with 1,3-dienes 254
 dideuterated, stereomutations 117
 effect of substituent on coupling 6
 isotopically substituted, couplings 4
 monosubstituted, nuclear magnetic resonance parameters 5, 9
  effect of medium 24
 multiply substituted, chemical shifts 15
  coupling constants 18
 oligomerization 247
 palladium catalysed oxidation 237–240
  kinetic isotope effect 237
  rate constant and complex formation constant 237
  reaction mechanism 238–240
 polymerization, radiation-induced 399
 proton magnetic resonance spectra 4
 radiolysis 403, 404
  ionic reactions 390–394
  mechanism 374, 390–394
   charge exchange 393
   electron transfer 391, 392
   from vinyl ion 393
   geminate neutralization 394
   intermediate complex ion 390, 391, 393
  radical reactions 398–400
  sensitized 377
 reduction 235
 substituted, additivity relations of coupling constants 18
 symmetrical disubstituted, differential proton and ^{13}C shifts 15
Ethylene-metal complexes, stability 220
Ethylene-propene mixture, oligomerization 253
Ethylene-propylene terpolymer 413, 424, 425, 463
 addition of non-conjugated dienes for vulcanization 425
 structure 424, 425
Ethylenic compounds, coupling constants 4
Ethyl vinyl ether, analysis by nuclear magnetic resonance spectra 8

Fermi contact term, determination of coupling constants 5
Fluoropropenes, H–H, H–F couplings 28
Franck-Condon principle, photochemical dimerization 314
 photochemistry 270, 271, 274
Fumaric acid, reduction 235

Gensler coupling 130, 131
GR-A—*see* Butadiene-acrylonitrile copolymer
Graft copolymers 426
GR-I—*see* Butyl rubber
Grignard reagents, allyl, proton magnetic resonance spectra 74, 75
 containing non-conjugated double bonds, cyclization 104
 nature of ions 41, 42
 stability from equilibrium studies 49, 51
 vinyl, geometry, configurational stability 54
  proton magnetic resonance spectra 54
  stability 49, 51
  stereochemical retention 163
GR-S—*see* Styrene-butadiene copolymer
γ-Gurjunene, selective hydrogenation 233
Gutta-percha, cyclization 441
 hydrochlorinated, nuclear magnetic

resonance spectra 439
hydrogenated, infrared spectra 462
hydrogenation 460
isomerization 449, 455, 458
  effect on rate of crystallization 458
maleic anhydride addition 497
oxidation 478
  chain scission mechanism 481, 482
peroxide vulcanization, crosslinking efficiency 466
physical properties 413, 416
pyrolysis 487
structure 415, 416

Heat of combustion, cyclopropane derivatives 552
1,3-Heptadiene polymers 420
1,6-Heptadiene-silver complex, stability 220
1-Heptene, palladium catalysed isomerization 230
2-Heptene, mass spectra, McLafferty rearrangement 332
Heptenes, isomerization, metal hydride addition-elimination mechanism 226, 228
Hevea 416, 418, 423, 424, 429
chlorinated, uses 436
chlorination 434, 435
cyclization 441–443, 445, 460
cyclized, structure 441–443
hydrochlorination 439
hydrofluorination 439
hydrogenation 460
  rate 461
γ-irradiated 458
isomerization 449, 455, 458, 466
  effect on rate of crystallization 458
  under vulcanizing conditions 460
maleic anhydride addition 497
mechanochemical degradation 489, 490
mastication 490
oxidation 477, 478, 482
  chain scission mechanism 480–482
  effect on physical properties 478
  peroxide vulcanizate 478, 482
  resistance to 482
  sulphur vulcanizate 478, 482
ozonization 416, 484–486
photochemical production of hydrogen 457
physical properties 413, 416
pyrolysis 487
reaction with aldehydes 498
structure 415, 416
vulcanization 463
  peroxide 464
  sulphur 468
    crosslinking efficiency 469, 470
    double bond shift 471
    function of accelerator 470
1,4-Hexadiene, production by oligomerization of butadiene 253
1,3-Hexadiene polymers 420
1,5-Hexadiene-silver complex, stability 220
1-Hexene, vicinal coupling constants 26
Hexenes, isomerization, metal hydride addition-elimination mechanism 228
mass spectra 331
  McLafferty rearrangement 332
selective reduction 236
Homoallylic cation 95
Horiuti-Polanyi mechanism, catalytic hydrogenation 176, 179, 188
interpretation of half-hydrogenated state 177
Hydrogenation 175–210, 230–236
catalysts—*see* Catalysts
catalytic, double bond migration 177, 342
selective 232, 236
chemical reduction 203–210
  by diimide 208–210
  by dissolving metals 203–206
    Birch reduction 203–206
    zinc in boiling acetic acid 203
  by metal hydrides 206–208
    cobalt hydrocarbonyl 207, 208
    lithium aluminium hydride 206, 207
    mechanism 208
    sodium borohydride 206
    triphenylstannane 207
cyclopropanes 588–592
-half, heterogeneous catalytic 195
heterogeneous catalytic 175–196
  adsorption states 175
  π-allylic complex intermediate 176, 181
  associative adsorption, half-hydrogenated state 176
  π-complex formation 176

Hydrogenation—*continued*
  isomerization on catalyst surface 178
  mechanisms 175–192
    deuterium exchange, intramolecular hydrogen shift 182
    half-hydrogenated state 184–186
    rate as a function of concentration of half-hydrogenated intermediates 180
    hydrogen pressure 179, 180
    rate-limiting step 184
  selective 192–196
    competitive adsorption 193
    heat of hydrogenation 193
  stereochemistry 175–192
    cis,trans addition 177–192
      effect of temperature 186
      effect of weight of catalyst 187
      half-hydrogenated state 183
      adsorption in a boat conformation 184
      deuterium exchange 181, 182
      directive effect of hydroxyl group 188
      effect of catalyst hindrance 186
      effect of hydrogen pressure 183, 184
      rate limiting step 185
  homogeneous catalytic 196–203, 232–234
    cis-addition 197
    decarbonylation, avoidance 233
    hydrido-transition metal complex 196
    isotope effect on rate 234
    selective 232, 233
  polymers containing C=C bond 460–463
    mechanism 462
  transition metal-alkene complexes as intermediates 230–236

IIR—*see* Butyl rubber
Infrared spectroscopy, cyclopropane derivatives 538–541
  stretching frequency C—H bond 539
  structure of hydrogenated polymers 462
  structure of polymers containing C=C bond 415, 417, 428, 429, 444, 446
  study of polymer oxidation 484
Ingolds' rule, rearrangements 141
Ion cyclotron double resonance, study of ion-molecule reactions 390
Ionization potentials 337–342, 348, 349, 361
  cyclopropane derivatives 530–532
    correlation with extended Hammett equation 531
Ion-molecule reactions, atom transfer 390
  condensation-type radiolysis of ethylene 393, 394
  high-order, ethylene radiolysis 394
  ion transfer 390
  long-chain, radiation induced polymerization 394
  radiolysis 369
  studies with ion cyclotron double resonance 390
  studies with mass spectrometer 390, 394
α-Ionone, photochemical reactions 145
β-Ionone, hydrogen transfer in 146
β-Ionylidene compounds, anionotropic rearrangement 152
Isobutene, carbonylation 246
  mass spectra, ionization potential 339
Isobutylene, polymerization, ion-injection techniques 395
Isobutylene-isoprene copolymer—*see* Butyl rubber
Isobutyl vinyl ether, analysis by nuclear magnetic resonance spectra 8
Isomerization, π-allyl complex mechanism 228–230
  catalysts—*see* Catalysts
  cis–trans, during cyclization of polydienes 446
    in polymers containing C=C bonds 449–460
    photochemical 277
  conjugated dienes 252
  during catalytic heterogeneous hydrogenation 178, 195
  during mass spectrometry 331
  during radiolysis 374, 397
  effect of electron donating power of phosphine 248

Subject Index 661

mechanism, ambident allylic cation 136
metal hydride addition-elimination mechanism 225–228
photochemical 275–297
  double bond migration, twisted alkene triplet 278
  electrocyclic reactions 284–297
    conrotatory cyclization 287
    internal conversion 288
    mechanisms, triplet state 287
    quantum yields 286
    ring opening 285, 289–291
  sigmatropic reactions 281–284
    1,3-shifts 281–283
    1,5-shifts 282–284
    1,7-shifts 281, 284
  Woodward-Hoffmann rules—see Woodward-Hoffmann rules
photoinduced, of polymers 449–452, 455–458
photosensitized 449
  triplet states 276
radiation-induced, of polymers 449–452
sigmatropic chlorine transfer 162
sigmatropic hydrogen transfer 167
sigmatropic shifts 132
solvent effects 143, 144
thermodynamics of 135
Isoprene, reduction 231
Isoprene-butadiene copolymer, microstructural analysis 417
Isoprene copolymers, epoxidation 493
  isomerization 454
  radiation vulcanization 474, 475
  see also Butyl rubber; Isoprene-butadiene copolymer
Isoprene homopolymers—see Polyisoprenes

Jesse effect, negative, radiolysis of mixtures 401
radiolysis 376

Ladder polymers 432, 443, 449
Langmuir-Hinshelwood mechanism, catalytic hydrogenation 176
Linalool, selective hydrogenation 232, 233
Δ^{12}-Lupene series, mass spectra, retro-Diels-Alder ions 352, 353

Maleic acid, reduction 235
Mass spectrometry 327–357
  allyl cation stability 44, 60

allylic fission 328, 331, 336
appearance potential 337–342, 348, 349, 353, 354
  calculation from heats of formation of ions and molecules 340, 341
  determination from ionization efficiency curves 338, 339
  for an excited state 341
double bond location 342–347
  by deuterium introduction 342
double bond migration, preferential 350
  under electron impact 342, 347, 350
electron-impact induced migration prior to fragmentation 333
explusion of alkyl radicals from cyclic alkenes 353–357
heat of formation of ions 340
hydrogen scrambling 334, 335
1,3 hydrogen shifts, in pentene 335
  in propene, deuterated 334
ion formation 379
ion lifetimes 380
ionization potentials 337–342, 348, 349
  determination by electron impact 338
  determination from ionization efficiency curves 338, 339
isomerization prior to fragmentation 331
  straight chain alkenes 328–330, 353
  studies of ion-molecule reactions 394
  tandem, studies of ion-molecule reactions 390
McLafferty rearrangement—see McLafferty rearrangement
partial, n-hexene isomers, abundance of ions 331
quasi-equilibrium theory of fragmentation 379, 380
retro-Diels-Alder rearrangement 332, 337, 347–353, 355, 356
  in structure determination 352, 353
statistical theory of fragmentation 380
McLafferty rearrangement 331–335, 348
  second, following hydrogen migration 334

Menthene, mass spectra, allylic fragmentation 336
  hydrogen migration caused by electron impact 335, 336
  rate of double bond migration 336
  retro-Diels-Alder reaction 351
Metal-hydride addition-elimination—see Addition-elimination, metal hydride
Methoxyethylene-rhodium complex, stability 220
Methyl acrylate, carbonylation 246
2-Methyl-1,3-Butadiene, condensation with ethylene 254
3-Methyl-5α-2-cholestene, selective reduction 232
Methylcyclohexene, deuterated, mass spectra 337
  mass spectra, retro-Diels-Alder fragmentation 332, 348–351
1-Methylcyclopentadiene, chemical shift 33
Methylcyclopropane, ionization potential 530
Methyl cyclopropyl ketone, rate of reaction with hydrogen 565, 566
Methyl-1-hexenes, mass spectra, McLafferty rearrangement 332
9-Methyl-trans-Δ^2-octalin, mass spectra 350
2-Methylpent-2-enyl thiolacetate, cis, trans isomerism investigated by nuclear magnetic resonance 34
Methyl vinyl sulphide, charging of d orbitals of sulphur by hyperconjugation 8
  nuclear magnetic resonance spectra 7
Molar refractivities, cyclopropanes 548, 549

Naphthoquinone, selective reduction 233
NBR—see Butadiene-acrylonitrile copolymer
Neoprene—see Polychloroprene
Nitrile rubbers—see Butadiene-acrylonitrile copolymer
Nitrocyclopropane, electron paramagnetic resonance study 547
  ultraviolet spectra 596
p-Nitrostilbene, stereomutation 123
Non-conjugated anions 40, 77–83, 101–106
  formation from non-conjugated carbonyl groups 101, 102
Non-conjugated cations 40, 77–101
Non-conjugated double bond, as neighbouring group 77
  cyclization in Grignard reagents 104
  participation by, in carbanion reaction, kinetic evidence 104
    molecular orbital calculations 81–83
    product evidence 104
    ring size 105
  in carbon radical reaction 81–83
  in carbonium ion reaction 78, 79, 83–85
    effect of solvent 83–86
    effect of substituent 85
    kinetic evidence 78
    molecular orbital calculations 81–83
    nature of intermediate and transition states 79
    product evidence 78
  in formation of anions 101
  3,4-, participation by, in carbonium ion reaction 95, 96
  4,5-, participation by, in carbonium ion reaction 93
  5,6-, participation by, in carbonium ion reaction 87–90
    effect of nucleophile 89
    ring size formed in 87
    stereochemical consequences 90
    substituent effect 88
  6,7-, participation by, in carbonium ion reaction 93
Norbornadiene, nuclear magnetic resonance spectra 33
  stereochemistry of carbonylation 247
Norbornadiene-metal complexes 219, 220
7-Norbornadienyl cation, geometry 98, 99
  proton magnetic resonance spectra 100
Norbornene, cis addition to double bond 240
  nuclear magnetic resonance spectra 33
  ^{13}C—H satellites 34
  selective solvent effects 34
5-Norbornene-2,3-dicarboxylic anhydride, reduction 235

7-Norbornenyl cation, geometry 98, 99
  proton magnetic resonance spectra 100
Norbornyl cation, structure 79, 80
Nuclear magnetic resonance spectra 2–35, 544–548
  carbanions 142
  ^{13}C—H satellites 5, 34
  chemical shift—*see* Chemical shift
  coupling constants—*see* Coupling constants
  cyclic compounds with more than one double bond 32
  geometry of ions 54, 61, 62, 74–77, 100, 101
  hyperconjugation 12
  medium effects 22
  reaction field estimation 24
  solvent effect 34
  structure determination, polymers 415, 417–419, 422, 444, 446
  vinylethers and sulphides 7
  study of rotational conformers about C—C bond 26
  vinyl derivatives containing Si, Hg, Al, Sn 9
Nuclear quadruple resonance, cyclopropane derivatives 544–548

Octalin, mass spectra 350
1-Octene, isomerization in presence of 1-hexene or undecene 228, 229
  selective reduction 232
Olefins—*see* Alkenes
Oligomerization 247–256
  π-allyl complexes, effect of aluminium halide 248
    effect of phosphine 248
  π-allyl nickel halide 248
  catalysts—*see* Catalysts
  cyclic 250–252
  cyclobutanation 255, 256
  dimerization 255, 256
  linear 249, 250
  mixed 253–255
    cyclic condensation 255
    linear condensation 253–255
d Orbitals of sulphur, charged by hyperconjugation 8
Organolithium reagent, stability from equilibrium studies 50
Organometallic compounds, of Groups IA and IIA, nature of ions 41
  proton magnetic resonance spectra 10, 11
  shifts explained in terms of d_π–p_π bonding 10
Oxetenes, Paterno–Buchi reaction 309
Oxidation 236–240
  to alkylidene diacetates 240
  to enol acetates 240
  palladium catalysed 237–240
    π-allylic complexes 239
    reaction mechanism 238–240
  polymers containing C=C bond 476–484
    chain scission and crosslinking 478
    chain scission mechanism 480, 481
Oximes, α,β unsaturated carbonyl compounds, nuclear magnetic resonance spectra 34
Oxy-Cope rearrangement 159
Ozonization, polymers containing C=C bond 484–486
  mechanism 486
  microstructural analysis 485

Paterno–Buchi reaction, intermolecular photoaddition 308–310
  mechanism 310
  oxetanes formation 308–310
Pauling hindrance, vitamin A_2 130
Pentacyanocobaltate, hydrogenation catalyst 197–199, 231, 232
1,3-Pentadiene, condensation with ethylene 254
  see also Piperylene
1,4-Pentadiene, stereochemistry of carbonylation 247
Pentadiene-butadiene copolymer 421
1,4-Pentadiene-silver complex, stability 220
Pentadienyl anion, rotational barrier 143
Pentadienyllithium, proton magnetic resonance 76
Pentadienyl radical, stabilization energy 136
1-Pentene, deuterated, mass spectra 335
  isomerization, metal hydride addition-elimination mechanism 226

1-Pentene—*continued*
  palladium catalysed 230
  mass spectra 331
    McLafferty rearrangement 332
  selective reduction 236
Phenylcyclopropane, ultraviolet spectra 532, 533
Phenylcyclopropane carboxylic acids, transmission of resonance effects 567
Phenylethylene-rhodium complex, stability 220
Photoaddition 297–312
  intermolecular 303–312
    carboxymethylation 303
    cleavage of C=C bond 303
    Paterno-Buchi reaction—*see* Paterno-Buchi reaction
    to aromatic compounds 311, 312
  intramolecular 297–302
  mechanism, intercrossing system 299
  to steroids 303, 304, 306
Photochemistry 267–316
  dimerization—*see* Dimerization, photochemical
  energy transfer, intersystem crossing 271–273
  flash photolysis 269
  fluorescence 271
  heterocyclic photoproducts, quantum yield 302
  'hot ground-state' reactions 272
  cis,trans isomerization 275–278
    see also Isomerization
  laser 269
  laws 269
  phosphorescence 271
  ring opening 285, 289–291
  sensitized 274–277
    diradical character of triplets 275
    energy transfer, non-vertical transitions 274
    production of twisted triplet 276
    heterogeneous 274
  see also Photoaddition; Ultraviolet absorption spectroscopy
Photocyclization of polymers containing C=C bond 456, 457
Photocycloaddition 304–308
Photoisomerization 120
Photolysis, double bond migration 457
  flash 269, 276

Piperylene polymers—*see* Polypiperylenes
Polyacetylene, properties, synthesis and uses 426, 427
Polyacrylonitrile, cyclization to a 'ladder' polymer 432, 433
Polyalkenes, double bond location by mass spectra 347
  cis,trans isomerization 125–132
  non-conjugated, isomerization 121
  see also individual polymers
Polyalkynes 426, 427
  oxidation 484
  see also individual polymers
Polybutadiene-2,3-d_2, isomerization 451
  accompanied by vulcanization 453
1,2-Polybutadiene, reaction with HCl 440
1,4-Polybutadiene, maleic anhydride addition 497
cis-1,4-Polybutadiene, reaction with chloral 499
  substituted 421
  use in place of natural rubber 416
Polybutadienes 416, 417
  bromination 437, 438
  chlorination 435, 436
  compared with neoprene 418
  cyclization 444–446
    cis,trans isomerization 446
  dichlorocarbene addition 495
  gem-dihalocyclopropane derivative 495
  double bond shift 454
  epoxidation 492
  hydrogenated, physical properties 463
  hydrogenation 461, 462
  infrared spectra 428
  isomerization 446, 449–455
    accompanied by sulphur vulcanization 452–454, 459, 460
    mechanism 455
    photoinduced 449–452, 454
      kinetic studies 450
      mechanism 451
    radiation-induced 449–452
      cyclization 451
      kinetic studies 450
      mechanism 451
      unsensitized 450
    thiylcatalysed 494
  mechanochemical degradation 490

oxidation 483, 484
peroxide-cured 468
photocyclization 456
physical properties 416, 417, 424, 429
reaction with sulphur 452, 453
structure 417
synthesis, yield of isomers 416
thermal degradation 488, 489
thiol addition 493, 494
treatment with dichlorocarbene 495
vulcanization 452
  peroxide 467, 468, 476
    crosslinking efficiency 467
  radiation 474–476
    chemical production of crosslinks 476
    crosslinking mechanism 475, 476
  sulphur, mechanism 473
Polychloroprenes 413, 418, 419
gem-dihalocyclopropane derivative 495
epoxidation 492
hydrogenation 461, 462
mechanochemical degradation 490
production 418
properties 418
structure by nuclear magnetic resonance spectra 418, 419
vulcanization 464, 474
  crosslinking by dicumyl peroxide 474
Polycyclopentadiene, microstructure by nuclear magnetic resonance study 422
synthesis 421, 422
Poly(2,3-dimethyl-1,3-butadiene), cyclization 446
Polyenes—*see* individual polymers; Polyalkenes
Polyethylene, radiation induced C=C bond formation 433
Polyisoprene-3-*d*, electron irradiation, double bond shifts 458
  isomerization mechanism 459, 460
cis-Polyisoprene, oxidation 482
  reaction with fluoral 499
  sulphur vulcanization 471
  treatment with dichlorocarbene 494
cis-1,4-Polyisoprene, hydrohalogenation 439
  physical properties 416
  see also Hevea

cis-1,4-Polyisoprene-3-*d*, hydrogenation 463
trans-1,4-Polyisoprene, hydrohalogenation 439
  physical properties 416
  see also Gutta-percha
3,4-Polyisoprene, cyclized, structure 444
Polyisoprenes 413–416
bromination 437
chloral addition 499
chlorination 435
cis,trans composition 415
  effect on physical properties 416
cyclization 439–444, 446, 447
dichlorocarbene addition 495, 496
gem-dihalocyclopropane derivatives 495
epoxidation 492, 493
hydrogenated, infrared spectra 462
hydrogenation 425, 461, 462
hydrogenation catalysts 461
hydrogen chloride addition 439, 440
isomerization 454–460
  mechanism 455, 459
  photosensitized 454
  thiyl-catalysed 494
cis,trans isomerization, during hydrogenation 462
maleic anhydride addition 496, 497
oxidation 477–483
  chain scission 478
  mechanism 480, 482
photocyclization 456
photolysis, double bond migration 457
pyrolysis 487–489
radiation-induced crosslinking 476
reaction with ethylenic compounds 497, 498
reaction with glyoxal 498
reaction with molecular oxygen 416
structures 415
vulcanization, peroxide 464–467, 476
  accompanied by cyclization 464, 465
  crosslinking efficiency 466
  radiation 474, 475
    chemical production of crosslinks 476
    crosslinking mechanisms 475, 476

Polyisoprenes—*continued*
  sulphur, accompanied by *cis,trans* isomerization 471
1,4-Polyisoprenes, cyclization mechanism 441, 442
Polymerization, radiation-induced, long-chain ion-molecule reactions 394
  see also Dimerization; Oligomerization
Polymers containing C=C bond 411–500
  chemical reactions 433–500
    addition reactions 493–500
    cyclization 441–449
    degradation 486–491
    epoxidation 492
    halogenation 434–440
    hydroformylation 499, 500
    hydrogenation 460–463
    hydrohalogenation 434–440
    cis,trans isomerization 449–460
    oxidation 476–484
    ozonization 484–486
      mechanism 486
    vulcanization 452, 463–476
  see also individual polymers
Polynorbornene, infrared spectra 429
Polypentenamers, infrared spectra 428
  production by polymerization of cyclopentene 428
  physical properties 429
Polypiperylenes 419–421
  hydrogenation 425, 461–463
  peroxide vulcanization, crosslinking efficiency 466, 467
Polyvinyl chloride, dehydrochlorination 431, 432
Polyvinylidene chloride, dechlorination 432
Poly-ynoic acids, stepped, synthesis 130
3-^{13}C-Propene, mass spectra, loss of methyl radical 334
Propene-ethylene mixture, oligomerization 253
Propene-nickel complex, structure 221
Propenes, deuterated, mass spectra 334, 335
  long range couplings between methyl and vinyl protons 13
  mass spectra, ionization potentials 338
  2-substituted, differential shieldings 16
  vicinal coupling constants 26
Propenylbenzene, production by isomerization of allyl benzene 230
Propylene, mass spectra, appearance potentials of fragment ions 339, 340
  microwave spectrum 26
  palladium catalysed oxidation 237
  radiolysis, liquid-phase 405
    photophysical processes 375
  reduction 235
  vacuum ultraviolet photochemistry 394, 395
Propylene-metal complexes, stability 220
Proton magnetic resonance spectra— *see* Nuclear magnetic resonance spectra
Pyrolysis, polymers containing C=C bond 487–489

Radiolysis 359–405
  condensed-phase, by kinetic spectroscopy 365
    mechanism, degradation spectrum of electrons 385–387
    by pulse radiolysis 365
  dosimetry 366–368
    by electromagnetic radiation 366
    by electron accelerators 362–366
      characteristics 363
      steady state kinetics 363
    excitation by secondary electrons 362
  gas-phase, by electron accelerator 364
    ion-pair energy of formation 366, 367
    ion-pair yield 366, 367
    mechanism, energy deposition along columns of dense ionization 384
    ionic intermediates 368
    track effects 385
    triplet states 371
  by high energy radiation 360
    energy deposition 362
  ionization efficiencies 376
  ionization potentials 361
  by ionizing radiation 360
  liquid-phase, mechanisms, effect of solvation 369
    neutralization 368

Subject Index

mechanisms, associative ionization 389
bimolecular stage 370, 388–401
  diffusion of radicals and spur size 400
  electron attachment reactions 397
  geminate neutralization 396, 397
  ion-molecular reactions 389–396
    see also Ion-molecule reactions
physical stage 370
primary unimolecular stage 370
  radical reactions 397–401
  reactions of excited parent ion 388
  scavenging of radicals 400
  spectroscopic and collision processes 388
complete product distributions 403
escape probability of ions 386, 387
excess electronic energies in ion formation 372
ion-electron separation 387
ionic polymerization as evidence for free ions 387
initial species 371
  geminate neutralization 386
  optimal approximation 371
  production 372
Jesse effect 376
material balance 403
observed products 388
optical approximation and oscillator strength 375
original products 388
physiochemical and chemical stages 370, 371
primary products 388
reactive intermediates 368–403
recapture of electrons 386
super-excited states 371, 376
survival of escaped electrons 387
time scale 370
  physical stage 370–373
ultimate product formation 371
unimolecular stage 370, 373–384
  chemical fate of molecule 381, 382
  dissociation 373
  electron attachment 381
  energy transfer function 381
  'hot' atoms 375
  isomerization 374
  optical approximation 382
  sub-excitation electrons 380, 381
  vibrational degradation 374
mixtures, electron density 382
  energy division between components 382, 383
  energy transfer 401
  mechanism, effect of scavengers 402
  preferential excitation 401
  stopping power, calculation 383
by penetrating radiation 360
stopping power of medium 368
by vacuum ultraviolet radiation 362
yields in 'G values' 366
Raman spectra, cyclopropane derivatives 541–544
Rearrangement catalysts—see Catalysts
Rearrangements 115–167
  aniontropic 151–154
  base-catalysed 141
  during mass spectrometry—see individual rearrangements
  mechanisms, double bond shift 165
  metal hydride addition-elimination 166
  prototropic, acid-catalysed 144
  retrodehydrative 153
  S_ni' process 149
  S_n2' process 149
  see also Isomerization; Stereomutation; individual rearrangements
Reduction—see Hydrogenation
Retro-Diels-Alder rearrangement—see under Mass spectrometry
Rideal mechanism, catalytic hydrogenation 176
Rubber, natural—see Hevea
Rubber hydrochloride, uses 439
Rubber ozonide, infrared spectra 485
Ruthenium(II) chloride, hydrogenation catalyst 196, 197, 235, 236

SBR—see Styrene-butadiene copolymer
Sorbic acid, reduction 231

Squalene, dichlorocarbene addition 495
  isomerization 458
  oxidation 478
Stereochemistry, carbonylation 247
  dynamic 222–224
    substituent effect 223
    transition metal-alkene complexes 222–224
    heterogeneous catalytic hydrogenation 175–192
Stereoisomers, equilibria between 116
Stereomutation, addition-elimination mechanism 122
  allylic resonance 117
  benzyl migration 134
  'forbidden' transitions 121
  nitric oxide catalysed 136
  photoactivated 119
  radical-catalysed 117
  sensitized 119
  triplet energy transfer, from carbonyl groups 120
  triplet intermediates 120
Stilbene, photoisomerization 276, 277
  triplet states 276
  radiation-induced isomerization 397
  reduction 231
  stereomutation 121, 124
    base-catalysed 122
Styrene, reduction 231
  substituted, vinylic, proton magnetic resonance spectra 34
Styrene-butadiene copolymer 413, 423, 424
  cyclization 445
  epoxidation 492
  general purpose elastomer 418
  hydrogenation 461, 462
  mechanochemical degradation 490
  microstructural analysis 417
  pyrolysis 489
  synthesis 421
  uses 423
  vulcanization, peroxide, crosslinking 468
    sulphur, mechanism 473

Tetraene polymers 422, 423
Tetraphenylbutadiene, isomerization 143
Thermal degradation—*see* Pyrolysis
Trienes, conjugated, polymerization 429, 430
Tris(triphenylphosphine)dichlororuthenium, hydrogenation catalyst 201
Tris(triphenylphosphine) halogenorhodium, hydrogenation catalyst 199, 200, 232–234
Tris(triphenylphosphine)hydridochlororuthenium(II), hydrogenation catalyst 236
Triterpenes, mass spectrometry and structure determination 352
Tropolones, electrocyclic photochemical isomerization 285
γ-Tropolones, production by photocycloaddition 305, 306

Ultraviolet absorption spectroscopy 269–271, 532–538
  cyclopropane derivatives 532–538
  diamagnetism 270
  Franck-Condon principle 270, 271
  mechanisms, single triplet transitions 270
    spin inversion (intersystem crossing) 270
    spin-orbital coupling 270
    triplet manifold 269
  paramagnetism 270
β,γ-Unsaturated acids, production by carbonylation of π-allyl complexes 244
α,β-Unsaturated esters, palladium catalysed oxidation 239
Unsaturated fatty acid esters, double bond location 345
Unsaturated polyesters 430, 431

Vinylacetylene, nuclear magnetic resonance spectra 8
Vinyl anion 40, 42, 49–56
  formation and reactions 56
  geometry 53
    hybridization 42
  spectral studies 53
  stability 43, 49–53
  stereochemistry of reactions 53
Vinyl cation 40, 43–49
  formation and reactions 48, 49
  geometry 48
    hybridization 42
  intermediate formed by loss of a nucleophilic group 47
  intermediate in electrophilic additions to alkynes 45

intermediate in solvolysis reactions of alkynes 47
stability 43–48
  effect of substituent 47
  from kinetic studies 45
  from mass spectral studies 44
Vinyl compounds, chemical shifts 9
  methyl substituent effect 13
  specific interaction of solvent with protons of 25
1-Vinylcyclohexanols, chemical shift 35
4-Vinylcyclohexene, production from butadiene 252
Vinylcyclohexene polymers 422
Vinylcyclopropane, conformations, s-trans, s-cis and gauche 525
  terminal addition 585
Vinyl derivatives, nuclear magnetic resonance spectra 57–59
Vinyl groups, bonded to Si, Hg, Al, Sn, chemical shifts 9
  coupling constants 11
Vinyl halides, ^{13}CH satellite spectra 5
  differential chemical shift 7
  palladium catalysed oxidation 239
Vinylic esters, rearrangements 157
Vinylic radical, stereochemical integrity 118
Vinylic silver compounds, synthesis 164
4-Vinylidenecyclopentene, long-range couplings 32
Vinyllithium reagent, infrared spectra 54
  cis,trans isomerization 53, 54
  proton magnetic resonance spectra 54
  stability from equilibrium studies 49, 50

synthesis 163
Vinylorganometallic compounds, synthesis and reactions 56
Vinylpotassium reagents, cis,trans isomerization 55, 56
Vinyl radical, geometry 48
Vitamin A analogues, synthesis 128
Vulcanization, accompanied by cis, trans isomerization 471
  peroxide 464–468
  polymers containing C=C bond 452, 463–476
    double bond shift 464
  radiation 474–476
  sulphur 464, 468–474
    accelerated 473
    polar mechanism 472, 473
    time of, effect on composition of polysulphide 469
    unaccelerated 468–473

Wittig rearrangements 128, 162
Woodward-Hoffmann rules, concerted cyclization 255
  photochemical isomerization 279–281, 294
    conrotatory and disrotatory modes of ring closure 279
    cyclobutene formation 287, 288
    electrocyclic reactions 280
    sigmatropic reactions 280, 281
    suprafacial and antarafacial migration 280

Zeise's salt, stability 219
Ziegler-type catalysts, homogeneous hydrogenation 202, 203
  polymerization 416, 421, 422, 424, 427, 428, 447
  rearrangements 148